Landslide Dynamics: ISDR-ICL Landslide Interactive Teaching Tools

Kyoji Sassa · Fausto Guzzetti
Hiromitsu Yamagishi · Željko Arbanas
Nicola Casagli · Mauri McSaveney
Khang Dang
Editors

Landslide Dynamics: ISDR-ICL Landslide Interactive Teaching Tools

Volume 1: Fundamentals, Mapping and Monitoring

A Programme of
the ICL for ISDR
ISDR

Springer

Editors
Kyoji Sassa
International Consortium on Landslides
Kyoto
Japan

Fausto Guzzetti
Research Institute for Hydrogeological Protection
National Research Council (CNR)
Perugia
Italy

Hiromitsu Yamagishi
Shin Engineering Consultant Co. Ltd.
Sapporo
Japan

Željko Arbanas
Faculty of Civil Engineering
University of Rijeka
Rijeka
Croatia

Nicola Casagli
Department of Earth Sciences
University of Florence
Florence
Italy

Mauri McSaveney
GNS Science
Lower Hutt
New Zealand

Khang Dang
International Consortium on Landslides
Kyoto
Japan

Additional material to this book can be downloaded from http://extras.springer.com.

ISBN 978-3-319-57773-9 ISBN 978-3-319-57774-6 (eBook)
https://doi.org/10.1007/978-3-319-57774-6

Library of Congress Control Number: 2017946030

Cover illustration: Bird's-eye view which based on the geomorphological map by photo interpretation around Mt. Unzen-Fugendake and Mt. Mayu-Yama (Inoue 1999, Unzen Restoration Office 2002) Inoue K (1999) Shimabara-Shigatsusaku Earthquake and topographic changes by Shimabara catastrophe. Journal of the Japan Society of Erosion Control Engineering 52(4):45–54 (in Japanese) Unzen Restoration Office of the Ministry of Land, Infrastructure and Transport of Japan (2002) The Catastrophe in Shimabara—1791–1792 eruption of Unzen–Fugendake and the sector collapse of Mayu-Yama. An English leaflet (23 pages)

Printed on acid-free paper

This Springer imprint is published by Springer Nature
The registered company is Springer International Publishing AG
The registered company address is: Gewerbestrasse 11, 6330 Cham, Switzerland

Foreword for the ISDR-ICL Landslide Interactive Teaching Tools

Disasters induced by natural hazards annually affect millions of people. Disasters may pose a serious threat to health, cultures, and development prospects. Disasters have been increasing in this period of climate change, urbanization, and overpopulation. Predictions are that several types of events arising from natural hazards will become more frequent, thus impacts of resulting disasters could become more devastating. Among these potentially catastrophic hazards are landslides.

Two major milestones, 2030 Agenda for Sustainable Development and the Paris Agreement on Climate change, were adopted in 2015. The same year in Sendai, Japan, Member States adopted the Sendai Framework for Disaster Risk Reduction 2015–2030. The Framework aims of assessing the global progress on disaster risk reduction, identifying its priority actions, and setting the role of stakeholders.

The key outcome relating to landslides from the Sendai Framework for Disaster Risk Reduction 2015–2030 is the *International Strategy for Disaster Risk Reduction—International Consortium on Landslides Sendai Partnerships 2015–2025 (ISDR-ICL Sendai Partnerships).*

The *International Consortium on Landslides, ICL,* actively supported by UNESCO, is a vital organization for landslide disaster risk reduction. Through international scientific communities, it conducts advanced research and builds capacities in landslide prevention and mitigation.

In line with the objectives of Sendai Partnership, ISDR-ICL is launching the Landslide Interactive Teaching Tools (LITT). This interactive open-access teaching toolkit aims to help academia, leading engineers, and practitioners to formulate necessary capacities and technical training for landslide disaster risk reduction. The set of tools includes methodologies and case studies on how to map, monitor, and forecast such extreme events. This open-access teaching tool could be periodically revised and updated by the community of users based on the reviews, comments, and new research findings.

Nowadays, more than ever, we need stronger scientific cooperation and broader results dissemination. In this spirit, I thank all the contributors to this initiative, and I pledge UNESCO's continuing support to deepening partnerships for education and resilience in societies across the world.

Flavia Schlegal
Assistant Director-General of UNESCO
for Natural Sciences

Preface: Aim and Outline of the ISDR-ICL Landslide Interactive Teaching Tools

Aim

The International Consortium on Landslides (ICL) proposed the ISDR-ICL Sendai Partnerships 2015–2025 for global promotion of understanding and reducing landslide disaster risk at a session of "Underlying risk factors" of the 3rd WCDRR on the morning of March 16, 2015. The partnership was proposed as a voluntary commitment to the World Conference on Disaster Risk Reduction, Sendai, Japan, 2015, and also as tools for implementing and monitoring the Post-2015 Framework for Disaster Risk Reduction and the Sustainable Development Goals. It was approved and signed by 16 global stakeholders in the afternoon of the same day in Sendai, Japan, and the Secretary-General Mr. Petteri Taalas of the World Meteorological Organization (WMO) signed it on April 16, 2016.

The Sendai partnerships acknowledge that

- At a higher level, social and financial investment is vital for understanding and reducing landslide disaster risk, in particular social and institutional vulnerability, through coordination of policies, planning, research, capacity development, and the production of publications and tools that are accessible, available free of charge and are easy to use for everyone in both developing and developed countries.

Landslide science and technologies have continuously been developed to be more reliable, precise, or cost-effective for landslide disaster risk reduction over the world. However, this scientific and technological progress has not been shared equally over the world. The gap between the available level of science and technologies and the practical use of those in many countries, regions, and communities is very wide. To fill this gap, ICL has created Landslide Interactive Teaching Tools, which are always updated and continuously improved, based on responses from users and lessons learned during their application. All textbooks gradually become outdated. To avoid this problem, ICL plans to upload the latest teaching tools in the WEB of Teaching Tools and print text tools periodically.

Landslide Dynamics
A landslide is a downslope movement of rock, soil, or both. Landslide disasters are caused by exposure to **hazardous motions of soil and rock** that threaten vulnerable human settlement in mountains, cities, coasts, and islands, as stated in the Sendai Partnerships. Understanding "Landslide dynamics" is the very basis of landslide disaster risk reduction.

Organizations Contributing Teaching Tools
Each teaching tool will be submitted by the teaching tool contributing organization as shown in the list of contributing organizations. Each organization has its own Teaching Tool Identifying Number consisting of telephone number of the country and the registered number within the country. The involvement of organization as well as individual researcher is adopted to keep quality and updating of each tool.

Outline

The teaching tools are classified into five major parts. The part number is included in each teaching tool identifier.

Teaching tools have fundamentals and 4 parts of tools. Each part has 3–5 subparts as below. However, the subparts do not appear in the content because of two levels of publication structure. Please refer the following subparts and pages.

Fundamentals

(1) Landslide Types: Description, illustrations and photos (page 1–38 in Vol. 1)
(2) Landslide Dynamics for risk assessment (page 1–79 in Vol. 2)

Part 1 Mapping and Site Prediction in Vol. 1

(1) Basic Mapping (5 tools: page 41–112)
(2) Site Prediction Using GIS (8 tools: page 113–195)
(3) Field Guidelines (2 tools: page 197–232)

Part 2 Monitoring and Early Warning in Vol. 1

(1) Remote Sensing Techniques for Landslide Monitoring (4 tools: page 235–295)
(2) Monitoring System Instrumentation (3 tools: page 297–340)
(3) Rainfall Threshold for Landslide Prediction (10 tools: page 341–447)
(4) Landslide Time Prediction from Pre-failure Movement Monitoring (6 tools: page 449–551)
(5) Guidelines for Landslide Monitoring and Early Warning Systems (3 tools: page 553–599)

Part 3 Risk Assessment in Vol. 2

(1) Numerical Modeling and Simulation (12 tools: page 83–275)
(2) Physical and Mathematical Modeling (4 tools: page 277–319)
(3) Laboratory Soil Testing for Landslide Analysis (4 tools: page 321–402)
(4) Analysis and Assessment of Landslides (4 tools: page 403–443)

Part 4 Risk Management and Country Practices in Vol. 2

(1) Landslide Risk Management (10 tools: page 447–597)
(2) Community Risk Management (5 tools: page 599–667)
(3) Country Practices (14 tools: page 669–831)

The teaching tools consist of three types of tools.

1. The first type is TXT-tools consisting of original texts with figures. The first edition includes two volumes of books.
2. The second type is PPT-tools consisting of PowerPoint files and video tools made for visual lectures.
3. The third type is PDF-tools consisting of already published reference papers/reports, guidelines, and others.

The second and the third types of tools are supplementary tools of the text tools (text books).
Each teaching tool has its own identifier. The identifier of each tool consists of three parts:

1. the number of the part of the tool box in which it appears (Parts 0–4);
2. the country telephone code and an assigned unique number for each contributing organization (for example 081-1 signifies Japan-ICL headquarters, and 081-3 signifies Japan—Erosion and Sediment Control Department, Ministry of Land, Infrastructure, Transport and Tourism);
3. the last part of the identifier is a consecutive number assigned to the teaching tool by its contributing organization.

Call for Contribution to the Interactive Teaching Tools

The International Consortium on Landslides calls for new tools to this Landslide Interactive Teaching tools.

New progress of Landslide Science and Technology, and case studies and country practices in the world are good information for practitioners and engineers in other countries and other fields as well. We plan to establish a new Editorial Manager for Landslide Teaching Tools to promote the capacity building for the contribution of new tools. The authors of teaching tools in this publication are also requested to improve their contributed tools and contribute new tools for other people in the world landslide community.

Those who are willing to contribute to the Landslide Teaching Tools are requested to contact ICL Secretariat and one of Editors of this teaching tools Vol. 1 and Vol. 2. All authors and editors of this teaching tools are highly appreciated for their useful contribution and editorial efforts.

Contacts

Editor-in-Chief

Kyoji Sassa: International Consortium on Landslides (ICL) (e-mail: sassa@iclhq.org)

Editorial Secretariat

- Kyoji Sassa: Secretary-General in ICL Secretariat, Kyoto, Japan (e-mail: secretariat@ichhq.org)
- Khang Dang: Assistant to Editor-in-Chief (e-mail: khangdq@gmail.com)
- Hendy Setiawan: Assistant to Editor-in-Chief (e-mail: sethiawan.hendi.45w@kyoto-u.jp)

Editors of Parts

Fundamentals

- Kyoji Sassa: International Consortium on Landslides (ICL), Japan (e-mail: sassa@iclhq.org)

Part 1: Mapping and Site Prediction

- Fausto Guzzetti : CNR—Consiglio Nazionale delle Ricerche, Istituto di Ricerca per la Protezione Idrogeologica, Italy (e-mail: F.Guzzetti@irpi.cnr.it)
- Hiromitsu Yamagishi: Shin Engineering Consultant Co. Ltd, Japan (e-mail: hiromitsuyamagishi88@gmail.com)

Part 2: Monitoring and Early Warning

- Željko Arbanas: University of Rijeka, Faculty of Civil Engineering, Croatia (e-mail: zeljko.arbanas@gradri.uniri.hr)
- Nicola Casagli: University of Florence, Department of Earth Sciences, Italy (e-mail: nicola.casagli@unifi.it)

Part 3: Risk Assessment

- Kyoji Sassa: International Consortium on Landslides, Japan (e-mail: Sassa@iclhq.org)
- Binod Tiwari: Civil & Environmental Engineering Department, California State University, USA, (e-mail: btiwari@fullerton.edu)

Part 4: Risk Management and Country Practices

- Mauri McSaveney: GNS Science, New Zealand (e-mail: m.mcsaveney@gns.cri.nz)
- Ko-Fei Liu: National Taiwan University, Chinese Taipei (e-mail: kfliu@ntu.edu.tw)
- Alexander Strom: Geodynamics Research Centre—branch of JSC "Hydroproject Institute", Russia (e-mail: strom.alexandr@yandex.ru)

Kyoji Sassa
Executive Director
International Consortium on Landslides

Contents

TXT-tool 0.001-2.1
Landslide Types: Descriptions, Illustrations and Photos 1
Lynn Highland and Peter Bobrowsky

Part I Mapping and Site Prediction

TXT-tool 1.081-2.1
Landslide Mapping Through the Interpretation of Aerial Photographs 41
Toyohiko Miyagi

TXT-tool 1.081-2.2
Landslide Mapping Through the Interpretation of Aerial Photographs and Topographic Maps 53
Eisaku Hamasaki and Toyohiko Miyagi

TXT-tool 1.081-3.1
Landslide Recognition and Mapping Using Aerial Photographs and Google Earth 67
Hiromitsu Yamagishi and Rigoberto Moncada

TXT-tool 1.039-1.1
Very-High Resolution Stereo Satellite Images for Landslide Mapping 83
Francesca Ardizzone, Federica Fiorucci, Alessandro Cesare Mondini and Fausto Guzzetti

TXT-tool 1.504-1.1
Landslide Inventory Educational Methodology Derived from Experiences in Latin America 95
Rigoberto Moncada and Hiromitsu Yamagishi

TXT-tool 1.052-1.1
GIS Using Landslides Inventory Mapping for Volcanoes 113
Gabriel Legorreta Paulín

TXT-tool 1.052-1.2
GIS Using Landslides Susceptibility Mapping Model for Volcanoes 123
Gabriel Legorreta Paulín, Michael Polenz and Trevor Contreras

TXT-tool 1.386-2.1
Landslide Susceptibility Assessment Method 133
Marko Komac and Jernej Jež

TXT-tool 1.081-6.1
A Comparative Study of the Binary Logistic Regression (BLR) and Artificial Neural Network (ANN) Models for GIS-Based Spatial Predicting Landslides at a Regional Scale 139
Jie Dou, Hiromitsu Yamagishi, Zhongfan Zhu, Ali P. Yunus and Chi Wen Chen

TXT-tool 1.386-2.2
Practical Approach to Assessing the Factors Influencing Landslide Susceptibility Modelling—The Case of Slovenia 153
Marko Komac

TXT-tool 1.084-3.1
Landslide Susceptibility Mapping at a Regional Scale in Vietnam 161
Quoc Hung Le, Thi Hai Van Nguyen, Minh Duc Do, Thi Chau Ha Le, Van Son Pham, Ho Khanh Nguyen and Thanh Binh Luu

TXT-tool 1.039-1.2
Bedding Attitude Information Through the Interpretation of Stereoscopic Aerial Photographs and GIS Modeling 175
Ivan Marchesini, Michele Santangelo, Federica Fiorucci, Mauro Cardinali, Mauro Rossi, Francesco Bucci and Fausto Guzzetti

TXT-tool 1.086-1.1
Distribution of Island-like Permafrost in the Lesser Khingan Mountains of Northeast China Using Landsat7 ETM+ Imagery 187
Chunjiao Wang and Wei Shan

TXT-tool 1.081-7.1
Investigating Landslides in the Field Using Google Earth and PowerPoint: A Case Study of Altos de Loarque in Honduras 197
Kiyoharu Hirota

TXT-tool 1.064-1.1
Field Guide for the Identification and Assessment of Landslide and Erosion Features and Related Hazards Affecting Pipelines 209
Chris Massey, Graham Hancox and Mike Page

Part II Monitoring and Early Warning

TXT-tool 2.039-3.1
Satellite Remote Sensing Techniques for Landslides Detection and Mapping 235
Nicola Casagli, Veronica Tofani, Andrea Ciampalini, Federico Raspini, Ping Lu and Stefano Morelli

TXT-tool 2.039-3.2
Ground-Based Remote Sensing Techniques for Landslides Mapping, Monitoring and Early Warning 255
Nicola Casagli, Stefano Morelli, William Frodella, Emanuele Intrieri and Veronica Tofani

TXT-tool 2.386-2.1
SAR Interferometry as a Tool for Detection of Landslides in Early Phases 275
Marko Komac, Tina Peternel and Mateja Jemec Auflič

TXT-tool 2.039-3.3
Ground-Based Radar Interferometry for Landslide Monitoring 287
Nicola Casagli, Filippo Catani, Chiara Del Ventisette and Guido Luzi

TXT-tool 2.062-1.1
A Landslide Monitoring and Early Warning System 297
Teuku Faisal Fathani and Dwikorita Karnawati

TXT-tool 2.007-1.1
Monitoring Alarm System of Landslide and Seismic Safety for Potentially Hazardous Objects 309
Alexander Ginzburg, Alexey Nikolaev, Valentina Svalova, Anatoliy Manukin and Vladimir Savosin

TXT-tool 2.007-1.2
Landslide and Seismic Monitoring System on the Base of Unified Automatic Equipment 327
Alexander Ginzburg, Alexey Nikolaev, Valentina Svalova, German Postoev and Andrey Kazeev

TXT-tool 2.039-1.1
Italian National Early Warning System 341
Mauro Rossi, Ivan Marchesini, Gabriele Tonelli, Silvia Peruccacci, Maria Teresa Brunetti, Silvia Luciani, Francesca Ardizzone, Vinicio Balducci, Cinzia Bianchi, Mauro Cardinali, Federica Fiorucci, Alessandro Cesare Mondini, Paola Reichenbach, Paola Salvati, Michele Santangelo and Fausto Guzzetti

TXT-tool 2.084-3.1
Rainfall Thresholds for Triggering Geohazards in Bac Kan Province (Vietnam) 351
Duc Ha Nguyen, Thi Hai Van Nguyen, Quoc Hung Le, Van Son Pham and Ho Khanh Nguyen

TXT-tool 2.039-1.2
Rainfall Thresholds for the Possible Initiation of Landslides in the Italian Alps 361
Maria Teresa Brunetti, Silvia Peruccacci, Michela Rosa Palladino, Alessia Viero and Fausto Guzzetti

TXT-tool 2.039-1.3
Topographic and Pedological Rainfall Thresholds for the Prediction of Shallow Landslides in Central Italy 371
Silvia Peruccacci, Maria Teresa Brunetti and Fausto Guzzetti

TXT-tool 2.039-4.1
FLaIR Model (Forecasting of Landslides Induced by Rainfalls) 381
Pasquale Versace, Giovanna Capparelli and Davide Luciano De Luca

TXT-tool 2.386-2.1
A System to Forecast Rainfall-Induced Landslides in Slovenia 391
Mateja Jemec Auflič, Jasna Šinigoj, Tina Peternel, Martin Podboj, Matija Krivic and Marko Komac

TXT-tool 2.886-1.1
Early Warning Criteria for Debris Flows and Their Application in Taiwan 405
Chyan-Deng Jan, Ji-Shang Wang, Yi-Chao Zeng and Feng-Hao Kuo

TXT-tool 2.081-5.1
High-Resolution Rainfall Prediction for Early Warning of Landslides 415
Ryo Onishi, Keigo Matsuda and Keiko Takahashi

TXT-tool 2.386-1.1
Intensity-Duration-Frequency Curves for Rainfall-Induced Shallow Landslides and Debris Flows Using Copula Functions 425
Nejc Bezak, Mitja Brilly, Mojca Šraj and Matjaž Mikoš

TXT-tool 2.039-1.5
An Algorithm for the Objective Reconstruction of Rainfall Events Responsible for Landslides 433
Massimo Melillo, Maria Teresa Brunetti, Silvia Peruccacci, Stefano Luigi Gariano and Fausto Guzzetti

TXT-tool 2.385-1.1
A Comprehensive Landslide Monitoring System: The Kostanjek Landslide, Croatia 449
Snježana Mihalić Arbanas, Martin Krkač, Sanja Bernat Gazibara, Marko Komac, Marin Sečanj and Željko Arbanas

TXT-tool 2.385-1.2
Landslide Comprehensive Monitoring System: The Grohovo Landslide Case Study, Croatia 465
Željko Arbanas, Snježana Mihalić Arbanas, Martina Vivoda Prodan, Josip Peranić, Sanja Dugonjić Jovančević and Vedran Jagodnik

TXT-tool 2.062-1.2
A Monitoring and Early Warning System for Debris Flows in Rivers on Volcanoes 479
Teuku Faisal Fathani and Djoko Legono

TXT-tool 2.380-1.1
Monitoring and Early Warning System of the Building Constructions of the Livadia Palace, Ukraine 491
Oleksander Trofymchuk, Iurii Kaliukh and Oleg Klymenkov

TXT-tool 2.039-4.2
LEWIS Project: An Integrated System for Landslides Early Warning 509
Pasquale Versacc, Giovanna Capparelli and Davide Luciano De Luca

TXT-tool 2.039-3.4
Methods to Improve the Reliability of Time of Slope Failure Predictions and to Setup Alarm Levels Based on the Inverse Velocity Method 537
Tommaso Carlà, Emanuele Intrieri, Federico Di Traglia, Giovanni Gigli and Nicola Casagli

TXT-tool 2.886-1.2
Guidelines for Landslide Monitoring Systems 553
An-Bin Huang and Wen-Jong Chang

TXT-tool 2.386-1.2
Practice Guidelines on Monitoring and Warning Technology for Debris Flows 567
Johannes Hübl and Matjaž Mikoš

TXT-tool 2.886-1.3
Debris Flow Monitoring Guidelines 587
Hsiao-Yuan Yin and Yi-Min Huang

List of PPT-tools and PDF-tools 601

TXT-tool 0.001-2.1 Landslide Types: Descriptions, Illustrations and Photos

Lynn Highland and Peter Bobrowsky

Abstract

Regardless of the exact definition used or the type of landslide under discussion, understanding the basic mechanics and processes of a typical landslide is critical. With an increasing awareness and mandate to fully understand new building site characteristics, improve existing critical infrastructure and appreciating the importance of an area's landslide history, typing landslides according to shared characteristics gives vital information about the future performance of a site, in relation to potential landslide hazards. We provide here, a basic primer for an understanding of the similarities and differences of the nature and physics of landslide movement. Type of material, speed of motion, slope angle, potential frequency of occurrence, weather and climatic influences, and man-made disturbances as well as other factors have a bearing on landslide motion, size and impact, yet we can generally group and categorize most landslides into more understandable groupings. Landslide typology is constantly evolving and becoming more exact, given the expanding tools of site investigation, improving computer and GIS modeling, and careful peer analysis.

Keywords

Landslide types • Landslide classification • Typology
Landslide materials • Landslide speed • Landslide process

L. Highland (✉)
Morrison, CO, USA
e-mail: highlandl2016@gmail.com

P. Bobrowsky
Sidney, Canada
e-mail: peter.bobrowsky@canada.ca

K. Sassa et al. (eds.), *Landslide Dynamics: ISDR-ICL Landslide Interactive Teaching Tools*, https://doi.org/10.1007/978-3-319-57774-6_1

Contents

1 Introduction 2
2 Basic Landslide Types 2
2.1 Falls 3
2.2 Slides 5
2.3 Spreads 10
2.4 Flows 17
2.5 Complex 36
3 Conclusion 36
References 38

1 Introduction

Geologists, engineering geologists, geotechnical engineers and other professionals often rely on unique and slightly differing definitions of landslides. This diversity in definitions reflects the complex nature of the many disciplines associated with studying landslide phenomena. For our purposes, landslide is a general term used to describe the downslope movement of soil, rock, and organic materials under the effects of gravity and also the landform that results from such movement (Cruden 1991) (see Fig. 1 for an example of one type of landslide, showing the commonly-accepted terms for the various parts of a landslide).

Varying classifications of landslides are associated with specific mechanics of slope failure and the properties and characteristics of failure types; these will be discussed briefly herein.

There are a number of other phrases/terms that are often used interchangeably with the term "landslide" including most frequently the terms mass movement and slope failure. One commonly hears such terms applied to all types and sizes of landslides.

Figure 1 shows the position and the most common terms used to describe the unique parts of a landslide. From there, we paraphrase the work of Highland and Bobrowsky (2008) with an update and improvement on the topics of landslide type schematics and example photographs.

2 Basic Landslide Types

A landslide is a downslope movement of rock or soil, or both, occurring on the surface of rupture —either curved (rotational slide) or planar (translational slide) rupture—in which much of the material often moves as a coherent or semi-coherent mass with little internal deformation. Notably, in some cases, landslides may also involve other types of movement, either at the inception of the failure or later, if properties change as the displaced material moves downslope.

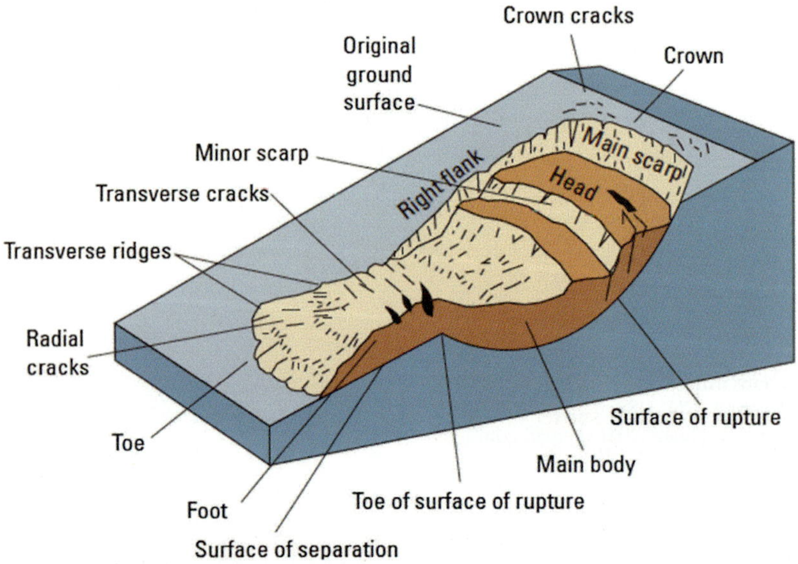

Fig. 1 This graphic illustrates commonly-used labels for the parts of a landslide. The image shows a rotational landslide that has evolved into an earthflow. (modified from Varnes 1978)

This section provides descriptions and illustrations of the various types of landslides. Understanding the characteristics of the specific type of landslide hazard in your area is vitally important to consider when planning or adopting appropriate mitigating action to lessen the risk of loss and damage. The type of landslide will determine the potential speed of movement, likely volume of displacement, distance of run-out, as well as the possible effects of the landslide and the appropriate mitigating measures to be considered.

Landslides can be classified into different types on the basis of the type of movement and the type of material involved (Cruden and Varnes 1996). In brief, material in a landslide mass is either rock or soil (or both); the latter is described as earth if mainly composed of sand-sized or finer particles and debris if composed of coarser fragments. The type of movement describes the actual internal mechanics of how the landslide mass is displaced: fall, topple, slide, spread, or flow. Thus, landslides are primarily described using two terms that refer respectively to material and movement (that is, rockfall, debris flow, and so forth). Landslides may also form a complex failure encompassing more than one type of movement (that is, rock slide—debris flow).

We define "type of movement" as synonymous with "landslide type." Each type of movement can be further subdivided according to specific properties and characteristics, and the main subcategories of each type are described elsewhere. Less common subcategories are not discussed here but are referred to in the source reference.

2.1 Falls

A fall begins with the detachment of soil or rock, or both, from a steep slope along a surface on which little or no shear displacement has occurred. The material subsequently descends mainly by falling, bouncing, or rolling.

2.1.1 Rockfall

Falls are abrupt, downward movements of rock or earth, or both, that detach from steep slopes or cliffs. The falling material usually strikes the lower slope at angles less than the angle of fall, causing bouncing. The falling mass may break on impact, may begin rolling on steeper slopes, and may continue until the terrain flattens, or encounters a structure or other obstacles.

Occurrence and relative size/range

Common worldwide on steep or vertical slopes—also in coastal areas, and along rocky banks of rivers and streams. The volume of material in a fall can vary substantially, from individual rocks and clumps of soil to massive blocks thousands of cubic meters in size.

Velocity of travel

Very rapid to extremely rapid, free-fall; bouncing and rolling of detached soil, rock, and boulders. The rolling velocity depends on slope steepness.

Triggering mechanism

Undercutting of slope by natural processes such as streams and rivers or differential weathering (such as the freeze/thaw cycle), human activities such as excavation during road building and (or) maintenance, and earthquake shaking or other intense vibration.

Effects (direct/indirect)

Falling material can be life-threatening. Falls can damage property beneath the fall-line of large rocks. Boulders can bounce or roll great distances and damage structures or kill people. Damage to roads, railroads, and highways is particularly intense: rockfalls can cause deaths to those in vehicles hit by rocks and can block highways and railroads.

Corrective measures/mitigation

Rock curtains or other slope covers, protective covers over roadways, retaining walls to prevent rolling or bouncing, explosive blasting of hazardous target areas to remove the source, scaling, removal of rocks or other materials from highways and railroads can be used. Rock bolts or other similar types of anchoring used to stabilize cliffs, as well as scaling, can lessen the hazard. Warning signs are recommended in hazardous areas for awareness. Stopping or parking under hazardous cliffs should be avoided.

Predictability

Mapping of hazardous rockfall areas has been completed in a few areas around the world. Rock-bounce calculations and estimation methods for delineating the perimeter of rockfall zones have also been determined and the information widely published. Indicators of imminent rockfall include terrain with overhanging rock or fractured or jointed rock along steep slopes, particularly in areas subject to frequent freeze-thaw cycles. Also, cut faces in gravel pits may be particularly subject to falls. Figures 2, 3, 4, 5 and 6 show a schematic and photographs of rockfall.

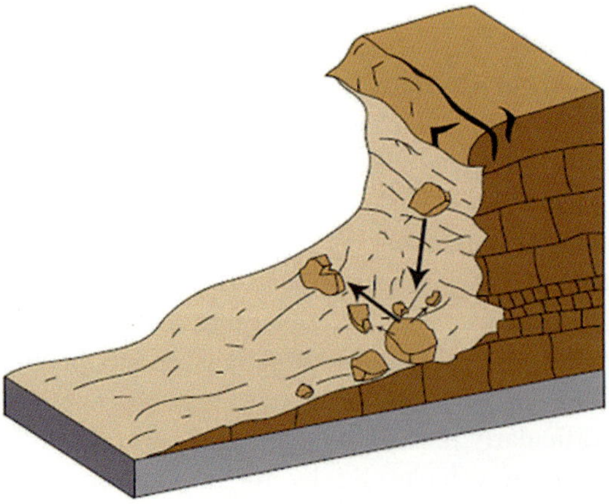

Fig. 2 Schematic illustration of rockfall. Note that rocks may roll and bounce at potentially great distances depending on a number of factors

2.1.2 Topple

A topple is recognized as the forward rotation out of a slope of a mass of soil or rock around a point or axis below the center of gravity of the displaced mass. Toppling is sometimes driven by gravity exerted by the weight of material upslope from the displaced mass. Sometimes toppling is due to water or ice collecting in cracks in the mass. Topples can consist of rock, debris (coarse material), or earth materials (fine-grained material). Topples can be complex and composite.

Occurrence

They are known to occur globally, often prevalent in columnar-jointed volcanic terrain, as well as along stream and river courses where the banks are steep.

Velocity of travel

They are extremely slow to extremely rapid, sometimes accelerating throughout the movement depending on distance of travel.

Triggering mechanism

Topples are sometimes driven by gravity exerted by material located upslope from the displaced mass and sometimes by water or ice occurring in cracks within the mass; also, vibration, undercutting, differential weathering, excavation, or stream erosion.

Effects (direct/indirect)

They can be extremely destructive, especially when failure is sudden and (or) the velocity is rapid. They can potentially block waterways or highways causing flooding and interference of travel.

Corrective measures/mitigation

In rock there are many options for the stabilization of topple-prone areas. Some examples for reinforcement of these slopes include rock

Fig. 3 This rockfall in North Carolina (USA), occurred on a highway through Pigeon River Gorge, on October, 2009, and caused traffic delays for weeks. Photo by North Carolina Department of Transportation

bolts and mechanical and other types of anchors. Seepage is also a contributing factor to rock instability, and drainage should be considered and addressed as a corrective means.

Predictability

Topples are not generally mapped for susceptibility; some inventory of occurrence exists for certain areas. Monitoring of topple-prone areas is useful; for example, the use of tiltmeters. Tiltmeters are used to record changes in slope inclination near cracks and areas of greatest vertical movements. Warning systems based on movement measured by tiltmeters could be effective. Figures 7, 8, 9 and 10 show a schematic and photos of topple.

2.2 Slides

A slide is a downslope movement of a soil or rock mass occurring on surfaces of rupture or on relatively thin zones of intense shear strain. Movement does not initially occur simultaneously over the whole of what eventually becomes the surface of rupture; the volume of displacing material enlarges from an area of local failure.

2.2.1 Rotational Landslide

A landslide on which the surface of rupture is curved upward (spoon-shaped) and the slide movement is more or less rotational about an axis that is parallel to the contour of the slope. The displaced entity may, under certain circumstances, move as a relatively coherent mass along the rupture surface with little internal deformation. The head of the displaced material may move almost vertically downward, and the upper surface of the displaced material may tilt backwards toward the scarp. If the slide is rotational and has several parallel curved planes of movement, it is called a slump.

Occurrence

Because rotational slides occur most frequently in homogeneous materials, they are the

Fig. 4 Large rocks brought down by the January 12, 2010 (Magnitude 7) earthquake. The earthquake caused many landslides and rockfalls. Photograph by Randy Jibson, U.S. Geological Survey

most common landslide occurring in "fill" materials.

Relative size/range

They are associated with slopes ranging from about 20°–40°. In soils, the surface of rupture generally has a depth-to-length ratio between 0.3 and 0.1.

Velocity of travel (rate of movement)

Velocity is extremely slow (less than 0.3 m or 1 ft every 5 years) to moderately fast (1.5 m or 5 feet per month) to rapid.

Triggering mechanism

Intense and (or) sustained rainfall or rapid snowmelt can lead to the saturation of slopes and increased groundwater levels within the mass; rapid drops in river level following floods, ground-water levels rising as a result of filling reservoirs, or the rise in level of streams, lakes, and rivers, which cause erosion at the base of slopes. These types of slides can also be earthquake-induced.

Effects (direct/indirect)

These can be extremely damaging to structures, roads, and lifelines but are not usually life-threatening if movement is slow. Structures situated on the moving mass also can be severely damaged as the mass tilts and deforms. The large volume of material that is displaced is difficult to permanently stabilize. Such failures can dam rivers, causing flooding.

Fig. 5 Jalalabad Road, Afghanistan—Rockfall on the Jalalabad Road which is a highway from Kabul to Surobi and Jalalabad, Afghanistan, February 28, 2007. Photo by Sven Dirks http://commons.wikimedia.org/wiki/File:Jalalabad_Road_rock_fall.jpg

Mitigation measures

Instrumental monitoring to detect movement and the rate of movement can be implemented. Disrupted drainage pathways should be restored or re-engineered to prevent future water buildup in the slide mass. Proper grading and engineering of slopes, where possible, will reduce the hazard considerably. Construction of retaining walls at the toe may be effective to slow or deflect the moving soil; however, the slide may overtop such retaining structures despite good construction.

Predictability

Historical slides can be reactivated; cracks at tops (heads) of slopes are good indicators of the initiation of failure. Figures 11, 12 and 13 show a schematic and photos of rotational landslides.

2.2.2 Translational Landslide

The mass in a translational landslide moves out, or down and outward, along a relatively planar surface with little rotational movement or backward tilting. This type of slide may progress over considerable distances if the surface of rupture is sufficiently inclined, in contrast to rotational slides, which tend to restore the slide equilibrium. The material in the slide may range from loose, unconsolidated soils to extensive slabs of rock, or both. Translational slides commonly fail along geologic discontinuities such as faults, joints, bedding surfaces, or the contact between

Fig. 6 A large rockfall due to the May, 2008 Wenchuan, China Earthquake. Rockfalls such as this one are often deadly, although rockfalls can involve small rocks, large boulders, and/or combinations of both, and may disrupt transportation routes for weeks or months. Photo by Dave Wald, U.S. Geological Survey

Fig. 7 Schematic illustration of a topple

Fig. 8 A Topple in the vicinity of Jasper National Park, British Columbia, Canada. Photo by G. Bianchi Fasani

rock and soil. In northern environments the slide may also move along the permafrost layer.

Occurrence

These are one of the most common types of landslides, worldwide. They are found globally in all types of environments and conditions.

Relative size/range

They are generally shallower than rotational slides. The surface of rupture has a distance-to-length ratio of less than 0.1 and can range from small (residential lot size) failures to very large, regional landslides that are kilometers wide.

Velocity of travel

Movement may initially be slow (5 feet per month or 1.5 m per month) but many are moderate in velocity (5 feet per day or 1.5 m per day) to extremely rapid. With increased velocity, the landslide mass of translational failures may disintegrate and develop into a debris flow.

Triggering mechanism

They are primarily triggered by intense rainfall, rise in groundwater within the slide due to rainfall, snowmelt, flooding, or other inundation of water resulting from irrigation, or leakage from pipes or human-related disturbances such as

Fig. 9 A topple in a granite outcrop, near the summit of Mount Evans, Arapaho National Forest, Colorado (USA). Photo by Lynn Highland, retired U.S. Geological Survey

undercutting. These types of landslides can be earthquake-induced.

Effects (direct/indirect)

Translational slides may initially be slow, damaging property and (or) lifelines; in some cases they can gain speed and become life-threatening. They can also dam rivers, causing flooding.

Mitigation measures

Adequate drainage is necessary to prevent sliding or, in the case of an existing failure, to prevent a reactivation of the movement. Common corrective measures include leveling, proper grading and drainage, and retaining walls. More sophisticated remedies in rock include anchors, bolts, and dowels, which in all situations are best implemented by professionals. Translational slides on moderate to steep slopes are very difficult to stabilize permanently.

Predictability

There is a high probability of repetitive occurrence in areas where they have occurred in the past, including areas subject to frequent strong earthquakes. Widening cracks at the head or toe bulge may be an indicator of imminent failure. Figures 14, 15 and 16 show a schematic and photos of a translational landslide.

2.3 Spreads

A spread is an extension of a cohesive soil or rock mass combined with the general subsidence of the fractured mass of cohesive material into softer underlying material. Spreads may result from liquefaction or flow (and extrusion) of the softer underlying material. Types of spreads include block spreads, liquefaction spreads, and lateral spreads.

2.3.1 Lateral Spreads

Lateral spreads usually occur on very gentle slopes or essentially flat terrain, especially where a stronger upper layer of rock or soil undergoes extension and moves above an underlying softer, weaker layer. Such failures commonly are accompanied by some general subsidence into the weaker underlying unit. In rock spreads, solid ground extends and

Fig. 10 A topple in sandstone, at Fisher Towers, near Moab, Utah (USA). Photo by Lynn Highland, retired U.S. Geological Survey

Fig. 11 Schematic of a Rotational landslide

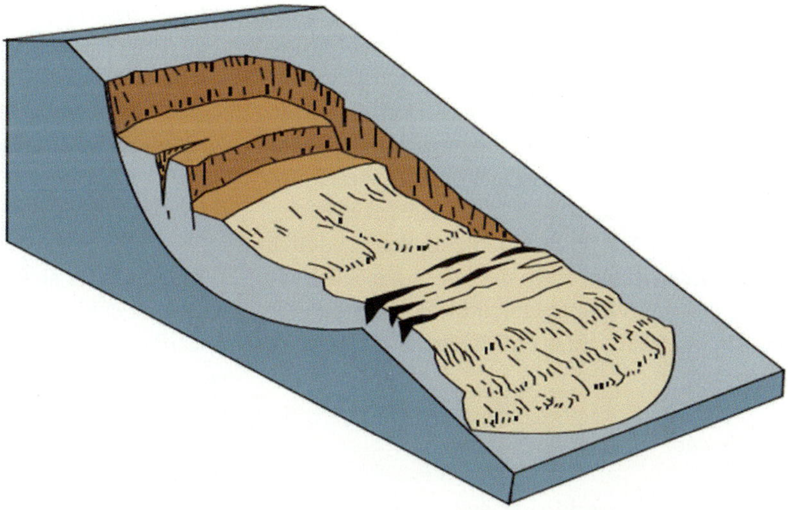

Fig. 12 This photo shows a rotational landslide triggered by rainfall in San Benito County, California (USA), in 1998. This landslide occurred as a reactivation of an old landslide in Pleistocene non-marine claystone, sandstone and conglomerate, and was located next to the San Andreas Rift Zone. Photo by Lynn Highland, retired U.S. Geological Survey

fractures, pulling away slowly from stable ground and moving over the weaker layer without necessarily forming a recognizable surface of rupture. The softer, weaker unit may, under certain conditions, squeeze upward into fractures that divide the extending layer into blocks. In earth spreads, the upper stable layer extends along a weaker underlying unit that has flowed following liquefaction or plastic deformation. If the weaker unit is relatively thick, the overriding fractured blocks may subside into it, translate, rotate, disintegrate, liquefy, or even flow.

Occurrence

They occur worldwide and known to occur where there are liquefiable soils. Common, but not restricted, to areas of seismic activity.

Relative size/range

The area affected may start small in size and have a few cracks that may spread quickly, affecting areas of hundreds of meters in width.

Velocity of travel

They may be slow to moderate and sometimes rapid after certain triggering mechanisms, such as an earthquake. Ground may then slowly spread over time from a few millimeters per day to tens of square meters per day.

Triggering mechanism

Triggers that destabilize the weak layer include: Liquefaction of lower weak layer by

Fig. 13 A photo of a rotational landslide, showing the Dainichi-san landslide triggered by the October 23, 2004 Mid-Niigata Prefecture earthquake (Sassa 2005)

Fig. 14 Schematic of a Translational landslide

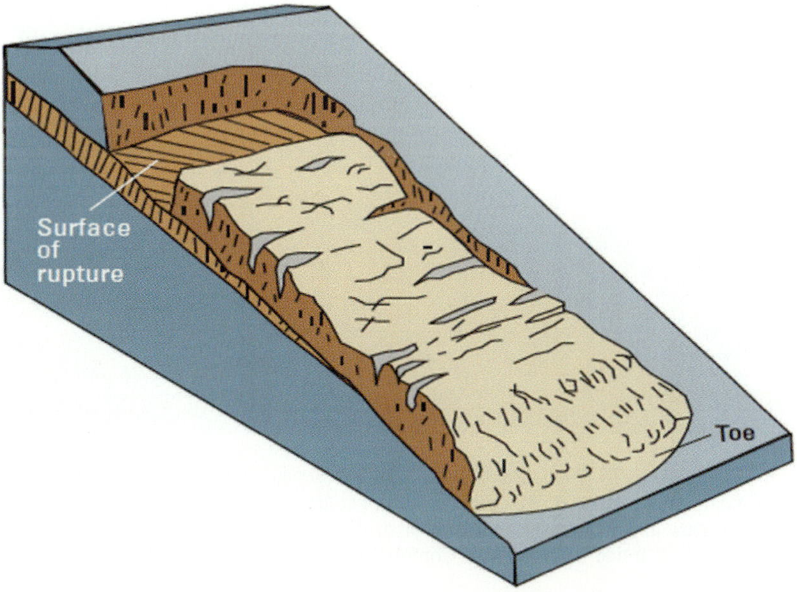

Fig. 15 The Minami-Aso landslide shows a translational landslide in a steep slope. The landslide was triggered by the Kumamoto Earthquake of 2016 in Japan (Dang et al. 2016). Photo taken from UAV by Khang Dang and Kyoji Sassa

Fig. 16 The June 14, 2008 Aratozawa landslide in Japan is an example of rotational movement in the upper area of the headscarp and translational movement in the central and lower part. It can be typed as a Complex Landslide, since it includes more than one type of movement (Miyagi et al. 2010). This photo was taken by Fukuoka H. in 2008. See Figs. 44 and 45 for thematic illustration and photo of a Complex landslide

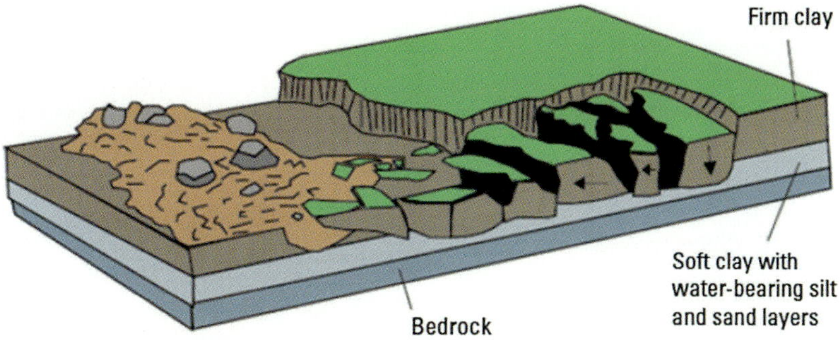

Fig. 17 Schematic of a Lateral Spread

Fig. 18 Lateral spreads at Hebgen Lake near West Yellowstone, Montana (USA), due to the effects of the Magnitude 7.3 Hebgen Lake earthquake, on August 18, 1959. Photo by R.B. Colton, U.S. Geological Survey

earthquake shaking; Natural or anthropogenic overloading of the ground above an unstable slope; Saturation of underlying weaker layer due to precipitation, snowmelt, and (or) ground-water changes; Liquefaction of underlying sensitive marine clay following an erosional disturbance at base of a riverbank/slope; Plastic deformation of unstable material at depth (for example, salt).

Fig. 19 The effects of lateral spreading in Christchurch, New Zealand as a result of the February 22, 2011 Christchurch Earthquake. *Photo* Wikipedia Commons. https://commons.wikimedia.org/wiki/File:25_Feb_River_Road.jpg

Effects (direct/indirect)

These can cause extensive property damage to buildings, roads, railroads, and lifelines. They can spread slowly or quickly, depending on the extent of water saturation of the various soil layers. Lateral spreads may be a precursor to earthflows.

Fig. 20 Lateral spreading caused by the magnitude 6.5 El Salvador, Central America earthquake of February 13, 2001. Area shown is on the eastern shore of Lago de Ilopango. Photograph by Anthony Crone, U.S. Geological Survey

Mitigation measures

Liquefaction-potential maps exist for some places but are not widely available. Areas with potentially liquefiable soils can be avoided as construction sites, particularly in regions that are known to experience frequent earthquakes. If high ground-water levels are involved, sites can be drained or other water-diversion efforts can be adopted.

Predictability

There is a high probability of recurrence in areas that have experienced previous problems. They are most prevalent in areas that have an extreme earthquake hazard as well as liquefiable soils. Lateral spreads are also associated with susceptible marine clays and are a common problem throughout the St. Lawrence Lowlands of eastern Canada. Figures 17, 18, 19 and 20 show a schematic and photos of lateral spread.

2.4 Flows

A flow is a spatially continuous movement in which the surfaces of shear are short-lived, closely spaced, and usually not preserved. The component velocities in the displacing mass of a flow resemble those in a viscous liquid. Often, there is a gradation of change from slides to flows, depending on the water content, mobility, and evolution of the movement.

2.4.1 Debris Flows

A debris flow is a form of rapid mass movement in which loose soil, rock and sometimes organic matter combine with water to form a slurry that flows downslope. They have been informally and inappropriately called "mudslides" due to the large quantity of fine material that may be present in the flow. Occasionally, as a rotational or translational slide gains velocity and the internal mass loses cohesion or gains water, it may evolve into a debris flow. Dry flows can sometimes occur in cohesionless sand (sand flows). Debris flows can be deadly as they can be extremely rapid and may occur without any warning.

Occurrence

Debris flows occur around the world and are prevalent in steep gullies and canyons; they can be intensified when occurring on slopes or in gullies that have been denuded of vegetation due to wildfires or forest logging. They are common in volcanic areas with weak soil.

Relative size/range

These types of flows can be thin and watery or thick with sediment and debris and are usually confined to the dimensions of the steep gullies that facilitate their downward movement. Generally the movement is relatively shallow and the runout is both long and narrow, sometimes extending for kilometers in steep terrain. The debris and mud usually terminate at the base of the slopes and create fanlike, triangular deposits called debris fans, which may also be unstable.

Velocity of travel

They can be rapid to extremely rapid (35 miles per hour or 56 km per hour) depending on consistency and slope angle.

Triggering mechanisms

Debris flows are commonly caused by intense surface-water flow, due to heavy precipitation or rapid snowmelt, that erodes and mobilizes loose soil or rock on steep slopes. Debris flows also commonly mobilize from other types of landslides that occur on steep slopes, are nearly saturated, and consist of a large proportion of silt- and sand-sized material.

Effects (direct/indirect)

Debris flows can be lethal because of their rapid onset, high speed of movement, and the fact that they can incorporate large boulders and other pieces of debris. They can move objects as large as houses in their downslope flow or can fill structures with a rapid accumulation of sediment

Fig. 21 Schematic illustration of a debris flow

Fig. 22 Heavy rains triggered a debris flow at Arapahoe Ski Area, Colorado (USA) July 28, 1999. Photo by Ed Harp, U.S. Geological Survey

and organic matter. They can affect the quality of water by depositing large amounts of silt and debris.

Mitigation measures

Flows usually cannot be prevented; thus, homes should not be built in or adjacent to steep-walled gullies that have a history of debris flows or are otherwise susceptible due to wildfires, soil type, or other related factors. New flows can be directed away from structures by means of deflection, debris-flow basins can be built to contain flow, and warning systems can be put in place in areas where it is known at what rainfall thresholds debris flows are triggered.

Fig. 23 A photo of the July 20, 2003 debri flow which occurred in Minamata City, Kyushu Island, Japan, resulting in 14 deaths and 15 houses destroyed (Sassa et al. 2004)

Evacuation, avoidance, and (or) relocation are the best methods to prevent injury and life loss.

Predictability

Maps of potential debris-flow hazards exist for some areas. Debris flows can be frequent in any area of steep slopes and heavy rainfall, either seasonally or intermittently, and especially in areas that have been recently burned or the vegetation removed by other means. Figures 21, 22, 23 and 24 show a schematic and images of debris flows.

2.4.2 Lahars (Volcanic Debris Flows)

The word "lahar" is an Indonesian term. Lahars are also known as volcanic mudflows. These are flows that originate on the slopes of volcanoes and are a type of debris flow. A lahar mobilizes the loose accumulations of tephra (the airborne solids erupted from the volcano) and related debris.

Occurrence

Lahars are found in nearly all volcanic areas of the world.

Relative size/range

Lahars can be hundreds of square kilometers or miles in area and can become larger as they gain speed and accumulate debris as they travel downslope; or, they can be small in volume and affect limited areas of the volcano and then dissipate downslope.

Fig. 24 Debris-flow damage to the city of Caraballeda, located at the base of the Cordillera de la Costan, on the north coast of Venezuela. In December 1999, this area was hit by Venezuela's worst natural disaster of the 20th century; several days of torrential rain triggered flows of mud, boulders, water, and trees that killed as many as 30,000 people (Photo by L.M. Smith, Waterways Experiment Station, U.S. Army Corps of Engineers)

Velocity of travel

Lahars can be very rapid (more than 35 miles per hour or 56 kilometres per hour) especially if they mix with a source of water such as melting snowfields or glaciers. If they are viscous and thick with debris and less water, the movement will be slow to moderately slow.

Triggering mechanism

Water is the primary triggering mechanism, and it can originate from crater lakes, condensation of erupted steam on volcano particles, or the melting of snow and ice at the top of high volcanoes. Some of the largest and most deadly lahars have originated from volcanic eruptions or venting which suddenly melts surrounding snow and ice and causes rapid liquefaction and flow down steep volcanic slopes at catastrophic speeds.

Effects (direct/indirect)

Effects can be extremely large and devastating, especially when triggered by a volcanic eruption and consequent rapid melting of any snow and ice—the flow can bury human settlements located on the volcano slopes. Some large flows can also dam rivers, causing flooding upstream. Subsequent breaching of these weakly cemented dams can cause catastrophic flooding downstream. This type of landslide often results in large numbers of human casualties.

Mitigation measures

No corrective measures are known that can be taken to prevent damage from lahars except for avoidance by not building or locating in their paths or on the slopes of volcanoes. Warning systems and subsequent evacuation work in some instances may save lives. However, warning

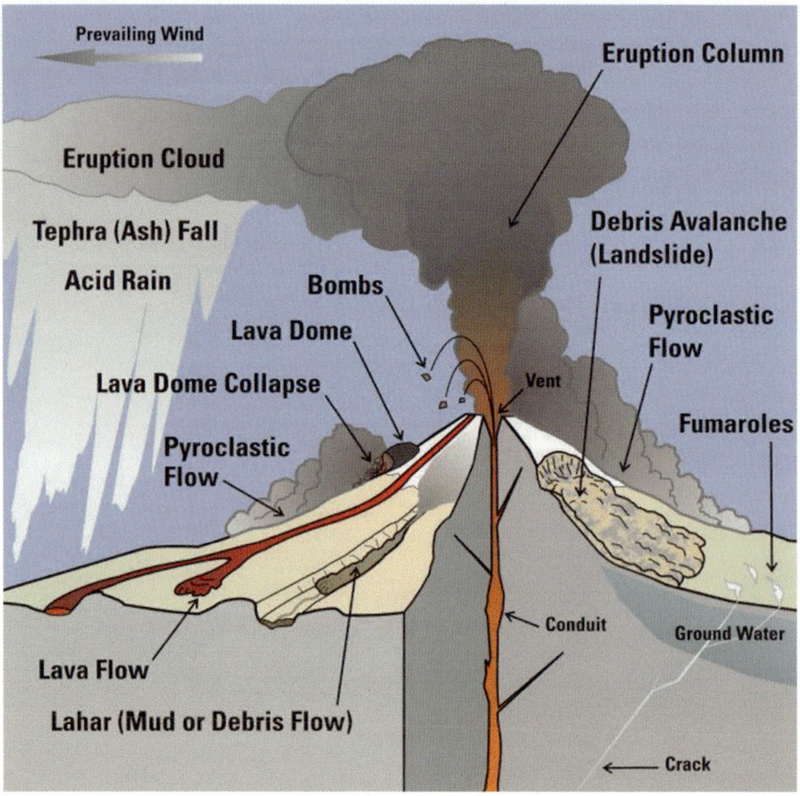

Fig. 25 Schematic illustration of volcano hazards, which includes lahars (lower *left* of illustration). Note that debris avalanches and pyroclastic flows can also occur. Lahars often flow in the same manner as debris and lava flows, and course along the same drainage paths. Illustration modified from U.S. Geological Survey

systems require active monitoring, and a reliable evacuation method is essential.

Predictability

Susceptibility maps based on past occurrences of lahars can be constructed, as well as runout estimations of potential flows. Such maps are not readily available for most hazardous areas. Figures 25, 26 and 27 show a schematic and photos of lahars.

2.4.3 Debris Avalanche

Debris avalanches are essentially large, extremely rapid, often open-slope flows formed when an unstable slope collapses and the resulting fragmented debris is rapidly transported away from the slope. In some cases, snow and ice will contribute to the movement if sufficient water is present, and the flow may become a debris flow and (or) a lahar.

Occurrence

Debris avalanches occur worldwide in steep terrain environments. They are also common on very steep volcanoes where they may follow drainage courses.

Relative size/range

Some large avalanches have been known to transport material blocks as large as 3 km in size, several kilometers from their source.

Fig. 26 On November 13, 1985, the Nevado del Ruiz stratovolcano in Tolima, Colombia, erupted, after 69 years of dormancy. As pyroclastic flows erupted from the volcano's crater, they melted the mountain's glaciers, sending four enormous lahars down its slopes at 50 kilometers per hour (30 miles per hour). The lahars picked up speed in gullies and coursed into the six major rivers at the base of the volcano; they engulfed the town of Armero, killing more than 20,000 of its almost 29,000 inhabitants. Photo by Darrell Herd, U.S. Geological Survey

Velocity of travel

Travel is rapid to extremely rapid; such debris avalanches can travel close to 100 m/s.

Triggering mechanism

In general, the two types of debris avalanches are those that are "cold" and those that are "hot." A cold debris avalanche usually results from a slope becoming unstable, such as during collapse of weathered slopes in steep terrain or through the disintegration of bedrock during a slide-type landslide as it moves downslope at high velocity. At that point, the mass can then transform into a debris avalanche. A hot debris avalanche is one that results from volcanic activity including volcanic earthquakes or the injection of magma, which causes slope instability.

Effects (direct/indirect)

Debris avalanches may travel several kilometers before stopping, or they may transform into more water-rich lahars or debris flows that travel many tens of kilometers farther downstream. Such failures may inundate towns and villages and impair stream quality. They move very fast and thus may prove deadly because there is little chance for warning and response.

Fig. 27 Photo of a lahar, caused by the 1982 eruption of Mount St. Helens, Washington, USA. Photo by Tom Casadevall, U.S. Geological Survey

Fig. 28 Schematic of a Debris Avalanche

Corrective measures/mitigation

Avoidance of construction in valleys on volcanoes or steep mountain slopes and real-time warning systems may lessen damages. However, warning systems may prove difficult due to the speed at which debris avalanches occur—there may not be enough time after the initiation of the event for people to evacuate. Debris avalanches cannot be stopped or prevented by engineering means because the associated triggering mechanisms are not preventable.

Fig. 29 A debris avalanche which buried a village in Guinsaugon, Southern Leyte, Philippines, in February, 2006 (Photo by University of Tokyo Geotechnical Team)

Predictability

If evidence of prior debris avalanches exists in an area, and if such evidence can be dated, a probabilistic recurrence period might be established. During volcanic eruptions, chances are greater for a debris avalanche to occur, so appropriate cautionary actions could be adopted. Figures 28, 29 and 30 show a schematic and photos of debris avalanches.

2.4.4 Earthflow

Earthflows can occur on gentle to moderate slopes, (see Fig. 33) generally in fine-grained soil, commonly clay or silt, but also in very weathered, clay-bearing bedrock. The mass in an earthflow moves as a plastic or viscous flow with strong internal deformation. Susceptible marine clay (quick clay) when disturbed is very vulnerable and may lose all shear strength with a change in its natural moisture content and suddenly liquefy, potentially destroying large areas and flowing for several kilometers. Size commonly increases through headscarp retrogression. Slides or lateral spreads may also evolve downslope into earthflows. Earthflows can range from very slow (creep) to rapid and catastrophic. Very slow flows and specialized forms of earthflow restricted to northern permafrost environments are discussed elsewhere.

Occurrence

Earthflows occur worldwide in regions underlain by fine-grained soil or very weathered bedrock. Catastrophic rapid earthflows are common in the susceptible marine clays of the St. Lawrence Lowlands of North America, coastal Alaska and British Columbia, and in Scandinavia.

Relative (size/range)

Flows can range from small events of 100 m^2 in size to large events encompassing several square kilometers in area. Earthflows in

Fig. 30 Oblique aerial view of the Hattian Bala, Pakistan debris avalanche showing the avalanche scarp, travel path and debris. The debris avalanche was triggered by the October 8, 2005 Pakistan Earthquake. As a result, the Karli stream drainage was blocked, causing the impoundment of a lake (seen in *lower left* area of photo). Photograph by USAID

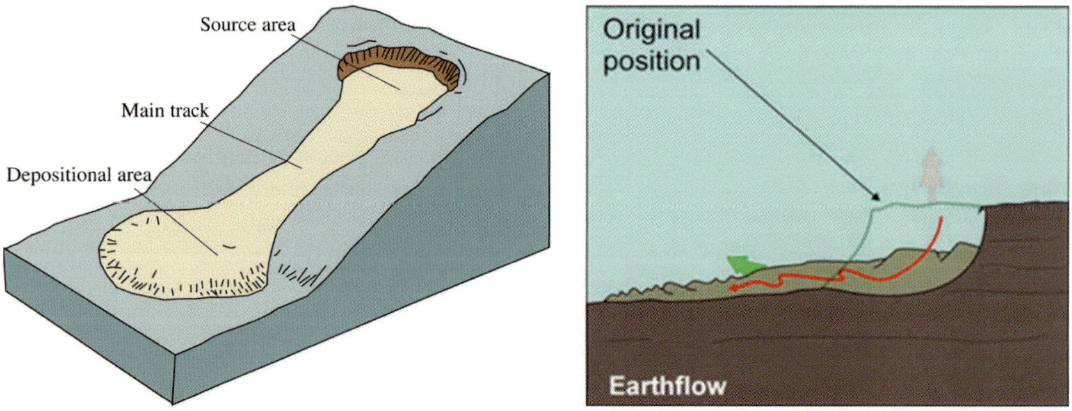

Fig. 31 Schematic illustrations of an Earthflow. The photo at *left* shows a schematic of a steep slope earthflow (see Fig. 35 for a photo), and the photo at *right* shows a cross-section/schematic of an earthflow on a shallow slope (see Fig. 34 for a photo of a shallow-slope earthflow)

Fig. 32 Photo the Slumgullion Landslide, Colorado, USA. This earth flow occurred approximately 700 years ago, dammed the Lake Fork of the Gunnison River, which caused flooding the valley, forming Lake Cristobal. Photo by Jeff Coe, U.S. Geological Survey

susceptible marine clays may runout for several kilometers. Depth of the failure ranges from shallow to many tens of meters.

Velocity of travel

They range from slow to very rapid.

Triggering mechanisms

Triggers include saturation of soil due to prolonged or intense rainfall or snowmelt, sudden lowering of adjacent water surfaces causing rapid drawdown of the ground-water table, stream erosion at the bottom of a slope, excavation and construction activities, excessive loading on a slope, earthquakes, or human-induced vibration.

Effects (direct/indirect)

Rapid, retrogressive earthflows in susceptible marine clay may devastate large areas of flat land lying above the slope and also may runout for considerable distances, potentially resulting in human fatalities, destruction of buildings and linear infrastructure, and damming of rivers with resultant flooding upstream and water siltation problems downstream. Slower earthflows may damage properties and sever linear infrastructure.

Corrective measures/mitigation

Improved drainage is an important corrective measure, as is grading of slopes and protecting the base of the slope from erosion or excavation. Shear strength of clay can be measured, and

Fig. 33 On April 27, 1993, a landslide severely damaged three homes near the town of LaFayette in the Tully Valley, 24 km (15 miles) south of Syracuse, New York. This landslide can be classified as a rapid slump-earth flow. Material involved consists of red lake clay deposits of glacial origin, covered by glacial till and colluvium of varying thickness. Photo by Gerry Wieczorek, U.S. Geological Survey

potential pressure can be monitored in suspect slopes. However, the best mitigation is to avoid development activities near such slopes.

Predictability

Evidence of past earthflows is the best indication of vulnerability. Distribution of clay likely to liquefy can in some cases be mapped and has been mapped in many parts of eastern North America. Cracks opening near the top of the slope may indicate potential failure. Figures 31, 32, 33, 34 and 35 show schematics and photos of an earthflow.

2.4.5 Creep

Creep is the informal name for a slow earthflow and consists of the imperceptibly slow, steady downward movement of slope-forming soil or rock. Movement is caused by internal shear stress sufficient to cause deformation but insufficient to cause failure. Generally, the three types of creep are: (1) seasonal, where movement is within the depth of soil affected by seasonal changes in soil moisture and temperature; (2) continuous, where shear stress continuously exceeds the strength of the material; and (3) progressive, where slopes are reaching the point of failure for other types of mass movements.

Occurrence

Creep is widespread around the world and is probably the most common type of landslide, often preceding more rapid and damaging types of landslides. Solifluction, a specialized form of creep common to permafrost environments, occurs in the upper layer of ice-rich, fine-grained soils during the annual thaw of this layer.

Fig. 34 The 1993 Lemieux Landslide was a rapid earth flow in sensitive marine clay near Ottawa, Canada. The headscarp retrogressed 680 m into level ground above the riverbank. About 2.8 million tons of clay and silt liquefied and flowed into the South Nation River valley, damming the river (Photo by G.R. Brooks, Geological Survey of Canada)

Relative size/range

Creep can be very regional in nature (tens of square kilometers) or simply confined to small areas. It is difficult to discern the boundaries of creep since the event itself is so slow and surface features representing perceptible deformation may be lacking.

Velocity of travel

Velocity is very slow to extremely slow. Usually less than 1 m (0.3 ft) per decade.

Triggering mechanism

For seasonal creep, rainfall and snowmelt are typical triggers, whereas for other types of creep there could be numerous causes, such as chemical or physical weathering, leaking pipes, poor drainage, destabilizing types of construction, and so on.

Effects

Because it is hard to detect in certain places because of the slowness of movement, creep is sometimes not recognized when assessing the suitability of a building site. Creep can slowly pull apart pipelines, buildings, highways, fences, and so forth, and can lead to more drastic ground failures that are more destructive and faster moving.

Corrective measures/mitigation

The most common mitigation for creep is to ensure proper drainage of water, especially for

Fig. 35 The 1995 La Conchita, California (USA) earthflow occurred after heavy rains. No one was killed or injured, as it moved relatively slowly. This landslide later reactivated in 2005, and moving much more rapidly, killed 10 people. Photo by Mark Reid, U.S. Geological Survey

Fig. 36 Schematic illustration of a slow-moving flow, called Creep

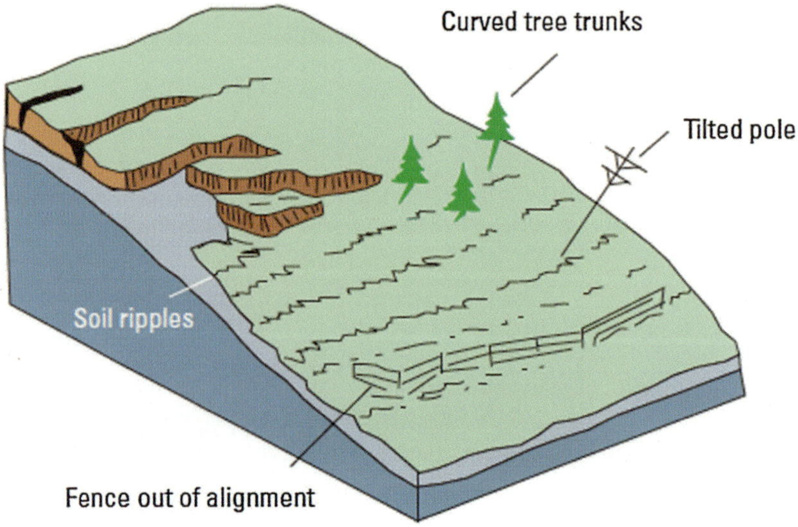

Fig. 37 An Illustration of creep and it's effect on trees and forests. Photo by NOAA/NGDC/Colorado Geological Survey, Public Domain http://www.ngdc.noaa.gov/hazardimages/picture/show/1551

the seasonal type of creep. Slope modification such as flattening or removing all or part of the landslide mass, can be attempted, as well as the construction of retaining walls.

Predictability

These are indicated by curved tree trunks, bent fences and (or) retaining walls, tilted poles or fences, and small soil ripples or ridges on the surface. Rates of creep can be measured by inclinometers installed in boreholes or by detailed surface measurements. Figures 36, 37, 38 and 39 show a schematic and photos of creep.

2.4.6 Flows in Permafrost

Failures in permafrost conditions involve the movement of fine-grained, previously ice-rich

Fig. 38 A photo showing the surface expression of Creep in a region in the Buzau subcarpathians in Romania. The slope is composed of Neogene clays and marls, and entirely covered with colluvial deposits. Photo by Dr. Dan Balteanu, Romanian Academy

soil and can occur on gentle slopes. Seasonal thaw of the upper meter of frozen ground melts ground ice and results in oversaturation of the soil, which in turn loses shear strength and initiates flows. Solifluction, a form of cold environment creep, involves very slow deformation of the surface and forms shallow lobes elongated downslope. Active layer detachments, also known as skinflows, involve rapid flow of a shallow layer of saturated soil and vegetation, forming long, narrow flows moving on the surface but over the underlying

Fig. 39 This is an earthflow on Mission Pass in the California (USA) coastal ranges. The lateral lines on the hillside show creep. Public domain: NOAA/NGDC, B. Bradley, University of Colorado. http://www.ngdc.noaa.gov/hazardimages/picture/show/1537

permanently frozen soil. This type of movement may expose buried ice lenses, which when thawed may develop into retrogressive thaw flows or possibly debris flows. Retrogressive thaw flows are larger features with a bimodal shape of a steep headwall and low-angle tongue of saturated soil. This type of feature will continue to expand through headscarp retrogression until displaced vegetation buries and insulates the ice-rich scarp.

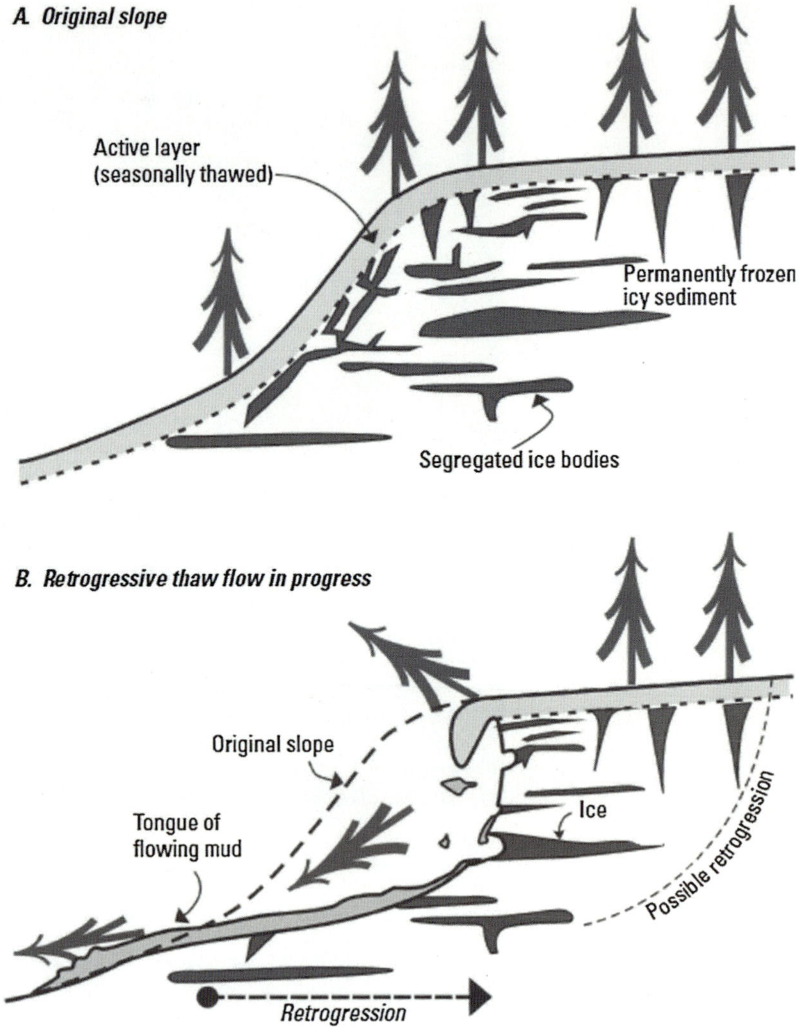

Fig. 40 Schematic illustration of a retrograde flow in permafrost

Occurrence

Flows are common in ice-rich permafrost soils in northern latitudes and high altitudes (cold environments).

Relative size/range

Flows are generally small but can increase in size through headscarp retrogression. They may evolve into a larger debris flow.7

Velocity of travel

Velocity is very slow (solifluction); slow (retrogressive thaw flow); rapid (active layer detachment).

Triggering mechanisms

May be triggered by above-average summer temperatures, frost wedges, wildfire, and anthropogenic disturbances to insulating peat

Fig. 41 Lake with a large, active retrogressive thaw slump. Image courtesy of Steve Kokelj Government of the Northwest Territories, Canada

layer. Such landslides are particularly likely in warming climates.

Effects (direct/indirect)

Damage to pipelines and roads and other structures can be severe.

Corrective measures/mitigation

Infrastructure designs that have minimal effect on the surface peat layer or temperature of the active layer and avoidance, when possible, of ice-rich soils when planning roads and other infrastructure, can reduce risk. Ice content of the upper soil can be readily tested.

Predictability

If ice-rich soil thaws, it will flow. In some areas, ice content has been mapped; in other areas, ice content can be estimated on the basis of specific mapped units shown on surficial geology maps. Figures 40, 41, 42 and 43 show a schematic and photos of permafrost-related flow.

Fig. 42 In 2004 a large chunk of earth collapsed along the upper Selawik River in Alaska as a result of permafrost failure, spewing mud and silt into the water and pushing the river against its far bank. Sediment from the collapse turned the clear upper Selawik River into an opaque, turbid stream. Climate change may make these more common as permafrost begins to melt, with severe impacts on water quality and wildlife

2.5 Complex

These are landslides that feature components of two or more of the basic types of landslides and can occur either simultaneously or at different times during the onset of slope failure. Figures 44 and 45 show a schematic and photo of a complex landslide.

The landslide involved a complex sequence of events–including rotation, translation and flow mechanisms, including transformation into a debris flow at the terminus of the landslide–as such, it can be referred to as a debris-avalanche flow (Wartman et al. 2016)

3 Conclusion

The development of the typing of landslides is ongoing and evolving, as new information is obtained, and some past landslides are re-evaluated. Classifications of natural phenomena are always evolving and there are differing professional opinions on the process. One recent effort is a publication by Hungr et al. (2014) which is published in its entirety in the Teaching Tools. The reader is encouraged to read this publication for both new information, and as an example of how the typing of landslides is

Fig. 43 Photograph of a retrogressive thaw flow, in the Northwest Territories, Canada. Wildfire has likely contributed to the size of the flow by means of damage to an insulating moss layer, resulting in the thickening of the active layer, which is thawing permafrost. Photo by Marten Geertsema, Ministry of Forests, British Columbia, Canada

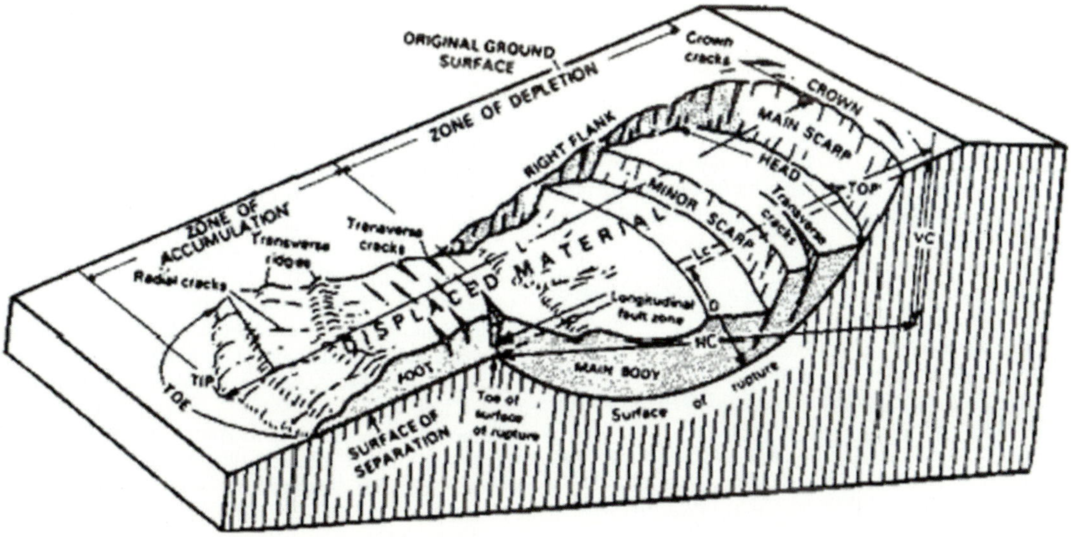

Fig. 44 Schematic illustration of a Complex landslide

Fig. 45 This photo shows a recent example of a complex landslide. This landslide occurred in northwest Washington (USA) on March 22, 2014. Landslide debris covered about 40 homes and other structures as well as nearly a mile of a highway, State Route 530. It also caused 43 fatalities in the community of Steelhead Haven near Oso, Washington

approached, as well as how the science of landslide typing continues to improve.

References

Cruden DM (1991) A Simple definition of a landslide. Bull Int Assoc Eng Geol 43:27–29

Cruden DM, Varnes DJ (1996) Landslide types and processes. In: Turner AK, Schuster RL (eds) Landslides investigation and mitigation, Chapter 3, Special Report 247. Transportation research board, US National Research Council, Washington, DC, pp 36–75

Dang K, Sassa K, Fukuoka H et al (2016) Mechanism of two rapid and long-runout landslides in the 16 April 2016 Kumamoto earthquake using a ring-shear apparatus and computer simulation (LS-RAPID). Landslides 13(6):1525–1534. doi:10.1007/s10346-016-0748-9

Highland LM, Bobrowsky P (2008) The landslide handbook: a guide to understanding landslides. U.S. Geological Survey Circular 1325, 129 p. http://pubs.usgs.gov/circ/1325/pdf/C1325_508.pdf

Hungr O, Leroueil S, Picarelli L (2014) The varnes classification of landslide types, an update. Landslides 11:167–194. doi:10.1007/s10346-013-0436-y

Miyagi T, Yamashina S, Esaka F, Abe S (2010) Massive landslide triggered by 2008 Iwate-Miyagi inland earthquake in the Aratozawa Dam area, Tohoku, Japan. Landslides 8:99–108. doi:10.1007/s10346-010-0226-8(Springer, June 2010)

Sassa K, Fukuoka H, Wang G, Ishikawa N (2004) Undrained dynamic-loading ring-shear apparatus and its application to landslide dynamics. Landslides. 1(1):7–19

Sassa K (2005) Landslide disasters triggered by the 2004 Mid-Niigata Prefecture earthquake in Japan. Landslides 2:135–142. doi:10.1007/s10346-0054-4 (Springer, May 2005)

Varnes DJ (1978) Slope movement types and processes. In: Schuster RL, Krizek RJ (eds) Landslides, analysis and control, special report 176: transportation research board. National Academy of Sciences, Washington, DC., pp 11–33

Wartman J, Montgomery DR, Anderson SA, Keaton JR, Benoit J, dela Chapelle J (2016) The 22 March 2014 Oso landslide, Washington, USA. Geomorphology 253:275–288. doi:10.1016/j.geomorph.2015.10.022 (15 Jan 2016)

Part I

Mapping and Site Prediction

TXT-tool 1.081-2.1 Landslide Mapping Through the Interpretation of Aerial Photographs

Toyohiko Miyagi

Abstract

Topography created by landslides is widely distributed. Landslide risk evaluation is very important for mitigating or preventing these natural disasters, and finding facts can be very difficult. We here describe the techniques for aerial photo interpretation and explain the results. Sometimes, we "can't see the forest for the trees." This manual will clearly and practically explain the following: (1) why stereo pair interpretation is necessary; (2) how to perform stereo pair interpretation with aerial photos etc. to determine the landslide area; (3) what can be learned using this method of interpretation; (4) how landslide motion characteristics can be ascertained once interpretation of landslide micro-topography is performed; (5) once this is ascertained, how to rationalize landslide surveys and countermeasures.

Keywords

Landslide dynamics · Aerial photo interpretation · Micro topography Risk · Reactivation

Contents

1 The Significance of Mapping Based on Photo Interpretation 41
2 Why Is Stereo Pair Interpretation Necessary? 42
3 Difficulties with Direct Sensing of Images and Their Solutions 44
4 How Is Stereo Pair Interpretation of Aerial Photos for Determining the Landslide Area?. 44
5 What Can Be Learned Through Aerial Photo Interpretation? 45
6 Examples of Landslide Topography and Distribution Maps 45
References 51

1 The Significance of Mapping Based on Photo Interpretation

First we wish to discuss the necessity of ascertaining landslide topography using stereo pair interpretation, such as aerial photo interpretation.

T. Miyagi (✉)
Faculty of Arts, Department of Regional Management, Tohoku-Gakuin University, 2-1-1, Tenjinsawa, Izumi, Sendai 981-3193, Japan
e-mail: miyagi@mail.tohoku-gakuin.ac.jp

K. Sassa et al. (eds.), *Landslide Dynamics: ISDR-ICL Landslide Interactive Teaching Tools*, https://doi.org/10.1007/978-3-319-57774-6_2

Sometimes, it is said, we "can't see the forest for the trees". This proverb is applicable to landslides. For example, cut slopes frequently collapse when building roads. If one supervises in the field, the engineer is sure to see such accidents. In order to find countermeasures, we must first consider why such surface landslides occur.

By investigating the causes of surface landslides, we can also investigate the dynamic characteristics of rock. If it rained heavily before the surface landslide, we should also take an account into consideration. By resolving issues of material characteristics related to rock deformation, an understanding of dynamic characteristics can lead to solutions to problems. Taking rainfall into consideration can lead to a lower evaluation of a site's safety as a rise in pore water pressure can trigger landslides. Of course, such perspectives are necessary in responding to surface landslides on site. However, from the perspective of landslide topography mapping, it should be emphasized that it is necessary to rank the reasons why the surface landslide occurred where it dose on the road slope.

In many cases it is difficult to accurately ascertain the overall form of massive landslides from field observation alone. It is impossible to know if the surface landslide occurring before one's eyes the toe of a large-scale landslide, spreading out across a huge slope behind it, or a small phenomenon occurring only in that location. These small surface landslides also sometimes occur due to fissures caused by large landslides.

This manual will clearly and practically explain the following: (1) why stereo pair interpretation is necessary; (2) how to perform stereo pair interpretation with aerial photos etc. to determine the landslide area; (3) what can be learned using this method of interpretation; (4) how landslide motion characteristics can be ascertained by interpretation of landslide micro topography; (5) After it is ascertained, how to rationalize landslide surveys and countermeasures.

For those with no experience in stereo pair interpretation, the relevance of these tasks may be difficult to do. The ability to observe and analyze topography can be developed through adequate methods, but the study of civil engineering alone imparts only a mechanical understanding and material properties; normally it does not prompt one to question why failures occur where they do. This is analogous to bandaging a cut on an arm, without realizing that the arm is actually broken. However, although mapping of landslide topography can be performed by interpretation of aerial photos, the research on using it for risk evaluation is still very much work in progress. In other words, this method is not yet in some respects. Thus, the development of this risk evaluation method and its application suitable to regional characteristics are also goals of this research. Below, the significance and weaknesses of aerial photo interpretation will be listed (Fig. 1).

2 Why Is Stereo Pair Interpretation Necessary?

It is necessary to ascertain the actual shape of the landslide breakdown region. Landslides occur when the landslide block slides over a portion of slope bounded by a contact surface, called the slip surface, while this landslide block remains on the slope. This landslide block is referred to as the landslide body. Its size can vary from a few meters on the extremely small size, up to several kilometers.

Much research and many reports exist about the evaluation of landslide hazard potential. In general, such research combines data from detailed geological maps, land gradient, and estimated weight of rainfall, and focuses on areas where much of the information overlaps, to render evaluations such as "landslides probably have a high chance of occurring here." Needless to say, such analysis and evaluations are important. However, each landslide is unique. Put another way, the largest characteristic of the slope fluctuations referred to as landslides is that the area of deformation caused by the landslide is clearly distinguished from the surrounding, immobile area. This being the case, if we are

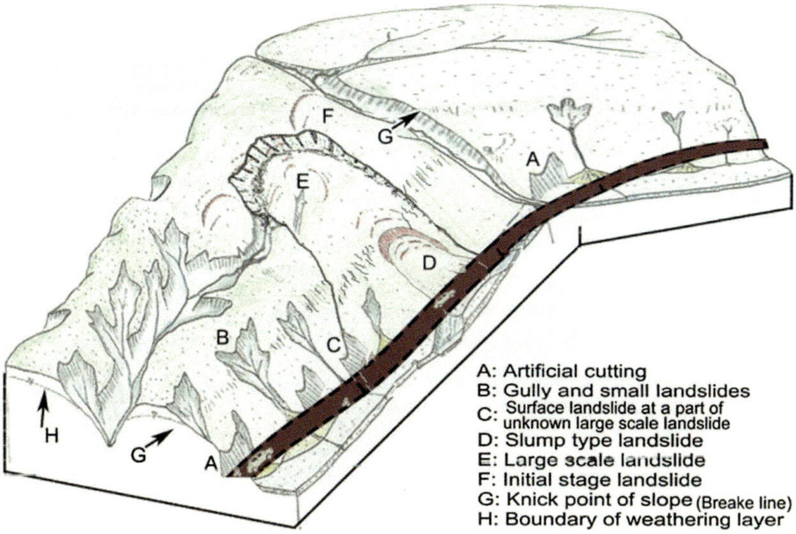

Fig. 1 Schematic illustration of the relationship between the trees (the surface landslide before one's eyes) and the forest (large-scale landslide, incorporating the surface landslides)

unable to determine the extent of the area deformed by the landslide, disaster countermeasures, and judgment of the probability of future disasters, will be impossible. Another characteristic of landslides is that even after the landslide has triggered fluctuations, the landslide body, causing failures within the landslide area, still remains, and has the potential to cause further failures. By ascertaining the actual shape of the landslide body, a host of disaster prevention measures, such as evaluations, countermeasures, and monitoring, can undertakes. Through photo interpretation, a distribution map of the landslide topography can be created, allowing the main scarp, which constitutes the actual shape of the generation area, to be realistically interpreted and illustrated. This in turn enables making contours of the landslide body, i.e. the area that broke free from the main scarp and moved, to be realistically illustrated, thus allowing to the determine the landslide topography.

Whole landslide topography cannot be ascertained from field investigation alone. The scale of landslide phenomena range from a few meters to several kilometers, and have drastic relief and depth. Thus, the visible phenomenon conceals the underlying phenomena. It is also usual for the cases in which we are unable to accurately ascertain the overall picture of the landslide fluctuation zone in the field, and for the positioning the phenomena which we are observing within the landslide overall. Before heading to the field, it is advisable to get a whole picture of the landslide from the air, and objectively ascertain the characteristics of its topography using vertical, somehow exaggerated, and relief images. Ascertaining the landslide's actual shape using photographic interpretation, and considering what part of the whole landslide the site constitutes, will lead to a rational understanding of it.

Ascertaining the micro topography of the landslide generation area: lead to performing risk evaluation of landslide topography using AHP (Analytic Hierarchy Process (Saaty 1980)), and to distinguishing from the non-landslide area. By ascertaining micro characteristics of the landslide area, indicators of the mode of movement, extent of crushing, and localized atypical deformation, can be discerned. Understanding the landslide's minor structures requires the use of high resolution data such as photos, and precise interpretation by an experienced technician.

3 Difficulties with Direct Sensing of Images and Their Solutions

Difficulty in confirming the landslide area: In surfaces covered with vegetation, such as tall trees, it will be difficult to confirm the area of landslide topography and micro-topography. If vegetation is scant or devastated, interpretation of the topography will be simple. On the other hand, if the surface is densely covered with a forest of tall trees, such as a tropical rain forest, interpretation will be difficult. This point is likely to be reviewed during the current Vietnam project.

Difficulty in understanding topography due to artificial land deformation: Landslide-prone areas tend to have abundant water, and often minor flatlands. Since such land is often used for habitation and agriculture, it tends to be subject to artificial land deformation. Artificial land deformation often produces topography similar to landslide micro topography. Therefore, caution is required (Fig. 2).

Difficulty in understanding landslide topography due to the motion characteristics of the landslide itself: depending on the characteristics of the rock and style of failure, a clear range of movement and main scarp might not be formed. When landslides occur particularly in schist regions of Japan, the entire slope often moves, therefore no clear landslide topography is established.

4 How Is Stereo Pair Interpretation of Aerial Photos for Determining the Landslide Area?

Taking aerial photographs: A camera is brought on board an aircraft, and consecutive elevated vertical photographs are taken. About 60% of the consecutive photographs should overlap. These

Fig. 2 Aerial photo interpretation performed using aerial photographs with a stereoscope of mirror type, using simple stereoscope, topography map, geological map, writing materials etc. The series of aerial photographs should be arranged as shown above, and examined from directly above. The gap between photos differs depending on the person; appropriate adjusting the image by moving the photos while a time leads to clearly seeing

overlapping portions will be used for suitable views. In a pair of two images, the images should be taken from slightly different perspectives; they should not be exactly the same. The difference in viewing angle between the two photos is what enables stereoscopic interpretation. The aerial photographs can be taken on roll negative film of 20 cm or 24 cm width, or with a digital camera. Due to the nature of cameras, which use central projection, because the image will be distorted around the edge. However, despite this distortion, interpretation can be performed adequately.

On the other hand, ortho images are central projection images to which processing has been applied to remove the distortions, recreating them as orthogonal projections. Consequently, they do not appear stereoscopic, even if two side-by-side orthogonal projection photographs are arranged and observed.

5 What Can Be Learned Through Aerial Photo Interpretation?

(1) A whole picture of the land can be observed. The images capture the whole land conditions at the time the aerial photograph was taken. You can observe land utility, the survey route, and of course the object of interpretation.

(2) Landform classification, land cover classification, and vegetation classification can be determined. This will be explained later, but land comprises a combination of the surface and the break line of the slope profile. Such topography is most easily observed in aerial photo interpretation.

(3) The area of land deformed by the landslide can be clearly observed. The main scarp as well as the landslide body can be ascertained by clearly determining the contours of the landslide topography. If this is unclear, it suggests that the landslide is inactive or suspended.

(4) Interpretation enables topography mapping. A landslide distribution map can be completed by transcribing the interpreted information into a topography map.
Moreover, you can estimate

(5) how landslide motion characteristics can be ascertained by that interpretation of landslide micro-topography is performed, and

(6) how to rationalize landslide surveys and countermeasures.

6 Examples of Landslide Topography and Distribution Maps

See Figs. 3, 4, 5, 6, 7, 8 and 9.

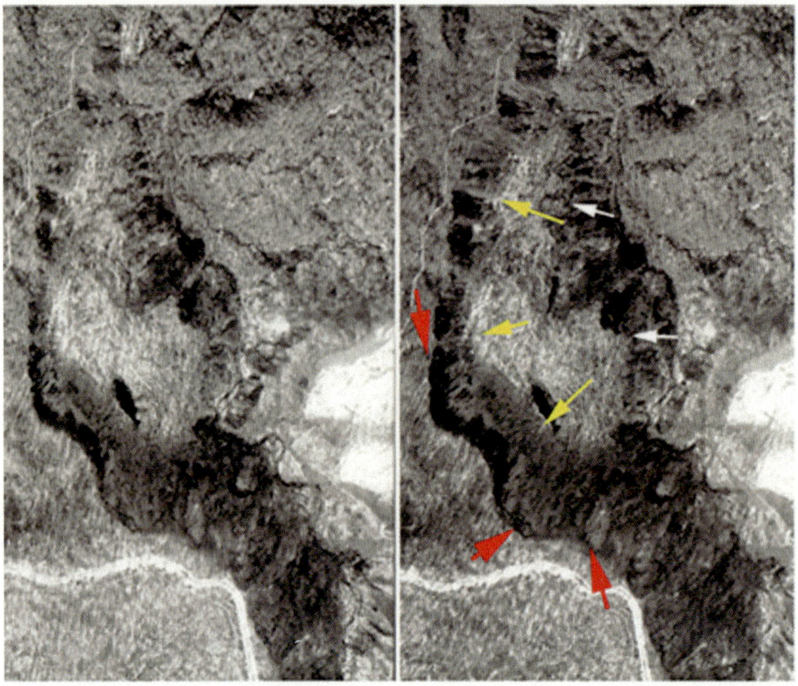

Fig. 3 An example of the appearance of a large-scale landslide topography by stereo pair aerial photographs. This is a typical large-scale block slump landslide (300 by 600 m and the AHP score: 85). *Red* main scarp; *yellow* boundary area of landslide body between main scarp (gentle slope created by talus accumulation is almost invisible), tip of landslide body (numerous small-scale surface landslides occurring) (Miyagi et al. 2004)

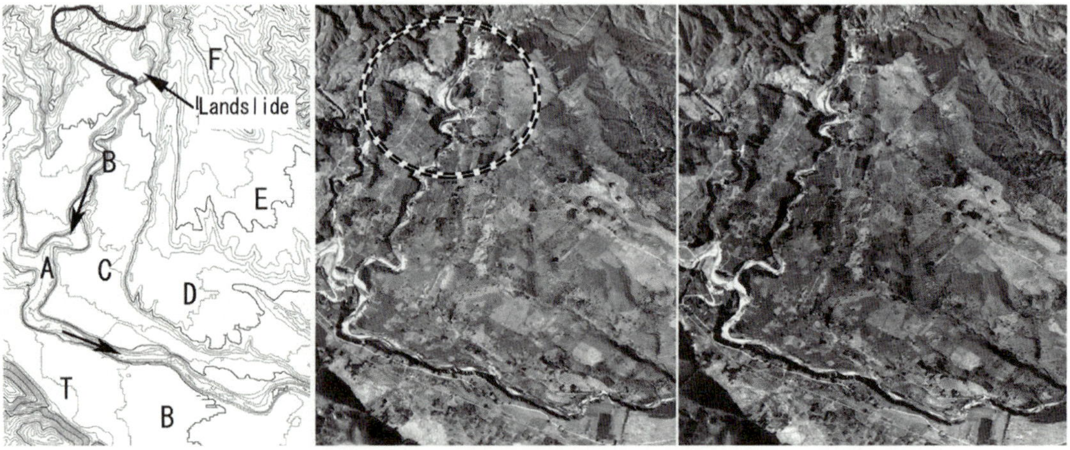

Fig. 4 Example of stereo pair image and the topo map (*left*). *A* river, *B–F* river terraces, *T* alluvial cone. A landslide is visible at the *upper part*. The landslide cut the river terraces (Miyagi et al. 2004)

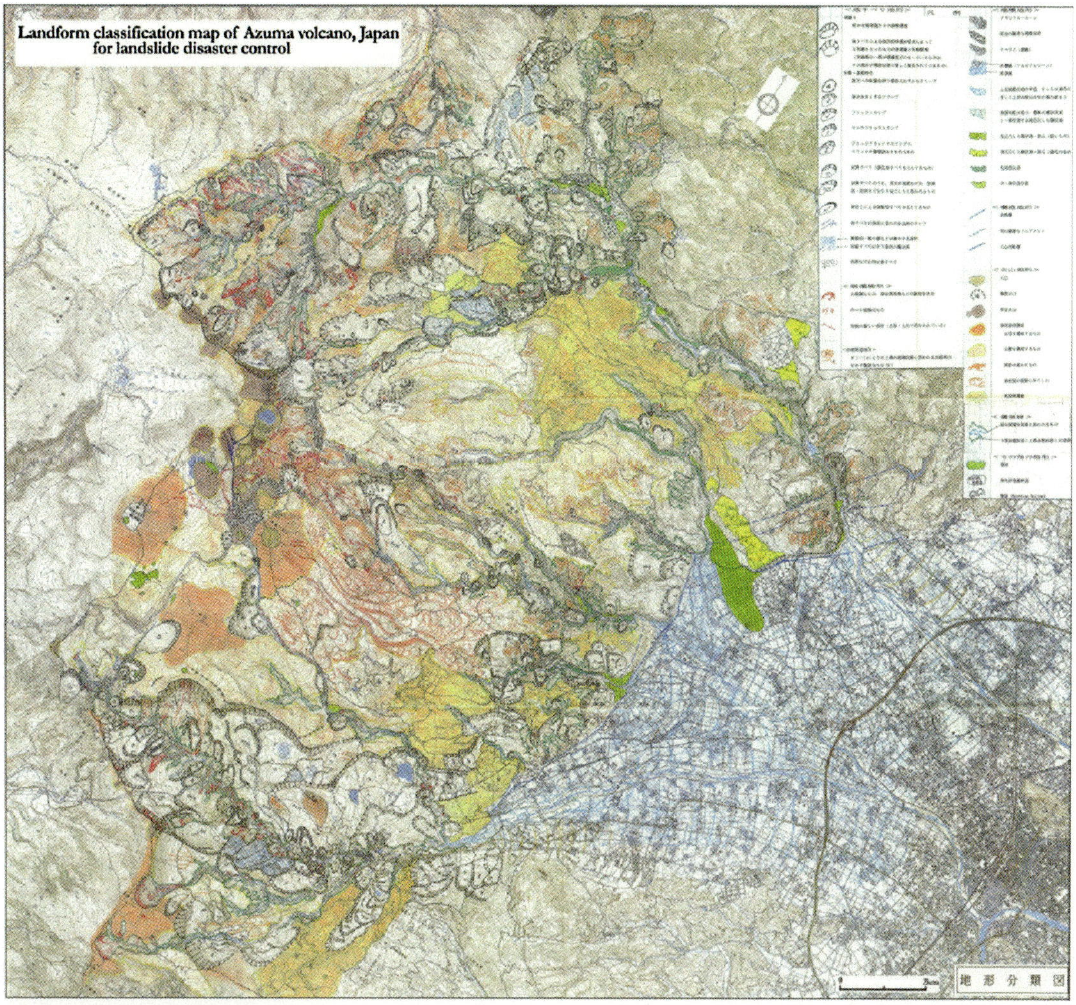

Fig. 5 Azuma volcanic region landform classification map (*top*) rendered as a 1:25,000 scale topography map. It depicts the landslide topography (the type of movement is also classified), volcanic slope, accumulation topography of the mountain foot, and a convex break line (corrosion line), allowing estimation of the landslide topography distribution, as well as the relationship between the landslide area and volcanic slope, and potential sediment yield per watershed. Aerial photograph: 1:20,000 scale monochrome adhesive photo (drawn by T. Miyagi)

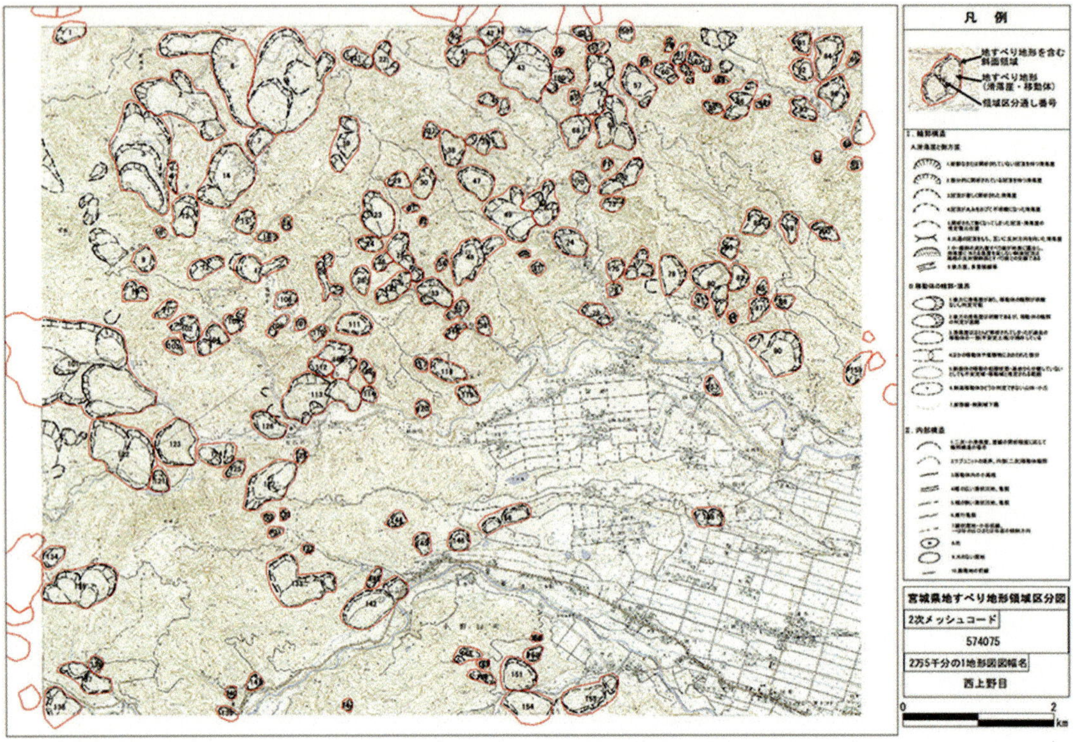

Fig. 6 Landslide topography distribution map (created at the request of the Japan Landslide Society). Depicts landslide topography derived from aerial photo interpretation in a 1:25,000 scale topography map. Here, maps are annotated as per the legend on the right. The minimum landslide fluctuation range has been determined, and the main scarp is indicated by *heavy lines* and *hatch marks*. The actual shape of the landslide body is indicated with *thin lines*. If large-scale internal structures, on a scale that permits interpretation, and if the landslide body are confirmed, these too are noted. The main scarp and its surroundings, the height difference at the top part of the slope, and fissures, probably suggest instability. They are noted with extreme caution. The aerial photographs used were 1:20,000 scale monochrome adhesive photos from the Nishikaminome region of northwestern Miyagi Prefecture. Each landslide topography map is assigned an ID number linked to relevant information (Japan Landslide Society and Miyagi Prefectural Office 2002). The creation of this distribution data began in the 1980s, and comprises a 1:50,000 scale map of recent information covering all of Japan (NIED 2015)

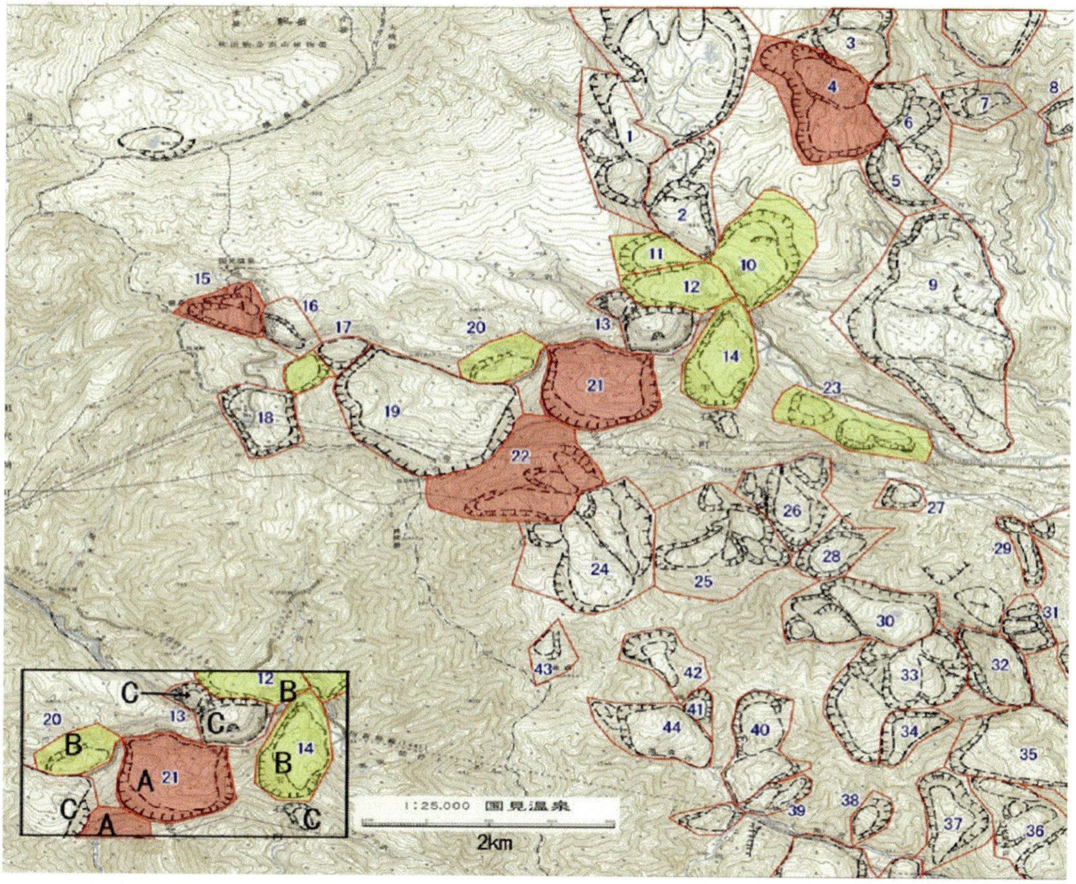

Fig. 7 Landslide topography distribution and risk level chart (stereo pair interpretation) was performed on the distribution map from Fig. 5 and the images recorded in the interpretation evaluation chart; landslide risk was evaluated using AHP. Hamasaki provides an explanation of this. The evaluation points obtained through interpretation do not constitute a rigorous scale (e.g., there is not a significant difference between a score of 60 and 65), and are divided into three general ranks. *A-rank areas*, which have an AHP score of 80 or higher, if synthesized with AHP evaluations in actual disaster countermeasure case studies, can be judged to be essentially active. *B-rank areas* have a score of 60 or higher; caution is necessary when surveying them. *C-rank areas* have been judged to be essentially unmoving, but caution is necessary as the landslide body itself has sustained damage from the landslide, and actions such as tearing might lower the landslide body's stability. Evaluation of this is done using aerial photo interpretation alone; use of estimated values is legitimate. It is very useful as a draft map for various plans, and is the basis for assigning priority to the implementation of field surveys (Miyagi et al. 2004)

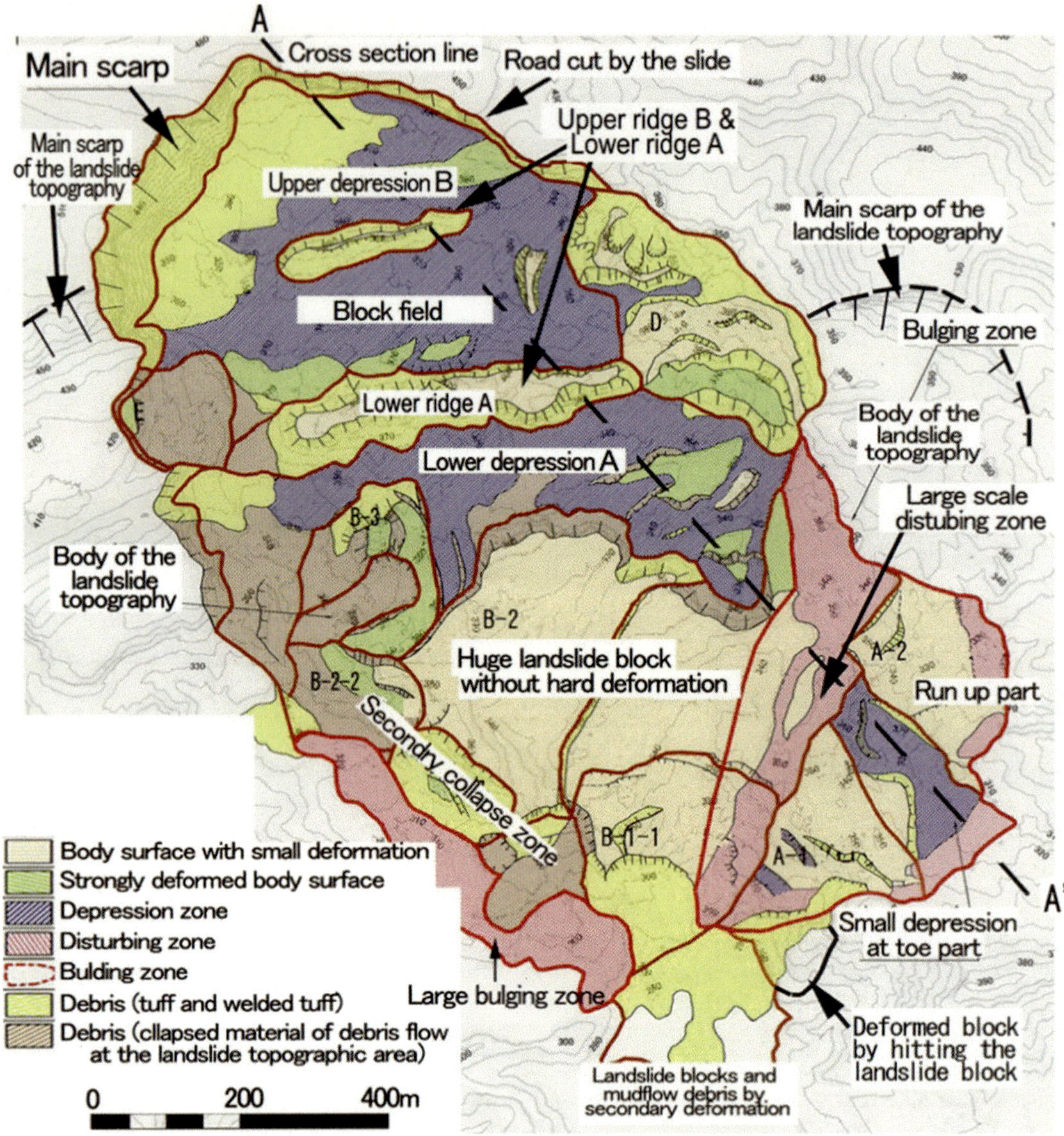

Fig. 8 Micro-topography landform classification map showing the internal state of the landslide topography. Land deformation caused by landslide action creates new micro-topographies. These or their adjacent topographies move together, giving rise to the possibility of new fluctuations. Aerial photograph scale: 1:8000, real color photograph (Miyagi et al. 2011)

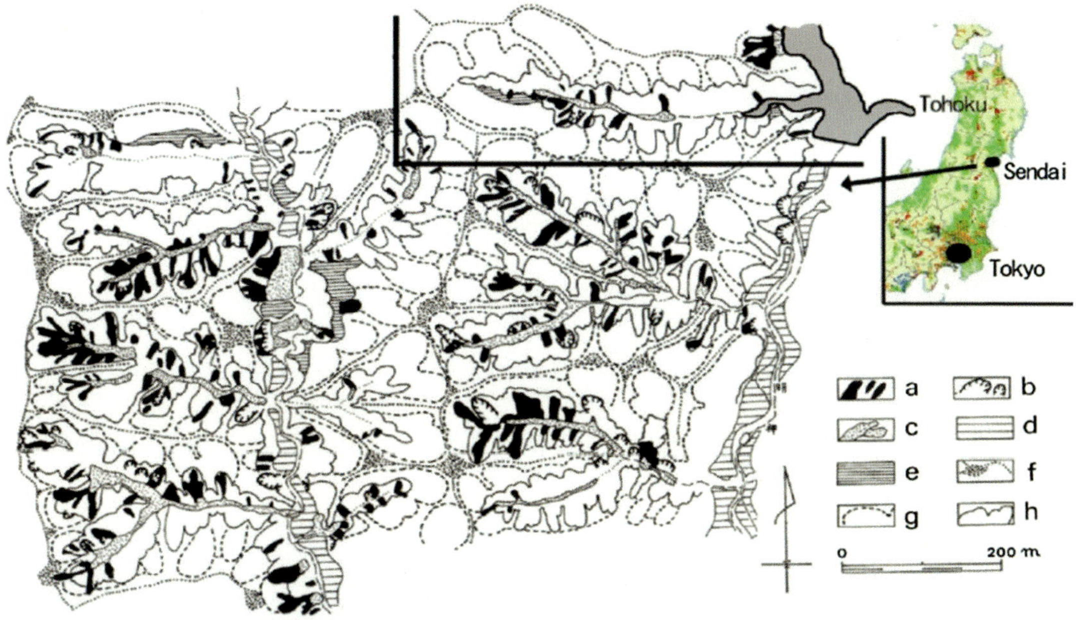

Fig. 9 Distribution map showing small-scale surface failures occurring in dissected valleys on hilly terrain, and distribution and convex break lines (corrosion line) of superficial landslides. Aerial photograph scale (1:10,000 adhesive); 91% of topographical sites of landslide occurrence were near the lower convex break line (Miyagi et al. 2004)

Acknowledgements The authors gratefully acknowledge Emeritus Professor Sassa, K. and Dr. Yamagishi, H. for all kind assistances.

References

Japan Landslide Society and Miyagi Prefectural Office (2002) Landslide topographic area mapping and the risk evaluation in Miyagi Prefecture. Mapps and Reports, Japan (in Japanese)

Miyagi T, Prasada GB, Tanavud C, Potichan A, Hamasaki E (2004) Landslide risk evaluation and mapping—manual of landslide topograhy and risk management. Report of the National Research Institute for Earth Science and Disaster Prevention No. 66, pp 75–137

Miyagi T, Yamashina S, Esaka F, Abe S (2011) Massive landslide triggered by 2008 Iwate-Miyagi inland earthquake in the Aratozawa dam area, Tohoku, Japan. Landslides 8:99–108

NIED: National Institute of Earth Science and Disaster Prevention Japan (1985–2015) Landslide topography mapping (1:50,000)

Saaty TL (1980) The analytic hierarchy process. McGraw-Hill Book Company, NewYork, p 265p

TXT-tool 1.081-2.2 Landslide Mapping Through the Interpretation of Aerial Photographs and Topographic Maps

Eisaku Hamasaki and Toyohiko Miyagi

Abstract

Processes of formation of topography and slopes are the result of a comprehensive combination of climatic change and crustal movement. When examining long time-scales, it is necessary to examine variations in internal and external stresses to understand the formative history of the topography. In this manual, the steps for understanding landslides will be described. The concept and frameworks for identifying unstable slopes (landslide topography) using aerial photos and topographic maps are introduced in details.

Keywords

Landslide topography • Aerial photo interpretation

Contents

1 **Introduction** 53
2 **Model of Landform Development** 54
3 **Learning the Characteristic Appearance and Micro-Landforms of Landslides** 56
3.1 Interior of the Landslide Body 57
3.2 Head Phenomena 59
3.3 Lateral Sections 59
3.4 Toe 59
4 **Transitions in Landslides from Beginning to Termination** 59
4.1 Definition of the Internal Micro-topography of Landslide Bodies and Interpretation 59
4.2 How to Make Topographic and Classical Maps Step by Step, and Clarify Landslide Outlines 63
5 **Summary** 63
References 66

E. Hamasaki (✉)
Advantechnology Co. Ltd, Aobaku, Sendai, Japan
e-mail: hamasaki@adovantechnology.co.jp

T. Miyagi
Department of Regional Management, Faculty of Arts, Tohoku-Gakuin University, 2-1-1, Tenjinsawa, Izumi, Sendai 981-3193, Japan
e-mail: miyagi@mail.tohoku-gakuin.ac.jp

1 Introduction

Many factors are related to both landform development and slope destabilization mechanisms (occurrence of landslides). When considering long time-scales, it is necessary to examine variations in

K. Sassa et al. (eds.), *Landslide Dynamics: ISDR-ICL Landslide Interactive Teaching Tools*, https://doi.org/10.1007/978-3-319-57774-6_3

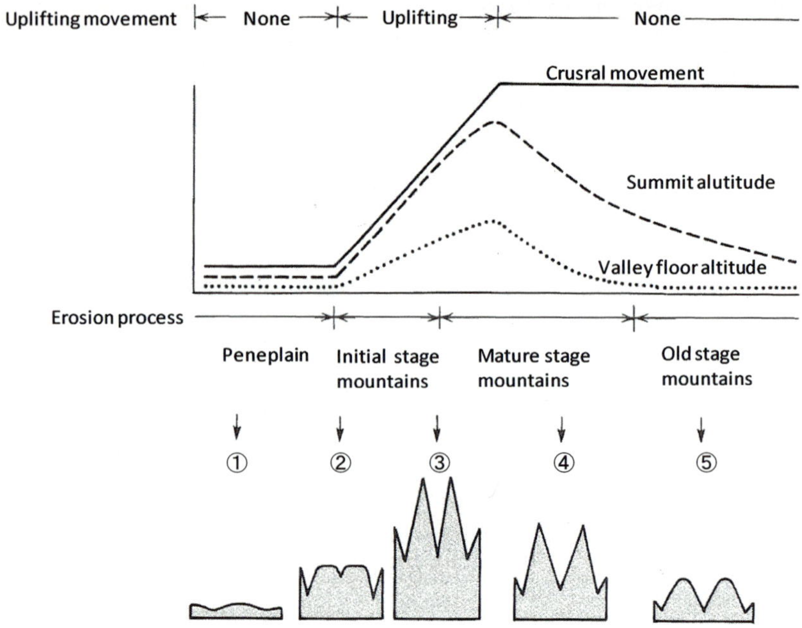

Fig. 1 Fluctuation of mountain topography due to uplift and erosion (Kaiduka 1977)

internal and external stresses to understand the formative history of the topography. Internal stress, including motions within the earth's interior, such as uplift (mountain formation) and submergence (basin formation), directly contribute to the formation of faults and folds (Fig. 1). On the other hand, external stress includes base-level changes (Fig. 2) caused by global climate change (glacial-interglacial) and by crustal movements (Fig. 1).

Processes of formation of topography and slope are the result of a comprehensive combination of climatic change and crustal movement (Fig. 3). From the view point of a short time scale, topography is essentially formed by a combination of rock strength, weathering processes, and erosion stress (Fig. 4). Only in extremely rare cases are erosion stresses kept constant. For example, temporary heavy rain may suddenly increase the strength of stream erosion. Or, if there are prolonged high temperatures and dry periods, erosion diminishes and weathering increases.

Under those conditions, the internal factors making slopes unstable (landslides) are physical changes—such as weathering, metamorphosis, and argillization—related to slip surface formation and inner stress fields. On the other hand, direct external factors are rainfall and snow melt (rise in pore water pressure), river channel erosion (impedance unloading), and seismic force (increased strain and decreased shearing strength). Of course, artificial changes can also be a contributing factor to landslides.

Given the above context, in this manual, the steps for understanding landslides will be described below.

2 Model of Landform Development

Steps in modelling landslides and landform development:

(1) Increase your understanding of the geomorphic development of the relevant slope: Consider the geomorphic development history in order to deepen understanding of landslides.
(2) Know the characteristic appearance and micro-topography of landslides: Know the appearance and micro topography landslides possess when examined close up.
(3) Know the transitions of landslide topography, from outbreak to cessation:

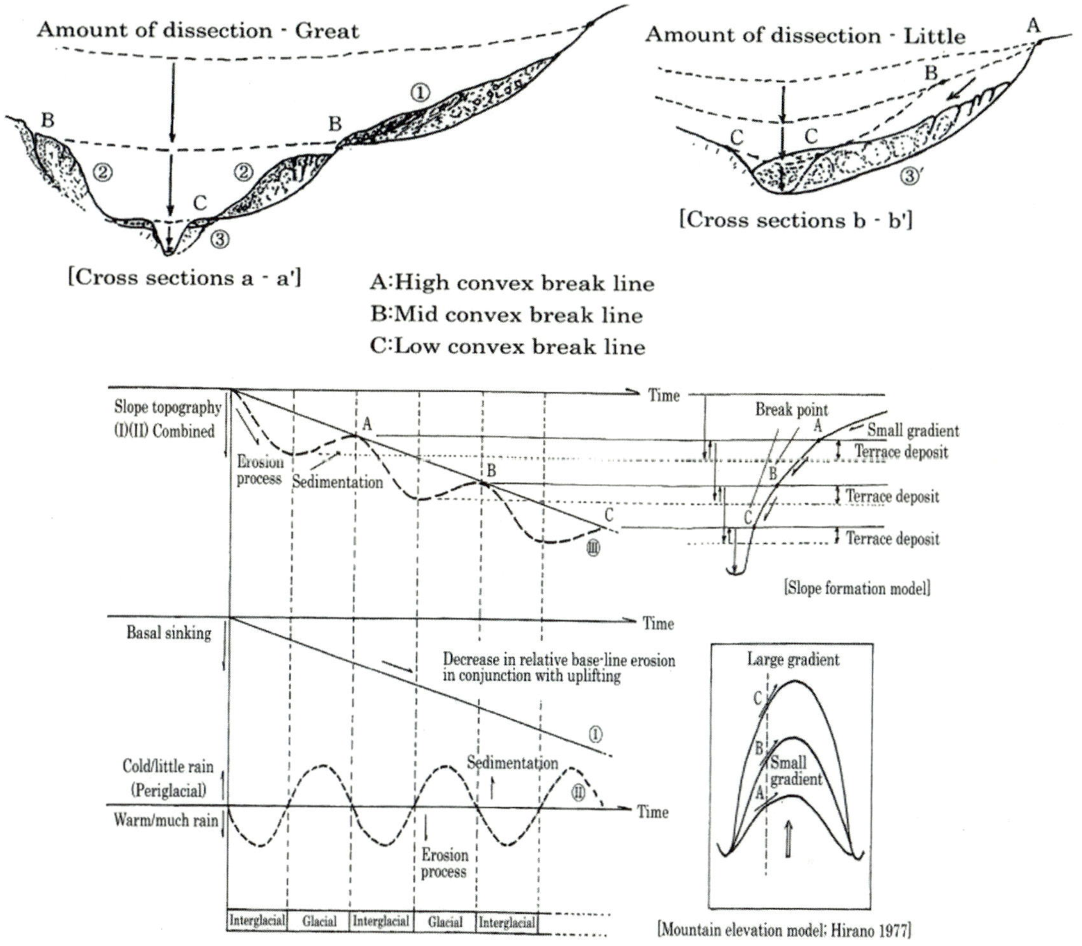

Fig. 2 Model of slope formation of uplifting mountainous areas (Kobayashi and Hamasaki 1988)

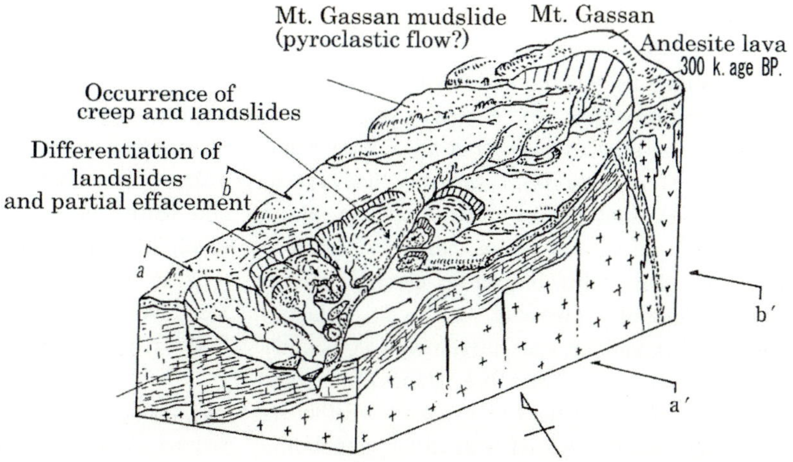

Fig. 3 Diagram of a landslide on the west slope of Mt. Gassan (Kobayashi and Hamasaki 1988)

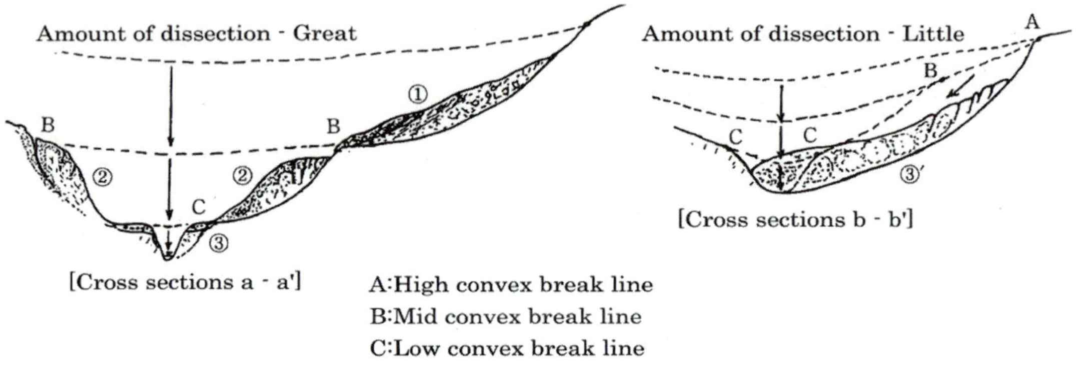

Fig. 4 Cross-section of a series of landslides. The landslides occurred in the numbered order: (*1*)–(*2*)–(*3*) (Kobayashi and Hamasaki 1988)

Understand the general topographical outline, from landslide outbreak to cessation.

(4) Landslide micro-topography definitions and interpretation.
(5) Method of topography boundary classification:
Learn topography classification methods, and then ideas and methods for understanding typical landslide topographies.
(6) Gain training in photographic interpretation to find landslide areas. Increase your amount of experience with landslide interpretation using real samples of aerial photographs.

Figures 5 and 6 illustrate these models.

In order to understand unstable slopes, such as landslides, it is important to deepen your understanding of the development of slope topography on location. Topography is essentially formed by a combination of rock strength, weathering processes, and erosion stress. Only in extremely rare cases is erosion always constant. i.e. over short periods. Temporary heavy rain may be followed by dry periods during which weathering processes make headway. Viewed over long timescales, base-level changes due to major global climatic change (glacial-interglacial) and crustal movements constantly alter erosion stress. Therefore, slopes form various topographies. The strength and weathering characteristics of bedrock can differ, depending on the type of rock, so in considering rock strength characteristics:

(1) When rock strength > erosion stress, there is little sediment runoff, and moderate slopes due to weathering are predominant.
(2) When rock strength < erosion stress, erosion topography progresses in stages, e.g., Stages 1–3.[1]
(3) When rock strength $\leq$ erosion stress, if an external force greatly exceeding the rock strength occurs during some kind of event (earthquake, sudden rainfall, snow melt), a high mass of the slope will become unstable (see Footnote 1).

3 Learning the Characteristic Appearance and Micro-Landforms of Landslides

Some characteristics are shown for interpretation of unstable slope such as landslides (Fig. 7). In order to understand an unstable slope (landslide), and learn about its internal structure and risk level, it is important to understand the characteristics of its exterior and micro-topography. Here we discuss the (1) interior of the landslide body, (2) head (3) side, and (4) toe.

[1]When erosion stress (from base-level sinking caused by bedrock rising and from precipitation), and geological conditions (strength of slope structure materials) are in dynamic balance, unstable slopes where large amounts of mass can move, i.e. landslides, are difficult to create.

Fig. 5 Gaining a deeper understanding of geomorphic development of the observed slopes

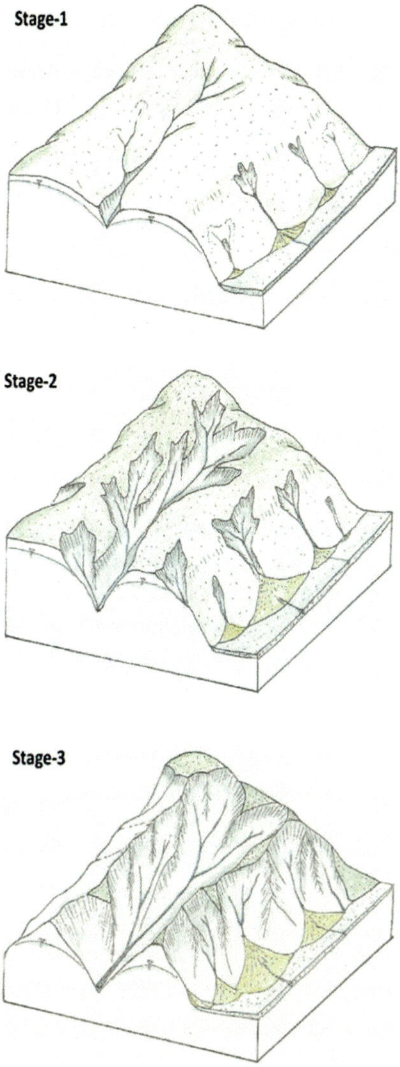

Original surfaces, which are found in places with higher than base-line erosion (e.g., formed from uplifted peneplains or pyroclastic flow surfaces from volcanic activity) are gradually eroded from the bottom (linear erosion from river water, etc.). Initially, they form small furrows such as rills and gullies.

The erosion area then spreads, with rills and gullies becoming v-shaped valleys. The eroded materials are swept downward, consequently they form depositional areas such as fans and alluvial cones with low gradients where flow energy is lost.

Moreover, when the base-level sinks, the original surface is lost, and steep topography, with predominant knife ridges and valley walls, is formed. From this stage, if basal level erosion does not fall temporarily, and becomes arrested, erosional forces weaken, moderate slopes develop due to weathering processes, and the topography will likely become rounded and hilly.

3.1 Interior of the Landslide Body

- Compared with its surroundings, the landslide body is not smooth, but will have distinct micro-relief features and be very rugged.
- In extremely newly formed landslides, there are clear fissures in the landslide body.
- Usually its slope will be gentle compared with its surroundings.
- In clayey landslides etc., the block has a high water retention potential; thus, there are often some lakes and bogs on the surface of the block.
- Within the landslide body, repeated small fluctuations lead to low channel density.
- Clay soils are easily utilized for rice paddies, etc.
- Many types of vegetation, such as reeds, cedar, and bamboo, are often overgrow the landslide. This is because the landslide includes soft soil and high groundwater levels that favour plant growth.
- Tongue fissures, flow mounds, and isolated mounds appear from time to time (Fig. 7).
- Volcanic mudflow and lava topography can also create tongue fissures, but they will not contain head scarp topography.

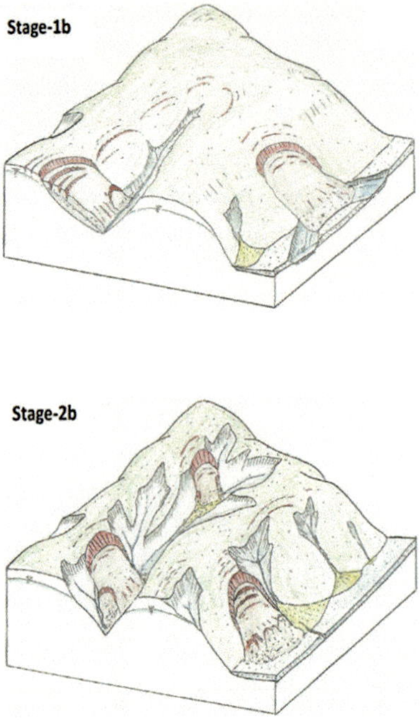

In places if bottom erosion has progressed slowly (Stage-1, 2), we may question the amount of sudden erosion that will occur due to sudden downpours and accelerated uplift of bedrock. We may also ask what effects sudden, large external forces, such as seismic movements, may have.

In such cases, dynamic equilibrium will be broken, and slope destabilization will suddenly increase. As a result, mass movements, such as landslides, are likely to occur over a wide area, including part of the erosion area.

Stage-1b shows some landslides occurring due to some large external forces or sudden downpours after the Stage-1 situation. On the other hand, **Stage-2b** shows more many landslides happen by similar external forces after Stage-2.

In this way, the landscape in landslide areas is established by the combination of linear erosion by fluvial processes and the processes of autonomous destruction by landslides.

Fig. 6 Deeper understanding of geomorphic development in the observed slope overall. *Stage b* is the development of a series of landslides

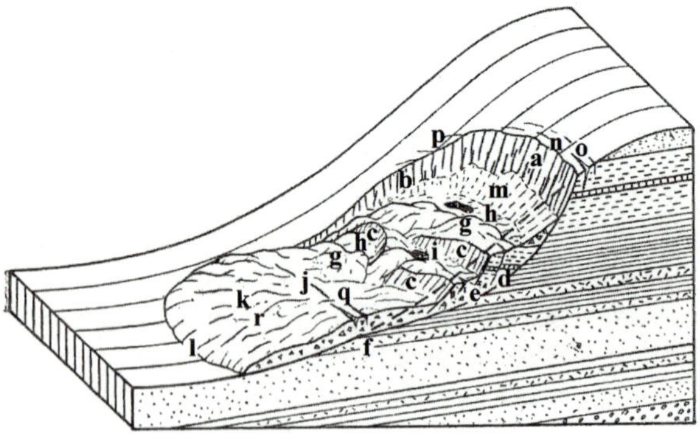

a.Main scarp, b.Lateral scarp, c.Secondary scarp,

d.Main slide surface, e.Secondary slide surface,

f.Toe part, g.Small prominence,h.Depression, i.pond (bog),

j.Uplifted area,k.Tongue, l.Tongue line, m.Talus, n.Crown,

o.Crown fissures,

p.Lateral (echelon) fissures,q.Fissures on uplifted area,

Notes

1) Not all of this micro topography forms with every landslide.
2) The shapes and sizes of elements of the micro-topography are extremely varied.

Fig. 7 Pattern diagram of landslide topography (Suzuki 1982)

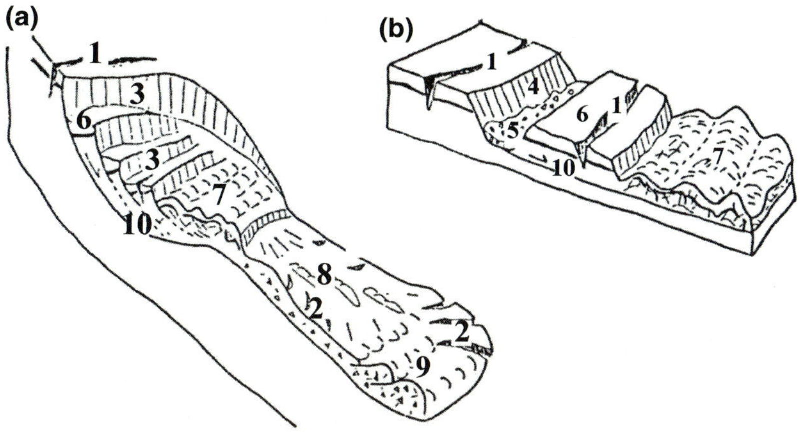

Fig. 8 Model of basic topographic units in landslide terrain (Kimata and Miyagi 1985)

3.2 Head Phenomena

The head part comprises a tension zone where scarps and detached scarps (depressions) form. Figure 8 shows a rotational landslide as A, a translational landslide.

A scarp may also appeared as a steep slope between the convex break line and the concave break line directly beneath it (Fig. 8—features 3 and 4). Consequently, if this scarp shape is found within a certain area, this scarp is the head part of landslide.

Rotational landslides typically have an arcuate scarp (Fig. 8—feature 3). On the other hand, translational landslides create a linear scarp and a depression zone in a separated head area. It is called a detached scarp (Fig. 8—feature 4, Fig. 12).

3.3 Lateral Sections

- Gullies and valley topography sometimes forms by erosion through cracks (Fig. 9—feature 7).
- Small failures may appear from the side to the toe (Fig. 9—feature 6).

3.4 Toe

- Small-scale landslides and slope failures often appear (Figs. 9 and 10).
- If the toe area pushes into a river, it can bury the river, creating a bend in the river channel.

4 Transitions in Landslides from Beginning to Termination

To identify an unstable slope, it is important to understand how the topography changes, from the start of the landslide to its termination. Landslide development and termination processes are not limited to a single pattern. For example, landslide displacement often increase or change the initial micro-topography structure by causing continual expansion. Or, if the landslide remains stopped for a long time, its micro-topography features, such as scarps and fissures, become gentle terrain, and displacement features tend to disappear. Understanding the transitions in landslide topography, based on case studies, is important for recognizing landslides and performing stability evaluations.

4.1 Definition of the Internal Micro-topography of Landslide Bodies and Interpretation

4.1.1 Pressure Ridges

Definition: Pressure ridges are micro-relief features that occurs perpendicular to the direction of movement in a place of intense compression stress due to differences in relative movement speeds in the landslide body.

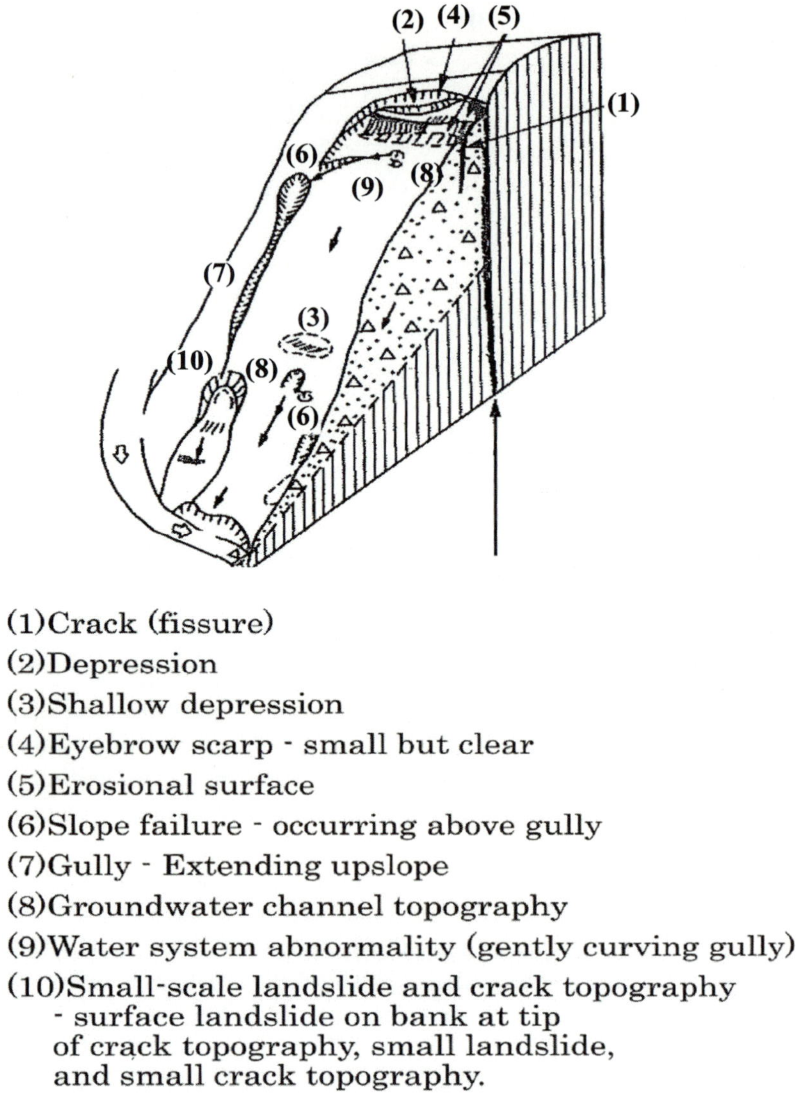

Fig. 9 Example of micro-landform features (Suzuki 1982)

Interpretation: As they are formed by small-scale thrusting, they are recognised ripples.

4.1.2 Flow Traces/Flow Mounds or Flow Hills

Definition: Flow mounds are a micro-relief feature running parallel to the direction of movement, which occurs when a part of the landslide body turns to muddy or colluvial soil, and begins to move like a fluid. Flow hills are relatively high points on micro-relief features formed by this movement (so-called flow peaks).

Interpretation: They can often recognized as gentle micro-relief features in photographs that have no clear directionality. However, they usually exhibit fluid movement, have a very small sub-scarp forming the top edge of their area, and have a movement area that assumes a long, elliptical shape (Fig. 10).

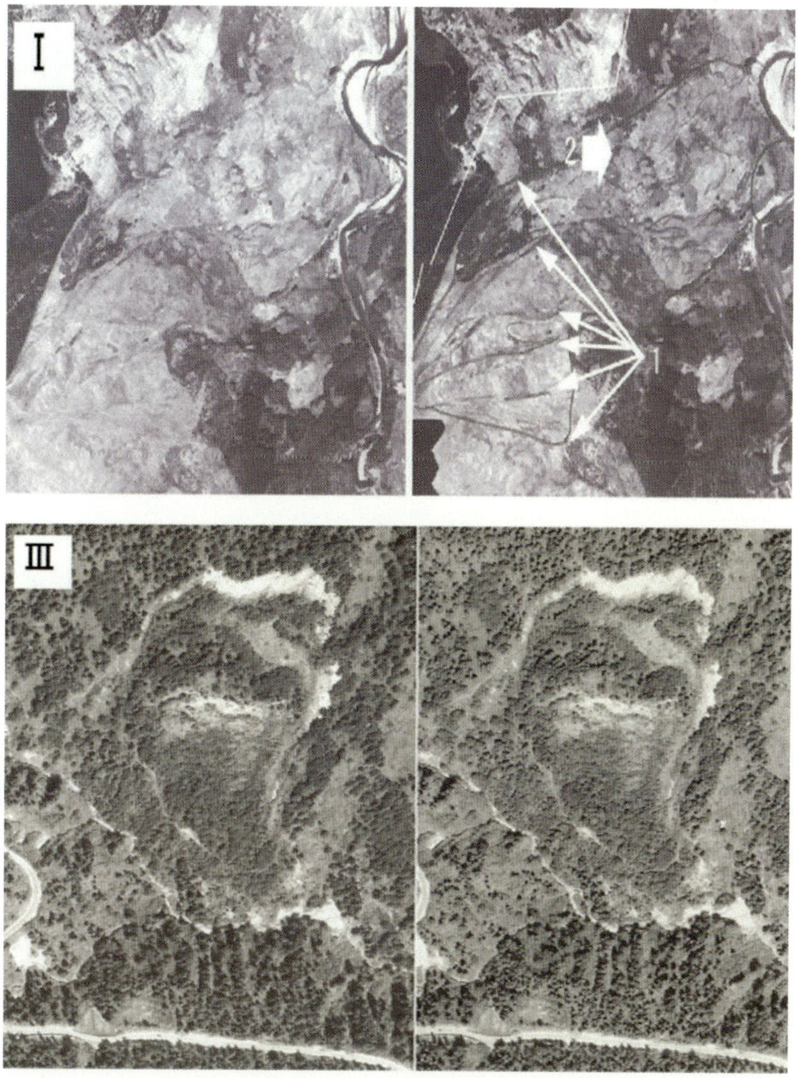

Fig. 10 Flow traces and pressure ridge (Miyagi et al. 2004) *Arrows 1* flow traces—the movement area assumes a long, narrow planar shape, and one part of it extends (obliquely) in the direction of movement, forming minute ridges and small talus topography. *Arrow 2* pressure ridge—protrusion topography that appears in the *lower part* of the movement area

4.1.3 Sub-scarp (Minor Scarp, Scarplet)

Definition: Scarp topography found in landslides in which fragmentation occurs.

Interpretation: After a small landslide occurs in the main landslide body, the sub-scarf exists there. Even though it has very similar shape to the main-scarp, it may not necessarily have a similar slip to the main-scarp (Figs. 11 and 12).

4.1.4 Detached Scarp/Fissured Depression

Definition: A steep scarp formed due to the movement caused by tension in the landslide body interior, and the relatively low-lying area interposed in that steep scarp. In such cases, the low-lying area is often an outcropping of the slip surface.

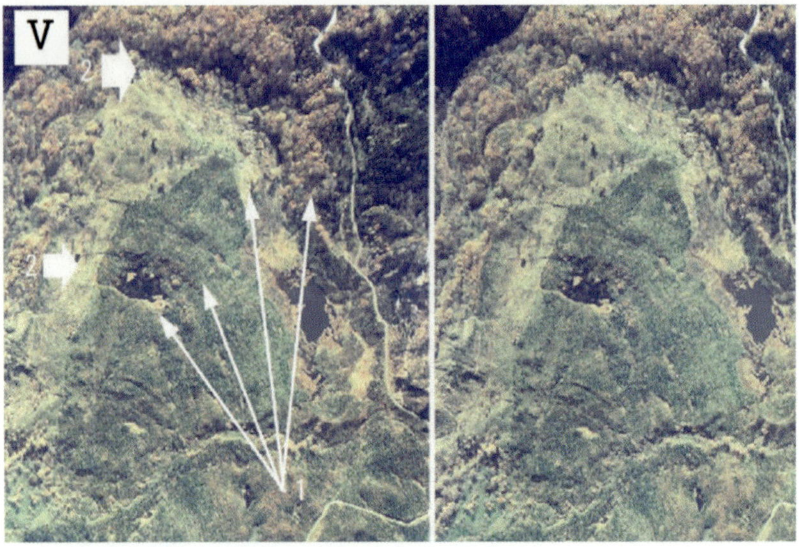

Fig. 11 Main scarp and sub-scarp. There is a large-scale sub-scarp in the central part of the landslide. Its *upper part* is formed of a large block, while in its *lower part* a characteristic colluvial landslide can be seen in low relief (Miyagi et al. 2004)

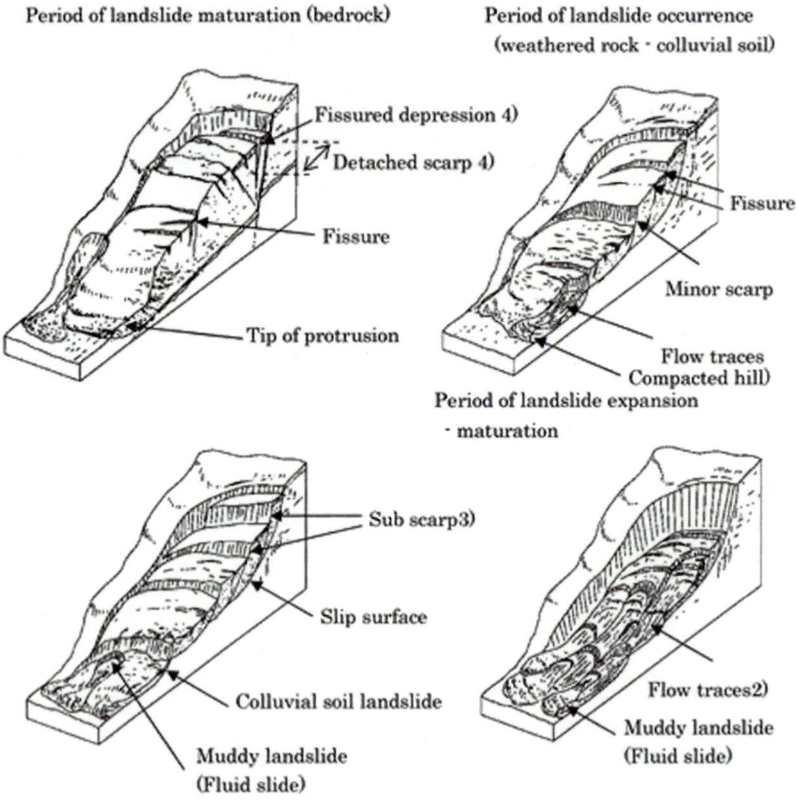

Fig. 12 Detached scarp and fissured depression. Detached scarps and fissured depressions frequently appear as dip slipping along the bedding plane of the stratum of sedimentary rocks. This shows the archetypal development of a detached scarp and fissured depression at the front of the main scarp and center of the landslide body. Ponds or bogs can be found in the depressions (Miyagi et al. 2004)

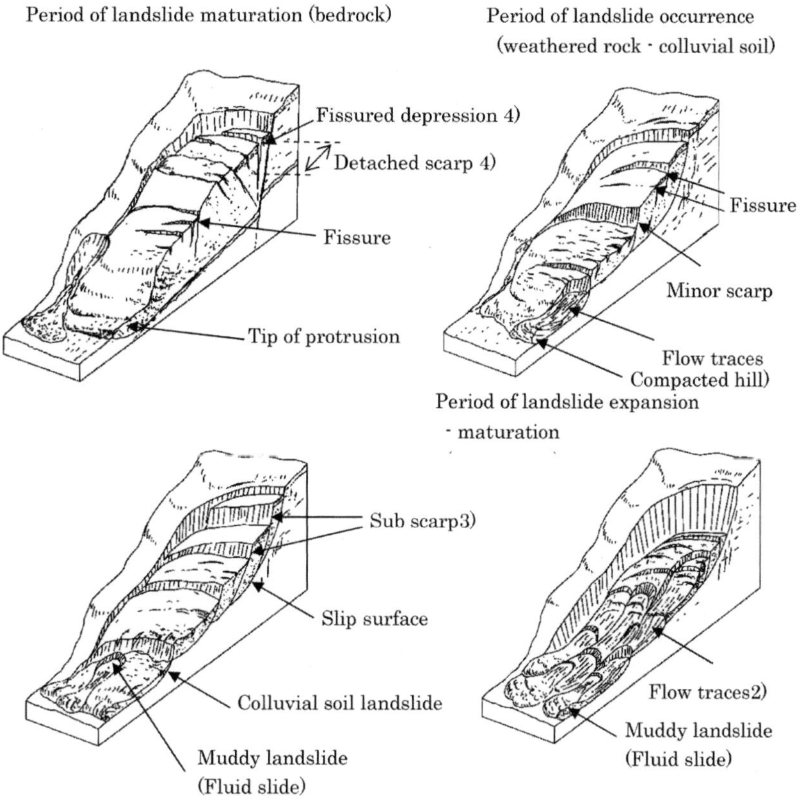

Fig. 13 Schematic diagram of landslide progress (Miyagi et al. 2004)

Interpretation: A steep scarp, similar to a sub-scarp, but clearly distinct from the main scarp in that it opposes a steep scarp.

With an understanding of the description so far, Fig. 13 shows a schematic diagram of landslide progress step by step.

4.2 How to Make Topographic and Classical Maps Step by Step, and Clarify Landslide Outlines

To clarify the landslide outlines on the topographic map, some techniques, such as drawing lines on the crest and stream network, are very useful. Figure 14a, b outline these techniques step by step.

5 Summary

(1) We need to understand that the processes of formation of topography and slope are the result of a comprehensive combination of climatic change and crustal movement. Furthermore it is important to consider the variations in internal and external stresses to understand the formative history of the topography, particularly when examining long time-scales.

(2) Especially, for landslides, it is important to consider the geomorphic development history as described above in order to understand the landslide mechanisms. Moreover, it is necessary to examine landslides and to know the mechanisms producing the micro topography on the landslides.

STEP ONE

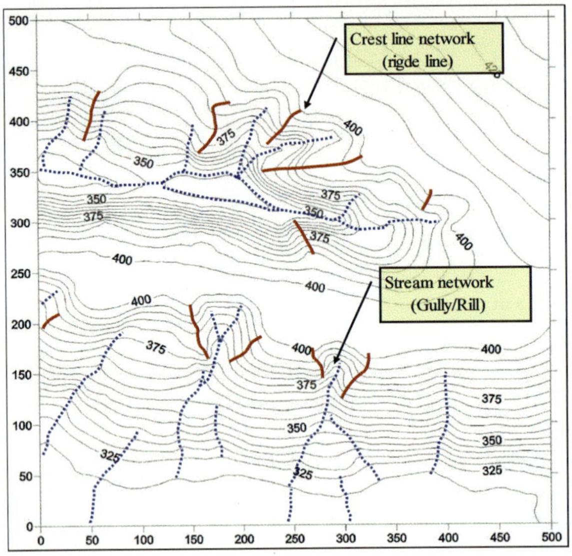

First, draw the stream network (valleys, runoff, gullies, rills) on the figure.

Also, add the crest line network (ridge lines).

When doing this, it is useful to refer to both the aerial photograph and topography map.

*This will make the catchment area for each area on the topography clear.

By outlining catchment areas, areas of active erosion will become clear.

*Channel density will reveal areas undergoing minute fluctuations.

*Suggest the existence of some weak lines, such as fault or lithological effects, in areas with broken ridges.

STEP TWO

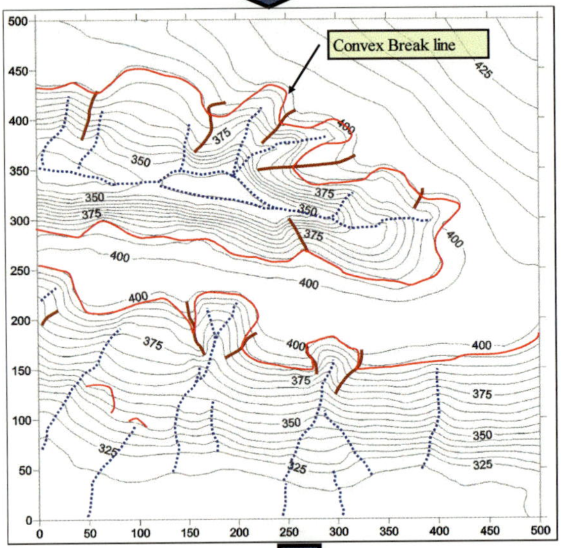

Next, add convex break lines.

*In general, the steep slope below the concave break line is often an area of active erosion, younger than the slope above it.

Fig. 14 Schematic diagram for landform mapping step by step

STEP THREE

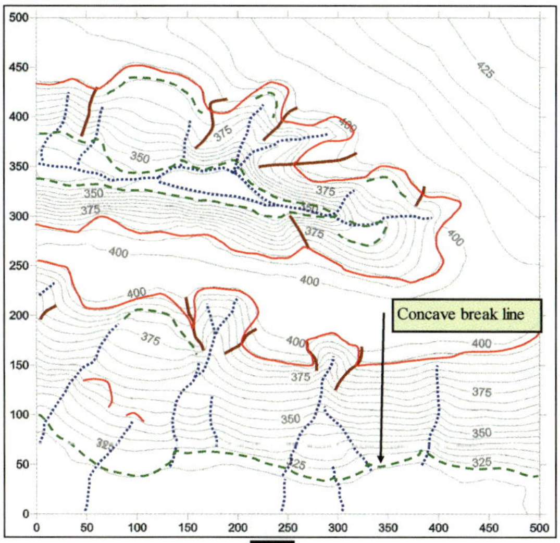

Then, add concave break lines.

In general, the gentle slope below the concave break line is often a depositional area, and younger than the above slope.

* Characteristics of landslides (unstable slopes)

- Distinct from surrounding topography.
- Covered in micro-relief.
- Have a tongue-shaped landform at the toe part.
- Low channel density.
- When there is an arced convex break line, and just beneath it the concave break line is almost parallel, there is a high probability of scarp due to land-sliding.

FINAL

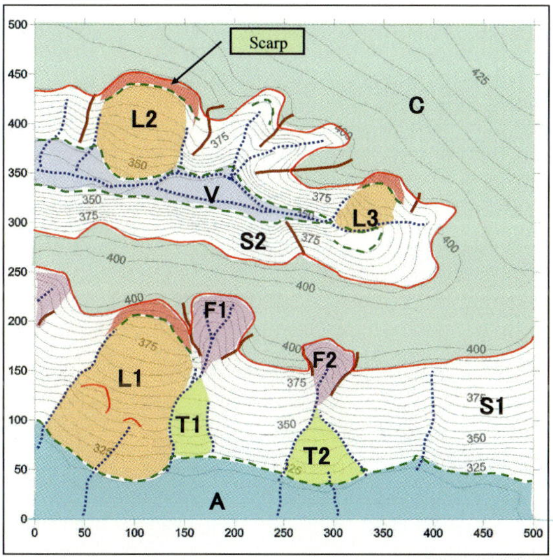

Lastly, color in the landforms, referring to the boundaries of landform units, such as the convex break lines, concave break lines, and stream networks.

When doing this, in areas with similar geological formations (internal factors), and areas subject to identical external stresses, divide the surface with consideration to creating a basically continuous topography.

Also, consider *1 (Step 2) with regard to stable and unstable surfaces.

*Left figure: example of landform classification

Classified into: crest gentle slope (C), valley side slope (S), alluvial plain (A), valley bottom (V), slope failure (F), alluvial cone - talus (T) and landslide (L)

Final 3D image

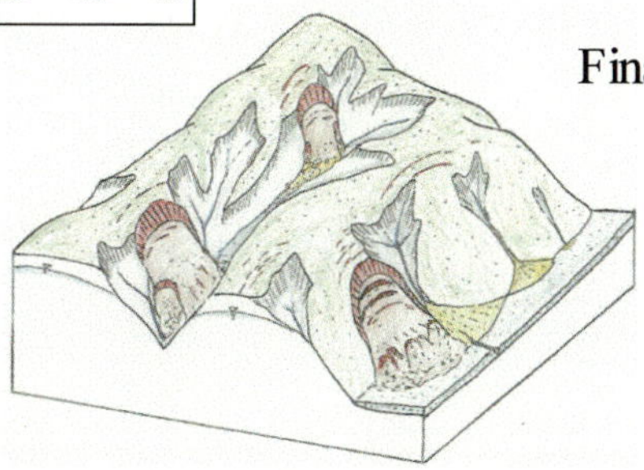

Fig. 14 (continued)

(3) This tool describes the several steps for understanding landslides using aerial photos and topographic maps.

Acknowledgments First of all, we would like to express Vietnam project leader and our supervisor, Prof. Kyoji Sassa, who give us some chances to study Vietnam Area, and many research experiences. And, we are grateful to Prof. Hiromitsu Yamagishi for helpful discussions. Finally, we wish to thank Haruna Ishikawa for the support in writing this tool.

References

Kaiduka S (1977) Japan of terrain—nature and origin, Iwanami Shoten, 215 p

Kimata R, Miyagi T (1985) The basic landform units composing of landslide area. J Jpn Landslide Soc 8:10–17

Kobayashi Y, Hamasaki E (1988) Landslide with transition of topography—case of west slope areas of Gassan Mt. Appl Geol Yamagata Jpn 8:10–17

Miyagi T, Prasada GB, Tanavud CI, Potichan A, Hamasaki E (2004) Landslide risk evaluation and mapping—manual of landslide topography and risk management. Report of the National Research Institute for Earth Science and Disaster Prevention No. 66, pp 75–137

Suzuki T (1982) Introduction of how to read of topographical map for construction engineers, vol 1–4. Kokonshoin

TXT-tool 1.081-3.1 Landslide Recognition and Mapping Using Aerial Photographs and Google Earth

Hiromitsu Yamagishi and Rigoberto Moncada

Abstract

To detect landslides using topographic maps and aerial photographs, it is necessary to know the topographic features characteristics of landslides i.e., the "landslide signature" (Pike in Math Geol 20(5), 1988) and how landslide topographic characteristics differ from the topography around a landslide. In this landslide teaching tool, representative topograhic features of landslides are discussed as a means to identify them in representations of terrain such as topographic maps and aerial photographs. Several examples and indications for this recognition tool are presented. Finally, the use of satellite images and terrain modeling in Google Earth to detect and map landslides is demonstrated as well.

Keywords

Landslide topographic criteria · Stereo viewing · Aerial photograph · Satellite image · Deep-seated landslide · Slope failure

Contents

1 **Introduction** ... 67
2 **Topographic Landslides Characteristics and Landslide Types** ... 68
3 **Detecting Landslides Using Topographic Maps and Aerial Photographs** ... 71
3.1 Use of Topographic Maps ... 71
3.2 Application of Aerial Photographs and Stereoscopy ... 73
4 **Satellite Images from Google Earth to Detect and Mapping Landslides** ... 77
5 **Conclusions** ... 81
References ... 81

H. Yamagishi (✉)
Shin Engineering Consultant Co.Ltd., 2-8-30, Sakaedori, Shiroishi-ku Sapporo 003-0021, Japan
e-mail: hiromitsuyamagishi88@gmail.com

R. Moncada
Universidad Tecnológica Centroamericana-UNITEC, Zona Jacaleapa, Tegucigalpa, Honduras
e-mail: rigoberto.moncada@unitec.edu; rigoberto.moncada.lopez@gmail.com

1 Introduction

A landslide is the movement of a mass of rock, earth or debris on a slope under the influence of gravity. Landslides are triggered by rainfall, earthquakes, volcanic eruptions, slope toe erosion by rivers and human activities, including

K. Sassa et al. (eds.), *Landslide Dynamics: ISDR-ICL Landslide Interactive Teaching Tools*, https://doi.org/10.1007/978-3-319-57774-6_4

e.g., construction works that involve terrain modification and vibrations.

Several mass movements and slope failures develop within and at the margins of already existing landslides, for example deep-seated landslides and also shallow landslides. So if landslide topography can be identified for a study area, sites with potential of developing new landslides can be defined as well. The method presented in this landslide teaching tool applies the situation explained above, in combination with disciplines that interpret terrain features by photogrammetry, stereoscopy and photogeology to generate information for landslide inventory. Among the disciplines mentioned, the use of stereoscopy is particularly important for landslide detection (Fig. 1). Inventory information can then be used for further decision-making or the elaboration of hazard and risk landslide mapping.

Both deep-seated landslides (and shallow landslides) are usually composed of a main scarp (scar part), and body or debris (flow part and deposit part). Since both types of landslides have the potential to alter topography appreciably, superficial alterations caused by them can be used as means of detection. Therefore, resources such as contour maps, aerial photographs, viewed using a stereoscope, and satellite images are ideal to observe them. It is important to mention that this detection functions only for mass movements that significantly modify surface topography, so not all kinds of landslides can be found or predicted using the proposed tools. However, the availability and low cost of terrain surface information and its applicability to large areas in short periods of time validates this tool. Desk and field studies elaborated with this tool can then be complemented by alternative landslide-detection techniques, such as underground sensors.

2 Topographic Landslides Characteristics and Landslide Types

This chapter presents criteria for identification and recognition of both deep-seated and shallow landslides (for this paper shallow landslide and slope failure will be used interchangeably), making use of topographic, geologic and photographic information. Figure 2 is a schematic model of mountain topography (Suzuki 1997). This figure shows a three-dimensional model of a mountain area including location of ridges, valleys, slopes and flat plateau that while not in scale show the changes that are to be identified in terrain. In particular, distinctions between the slopes or knick points are an important factor for finding landslides. With enough practice it is possible to visualize and imagine 3D models of terrain in order to find and understand the shapes

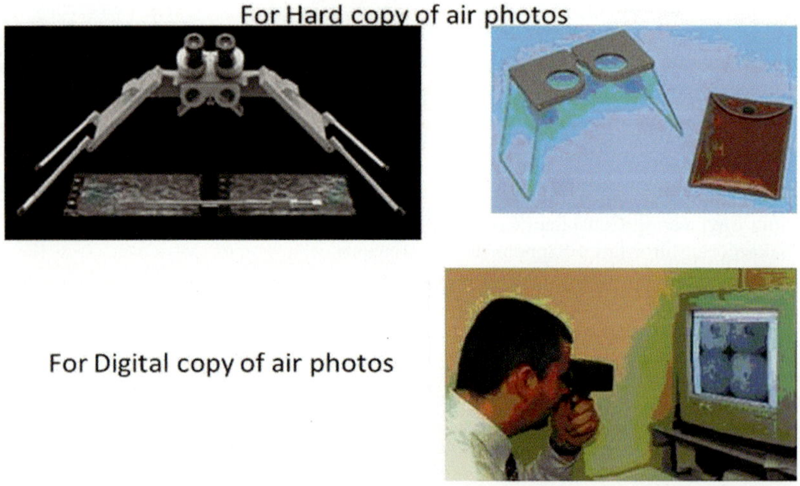

Fig. 1 Instruments for the stereoscopic vision of hard copy (*up*) and digital (*bottom*) stereoscopic aerial photographs

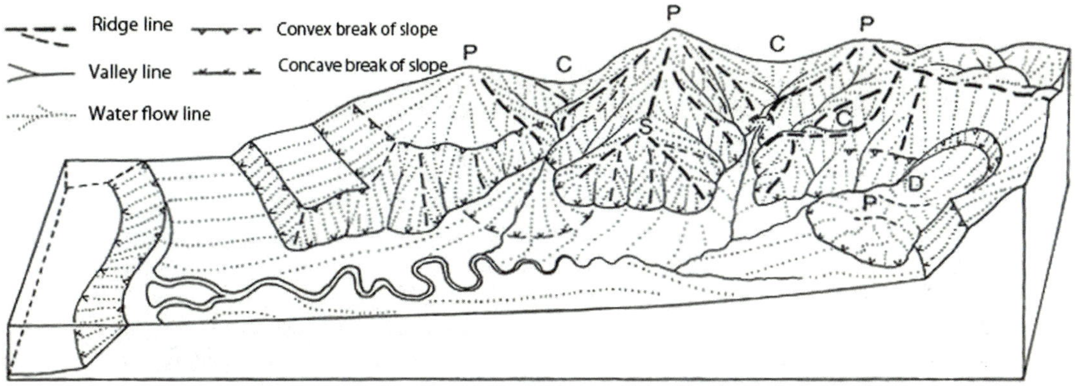

Fig. 2 Schematic model of mountain slopes (after: Suzuki 1997). *P* peak, *D* enclosed depression, *C* col (saddle), *S* spur

of landslides from two dimensional topographic maps and stereo-paired photographs. Use of software that models topography in three dimensions such as Google Earth can also be very useful to attain this visualization skill for landslide detection.

Deep-seated landslides are of great interest for the identification of sites where additional landslides can develop, as they tend to present more threatening and plentiful instability conditions in comparison to most slope failures. Thus, they will be focus of a more thorough description regarding their components and classification in order to support a better detection procedure. By recognizing accurately the main features and types of these landslides, it is possible to find them using available surface information in less time and with more consistency.

The following topographic characteristics are typical of deep-seated landslides Fig. 3: (1) main and minor scarps, (2) cracks (3) mounds or ridges, (4) hollows or ponds, (5) alterations/meanders of the drainage system (streams, rivers, etc.). Scarps are significant features that form part of deep-seated landslides and their equivalent deposits Fig. 4. The main scarp can result in variable landslide shapes: an amphitheatre-shape for a rotational slump or a straight and planar failure surface for a planar slide. The landslide body and toe can present hollows or ponds, mounds and cracks (fissures) showing variable shapes and locations. In case of a planar type, hollows are graben-shaped (depressions). Figure 3 schematizes the typical form of a deep-seated landslide (Cruden and Varmes 1996; Highland and Bobrowsky 2008).

Deep-seated landslides can be classified into the following main types: (a) Rotational type, (b) Translational type, (c) Debris avalanche, and (d) Earth flow. (Fig. 4, Highland and Bobrowsky 2008). Based on the new classification version proposed by Hungr et al. (2013), translational slides will be referred from now on as planar slides.

Rotational type is defined as a landslide in which the rupture surface is curved upward in cross section and the movement is rotational along the axis parallel to the slope surface. The head moves vertically downward, and the displaced materials tilt backward toward the scarp. Therefore, the portion in front of the scarp is regarded as a tensional zone; on the contrary, the marginal portion is a compression zone (Fig. 4a). Planar type is a landslide that moves down and outward, along a relatively planar surface (Fig. 4b). This type of slide may progress over considerable distances if the surface of rupture is sufficiently inclined, in contrast to rotational slides, which tend to restore their slide equilibrium.

Debris avalanche (Fig. 4c) is essentially large, extremely rapid and, often, open-slope flows formed when an unstable slope collapses and the resulting fragmented debris is rapidly transported away from the slope. In some cases, snow and ice will contribute to the movement if sufficient water is present. Finally, earth flow (Fig. 4d) can occur on gentle to moderate slopes, generally in

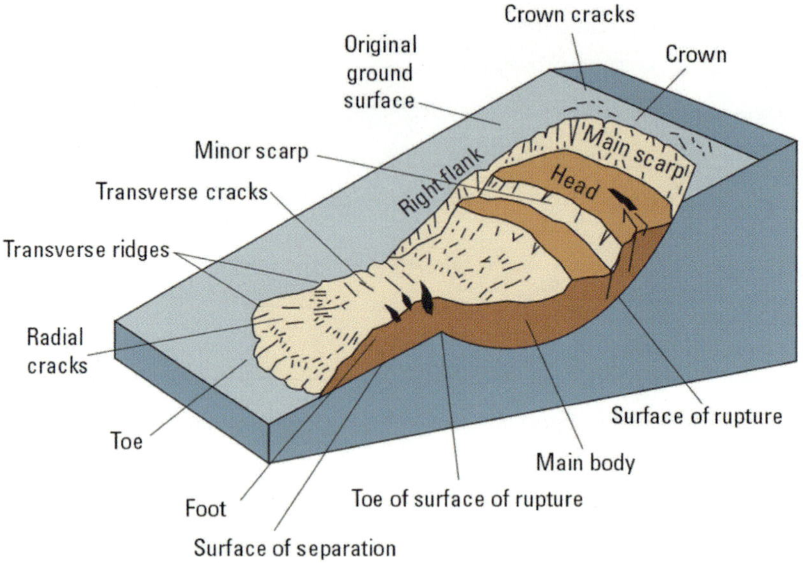

Fig. 3 Schematic model of typical deep-seated landslide (Cruden and Varnes 1996; Highland and Bobrowsky 2008)

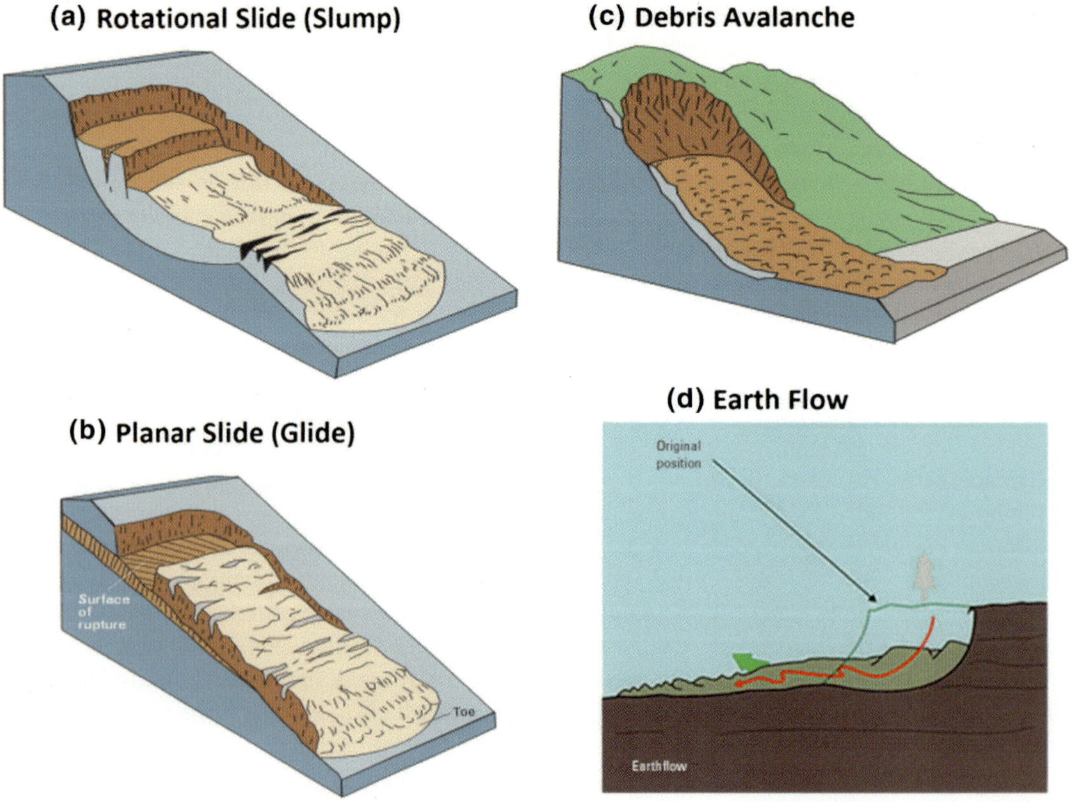

Fig. 4 Schematic types of deep-seated landslides (Varnes 1978; Highland and Bobrowsky 2008)

Fig. 5 Slope failures caused by heavy-rainfalls on July 14, 2004 in Niigata, Japan

fine-grained soils, commonly clay or silt, but also in very weathered, clay-bearing bedrock. The mass in an earth flow moves as a plastic or viscous flow with strong internal deformation.

While focus on identification has been on deep-seated landslides (as they can be spots for development of new mass movements), shallow landslides must be considered as well. Zones with numerous shallow landslides might indicate problematic conditions that can result in even more additional movements and even an individual slopes failure in a specific place can be an indicator of possible new landslides.

Slope failures tend to be triggered by heavy rainfall and high intensity earthquakes. They are characterized by shallow depth up to several meters and many failures occurring at once. In addition, they are sometimes associated with mud flows and debris flows as deposit parts. Mainly, slope failures are classified into (Fig. 5a) and spoon(concave) types (Fig. 5b).

3 Detecting Landslides Using Topographic Maps and Aerial Photographs

3.1 Use of Topographic Maps

After introducing the basic topographic features that allow identification of landslides, and with a general idea of its classification, topographic maps and aerial photographs can be used to locate them. The effectiveness and scale of this methodology will depend on the quality and scale of the input information utilized. Regional topographic maps or aerial photographs can be useful to find very large scale landslides. But if local or small scale landslides or slope failures are to be found, information with scales around 1:5000–1:10,000 should be available.

Topographic maps with detailed contour lines are originated from air photographs, therefore, if there are topographic maps, at first it is recommended to read the map and find large scale landslides and then smaller ones if possible. Important indications for landslide detection to be searched in topographic maps include: (1) Sudden changes in slope, which can be the location of a scarp, especially if it has a semi-circular or amphitheatre shape. (2) Changes in concavity of the contour lines in a slope, which can indicate a transition from a scarp (concave towards downhill) to a landslide body or deposit (convex towards downhill). (3) Formation of mounds, hollows or ponds in an area. (4) Location of very steep areas close to very flat ones, in intermittent patterns.

The characteristics above will not necessarily guarantee the identification of a landslide, but they can be good indicators for potential sites, which then should be confirmed through field work. This way of elaboration of a landslide inventory map can be reliable and feasible.

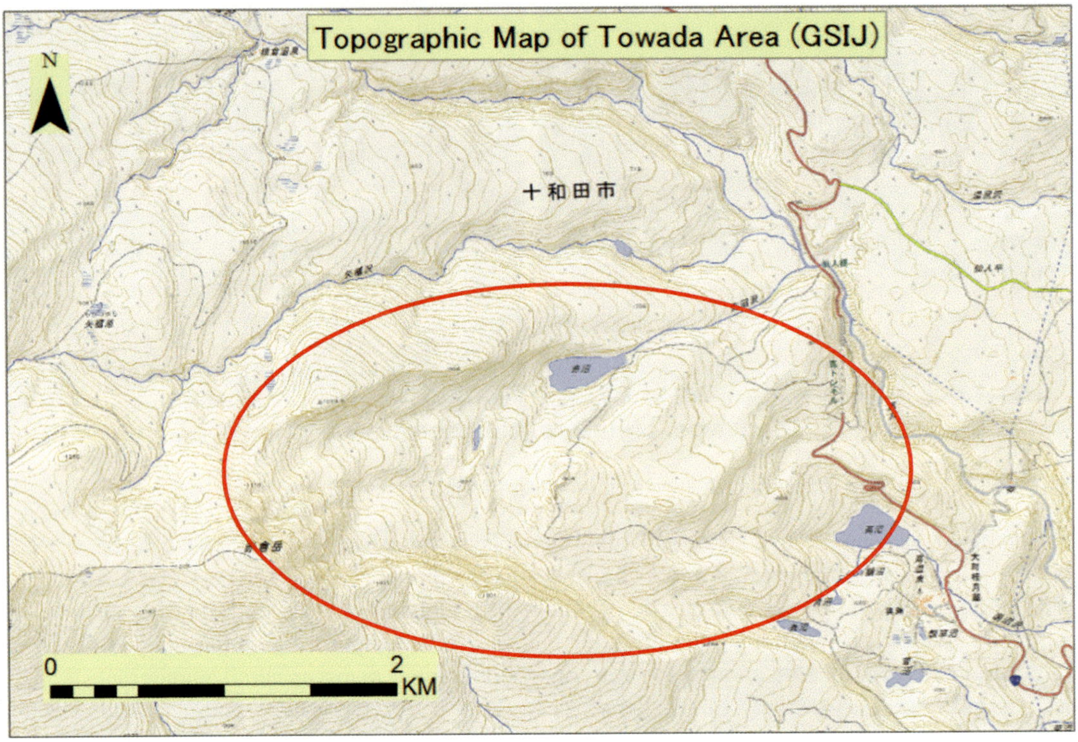

Fig. 6 Topographic map of Towada Area including typical large scale deep-seated landslides

Figures 6 and 7 show the implementation of the features explained above to find landslides, in this case, within the volcanic area of Towada in Akita, Japan. Figure 6 shows the original contour map and Fig. 7 is the same contour map with different types of scarps identified. As it can be observed, typical deep-seated landslides are outlined by the steep scars of the main scarp (amphitheatre-shaped) and flank scarps, and the body that includes mounds and hollows or ponds. Changes in concavity of the contours can also be observed between the zone near the scarp and close to the landslide toe.

Other examples of landslide detection using topographic maps can be seen in Figs. 8 and 9. The base topographic map in this case (1:25,000) is that of the Yoichi area in Hokkaido, Japan. Just as in the first example, it was provided online by the Geographical Survey Institute of Japan (GSIJ). The first landslide map (Fig. 8) belongs to the National Research Institute for Earth Science and Disaster Prevention (NIED) and the second one (Fig. 9) belongs to the Hokkaido Landslide Digital Map by Yamagishi (2012). Both of them applied principles of topographic characteristics in the base map to find probable locations of landslides. Differences between them can be attributed to the criteria of specialists and their particular observations during site inspections.

For identification of shallow landslides, topographic maps can also be used but scale must be larger. Figures 10 and 11 show a map of Izumozaki in Niigata, Japan, where slope failures were inventoried after intense precipitation events in 2004. Figure 10 displays the full scope of the study area and Fig. 11 illustrates how the level of detail a 1:10,000 contour map allows detection of small scale landslides. Changes in contour concavity and slope were used as important indicators to locate flows or other types of mass movements induced by heavy rainfalls.

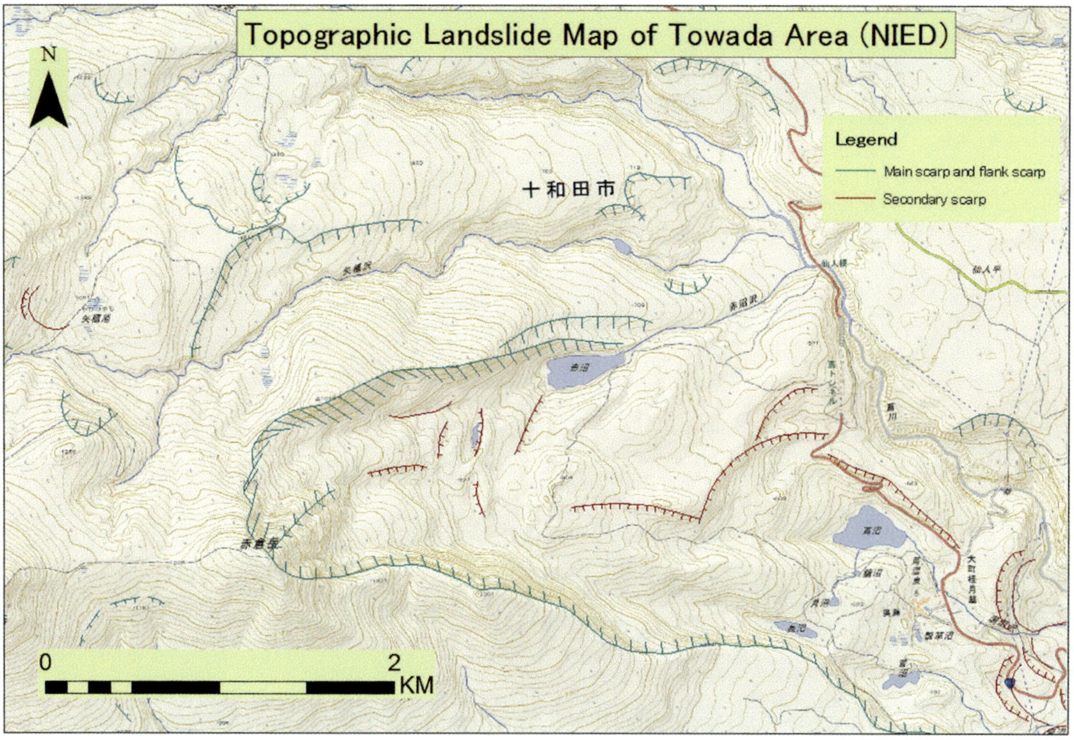

Fig. 7 Topographic map of Towada area in Akita, Japan (1:25,000) with large scale landslides indicated

3.2 Application of Aerial Photographs and Stereoscopy

Stereo paired aerial photographs, while requiring experience in photogeology and air photo interpretation, can provide much more information on landslide features than topographic maps. In addition of perceiving the same topographic variations of contour maps, aerial photographs in stereo pairs allow observation of vegetation cover, soil and rock type, alterations in small objects such as streams, fences, roads or other kind of infrastructure and terrain texture that can provide important indications whether a specific zone is a potential landslide site.

Another important advantage in the use of a stereoscope to view aerial photographs, such as the mirror stereoscope (up and left in Fig. 1) or pocket stereoscope (up and right in Fig. 1), is that stereo pairs tend to appreciably exaggerate elevation differences. Thus, important details for landslide detection that rely of changes in slopes, formation of hollows/ponds or steep scarps can be emphasized by this methodology. Of course, this distortion or exaggeration must be taken into account to also avoid misrepresentation of terrain.

Thus, if available, use of aerial photography is greatly encouraged. Its main drawbacks can be having resolutions unsuitable for the precision required by the landslide study, high cost and complexity for their elaboration and lack of availability for target zones that have limited or minimal geographical studies (unpopulated areas).

Examples of stereo pairs used to find landslides are presented in Figs. 12 and 13, with the major landslide identified circled in red. Figure 12 also shows a topographic map for comparison. While it is obvious for both cases that a topographic map would have achieved the

Fig. 8 Landslide inventory map of Yoichi, Hokkaido. Developed by NIED and provided from website of Shin Engineering Consultant Co. Ltd

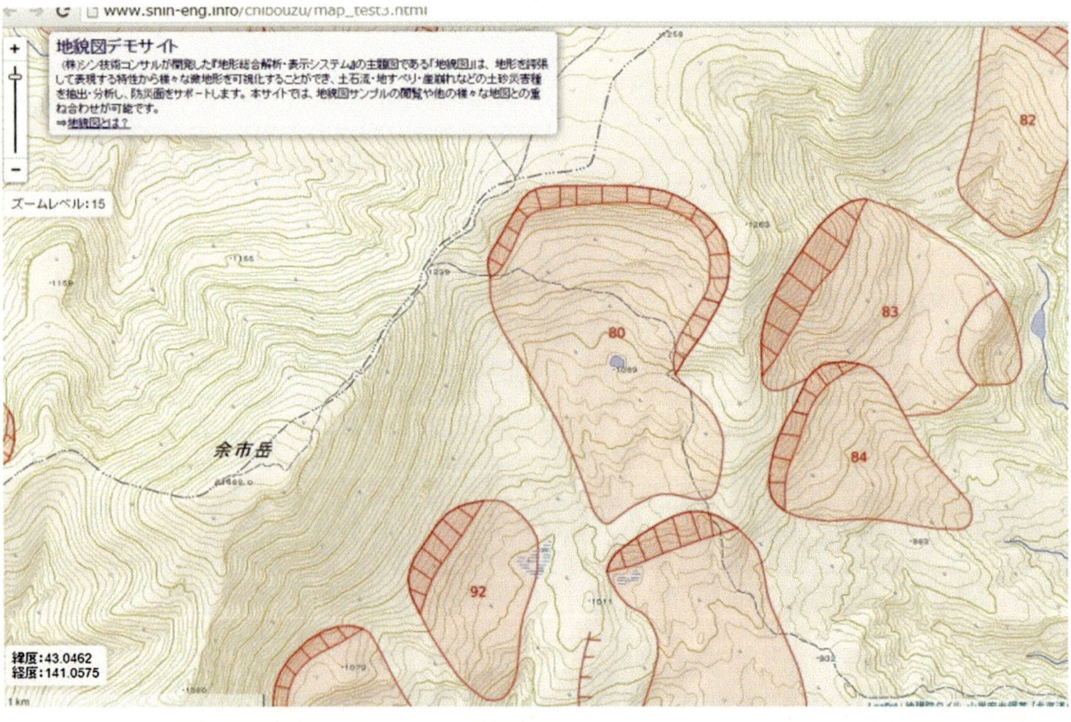

Fig. 9 Landslide inventory map of Yoichi, Hokkaido. Published by Yamagishi (2012) and provided from website of Shin Engineering Consultant Co. Ltd

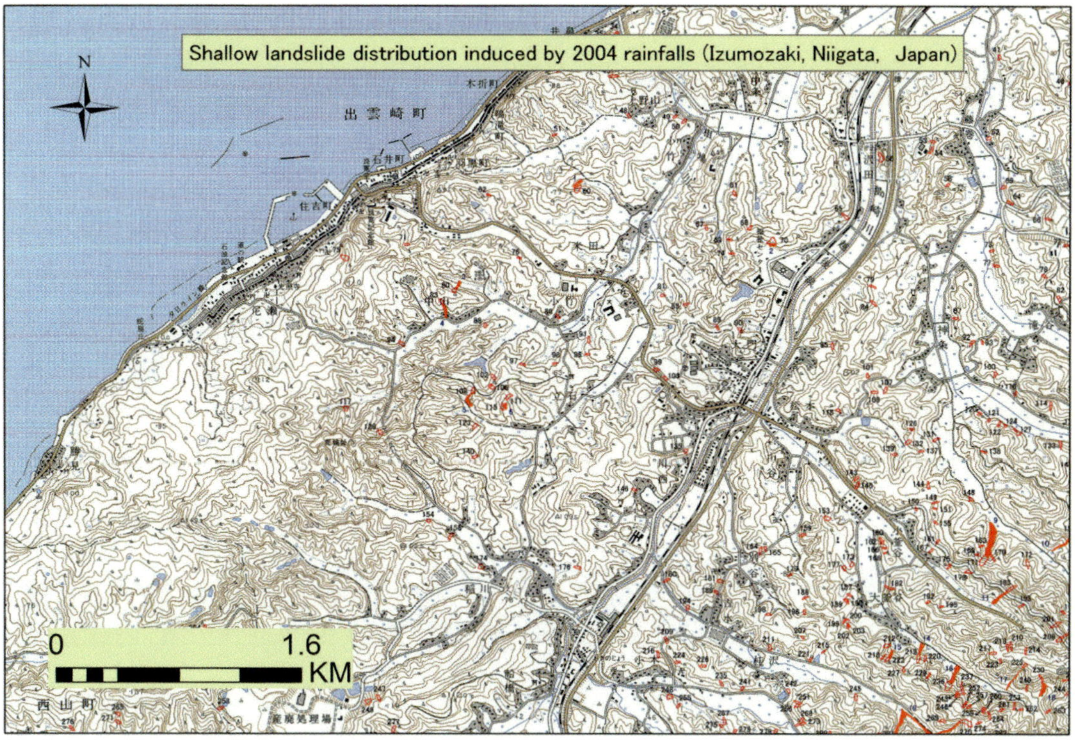

Fig. 10 Map of shallow landslide distribution for Izumozaki, Niigata, Japan after intense precipitation on July, 2004 (provided by Asia Air Survey Co. Ltd)

same results for identifying large scale landslides, this may not be applicable for other details. Additional information present in aerial photographs provides a more thorough insight of local conditions and facilitates identification of smaller landslide blocks within large landslides, which is something more complicated to achieve in topographic maps. Moreover, given that the stereoscopic effect requires overlapping between photographs (60% or more in the same flight line), several images of the same area with variations in illumination and position are available. Different points of view of the same site can assist in landslide detection.

With adequate scale and sufficient spatial resolution, aerial photographs can provide trustworthy data for elaboration of high quality local scale inventory maps. Individual landslide blocks, potential areas of rock creep and subtle slope failures can be easily perceived by the trained eyes. Therefore, landslides inventoried can also be characterized and classified in a preliminary way for hazard mapping with aerial photographs, before being fully confirmed by a field study.

Figure 14 shows an aerial photograph, scale 1:9000, with landslides identified for a site in Tegucigalpa City, Honduras. With assistance of a mirror stereoscope that includes magnifying lens, greater precision in defining the extents, level of certainty and direction of the landslides were achieved. This convenient scale and quality of images permitted interpretation of aerial photographs to develop a very complete landslide inventory study for the city of Tegucigalpa. As a result, 400 km^2 of metropolitan area were examined with high accuracy and the inventory map, created in a GIS database (.shp files), presented locations for more than 1500 potential landslides. Figure 15 shows a section of this study area and the landslide shapes generated. It can be appreciated that slope failures as small as a few square meters could be included in the inventory as well as large deep-seated landslides.

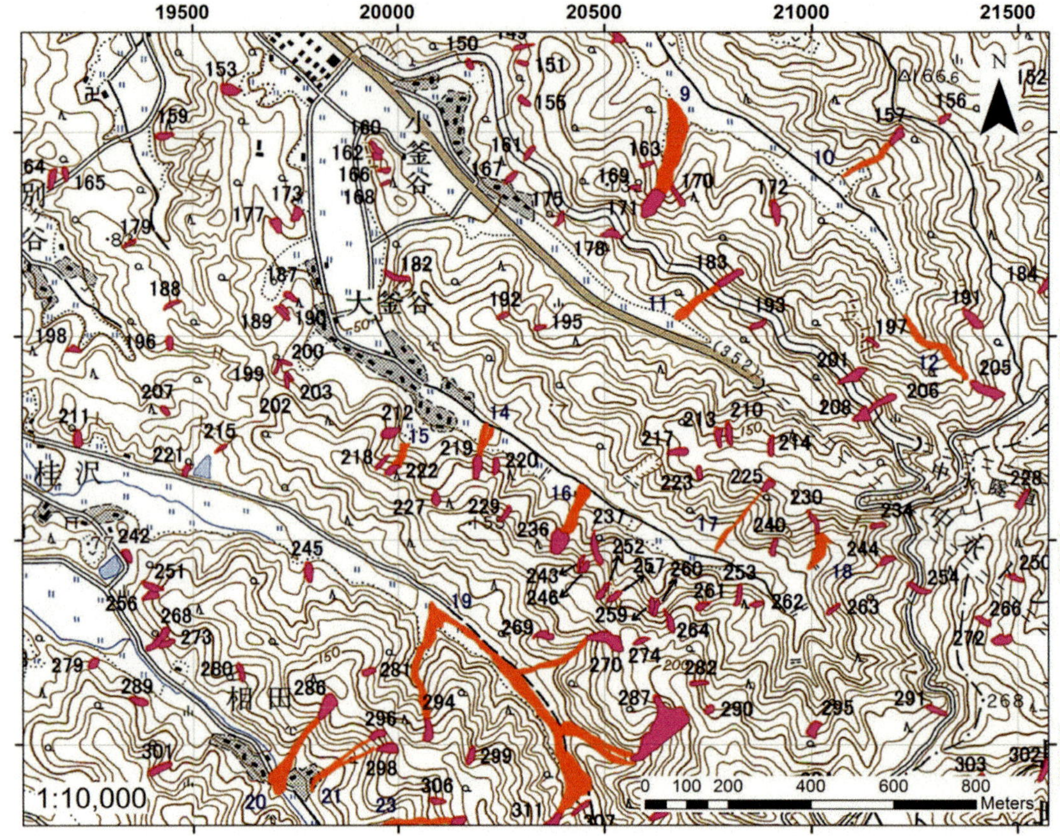

Fig. 11 Detailed map of shallow landslide distribution in Izumozaki, Niigata, Japan after intense precipitation on July, 2004. Notice that larger-sized shallow landslides are associated with mudflows, and they occurred along the valleys (provided by Asis Air Survey Co. Ltd)

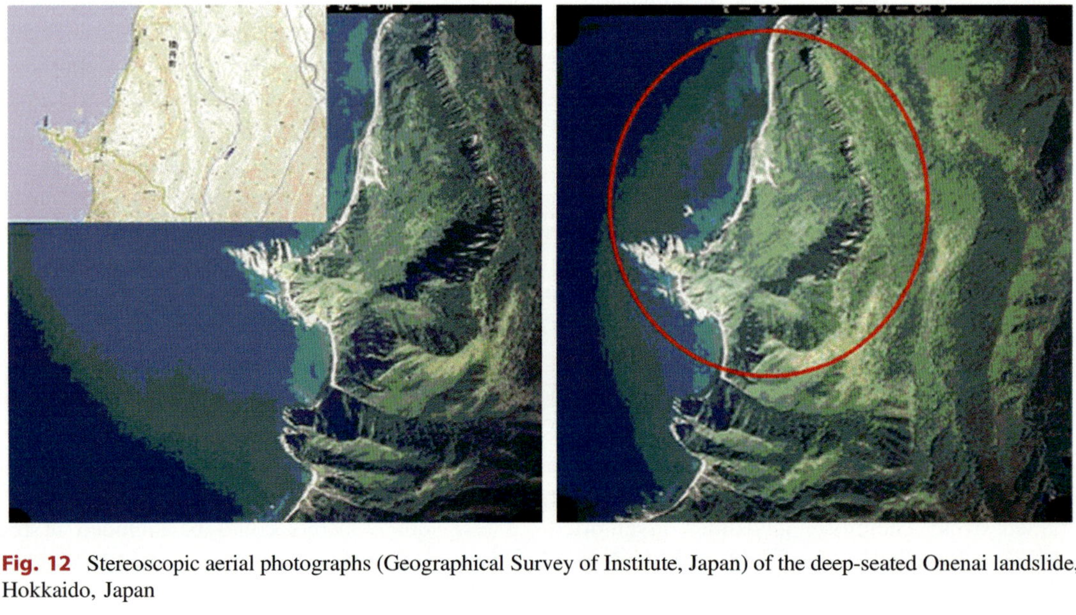

Fig. 12 Stereoscopic aerial photographs (Geographical Survey of Institute, Japan) of the deep-seated Onenai landslide, Hokkaido, Japan

Fig. 13 Stereoscopic aerial photographs (digital aerial photos by JICA Honduras) of deep-seated landslide close to Choluteca River in Tegucigalpa, Honduras

Fig. 14 Landslide identification in aerial photographs of Tegucigalpa, Honduras

4 Satellite Images from Google Earth to Detect and Mapping Landslides

As mentioned above, landslides are easily found using stereo-paired photographs and, to some extent, with detailed topographic maps. But if such photographs and topographic maps are not available, it is possible to find landslides using recent satellite images such as those of Google Earth.

In addition to its worldwide database, Google Earth allows images to be seen in 3D with terrain exaggeration to draw attention to scarps and other features and, therefore, large-scale

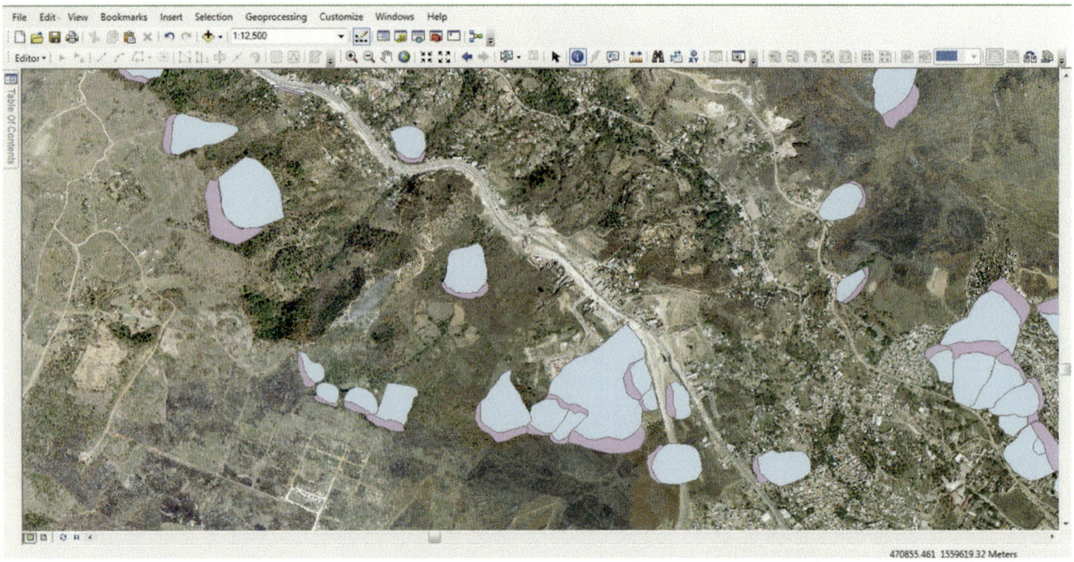

Fig. 15 Use of GIS and digital aerial photograph for elaboration of a digital landslide inventory map based on interpretation of aerial photographs (20 cm resolution) of Tegucigalpa, Honduras; provided by Japan International Cooperation Agency (JICA) Honduras

Fig. 16 Identification of deep-seated landslides using Google Earth for a very large valley in northeastern Afghanistan

landslides are easily identified. Other important tools that can be used to assist in landslide identification, include time series satellite images and profile views. These two options can help track slow mass movements over time and predict future positions or confirm possible landslide topography, respectively.

Figure 16 is a Google Earth image showing landslides associated with lakes in northeast Afghanistan. This Google Earth image shows many deep-seated landslides. It can be interpreted in the following way: the red-colored drawn polygon (debris of the landslide) blocked the valley, as a result the pond behind it increased in water level, and then in future, there is concern that the lake will overflow the debris, resulting in flooding. On the other hand, the yellow-colored polygon (scarp of the landslide) from the left hand scarp, formed the pond by slumping movement, and in this case there is no concern of water level increasing, thus, this slide is probably more stable than the red-colored one. Other landslide body can be observed highlighted in beige.

With assistance of its 3D view and acceptable to high quality of satellite images, it was possible to attain a dependable level of identification as good as that of aerial photographs, at least for deep seated landslides. Figure 17 shows additional landslides identified for the northeast mountains of Afghanistan, were both landslide scarp and body were easily distinguishable.

On the other hand Google Earth can also be used to find slope failures. Figure 18a from Google Earth and Fig. 18b (Digital Air Photo of 2013) show slope failures along the Choluteca River in Tegucigalpa City, Honduras. Comparing both pictures, it is clear that Google Earth can be used as a substitute in case aerial photographs are not available, since the difference in quality tends to be similar. Only if fine details

Fig. 17 More detailed identification of deep-seated landslides with scarps using Google Earth for the mountains in northeastern Afghanistan (Green-colored and Orange-colored polygons are scarps and bodies of landslides, respectively)

(a) (b)

Fig. 18 Failure at steep slope in Tegucigalpa, Honduras. **a** Air photo from Google Earth 2013, **b** Air photo of digital aerial photo from the same year. *Red areas* in (**b**) are failures on steep slopes at Tegucigalpa city

of texture and landslide classification or precise photogrammetric procedures are required, then aerial photographs become mandatory for a study area.

For implementation of landslide inventory and its respective education in different areas of the world, Google Earth is a very valuable tool. After defining the target area, terrain can be inspected both in 2D and 3D for identification of potential sites. Then, areas of interest are examined carefully and their general limits are marked with a polygon. Afterwards, specific polygons can be created for scarps, main body of deep-seated landslides, and slope failures,

Fig. 19 Shallow landslides identified on Google Earth in Santiago, Chile

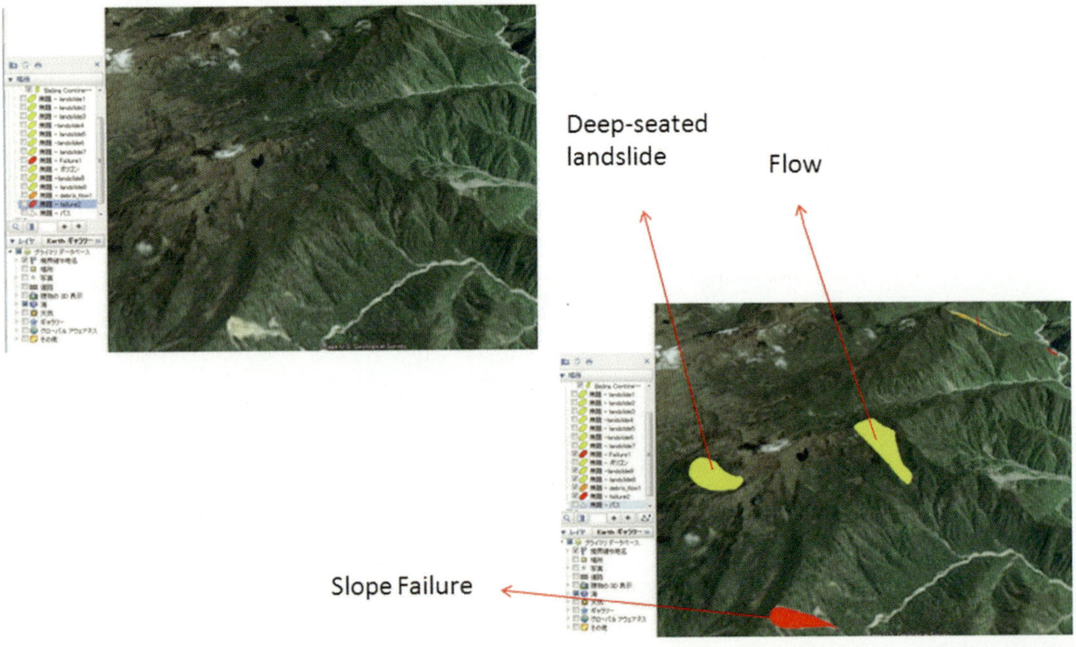

Fig. 20 Different types of mass movements identified on Google Earth in Quito, Ecuador

etc. Also, evolution of the landslide over time can be analyzed using the time series mode if historical images are available. Finally, a profile of terrain can be created to confirm additional landslide characteristics. This basic procedure is performed for each specific site and a kml shape compatible with GIS can be created for sharing and comparing findings. Results for landslide identification for other example study sites can be found in Figs. 19 and 20. While some landslides are evident from an initial observation, some require more careful analysis for correct identification.

5 Conclusions

In this landslide teaching tool (LTT), the generalities of how to elaborate landslide inventory using topographic maps, aerial photographs, and satellite images from Google Earth are described. In order to do it, it is required to know the models and features of deep-seated and shallow landslides and to develop training in stereo-pair viewing and aerial photo interpretation. It is also important to recognize the main types of landslides, such as deep-seated and shallow landslides. Both types of landslides are necessary to be detected for a complete assessment of a target area.

Basic indications for using topographic maps, aerial photographs, and Google Earth images for landslide recognition and mapping are presented and discussed, with their level of applicability examined. Finally, application examples of these criteria are illustrated for different sites and information sources. These examples confirm and demonstrate that the objectives of the LTT are attainable and feasible with the correct procedure and criteria.

Comparing the different types of information sources discussed, some points are relevant to emphasize. For example, while all sources of terrain information used in this document are important and vital for landslide identification, some have additional advantages. Use of aerial photography is encouraged and preferred to guarantee the greatest amount of detail and spatial resolution when identifying landslides. However, high quality aerial photographs are not always available and tend to be very costly. On the other hand, topographic maps, while not providing as many details, are very valuable for sites covered with vegetation or infrastructure that can obscure observation in aerial photographs. But these contour maps must have a large scale (around 1:10,000), otherwise they are only useful to find major landslides or scarps. An alternative for obtaining topographic maps when not available is to create them by GIS software from 30 m DEM from USGS (EARTHDATA) and ALOS Global Digital Surface Model "ALOS World 3D", but this is not the subject of this paper.

If contour maps and aerial photographs are not accessible, it was confirmed that Google Earth can be a very effective replacement. Unfortunately, if very detailed studies are required, Google Earth may not offer high resolution images for unpopulated areas, some which can be important landslide study sites. The other useful tools are now developing, such as three-dimensional models and variable image systems using GIS (Geographical Information System) software.

It is evident that integration of different terrain feature sources is necessary; both to compensate for limitations of each individual geographic information source and to have different points view that can strengthen the identification process. This is important since this detection process depends on qualitative or subjective interpretation of terrain based on experience. Going further, interpretation by different trained individuals of the same area is encouraged in order to enhance landslide identification.

Acknowledgments We thank JICA office in Honduras and the JICA experts (Drs. Yagi, H., Sato, G., and Hirota, K.) for their contribution and cooperation during 2011–2016 JICA Project of Landslide training in Tegucigalpa City.

References

Cruden DM, Varnes DJ (1996) Landslide types and processes, vol 247. Special Report, Transportation Research Board, National Academy of Sciences, pp 36–75

Hungr O, Leroueil S, Picarelli L (2013) The Varnes classification of landslide types, an update. Landslides 11(2):167–194
Highland LM, Bobrowsky P (2008) The landslide handbook—a guide to understanding landslides. US Department of the Interior, US Geological Survey, Circular 1325, 129p
NIED (2013) Digital archive for landslide distribution maps. National Research Institute for Earth Science and Disaster Prevention, Japan. Website: http://dil-opac.bosai.go.jp/publication/nied_tech_note/landslidemap/gis.html
Pike RJ (1988) The geometric signature: quantifying landslide-terrain types from digital elevation models. Math Geol 20(5):491–511. doi:10.1007/BF00890333
Shin Engineering Consulant Co. Ltd. (2013) Digital archive for landslide distribution maps. Chibouzu. Shin Engineering Co. Ltd, Japan. Website: http://www.shin-eng.info/chibouzu/map.html
Suzuki T (1997) Introduction to map reading for civil engineering. Geomorphological Basis for map reading, vol 1. KokonShoin, 200p
Varnes DJ (1978) Slope movement types and processes. In: Schuster RL, Krizek RJ (eds) Landslides-analyses and control special report 176. National Academy of Sciences, Washington, D.C., pp 11–34
Yamagishi H (2012) Hokkaido Landslide Digital Map (+DVD), 102p. Hokkaido University Press, Sapporo

TXT-tool 1.039-1.1
Very-High Resolution Stereo Satellite Images for Landslide Mapping

Francesca Ardizzone, Federica Fiorucci,
Alessandro Cesare Mondini and Fausto Guzzetti

Abstract

Landslides are common phenomena in mountainous countries, and play an important role in the evolution of landscapes. They also represent a serious hazard in many areas of the world. Acquiring systematic information on the type, abundance, and distribution of landslides, and preparing landslide inventory maps is of fundamental importance to mitigate landslide risk. Landslide inventory maps are essential for evaluating landslide hazard, vulnerability and risk, and for studying the evolution of landscapes dominated by mass-wasting processes. Landslide maps, including geomorphological, event, seasonal, and multi-temporal inventory maps, can be prepared using different techniques. We present the results of an experiment aiming at testing the possibility of using very high resolution, stereoscopic satellite images to map rainfall-induced shallow landslides. Three landslide inventory maps were prepared for the Collazzone study area, Umbria, Italy. Two of the maps were prepared through the visual interpretation of stereoscopic satellite images, and cover the periods January–March 2010, and March–May 2010. The third inventory map shows landslides occurred in the period January–March 2010, and was obtained through reconnaissance field surveys. We describe the statistics of landslide area for the three inventories, and compare quantitatively two of the landslide maps.

Keywords

Landslide · Inventory map · Remote sensing · Satellite image · GIS

F. Ardizzone (✉) · F. Fiorucci · A.C. Mondini · F. Guzzetti
Consiglio Nazionale delle Ricerche, Istituto di Ricerca per la Protezione Idrogeologica, Via della Madonna Alta 126, 06128 Perugia, Italy
e-mail: francesca.ardizzone@irpi.cnr.it

F. Fiorucci
e-mail: federica.fiorucci@irpi.cnr.it

A.C. Mondini
e-mail: alessandro.mondini@irpi.cnr.it

F. Guzzetti
e-mail: fausto.guzzetti@irpi.cnr.it

K. Sassa et al. (eds.), *Landslide Dynamics: ISDR-ICL Landslide Interactive Teaching Tools*, https://doi.org/10.1007/978-3-319-57774-6_5

Contents

1 **Introduction** ... 84
2 **Study Area** ... 84
3 **Materials and Methods** ... 85
3.1 VHR Stereoscopic Satellite Images ... 85
3.2 Hardware and Software Visualization System. 85
3.3 Visual Interpretation Criteria ... 87
3.4 Field Surveys ... 88
3.5 Rainfall Conditions ... 88
4 **Results** ... 88
4.1 Analysis of the Landslide Inventories ... 89
5 **Discussion and Conclusion** ... 91
References ... 93

1 Introduction

A landslide inventory is the simplest form of landslide map (Guzzetti et al. 2000), and shows the location, the date of occurrence, and the types of mass movements that have left discernible traces in an area. A landslide map is essential for geomorphological and ecological studies, and to evaluate landslide hazard, vulnerability, and risk. Landslide maps can be classified into different types based on variable causes: geomorphological, event, seasonal, and multi-temporal inventory maps (Cardinali et al. 2001; Guzzetti et al. 2006), therefore they are traditionally prepared exploiting different techniques, such as the visual interpretation of stereoscopic aerial photographs (Bucknam et al. 2001; Cardinali et al. 2001), reconnaissance field survey (Dapporto et al. 2005; Cardinali et al. 2006; Santangelo et al. 2010), and analysis of archive information on historical landslide events (Taylor and Brabb 1986; Guzzetti 2000; Salvati et al. 2013; Devoli et al. 2007; Damm and Klose 2014; Zêzere et al. 2014; Taylor et al. 2015).

The availability of high resolution (HR) and very-high resolution (VHR) satellite images, and improved digital visualization and analysis techniques, has encouraged investigators to exploit satellite images to detect and to map landslides (e.g., Mantovani et al. 1996; Saba et al. 2010; Murillo-Garcia et al. 2015) and erosion features (Fiorucci et al. 2015). Stereoscopic and 3-dimensional (3-D) models obtained from HR and VHR images can be examined visually to detect individual landslides, or groups of landslides, and to prepare landslide inventory maps (Alkevli and Ercanoglu 2010).

Guzzetti et al. (2012), in a recent review of the literature on landslide inventories, have shown that one of the most promising approaches exploits VHR optical, monoscopic and stereoscopic satellite images, analysed visually or through semi-automatic procedures. The authors also conclude that a combination of satellite, aerial, and terrestrial remote sensing data represents the optimal solution for landslide detection and mapping, in different physiographic, climatic, and land cover conditions.

We present the results of an experiment to test the possibility of using VHR stereoscopic satellite images, and innovative 3-D visualization technology, to prepare landslide inventory maps. The experiment was conducted in Umbria, Italy, where two inventories were prepared exploiting multiple sets of stereoscopic satellite images. A third inventory was obtained through a reconnaissance field survey conducted after a rainfall period that resulted in landslides. We measured the degree of matching between two inventories, adopting the method proposed by Carrara et al. (1992) and Ardizzone et al. (2002).

2 Study Area

The study area is in Umbria region, Italy (Fig. 1), and extends for about 90 km^2, including 78.9 km^2 of hilly terrain. Sedimentary rocks, Cretaceous to Recent in age, crop out in the area, and comprise recent fluvial deposits, continental gravel, sand and clay, travertine, layered sandstone and marl, and thinly layered limestone (Servizio Geologico Nazionale 1980). The soil texture is medium-fine, and soil thickness ranges from a few decimetres to more than one meter. Arable land, forests, urban areas, pastures, vineyards and orchards represent the land cover. Farming in the area favours the development of slope instabilities, including landslides and channelled and sheet erosion. The climate is Mediterranean, with most of the precipitation falling from October to December, and from February to May. The historical rainfall time series (1951–2013) for the Todi rain gauge (283 m of elevation, Fig. 1) shows that the average annual rainfall is 841.1 mm, and that the maximum and minimum average monthly precipitation levels were in November (116 mm) and July (38 mm), respectively.

Landslides are abundant in the area, and are caused by meteorological triggers i.e., prolonged rainfall and rapid snowmelt events. Mass movements in the area include soil slides and flows (shallow landslides), deep-seated slides, and compound failures (Guzzetti et al. 2006; Galli et al. 2008). Shallow landslides are mainly represented and occur primarily on cultivated or abandoned areas, and are rare in the forested terrain. In the cultivated areas, mechanical ploughing and harrowing obliterate landslides features. For this reason, the lifetime of the individual shallow landslides rarely exceeds a few seasons, although reactivations and new slope failures are common where previous landslides have occurred (Fiorucci et al. 2011).

3 Materials and Methods

For the hilly portion of the study area we prepared three inventory maps using two different methods: (1) the visual interpretation of VHR stereoscopic satellite images, and (2) reconnaissance geomorphological field mapping. In this section, we first describe the satellite images, the hardware and software visualization technology, and the interpretation criteria used to prepare the landslide maps. Next, we discuss the production of the reconnaissance geomorphological field mapping, and we provide information on the rainfall history that has resulted in the many landslides.

3.1 VHR Stereoscopic Satellite Images

For our experiment, we used VHR panchromatic images taken (1) by the GeoEye-1 satellite on 12 August 2009 and on 27 May 2010, and (2) by the WorldView-1 satellite on 8 March 2010. GeoEye-1 was launched in September 2008, and flying at an altitude of 681 km it captures images at 0.41-m panchromatic (black and white, resampled at 0.50-m) and 1.65-m multispectral (resampled at 2-m) resolution. WorldView-1 was launched in September 2007, and operating from an altitude of 496 km, it takes images at 0.50-m panchromatic and 2-m multispectral resolution. Table 1 lists the main characteristics of the stereoscopic images taken by the two satellites, and used to detect and map landslides in our study area.

Rational Polynomial Coefficients (RPCs) for the satellite images were available to us. RPCs provide a representation of the ground-to-image geometry, allowing for photogrammetric processing. We used the RPCs to generate 3-D models of each pair of stereoscopic satellite images. We estimated the accuracy of operation using eight checkpoints. Horizontal accuracy of the models, measured by the root mean square error (RMSE), was 2.2 m for the GeoEye-1 and 1.12 m for the WorldView-1 images. Vertical accuracy was 2.0 m for GeoEye-1, and 1.9 m for WorldView-1. The Bi-dimensional RMSE was 2.9 m for GeoEye-1, and 2.2 m for WorldView-1. The obtained values are within the accuracy of a 1:10,000 topographic map, and hence well suited for geomorphological and multi-temporal landslide inventory maps.

3.2 Hardware and Software Visualization System

We used ERDAS IMAGINE® and Leica Photogrammetry Suite (LPS) software for block orientation, and Stereo Analyst for ArcGIS® software for image visualization and landslide mapping. To obtain 3-D views of the VHR satellite images, we used the StereoMirrorTM hardware technology (Fig. 2).

Using Stereo Analyst for Arc ArcGIS®, we studied the 3-D views of the topographic surface prepared with the oriented satellite images, and we collected 3-D geographical information on landslides. The software allowed the preparation of multiple sets of oriented images, all having the same reference system. To map the

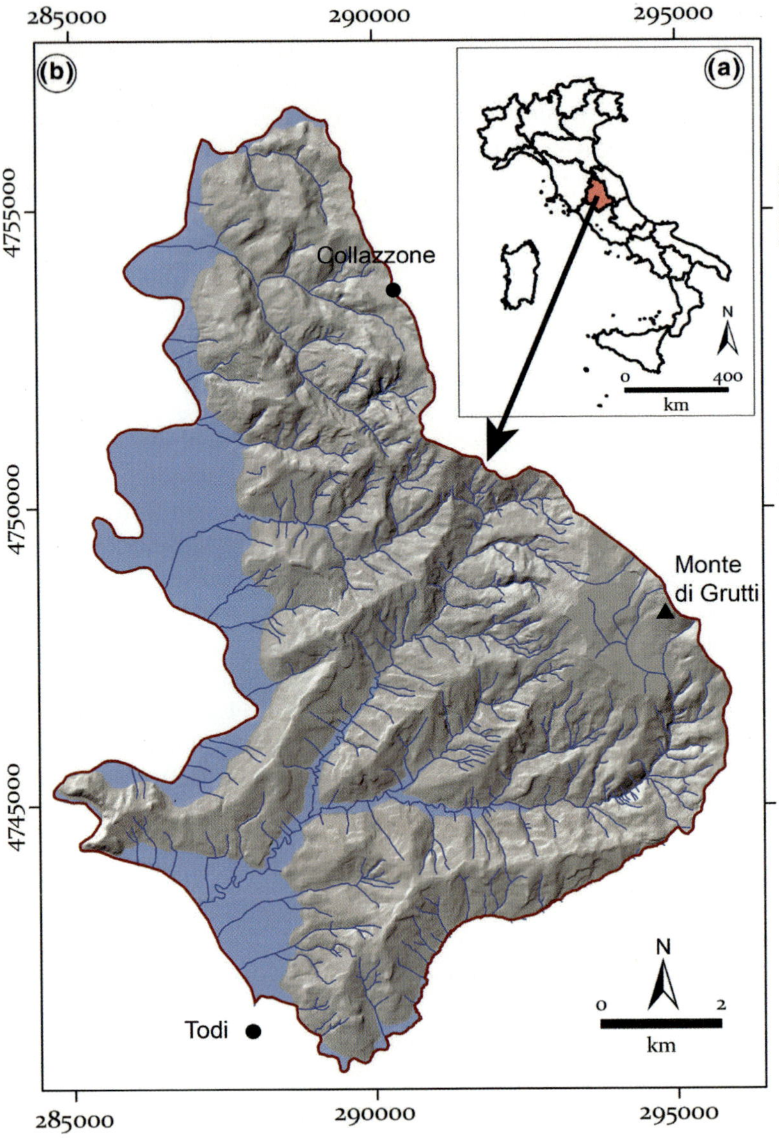

Fig. 1 Location and morphology of the Collazzone study area. **a** Italy territory. Umbria region is represented by the *red polygon*. **b** Collazzone study area. The coordinate system is UTM ED50 zone 33 (EPSG 23033)

Table 1 Characteristics of the stereoscopic satellite images acquired by the GeoEye-1 and the WorldView-1 satellites, and used in our work

	Date	Overlap (%)	Azimuth angle (°)	Elevation angle (°)
GeoEye-1	12/08/2009	90	45.18	86.18
GeoEye-1	12/08/2009	90	189.90	62.3
GeoEye-1	12/08/2009	90	349.99	61.97
GeoEye-1	12/08/2009	90	244.90	13.80
WorldView-1	08/03/2010	70	77.30	57.90
WorldView-1	08/03/2010	70	141.1	52.70
GeoEye-1	27/05/2010	95	8.41	72.26
GeoEye-1	27/05/2010	95	199.51	71.99

landslides in real-world 3-dimensional geographical coordinates, a 3-D floating cursor was used. When using a floating cursor, it is important that the cursor stays on the topographic surface. The software facilitates the task with an automatic terrain-following mode. A 3-D floating cursor consists of an independent cursor that is displayed on the left image and an independent cursor that is displayed on the right image of the stereo-pair. When images are not viewed in stereo, the 3-D floating cursor appears as two separate cursors. However, when viewed in stereo, the two cursors fuse to create the perception of a 3-D floating cursor. The manual adjustment of the position of the 3-D floating cursor requires the continuous attention of the operator. We used image correlation to obtain elevation information using the floating cursor (Murillo-Garcia et al. 2015). The StereoMirrorTM technology allows for the 3-D visualization of large areas. For our experiment, the area covered by the GeoEye-1 stereoscopic images was 13 km × 17 km (160 km^2), and the area covered by the WorldView-1 stereoscopic images was 24 km × 24 km (576 km^2). The ability to analyse in 3-D multiple sets of stereoscopic images for a large area proves important for landslide mapping, chiefly for multi-temporal mapping. The hardware and software technology simplified the acquisition of landslide information from stereoscopic images. Further, the landslide and morphological information was obtained in 3-D and stored directly in a GIS database, reducing the acquisition time and the problems (and errors) associated with the manual digitization of the landslide information and the construction of the geographical database (Galli et al. 2008).

3.3 Visual Interpretation Criteria

To recognize the landslides in the digital stereoscopic satellite images, we used the same interpretation criteria commonly adopted by geomorphologists to identify landslides on stereoscopic aerial photographs. The interpreter detects and classifies a landslide based on experience, and on the visual analysis of a set of features or characteristics that can be identified on the images (Fig. 3). These include: shape, size, tone, colour, mottling, texture, pattern of objects, and site topography (Ray 1960; Allum 1966; Rib and Liang 1978; van Zuidan 1985). Shape refers to the form of the topographic surface. Size describes the areal extent of an object. Colour, tone, mottling, and texture depend on the light reflected by the surface, and can be used to infer rock, soil and vegetation types. Mottling and texture are measures of terrain roughness and

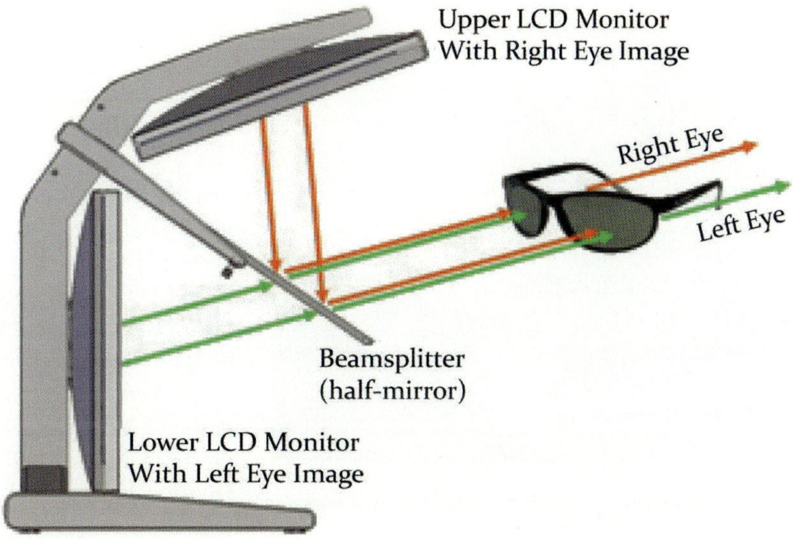

Fig. 2 Sketch of the StereoMirrorTM technology (http://www.planar.com/)

can be used to identify surface types and the size of debris. Pattern is the spatial arrangement of objects in a repeated or characteristic order or form. Site topography indicates the position of a place with reference to its surroundings and reflects morphometric characters. Because of the vertical exaggeration of stereoscopic vision, shape is a very useful characteristic for the identification and classification of the landslide from aerial or satellite stereoscopic images. Stereoscopic vision allows the recognition of landslides on aerial photographs or satellite images, particularly where landslides do not show evident morphological signs (Fig. 3).

3.4 Field Surveys

Following rainfall events in December 2009 (Fig. 4), we performed reconnaissance field surveys to identify and map rainfall-induced landslides in the study area. Four geomorphologists searched an area of about 90 km^2 in four days between January and March 2010. They drove and walked along roads, stopped where single or multiple landslides were identified, and at viewing points to check the slopes. Landslides were mapped in the field at a 1:10,000 scale, using the geomorphological and topographic information available on site. Single and pseudo-stereoscopic colour photographs of landslides were taken with digital hand-held cameras. The cartographic and photographic information obtained in the field was used in the laboratory to map visually the individual landslides on 1:10,000 scale ortho-photographs. The landslide and morphological information was then stored digitally in a GIS database.

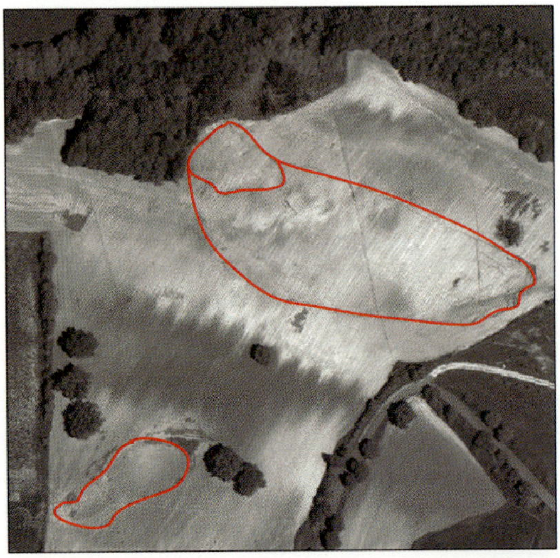

Fig. 3 Example of landslides on satellite stereo image in the Collazzone study area

3.5 Rainfall Conditions

To investigate the rainfall conditions that have resulted in landslides in the study area in the period from September 2009 to May 2010, we used rainfall measurements obtained by the Todi rain gauge, located 3 km south of the study area (Fig. 1). In autumn 2009, the rain gauge measured 194 mm of rain, with 34.37 mm in four days between 14 and 17 September 2009. In the 2009–2010 winter, the rain gauge recorded 323.84 mm, with 170.86 mm in 22 days from 19 December 2009 to 9 January 2010. In the spring of 2010, the rain gauge measured a cumulated rainfall of 205 mm, with 132.34 mm in 17 days, 3–19 May 2010 (Fig. 4). Inspection of the rainfall record of Fig. 4 indicates that most of the landslides mapped in this inventory occurred presumably in the period between the end of December 2009 and the first decade of January 2010, with a second wet period in the month of May when rainfall probably caused landslides that were identified in the GeoEye image of 27 May.

4 Results

For the Collazzone hilly study area, we prepared three landslide inventory maps using the hardware and software visualization system described in Sect. 3.2, and applying the visual interpretation criteria (Sect. 3.3) to the VHR satellite images illustrated in Sect. 3.1 (Fig. 5). Two maps were obtained through the visual interpretation of VHR stereoscopic satellite images taken on 12 August 2009 and 8 March 2010 by the GeoEye-1 and the World-View-1 satellites, (MAP A), and on 27 May 2010 by the GeoEye-1

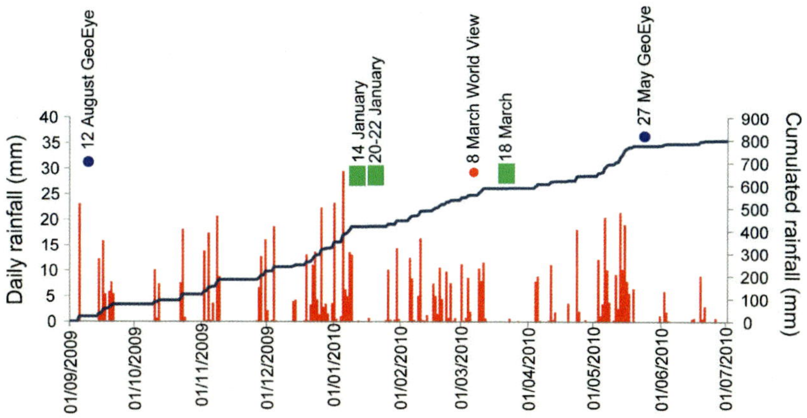

Fig. 4 Rainfall in Todi rain gauge (see Fig. 1). Rainfall conditions in the period 1 September 2009–1 July 2010. *Bars* show cumulated daily rainfall. *Blue line* shows total cumulate rainfall. *Green squares* show dates of field surveys. *Red dot* shows date of WorldView-1 image. *Blue dots* show date of GeoEye-1 images (Table 1)

satellite (MAP B). The third map (MAP C) was obtained through a reconnaissance field survey conducted between 14 of January and 18 of March 2010.

To detect and map the rainfall-induced shallow landslides on 3-D digital representations of the stereoscopic satellite images, we adopted a multi-temporal approach (Fiorucci et al. 2011), and we recognized new landslides by comparing images from different dates.

To prepare MAP A (Fig. 6), we compared visually the GeoEye-1 images taken on 12 August 2009 and the World-View-1 images taken on 8 March 2010. Thus, MAP A covers the period August 2009–March 2010.

To obtain MAP B (Fig. 6), we compared visually the World-View-1 images taken on 8 March 2010 with the next images obtained on 27 May 2010 by the GeoEye-1 satellite. Therefore, MAP B covers the period March–May 2010, with most of the landslides presumably occurring in the first half of May (Fig. 4).

To obtain MAP C (Fig. 6), a reconnaissance field survey was carried out on 14, 20 and 22 January, and 18 March, 2010 with most of the landslides in the period between the end of December 2009 and the first decade of January 2010.

MAP A shows 159 shallow landslides (Fig. 6) ranging in size from 3.7×10^1 to 1.51×10^4 m^2, for a total landslide area A_{LT} = 2.94×10^5 m^2, that corresponds to 0.37% of the hilly portion of the study area, an average of 2.0 landslides per square kilometre (Table 2). MAP B shows 55 shallow landslides (Fig. 6), with individual landslides in the range from 2.98×10^1 to 1.16×10^4 m^2, for a total landslide area $A_{LT} = 8.52 \times 10^4$ m^2. This corresponds to 0.11% of the hilly portion of the study area, an average of 0.7 landslides per square kilometre.

The reconnaissance inventory shown in MAP C shows 76 landslides (Fig. 6), ranging in area from 4.59×10^1 to 1.89×10^4 m^2, for a total landslide area $A_{LT} = 1.37 \times 10^5$ m^2, that correspond to 0.17% of the hilly portion of the study area, and an average of 0.04 landslides per square kilometre (Table 2).

4.1 Analysis of the Landslide Inventories

We analysed the landslide inventory maps for studying the statistics of landslide area, and we compared quantitatively MAP A and MAP C that show landslides presumably occurring in the same period. For each landslide map, the planimetric area was obtained in a GIS. Table 2 lists summary statistics for the mapped landslides in the three inventories. We estimated the frequency density distribution of the landslide areas for the

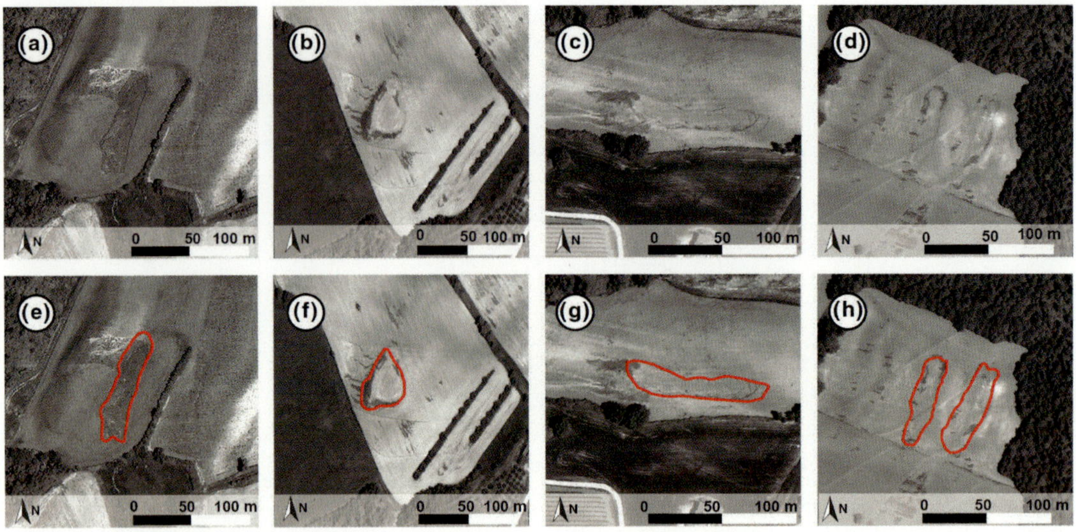

Fig. 5 Examples of landslides recognized on the GeoEye-1 satellite image taken on 27 May 2010 are shown in **a–d**. They are shown *encircled* in **e–h**, respectively

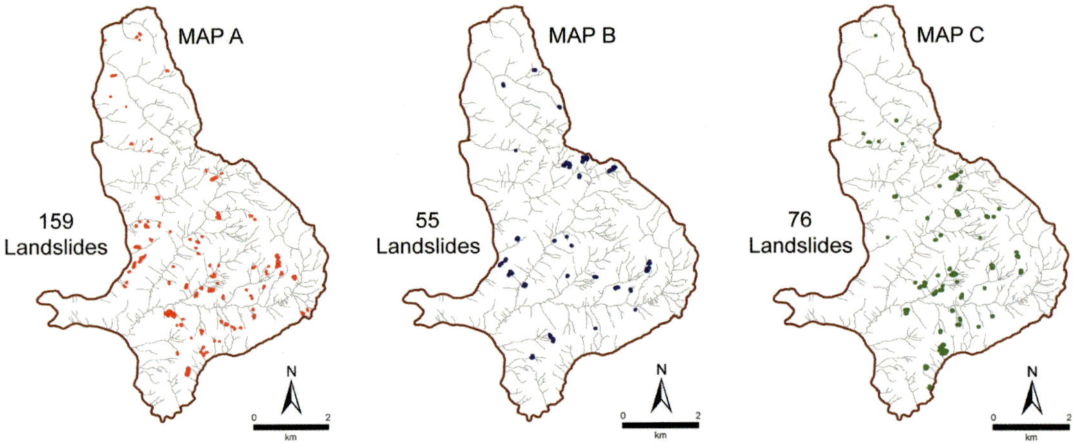

Fig. 6 Landslide inventory maps for the Collazzone study area. *MAP A* shows landslides mapped through the visual interpretation of the WorldView-1 stereoscopic images taken on 8 March 2010. *MAP B* shows landslides mapped through the visual interpretation of the GeoEye-1 stereoscopic images taken on 27 May 2010. *MAP C* shows the landslide inventory prepared through field surveys in the period January–March 2010

three maps using the Inverse Gamma (IG) and the Double Pareto (DP) functions (Stark and Hovius 2001; Malamud et al. 2004) (Fig. 7). The two distributions provide similar results within the three inventories (Fig. 7 and Table 3), with the standard error (ε in Table 3) largest for MAP B, the map with the least number of landslides. Figure 7 shows a distinct "rollover" in the Inverse Gamma distribution of landslide area for the three maps, while a problem exists

Table 2 Descriptive statistics for the three landslide inventory maps available for the Collazzone study area

	MAP A	MAP B	MAP C
Area covered (km^2)	90.0	90.0	90.0
Hilly area (km^2)	78.9	78.9	78.9
N_L (#)	159	55	76
Min A_L (m^2)	3.7×10^1	2.98×10^1	4.59×10^1
Max A_L (m^2)	1.51×10^4	1.16×10^4	1.89×10^4
Mean A_L (m^2)	1.86×10^3	1.55×10^3	1.81×10^3
Median A_L (m^2)	1.01×10^3	9.78×10^2	9.82×10^2
St. Dev. A_L (m^2)	2.5×10^3	2.05×10^3	2.68×10^3
A_{LT} (m^2)	2.94×10^5	8.52×10^4	1.37×10^5
$\delta(A_L)$ # km^{-2}	2.0	0.7	0.04

with the estimation of the Double Pareto function for MAP B and MAP C, probably due to the limited number of landslides in these two maps.

To evaluate the geographical mismatch between MAP A and MAP C we used the Error index proposed by Carrara et al. (1992), and the corresponding Matching index proposed by Galli et al. (2008). The two indices quantify the degree of similarity between the different landslide maps, using:

$$E = \frac{(A \cup C) - (A \cap C)}{(A \cup C)} \quad 0 \leq E \leq 1 \qquad (1)$$

$$M = 1 - E \quad 0 \leq M \leq 1 \qquad (2)$$

We applied Eqs. (1) and (2) to the entire study area, and to the individual homologous landslides in the two inventories (MAP A and MAP C). Geographical union ($\cup$) and intersection ($\cap$) of MAP A and MAP C were obtained in a GIS. Geographical union of MAP A and MAP C was 3.73×10^5 m^2, and the landslide area common to both inventories was 5.89×10^4 m^2. The error index E was equal to 0.84, and the corresponding match index M was equal to 0.16. These values indicate a significant discrepancy between the two inventories, as indicated in Galli et al. (2008). E indices (and M indices) for individual homologous landslides range from E = 0.35 to E = 0.90 (mean = 0.65, σ = 0.15), and M = 0.10 to M = 0.65 (mean = 0.35, σ = 0.15). The figures are similar to the values obtained by Santangelo et al. (2010) for a nearby area.

5 Discussion and Conclusion

In the literature, most of the attempts to detect and map landslides using satellite imagery (Mantovani et al. 1996; Singhroy 2005) consisted in the recognition and mapping of slope failures that have left easily recognizable morphological signs, essentially evident changes in land cover (e.g., from dense forest to exposed soil and rock). In our experiment, VHR stereoscopic satellite images were used to detect and map landslides that did not result in distinct morphological or land cover signatures that could be recognized easily using aerial photographs or field surveys. We consider the use of VHR satellite image in geographical areas where landslides are cancelled by human activity and where the signatures of failures are subtle to be an efficient tool to detect and map landslides. This is an advancement compared to previous results.

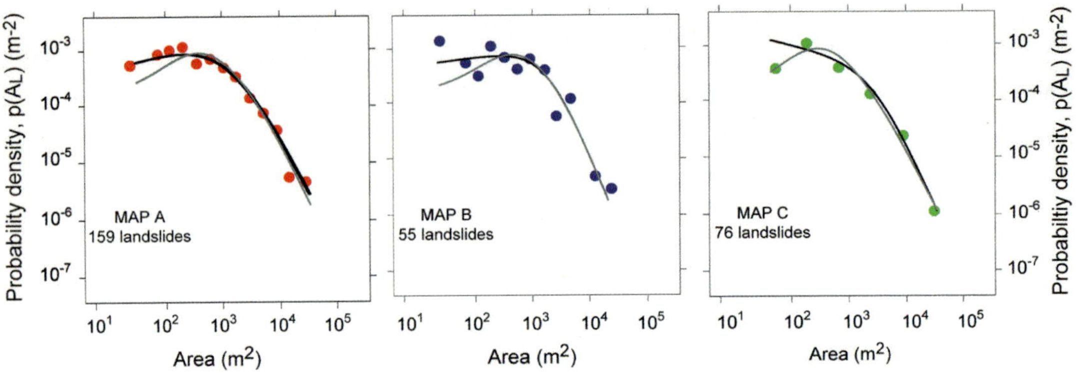

Fig. 7 Statistics of landslide size. The *graphs* show the probability density of landslide area, p(AL) for the three maps. *Coloured dots* are frequency values calculated by means of histogram estimation of logarithm of the data. *Black lines* are double Pareto, and *grey lines* are inverse Gamma models of p(AL) obtained through maximum likelihood estimation

Table 3 Parameters estimated for the double Pareto (Stark and Hovius 2001) and the inverse Gamma (Malamud et al. 2004) models for the probability density of landslide area, $p(A_L)$

	IG		DP	
	$\alpha + 1$ (ε)	$\bar{A}_L$ (m^2)	$\alpha + 1$ (ε)	$\bar{A}_L$
MAP A	2.45 (0.25)	325	2.24 (0.28)	201
MAP B	3.06 (0.65)	393	2.79 (0.55)	270
MAP C	2.20 (0.28)	250	2.49 (0.26)	270

The hardware and software used in the experiment allowed for the effective 3-D visualization of a large territory (>100 km^2), and simplified the acquisition of 3-D landslide information and storage in a GIS database. Advantages of the system include: (i) the ability to examine a large study area, (ii) the possibility to dynamically zoom in and out and to change the image contrast during the visual inspection, facilitating the mapping of landslides of different sizes, and (iii) the ability to digitize in 3-D during the image interpretation phase, reducing the time and errors associated with the acquisition of landslide information in digital format (Galli et al. 2008).

The landslide inventory produced by visual interpretation of stereoscopic satellite images (MAP A) showed 109% more landslides, covering 114% more landslide area than the reconnaissance landslide inventory (MAP C). We attribute the difference to the possibility of recognizing slope failures in the VHR stereoscopic images. The spatial resolution of the satellite images, 0.5 × 0.5 m for GeoEye-1 and WorldView-1, are adequate to detect and map the smallest landslides in the study area. Further, the vertical exaggeration of stereoscopic vision allowed for the detection of small (faint) morphological features related to landslides.

The area covered by the satellite stereoscopic images (>100 km^2) is comparable to the area covered by small-scale aerial photographs (1:50,000 and 1:75,000), and allows the obtaining of enlargements up to 1:10,000 scale. The use of Rational Polynomial Coefficient information to generate the 3-D models allowed us to obtain accuracy levels comparable to the accuracy of 1:10,000 topographic base maps.

We estimated the probability density of landslide area $p(A_L)$ for the different landslide maps (Fig. 7 and Table 3). Inspection of the probability density curves reveals similarities in the density distribution of the landslide area, p (A_L). All the empirical distributions exhibit the same general shape, with the density for large and very large landslides obeying a negative power law trend. The Double Pareto function revealed estimation problems for MAP B and MAP C, due to the reduced number of landslides in the two maps.

The geographical mismatch between landslide maps computed using the method proposed by Carrara et al. (1992) and Ardizzone et al. (2002) revealed a discrepancy between the two inventories (MAP A and MAP C). The large error index, E = 0.84 depends chiefly on differences in the number of the mapped landslides in the two inventories, while the mismatch between homologous landslides is smaller, E = 0.65.

The experiment showed that VHR stereoscopic satellite and adequate 3-D viewing technology facilitates the production of accurate landslide inventory maps, even in areas where slope failures have left subtle signatures. This is an important step towards the systematic production of multi-temporal landslide inventories, seasonal landslide maps, and event landslide inventory maps. GeoEye-1 and WorldView-1 satellite sensors have revisiting times of only a few days (2–8 days for GeoEye-1, and 1.5–5 days for WorldView-1) allowing for the potential production of multi-temporal inventories and their rapid and frequent update. This is a mandatory step for landslide hazard and risk assessment, and for erosion and landscape evolution studies.

Acknowledgments Work conducted in the framework of the ASI MORFEO project. FF is supported by a grant of Umbria Region.

References

Alkevli T, Ercanoglu M (2010) Assessment of ASTER satellite images in landslide inventory mapping: Yenice-Gökçebey (Western Black Sea Region, Turkey). DOI, Bull Eng Geol Environ. doi:10.1007/s10064-011-0353-z

Allum JAE (1966) Photogeology and regional mapping. Institute of Geological Sciences, Photogeological Unit. Pergamon Press, Oxford, p 107

Ardizzone F, Cardinali M, Carrara A, Guzzetti F, Reichenbach P (2002) Impact of mapping errors on the reliability of landslide hazard maps. Nat Hazards Earth Syst Sci 2(1/2):3–14

Bucknam RC, Coe JA, Chavarria MM, Godt JW, Tarr AC, Bradley LA, Rafferty S, Hancock D, Dart RL, Johnson ML (2001) Landslides triggered by Hurricane Mitch in Guatemala—inventory and discussion. U.S. Geological Survey Open File Report 01-443

Cardinali M, Ardizzone F, Galli M, Guzzetti F, Reichenbach P (2001) Landslides triggered by rapid snow melting: the December 1996–January 1997 event in Central Italy. In: Proceedings of 1st EGS Plinius conference, 14–16 Oct 1999. Bios Publisher, Cosenza, pp 439–448

Cardinali M, Galli M, Guzzetti F, Ardizzone F, Reichenbach P, Bartoccini P (2006) Rainfall induced landslides in December 2004 in south-western Umbria, central Italy: types, extent, damage and risk assessment. Nat Hazards Earth Syst Sci 6:237–260

Carrara A, Cardinali M, Guzzetti F (1992) Uncertainty in assessing landslide hazard and risk. ITC J 2:172–183

Damm B, Klose M. (2014) Landslide database for the Federal Republic of Germany: a tool for analysis of mass movement processes. In: Sassa K, Canuti P, Yin Y (eds) Landslide science for a safer geoenvironment, vol 2. Springer International Publishing, Switzerland, pp 787–792

Dapporto S, Aleotti P, Casagli N, Polloni G (2005) Analysis of shallow failures triggered by the 14–16 November 2002 event in the Albaredo valley, Valtellina (Northern Italy). Adv Geosci 2:305–308

Devoli G, Strauch W, Chavez G, Hoeg K (2007) A landslide database for Nicaragua: a tool for landslide-hazard management. Landslides 4(2):163–176

Fiorucci F, Cardinali M, Carlà R, Rossi M, Mondini AC, Santurri L, Ardizzone F, Guzzetti F (2011) Seasonal landslide mapping and estimation of landslide mobilization rates using aerial and satellite images. Geomorphology 129:59–70

Fiorucci F, Ardizzone F, Rossi M, Torri D (2015) The use of stereoscopic satellite images to map Rills and Ephemeral Gullies. Remote Sens 7(10):14151–14178

Galli M, Ardizzone F, Cardinali M, Guzzetti F, Reichenbach P (2008) Comparing landslide inventory maps. Geomorphology 94:268–289

Guzzetti F (2000) Landslide fatalities and evaluation of landslide risk in Italy. Eng Geol 58:89–107

Guzzetti F, Cardinali M, Reichenbach P, Carrara A (2000) Comparing landslide maps: a case study in the upper Tiber River Basin, Central Italy. Environ Manage 25(3):247–363

Guzzetti F, Galli M, Reichenbach P, Ardizzone F, Cardinali M (2006) Landslide hazard assessment in the Collazzone area, Umbria, central Italy. Nat Hazards Earth Syst Sci 6:115–131

Guzzetti F, Mondini AC, Cardinali M, Fiorucci F, Santangelo M, Chang KT (2012) Landslide inventory maps: new tools for an old problem. Earth Sci Rev 112(1):42–66

Malamud BD, Turcotte DL, Guzzetti F, Reichenbach P (2004) Landslide inventories and their statistical properties. Earth Surf Proc Land 29:687–711

Mantovani F, Soeters R, van Westen C (1996) Remote sensing techniques for landslide studies and hazard zonation in Europe. Geomorphology 15:213–225

Murillo-García FG, Alcántara-Ayala I, Ardizzone F, Cardinali M, Fiourucci F, Guzzetti F (2015) Satellite stereoscopic pair images of very high resolution: a step forward for the development of landslide inventories. Landslides 12(2):277–291

Ray RG (1960) Aerial Photographs in geological interpretation and mapping. Geological survey professional paper 373, Washington, USA

Rib HT, Liang T (1978) Recognition and identification. In: Schuster RL, Krizek RJ (eds) Landslide analysis and control. Transportation research board special report 176. National Academy of Sciences, Washington D.C., pp 34–80

Saba SB, van der Meijde M, van der Werff H (2010) Spatiotemporal landslide detection for the 2005 Kashmir earthquake region. Geomorphology 124:17–25

Salvati P, Marchesini I, Balducci V, Bianchi C, Guzzetti F (2013) A new digital catalogue of harmful landslides and floods in Italy. In: Margottini C, Canuti P, Sassa K (eds) Landslide Science and practice. Proceedings of the second world landslide forum, Rome, 19–25 Sept 2011, vol. 3, pp 409–414 (Spatial analysis and modelling)

Santangelo M, Cardinali M, Rossi M, Mondini AC, Guzzetti F (2010) Remote landslide mapping using a laser rangefinder binocular and GPS. Nat Hazards Earth Syst Sci 10:2539–2546

Servizio Geologico Nazionale (1980) Carta Geologica dell'Umbria. Map at 1:250,000 scale (in Italian)

Singhroy V (2005) Remote sensing of landslides. In: Glade T, Anderson M, Crozier MJ (eds) Landslide hazard and risk. Wiley Press, New York, pp 469–492

Stark CP, Hovius N (2001) The characterization of landslide size distributions. Geophys Res Lett 28:1091–1094

Taylor F, Brabb EE (1986) Map showing landslides in California that have caused fatalities or at least $1,000,000 in damages from 1906 to 1984. U.S. Geological survey miscellaneous field studies map, MF-1867

Taylor FE, Malamud BD, Freeborough K, Demeritt D (2015) Enriching Great Britain's National landslide database by searching newspaper archives. Geomorphology 249:1–17

van Zuidam RA (1985) Aerial photo-interpretation in terrain analysis and geomorphologic mapping. Smits Publishers, The Hague, Netherlands, ITC, p 442

Zêzere JL, Pereira S, Tavares AO, Bateira C, Trigo RM, Quaresma I, Santos PP, Santos M Verde J (2014) DISASTER: a GIS database on hydro-geomorphologic disasters in Portugal. Nat Hazards 72(2):503–532

TXT-tool 1.504-1.1 Landslide Inventory Educational Methodology Derived from Experiences in Latin America

Rigoberto Moncada and Hiromitsu Yamagishi

Abstract

This document provides details and summarizes experiences of an educational methodology program for landslide inventory mapping and characterization. Different application cases are described to show the implementation of the educational tool. The methodology utilizes multiple resources, including aerial photographs, satellites images, digital elevation models (DEM), and geographic information systems (GIS). These tools allowed trainees, even beginners, to quickly start developing their own inventory maps. In first place, application cases of this educational methodology are introduced. These include projects developed by Japan International Cooperation Agency (JICA) in Honduras, a virtual seminar instructed by the Pan-American Center for Geographical Studies and Research (CEPEIGE) of Ecuador and intensive training activities with government officials executed with the United Nations Education, Science and Culture Organization (UNESCO) ENHANS project planned for Chile, Ecuador and Peru. A general educational methodology was extracted from these activities and the current document describes its core aspects so it can be replicated for similar learning endeavors. This consolidated program for landslide education was established in order to teach basic tools for landslide identification and characterization at local scales and was proposed for academic staff and professional specialists. It focused on a combination of stereoscopic interpretation and topographic feature analysis of aerial photographs, satellite images, field surveys and application of GIS tools. Objectives, procedures, evaluation methods, profile of the participants and instructors and general requirements are established to enable interested parties to use this document as a reference for organization of related landslide inventory training. This type of

R. Moncada (✉)
Faculty of Engineering and Architecture, Universidad Tecnológica Centroamericana, Zona Jacaleapa, Tegucigalpa, Honduras
e-mail: rigoberto.moncada@unitec.edu; rigoberto.moncada.lopez@gmail.com

H. Yamagishi
Shin Engineering Consultant Co.Ltd., 2-8-30, Sakaedori, Shiroishi-ku Sapporo 003-0021, Japan
e-mail: hiromitsuyamagishi88@gmail.com

K. Sassa et al. (eds.), *Landslide Dynamics: ISDR-ICL Landslide Interactive Teaching Tools*, https://doi.org/10.1007/978-3-319-57774-6_6

program can be a fundamental and low-cost input for other landslide studies. Subsequent educational activities will have to include evaluation guidelines for preparation of trainers in order to attain a continuous socialization of the knowledge and skills required for landslide mapping.

Keywords
Aerial photograph interpretation · DEM · GIS · Landslide education
Landslide inventory mapping · Stereoscopy

Contents

1 Introduction 96
2 Application of Landslide Inventory Education Methods in Latin America 97
2.1 Hazard Geology Focusing on Landslides in Tegucigalpa 97
2.2 CEPEIGE Course 98
2.3 UNESCO ENHANS Project Training 100
3 Landslide Inventory Methodology Developed from Application Cases 102
3.1 Learning Objectives 102
3.2 Explanation of Procedure 103
3.3 Evaluation of Participants 107
3.4 Participant Profile/Instructor Profile/Organization Profile 107
3.5 Tools and Data Required 110
4 Conclusions and Recommendations 110
References 111

1 Introduction

In order to develop a detailed assessment of landslide hazard and risk for a specific study area, it is first necessary to identify the locations of mass movements on an inventory map. These location maps provide a clear picture of the geographic distribution and density of landslides in a zone and indicate the general characteristics of different parts of each landslide identified (such as scarps, cracks and main body). Combined with geologic and hydrologic information, they can reveal sections of a study region that are more prone to mass movements and should be subject of more specialized research. Thus, they become an essential input for subsequent analyses and designs related to landslide prevention and countermeasure design.

However, landslide inventory maps can be difficult to produce. For example, mapping can be hindered by the modification of topography due to weathering or infrastructure development, presence of vegetation cover and/or prevalence of rugged terrain, all of which either erase distinctive landslide features or impede access to zones for their examination. In addition, identifying features in large areas can be time- and resource-consuming. Also, in situ and remote landslide recognition and classification require specialized skills to develop reliable inventory maps.

To address all of these complications, a landslide inventory educational methodology has been formulated. Use of remote sensing, stereoscopy and geographic information systems applied in this methodology allow coverage of considerable extents of territory, including places with difficult access. These tools also provide a more general view of landslide sites in a way that can bypass obstructions such as vegetation and exogenous processes and still allow visualization of landslide features. Combining new tools with fundamental skills of geologic, geotechnical and topographic surveying, objects detected by desk work can be fully confirmed on site. By designing a program that teaches the necessary proficiency for creating a landslide inventory, the knowledge gap can be closed and implementation of these types of maps becomes technically and economically feasible. This is especially important for landslide-threatened communities with limited resources, which can be assisted to apply inventory methods independently and reduce their reliance on external specialists. Therefore this methodology is targeted at government officials, community leaders, professional engineers, academic researchers and any representatives concerned with disaster prevention.

2 Application of Landslide Inventory Education Methods in Latin America

The projects "Hazard Geology Focusing on Landslides in Tegucigalpa" and "Strengthening and Capacity Building of Professional Techniques for the Control and Mitigation of Landslide in Tegucigalpa Metropolitan Area" developed by Japan International Cooperation Agency (JICA) experts in Tegucigalpa, Honduras as described by Yamagishi et al. (2014), is one of the earliest applications of this specific inventory methodology in Latin America. Yamagishi (2014), Hirota (2015) and Sato (2015) explain some of the first examples of application in the Honduran projects and provide details of successive steps after inventory, such as hazard assessment and detailed geologic mapping. For this document, applications of this methodology in other countries of Latin America are described. One important distinction of this type of methodology is combining both traditional techniques such as stereoscopy and modern GIS packages, exploiting all available resources for executing the inventory.

Authors such as Guzzeti et al. (2012), Burchfiel (2012), Colombo et al. (2005) and Flentje et al. (2011) provide explanations about landslide inventory development, relationship with hazard and risk maps and applications using GIS and other information technologies. This document, instead, focuses on the educational approach of the landslide inventory topic, first presenting successful cases and then synthesizing the most important details for an educational program that can be applied, even with conditions of limited resources, for Latin America and other regions of the world.

2.1 Hazard Geology Focusing on Landslides in Tegucigalpa

To better understand the antecedents of the specific methodology discussed in this paper, the first application is briefly described. It was developed in the project "Hazard Geology Focusing on Landslides in Tegucigalpa" executed by JICA, the Japanese Society for the Promotion of Science (JSPS) and the Engineering Polytechnic University (UPI) of Honduras from 2012 to 2014. During the duration of this project both professional engineers (from disaster prevention agencies and universities) and undergraduate students were trained in aerial photograph interpretation and contour map analysis in order to develop an inventory map for the city of Tegucigalpa. They also learned how to transfer their inventory to digital format, editable in GIS. The map generated was limited to inventory only and has been used as input for the subsequent project "Strengthening and Capacity Building of Professional Techniques for the Control and Mitigation of Landslide in Tegucigalpa Metropolitan Area" (2014–2016). See Fig. 1 for inventory activities and Fig. 2 (left) for a sample of this inventory map.

In addition of creating a landslide inventory map for all of Tegucigalpa, the learning process developed during this project formed the basis for resulting applications and improvements in the landslide inventory educational methods focused on aerial photographs, contour maps and GIS applications. Specifically, students were guided and corrected by the Japanese project experts in order to develop their own methodological document which narrated all the steps for their inventory mapping. The document was called: "*Manual Para Elaboración de Mapa de Inventario de Deslizamientos de Tierra, Caso de aplicación: Ciudad de Tegucigalpa*"; in English: "Landslide Inventory Map Elaboration Manual, Application Case: Tegucigalpa City" (Grupo de

Fig. 1 Training activities developed during JICA-JSPS-UPI project

Fig. 2 Landslide inventory map developed for Tegucigalpa-North East Zone (*left*) and cover of landslide inventory map elaboration manual, application case: Tegucigalpa city (*right*)

Mapeo de Deslizamientos de Tierra de Tegucigalpa 2014). Figure 2 (right) shows the cover of this document. This manual was and is currently being used as reference text for the training activities described next.

2.2 CEPEIGE Course

The first example outside of Honduras regarding application of this methodology is a virtual course organized by the Pan-American Center for Geographical Studies and Research (CEPEIGE) of Ecuador in 2014. The title of this course was "XLI International Course of Applied Geography: Geomorphology and Landslide Risk Management in Latin America". The international course was integrated by different training components, including risk management and basic landslide concepts, and the inventory methodology was applied in the topic "Identification, Cartography and Monitoring of Unstable

Terrain", which was imparted from September until December of 2014.

This international course was aimed for professional civil engineers, geologists, geophysicists, geographers and university professors and researchers of different Latin American countries such as Ecuador, Argentina, Peru, Colombia, Chile, Mexico and Dominican Republic. As a result of having participants well-versed in the skills necessary for inventory mapping, training of this component was divided into two phases: 1. Inventory Theory and 2. Research Project.

First, participants were trained in topics such as landslide classification and typology, landslide maps, topographic feature interpretation, inventory techniques, GIS applications and landslide block monitoring. During these initial sessions, participants applied their knowledge using information of their respective localities. Then, for their final research project, participants chose a landslide site in their vicinity, proceeded to collect all necessary data, performed inventory analysis and put into practice all the skills acquired in the course to make a landslide inventory report, inventory map and a preliminary hazard analysis to deliver as a final product. Figures 3 and 4 show examples of landslide inventory maps generated by different participants. Final report and mapping delivered by course students demonstrated the effectiveness of the training methodology for inventory, even with limited resources and time.

With assistance of virtual education platforms it was possible to further develop participants' knowledge and skills using comment boards and videoconferences for more complete feedback with the instructors (Fig. 5). Flexibility provided by this virtual modality permitted access to this education to a greater number and diversity of participants, and during a longer period of time, that allowed developing each learning objective fully. On the other hand, limitations of this approach were: I. a moderate percentage of participants did not complete their course and II. instructors were not able to assist participants in their inventory confirmation visits at each site.

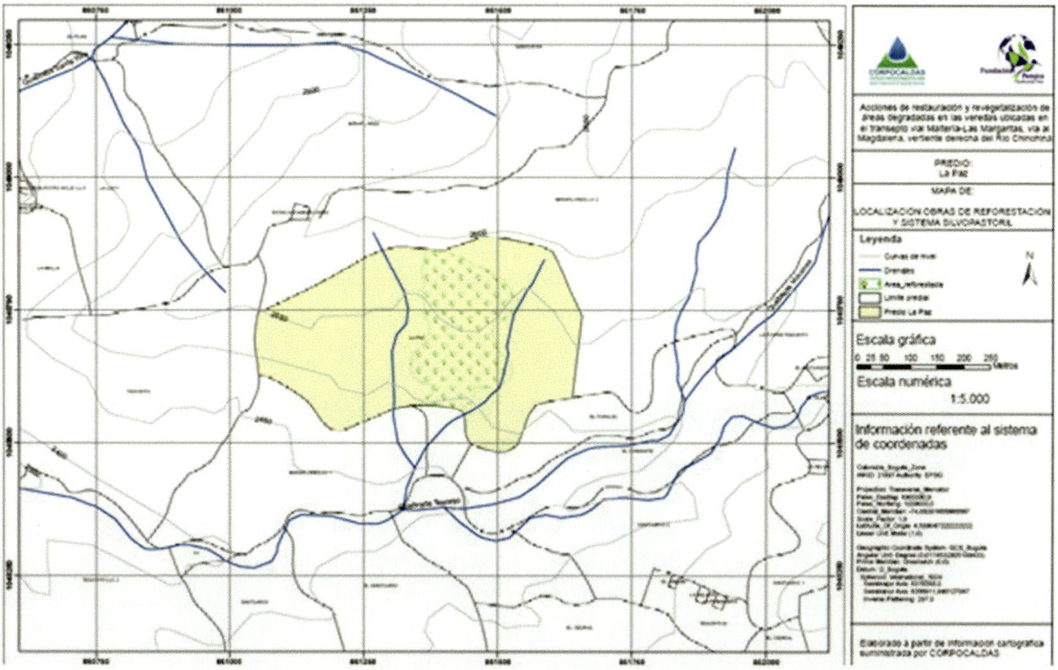

Fig. 3 Landslide feature identified over topographic map

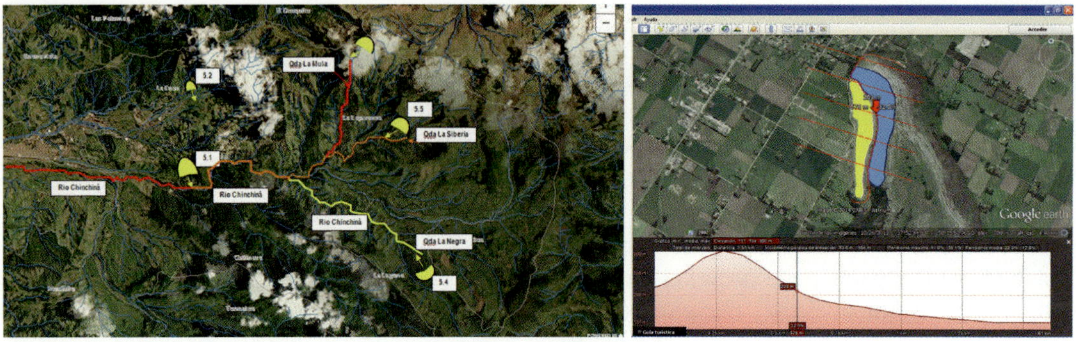

Fig. 4 Landslide inventory maps examples over satellite images

Fig. 5 Screen capture of videoconference activity

2.3 UNESCO ENHANS Project Training

A second experience, outside of the JICA project, that has been applied for inventory education in Latin America is that of the landslide inventory component of United Nations Education, Science and Culture Organization (UNESCO) project called Extreme Natural Hazards and Societal Implications (ENHANS). Particularly speaking, the project involved was that directed to the South American countries of Chile, Ecuador, Peru, Uruguay and others from 2015 to 2017. The purpose of ENHANS project was the preparation of key officials belonging to disaster management-related institutions in topics such as hazard assessment, risk assessment and resilience. For hazard assessment, the landslide component focused on inventory mapping to, at least, attain preliminary hazard mapping with aid of geology specialists in the target countries (except for Uruguay).

The key members selected for this training had a similar profile to those of the virtual course mentioned before (geologists, geographers, geophysicists, hydrologists, etc.), but included more government officials and disaster reduction management specialists instead of academic researchers.

Initially ENHANS Project started with introductory meetings at each of the target countries in order to present the project to stakeholders.

Fig. 6 Preparatory ceremonies of UNESCO ENHANS project in target countries, for example Santiago (*left*) and Quito (*right*)

Landslide component was introduced in the meetings at Santiago (Fig. 6 left) and Quito (Fig. 6 right) on April of 2015 and at Lima on September of 2015. This way the landslide component established the scope of the inventory methodology, the general details of how it would be implemented and the benefits it could provide for the work of officials and decision-makers in target countries.

Due to time constraints, training in the landslide inventory component was limited to one week for a classroom learning session at each of the target countries. In the case to be exposed next, that of inventory instruction in Peru, training was developed in November of 2015 and additional sessions are expected by the end of 2016 or beginning of 2017. So, the landslide inventory workshop was adapted to a shorter amount of time and, with the opportunity to execute site inspections, it was done this way: First, landslide inventory concept and techniques (such as stereoscopy) were introduced and preliminary applied to a nearby target area (Figs. 7 and 8). For this training the neighboring site of Chosica was selected and some additional maps were developed for the region of Huancavelica. Then, initial inventory data identified remotely was used to define a field inspection activity at Chosica and was executed at the middle of the workshop. Feedback provided by this field inspection was used to enhance inventory and provided some initial insight for hazard mapping. Finally, all inventory information was combined with GIS and digital elevation model (DEM) data to generate landslide location maps. An important advantage of working with government officials is that they must adhere to regulations and a Multinational Mass Movement Hazard Assessment Guide for the Andean Region (Proyecto Multinacional Andino 2007) could be utilized to generate standardized mapping, symbology and inventory.

For UNESCO Project case, the complete extent of methodology was also divided in two phases, albeit differently than CEPEIGE course due to a vital objective of the project: preparation

Fig. 7 Explanation of general landslide inventory principles

Fig. 8 Use of stereoscopes during training

of key officials who then can replicate this methodology in their respective communities. So the first phase consisted in preparing a leading group of officials in landslide inventory mapping (as explained before) and for a second phase, this leading group will be supervised by the instructors so they can teach using the same methodology (adapted to their necessities) to new officials and then continue this replication autonomously.

Training has yielded successful results with participating officials being able to generate their inventory maps (see Figs. 9, 10 and 11), refined by field visits (Fig. 12) and establishing a draft for future pilot workshops initially assisted by specialized instructors as guides.

3 Landslide Inventory Methodology Developed from Application Cases

As a result of the necessity identified and the experiences described in this document a general landslide inventory methodology was derived. It can be used as a base model to be duplicated and adapted, taking into account the conditions in Latin American countries. This way landslide inventory mapping can be executed in more places, by more officials, researchers and professionals and at different scales (local and regional) allowing the significant improvement of landslide disaster management. The details of this methodology are explained next.

3.1 Learning Objectives

The main objective for the methodology is for all participants to be able to generate by the end of the training a landslide inventory map that accurately reflects field conditions, by correct interpretation of landslide topography. More specifically, participants must attain the following:

- Learn to use the tools related with airphoto interpretation and GIS that allow landslide topography identification.
- Recognize the basic principles of landslide classification.

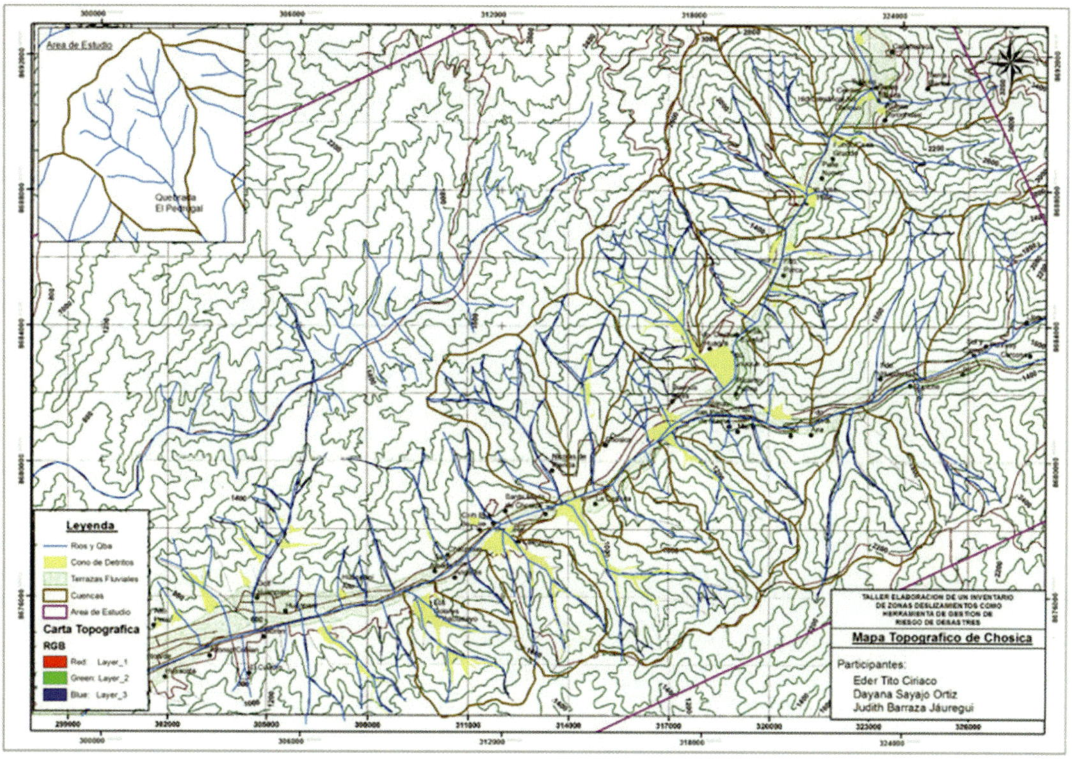

Fig. 9 Landslide inventory map over topographic maps of Chosica

- Apply identification and classification tools to generate a landslide inventory map both individually and in group.
- Validate and enhance inventory through field visits of landslide sites.
- Develop the preliminary basis of the creation of hazard, vulnerability and risk maps for landslides.
- Acquire the capabilities to share the inventory methodology with collaborators.

3.2 Explanation of Procedure

The procedure necessary to reach the objectives is summarized in the outline of Fig. 13. Two main phases apply this same procedure: an initial course to develop basic skills and a pilot course taught by the same participants that graduated from the initial course that allows them to apply these skills in the socialization of knowledge as new trainers. In turn, this pilot course can be the initial course for other completely new participants. In both cases, training starts with a previously selected area of interest and participants must begin by identifying and collecting all required information to develop the mapping method (see Data section). This is a critical step in the process since the quality of the map will greatly depend on the quality of the input information, so this step must be prepared fully before continuing with the rest of the activities.

Next, basic principles of landslide classification, triggers, most common locations and explanation of how landslides affect topography must be introduced or reinforced to participants. Then, depending on the information available for the target area, the methodology has to be adapted in case some information is not available. Ideally, aerial photographs of the study area should be obtained and, while pictures in 1:10,000 are recommended, other scales such as 1:25,000 or even 1:50,000 (as a last option) can

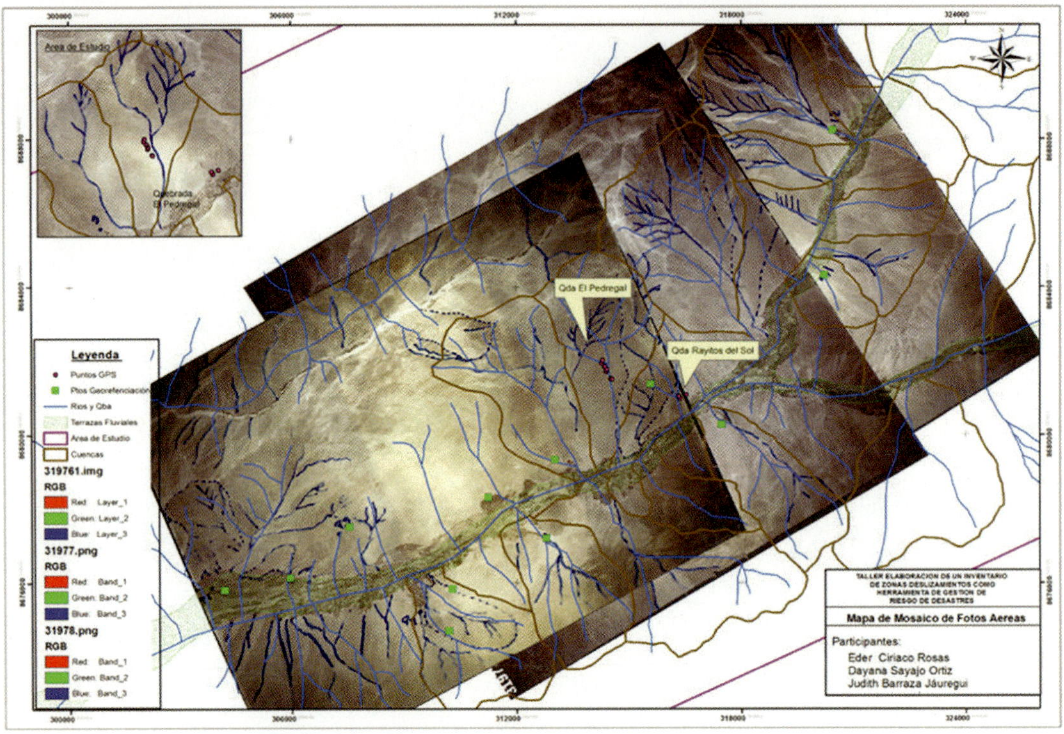

Fig. 10 Landslide inventory map of Chosica with aerial photographs used as data source

be used if they are the only photographs available. The great advantage of using aerial photography is that the stereoscopic effect can be used to perceive terrain abnormalities with more ease due to exaggeration of depth. Thus landslides detection can be enhanced in comparison to other methods. To complement the interpretation from air photographs or even to compensate for the lack of data or tools to use stereoscopy, it is also important to learn how to interpret topographic maps and satellite images, identifying important accidents that lead to landslide location. Local scale topographic maps can be of great use to recognize minor alterations in terrain that indicate landslide development. The indispensable component of landslide interpretation is applying geology and geomorphology principles to read topographic maps and stereo paired photographs by identifying details in contour lines and changes in depth and texture in the photos in order to reveal typical landslide topography.

Other option is the application of Digital Elevation Models to generate both contour maps and 3D models that can aid landslide identification. Free online software like Google Earth®, even with limited resolution in some areas, is an excellent source of satellite imagery that can be used during the inventory process and, with functions such as 3D viewing, terrain profiles and time series photographies, can become a very valuable resource for inventory as well. Participants, oriented by the instructor, must combine all these tools to find location of landslide features, indicating main body, scarps and cracks. Initially they develop their work individually.

Once landslide inventory has been performed in desk and, before integrating it into a map, these features must be verified in the field. Use of GPS (Global Positioning System) equipment and existing cartography plays a very important role in this aspect and should also be verified with participants prior to in situ work. It is also necessary to prepare beforehand standardized field

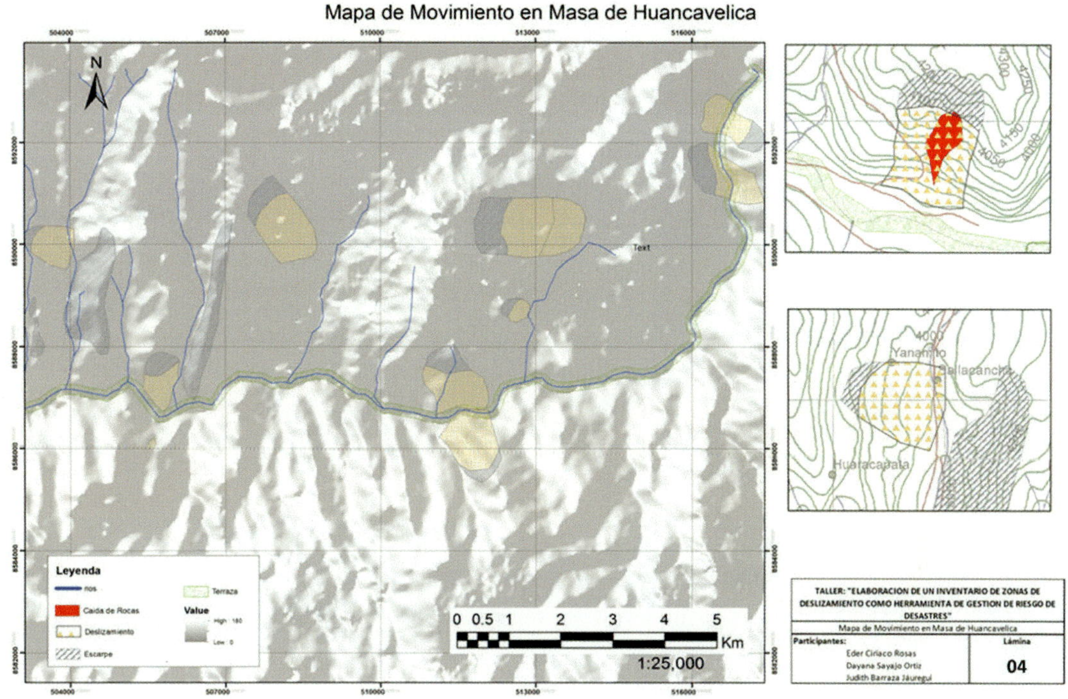

Fig. 11 Landslide inventory map of site in Huancavelica over DEM

inventory formats and map symbols, such as those defined by "Proyecto Multinacional Andino" in 2007, for managing inventory and sharing the maps developed with a greater audience. In site, the preferred course of action is to verify one by one each of the separate landslide features identified, confirming their existence and improving details such as the existence of cracks, scarps and individual blocks. Enough time must be allocated for a complete survey, which is also essential to eliminate features wrongly identified, improve those confirmed in the field or add new features not visualized during desk work. Thus, the use of GPS to compare in situ and desk-identified landslides is of great convenience for this depuration of the initial identification. For reference, a field visit report must be elaborated with detailed information of activities and geologic characterization, if possible.

Then, landslides confirmed should be converted to vectorial-format shapefiles within the GIS software. Information from DEM, digital contour maps, Google Earth® and GPS is relatively easy to import, while landslides found in air photographs and printed topography must be scanned and georeferenced to then be digitalized. However, printed data can allow participants to find landslides that digital data may not reveal.

Once all information is digitized, the next step is for each participant to consolidate all their information in a single version of their individual landslide inventory, uniting or altering the different landslides found from different information sources and giving priority to field results. Given that the participants are usually beginners in the inventory process, a very convenient way to reduce mistakes from their individual results is to then form groups (4–6 participants) to generate a consolidated version of the inventory. That is because participants will interpret landslides in different ways and even in different places, so the idea is to analyze coincidences and criticize the mapping of all group members. Then, based on the level of experience of each member a weighted average can be assigned to each of the different landslide features identified by each member so only those with greater total weight are used in the final group map.

Fig. 12 Field visit at Chosica during training

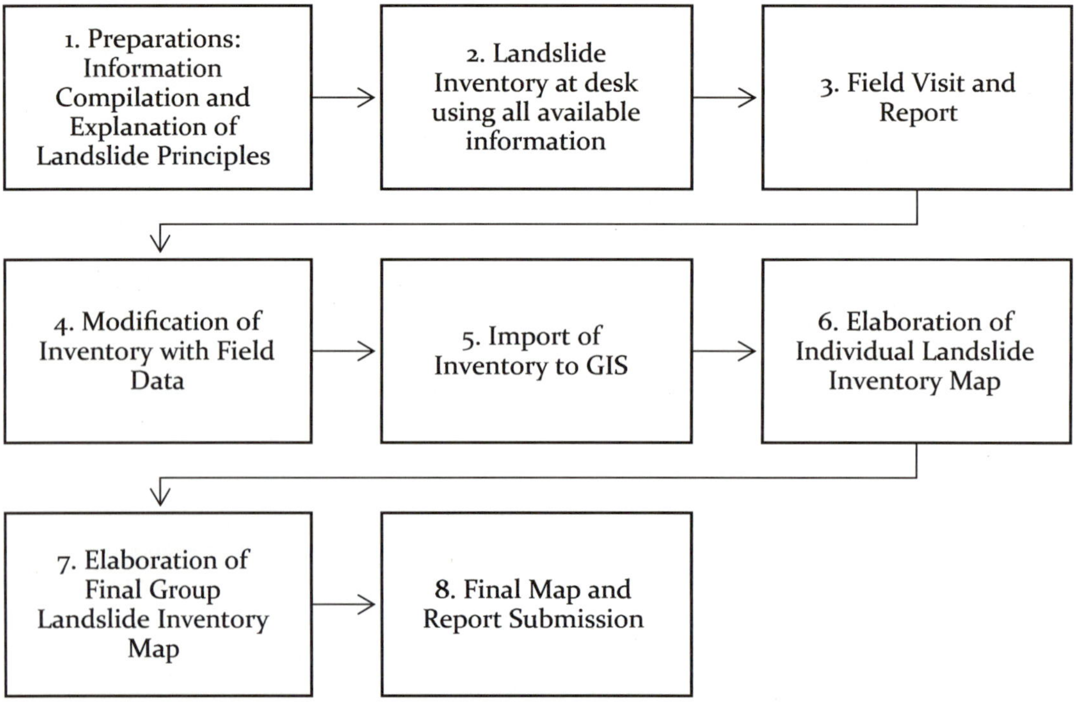

Fig. 13 Summary of landslide inventory procedure

Finally, with aid of the GIS, a final group inventory map is developed and presented to all participants of the course. Additional details can be included in the database of the shapefile that can assist hazard mapping in the future. Maps developed by each group are reviewed and evaluated by the instructor of the course, recommendations are summarized and feedback is provided to each group and participant. At the end of the course the most dedicated and proactive members are nominated as representatives and future instructors for socialization of knowledge. Participants must also write a final group report of all their training activities.

For the pilot course the process is similar to that of the initial course: the learning contents for participants are exactly the same. But the most important difference is that the instructors of this pilot course are the representatives chosen in phase 1. During this first pilot, experienced/senior instructors will supervise the training provided by the new/junior instructors and assess them in the application of the methodology and ensure quality of the products. Senior instructors then evaluate junior instructors so they can improve in subsequent training sessions. New trainers are also identified and the process is repeated over and over again. Eventually junior instructors can also be evaluated and promoted to senior instructors.

This methodology can be designed to be imparted both in personal and virtual environment. Ideally, doing it in person is preferred for the following reasons: Easier and more fluid responses to questions and feedback, stronger social interaction for identification of potential junior instructors and possibility of instructors to go with participants to their landslide sites. However, if adequate follow-up and evaluation is implemented to each participant and videoconferences and multimedia are used frequently for feedback, virtual training can also be used. This was demonstrated in the CEPEIGE experience explained in this article. Mixed/hybrid education modalities can also be applied depending of the convenience of participants and instructors.

3.3 Evaluation of Participants

An important result of the course is the preparation of participants who have developed their own inventory maps with the tools available (see tools and data section) and were able to use teamwork to achieve a map that closely resembles the landslide inventory reality in a way that is useful for hazard mapping, prevention activities and countermeasures. As a result, the expected products to be evaluated are the participants′ individual and group inventory maps and reports. For this assessment, the rubric defined in Table 1 shows the expected results of the training. Participants can also demonstrate their mastery of the training in alternative ways. For example, students of Tegucigalpa Landslide Mapping Group, civil engineering and geology technician undergraduates, were able to produce an understandable landslide inventory manual to show their proficiency. However, for a wider application of this methodology the use of a standardized rubric is recommended. For application of this rubric, a minimum score of 60 out of 100 is necessary to complete the training successfully and 80 is the minimum score for participants to be chosen as a junior instructors. For pilot courses, junior instructors should be evaluated by senior instructors, as well. A rubric for evaluating junior instructors is also being developed.

3.4 Participant Profile/Instructor Profile/Organization Profile

In order to train participants capable of attaining the minimum evaluation criteria for approval of the training and for junior instructor promotion, careful selection is necessary. Therefore, a specific profile has been identified. The participants should have understanding of topography and possess the ability of interpreting contour maps and aerial photographs (with or without stereoviewer). Moreover, they should be already familiar with mapping and geography topics,

Table 1 Participant rubric

Criteria	Global weight	Individual weight for each criteria					
		0%	20%	40%	60%	80%	100%
Quality of inventory maps (individual and group maps)	40	No map is developed	Map identifies few landslide features, and they do not match contour maps and aerial photograph stereopairs. No GIS tools are applied	Map identifies several landslide features, but most do not match contour maps and aerial photograph stereopairs. Few GIS tools are applied	Map identifies several landslide features correctly by using contours and stereopairs but still has some mistakes. Some GIS tools are applied	Map identifies most landslides accurately, with adequate interpretation of contour lines and stereopairs but the map does not apply all available GIS tools	Map identifies landslides accurately with correct interpretation of contour maps and stereopairs. It also uses all available GIS tools
Participation in group activities and final group report	20	No group map and report is developed with involvement of the participant. No report submitted	Participant is involved minimally in group map and report elaboration. Report incomplete	Participant's individual map contains few important features and has several mistakes. Minimal collaboration in making the map and report. Only basic information in final report	Participant's individual map contains some important features but has many mistakes. Some collaboration in making the report and map. Adequate description in report	Participant's individual map contains several important features but some mistakes. Full collaboration in making the group map and report. Complete information in report	Participant's individual map is an essential part of the group map. Full collaboration in making the group map. Fully detailed and high quality report
Field inspection and report	25	No field inspection is performed	Minimal field inspection is done, but not useful for the map, mainly only site pictures	Some field inspection is done, but only for a few areas of the target site and information reported is poor	Some significant features are visited, registered and georeferenced in the field. A very basic field report is delivered	Most significant features are visited, registered, inventoried and georeferenced in the field. A general field report is delivered	All significant features are visited, registered, inventoried and georeferenced in the field. A complete field report is delivered

(continued)

Table 1 (continued)

Criteria	Global weight	Individual weight for each criteria					
		0%	20%	40%	60%	80%	100%
Attitude/involvement	15	Two or less sessions attended, null involvement and no results presented	Very incomplete results presented and little to none interaction with other members	Late submission of products and scarce interaction with everyone involved	Some products delivered late, some interaction and assistance to other participants and at least weekly communication with instructor	Most products delivered in time, interaction and assistance to other participants and regular communication with instructor	Submission of all work and information in time, assistance to other participants, proposals for new learning and constant communication with instructor

also with geology, geohazards and basic principles of slope stability. This way all spatial information applications, remote sensing and landslide identification and typology are easy to grasp and constitute, for many, a review of information they already know.

As a result, this type of training is recommended for professional civil engineers, geologists, geophysicists, geographers and disaster management specialists. This profile is also necessary because it is intended that a percentage of these members will be selected as junior instructors. Finally, participants must belong to or work as consultants with an institution or organization that applies this knowledge in a non-profit or educational mode as specified below.

For junior instructors it is highly recommended for them to have experience in teaching, at least in the modality (online, classroom, etc.) they would be interested in developing their own course. Other requirements for junior instructors include high assessment score in the initial course, interest and disposition during course development and an educational profile as those of general participants. For senior instructors it is necessary for them to have at least three years of experience in landslide topics, two years of experience in educational activities, a Master's degree (or Ph.D.) in subjects related to soil mechanics, rock mechanics, geology and geography and all the qualifications of the general participant. In addition, sufficient proficiency in the execution of the teaching methodology must be demonstrated by them. It is also recommended that senior instructors should have a GIS Expert qualification if possible.

Organizations targeted by this type of training include government agencies dedicated to disaster prevention, response and mitigation, municipal governments, geology/civil engineering-related research centers, universities and NGOs with disaster-related projects.

3.5 Tools and Data Required

As mentioned in the procedure section, before the beginning of any training activity it is mandatory to fully assess the advantages and limitations of the resources of the target group. The following equipment must be available for effective training: large mirror stereoscopes (or at least medium or pocket stereoscopes), PC with all necessary software for GIS, including Google Earth for use of satellite images, GPS and image processing software, internet connection, printer, scanner, projector, screen, explorer GPS, digital camera and consumables such as pencils, paper, clear paper, masking tape, clear tape, etc.

Under ideal conditions information of the target area to be worked with should be: historical landslide registries or news articles, landslide sites point inventory with coordinates, other previous landslide inventory maps, field visit reports, topographic maps, aerial photographs, digital elevation maps, geologic maps, soil maps and geomorphologic maps, satellite images from Google Earth or Landsat, etc. Minimum requirements for information are regional topographic maps (ideally, 1:10,000 or at least 1:50,000), general aerial photographs (ideally, 1:10,000 or at least 1:50,000), digital elevation models (ideally 5 meters or less, but at least 30 m resolution, which is the resolution available worldwide with ASTER GDEM) and regional geologic maps. Where aerial photographs are not available, satellite imagery should be used with the most detailed topographic maps and DEM available to compensate.

4 Conclusions and Recommendations

Training methodology, in all of the cases presented, was able to successfully teach participants how to use available information for the development of landslide inventory maps. So both conventional classroom education, as that in UNESCO ENHANS Project, and virtual classroom education, as that of the CEPEIGE landslide inventory module, can be applied to impart this methodology. Selection of the appropriate participants for the training is vital for its success. For the two cases exposed, both groups had already significant background and experience so training could be imparted in a week or two when time was very limited. If less favourable profiles are among those selected, training should take more time, but can also be successfully implemented like in the case of JICA projects in Honduras, such as JICA_JSPS project which was able to generate a complete methodology manual with undergraduate trainees. One of the most important aspects of this training is the capacity to execute an adequate field visit and report of the landslide sites. Hence, instructors are encouraged to visit the field directly with participants. For virtual courses, it is recommended a more rigorous inspection of field reports by instructors. In addition to landslide inventory maps, developing the capacities to form new trainers is a fundamental point of the proposed methodology. With adequate selection of junior instructors and thorough evaluation of them, a replicable model can be created. If an editable web database is available, such as that of the JICA project in Honduras, replication of this methodology can help to complete landslide information voids much faster, involving the collaboration of a trainee network. Other important advantage of this methodology is that maps can be worked at a local scale, with a detailed inventory. If new instructors are able to work together with community leaders and local disaster professionals, this model can be applied to communities difficult to access by government authorities and help them strengthen their landslide hazard component independently. Important aspects to continue improving of this methodology include: evaluation of junior instructors, customized methodologies for different data and equipment available depending on target region and design of a simplified version for local emergency committee representatives with few or nonexistent engineering or scientific background.

Acknowledgments We thank the following people and institutions for their collaboration and support that made possible the execution of activities related to this document: Akihiko Yamada and Hiroshi Nishiki from JICA office in Honduras, JICA experts Dr. Hiroshi Yagi, Dr. Go Sato and Dr. Kiyoharu Hirota, Eng. Luis Eveline from UPI, CEPEIGE Academic Director Dr. Fernando Vera and Dr. Alexandros Makarigakis and Dr. Soichiro Yasukawa from UNESCO ENHANS Project.

References

Burchfiel P (2012) Visualizing landslide hazards: methods for empowering communities in Guatemala through hazard mapping. Master's theses. Paper 4187

Colombo A, Lanteri L, Ramasco M, Troisi C (2005) Systematic GIS-based landslide inventory as the first step for effective landslide-hazard management. Landslides 2(4):291–301

Flentje P, Stirling D, Chowdhury RN (2011) Landslide inventory, susceptibility, frequency and hazard zoning in the Wollongong and wider Sydney Basin Area. Semin Landslide Risk Manag Aust Geomech J, vol 46

Grupo de Mapeo de Deslizamientos de Tierra de Tegucigalpa (2014) Manual Para Elaboración de Inventario de Deslizamientos de Tierra. Caso de aplicación: Ciudad de Tegucigalpa. Agencia de Cooperación Internacional del Japón (JICA). Tegucigalpa. ISBN: 978-99926-701-1-8

Guzzetti F, Mondini AC, Cardinali M, Fiorucci F, Santangelo M, Chang K-T (2012) Landslide inventory maps: new tools for an old problem. Earth Sci Rev 112 (1–2):42–66. doi:10.1016/j.earscirev.2012.02.001

Hirota K (2015) Landslide and geology education in Honduras, as activities of Japan International Cooperation Agency senior volunteer. J Jpn Landslide Soc 52(4):36–42

Proyecto Multinacional Andino: Geociencias para las Comunidades Andinas (2007) Movimientos en Masa en la Región Andina: Una guía para la evaluación de amenazas. Servicio Nacional de Geología y Minería, Publicación Geológica Multinacional, No. 4, 432 P., 1 CD-ROM

Sato G (2015) Investigation and mitigation of landslide in Republic of Honduras as Japan's International Cooperation Program. J Jpn Landslide Soc 52(4):1–7

Yamagishi H (2014) Integrated disaster prevention maps and landslide hazard maps using GIS—examples of Shikoku and Central America Honduras. J Jpn Landslide Soc 51(2):24–29

Yamagishi H, Yagi H, Sato G (2014) Landslide hazard mapping of Tegucigalpa, Honduras-capacity building by JICA-JSPS project of 2012–2014. In: Proceedings of World landslide forum 3, 2–6 June 2014, Beijing, pp 712–715

TXT-tool 1.052-1.1 GIS Using Landslides Inventory Mapping for Volcanoes

Gabriel Legorreta Paulín

Abstract

This tool is a systematic method for carrying out a landslide inventory on Quaternary volcanoes. The goal of this tool is to provide the foundation for mapping landslides in volcanic areas using Geographic Information System (GIS). The mapping methodology uses aerial photographs, fieldwork, and adaptation of the Landslide Hazard Zonation Protocol of the Washington State Department of Natural Resources, USA to address the distribution, type, and abundance of landslides. The training area is along the Río La Carbonera watershed on the SE flank of Pico de Orizaba volcano, Mexico. The area is predisposed to landslides due to physiographic conditions (loose volcanic materials and high seasonal rainfall) and decades of deforestation.

Keywords

GIS • DEM • Landslide • Mapping • Inventory • Pico de Orizaba volcano

Contents

1	**Introduction**	113
2	**Methodology**	114
2.1	Step 1: Selection of Study Area	114
2.2	Step 2: Literature Review and Data Gathering	116
2.3	Step 3: GIS Base Map	116
2.4	Step 4: Selection of Aerial Photographs	117
2.5	Step 5: Field Work	118
2.6	Step 6: Landslide Inventory Map	119
3	**Conclusion**	120
References		120

1 Introduction

To portray past and current landslide occurrence, an heuristic landslide inventory approach based on Geographic Information Systems (GIS) applications can be used (Fig. 1). All heuristic landslides mapping has advantages and limitations based on the researcher's experience and the final user needs, and they can be applied in many creative ways. However, common tools and methods can be standardized. This teaching tool is based on the on-going research project (Grant PAPIIT, no. IN102115 and IPL project #187) of the Institute of Geography at the National Autonomous University of Mexico (UNAM) to conduct a landslide inventory in volcanic terrains using GIS. The Río

G.L. Paulín (✉)
Instituto de Geografía, Universidad Nacional Autónoma de México, Mexico City, Mexico
e-mail: legorretag@hotmail.com

K. Sassa et al. (eds.), *Landslide Dynamics: ISDR-ICL Landslide Interactive Teaching Tools*, https://doi.org/10.1007/978-3-319-57774-6_7

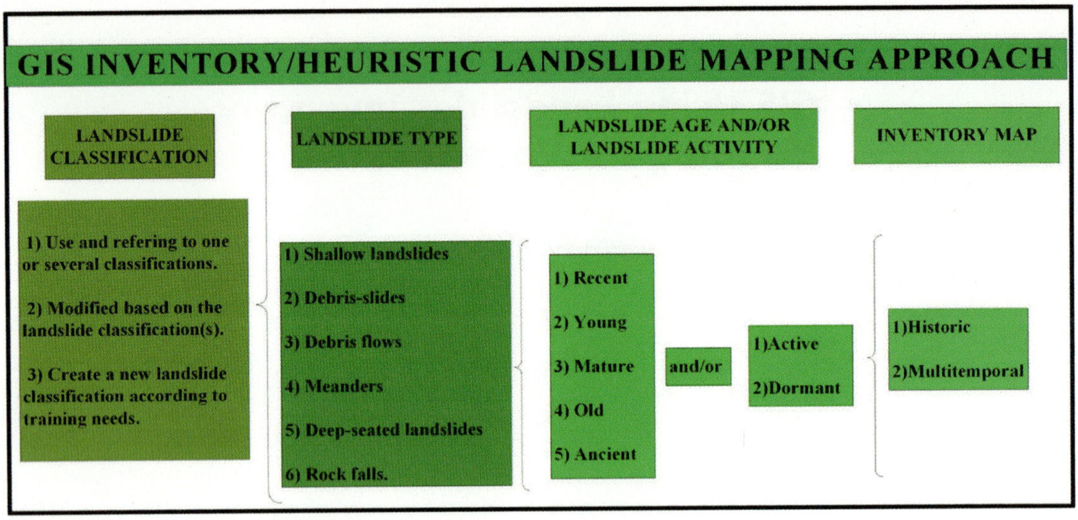

Fig. 1 Landslide mapping approach. In this tool, a modification of the Landslide Hazard Zonation Protocol of the Washington State Department of Natural Resources, USA is used

La Carbonera watershed on the southeastern flank of Pico de Orizaba volcano has been selected as a case study area. The study area is used to show a systematic methodology for landslide inventory mapping in volcanic areas that have scarce information on landslide distribution, type, and abundance. The watershed has been studied using aerial photographs, fieldwork, and adaptation of the Landslide Hazard Zonation Protocol of the Washington State Department of Natural Resources (DNR), USA. More than 230 landslides of six types were recognized: shallow landslides, debris-slides, debris flows, meanders, deep-seated landslides, and rock falls. The techniques and their implementation for each level in a GIS-based technology is presented and discussed.

2 Methodology

The methodology encompasses six main steps to build an GIS historic or a multitemporal landslide inventory map (Fig. 2).

2.1 Step 1: Selection of Study Area

In order to standardize analysis and comparison, we use data on existing natural watersheds. A watershed is territorial area bounded by ridges or divides (upper boundaries) where all runoff (streams and/or rivers) converge and flow into a common point called the exit point of the watershed (Fig. 3). The watershed can be endorheic (when it does not have a direct outflow to the sea, but converges instead into lakes or continental flat areas) or exorheic (when it has a direct outflow to the sea). The larger river systems, can be divided into several sub-watersheds to limit the watershed analysis to a maximum square kilometer area that reasonably could be assessed in a limited time period or with available economic resources (Fig. 3).

To study a watershed, it may be necessary to spend a day with a forester or a geologist who knows the study area and/or with recent aerial photographs of the area to establish a generalized characterization of local landslide processes within the context of the area's topography, geology, and land-use cover.

This work uses as an example the Río La Carbonera watershed, located on the southeastern flank of Pico de Orizaba volcano (the highest stratovolcano in Mexico, 5675 m a.s.l.) within Veracruz and Puebla states (Fig. 4). The watershed covers 71.9 km^2. The watershed is characterized by hilly and steep terrain, with elevations

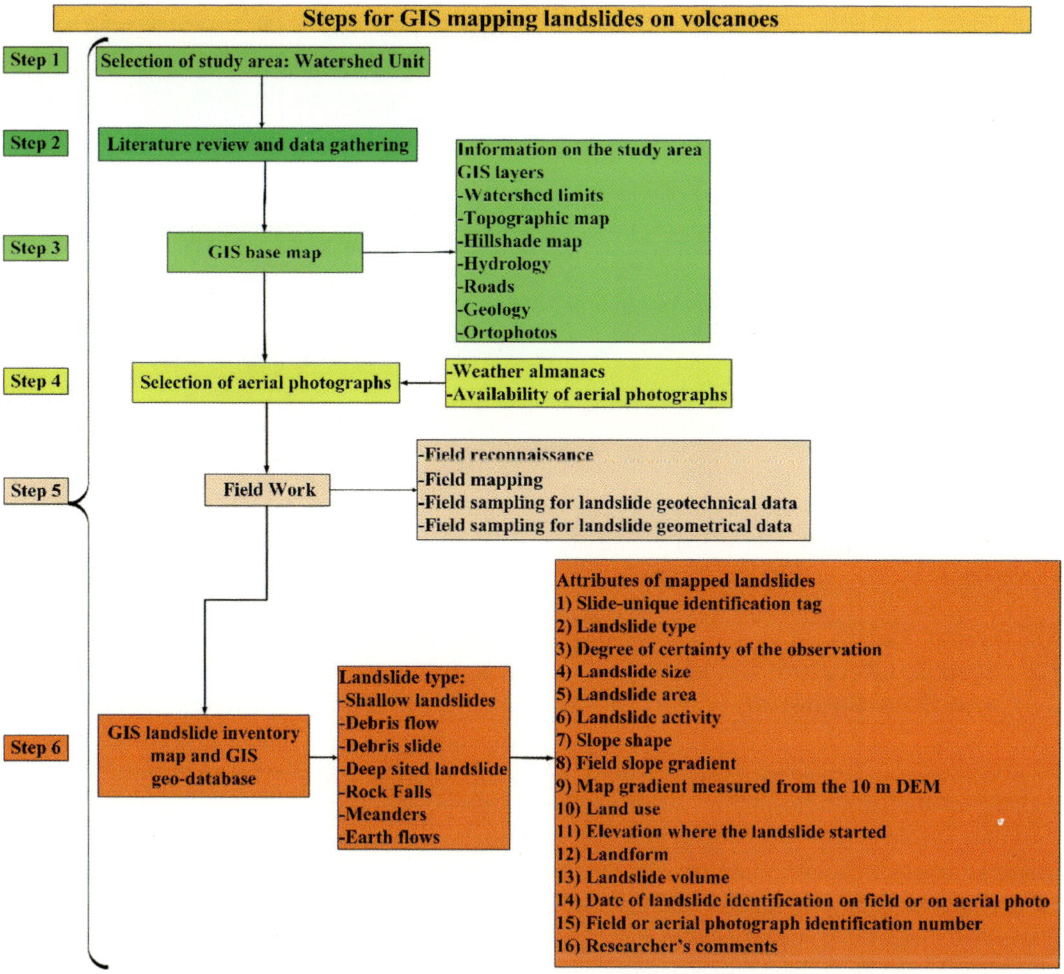

Fig. 2 Methodology for mapping landslides on volcanoes

Fig. 3 **a** Endorheic watershed; **b** Exorheic watershed

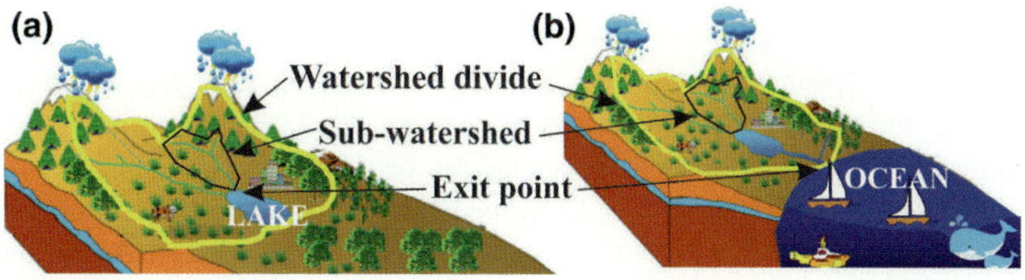

from 1224 to 3643 m a.s.l. and slopes between 3° (inner valleys of relatively flat plains) and 61° (mountainous terrain). It is a tributary of the Río Blanco, which flows into the Gulf of Mexico. The stream system of Río La Carbonera watershed erodes in the middle upper portion Tertiary and Quaternary lavas, pyroclastic flows, and fall deposits and in the middle lower portion a weathered and highly dissected, folded and faulted Cretaceous basement of limestone and shale. The study area is prone to landslides due to its large area of collapsible weathered or

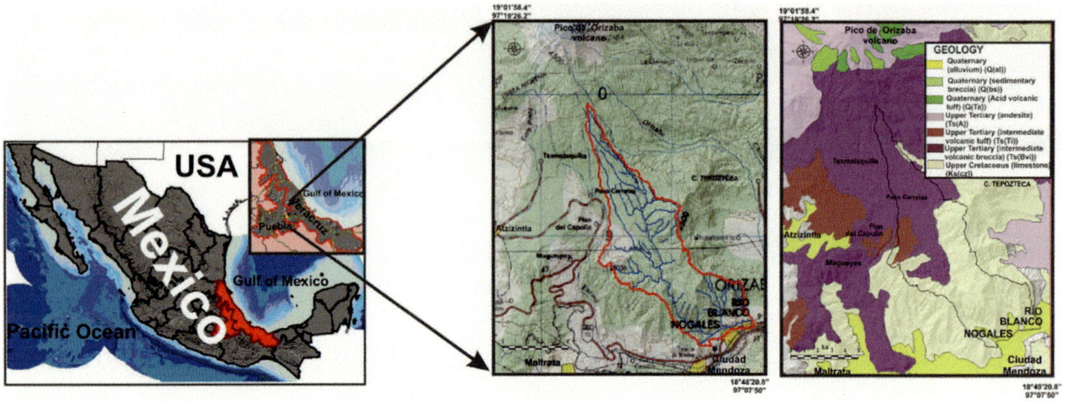

Fig. 4 Study area and geology

disjointed volcanoclastic and sedimentary material at high altitudes and the region's high seasonal rainfall.

2.2 Step 2: Literature Review and Data Gathering

To be successful in the analysis, interpretation, and mapping of the landslides in the watershed, the researcher should collect and study as much background information about the basin as possible. The information must provide context to establish a generalized characterization of landslide processes within the watershed. Background information could include topography, geology, land use, climate (including data on large storm events that potentially trigger landslides), aerial photos (chosen by the year in which unusual storms occur as an attempt to map landslides triggered by unusual rains within the watershed), hydrology, pre-existing landslide maps, geo-databases (including pertinent landslide attributes such as landslide location, landslide type, size, age, etc.), and many documents, including reports.

2.3 Step 3: GIS Base Map

All background map information must be digitized, georeferenced, and incorporated into the GIS system. All paper maps are converted to raster format, at the same pixel size resolution as the digital elevation model (DEM). The problem of scaling the thematic maps to the same resolution as the DEM is solved by generalization of a cartographic and geographic database (Slocum et al. 2005; Fisher 2004; Tate and Atkinson 2001; Schneider 1997; Dent 1985). The cartographic generalization is conducted by a heuristic process of aggregation. The aggregation process involves grouping pixels and representing them as a unit. The purpose of the generalization is to conceive a single measurement for a group of pixels in order to characterize a region (Fig. 5). Although there are different aggregation methods (such as measures of central tendency, maximum likelihood, etc.), there is no standardized procedure.

The characteristics of catchment area, stream length, stream patterns, stream orders, and drainage density can be generated from the DEM and incorporated as background information. By retrieval and on-off switching of the layers in the GIS, a base map is created to assist in the digitizing of landslides (Fig. 6).

The background information in this example includes roads, topographic and geologic paper maps at a scale of 1:50,000, and paper maps of land use, climate, and hydrology at a scale of 1:250,000. Here, all paper maps were converted to a 10 m pixel size raster format. Background information also includes orthophotos at 1:10,000 as well as a 10 m digital elevation model and its derived slope angle, slope curvature, and contributing area thematic maps that were constructed from the 1:50,000 topographic maps.

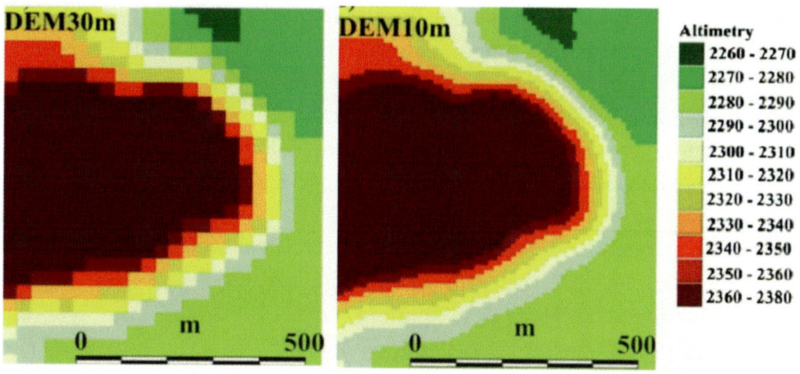

Fig. 5 Pixel size generalization from 30 m resolution to 10 m (Paulin et al. 2010)

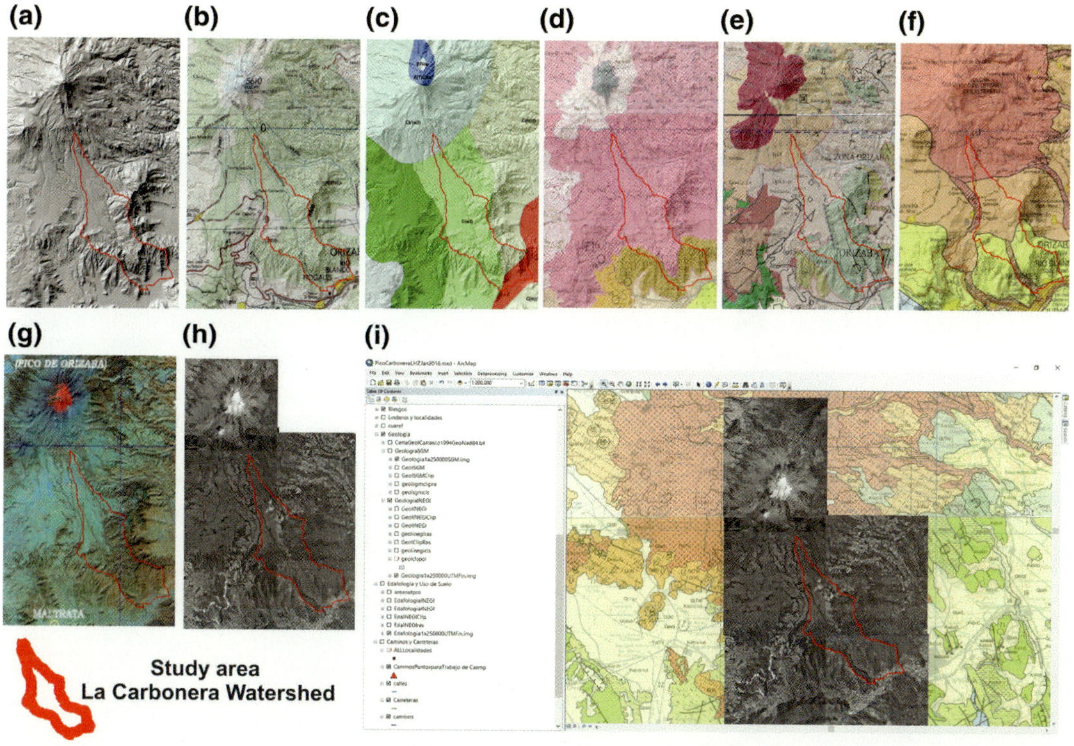

Fig. 6 Background information. **a** DEM (1:50,000); **b** Topography(1:50,000); **c** Climate (rain and temperature) (1:250,000); **d** Land use (1:250,000); **e** Geology (1:50,000); **f** Volcanic risk map (1:250,000);

2.4 Step 4: Selection of Aerial Photographs

Landslides in a study area can be identified by using a set or sets (multi-temporal) of aerial photographs to inventory historic landslides. Aerial photographs should be the most complete and relevant time sets based on climate almanacs, storm records, and availability of photos. For this teaching tool, landslide data are collected from two sets of aerial photographs and fieldwork to create a multi-temporal landslide inventory map. The aerial photographs cover a 14-year period from 1994 at a scale of 1:20,000 to 2008 at a scale of 1:10,000. The sets are analyzed with a mirror stereoscope (3× magnification). During aerial photographic interpretation, landslides can be mapped and classified by using the following approaches: (1) use and refer to one or several landslide classifications (e.g., Varnes 1958; Carson and Kirby 1972; Varnes 1978; Cruden and Varnes 1996) or (2) modifications based on the landslide classification(s) (e.g., Washington State Department of Natural Resources (DNR), Forest Practices Division 2006), and (3) create a new landslide classification according to training needs (Shan et al. 2013). For aerial photographic interpretation, landslides are mapped and classified into shallow landslides, debris flows, debris slides, deep-seated landslides, earthflows, rock falls, and meanders according to the landslide hazard zonation protocol (2006) of Washington State DNR Forest Practices Division, Cruden and Varnes (1996), and Wieczorek (1984). During stereo aerial photographic analysis, all landslides are mapped by "heads-up" digitizing in ArcGIS. "Heads-up" digitizing involves mapping on photo transparencies and digitizing right onto the screen into GIS at the same scale as the photos. Mapped landslides are interpreted and documented in the geo-dataset.

2.5 Step 5: Field Work

In parallel to the landslide mapping into the GIS, preliminary field reconnaissance is conducted in order to provide the essential feedback to

Fig. 7 Field work for landslide mapping and to assess and describe landslide distribution. **a** Field reconnaissance of the landslide flow zone and deposit. **b** Reconnaissance of landslide headscarp. **c** Statigraphic column description of landslide headscarp

improve the mapping of landslides. Subsequent field verification is necessary to resolve the differences between the sources of mapping, improve the accuracy of locations of landslides, and add detail or omitted landslides to the map (Fig. 7). The amount of field verification is often held to be between 15 and 25% of all mapped landslides, depending on project needs and budget. This allows a high statistical degree of confidence in the landslide assessment.

2.6 Step 6: Landslide Inventory Map

All landslides should be digitized into a GIS, and a spatial geo-database of landslides is constructed from a standardized GIS database. Pertinent attributes of mapped landslides are recorded on the GIS geo-database. These include (1) Slide-unique identification tag (2) landslide type, (3) level of certainty of the observation, (4) photo identification date, (5) landslide size, (6) landslide activity (Fig. 8a, b), (7) landslide parts (head, flow zone, deposit; Fig. 8c, d, e), (8) slope shape, (9) slope gradient, (10) map gradient measured from the 10 m digital elevation model (DEM), (11) land use, (12) elevation where the landslide occured, (13) identification number of the aerial photograph where the landslide occured, (14) landslide area, (15) landslide volume, and (16) researcher's comments (Fig. 9).

Fig. 8 **a** Landslide activity during wet season (active); **b** Landslide activity during dry season (dormant); **c** Aerial photograph from 2008 at a scale of 1:10,000; **d** Landslide polygon divided into three parts (head scarp, flow zone, and deposit); **e** Landslide deposit

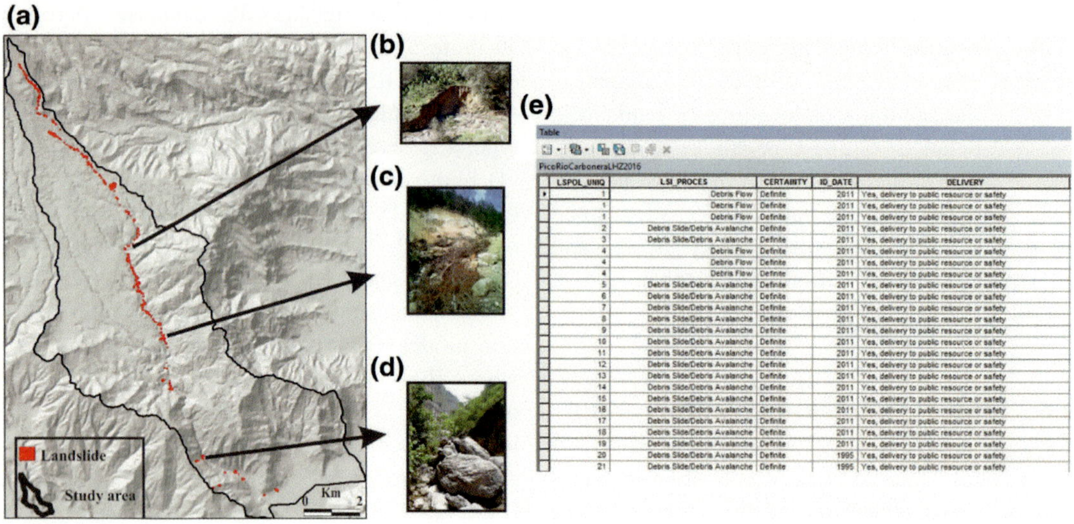

LSPOL_UNIQ	LSI_PROCES	CERTAINTY	ID_DATE	DELIVERY
1	Debris Flow	Definite	2011	Yes, delivery to public resource or safety
1	Debris Flow	Definite	2011	Yes, delivery to public resource or safety
1	Debris Flow	Definite	2011	Yes, delivery to public resource or safety
2	Debris Slide/Debris Avalanche	Definite	2011	Yes, delivery to public resource or safety
3	Debris Slide/Debris Avalanche	Definite	2011	Yes, delivery to public resource or safety
4	Debris Flow	Definite	2011	Yes, delivery to public resource or safety
4	Debris Flow	Definite	2011	Yes, delivery to public resource or safety
4	Debris Flow	Definite	2011	Yes, delivery to public resource or safety
5	Debris Slide/Debris Avalanche	Definite	2011	Yes, delivery to public resource or safety
6	Debris Slide/Debris Avalanche	Definite	2011	Yes, delivery to public resource or safety
7	Debris Slide/Debris Avalanche	Definite	2011	Yes, delivery to public resource or safety
8	Debris Slide/Debris Avalanche	Definite	2011	Yes, delivery to public resource or safety
9	Debris Slide/Debris Avalanche	Definite	2011	Yes, delivery to public resource or safety
10	Debris Slide/Debris Avalanche	Definite	2011	Yes, delivery to public resource or safety
11	Debris Slide/Debris Avalanche	Definite	2011	Yes, delivery to public resource or safety
12	Debris Slide/Debris Avalanche	Definite	2011	Yes, delivery to public resource or safety
13	Debris Slide/Debris Avalanche	Definite	2011	Yes, delivery to public resource or safety
14	Debris Slide/Debris Avalanche	Definite	2011	Yes, delivery to public resource or safety
15	Debris Slide/Debris Avalanche	Definite	2011	Yes, delivery to public resource or safety
16	Debris Slide/Debris Avalanche	Definite	2011	Yes, delivery to public resource or safety
17	Debris Slide/Debris Avalanche	Definite	2011	Yes, delivery to public resource or safety
18	Debris Slide/Debris Avalanche	Definite	2011	Yes, delivery to public resource or safety
19	Debris Slide/Debris Avalanche	Definite	2011	Yes, delivery to public resource or safety
20	Debris Slide/Debris Avalanche	Definite	1995	Yes, delivery to public resource or safety
21	Debris Slide/Debris Avalanche	Definite	1995	Yes, delivery to public resource or safety

Fig. 9 **a** Landslide inventory map, **b** Meander landslide type, **c** Deep seated landslide type, **d** Rock fall landslide type, **e** Landslide geo-database

3 Conclusion

In this tool, we briefly introduce a methodology for landslide inventory maps in unstable volcanic terrains. The mapping at Río La Carbonera watershed is an attempt to identify and select the best available thematic maps and to flag potential landslide areas in volcanic regions. The researchers must acknowledge the technical limitations of the landslide inventory map, whose quality depends on the experience of the investigators, the goals of the project, the complexity of the study area, and the available information. Despite the limitations, the author suggests this methodology has the potential to be the foundation for handling and supporting mapping of slope instability in areas with sparse information.

Acknowledgements This tool is supported by PAPIIT # IN102115 and the International Programme on Landslides (IPL project # 187).

References

Carson MA, Kirkby MJ (1972) Hillslope form and process. Cambridge University Press, Cambridge.

Cruden DM, Varnes DJ (1996) Chapter 3, Landslide types and process, pp 36–75. In: Turner AK, Schuster RL (eds) Landslides: investigation and mitigations: transportation research board, special report 247. Academy press, Washington, D.C., p 673

Dent DB (1985) Cartography thematic map design. Wm. C. Brown Publishers, Dubuque, Iowa

Fisher P (2004) Developments in spatial data handling. Springer, Berlin Heidelberg, New York

Paulin GL, Bursik M, Lugo-Hubp J, Orozco JZ (2010) Effect of pixel size on cartographic representation of shallow and deep-seated landslide, and its collateral effects on the forecasting of landslides by SINMAP and multiple logistic regression landslide models. Phys Chem Earth Parts A/B/C 35(3):137–148

Shan W, Guo Y, Wang F, Marui H, Strom A (eds) (2013) Landslides in cold regions in the context of climate change. Springer Science & Business Media, Berlin.

Slocum TA, Mcmaster RB, Kessler FC, Howard HH (2005) Thematic cartography and geographic visualization, 2nd edn. Prentice Hall, Upper Saddle River, NJ

Schneider M (1997) Spatial data types for database system. Springer, Berlin Heidelberg, New York

Tate NJ, Atkinson PM (2001) Modelling Scale in Geographical Information Science. Wiley, West Sussex, England

Varnes DJ (1958) Landslide type and process. In: Eckel EB (ed) Landslide and engineering practice, Special report no 29. Highway Research Board, pp 20–47

Varnes DJ (1978) Slope movement types and processes. In: Schuster RL, Raymond J, Krizek RJ (eds) Landslides—analyses and control special report 176. National Academy of Sciences, Washington, DC

Washington State Department of Natural Resources (DNR), Forest Practices Division (2006). Landslide hazard zonation (LHZ) mapping protocol, version 2.0. Accessed at: http://www.dnr.wa.gov/BusinessPermits/Topics/LandslideHazardZonation/Pages/fp_lhz_review.aspx

Wieczorek GF (1984) Preparing a detailed landslide inventory map for hazard evaluation and reduction. Bull Assoc Eng Geol 21:337–342

TXT-tool 1.052-1.2
GIS Using Landslides Susceptibility Mapping Model for Volcanoes

Gabriel Legorreta Paulín, Michael Polenz
and Trevor Contreras

Abstract

This tool is a systematic methodology for making landslide susceptibility maps using GIS technology. The main goal of this tool is to provide the foundation for mapping landslides susceptibility per landforms in areas where there is little information. Derivation of landform units used classification of morphometric parameters, expert knowledge, field verification, and adaptation of the Landslide Hazard Zonation Protocol of the Washington State Department of Natural Resources, USA. For each landform the landslide area rate (LAR) and the landslide frequency rate (LFR) is calculated in order to determine the landslide susceptibility. The training area is along the stream system of Río La Carbonera watershed on the SE flank of Pico de Orizaba volcano, Mexico.

Keywords

Landslide inventory · Landslide susceptibility · Modeling · Landforms
Pico de Orizaba volcano

Contents

1 **Introduction** 123
2 **Study Area** 124
3 **Methodology** 125
4 **Stage 1: Landslide Inventory Map** 125
5 **Stage 2: Mapping Landforms** 125
6 **Stage 3: Landslide Susceptibility** 128
7 **Conclusions** 130
References 131

1 Introduction

There are a number of approaches to generating landslide inventories and susceptibility maps based on geographic information systems (GIS) (Fig. 1). Each approach embedded in the GIS to map landslide susceptibility has its own advantages and limitations (based on the goals,

G.L. Paulín (✉)
Instituto de Geografía, Universidad Nacional Autónoma de México, Mexico City, Mexico
e-mail: legorretag@hotmail.com

M. Polenz · T. Contreras
Division of Geology and Earth Resources, Washington State, DNR, Washington, USA
e-mail: MICHAEL.POLENZ@dnr.wa.gov

T. Contreras
e-mail: Trevor.Contreras@dnr.wa.gov

K. Sassa et al. (eds.), *Landslide Dynamics: ISDR-ICL Landslide Interactive Teaching Tools*, https://doi.org/10.1007/978-3-319-57774-6_8

time, and resources of the project), and they can be applied in many creative ways. This teaching tool is based on the on-going research project (Grant PAPIIT, no. IN102115 and IPL project #187) from the Institute of Geography at the National Autonomous University of Mexico (UNAM) to map landslide susceptibility in volcanic terrains by using GIS. The Río La Carbonera watershed on the southeastern flank of Pico de Orizaba volcano, Mexico, has been selected as a case study area. This tool is a systematic methodology for landslide susceptibility mapping in volcanic areas with scarce information. The watershed has been studied using aerial photographs, fieldwork, and adaptation of the Landslide Hazard Zonation Protocol of the Washington State Department of Natural Resources, USA. More than 230 landslides of six types were recognized: shallow landslides, debris-slides, debris flows, meanders, deep-seated landslides, and rock falls. This analysis divides the watershed into 11 mass-wasting landforms, based on gravitational processes, and includes inner gorges, bedrock hollows, deep-seated landslides, meanders, and flats. These landforms are called rule-identified landforms (Fig. 4) and they are defined by rules (slope gradient and shape, lithology, landslide density and land use) adopted by the Washington State Department of Natural Resources to address landslide hazards (Washington State Department of Natural Resources 2006). For example, inner gorges and bedrock hollows have slopes generally greater than 70%. In this tool there are other areas that do not meet the rule-identified landform definitions (e.g., slope gradient < 70%). These areas are called non-rule-identified landforms (Fig. 4), such as low gradient hillslopes, moderate gradient hillslopes, high gradient hillslopes, rockfalls amphitheaters, non-rule-identified inner gorges, and non-rule-identified bedrock hollows. For each landform the landslide area rate (LAR) and the landslide frequency rate (LFR) (Corominas and Moya 2008) are calculated, as well as their slope stability susceptibility ratings. The technique and its implementation in a GIS is presented and discussed.

This tool uses an heuristic approach (options follow by this tool are underlined in Fig. 1) for mapping landslides, by considering the age and activity of landslides and applying a semi-quantitative approach for landslide susceptibility. The landslide susceptibility is derived from values that correspond to the area (landslide area rate) and number (landslide frequency rate) of landslides in each landform (Washington State Department of Natural Resources 2006).

2 Study Area

The Río La Carbonera watershed is selected for this teaching tool. The watershed has physiographic conditions that make it prone to landsliding. The area has high precipitation during the wet season, a high degree of weathering, steep slopes, weak rock types, and associated faults. Lying in the southeastern flank of Pico de Orizaba volcano, the area is representative of typical volcanic landscapes. The watershed is a tributary of the Río Blanco, which flows into the Gulf of Mexico within Veracruz state, and its area is 71.9 km^2, with an elevation range from 1224 to 3643 m a.s.l. and hillslopes ranging from 0° (inner valleys of relatively flat plains) to 68° (mountainous terrain). In the study area, the climate is subtropical semi-cold (Cb′(w)) at 3000–4400 m a.s.l., and subtropical temperate, subhumid (C(w1), C(w2)), and C(m)(f) and tropical semi-warm, humid (A)C(m)(f)) at <3000 m a.s.l. (García 2004). The mean annual precipitation is 1000–1100 mm/year at >4000 m a.s.l. and 927 mm/year at <1500 m a.s.l. (Palacios et al. 1999), with most falling as rain during seasonal storms between May and November. In combination with the intense precipitation, the geology in the area has a great influence on the topography and current landforms. The volcanic materials overlay weathered Cretaceous sedimentary

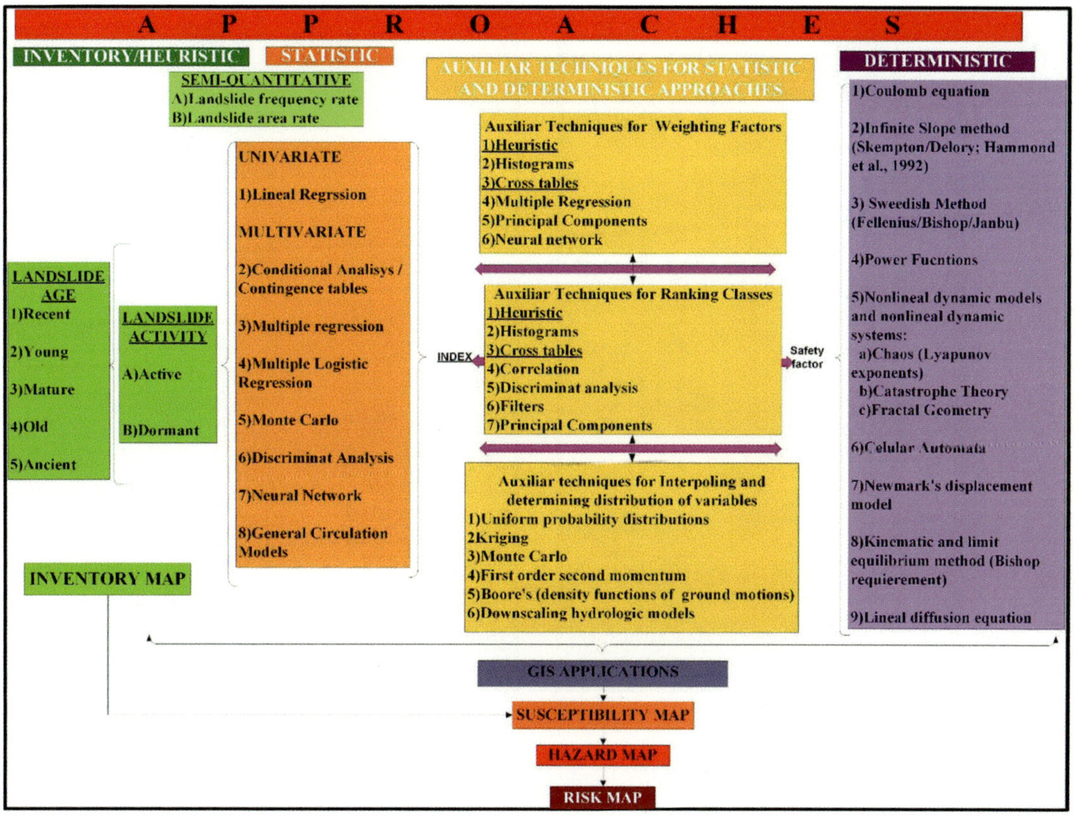

Fig. 1 Landslide susceptibility mapping modeling approaches

rocks. These influences are expressed in the patterns of resistance to erosion processes and the types of mass movement. In 2015, more than 230 landslides have been mapped (Fig. 2).

3 Methodology

The methodology encompasses three main stages of analysis (Fig. 3) to assess landslide susceptibility. Stage 1 is building a landslide inventory map; Stage 2 is building a landform map; and Stage 3 is calculating the landform landslide susceptibility.

4 Stage 1: Landslide Inventory Map

For the first stage, landslide data is collected from two sets of aerial photographs and during fieldwork to create a landslide inventory map. During aerial photographic interpretation, landslides are mapped and classified into shallow landslides, debris-slides, debris flows, meanders, deep-seated landslides, and rock falls, based on classifications by the Washington State Department of Natural Resources (2006), Cruden and Varnes (1996), and Wieczorek (1984). All landslides are digitized into a GIS, and a spatial geo-database of landslides was constructed from standardized GIS database.

5 Stage 2: Mapping Landforms

Once the locations of landslides are mapped and evaluated, similar potential landslide areas are grouped into individual landforms. Definition and delineation of landforms for landslide susceptibility zonation is conducted worldwide through various approaches dictated by the scale of the analysis, the requirements of the users, and the time and funding available. These approaches

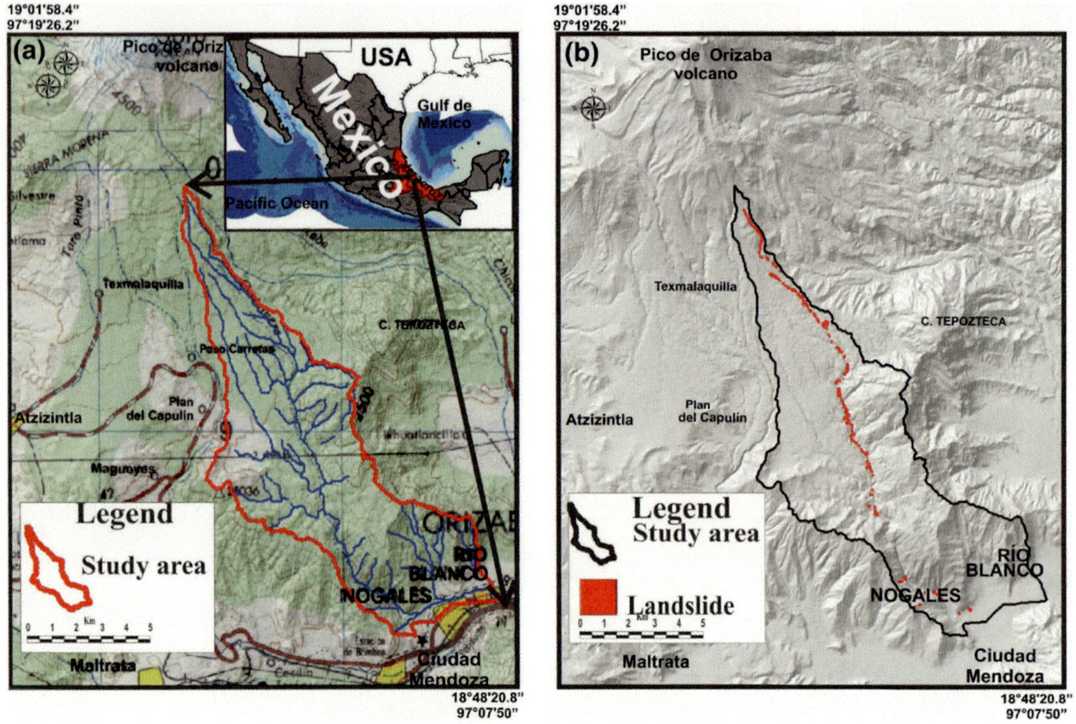

Fig. 2 Study area **a** location in Mexico, **b** landslide distribution along the valley

include visual identification based on expert knowledge and empirical evidence, as well as mapping with geomorphologic parameters, first and second derivatives from the digital elevation models, filter techniques, and statistical analysis (Pennock et al. 1987; Pike and Dikau 1995; Washington State Department of Natural Resources 2006; Evans 2012; Jasiewicz and Stepinski 2013). In this tool, specific landforms that exist across the study area are defined by rules adopted by Washington Forest Practices Protocol to address landslide hazards (Washington State Department of Natural Resources 2006). Their differentiation is based on slope gradient and shape, lithology, landslide density and sensitivity to forest practices. These landforms are called rule-identified landforms:

(1) Inner gorges are stream systems that are usually highly incised. They comprise areas with high gradient slopes (slopes are generally greater than 70%) embedded within bedrock hollows and convergent headwalls. Inner gorges can be either asymmetrical or symmetrical and may occur intermittently in lateral extent (Fig. 4).
(2) Bedrock hollows/Headwall (also called colluvium-filled bedrock hollows, swales, bedrock depressions, or simply hollows). These features are found primarily on convergent slopes but can also be found on planar slopes. They are often found up-gradient from inner gorges and on steep slopes >70%. The union of two or more bedrock hollows form a convergent headwall (Fig. 4).
(3) Deep-seated landslides (DSL). This landform encompasses active head and lateral scarps of deep-seated landslides (Figs. 4 and 5).
(4) Meanders occur along river valleys. The slope failures occur through stream

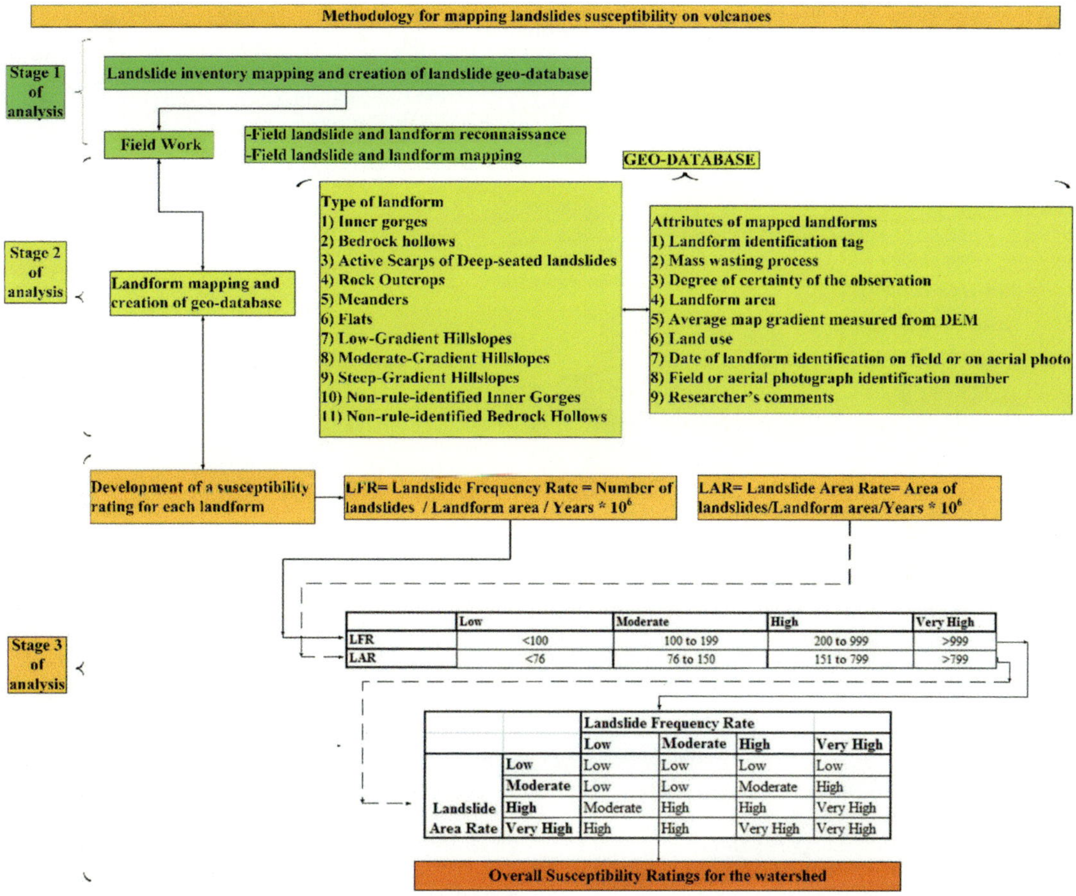

Fig. 3 Flow chart for modelling of landslide susceptibility

undercutting of outer edges of meanders along valley walls or high terraces of an unconfined meandering stream (Figs. 4 and 6).

5) Flats include slope forms that are mostly gentler than 10%. This category consists predominantly of lowland landforms such as valley bottoms, floodplains, and flat terrace surfaces (Fig. 4).

The aerial photos, the landslide inventory, and GIS layers are also used to identify other areas that do not meet the rule-identified landform definitions (e.g., having a slope gradient < 70%). These areas are called non-rule-identified landforms:

(6) Low-gradient hillslopes include all slope forms (convergent, divergent, and planar) and gradients between 11 and 40% (Fig. 4).

(7) Moderate-gradient hillslopes include all slope forms (convergent, divergent, and planar) that range between 41 and 70% (Fig. 4).

(8) Steep-gradient hillslopes, include all slope forms (convergent, divergent, and planar) with gradients of >70%. This category contains other high-hazard landforms (e.g., inner gorges and bedrock hollows) that are the initiation points for many debris slides and debris flows (Fig. 4).

(9) Non-rule-identified inner gorges have characteristics similar to those of rule-identified inner gorges but involve convergent slopes of 30–70%. They appear as gentle-walled canyons or gullies and seem to have been eroded by stream action but with evidence of mass wasting along

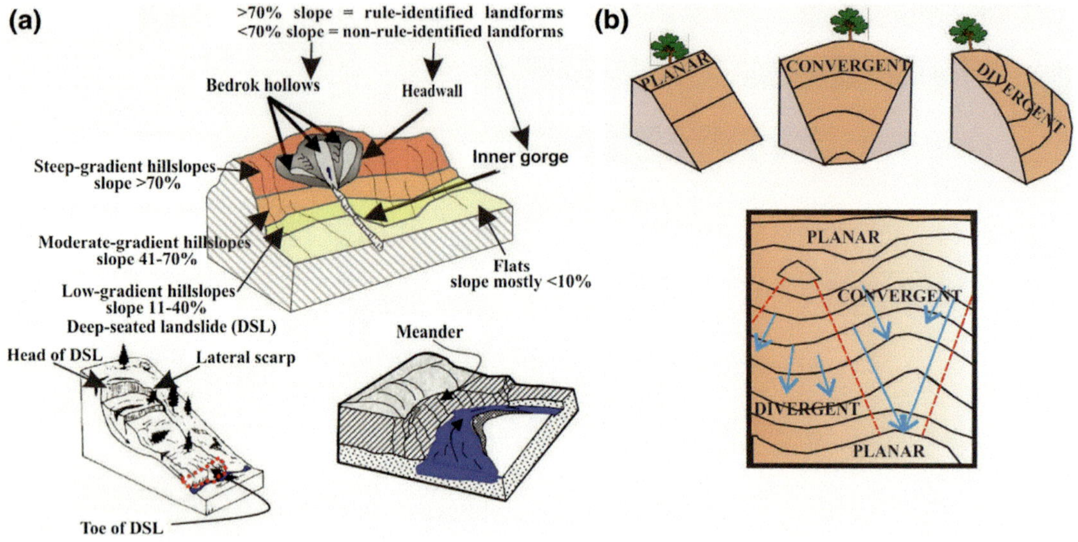

Fig. 4 **a** Rule-identified landforms: inner gorges, bedrock hollows/headwall, deep-seated landslide, meander, and flats (Washington State Department of Natural Resources 2006) and non-rule-identified landforms: Low-gradient hillslopes, Moderate-gradient hillslopes, and steep-gradient hillslopes. **b** Hillslope shape: planar, convergent, and divergent

Fig. 5 Deep-seated landslides

their sidewalls. They may be either symmetrical or asymmetrical in profile (Fig. 4).

10) Non-rule-identified bedrock hollows have characteristics similar to those of rule-identified bedrock hollows, but involve convergent slopes of <70%. This landform is the initiation point for many debris slides and debris flows (Fig. 4).

(11) Rockfalls amphitheaters can occur along the channel where lava flows form steep steps (Fig. 7).

Both rule- and non-rule-identified landforms are mapped into GIS as part of a landform polygon feature (Fig. 8).

6 Stage 3: Landslide Susceptibility

The third stage determines the landslide susceptibility for each landform. For both rule- and non-rule-identified landforms, a semi-quantitative overall hazard rating is derived

Fig. 6 Meader landslide. The *blue line* and *arrows* represent an old landslide debris deposit and landslide movement and the *red line* and *arrow* represent a new landslide debris deposit and movement

Fig. 7 Rockfall along *u-shape* amphitheatre. The rock fall deposit is on *top* of a Talus cone deposit

from values that correspond to the number and area of landslides within each landform, normalized for a period of time spanned by the aerial photographs used, and to the area of each landform (Fig. 9, step 1). Because the resulting numbers are always very small fractions, the values are then multiplied by one million and rounded in order to provide whole easily-usable numbers (Washington State Department of Natural Resources 2006). This procedure is restricted to shallow landslides (including debris flow and debris slides) that deliver sediment. Qualitative ratings of the landslide area rate and the landslide frequency rate are entered into a matrix

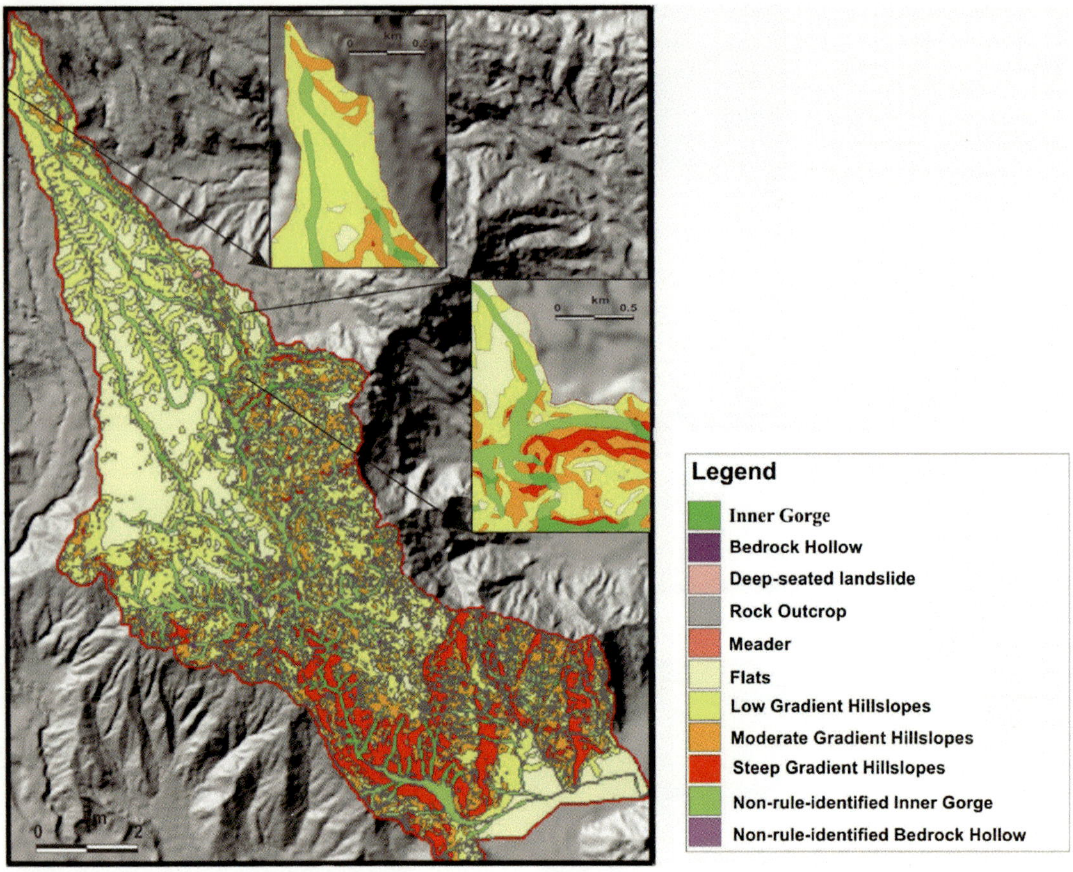

Fig. 8 Landforms map (GIS) of the study area

to determine the landslide susceptibility for each landform. As a result, each landform is assigned a susceptibility semi-quantitative rating of Low, Moderate, High, or Very High (Fig. 9, step 2). The overall landslide susceptibility per landforms is assigned by combining the landslide area rate and landslide frequency rate rating values. For example, if a landform has a low landslide frequency rate and very high landslide area rate, the overall susceptibility is high (Fig. 9, step 3).

7 Conclusions

With this tool, we briefly introduce a methodology for mapping landslide susceptibility in volcanic terrains. The mapping at Río La Carbonera watershed is one of the examples of flagging potential landslide areas by modeling landslide susceptibility in volcanic regions. The researcher must acknowledge the technical limitations of the landslide inventory map (Fig. 2b), landform map (Fig. 8), and the landslide susceptibility map (Fig. 9b), based on the experience of the investigators, the goals of the project, the complexity of the study area, and the available information. Despite the limitations, the methodology introduces a foundation for handling and supporting assessments of slope instability in areas with little information. For example, the semi-quantitative rating for volcanic landforms will provide the basis for landslide susceptibility comparison with other volcanic watersheds.

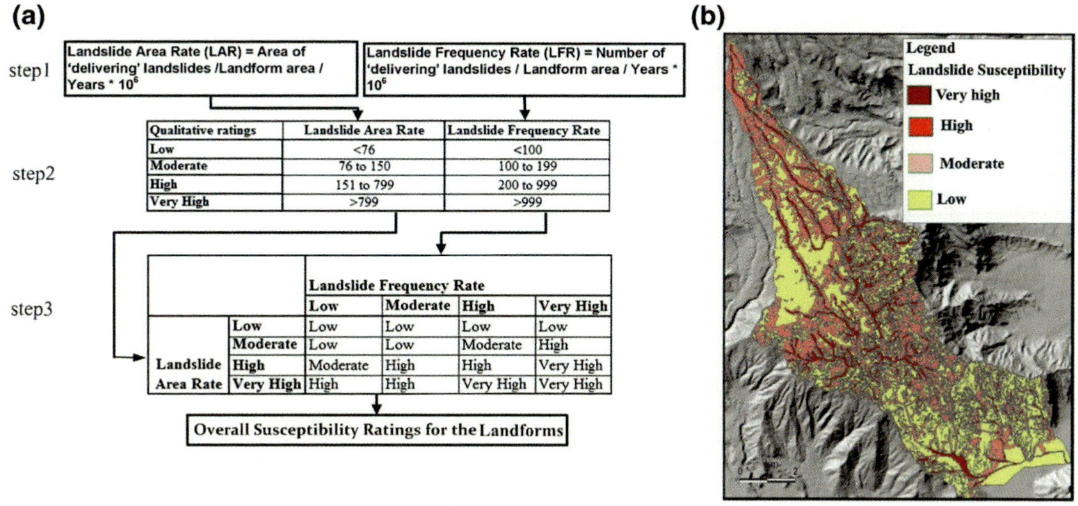

Fig. 9 **a** Procedure for producing a semi-quantitative overall landslide susceptibility, **b** landslide susceptibility map based on the landslide area rate and landslide frequency rate

8. Acknowledgements This tool is supported by PAPIIT #IN102115 and the International Programme on Landslides (IPL project #187).

References

Corominas J, Moya J (2008) A review of assessing landslide frequency for hazard zoning purposes. Eng Geol 102:193–213

Cruden DM, Varnes DJ (1996) Landslide types and process (Chapter 3, pp. 36–75). In: Turner AK, Schuster RL (eds) Landslides: investigation and mitigations: transportation research board, vol Special Report 247. Academy Press, Washington, D.C., p 673

Evans IS (2012) Geomorphometry and landform mapping: what is a landform? Geomorphology 137(1):94–106

García E (2004) Modificaciones al sistema de clasificación climatic de Köppen. Instituto de Geografía, UNAM. Serie Libros #6, p 90

Jasiewicz J, Stepinski T (2013) Geomorphons—a pattern recognition approach to classification and mapping landforms. Geomorphology 182:147–156

Palacios D, Parrilla G, Zamorano JJ (1999) Paraglacial and postglacial debris flows on Little Ice Age terminal moraine: Jamapa glacier, Pico de Orizaba (Mexico). Geomorphology 28:95–118

Pennock D, Zebarth B, De Jong E (1987) Landform classification and soil distribution in hummocky terrain, Saskatchewan, Canada. Geoderma 40:297–315

Pike RJ, Dikau R (1995) Advances in geomorphometry. Z Geomorphol NF Suppl Bd 101:238

Washington State Department of Natural Resources (2006) Landslide hazard zonation (LHZ) mapping protocol, version 2.0. Washington State Department of Natural Resources, Forest Practices Division. Accessed at: http://www.dnr.wa.gov/BusinessPermits/Topics/LandslideHazardZonation/Pages/fp_lhz_review.aspx

Wieczorek GF (1984) Preparing a detailed landslide inventory map for hazard evaluation and reduction. Bull Assoc Eng Geol 21:337–342

TXT-tool 1.386-2.1 Landslide Susceptibility Assessment Method

Marko Komac and Jernej Jež

Abstract

A method was developed for the susceptibility estimation of different landslide types, and tested in the Bovec municipality, NW Alpine Slovenia. The purpose of the susceptibility maps, at the 1:25,000 scale, is to be used in spatial planning by the local municipality to minimize potential landslide hazard. The requirements taken into account were the following: expert approach, GIS tools support, reasonable production time, and the product being simple to read and understand. The elaboration of the final products comprises four phases, the first three of which are desktop phases: (1) synthesis of archive data, (2) probabilistic model of landslide susceptibility, (3) compilation of phases 1 and 2 into the final map at the 1:25,000 scale. The fourth phase consists of field reconnaissance of areas considered most hazardous.

Keywords

Landslides · Methodology · Susceptibility · Assessment · Large-scale Geohazards

Contents

1 **Introduction** 133
2 **Landslide Susceptibility Map Development Method** 134
3 **Results and Discussion** 137
4 **Conclusions** 138
References 138

M. Komac (✉)
Faculty of Civil and Geodetic Engineering, University of Ljubljana, Ljubljana, Slovenia
e-mail: m.komac@telemach.net

J. Jež
Geological Survey of Slovenia, Ljubljana, Slovenia
e-mail: jernej.jez@geo-zs.si

K. Sassa et al. (eds.), *Landslide Dynamics: ISDR-ICL Landslide Interactive Teaching Tools*, https://doi.org/10.1007/978-3-319-57774-6_9

1 Introduction

Well-understood landslide hazard (i.e. the possibility of landslide occurrence in space, time, size and its potential impact on humans) to inhabitants and infrastructure at the local, municipal and regional levels, plays a key role in spatial planning and in civil engineering infrastructure planning. Defining potential landslide source areas is the first step in the process. By way of detailed geological mapping, analyses of existing geological data and the implementation of modern monitoring methods (i.e. remote sensing) it is possible to define landslide susceptible areas (Guzzetti et al. 1999; Van Westen 2000; Crozier and Glade 2005). The approach itself is scientific, and the results are applicable in direct risk aversion (preserving human lives and infrastructure) as well as in long-term spatial planning with the goal to minimize unwanted costs resulting from wrong human decisions and unnecessary interventions, often due to lack of understanding. The method's added value lays in its scientific base, and in the fact that it helps solve socio-economic challenges in areas prone to landslides. In Slovenia, in the period between 1994 and 2008, the total damage from slope wasting processes totalled more than 100 million euros (STAT 2009), not counting remediation costs, especially long-term remediation costs of larger occurrences (e.g. Log pod Mangartom, Slano blato, Macesnikov plaz, Koseč, Ciprnik) that pose danger to human lives or infrastructure. Systematic use of this method in spatial planning is expected to reduce many of these costs.

2 Landslide Susceptibility Map Development Method

A method to create landslide susceptibility maps was developed. In this context, the term landslides includes not only landslides in the strict sense, but also rockfalls and debris flows. The results outline the potential source areas of these processes. The method aims to produce a product designed to assist local communities and municipalities in spatial planning and help them avoiding potential landslide source areas. The method is based on expert work, and was developed progressively starting with small-scale maps of 1:250,000 (landslide susceptibility map, Komac and Ribičič 2006 (Fig. 1); debris flow susceptibility map, Komac et al. 2009). These maps provide a general overview of landslide susceptibility in Slovenia, which is useful for a synoptic overview that helps pinpoint the most susceptible areas or municipalities.

Consequently, more detailed susceptibility models at the scale of 1:25,000 were designed to address the use of spatial planning models at the municipal level. Development and validation of models were based on an extensive landslide database (Komac and Hribernik 2015) compiled and standardized at the national level and on analyses of landslide spatial occurrence. Finally, the process of model development was divided into four phases (Fig. 2), which are described below.

The first phase (Phase 1) comprises the compilation and organization of archived geological data (e.g. maps, reports, observation points, etc.) into a thematic landslide map at the 1:25,000 scale (Bavec et al. 2012), and into a collection of auxiliary data useful in the following phases. Specifically, Phase 1 consists of (1.a) terrain overview with the local representatives, and collection of more detailed data available from local archives, (1.b) gathering of geological data and information on landslide occurrences from general and long-term archives (i.e. Geological Survey, consultant companies, universities, press), (1.c) compilation of auxiliary data on spatio-temporal factors known to influence the occurrence of the different landslide types (also known as natural factors; i.e. digital elevation model, river network, land cover, populated areas, infrastructures, roads, pipelines, electric lines), (1.d) compilation of landslide data into a single, standardized database (or entering it into an existing database), and (1.e) expert evaluation of landslide susceptibility of individual lithologic or engineering geological units.

The second phase (Phase 2) comprises the spatial occurrence analysis of each landslide process type, with relation to each spatio-temporal factor's value with the pixel resolution that matches the scale of the landslide map obtained from Phase 1. Phase 2 encompasses two parts (Komac 2012). The first part (2.a) includes the assessment of influence within the spatio-temporal factors, including: (2.a.i) reclassification of spatio-temporal factors into classes based on literature or expert decision, (2.a.ii) analysis of landslide spatial occurrences within the classes for each individual spatio-temporal factor in order to determine the significance of

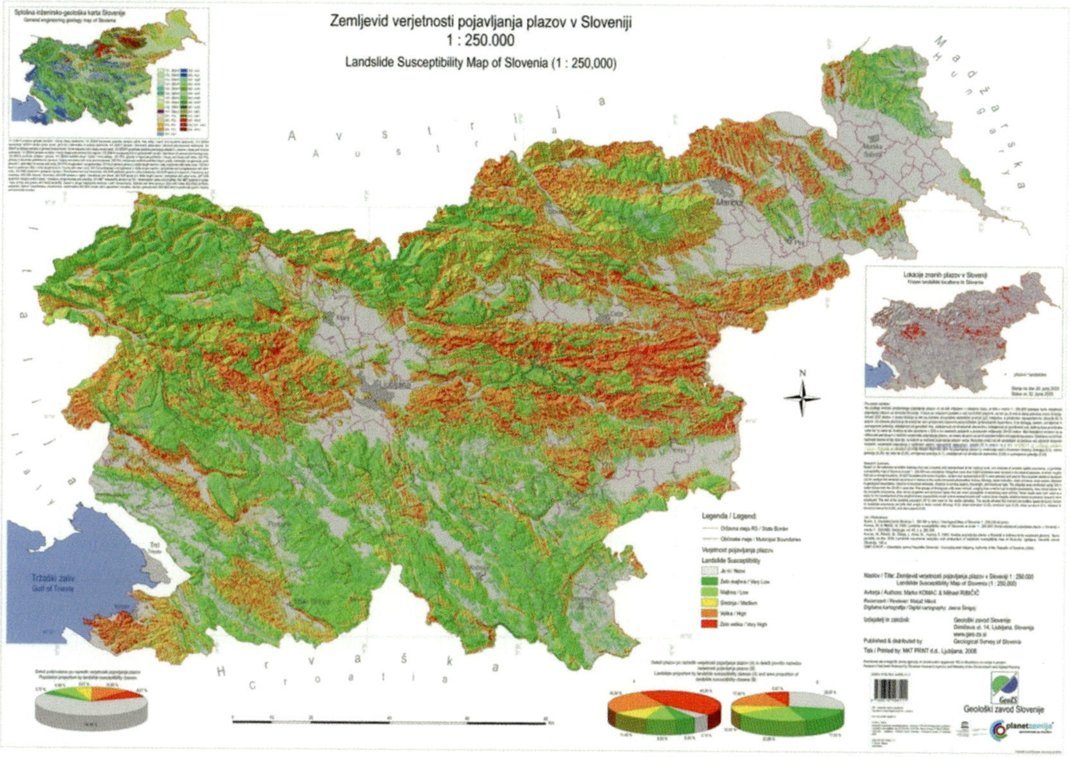

Fig. 1 Landslide susceptibility map of Slovenia, at the scale of 1:250,000 (from Komac and Ribičič 2006)

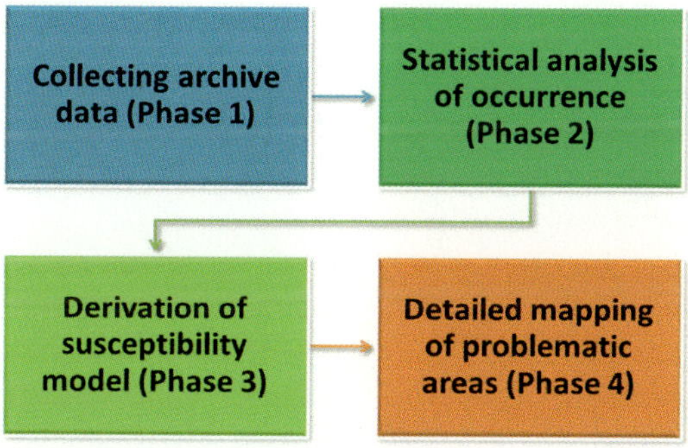

Fig. 2 A logical scheme of the four phases of the method to derive landslide susceptibility maps

the factor analyzed as a whole and to determine the significance of the classes within each (significant) factor. In the described methodology, a Chi-square (χ^2) statistical test on two thirds of the landslide occurrences was used to determine the importance of every class and the factor itself, (2.a.iii) reclassification (or, more precisely, reordering) of classes within each of the significant factors in accordance with the results of the class significance analysis (step 2.a.ii.), and (2.a.iv) normalization of newly ordered classes within the same value range (i.e. 0–1) with the purpose of eliminating the fake importance of a factor that could be brought into the model development when factors with a larger number of classes could have more influence on the model due to the maximum number being larger than the factors with less classes. The second part (2.b) includes the assessment of contribution of each spatio-temporal factor to the occurrence of a specific landslide type (determine the weight of a factor in the linear weighted model) and development of a statistically based model for each landslide type. The assessment can be done using values determined/defined in previous assessments, or it can be carried out specifically for the given case. In the described method, a quasi-Monte Carlo approach was used with several tens of thousands trials of different factor weights used to calculate numerous models. The quality of each model was assessed with the remaining one third of the landslide occurrences for every one of the 100 equal area classes of the derived model.

The third phase (Phase 3) consists of the calculation of the final susceptibility model for each landslide type, and the production of the susceptibility map, at the 1:25,000 scale, outlining the most problematic areas, which will be the focus of the next phase (Phase 4). Phase 3 includes the following: (3.a) Based on the inputs of the normalized spatio-temporal factors from Phase 2.a.iv., and well as on the spatio-temporal factors weights acquired in the phase 2.b., a linear weighted susceptibility model (raster-based map) for a given landslide type was developed. Secondly (3.b), the developed susceptibility model was fine-tuned with in situ verification, meaning the modified reclassification of units within the factor was introduced and the model was re-run. Finally (3.c), the susceptibility model was simplified by being divided into 6 susceptibility classes to be clearly understood by the non-expert end users.

The fourth and final phase (Phase 4) consisted in a detailed geological and landslide mapping of selected problematic areas, at the scale of 1:5000 or 1:10,000, depending on the availability of resources. Specifically, Phase 4 consisted of (4.a) choosing the most problematic areas (this was done in agreement with the municipality or the local community), (4.b) detailed field mapping, and (4.c) producing a detailed map at the scale of 1:5000, once again applying Phases 1–3 described above.

Based on an extensive literature search, past expert decisions, and correcting the results while developing the method, the factors (original, and their derivatives) that are listed in Table 1 were identified as those playing a significant role in the construction of the susceptibility models for the three considered landslide types. The conceptual frameworks for the landslide, debris flow and rockfall susceptibility models are exemplified in Fig. 3.

Table 1 Spatio-temporal factors used to model the three considered landslide types

Factor	Source	Rockfall	Landslide	Debris flow
Lithology	Archive/in situ	X	X	X
Strata dipping (inclusion depends on the model scale)	Archive/in situ	X	X	–
Structural elements	Archive/in situ	–	X	–
Fault zones	Archive/in situ	X	X	X
Slope (slope inclination)	DEM derivate	X	X	X
Curvature (slope concavity)	DEM derivate	–	X	X
Aspect (slope orientation)	DEM derivate	–	X	–
Land use/land-cover type	Archive/in situ	–	X	–
River net	Survey agency	–	–	X
DEM	Survey agency	–	–	X
48-h rainfall	Survey agency	–	–	X
Stream energy potential	DEM and river net derivate (Komac et al. 2009, 2010)			X
Energy potential	DEM derivate (Komac et al. 2009, 2010)			X

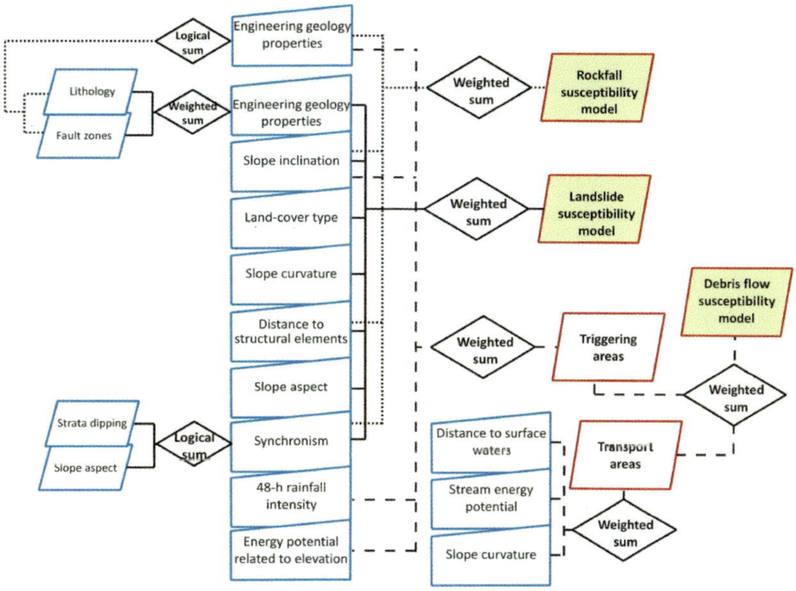

Fig. 3 Conceptual framework for the landslide susceptibility model. The processes marked with full lines represent the landslide susceptibility model derivation, the processes marked with *dashed lines* represent the debris-flow susceptibility model derivation, and the processes marked with *dotted lines* represent the rockfall susceptibility model derivation

In addition to the references listed in Table 1, there is abundant literature available on the natural factors influencing landslide occurrence (Guzzetti and Cardinali 1990; Cruden and Varnes 1996; California Department of Conservation, Division of Mines and Geology 1997; Guzzetti et al. 1999; Van Westen 2000; Crozier and Glade 2005).

3 Results and Discussion

Based on the experience from the methodology development process and by comparing the expert evaluation (Phase 1) of the individual lithological or engineering geological units to pure statistical approach (Phase 2), the results revealed the necessity of combining the two phases. This is due to several factors that hinder the successful application the method, including the following:

1. In some places, the number of landslide occurrences is insufficient for their statistical modelling.
2. Not rarely, there are variations in the lithological units or formations that contribute to local geological engineering property discrepancies. Such examples are lithologically heterogeneous units (e.g. clastic rocks with limestone intercalations). In such cases, the quickly changing mechanical properties of rocks could be neglected at larger scales, but still, the influence on landslide occurrence is significant.
3. The synchronism of strata and slope dipping plays a very important role in landslide occurrence, and it is rather difficult (and demanding in terms of resources) to model strata dipping on a larger area in high detail.
4. Statistical results can sometimes be misleading, e.g. elevation impact on landslide occurrence. Analysts need to be very careful, as usually the correlation between the slopes and the elevation is positive—the higher the elevation the steeper the slope (which holds true at least within a specific elevation range). Nevertheless, the influence of elevations on landslide occurrence is purely coincidental.
5. On the other hand, pure in situ expert assessment will produce overly general conclusions, and will miss any indications of joint influence of two or more factors that result in landslide occurrence.

4 Conclusions

There are many approaches to assessing slope mass wasting processes, and the method presented here is one of them. The method presented is based on extensive desktop statistical modelling, and in situ mapping, which makes it appropriate for large-scale applications. Despite the fact that merging the desktop approach with the in situ mapping and verification approach calls for more resources (both time and funds), the long-term benefits are clear, and directly applicable to spatial planning.

Acknowledgements The authors thank the Geological Survey of Slovenia, the Slovenian Research Agency, the Ministry for the Environment and spatial planning of Slovenia, and the Administration of the Republic of Slovenia for Civil Protection and Disaster Relief with the Ministry of defence of the Republic of Slovenia, for funding the mentioned research projects in the period from 2004 to 2013.

References

Bavec M, Čarman M, Durjava D, Jež J, Krivic M, Kumelj Š, Požar M, Komac M, Šinigoj J, Rižnar I, Jurkovšek B, Trajanova M, Poljak M, Celarc B, Demšar M, Milanič B, Mahne M, Otrin J, Čertalič S, Štih J, Hrvatin M et al (2012) Izdelava prostorske baze podatkov in spletnega informacijskega sistema geološko pogojenih nevarnosti zaradi procesov pobočnega premikanja, erozijskih kart ter kart snežnih plazov—pilotni projekt: sumarno poročilo (in Slovene). Geološki zavod Slovenije, Ljubljana, p 40

California Department of Conservation, Division of Mines and Geology (1997). Factors affecting landslides in forested terrain, Note 50. Online: http://www.consrv.ca.gov/cgs/information/publications/cgs_notes/note_50/Documents/note50.pdf

Crozier MJ, Glade T (2005) Landslide hazard and risk: issues, concepts and approach. In: Glade T, Anderson MG, Crozier MJ (eds) Landslide risk assessment. Wiley, New York, USA, pp 1–40

Cruden DM, Varnes DJ (1996) Landslide types and processes. In: Turner AK, Schuster RL (eds) Landslides—investigation and mitigation: transportation research board, Special report no. 247. National Research Council, National Academy Press, Washington, D.C., USA, pp 36–75

Guzzetti F, Cardinali M (1990) Landslide inventory map of the Umbria region, Central Italy. In: Cancelli A (ed) Proceedings ALPS 90 6th international conference and field workshop on landslides, Ricerca Scientifica ed Educazione Permanente, Università degli Studi di Milano, Milano, Italy, pp 273–284

Guzzetti F, Carrara A, Cardinali M, Reichenbach P (1999) Landslide hazard evaluation: a review of current techniques and their application in a multi-scale study central Italy. Geomorphology 31:181–216

Komac M (2012) Regional landslide susceptibility model using the Monte Carlo approach—the case of Slovenia. Geol Q 56(1):41–54

Komac M, Hribernik K (2015) Slovenian national landslide database as a basis for statistical assessment of landslide phenomena in Slovenia. Geomorphology, ISSN 0169555X. Accepted Manuscript, doi:10.1016/j.geomorph.2015.02.005

Komac M, Ribičič M (2006) Landslide susceptibility map of Slovenia at scale 1:250,000. Geologija 49(2):295–309

Komac M, Kumelj Š, Ribičič M (2009) Debris-flow susceptibility model of Slovenia at scale 1: 250,000. Geologija 52(1):87–104

Komac M, Kumelj Š, Ribičič M, Mikoš M (2010) Debris-flow susceptibility model of Slovenia at scale 1:250,000. In: CHEN SC (ed) INTERPRAEVENT 2010 international symposium in Pacific Rim, Taipei, Taiwan. Symposium proceedings: torrents, floods and storms: mass movements, landslide and slope stability: disaster and risk management. International Research Society INTERPRAEVENT, Taipei, pp 786–796

STAT (2009) Statistical information portal. Statistical Office of Slovenia, Slovenia

Van Westen CJ (2000) The modeling of landslide hazards using GIS. Surv Geophys 21:241–255

TXT-tool 1.081-6.1
A Comparative Study of the Binary Logistic Regression (BLR) and Artificial Neural Network (ANN) Models for GIS-Based Spatial Predicting Landslides at a Regional Scale

Jie Dou, Hiromitsu Yamagishi, Zhongfan Zhu,
Ali P. Yunus and Chi Wen Chen

Abstract

This teaching tool is to present how to generate the landslide susceptibility maps using binary logistic regression (BLR) and artificial neural network (ANN) methods at a regional scale. The study area is one of most landslide-prone areas in Japan. First, the landslide inventory data from the National Research Institute for Earth Science and Disaster Prevention (NIED) were randomly partitioned into two parts: training and testing data. Then, 10 m DEM data and geology map were analyzed to extract the landslide predisposing factors. Next, the susceptibility maps were produced in a geographic information system (GIS) environment. Then, the receiver operating characteristics (ROC) was used to assess the model accuracy. Validation results show that both of two methods can be obtained with acceptable results. The maps can provide useful information for the future planning of hazard mitigation.

Keywords

Landslides susceptibility · Logistic regression (LR)
Artificial neural network (ANN) · Predisposing factors
Hazard map · Osado of Sado Island

Contents

1 **Introduction** ... 140
2 **Case Study and Data** ... 140
2.1 Case Study Area ... 140
2.2 Landslide Inventory ... 141
2.3 Landslide Predisposing Factors ... 143
3 **Methods** ... 143

J. Dou (✉) · C.W. Chen
Center for Spatial Information Science, University of Tokyo, Kashiwa, Japan
e-mail: douj888@gmail.com

H. Yamagishi
Shin Engineering Consultant Co.Ltd., 2-8-30, Sakaedori, Shiroishi-ku Sapporo 003-0021, Japan

Z. Zhu
College of Water Sciences, Beijing Normal University, Beijing, China

A.P. Yunus
Department of Geology, Aligarh Muslim University, Aligarh, India

C.W. Chen
National Science and Technology Center for Disaster Reduction, Taipei, Taiwan

K. Sassa et al. (eds.), *Landslide Dynamics: ISDR-ICL Landslide Interactive Teaching Tools*, https://doi.org/10.1007/978-3-319-57774-6_10

3.1 Binary Logistic Regression (BLR) 143
3.2 Back-Propagation for Feed-Forward Artificial Neural Network (ANN) 145
4 Implementation of Landslide Susceptibility Mapping 146
4.1 Landslide Susceptibility Mapping Using BLR Mode 146
4.2 Landslide Susceptibility Mapping Using ANN Model 147
4.3 The Validation of the Landslide Susceptibility Models 150
5 Conclusion 150
References 150

1 Introduction

Japan is subject to several types of natural disasters that include earthquakes, tsunamis, volcanic eruptions, and typhoons. In addition, it is also susceptible to numerous landslides, owing to geologic settings called a subduction zone. Therefore, most of the rocks have been affected by faulting and folding considerably. In addition, Japan is located at temperate monsoon climate zone, and then much rains are documented in the rain seasons association with typhoon. This trend is expected to continue and may increase in the future with the influence of unplanned urbanization and economic development, deforestation. To mitigate from landslide disasters, susceptibility, hazard, and risk must be prepared (Turner and Schuster 1996). Hence, predicting landslide has become an important and challenging issue in the risk mitigation research (Chang and Chao 2006; Dou et al. 2014, 2015a, 2015b; Zhu et al. 2015, 2016).

The landslide susceptibility maps (LSM) delineate the earth's surface into zones of varying degrees of stability based on the evaluated significance of the conditioning factors inducing instability. LSM plays an important role in risk mitigation. LSM also provide an important basis for the measure aimed at decreasing the risks associated with landslides (Ayalew et al. 2005; Dou et al. 2014).

In literature, binary logistic regression (BLR) and artificial neural network (ANN) methods are considered to be the two of the most commonly used methods for the assessment of probability of landslide occurrence at medium and regional scales. The advantage of BLR over other multivariate analysis methods is that it is independent on data distribution and can handle a variety of data sets such as continuous, categorical, and binary data. One of the significant advantages of ANN is that it imposes less restrictive requirements with respect to the available information concerning the character of the relationships between the processed data and the type of distribution. It has robustness to deal with noisy or incomplete patterns and high fault tolerance compared with LR (Shi et al. 2012). Furthermore, ANN is a complex and flexible non-linear statistical method as Kanungo et al. (2006) noted the method can obtain high prediction accuracy for a classification, particularly with adequate samples. In this study, we use these two methods to produce LSM.

The outline of the teaching tool is shown in Fig. 1. We first prepare the predisposing factors, such as slope, aspect, etc. Then the BLR and ANN model are used to create landslide susceptibility maps. Finally, the accuracy is estimated using the Receiver Operating Characteristics (ROC) and historical data. This tool provides a common useful way to create landslide hazard maps for future planning of risk mitigation.

2 Case Study and Data

2.1 Case Study Area

The study area is located in a mountainous region of Sado Island of Niigata Prefecture, in the Japan Sea (Fig. 2). Sado Island is composed of two parallels, elongated ridges approximately northeast (NE) to the southwest (SW) as shown in Fig. 2, which are locally named Osado and Kosado and cover approximately 856 km^2. In this tool, we concentrate on the Osado Island. Climate is humid, subtropical with warm summers and cold winters. The average temperature is 13.7 °C and the annual precipitation is

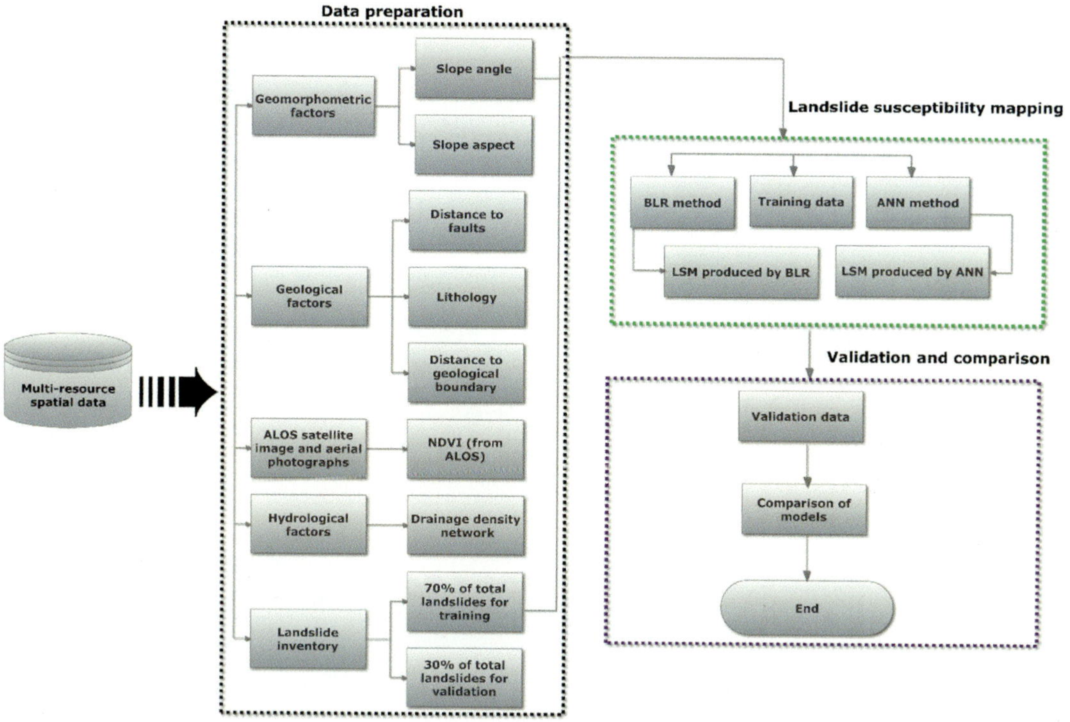

Fig. 1 Flowchart shows overall methodology adopted for this tool

approximately 1550 mm (Yamagishi 2008). The elevation varies from sea level to 1172 m with a mean of 333 m. The highest point of the island is the peak of Mt. Kimpoku in Osado. The geology of the island is Neogene terrestrial and marine volcanic: rhyolitics (dacitics) and andesitics, associated with pyroclastics and rhyolitic intrusives. Most of the coastal slopes are rugged rocky, and some shores are recently formed semi-consolidated and unconsolidated sand and gravel. Consisting principally of mountains, hills, and upheaved benches and active faults, study area is prone to landsliding and debris disasters (Yamagishi 2008). In the Osado area, the landslides are triggered mostly by rainfall and partly by snow melting. The landslide types are mostly deep-seated, translational and rotational slides as shown in Fig. 3, excluding the rock falls. Most of the deep-seated landslides are inventoried in the rhyolitics and dacitics. The landslide susceptibility may also be influenced by rainfall, snow melting, geology, slope aspect, and slope angle.

2.2 Landslide Inventory

The first step is to build the landslide inventory map. In this tool, a total of 825 known landslides (Fig. 4) was first obtained for the model development; these landslides were interpreted by the landslide experts at the National Research Institute for Earth Science and Disaster Prevention (NIED), Japan. NIED has been producing these landslide inventory since the year 2000 from the interpretation of multiple aerial photographs. The landslides are depicted as boundary polygons in GIS shape file format and are available at NIED archives. It is observed from the landslide inventory map that most landslide areas are

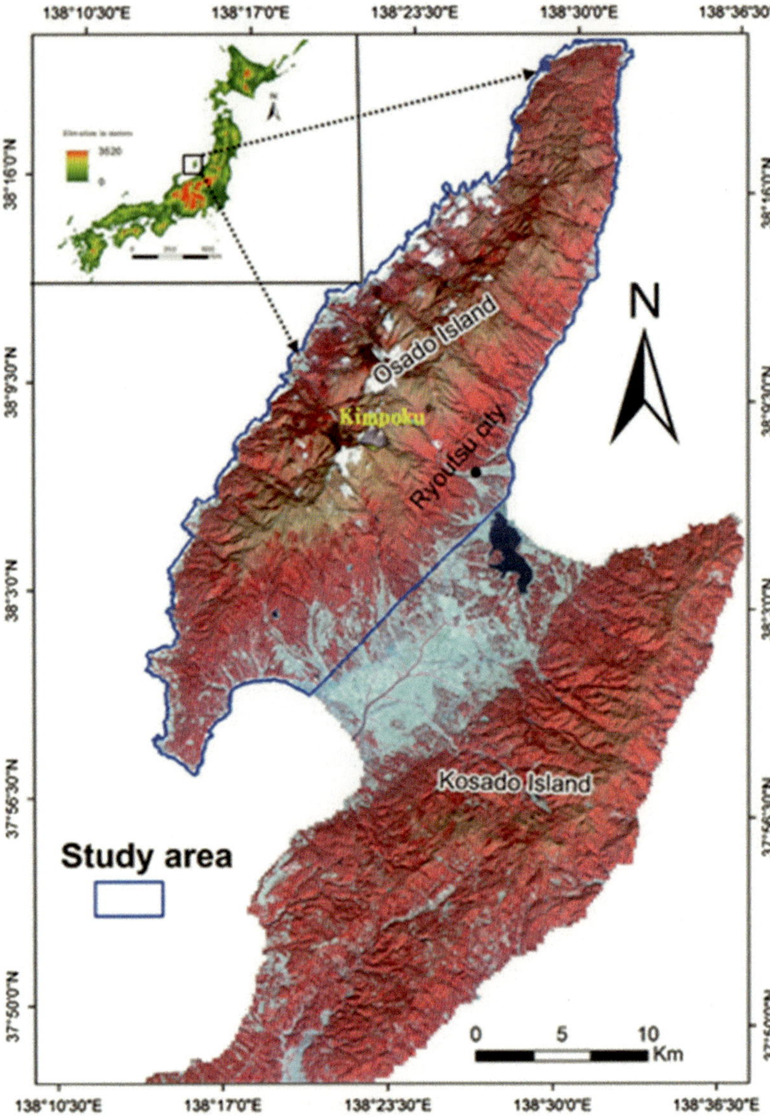

Fig. 2 False color composite 3D view of the study area

greater than 0.01 km^2. The total area of landslides is about 57 km^2, and accounts for approximately 15% of the Osado area. The areas of all the mapped landslides display a frequency distribution that can be described by a power law for approximately three orders of area with a good fit ($R^2 = 0.95$) (Fig. 5 in Osado). In this study, the landslide distribution was examined only for medium to large landslides (2.34×10^3–8.48×10^5 m^2) with a significant rollover value (1.13×10^4 m^2). The power law can be used to examine and describe the dominant landslide areas (Guzzetti et al. 2002). In this tool, we show that the landslide inventory from NIED offers further strong support for the theory that landslide frequency-area statistics are power law for landslides that exceed a smallest length scale. The exponent of the power law (1.6) indicates that the large landslide areas are dominant.

Different sampling strategies are available to construct the reliable landslide susceptibility maps. Several previous researches preferred to use 'points' to represent the spatial location of

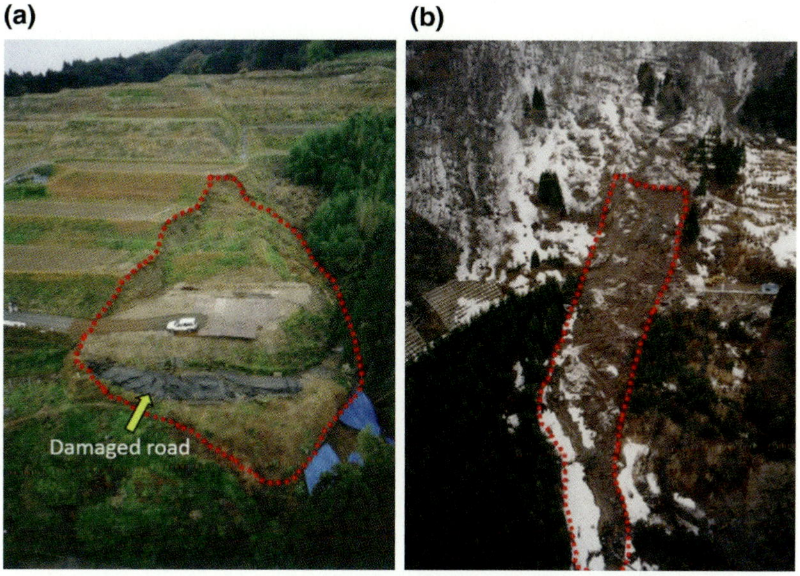

Fig. 3 Examples of identified major landslide types in the Kosado Island: **a** rotational slide that severely damaged the agricultural roads, **b** translational slide in the hilly terrains that carved the dense vegetation

landslides (Tien Bui et al. 2012). Dai and Lee (2003) delineated only the source areas during the landslide susceptibility assessments and excluded both the transport and the deposition zones of existing landslides. Comparisons of such sampling strategies are however beyond the purpose of this research. Here, we adopted one of the most popular methods- the polygon of landslide to represent the spatial location (Yalcin et al. 2011). For building models, the landslide inventory was randomly partitioned into two groups: a training dataset (70%) and a validation dataset (30%).

2.3 Landslide Predisposing Factors

Landslides occurrence are influenced by the interaction of topographic, hydrological and geological factors (Costanzo et al. 2012; Dou et al. 2014), therefore, the selection of the predisposing factors is considered to be a fundamental step in the susceptibility modeling.

According to the previous studies (Ayalew et al. 2005; Dou et al. 2015d; c) in this study, these six predisposing factors including slope angle, slope aspect, drainage density network, distance to the geologic boundaries, distance to fault and lithology (Fig. 6) have high relationship with landslide occurrence. Hence, we extract these six predisposing factors to produce the landslide susceptibility mapping.

Additionally, the source data for the landslide predisposing factors may vary in their scale and affect the accuracy of landslide susceptibility models. To be commensurate with the diversity of the data source and difference in the scales, we converted all the factors to a raster format with a resolution of 10 m.

3 Methods

3.1 Binary Logistic Regression (BLR)

Binary logistic regression (LR) is one of the most frequently used multivariate analysis methods for creating landslide susceptibility mapping models. The dependent variable in the LR method is a function of the probability and can be computed as follows (Lee and Pradhan 2006):

$$P(Y=1|x) = \frac{\exp(\sum bx)}{1+\exp\left(\sum bx\right)}x \qquad (1)$$

where P is the estimated probability of landslide occurrence and ranges from 0 to 1; Y is an indicator variable, X is the independent

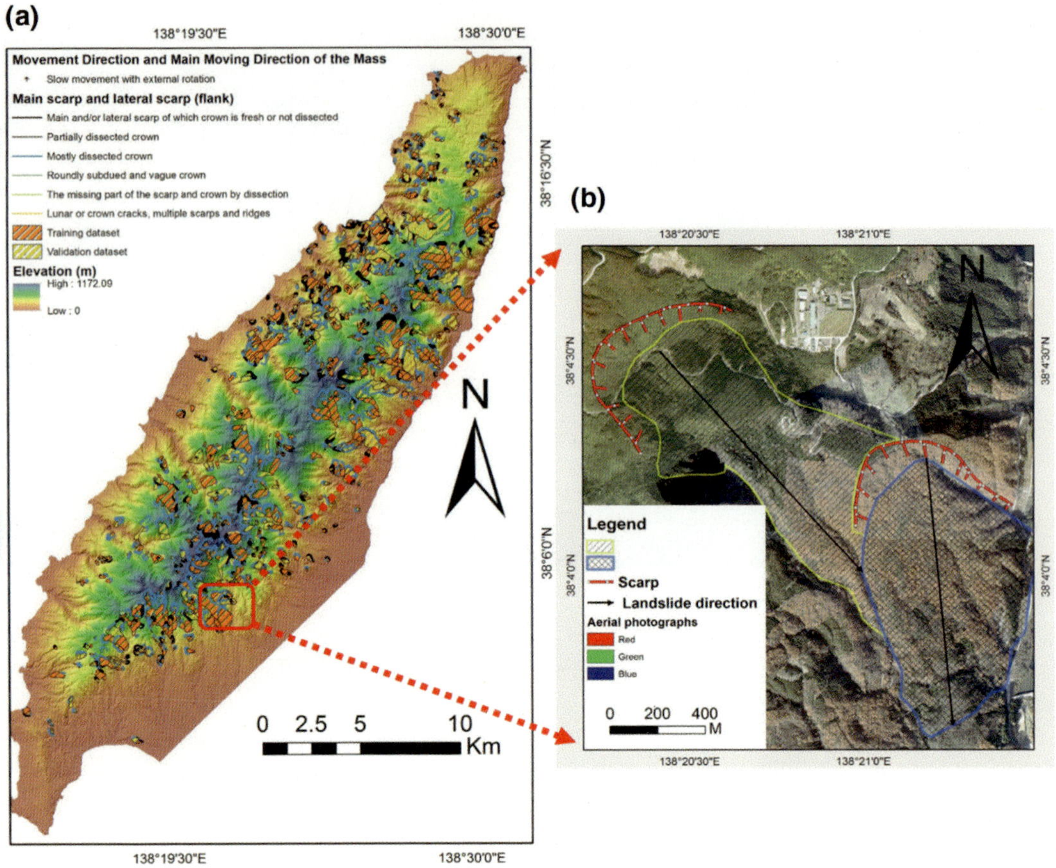

Fig. 4 **a** Landslide inventory map for the study area randomly divided into two groups overlaid the shaded relief (10 m DEM): training dataset and validation samples: **b** the enlargement of a landslide location overlaying the aerial ortho-photographs provided by the Midori Net Niigata and Sado city acquired in 2005. The *yellow line* is the primary slide and blue line is secondary one

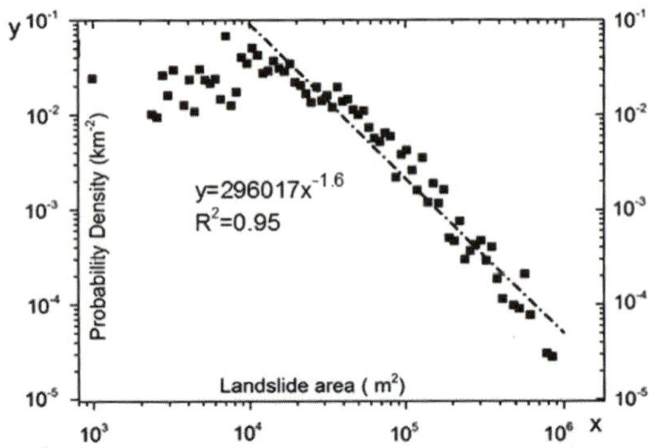

Fig. 5 Probability distribution of landslide areas in the Osado Island

variables, $X = (x_0, x_1, x_2, \ldots x_n), x_0 = 1$; b is regression coefficient.

To linearize the mentioned method as well as remove the 0/1 boundaries for the original dependent variable, the estimated P(probability) is transformed by the following formula:

$$P' = \ln\left(\frac{P}{1-P}\right) \tag{2}$$

The alteration is referred to as the logit transformation. Theoretically, the logit transformation of binary data can ensure that the

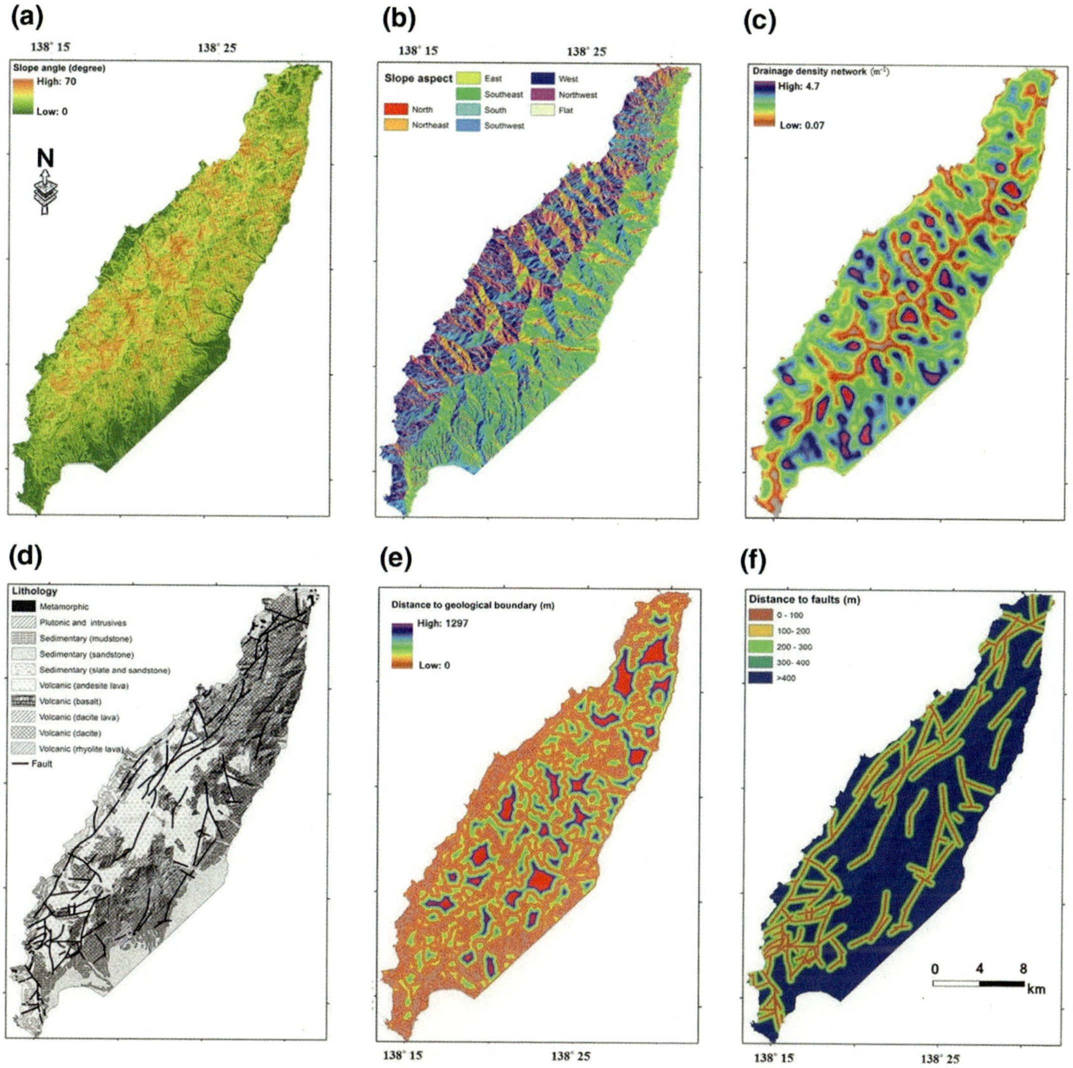

Fig. 6 Landslide predisposing factors: **a** slope angle, **b** slope aspect, **c** drainage density, **d** lithology, **e** distance to geological boundaries, and **f** distance to faults

dependent variable is continuous and the logit transformation is boundless. Moreover, it can ensure that the probability surface will be continuous within the range [0, 1]. Using the logit transformations, the standard linear regression models can be obtained as follows:

$$P' = \ln\left(\frac{P}{1-P}\right) = b_0 + b_1x_1 + b_2x_2 + \cdots + b_nx_n + \varepsilon \quad (3)$$

where, b_0 is the constant or intercept of the formula, $b_1, b_2, \ldots b_n$ represents the slope coefficients of the independent parameters, $x_1, x_2, \ldots x_n$ in the logistic regression and ε is standard error.

3.2 Back-Propagation for Feed-Forward Artificial Neural Network (ANN)

ANN has also been successfully implemented for evaluating landslide susceptibility by several researchers (Zare et al. 2013; Dou et al. 2015e). Our tool focuses on a particular type of ANN model, known as a back-propagation neural network. Back-propagation neural network (BPNN) algorithm is used in the feed-forward ANN. This is typically used to train the network among the different types of ANN models including radial basis function (RBF), general regression neural networks (GRNN), and probabilistic neural networks (PNN). The BPNN algorithm is simply a gradient-descent algorithm (also called a generalized delta rule) that uses to minimize the total error or mean error of target computed by the neural network. This algorithm is a neural network that is composed of three layers, input, hidden, and output. The structure of a typical three layer BPNN is displayed in Fig. 7. The input layer propagates components of a special input vector after weighting synaptic weights to each node in the hidden layer. Each hidden layer computes outputs corresponding to these weighted sums through a non-linear/linear function, e.g., log-sigmoid, purelin, or tan-sigomid (Prasad et al. 2012). The details can be referred to the reference (Dou et al. 2015d).

4 Implementation of Landslide Susceptibility Mapping

4.1 Landslide Susceptibility Mapping Using BLR Mode

In this tool, the forward stepwise logistic regression approach was used to incorporate predictor variables with a main contribution to the presence of landslides, using the SPSS 20 software. In the training dataset 578 landslides represented the presence of landslide points and were assigned the value 1. In agreement with the equal proportions of landslide and non-landslide, the same number of non-landslide points were

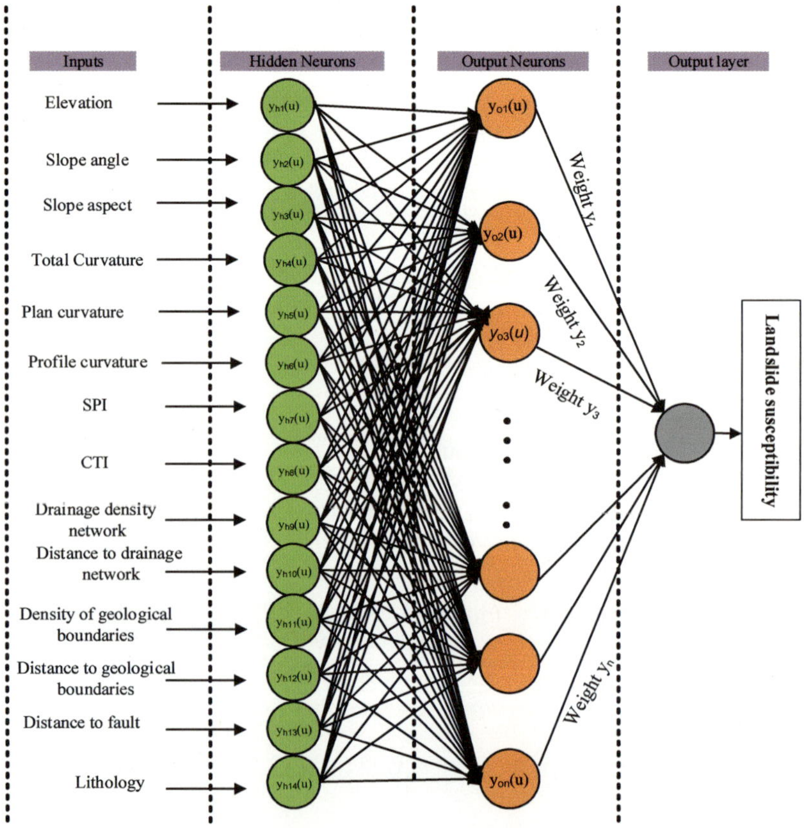

Fig. 7 An example of structure of a typical three-layer feed forward BPNN

Table 1 Coefficients, statistics of the factors and the multi-collinearity diagnosis indexes for variables used in the logistic regression equation

Predisposing factors	Coefficient	S.E.	Wald	df	P-value	Exp (B)	Collinearity statistics	
							Tolerance	VIF
Slope angle	1.209	0.193	39.364	1	0	3.351	0.977	1.023
Slope aspect	0.283	1.574	0.032	1	0.085	1.327	1	1
Drainage density	0.123	0.429	0.082	1	0.077	1.131	0.999	1.001
Lithology	0.879	1.564	1.442	1	0.023	0.153	0.994	1.006
Distance of geological boundaries	0.853	0.571	0.013	1	0.093	2.347	0.973	1.028
Distance to faults	−0.441	0.888	0.247	1	0.061	0.643	0.988	1.012
Constant	0.791	0.121	42.699	1	0.000	2.206		

randomly sampled from the landslide-free area and assigned the value 0.

The result is shown in Table 1. It shows that all the predisposing factors have a *P*-value less than 0.1, indicating a statistical correlation between factors and the susceptibility of landslides at the 90% confidence level. Additionally, it is necessary to examine the effect of correlation because logistic regression is sensitive to collinearity among the independent variables. The variance inflation factor (VIF) and tolerance (TOL) are widely used indexes of the degree of multi-collinearity. A VIF value greater than or equal to 5 and a TOL value less than 0.2 indicates a serious multi-collinearity problem (O'brien 2007). In this tool, both of these indexes were calculated (Table 1), the maximum VIF and minimum TOL were 1.028 and 0.973, respectively. Therefore, there is no multi-collinearity between these variables in the study.

Lastly, the regression coefficients of the predictors were imported to generate the landslide susceptibility map (Fig. 8) in GIS. The map of classes are also applied the natural break classification to divide the boundaries of each class.

4.2 Landslide Susceptibility Mapping Using ANN Model

In this tool, the three layers of the BPNN were simulated using the Matlab R2012a software. To select the number of hidden neurons, we randomly set the neurons and run the 10 times of network to choose the average value of R^2 for reducing the effect of initial value in the BP network. From Fig. 9, when the number of hidden neurons is 10 that R^2 has the highest values (0.96) as shown in Table 2. Hence, the numbers of hidden neurons were selected as 10. The structures of the networks required in the BPNN were constructed as 6-10-1 (input-hidden-output). The weights between the layers were obtained by training the BPNN network to compute the contribution of each landslide-conditioning factor. To test the stability of the BPNN model, it was iterated ten times, each with a random set of landslide data selected from the entire data pool. The results were found to be similar. The standard deviation was 0.0032. Hence, the random samples did not have a significant influence on the results. In this case, the average values were computed only to interpret the results.

When the BPNN reached the minimum RMSE values (0.001), the entire study area was fed into the network to assess the landslide susceptibility mapping. The weights for minimum error were recorded in the process and weights of each factor were determined for the study area. The set of landslide index values obtained in each pixel were then converted to raster in GIS environment. We used the six factors (slope angle, slope aspect, drainage density network,

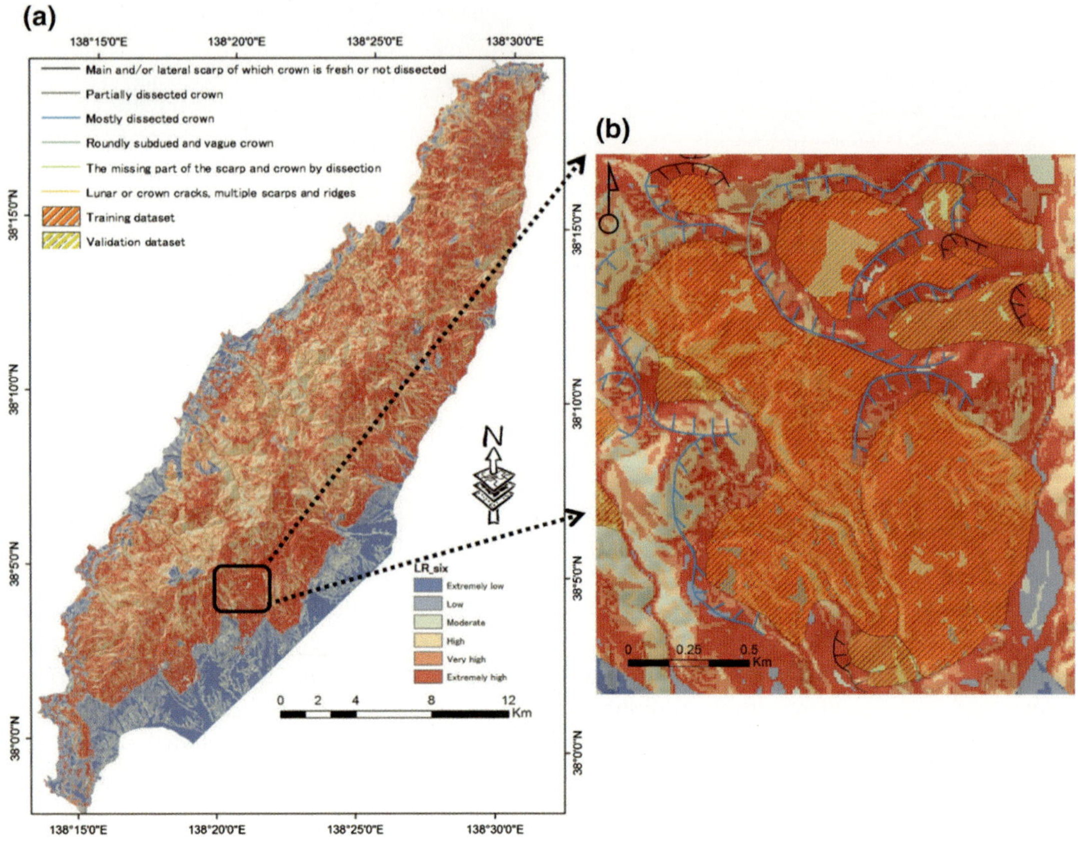

Fig. 8 LSM maps produced by the BLR method. Maps indicate the spatial probability of landslide occurrence in six classes (**a**). The right map is enlargements of the LSM maps (**b**). Referred to the Fig. 4b, we can see that the landslides locates in the extremely high areas

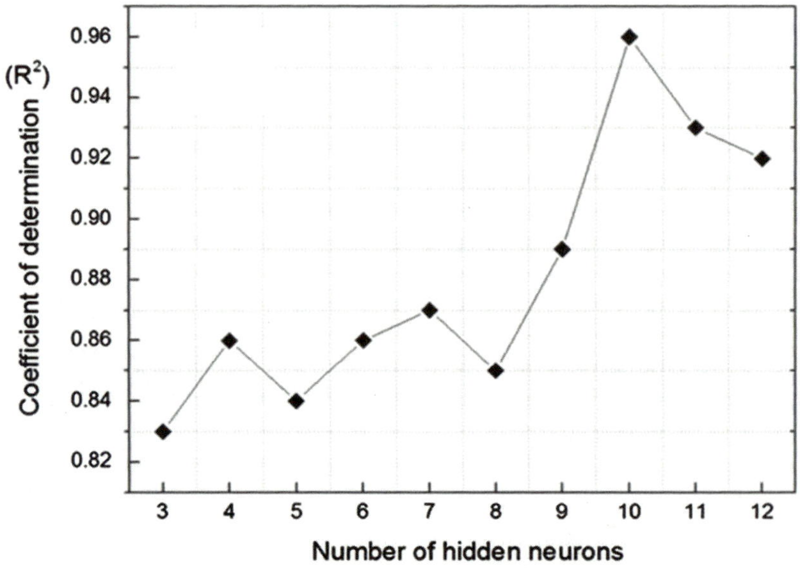

Fig. 9 Number of hidden neurons influences the back propagation network: coefficient of determination R^2 has the largest value (0.96) when the number of hidden neurons is 10

Table 2 Results of 10 times running the back propagation network model for selection of the number of hidden neurons

Number of hidden neurons	Coefficient of determination (R^2)		
	Minimum	Maximum	Average
3	0.67	0.93	0.83
4	0.73	0.95	0.86
5	0.68	0.92	0.84
6	0.64	0.91	0.86
7	0.76	0.94	0.87
8	0.69	0.92	0.85
9	0.63	0.94	0.89
10	0.84	0.97	0.96
11	0.78	0.96	0.93
12	0.74	0.95	0.92

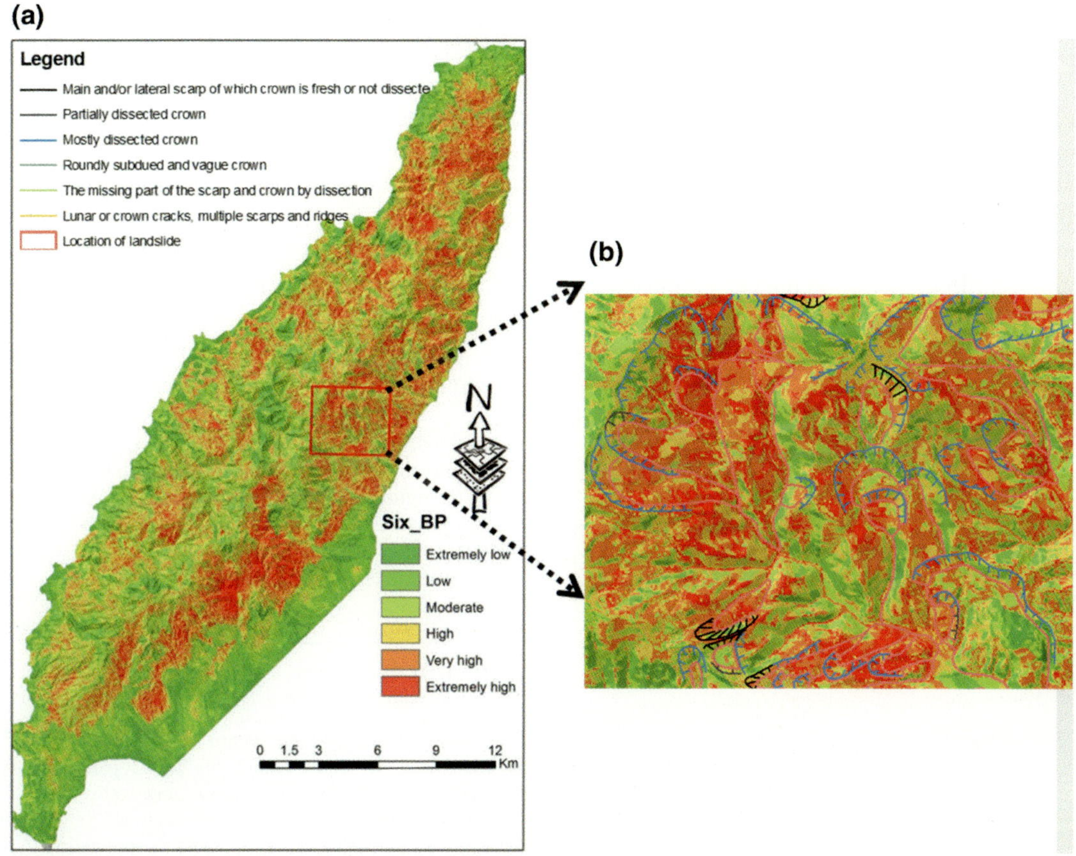

Fig. 10 LSM maps produced by the BPNN method. Maps indicate the spatial probability of landslide occurrence in six classes. The right map is enlargements of the LSM maps. At the right map, the black is the main scarp and the blue line is the almost dissected crown

distance from the geologic boundary, distance to fault and lithology) to produce the landslide susceptibility mappings as shown in Fig. 10.

4.3 The Validation of the Landslide Susceptibility Models

It is critical to verify the accuracy of any prediction model. Without the validation, the predicted model and LSM maps lack in the necessary scientific significance (Chung and Fabbri 1993). LSM results can be validated using the known landslide locations. For the verification, the total landslides were divided into two groups, training data, and validation data. We verified the accuracy of the BLR and BPNN model using a receiver operating characteristic (ROC) curve.

A ROC plot of sensitivity (true positive rate is the portion of false positives out of the total actual positives) and 1-specificity (false positive rate is the portion of false positives out of the total actual negatives) was made for the model validation (Swets 1988). The area under the ROC curve (AUC) can characterize the quality of a forecasting system by describing the system's ability to correctly predict the occurrence or non-occurrence of a predefined event (Swets 1988). The AUC value varies from 0.5 to 1.0. An ideal model has an AUC value of 1.0. For the proposed model, the AUC value (0.82) was higher using BPNN than 0.78 using BLR model (Fig. 11). Moreover, the AUC is significantly different from 0.5 because the p-value is 0.00 implying that the BPNN classified the group significantly better than by chance.

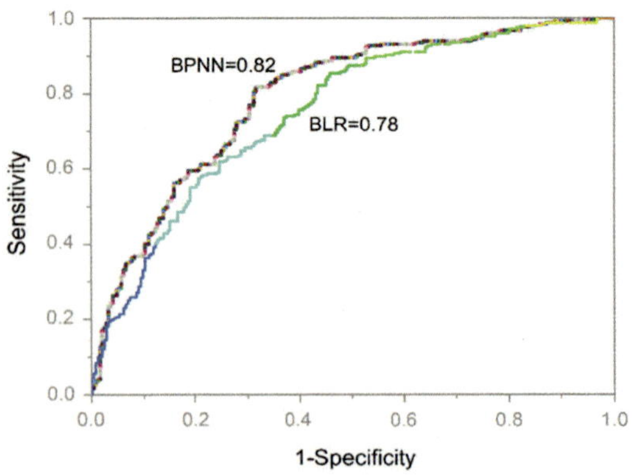

Fig. 11 ROC curves for landslide susceptibility maps produced using the BLR and BPNN methods

5 Conclusion

Landslide susceptibility mapping is essential to describe the propensity of a landslide in the area. The LSM maps were then produced by applying both the BLR and BPNN methods for the predisposing factors. Both methods can achieve the acceptable results. In addition, we noticed that the maps prepared from using BPNN can obtain a bit higher accuracy than BLR. This result is well agreement with the other researchers. This tool may be used to tackle similar issues of landslide hazard mitigation in other susceptible areas in any landslide-prone countries.

Additionally, the results of such studies can provide helpful information for the disaster managers, and for the decision makers in the landslide-prone area. These maps can be useful for them to select the suitable spatial locations to implement reconstruction and urban development landslide prevention and mitigation within the hilly terrain (Hadji et al. 2013). The urban planners can use such maps to avoid development in landslide threatened areas; the practice will represent the most efficient and economic way to decrease future damages and loss of lives.

Acknowledgements We would like to express our deep appreciation to Midori NET Niigata and Sado City for providing the ortho photographs of Sado Island and the NIED for providing the landslide data. Here, Dou highly appreciates Dr. Takashi Ougchi's and Dr. Yuichi S. Hayakawa's guidance and support from the University of Tokyo.

References

Ayalew L, Yamagishi H, Marui H, Kanno T (2005) Landslides in Sado Island of Japan: part I. Case studies, monitoring techniques and environmental considerations. Eng Geol 81:419–431

Chang T-C, Chao R-J (2006) Application of back-propagation networks in debris flow prediction. Eng Geol 85:270–280. doi:10.1016/j.enggeo.2006.02.007

Chung C-J, Fabbri AG (1993) The representation of geoscience information for data integration. Nonrenewable Resour 2:122–139. doi:10.1007/BF02272809

Costanzo D, Rotigliano E, Irigaray C et al (2012) Factors selection in landslide susceptibility modelling on large scale following the gis matrix method: application to the river Beiro basin (Spain). Nat Hazards Earth Syst Sci 12:327–340. doi:10.5194/nhess-12-327-2012

Dai FC, Lee CF (2003) A spatiotemporal probabilistic modelling of storm-induced shallow landsliding using aerial photographs and logistic regression. Earth Surf Process Landforms 28:527–545. doi:10.1002/Esp.456

Dou J, Oguchi T, Hayakawa YSS et al (2014) Susceptibility mapping using a certainty factor model and its validation in the Chuetsu Area, central Japan. Landslide Sci a Safer Geoenvironment 2:483–489. doi:10.1007/978-3-319-05050-8_65

Dou J, Chang K, Chen S et al (2015a) Automatic case-based reasoning approach for landslide detection: integration of object-oriented image analysis and a genetic algorithm. Remote Sens pp 4318–4342. doi:10.3390/rs70404318

Dou J, Paudel U, Oguchi T et al (2015b) Shallow and Deep-seated landslide differentiation using support vector machines: a case study of the Chuetsu Area. Japan 26:227–239. doi:10.3319/TAO.2014.12.02.07(EOSI)1

Dou J, Tien Bui D, Yunus Ap et al (2015c) Optimization of causative factors for landslide susceptibility evaluation using remote sensing and gis data in parts of Niigata, Japan PLoS One 10:e0133262. doi:10.1371/journal.pone.0133262

Dou J, Yamagishi H, Pourghasemi HR et al (2015d) An integrated artificial neural network model for the landslide susceptibility assessment of Osado. Nat Hazards. doi:10.1007/s11069-015-1799-2

Dou J, Yamagishi H, Pourghasemi HR et al (2015e) An integrated artificial neural network model for the landslide susceptibility assessment of Osado Island, Japan. Nat Hazards 78:1749–1776. doi:10.1007/s11069-015-1799-2

Guzzetti F, Malamud BD, Turcotte DL, Reichenbach P (2002) Power-law correlations of landslide areas in central Italy. Earth Planet Sci Lett 195:169–183. doi:10.1016/S0012-821X(01)00589-1

Hadji R, Boumazbeur AE, Limani Y et al (2013) Geologic, topographic and climatic controls in landslide hazard assessment using GIS modeling: a case study of Souk Ahras region, NE Algeria. Quat Int 302:224–237. doi:10.1016/j.quaint.2012.11.027

Kanungo DP, Arora MK, Sarkar S, Gupta RP (2006) A comparative study of conventional, ANN black box, fuzzy and combined neural and fuzzy weighting procedures for landslide susceptibility zonation in Darjeeling Himalayas. Eng Geol 85:347–366. doi:10.1016/j.enggeo.2006.03.004

Lee S, Pradhan B (2006) Landslide hazard mapping at Selangor, Malaysia using frequency ratio and logistic regression models. Landslides 4:33–41. doi:10.1007/s10346-006-0047-y

O'brien RM (2007) A caution regarding rules of thumb for variance inflation factors. Qual Quant 41:673–690. doi:10.1007/s11135-006-9018-6

Prasad R, Pandey a, Singh KP et al (2012) Retrieval of spinach crop parameters by microwave remote sensing with back propagation artificial neural networks: a comparison of different transfer functions. Adv Sp Res 50:363–370. doi:10.1016/j.asr.2012.04.010

Shi H-Y, Lee K-T, Lee H-H et al (2012) Comparison of artificial neural network and logistic regression models for predicting in-hospital mortality after primary liver cancer surgery. PLoS ONE 7:e35781. doi:10.1371/journal.pone.0035781

Swets JA (1988) Measuring the accuracy of diagnostic systems. Science 240:1285–1293

Tien Bui D, Pradhan B, Lofman O, Revhaug I (2012) Landslide susceptibility assessment in Vietnam Using support vector machines, decision tree, and naïve bayes models. Math Probl Eng 2012:1–26. doi:10.1155/2012/974638

Turner AK, Schuster RL (eds) (1996) Landslide; investigation and mitigation, special report 247, Transportation Research Board, National Research Council, National Academy Press, Washington DC, 673p

Yalcin A, Reis S, Aydinoglu AC, Yomralioglu T (2011) A GIS-based comparative study of frequency ratio, analytical hierarchy process, bivariate statistics and logistics regression methods for landslide susceptibility mapping in Trabzon, NE Turkey. CATENA 85:274–287. doi:10.1016/j.catena.2011.01.014

Yamagishi H (2008) GIS mapping of landscape and disasters of Sado Island, Japan. In: the international archives of the photogrammetry, remote sensing and spatial information sciences. Vol. XXXVII. Part B7. Beijing 2008. pp 1429–1432

Zare M, Pourghasemi HR, Vafakhah M, Pradhan B (2013) Landslide susceptibility mapping at Vaz Watershed (Iran) using an artificial neural network model: a comparison between multilayer perceptron (MLP) and radial basic function (RBF) algorithms. Arab J Geosci 6:2873–2888

Zhu Z, Yu J, Wang H et al (2015) Fractal dimension of cohesive sediment flocs at steady state under seven shear flow conditions. Water 7:4385–4408. doi:10.3390/w7084385

Zhu Z, Wang H, Yu J, Dou J (2016) On the kaolinite floc size at the steady state of flocculation in a turbulent flow. pp 1–16. doi:10.1371/journal.pone.0148895

TXT-tool 1.386-2.2 Practical Approach to Assessing the Factors Influencing Landslide Susceptibility Modelling—The Case of Slovenia

Marko Komac

Abstract

This paper introduces and explains a landslide susceptibility assessment undertaken in Slovenia at two geographical scales: 1:25,000 for municipal spatial planning, and 1:250,000 for national strategic spatial planning and synoptic exposure assessment. The approach is designed to help analysts decide which influencing factors to include in the modelling of landslide susceptibility. Inclusion of too many factors may represent a challenge from the resources perspective, whereas too few factors may reduce the quality of the model. It explains the method used for the cases of Slovenia, and how to approach to modelling; which weight values to use for a specific factor if a weighting method is chosen; and how to assess the modelling errors. Explanations, guidance and discussion of potential drawbacks of the approach are given.

Keywords

Landslides · Susceptibility · Assessment · Methodology · Factors Slovenia

Contents

1 **Introduction** 153
2 **Methodology** 154
3 **Results and Discussion** 157
4 **Conclusions** 159
References 159

1 Introduction

Assessing the relevant influencing factors that govern landslide spatial and temporal occurrence, and consequentially developing effective and accurate models of landslide susceptibility can be a difficult and error prone process. Because of high landslide occurrence in Slovenia an assessment has been made to provide experts and decision-makers with information on potential future spatial occurrence. First, analyses were made at the municipal level at 1:25,000 scale; these used the Bovec municipality in a NW Slovenia as a test case,

M. Komac (✉)
Faculty of Civil and Geodetic Engineering, University of Ljubljana, Ljubljana, Slovenia
e-mail: m.komac@telemach.net

K. Sassa et al. (eds.), *Landslide Dynamics: ISDR-ICL Landslide Interactive Teaching Tools*, https://doi.org/10.1007/978-3-319-57774-6_11

because of the presence of many types of landslides including rock-falls and debris-flows (Komac 2005; Bavec et al. 2005). A second stage of the landslide susceptibility assessment was conducted at 1:250,000 scale, to evaluate the national landslide exposure in Slovenia (Komac and Ribičič 2006). The detailed analyses were undertaken across an area of 367.3 km^2, and the country-wide analyses relate to an area of approximately 20,000 km^2. Despite the fact that there are common assessment and model similarities, the two approaches and the related models are not identical. Some important differences exist e.g., due to the quality of the data used for modelling landslide susceptibility. These will be presented in the following chapters, focusing on landslides (excluding rock-falls and debris-flows).

2 Methodology

For both landslide susceptibility assessment approaches the Chi-square (X^2) statistical analysis method was used as it is has been proved to be useful for the continuous and normally distributed nominal (classified) variables (factors) (Davis 1986; Stančič and Veljanovski 2000; Lineback Gritzner et al. 2000; Komac 2012) that govern the spatial occurrence of landslides. To effectively assess the landslide occurrence in a study area, by different classes of different spatio-temporal factors, the analysis requires the collection of reliable landslide population. Such a population needs to be divided into two groups, a "learning" ("training") set used for model development, and a testing ("evaluation" or "validation") set used for model evaluation (Fig. 1a). The division between the two varies from author to author, from 90%/10%, to 60%/40%, in favour of the learning set. In this work, a division in approximately 2/3 for learning set, and 1/3 for testing set was used. When dividing the landslide population into the two sets, it is important to ensure the representativeness of the two sets from the lithological units' aspect, to assure that the division between the two sets is as close to the selected ratio (learning/testing) for all units (or as many as possible). This will avoid

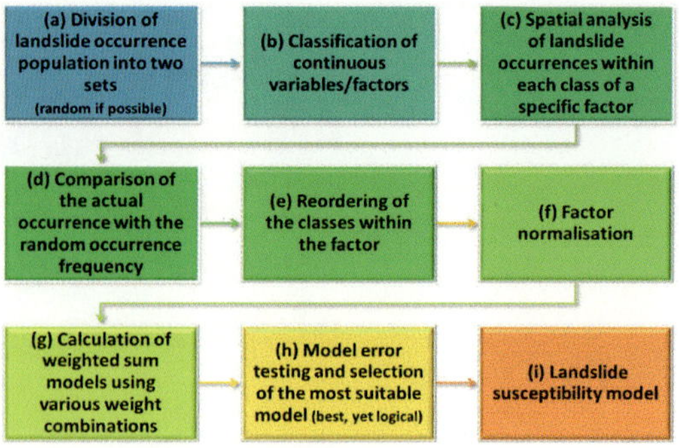

Fig. 1 Landslide susceptibility assessment approach

missing units, and associated biased results. If there are not enough landslide occurrences (insufficient population) within a unit, the expert knowledge classification based on the engineering-geological properties of this unit is needed. The unit could either be joined with another similar engineering-geological unit, or its susceptibility could be defined by an expert rather than by a statistical calculation from an insufficient population.

Nominal factors (variables) are already divided into classes, whereas continuous variables are not, and have to be classified prior to the Chi-square analysis (Fig. 1b). If the number of classes defined is too small, the result of the statistical analysis will be too general and of little use for susceptibility modelling. Conversely, if too many classes are used, the result of the Chi-square analysis will falsely indicate a specific factor to be influential, despite the fact that its impact may be negligible, or irrelevant. Here, an approach based on the expert knowledge would be the most advisable. For example, by comparing the analytical results and the field experience in the case of Slovenia, the most representative slope angle classification of values varying from 0° to 90°, would be a classification into classes of 3° to 5°. By placing the classification into classes of 1° the result would produce too many variables. The classification into classes by 10° would result in too few variables, resulting in general conclusions. When classifying the proximity factors (i.e., proximity to drainage lines), the class span should not be more precise than the resolution of the data—the pixel size of the image for the raster based analyses, or in the case of vector based approaches the class span should not be more precise than the resolution on which the analysis is based. Otherwise analyses would make no sense due to the limitations of the source data resolution.

After the classification is undertaken the spatial analysis of landslide occurrences within each class (zone) of a specific factor is performed to assess the frequency of occurrence (Fig. 1c). To accommodate the basic parameters of the Chi-square analysis, the random occurrence for each class (based on its area size) is calculated, and in case of the Chi-square analysis the two values, observed (actual occurrences) and expected (random occurrences), are compared (Fig. 1d). The later step is usually automated in most statistical software. The Chi-square analysis results in an overall score of the factor's influence on landslide spatial occurrence. It also indicates which classes within the factor are influencing the occurrence more or less, or even hinder it.

The next step comprises re-ordering the classes within the factor to accommodate their influence on the landslide occurrence (Fig. 1e). In practice this means that based on the results of the X^2 univariate analyses, the classes within each factor were ordered (ranked) according to the statistical landslide occurrence probability. Where obvious discrepancies of class ranking occurred, an expert decision was made to correct the error. Before selecting the relevant factor for susceptibility modelling, or more precisely the model calculation for which the weighted sum model was selected, all the factors included in the model have to be normalised to the same range of values (i.e., between 0 and 1) in order to eliminate the class number influence (Fig. 1f). For the factor normalisation phase, the following equation was used (Komac 2012):

$$NV = \frac{OV - min}{max - min}, \tag{1}$$

where NV is the new normalised value, OV is the old (nominal) value, and the *max—min* represents the difference between the maximum (max) and the minimum value of the variable. In the represented case, the normalised values ranged from 0 to 1. Table 1 lists an example of the steps from "c" to "f" in Fig. 1.

In the case used, the weighted sum model was chosen to calculate landslide susceptibility for each individual pixel (cell) for the study area. For the susceptibility model calculation, the following equation was used (Vooght 1983):

$$LSUSC_R = \sum_{j=1}^{n} w_j \times f_{ij}, \tag{2}$$

where, $LSUSC_R$ is the standardised relative landslide susceptibility (in the range 0–1) at a given location, w_j is the weighting for the given factor (variable), and f_{ij} is the value of a continuous or discrete variable at a given location.

To minimise the error of the resulting susceptibility model, a pseudo Monte-Carlo approach was chosen, where multiple models were calculated using different weighting combinations (Fig. 1g). The medium weighting values for the factors used were adopted from (Komac 2005a, b, 2012.) The total range of the weighting values (minimum weighting value—*min*, maximum weight value—*max*, and the step value between them—*K*) were defined by an expert, using trials and past experience. The selection of the weighting values for each of the variables included in the susceptibility model calculation was not completely random. Weighting values were chosen following the same procedure—first factor was assigned a specific weighting value from the values range for that specific factor, then the rest of the weightings were distributed among remaining factors, following their sequence until the total weightings' sum was 100%; this took into consideration the constraints defined by *min*, *max* and *K*. The same combination of weightings was never repeated. The algorithm is exemplified in Fig. 2, and some of the weights used are listed in Table 2.

After the calculation of each standardised relative susceptibility model, values were

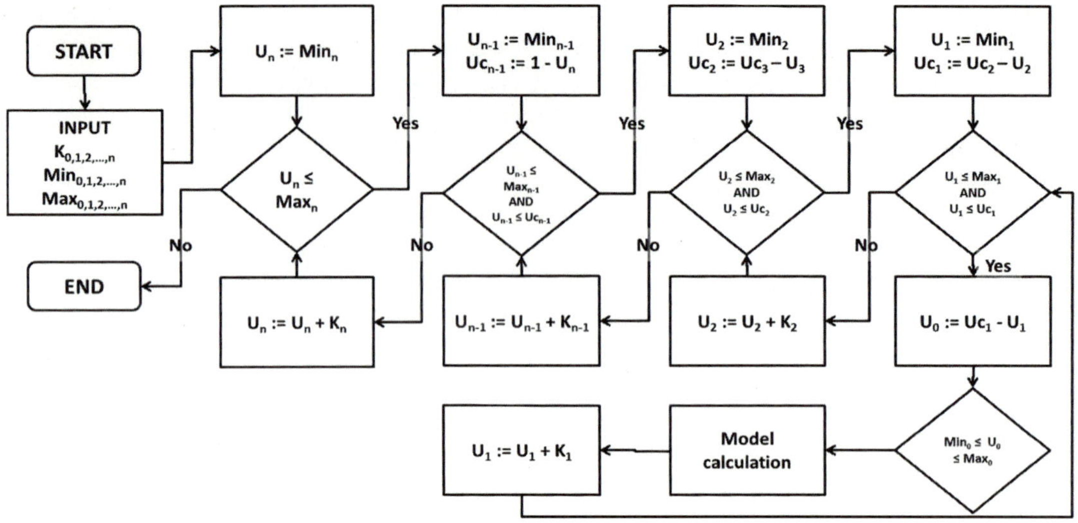

Fig. 2 Pseudo Monte-Carlo process used for an automatic calculation of susceptibility models (*K* step; *Min* minimum weighting value; *Max* maximum weight value; *U* current weighting; *Uc* difference between 1 and the cumulative sum of weightings)

reclassified into 100 classes (landslide susceptibility scores) based on equal area approach, each class occupying 1% of the total study area. As a result, the study area is divided into classes that have equal area, and these classes contain all pixels with landslide susceptibility scores within that specific percentage of the overall score distribution. The class with the lowest landslide susceptibility score was classified as 1, and the class with the highest landslide susceptibility score was classified as 100 (LSUSCreclass = 1–100). As a result of this approach, all the model results can be compared, as they all have the same (re)classification basis.

The next step was testing the landslide susceptibility model. This was achieved using the occurrence of landslides from the testing set, to which additional randomly selected locations where landslides cannot occur (also defined as "no_landslide" locations) were added, within the classes of the reclassified model. We assumed that for derived susceptibility prediction models the classes with low landslide susceptibility scores ($LSUSC_{reclass}$ = 1–50) represent areas that are resistant to landslide occurrences. Thus it is expected that within the area of these classes, all or a majority of the test locations where landslides cannot occur ("no_landslides") should be located when performing the model quality assessment. Conversely, the other half of the landslide susceptible classes ($LSUSC_{reclass}$ = 51–100) were defined as landslide prone (i.e., susceptible to landslides) (Fig. 1h). As a result, the majority of landslides should occur in the classes with large susceptibility scores. To correctly evaluate the susceptibility models obtained, the Cohen's kappa (k) index (Cohen 1960) was used:

$$\kappa = \frac{P_C - P_E}{1 - P_E}, \quad (3)$$

where: P_C represents the proportion of correctly classified control points (either "landslides" located at a landslide susceptible area or "no_landslides" located at a landslide area, resistant to landslide occurrences); P_E represents the proportion of hypothetical probability of correctly classified control points.

For the selection of the final landslide susceptibility model and with the purpose to minimise ambiguities in the result, an averaging of weighting values of several best models (e.g., best 10, 25, or best 5%, 10%) was applied. Komac (2012) showed that in addition to the best model, the average weightings of the best 10 models resulted

Table 1 Example of Chi-square analysis for the case of landslide occurrence within slope classes

#	Slope classes	O	E	O-E	$(O\text{-}E)^2/E$	CI	RCL	NORM
1	0–5	121	600.56	−479.56	382.94	−382.94	1	0.0625
2	5–8	146	237.76	−91.76	35.41	−35.41	2	0.125
3	8–11	255	234.78	20.22	1.74	1.741	10	0.625
4	11–14	326	213.61	112.39	**59.14**	59.14	13	0.8125
5	14–17	334	187.72	146.28	**113.98**	113.98	15	0.9375
6	17–20	295	157.55	137.45	**119.91**	119.91	16	1
7	20–23	228	130.02	97.98	**73.84**	73.84	14	0.875
8	23–26	178	103.73	74.28	**53.19**	53.19	12	0.75
9	26–29	99	82.69	16.31	3.21	3.22	11	0.6875
10	29–32	69	66.41	2.6	0.10	0.10	9	0.5625
11	32–35	39	51.78	−12.73	3.16	−3.16	3	0.1875
12	35–38	25	35.55	−10.55	3.13	−3.13	4	0.25
13	38–41	15	20.54	−5.51	1.49	−1.49	6	0.375
14	41–44	10	11.720	−1.72	0.25	−0.25	7	0.4375
15	44–47	7	7.181	−0.18	0.05	−0.05	8	0.5
16	47–90	9	14.412	−5.41	2.03	−2.03	5	0.3125
	Sum	2156	2156	0.0	853.53	–	–	–

Total $X^2 = 853.5316$, df = 15 $p < 0.00$; *O* Observed landslides; *E* Expected landslides (E = total landslide number × class area %); *CI* class influence (value of $(O\text{-}E)^2/E$ including the impact: if prone "+", if averse "−"; *RCL* class reordering values due to the CI value; *NORM* calculated normalization values (Eq. 2). The bold numbers represent statistically significant values derived with the Chi-square analysis—this indicates that these classes have a positive influence on landslide occurrence

in the same success rate (or the same error) as the model defined by an expert, followed by the average weightings from the best 5% of all models (or 7808 models). In this case, the latter was chosen as the final landslide susceptibility model (Fig. 1i). The result is shown in Fig. 3 (after Komac 2012), where six descriptive classes of landslide susceptibility were defined (1—None; 2—Very Low; 3—Low; 4—Moderate; 5—High; 6—Very High) based on natural breaks or on Jenks' optimisation technique (Jenks 1967) in the value distribution to maximise the between-class and minimise the within-class differences.

3 Results and Discussion

During the development of the methodology the following lessons were learned:

1. Synchronicity of strata dip and the slope aspect proved a useful parameter, but its modelling was very demanding due to the lack of representative data at a satisfactory spatial sampling frequency (for larger scales), and also because of cumbersome computing of strata dipping within units and inadequate geological boundary constraints.
2. When dividing the landslide population into learning and testing sets, it is important to ensure the representativeness of the two sets for each of the lithological units is as close as possible to the selected splitting proportions. In the case of insufficient landslide occurrences within a specific unit, the entire unit could be either merged with another unit of similar engineering-geological characteristics, or its susceptibility could be defined by an expert rather than by a statistical model based on insufficient information.

Table 2 Spatio-temporal factors that have been used to model landslides at the scale 1:250,000 (Komac 2005b, 2012) and at the scale 1:25,000 (Komac 2005a)

Factor	Weighting parameters 1:250,000	Weighting parameters 1:25,000 (a)	Weighting parameters 1:25,000 (b)	Weighting parameters 1:25,000 (c)
Lithology (engineering geological class)	Min = 0.2 Max = 0.6 K = 0.02	Min = 0.2 Max = 0.4 K = 0.01	Min = 0.0 Max = 1 K = 0.05	Min = 0.0 Max = 1 K = 0.2
Strata dipping (and synchronism with the slope aspect classified into classes of 22.5°)	–	Min = 0.1 Max = 0.2 K = 0.01	Min = 0.0 Max = 1 K = 0.05	Min = 0.0 Max = 1 K = 0.2
Proximity to structural elements (m)	Min = 0.0 Max = 0.2 K = 0.02	–	–	Min = 0.0 Max = 1 K = 0.2
Fault zones (presence /absence)	–	Joined with structural zones	Joined with structural zones	Joined with structural zones
Slope (slope inclination; in °)	Min = 0.2 Max = 0.6 K = 0.02	Min = 0.2 Max = 0.4 K = 0.01	Min = 0.0 Max = 1 K = 0.05	Min = 0.0 Max = 1 K = 0.2
Curvature (slope concavity; unitless)	Min = 0.0 Max = 0.3 K = 0.02	–	–	Min = 0.0 Max = 1 K = 0.2
Aspect (slope orientation; in °)	Min = 0.0 Max = 0.2 K = 0.02	–	–	Min = 0.0 Max = 1 K = 0.2
Land use /land-cover type (class)	Min = 0.0 Max = 0.4 K = 0.02	Min = 0.1 Max = 0.2 K = 0.01	Min = 0.0 Max = 1 K = 0.05	Min = 0.0 Max = 1 K = 0.2
# models calculated	156,169	900	1675	322

Min minimum weighting value; *max* maximum weighting value; *K* step value between Min and Max

3. When classifying continuous factors (variables), the number of selected classes needs to be selected carefully, and on the basis of expert knowledge to avoid erroneous or misleading statistical results.
4. A good approach to avoid model overfitting is to use the average weighting values of several best models (i.e., 10, 25, 50, 100 etc.) and to also use (again!) expert knowledge support—i.e., relatively unstable lithological unit (e.g., shale) can be stable on a slope with small angles and the opposite, the relatively stable rock (e.g., non-stratified limestone) can be stable at high slope angles, so both factors should be used. On the other hand, elevation is misleading, as high altitudes do not necessarily mean steep slopes and may not be associated to high landslide occurrence.
5. Representation of the final results needs to be simple and well designed in order that they may be understood by non-geoscience decision makers, politicians and the general public.

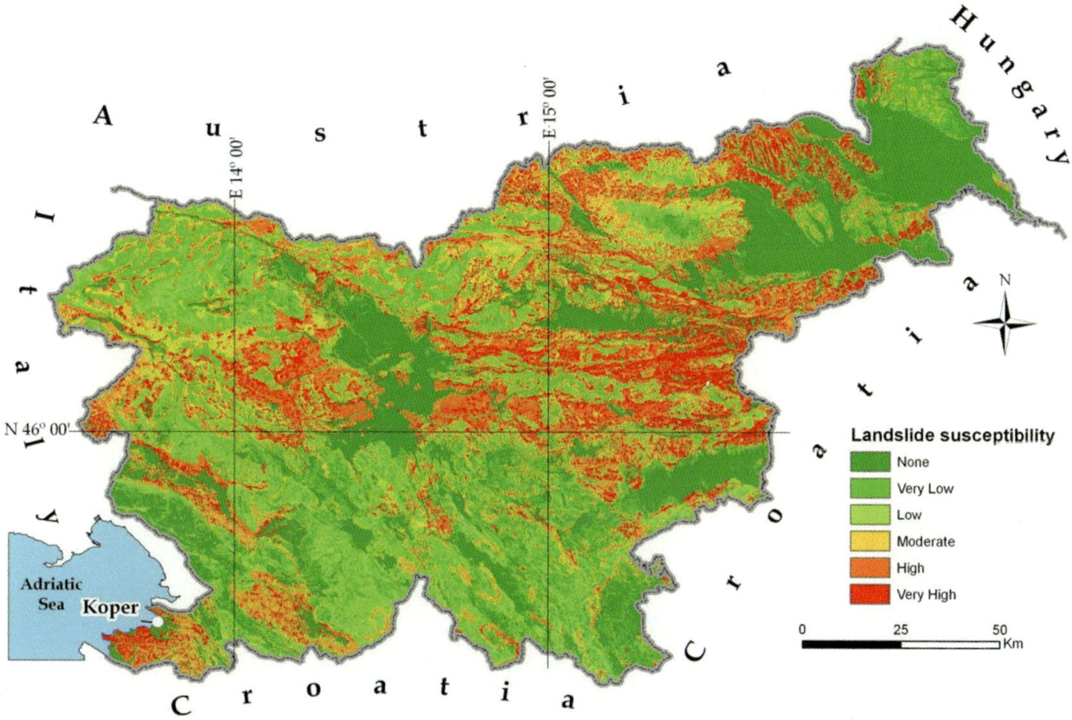

Fig. 3 Landslide susceptibility map of Slovenia derived from the average of best 5% (7808 models) of all models calculated in the trial performed by Komac (2012) (original scale is 1:250,000)

4 Conclusions

Landslide susceptibility assessment is the first step in the delineation and understanding of landslide hazard and risk. The process is thus very important as it represents the foundation on which all further steps are built. An informative and simple description is given on the steps in landslide susceptibility assessment and also potential pitfalls in the process. In conclusion some practical recommendations are presented that became apparent while developing this methodology.

Acknowledgements The author would like to thank the Geological Survey of Slovenia, the Slovenian Research Agency, the Ministry for the Environment and Spatial Planning of Slovenia and the Ministry of Defence of the Republic of Slovenia, Administration of the Republic of Slovenia for Civil Protection and Disaster Relief, for funding research projects from 2004 to 2013.

References

Bavec M, Budkovič T, Komac M (2005) Geohazard—geološko pogojena nevarnost zaradi procesov pobočnega premikanja. Primer občine Bovec: estimation of geohazard induced by mass movement processes. The Bovec municipality case study. Geologija, vol 48, no 2, pp 303–310. doi:10.5474/geologija.2005.025

Cohen J (1960) A coefficient of agreement for nominal scales. Educ Psychol Meas 20(1):37–46

Davis JC (1986) Statitics and data analysis in geology. Wiley, New York

Jenks GF (1967) The data model concept in statistical mapping. Int Yearb Cartography 7:186–190

Komac M (2005a) Verjetnostni model napovedi nevarnih območij glede na premike pobočnih mas—primer občine Bovec: probabilistic model of slope mass movement susceptibility—a case study of Bovec municipality. Slovenia Geologija 48(2):311–340

Komac M (2005b) Landslide susceptibility assessment based on the analysis of satellite data and other spatial data. Ljubljana, Geological Survey of Slovenia, p 232

Komac M (2012) Regional landslide susceptibility model using the Monte Carlo approach—the case of Slovenia. Geological Quarterly 56 (1):41–54

Komac M, Ribičič M (2006) Landslide susceptibility map of Slovenia at scale 1:250,000. Geologija 49(2):295–309

Lineback Gritzner M, Marcus WA, Aspinall R, Custer SG (2000) Assessing landslide potential using GIS, soil wetness modelling and topographic attributes, Payette River, Idaho. Geomorphology 37:149–165

Stančič Z, Veljanovski T (2000) Understanding Roman settlement patterns through multivariate statistics and predictive modelling. In: Lock G (ed) Beyond the map, IOS Press, The Netherlands, pp 147–156

Voogd H (1983) Multicriteria evaluation for urban and regional planning. Pion Ltd., London

TXT-tool 1.084-3.1 Landslide Susceptibility Mapping at a Regional Scale in Vietnam

Quoc Hung Le, Thi Hai Van Nguyen, Minh Duc Do, Thi Chau Ha Le, Van Son Pham, Ho Khanh Nguyen and Thanh Binh Luu

Abstract

This tool deals with a procedure for landslide susceptibility mapping at a regional scale. It is applied to the upper Lo River catchment in Vietnam, where data on spatial distribution of historic landslides and environmental factors are available, although they are limited. The available data include an inventory map and six landslide factor maps related to elevation, slope gradient, drainage density, fault density, types of weathering crust, and types of land cover. The Analytical Hierarchy Process method is used together with expert knowledge to define the weights of each factor and each factor class in order to determine the relative importance of the six landslide factors and their classes within the landslide susceptibility analysis. The Weighted Linear Combination method is used to assign the above-defined weights to all factor maps in raster format in order to compute the landslide susceptibility. As the approach is able to integrate expert knowledge in the weighting of the input factors, the actual study shows that the combination of Analytical Hierarchy Process and Weighted Linear Combination methods is suitable for landslide susceptibility mapping in large mountainous areas at medium scales, particularly for areas lacking detailed input data.

Q.H. Le (✉)
Vietnam Institute of Geosciences and Mineral Resources, No. 67 Chien Thang Street, Van Quan ward, Ha Dong district, Hanoi City, Vietnam
e-mail: le.quoc.hung@vigmr.vn; hunglan@gmail.com

T.H.V. Nguyen · V.S. Pham · H.K. Nguyen · T.B. Luu
Department of Remote Sensing and Geomatics, Vietnam Institute of Geosciences and Mineral Resources, No. 67 Chien Thang Street, Van Quan ward, Ha Dong district, Hanoi City, Vietnam

M.D. Do
VNU University of Science, Vietnam National University, Hanoi, 334 Nguyen Trai, Thanh Xuan, Hanoi, Vietnam
e-mail: ducdm@vnu.edu.vn

T.C.H. Le
Central Vietnam Institute for Water, 132 Dong Da, Hai Chau, Da Nang, Vietnam
e-mail: lechauha@gmail.com

K. Sassa et al. (eds.), *Landslide Dynamics: ISDR-ICL Landslide Interactive Teaching Tools*, https://doi.org/10.1007/978-3-319-57774-6_12

Keywords
Landslide susceptibility · Geographical information system (GIS) Analytical hierarchy process (AHP) · Weighted linear combination (WLC)

Contents

1 Introduction ... 162
2 Study Area ... 162
3 Methodology ... 164
3.1 Available Data ... 164
3.2 General Overview of the AHP and WLC Methods ... 164
3.3 The Workflow of Landslide Susceptibility Mapping for the Study Area ... 167
4 The Final Landslide Susceptibility Map ... 170
5 Conclusion ... 173
References ... 173

1 Introduction

A landslide is defined as "the movement of a mass of rock, earth or debris down a slope" (Cruden 1991). The term "landslide" encompasses five types of slope movement: fall, topples, slide, spread, and flows (Cruden and Varnes 1996). This tool deals with the fall and slide types, which can be shallow or deep-seated landslides.

Landslide susceptibility mapping (LSM) is the task of ranking areas with different degrees of landsliding potential by assessing all critical factors (landslide factors) that contribute to the re-occurrences of inventoried landslides in the past (Chalkias et al. 2014). LSM can provide a basic tool for decision-makers to make appropriate development plans (Gorsevski et al. 2006a; Feizizadeh et al. 2013). The process of LSM depends largely on the data availability, the scale of investigation, and the analysis methods (Fell et al. 2008). With the advanced technology development in GIS and remote sensing, more sophisticated and accurate spatial models have been increasingly used worldwide, especially for the LSM, as reviewed by Gorsevski et al (2006a).

Many studies on LSM have been conducted in mountainous areas in Vietnam by analysing the complex interactions among controlling factors (Saro and Dan 2005; Bui et al. 2012b), or using modeling approaches (Hung et al. 2005; Saro and Dan 2005; Long and De Smedt 2008, 2012; Bui et al. 2011, 2012a). However, LSM in Vietnam is still a challenge for scientists because the required data are not always available or, if available, they are of poor quality, which is a common problem worldwide, as remarked by Malczewski (2000), van Westen et al. (2006), and Fell et al. (2008).

Among several GIS-multicriteria decision analysis methods, the Analytical Hierarchy Process (AHP) and Weighted Linear Combination (WLC) have been considered the simplest approaches for LSM at a regional scale. These two methods are able to integrate expert knowledge in the weighting of the input factors. Therefore, they can be used in combination to carry out a LSM in a large area where data on spatial distribution of historic landslides and environmental factors are available although they are limited (Hung et al. 2015). These two methods can be applied to the upper Lo River catchment in northern Vietnam as a case study area.

2 Study Area

The upper Lo River catchment is located in Ha Giang Province (Fig. 1), a tectonically active area in northern mountains of Vietnam (Khien et al. 2012). The catchment covers an area of approximately 4528 km^2 with inclined terrain.

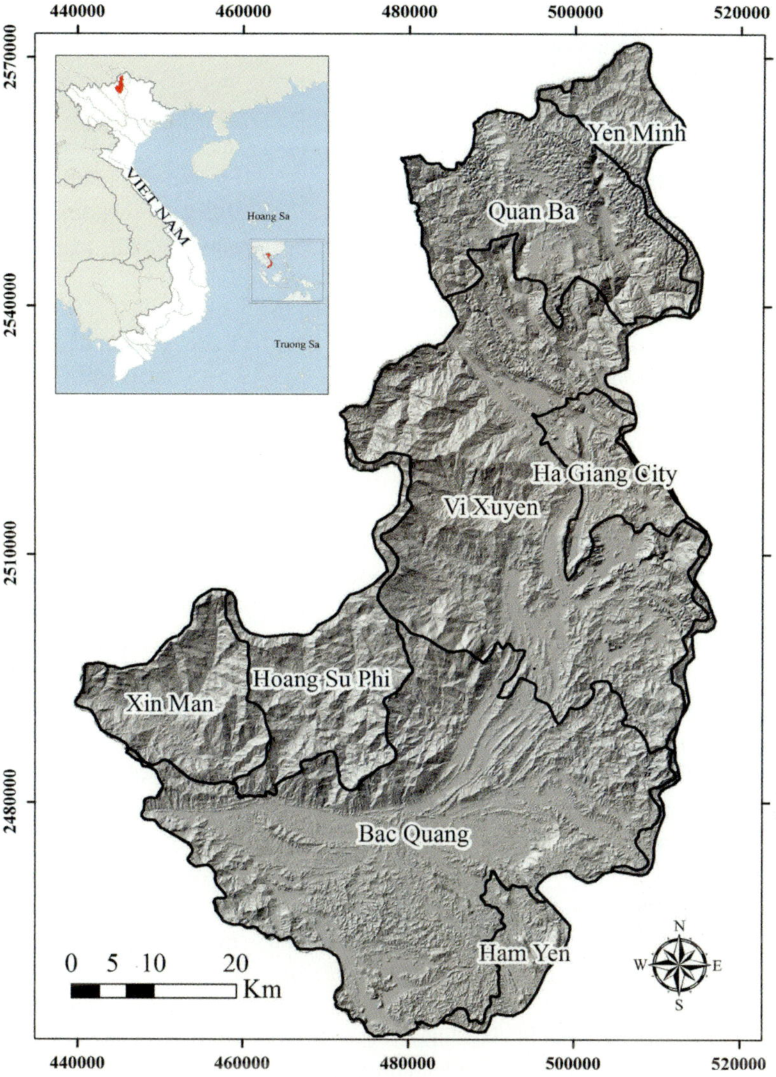

Fig. 1 A *shaded relief image* showing the surface morphology of the study area. This image is based on the 20 m-DEM, which is interpolated from the 1:50.000 topographic maps. The *black lines* indicates the boundaries of the administrative districts in the upper Lo River catchment

The Lo River is the main channel system in these regions: it originates from China and flows to the territory of Vietnam. The Lo River and its tributaries form a rather dense drainage network, with an average density of approximately 1 km/km^2. Land cover in the upper Lo River catchment varies with the topography, the weathering thickness of the substrate, and human activities, which have an impact on the distribution of different types of forest and plantation.

The upper Lo River catchment is characterized by a tropical climate with four seasons. In the study area, rainfall is considered to be the main trigger for a number of disastrous events, including landslides (Khien et al. 2012). According to the 1976–2014 rainfall record database of the National Centre for Hydro-meteorological Forecasting of Vietnam, the upper Lo River catchment has an average annual rainfall ranging from 2500 to 3200 mm/year, in which 90% of the total rainfall occurs in the summer (from May to October every year). Under extreme weather conditions, such as rains with high intensity or long duration, landslides often occur on natural slopes with highly weathered layers. The thicker the weathered layer is, the higher the volume of the landsliding mass will be. Field observations show that translational, rotational slides and rock fall are the most common types of landslides in the upper Lo River catchment.

3 Methodology

3.1 Available Data

In this tool, a spatial database was constructed in a GIS environment (e.g., ArcGIS 9.2) that includes a landslide inventory map and six landslide factor maps. The maps are briefly described below:

An inventory of 216 historic landslides was mapped only by field surveys in 2010 and 2011 due to the lack of aerial photographs for the whole area. This inventory includes 110 deep-seated and 105 shallow landslides. The landslides were mostly found in the central parts of the upper Lo River catchment, especially densely populated areas such as Ha Giang City, Vi Xuyen Town, and some surrounding communities. This landslide inventory map was used in the final stage to validate the reliability of the resulting map.

A landslide in any location usually has several controlling factors (so-called landslide factors), which determine the initial favourable conditions for landslide occurrence, but only one triggering factor, which determine the timing of landsliding (Ladas et al. 2007). In the upper Lo River catchment, heavy rainfall is the main landslide triggering factor; however, detailed rainfall data and maps were not available. In this tool, six landslide factors in the upper catchment at a scale of 1:100,000 were compiled from different available sources (Fig. 2). The *Elevation map* (a) was derived from a Digital Elevation Model (DEM) with a ground resolution of 20 × 20 m, which was interpolated from 1:50,000 scale topographic maps. The *Slope map* (b) and the *Drainage density map* (c) were derived from the same DEM and combined with the river system that was extracted from 1:50,000 scale topographic maps. The *Fault density map* (d) was extracted from the 1:200,000 scale geological maps; and the *Land cover map* (e) was extracted from the 1:100,000 scale forest maps, which had been constructed in 2010. The *Weathering crust map* (f) was produced from the 1:200,000 scale geological maps and field surveys. For the later analysis of landslide susceptibility, the main attributes of each factor maps were grouped into different classes using Jenks Natural Break classification in ArcGIS 9.2.

3.2 General Overview of the AHP and WLC Methods

The AHP, as first introduced by Saaty (1980), is based on three principles: decomposition, comparative judgment, and synthesis of priorities (Malczewski 1999). In landslide susceptibility mapping, AHP is applied to weight and rank the influence (the relative importance) of each landslide factor and its classes, based on the occurrences of landslides in the study area. The great advantage of this approach is that it re-arranges the complexity of the data set by a hierarchy, with a pairwise comparison between two landslide factors or between two classes within one landslide factor. The AHP also allows the validation of the pair consistency using eigenvector values, which express the degree of importance in the relationship between factors or between classes within one factor. From eigenvector values, one consistency value is determined, which is used to recognise the inconsistency or transitivity between two factors. The transitivity of factors in AHP is understood as, for example, factor A is more preferred than factor B, and factor B is more preferred than factor C, then factor A should be more preferred than factor C. From that, the consistency index CI, random index RI, and consistency ratio CR are calculated in order to validate the consistency of the comparison (Saaty 2000). All these indices and ration values are arranged in a range from 0 to 1. The CR is a ratio between the matrix's consistency index and a random index. The random index is the average consistence index obtained by generating large numbers of random matrices. If CR is less than 0.1 the consistency of the model is acceptable, if it is greater than 0.1 the

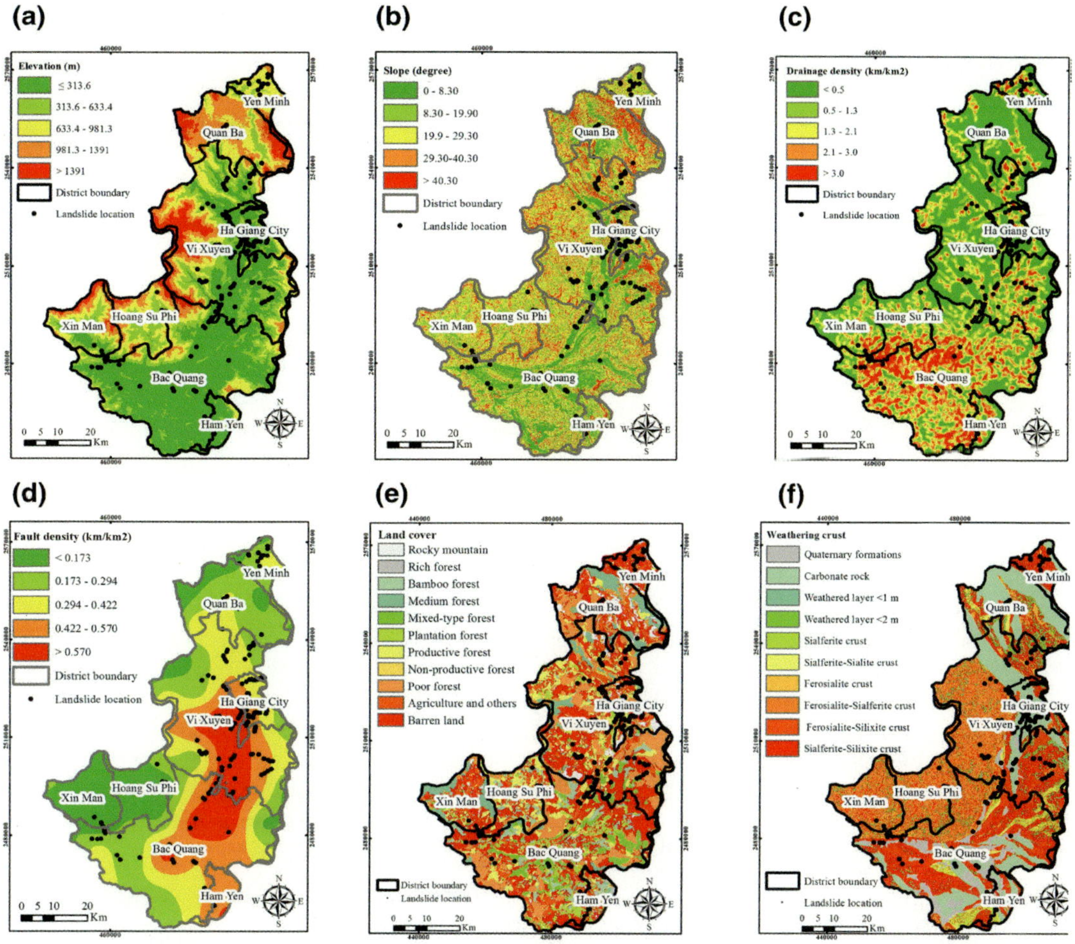

Fig. 2 Landslide factor maps: **a** elevation (*E*); **b** slope (*S*); **c** drainage density (*D*); **d** fault density (*F*); **e** land cover (*L*); **f** weathering crust (*W*)

pairwise comparison needs to be re-calculated. In landslide susceptibility mapping, AHP is applied to weight and rank the influence (the relative importance) of each landslide factor and its classes based on the occurrences of landslides in the study area. As adapted by Rajput and Shukla (2014), the AHP method consists of the following steps:

(1) Decomposition of the complex problem into smaller ones.
(2) Construction of a decision matrix and determination of the priority score using a 9-point scale for pairwise comparisons as described in Table 1.
(3) Execution of comparative judgment with the element in Table 1.
(4) Normalization of the comparison matrix by dividing each column by the sum of the entries of that column.
(5) Calculation of the eigenvector value of the normalized matrix to obtain the relative weight of the criteria. To calculate weights for each compared factor using the AHP approach, the comparison matrix means the weight matrix. Therefore, eigenvector values indicate weighted values of comparison factors.
(6) Checking the consistency of the comparison using the consistency index (CI) and

Table 1 Adopted scale of absolute numbers for pairwise comparison (Saaty 2008)

Intensity of importance	Definition	Explanation
1	Equal importance	Two activities contribute equally to objectives
2	Weak or slight	
3	Moderate importance	Experience and judgment slightly favour one activity over another
4	Moderate plus	
5	Strong importance	Experience and judgment strongly favour one activity over another
6	Strong plus	
7	Very strong or demonstrated importance	An activity is very strongly favoured over another; its dominance demonstrated in practice
8	Very, very strong	
9	Extreme importance	The evidence favouring one activity over another is of the highest possible order of affirmation
Reciprocals of above	If activity i has one of the above non-zero numbers assigned to it when compared with activity j, then j has the reciprocal value when compared with i	A reasonable assumption
Reciprocals of above	If activity i has one of the above non-zero numbers assigned to it when compared with activity j, then j has the reciprocal value when compared with i	A reasonable assumption

Note The determination of the priority score using a 9-point scale for pairwise comparisons as follows: one end of the scale is given a label with an expression and other end is given a label with an opposite expression:

Very high Susceptibility	9	8	7	6	5	4	3	2	1	Insusceptibility

Table 2 List of equations adopted in this study

Equation Nr.	Equation expression	Explanation of parameters
Equation 1	$CI = \frac{(\lambda_{\max} - n)}{n-1}$	CI: consistency index n: number of elements to be compared λ_{max}: maximum eigenvector value
Equation 2	$CR = \frac{CI}{RI}$	CR: consistency ratio that should be lower than 0.1, otherwise the pair comparison needs to be re-calculated RI: random index (Table 3)
Equation 3	$LSI = \sum_{j}^{n} W_j w_{ij}$	LSI: landslide susceptibility index W_j: weight of landslide factor j w_{ij}: weight of class i in landslide factor j n: number of landslide factors

Table 3 Random indices (RI) for a matrix of n elements (Saaty 1977)

n	1	2	3	4	5	6	7	8	9	10	11	12	13	14	15
RI	0	0	0.58	0.9	1.12	1.24	1.32	1.41	1.45	1.49	1.51	1.48	1.56	1.57	1.59

consistency ratio (CR) as calculated following Eqs. (1) and (2) in Table 2, and using the random index (RI) as given in Table 3. It should be noticed that CR must be lower than 0.1 to accept the computed weights; otherwise, the pair comparison needs to be recalculated.

(7) Using the resulting evaluation scores to order the decision alternatives from the most to the least desirable.

The WLC, as first introduced by Voogd (1983), is one of the most often used decision models in GIS to derive composite maps for landslide susceptibility assessment and mapping (Malczewski 2000; Ayalew et al. 2004). After the relative weights are generated by other methods such as AHP, the weights are aggregated by the WLC to form a single score of evaluation (Gorsevski et al. 2006b). In the spatial database prepared for the study, each thematic map, which represents a landslide factor, comprises a number of classes according to different homogeneous areas distributed in the territory. Using the WLC method, the classes of the landslide factors are standardized to a common numeric range and then combined by means of weighting (Ladas et al. 2007). After the relative weights are generated by other methods such as the AHP, the weights are aggregated by the WLC to form a single score of evaluation (Gorsevski et al. 2006b). Each criterion is multiplied by its weight from the pairwise comparison, and the results are summed up to form the final score, as expressed by Eq. (3) in Table 2. There are six steps involved in the WLC procedure (Malczewski 2000) including:

(1) Defining the set of landslide factors, which depend largely on the availability of georeferenced data in digital form.
(2) Defining the set of factor classes (feasible alternatives), into which each landslide factor is classified.
(3) Generating landslide factors and their classes as thematic maps in GIS.
(4) Assigning weights to thematic maps, in which weights are generated by the AHP method.
(5) Combining maps and weights to produce a new combined map using Eq. (3) in Table 2.
(6) Classifying the values (combined weights) of the new combined map into landslide susceptibility categories (the alternatives) to establish a landslide susceptibility zonation map.

3.3 The Workflow of Landslide Susceptibility Mapping for the Study Area

The procedure of applying the combination of AHP and WLC methods for mapping in the upper Lo River catchment is summarized in Fig. 3. The AHP method is used to define weights for the landslide factors and for the classes of each factor based on expert knowledge. Then, the WLC method is used to compute weighted factor maps to assess the landslide susceptibility using a landslide susceptibility index (LSI). The LSI is calculated by summation of the weighted value of each factor multiplied by the weighted value of each class of the landslide factors, as expressed by Eq. (3) in Table 2.

The weighting and ranking scale is defined in a range of 0–1 that is based on expert knowledge about the natural features that are distributed over the whole region. Six landslide factors are evaluated using pairwise comparison in the AHP method. The weights are presented by the eigenvalues as given in Table 4, in which the slope factor has the highest eigenvalue (0.3310) while the elevation factor has the lowest value (0.0463). From the results of pairwise

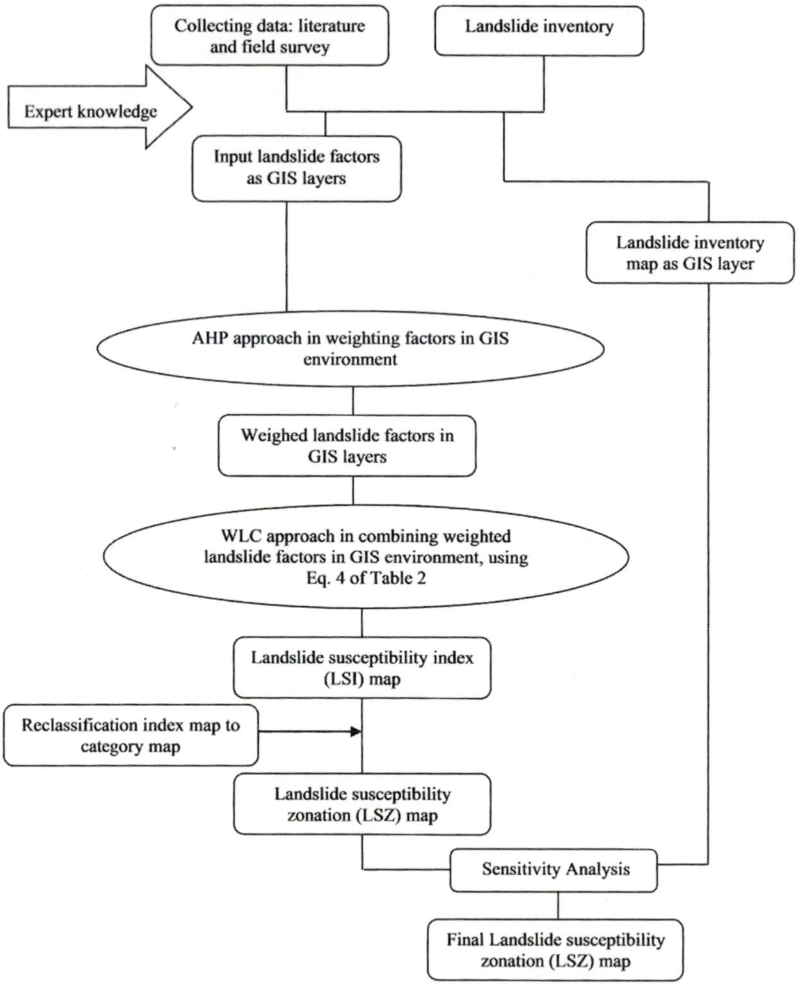

Fig. 3 Procedures for landslide susceptibility mapping (LSM) using the combination of AHP and WLC methods

Table 4 Pairwise comparison matrix, weights, eigenvector values and consistency ratio (CR) of the landslide factors

Landslide factors	(1)	(2)	(3)	(4)	(5)	(6)	Eigenvector value
(1) Elevation	1						0.0463
(2) Drainage density	2	1					0.0705
(3) Fault density	3	2	1				0.1116
(4) Land cover	4	3	2	1			0.1785
(5) Weathering crust	5	4	3	2	1		0.2621
(6) Slope	5	4	3	2	2	1	0.3310
							CR = 0.0218

comparison, the eigenvalues are assigned as weighting values Wi corresponding to individual landslide factors. The obtained consistency ratio (CR) of 0.0218 indicates an adequate degree of consistency in the comparison; thus, all values are taken into the WLC model in the GIS environment. From the results of these pairwise comparisons, as given in Table 5, the eigenvalues are assigned as weighting values wji, corresponding to classes of each landslide factor.

Table 5 Pairwise comparison matrices, weights, eigenvector values and consistency ratios (CR) of all classes of each factor

Landslide factors	(1)	(2)	(3)	(4)	(5)	(6)	(7)	(8)	(9)	(10)	(11)	Eigenvector value
Elevation (m)												
(1) $\leq$313.6	1											0.0694
(2) 313.6–633.4	2	1										0.1198
(3) 633.4–981.3	3	2	1									0.2121
(4) 981.3–1391	5	3	2	1								0.3865
(5) $\geq$1391	3	2	1	1/2	1							0.2121
												CR = 0.0039
Fault density (km/km^2)												
(1) 0.173	1											0.0419
(2) 0.173–0.294	3	1										0.0836
(3) 0.294–0.422	5	3	1									0.1772
(4) 0.422–0.570	6	4	2	1								0.2728
(5) >0.570	7	5	3	2	1							0.4245
												CR = 0.0307
Drainage density (km/km^2)												
(1) <0.5	1											0.0514
(2) 0.5–1.3	2	1										0.0780
(3) 1.3–2.1	4	3	1									0.1749
(4) 2.1–3.0	5	4	2	1								0.2717
(5) >3.0	6	5	3	2	1							0.4241
												CR = 0.0219
Land cover												
(1) Rocky mountain	1											0.0163
(2) Rich forest	2	1										0.0225
(3) Bamboo forest	3	2	1									0.0338
(4) Medium forest	3	2	1	1								0.0338
(5) Mixed-type forest	3	2	1	1	1							0.0338
(6) Plantation forest	4	3	2	2	2	1						0.0520
(7) Productive young forest	5	4	3	3	3	2	1					0.0756
(8) Non-productive young forest	6	5	4	4	4	3	2	1				0.1064
(9) Poor forest	7	6	5	5	5	4	3	2	1			0.1473
(10) Agriculture and other lands	8	7	6	6	6	5	4	3	2	1		0.2024
(11) Barren land	9	8	7	7	7	6	5	4	3	2	1	0.2761
												CR = 0.0286
Weathering crust												
(1) Quaternary formations	1											0.0159
(2) Carbonate rock	2	1										0.0194
	2	1	1									0.0241

(continued)

Table 5 (continued)

Landslide factors	(1)	(2)	(3)	(4)	(5)	(6)	(7)	(8)	(9)	(10)	(11)	Eigenvector value
(3) Bedrock and slightly weathered rock or weathered layer <1 m												
(4) Bedrock and slightly weathered rock or weathered layer <2 m	3	2	2	1								0.0349
(5) Sialferite crust	7	6	5	4	1							0.0950
(6) Sialferite-Sialite crust	7	6	5	4	1	1						0.0950
(7) Ferosialite crust	8	7	6	5	2	2	1					0.1413
(8) Ferosialite-Sialferite crust	8	7	6	5	2	2	1	1				0.1413
(9) Ferosialite-Silixite crust	9	8	7	6	3	3	2	2	1			0.2165
(10) Sialferite-Silixite crust	9	8	7	6	3	3	2	2	1	1		0.2165
												CR = 0.0105
Slope (°)												
(1) ≤8.3	1											0.0472
(2) 8.3–19.9	2	1										0.0713
(3) 19.9–29.3	4	3	1									0.1565
(4) 29.3–40.3	5	4	2	1								0.2362
(5) >40.3	7	6	4	3	1							0.4888
												CR = 0.0314

All CR values are smaller than 0.1, thus, indicated that the weights of all factor classes are accepted.

Using the WLC method, Eq. (3) as given in Table 2 is applied to all landslide factors to produce the landslide susceptibility index (LSI) map (Fig. 4) using the following expression:

$$\begin{aligned} \mathrm{LSI} = 0.0463 * \mathrm{E} + 0.0705 * \mathrm{D} + 0.1116 \\ * \mathrm{F} + 0.1785 * \mathrm{L} + 0.2621 * \mathrm{W} + 0.3310 \\ * \mathrm{S} \end{aligned}$$

in which variables E, D, F, L, W, and S are abbreviations of the landslide factors: Elevation (E), Drainage density (D), Fault density (F), Land cover (L), Weathering crust (W), and Slope (S), respectively. LSI represents the relative susceptibility to a landslide occurrence; therefore, the higher the LSI, the more susceptible the area is to landslides. The LSI values are normalized to the range 0–1 for consistency in comparison and classification across all factors.

4 The Final Landslide Susceptibility Map

The LSI map is then reclassified by splitting the cumulative frequency histogram of LSI values into four classes using method of Jenks Natural Break classification in ArcGIS 9.2 (Fig. 5 and Table 6). The produced map is a landslide susceptibility zonation (LSZ) map, which is then used as the final map (Fig. 6). In the LSZ map, the surfaces of the study area are classified into four landslide susceptibility zones, namely "low," "moderate," "high," and "very high," that account for 21.57, 37.46, 29.21, and 11.75% of the total areas, respectively (Table 7).

To validate the final map, as well as the overall methodology, the LSZ map is then

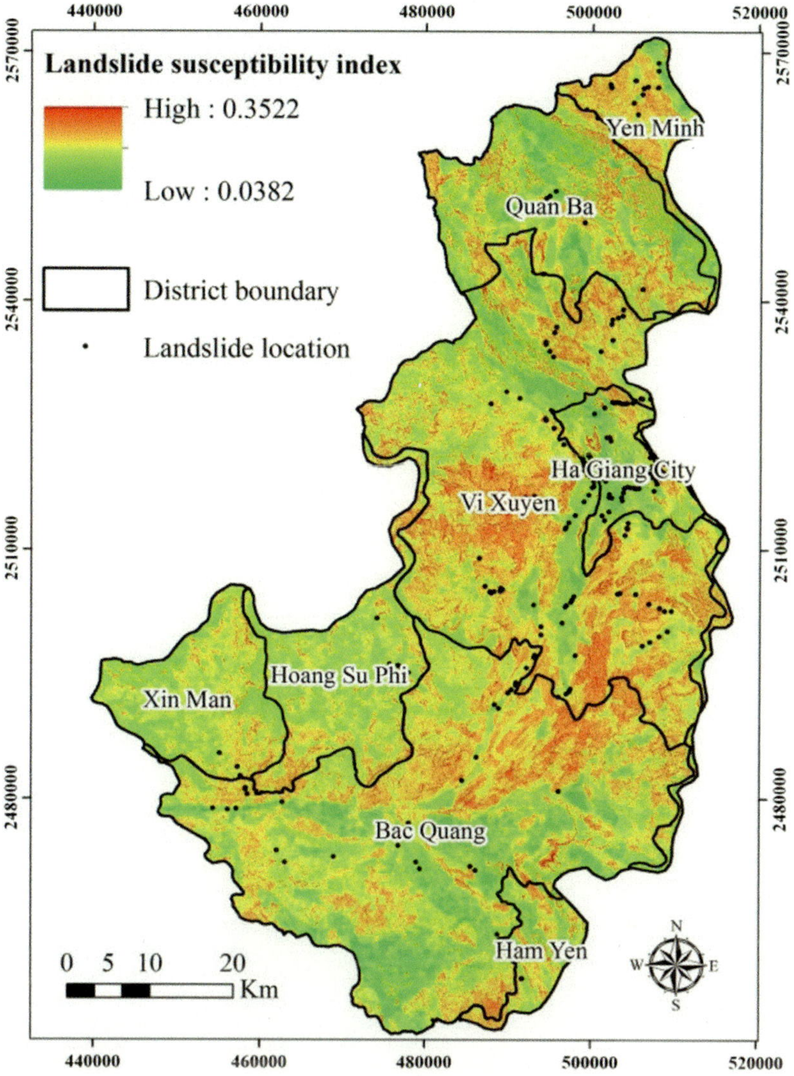

Fig. 4 The landslide susceptibility index (LSI) map is produced by applying Eq. (3) as given in Table 2 to all landslide factors

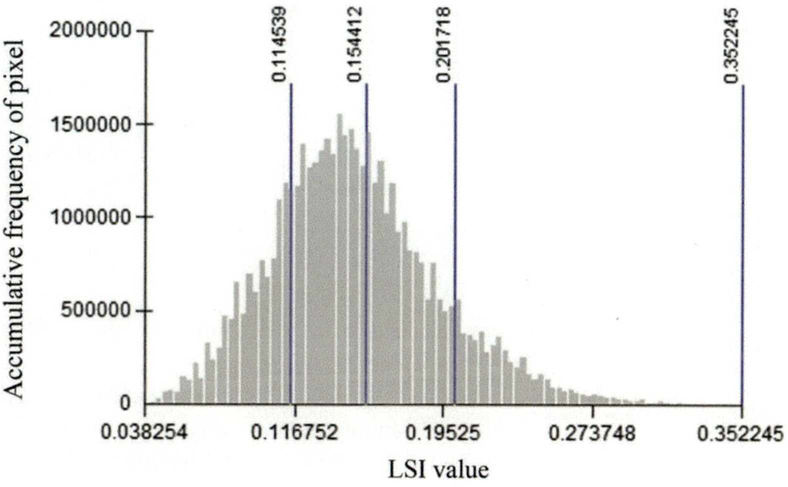

Fig. 5 The cumulative frequency histogram of LSI values is split into four classes using the method of Jenks natural break classification in ArcGIS 9.2

Table 6 Reclassification of LSI values to produce the landslide susceptibility zonation map

Cumulative frequency of LSI values (%)	Susceptibility index (LSI)	Landslide susceptibility classes
22	LSI < 0.11582	Low
59	0.11582 < LSI < 0.1564	Moderate
88	0.1564 < LSI < 0.2032	High
100	LSI > 0.2032	Very high

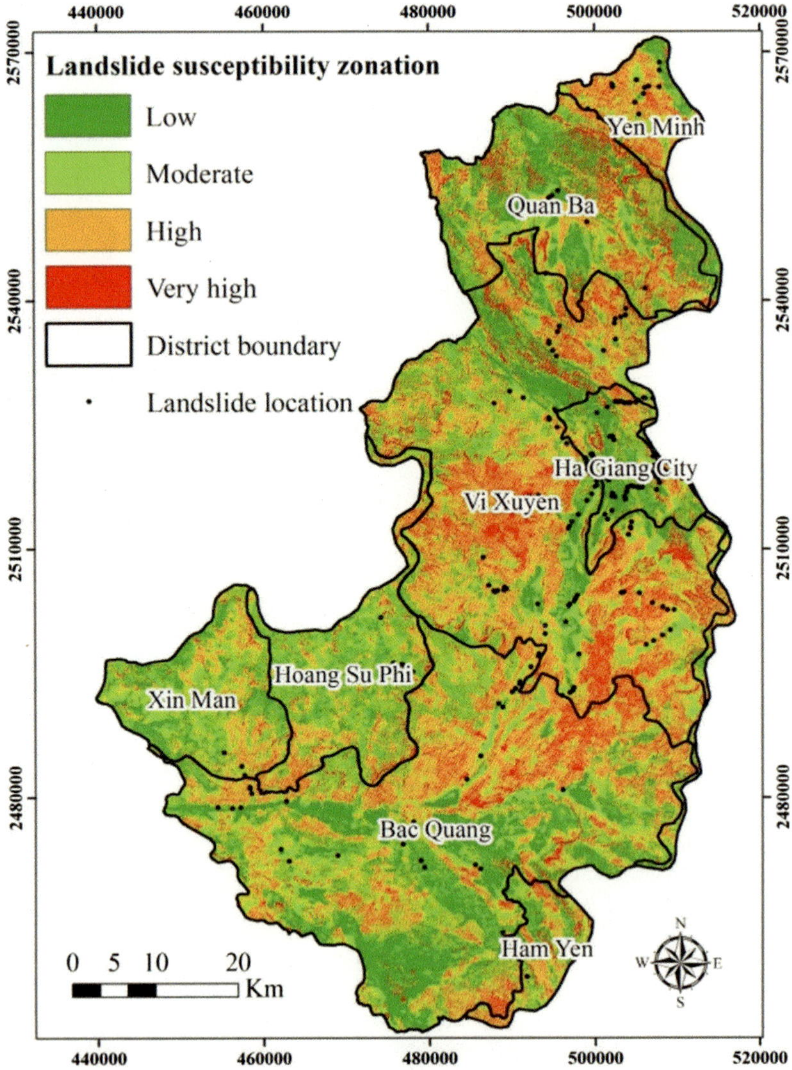

Fig. 6 The final susceptibility map of the upper Lo River catchment landslide susceptibility zonation (LSZ) map

overlaid with the observed landslides in the inventory map. As presented in Table 7, out of 216 observed landslides, 50 landslides (~23.15%) fall within the low-susceptibility zone, 63 landslides (~29.17%) fall within the moderate-susceptibility zone, 83 landslides (~38.43%) fall within the high-susceptibility zone, and 20 landslides (~9.26%) fall within the very-high-susceptibility zone. The results are in accordance with the occurrences of the observed

Table 7 Summary of the distribution of the predicted landslide susceptibility classes in the landslide susceptibility zonation map compared with the observed landslides in the landslide inventory map

Landslide susceptibility class	Predicted landslide susceptibility classes		Observed landslides	
	Area (km^2)	Percentage (%)	Number	Percentage (%)
Low	977	21.57	50	23.15
Moderate	1696	37.46	63	29.17
High	1323	29.21	83	38.43
Very high	532	11.75	20	9.26
Total	4528	100	216	100

landslides, in which 47.69% of observed landslides are located in the two most susceptible zones (very-high-susceptibility zone and high-susceptibility zone) that cover 40.96% of the total area. This simple type of validation based on spatial cross-checking of the mapping results serves as a first indicator for the plausibility of the landslide susceptibility map. A true validation of the overall methodology, however, is only supported to some extent by this point.

The validation is in accordance with the fact that many existing landslides were found inside or close to areas related to human activities such as settlements, transportation routes, terraced fields, mining sites, deforested land, and barren land; therefore, the occurrences of landslides in the moderate- or low-susceptibility zones are attributed to local impacts. Those susceptibility zones are often characterized by natural features such as low land, gentle slopes, low density of drainage and faulting, and loose sediment. Therefore, those susceptibility zones are theoretically favourable for human settlements and unfavourable for landsliding.

5 Conclusion

The frequency of landslides in the upper Lo River catchment in northern Vietnam has increased with the rapid urbanization and economic development of the catchment within the last 10 years. A regional landslide susceptibility map is required to provide a foundation for long-term land use planning that includes landslide mitigation measures in the region. The lack of historical landslide data in the region has made the landslide susceptibility assessment difficult. In this tool, among many available techniques worldwide, the integrated analytical hierarchy process (AHP) and weighted linear combination (WLC) approach in landslide susceptibility mapping was applied because this approach is able to integrate expert knowledge in the weighting of the input factors, which is suitable for areas lacking detailed input data. The results reveal the effectiveness of the combination of AHP and WLC for mapping at medium scales in a large mountainous region, which has limited data on spatial distribution of landslides and environment factors. Therefore, this approach can quickly result in the final maps in order to provide the local authority and community with better strategies for disaster mitigation and management. However, this approach still depends very much on the experience and knowledge of individual experts.

Acknowledgments This work is part of the research project "Building up an integrated system for assessing natural disasters in mountainous areas of Vietnam—A case study in upstream catchment of Lo River", project code: 105.11.50.09, funded by the National Foundation for Science and Technology Development (NAFOSTED), Vietnam Ministry of Science and Technology (MOST).

References

Ayalew L, Yamagishi H, Ugawa N (2004) Landslide susceptibility mapping using GIS based weighted linear combination, the case in Tsugawa area of Agano River. Niigata prefecture, Japan. Landslides 1:73–81. doi:10.1007/s10346-003-0006-9

Bui D, Lofman O, Revhaug I, Dick O (2011) Landslide susceptibility analysis in the Hoa Binh province of

Vietnam using statistical index and logistic regression. Nat Hazards 59:1413–1444. doi:10.1007/s11069-011-9844-2

Bui DT, Pradhan B, Lofman O, Revhaug I, Dick OB (2012a) Landslide susceptibility assessment in the Hoa Binh province of Vietnam. A comparison of the Levenberg-Marquardt and Bayesian regularized neural networks. Geomorphology 171:12–29. doi:10.1016/j.geomorph.2012.04.023

Bui DT, Pradhan B, Lofman O, Revhaug I, Dick OB (2012b) Landslide susceptibility mapping at Hoa Binh province (Vietnam) using an adaptive neuro-fuzzy inference system and GIS. Comput Geosci 45:199–211. doi:10.1016/j.cageo.2011.10.031

Chalkias C, Ferentinou M, Polykretis C (2014) GIS-based landslide susceptibility mapping on the Peloponnese Peninsula, Greece. Geosciences 4:176–190. doi:10.3390/geosciences4030176

Cruden DM (1991) A simple definition of a landslide. Bull Int Assoc Eng Geol 43:27–29

Cruden DM, Varnes DJ (1996) Landslide types and processes. In: Turner AK, Schuster RL (eds) Landslides investigation and mitigation. Transportation research board, US National Research Council. Special report 247, Washington, DC, pp 36–75 (Chapter 3)

Feizizadeh B, Blaschke T, Nazmfar H, Rezaei Moghaddam MH (2013) Landslide susceptibility mapping for the Urmia Lake basin. Iran: a multi-criteria evaluation approach using GIS. Int J Environ Res 7:319–336

Fell R, Corominas J, Bonnard C, Cascini L, Leroi E, Savage WZ (2008) Guidelines for landslide susceptibility, hazard and risk zoning for land use planning. Eng Geol 102:85–98. doi:10.1016/j.enggeo.2008.03.022

Gorsevski PV, Gessler PE, Boll J, Elliot WJ, Foltz RB (2006a) Spatially and temporally distributed modeling of landslide susceptibility. Geomorphology 80:178–198

Gorsevski PV, Jankowski P, Gessler PE (2006b) An heuristic approach for mapping landslide hazard by integrating fuzzy logic with analytic hierarchy process. Control Cybern 35:121–146

Hung LQ, Batelaan O, San DN, Van TT (2005) GIS—remote sensing application of landslide hazard mapping—case study Thua-Thien-Hue province, Vietnam. In: Oluic M (ed) New strategies for European remote sensing. Millpress, Rotterdam, Netherlands, pp 627–634 (Proceedings of the 24th symposium of the European Association of Remote Sensing Laboratories, Dubrovnik, Croatia, 25–27 May 20)

Hung LQ, Van NTH, Duc DM, Ha LTC, Van Son P, Khanh NH, Binh LT (2015) Landslide susceptibility mapping by combining the analytical hierarchy process and weighted linear combination methods: a case study in the upper Lo River catchment (Vietnam). Landslides, pp 1–17. doi:10.1007/s10346-015-0657-3

Khien NX, Hung LQ, Long NT (2012) Report of project "Assessment of status geohazards in four mountainous provinces in North of Viet Nam: Ha Giang, Bac Kan, Cao Bang, Tuyen Quang. Identification of sources, prediction and propose preventive measures and consequences mitigation". Sponsored by Ministry of Natural Resources and Environment (in Vietnamese). Center for Information and Archives of Geology, Hanoi, 318 p

Ladas I, Fountoulis I, Mariolakos I (2007) Using GIS and multicriteria decision analysis in landslide susceptibility mapping—a case study in Messinia prefecture area (SW Peloponnesus, Greece). Bull Geol Soc Greece XXXX:1973–1985

Long NT, De Smedt F (2008) Landslide susceptibility mapping of the mountainous area in A Luoi district, Thua Thien Hue province, Vietnam. Vrije University of Brussel, Dissertation. 255 p

Long NT, De Smedt F (2012) Application of an analytical hierarchical process approach for landslide susceptibility mapping in A Luoi district, Thua Thien Hue province, Vietnam. Environ Earth Sci 66:1739–1752. doi:10.1007/s12665-011-1397-x

Malczewski J (1999) GIS and multicriteria decision analysis. Wiley, New York, p 408

Malczewski J (2000) On the use of weighted linear combination method in GIS: common and best practice approaches. Trans GIS 4:5–22

Rajput V, Shukla AC (2014) Decision-making using the analytic hierarchy process (AHP). Int J Sci Res 3:135–136

Saaty TL (1977) A scaling method for priorities in hierarchical structures. J Math Psychol 15:234–281. doi:10.1016/0022-2496(77)90033-5

Saaty TL (1980) The analytic hierarchy process: planning, priority setting, resource allocation. McGraw-Hill, New York, p 287

Saaty TL (2000) Fundamentals of decision making and priority theory with the analytic hierarchy process. Analytic Hierarchy Process Series, vol 6. RWS Publications, Pittsburg, p 477

Saaty TL (2008) Decision making with the analytic hierarchy process. Int J Serv Sci 1:83–98

Saro L, Dan NT (2005) Probabilistic landslide susceptibility mapping in the Lai Chau province of Vietnam: focus on the relationship between tectonic fractures and landslides. Environ Geol 48:778–787. doi:10.1007/s00254-005-0019-x

van Westen CJ, van Asch TWJ, Soeters R (2006) Landslide hazard and risk zonation: why is it still so difficult? Bull Eng Geol Environ Off J Int Assoc Eng Geol Environ IAEG 65:67–184. doi:10.1007/s10064-005-0023-0

Voogd H (1983) Multicriteria evaluation for urban and regional planning, 1st edn. Pion Ltd., Princeton University, London, p 387

TXT-tool 1.039-1.2 Bedding Attitude Information Through the Interpretation of Stereoscopic Aerial Photographs and GIS Modeling

Ivan Marchesini, Michele Santangelo, Federica Fiorucci, Mauro Cardinali, Mauro Rossi, Francesco Bucci and Fausto Guzzetti

Abstract

Information on bedding attitude is necessary to define the geological and structural setting of an area and is relevant for the assessment of landslide susceptibility at different scales and in different physiographic environments. Bedding planes are a known factor controlling the type, abundance, and pattern of landslides. Where layered rocks crop out, the geometrical relationships between the attitude of the bedding and the geometry of the terrain is crucial to understand landslide phenomena. Obtaining information on bedding attitude for large areas through field surveys is time consuming, and resource intensive, hampering the possibility of quantitative investigations on the control of bedding planes on landslides. We describe a GIS-based method to extract information on bedding planes from the analysis of information captured through the visual interpretation of stereoscopic aerial photographs and a digital representation of the terrain. Results show that our procedure provides bedding attitude information in good agreement with bedding attitude data acquired during field surveys.

Keywords

Bedding attitude · GIS · Aerial photo-interpretation · Landslides

I. Marchesini (✉) · M. Santangelo · F. Fiorucci · M. Cardinali · M. Rossi · F. Bucci · F. Guzzetti
Research Institute for Geo-Hydrological Protection, CNR, via Madonna Alta 126, 06128 Perugia, Italy
e-mail: Ivan.Marchesini@irpi.cnr.it

M. Cardinali
e-mail: Mauro.Cardinali@irpi.cnr.it

M. Rossi
e-mail: Mauro.Rossi@irpi.cnr.it

F. Guzzetti
e-mail: Fausto.Guzzetti@irpi.cnr.it

K. Sassa et al. (eds.), *Landslide Dynamics: ISDR-ICL Landslide Interactive Teaching Tools*, https://doi.org/10.1007/978-3-319-57774-6_13

Contents

1 **Introduction** 176
2 **Bedding Trace Mapping** 177
3 **Bedding Attitude Estimation Algorithm** 178
4 **Case Study** 179
5 **Results** 181
6 **Discussion** 182
7 **Conclusions** 184
References 184

1 Introduction

Investigators have shown the importance of lithology and structure to control the type, abundance, and pattern of landslides (Fookes and Wilson 1966; Zaruba and Mencl 1969; Varnes 1978; Crozier 1986; Koukis and Ziourkas 1991; Guzzetti et al. 1996, 2008; Günther 2003; Grelle et al. 2011; Santangelo et al. 2015). The presence and abundance of discontinuities in the rock mass, including bedding planes and foliation (Guzzetti et al. 1996), and faults, joints, cleavages, and other fracture systems (Günther 2003; Goudie 2004), are known geological factors controlling the stability of slopes (Guzzetti et al. 1996; Günther 2003; Grelle et al. 2011).

Where hard rocks crop out, beddings, joints, and fracture systems condition the formation and volume of rock falls, topples, rock slides, and rock avalanches (Goodman and Bray 1976; Cruden 2003; Günther 2003). Where layered rocks crop out, the geometrical relationships between the slopes and the discontinuities in the slopes favour (e.g., along cataclinal slopes) or limit (e.g., along anaclinal slopes) the formation of deep-seated slides (Guzzetti et al. 1996), and determine the area and volume of the landslides (Katz and Aharonov 2006). At the regional scale, the presence of structurally complex rocks characterised by abundant and intricate fabrics is a known condition that controls the location, type and abundance of landslides (Esu 1977; Grelle et al. 2011; Santangelo et al. 2013). At the local scale, the superposition of rocks characterised by different permeability, a typical setting in layered sedimentary and volcanic rocks, can generate hydrogeological conditions that favour or limit slope instability (Carrara et al. 1992; Guzzetti et al. 1996).

In landscapes dominated by layered rocks, defining the geometrical relationships between the attitude (strike and dip) of bedding planes and the attitude (gradient and aspect) of the terrain is crucial to understand landslide phenomena, to explain their types, patterns and abundance (Guzzetti et al. 1996), and to determine landslide susceptibility at different geographical scales using heuristic (Ruff and Czurda 2008), statistical (Carrara et al. 1992; Thiery et al. 2007; Rossi et al. 2010), physically-based (Günther 2003), or combined (Frattini et al. 2008) methods.

Geometrical information on the attitude of bedding planes is most commonly obtained through geological field mapping (Clegg et al. 2006; De Donatis and Bruciatelli 2006; Bodien and Tipper 2013). The method is time consuming, particularly where large areas (hundreds to thousands of square kilometres) have to be covered by the survey. This hampers the possibility of using bedding and structural data for regional landslide modelling. Geological mapping is also conditioned by practical and operational constrains (e.g., inaccessible areas, poor visibility, limited number of outcrops) that can hamper, locally severely, the reliability and representativeness of the geological information (Grelle et al. 2011). In addition, bedding measurements are typically point measurements and, as such, they are conditioned by local anomalies (e.g., minor folds, minor faults, mass movements), and the single measurements may not represent the general geological trend that controls the presence (or the absence) of landslides in an area (Guzzetti et al. 1996; Grelle et al. 2011; Marchesini et al. 2013).

Photo-geological techniques can be used to obtain information on bedding attitude, in combination or as an alternative to geological field mapping. This is achieved through the visual interpretation of stereoscopic aerial photographs (Rib and Liang 1978; Antonini et al. 2002; Brardinoni et al. 2003; Weirich and Blesius 2007), or satellite images of sufficient ground resolution. Use of remote sensing imagery (aerial photographs and satellite images) emphasises spatial continuity or repetition of geological and morphological features, facilitating their recognition and mapping (Rib and Liang 1978). Unlike field surveys that obtain bedding information at single locations (point measurements), the photo-geological techniques allow for the detection and mapping of the general trend of the geological structures, including bedding planes. This ability is particularly important for defining and mapping bedding attitude domains i.e., portions of terrain characterised by a similar attitude, or by similar geometrical relationships between the attitude and the local terrain slope (Cardinali et al. 2001).

In this work, we describe a method implemented in a Geographical Information System (GIS), to extract quantitative information on bedding planes from the analysis of information captured through the visual interpretation of stereoscopic aerial photographs (Aerial Photo Interpretation—API).

2 Bedding Trace Mapping

Where layered rocks crop out, individual layers or groups of layers intersecting at an angle the topographic surface leave discernible linear signatures that can be recognized and mapped, in the field or through the interpretation of remote imagery (e.g., aerial photographs, very-high resolution satellite images) (Fig. 1). In this work, we give the name "bedding trace" to the linear signatures left by layered rocks on the topographic surface. Visual evidence of a bedding trace on the remote imagery depends on multiple factors, including (i) the type, mechanical, and hydrological characteristic of the rocks, (ii) the contrast in resistance to erosion of the different rock layers, (iii) the attitude of the bedding planes, (iv) the steepness of the slope and the complexity of the topography, and (v) the presence (or absence) of vegetation that may emphasize or hide individual beddings. Figure 1 shows examples of bedding traces, and of their

Fig. 1 Examples of bedding traces in different lithological settings in Umbria, Central Italy. All images taken from Google Earth®. *Scale bars* show approximate scale of the images. **a** Heterogeneous, hard- and soft-layered rocks. **b** Homogeneous, hard-layered rocks. **c** Heterogeneous, soft continental deposits. **d** Heterogeneous, hard- and soft-layered rocks with vegetation. From Santangelo et al. (2015)

visual signatures on aerial photographs in different geological settings in Umbria, Central Italy.

3 Bedding Attitude Estimation Algorithm

The orientation (or attitude) of a bedding plane is determined by the strike and the dip of the geometrical plane that describes the bedding plane. The strike line is the line that represents the intersection of the bedding with a horizontal plane. The strike angle (i.e., the azimuth) is the angle made by the strike line to a reference point, most commonly the North, N. The dip direction is 90° off the strike angle, and the angle of dip (inclination) is the angle between the dip direction to the horizontal. Figure 2 shows how the availability of a bedding trace derived through API (i) allows us to infer the bedding surface (i.e., the flat surface that locally approximates the bedding plane), and (ii) permits exploiting the bedding surface to estimate the bedding attitude, i.e., the dip direction and inclination of the bedding plane.

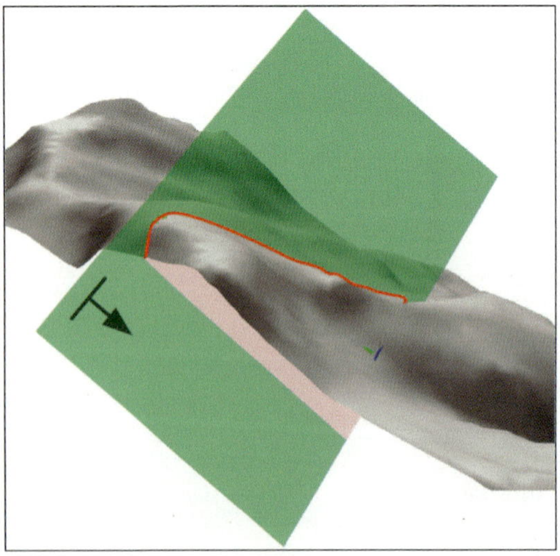

Fig. 2 Three-dimensional visualization of an example of (i) a bedding trace (*red line*), (ii) the bedding surface (*green plane*) inferred from the trace, and (iii) the dip direction and inclination values, represented by means of the symbol, of the estimated bedding attitude (*symbol with arrow*). From Marchesini et al. (2013)

To determine the attitude of the bedding planes, starting from bedding trace information, we implemented a GIS tool in GRASS GIS environment (GRASS Development Team 2016). The script adopts a stepwise approach (Fig. 3) that exploits information on the geographical location of a bedding trace, including latitude, longitude, and elevation of individual points along the bedding trace, to define strike and dip values that determine univocally the orientation of the bedding trace. First, the trace is drawn on a base map in geographical or projected coordinates (Fig. 3a). Next, the script draws the unique segment joining the two end points of the line representing the bedding trace (Fig. 3b). The orientation of the segment is determined by the position of the two end points of the trace, and is independent of (and does not represent, necessarily) the strike of the bedding. The bedding trace and the new added segment encompass a polygon. Then, the script generates a set of equally spaced points along the combined line (bedding trace plus added segment) representing the boundary of the polygon (Fig. 3c), and for each point the script obtains the terrain elevation from the available DEM. The "nn" library (Sakov 2012) implemented in the GRASS GIS add-on module "r.surf.nnbathy" is then used to interpolate geographically the elevation values measured at the single points. In particular, points are linearly interpolated exploiting the triangulation option of the "nn" library. The result is a raster surface (a sort of rasterized TIN model) that describes the modelled bedding plane in 3D (Fig. 3d).

Next, values for the local slope and the local aspect are computed for each grid cell in the modelled 3D bedding surface, and descriptive statistics are calculated, including the mean and the standard deviation of the dip angle, and the mean, the angular standard deviation, and the circular variance of the aspect (Davis 1990; Butler 1992; Nichols 2009). The latter values measure the uncertainty associated to the geometrical definition of the orientation of the bedding (strike and dip), and provide a quantitative evaluation of the quality of the orientation measurements obtained by fitting a plane through a

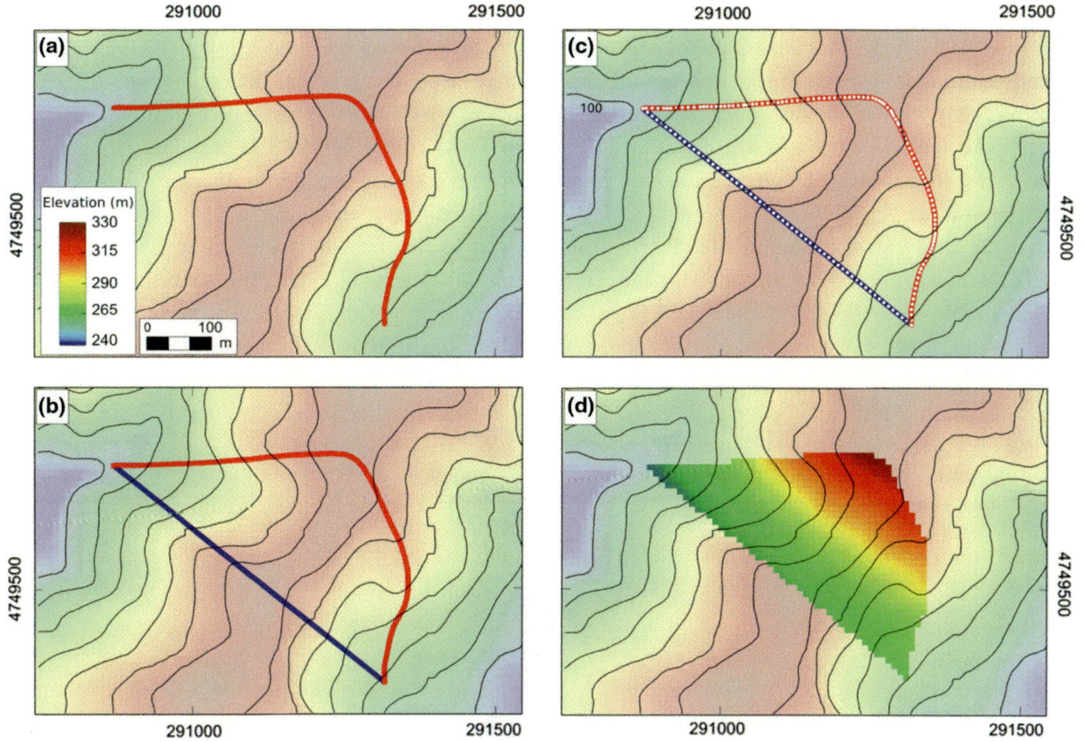

Fig. 3 Scheme of the method to determine the attitude of a bedding plane from a bedding trace. **a** Bedding trace (*red line*) is drawn on a base map. **b** End nodes of the line representing the bedding trace (*red line*) are joined with a single segment (*blue line*) to form a polygon. **c** Equally spaced points (*white dots*) are selected along the boundary of the polygon, and their elevation is obtained from the DEM. **d** Elevation values along the trace are interpolated to obtain a 3D model of the bedding plane. *Faint shades of colour* show terrain elevation values. *Solid shades of colour* show elevation of the interpolated bedding plane. Modified from Santangelo et al. (2015)

set of 3D points. With this respect, the method is an alternative to the method proposed by Fernández (2005). Lastly, the computed values, and information on the polarity (normal or reversed) of the bedding plane, are associated to a point located in proximity to the bedding trace, and stored in the GIS (Marchesini et al. 2013).

4 Case Study

We tested the method for the semi-automatic extraction of quantitative information on the orientation of bedding planes in the Collazzone area, Umbria, central Italy (Fig. 4). This is a portion of a larger area studied by Guzzetti et al. (2006) to map landslides and ascertain landslide hazard, and by Guzzetti et al. (2009) to map landslides and determine landslide mobilisation rates. Our study area covers 50.5 km^2 of hilly terrain, with elevations in the range between 145 and 380 m (mean = 236 m, standard deviation = 45.2 m), and terrain gradient in the range from 0° to 52° (mean = 10.1°, standard deviation = 5.6°). In the area, landscape is controlled by lithology and the attitude of bedding planes, which also control the shape and the extent of the slope failures. Valleys oriented N–S are shorter, asymmetrical, and parallel to the main direction of the bedding planes. The W-facing slopes are longer and less steep than the E-facing slopes, which are shorter and steeper. Valleys oriented E–W are longer, symmetrical, and mostly perpendicular to the direction of the bedding planes. Sedimentary rocks crop out in the area, including continental gravel, sand, and clay,

Plio-Pleistocene in age (Conti and Girotti 1977; Barchi et al. 1991). The Plio-Pleistocene continental sediments are arranged in a monocline dipping gently towards the west, the result of extensional tectonic active in the Central Apennines (Ambrosetti et al. 1987; Martini and Sagri 1993; Barchi et al 2001).

For the study area, a Digital Elevation Model (DEM) with a ground resolution of 10 m × 10 m was available to us (Fig. 4). The DEM was prepared through the interpolation of 5- and 10-m contour lines shown on topographic base maps (Guzzetti et al. 2006).

We obtained information on the geographical distribution of the bedding traces in the study area through the visual interpretation of 1:33,000 scale, black and white aerial photographs taken in 1954 (red lines in Fig. 4). The geographical information on the bedding traces was obtained in three steps, adopting a procedure commonly used to acquire digital geomorphological information from stereoscopic aerial photographs (Cardinali et al. 2001; Antonini et al. 2002). First, the individual traces were identified visually on the aerial photographs, and drawn on a transparent plastic sheet superimposed on the

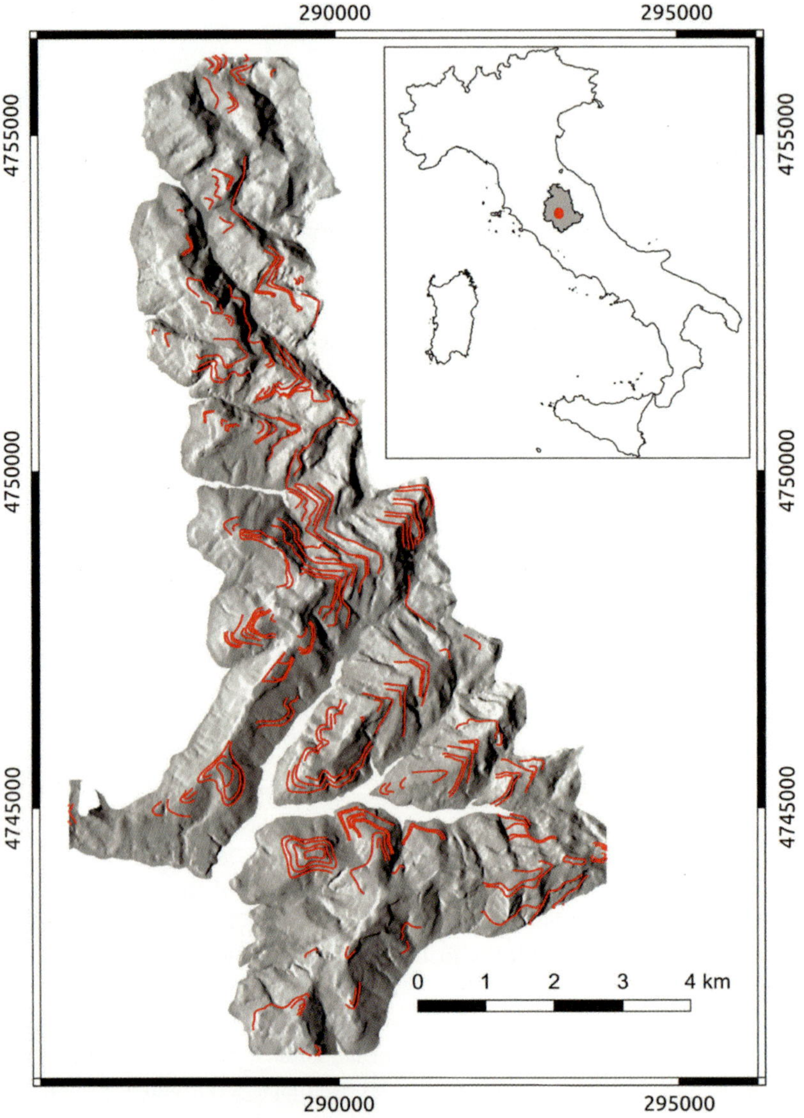

Fig. 4 Shaded relief image for the Collazzone study area, Umbria, Central Italy, obtained from a 10 m × 10 m DEM. *Red lines* show bedding traces obtained from the visual interpretation of stereoscopic aerial photographs. UTM zone 33, datum ED50 (EPSG:23033)

stereoscopic aerial photographs. Second, the individual mapped features were transferred visually onto non-deformable plastic sheets superimposed on 1:10,000 scale topographic base maps. In this phase, care was taken to accurately locate the bedding traces with respect to the local topography shown on the base maps. Third, the information drawn on the non-deformable plastic sheets was scanned (at 300 dot-per-inch (dpi), corresponding to a resolution of 0.84 m on the ground), geo-referenced, transformed into vector (line) format, and stored in a dedicated geo-database (Marchesini et al. 2013). Fourth, the position of the mapped bedding traces was checked on a digital stereo model obtained from a stereoscopic GeoEye satellite image, to improve their positional accuracy. The stereo model was obtained using ERDAS IMAGE® Leica Photogrammetry Suite SW, and the 3D visualisation was obtained using a PLANAR StereoMirror™ hardware (Ardizzone et al. 2013). This technology allows for the digitisation of 3D vectors on the 3D stereo model, which reduces the errors associated to the manual drawing of the interpreted geomorphological features, including the bedding traces.

Finally a run of the script (Sect. 3) was executed, using the bedding trace dataset and the DEM, to produce the bedding attitudes map.

5 Results

Figure 5 shows a detail of the modelled bedding attitude map. In the figure, bedding attitudes are represented by oriented symbols, scaled to the inclination value. The smaller the symbol the more horizontal is the bedding plane.

Bedding inclination spans the range from 1° (small bedding symbol) to 20° (large bedding symbol). The angular standard deviations, calculated by the algorithm for each bedding attitude, are

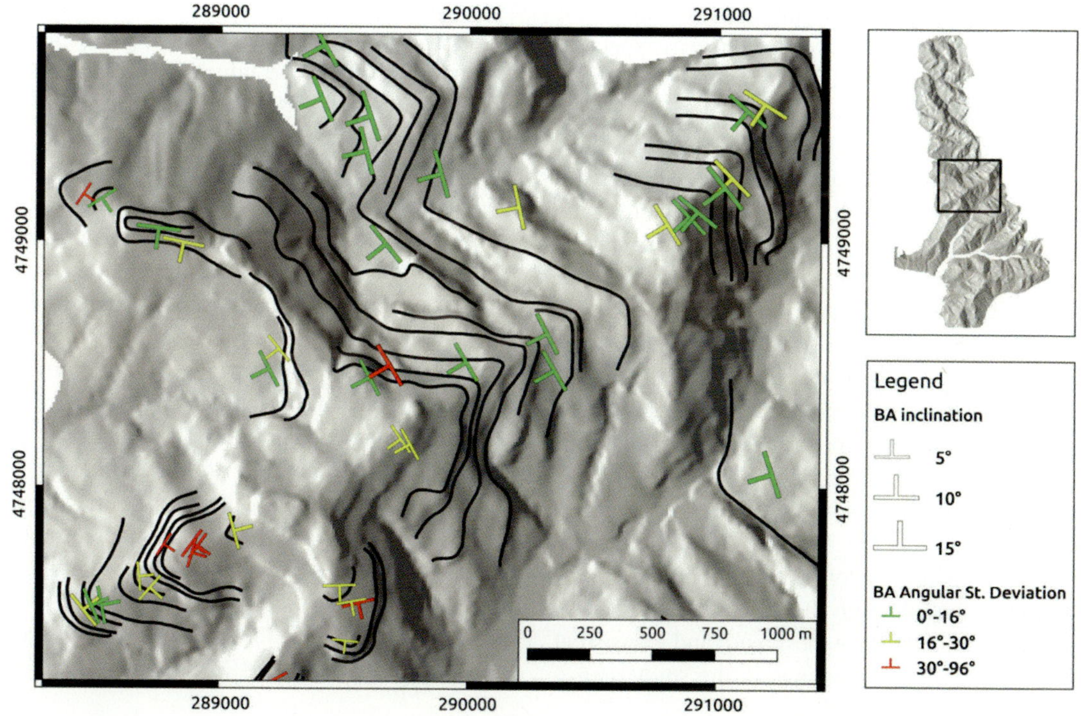

Fig. 5 Bedding attitudes represented using oriented symbols (according to dip direction), different dimensions (proportional to the inclination value), and different colours (according to three different levels of dip direction angular variance). From Marchesini et al. (2013)

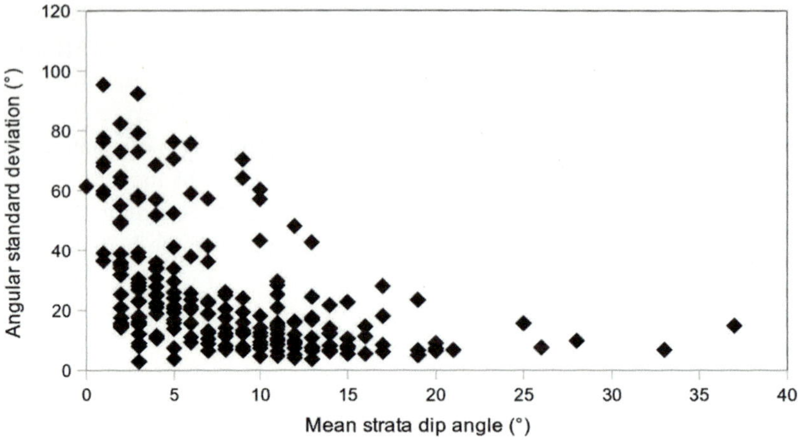

Fig. 6 The graph shows the bedding attitude angular standard deviation against the mean inclination of the attitude. The angular standard deviation of the dip direction is particularly high in correspondence of low values of inclination (dip angle). From Marchesini et al. (2013)

lower than 16° for half of the bedding attitude dataset, and lower than 30° for 75% of the bedding attitude data. Most of the bedding attitudes that have high values of angular standard deviation are characterized by low values of the mean inclination (i.e. the bedding attitudes are near horizontal, see chart in Fig. 6). We observe that this was expected, since, even during the field surveys, the dip direction is not always recognizable for nearly-horizontal beddings. Bedding inclination values are less scattered, i.e. the standard deviation of the values of inclination estimated by the algorithm for each bedding attitude is generally low. About 75% of the values of the dip standard deviation are smaller than 2°, with a maximum value of 8°.

Figure 7 shows a visual comparison of the field measured and modelled (API + GIS algorithm) bedding attitude data, for a portion of the study area. Visual inspection of this image confirms a fair agreement of the bedding attitude data collected by these two different methods (i.e., the field survey and the API + GIS algorithm procedure). Moreover, the bedding attitudes obtained through the API + GIS algorithm procedure are more widespread and consistent with the dominant structures in the study area.

6 Discussion

With reference to Fig. 7, we believe that a perfect match between modelled and measured data could not have been achieved. Indeed, since the field survey data represent local bedding information, they cannot be completely comparable with those obtained from bedding traces which, in turn, represent a more general setting of inclination and dip direction. Essentially, we believe that, from a quantitative point of view, neither the field data nor the bedding traces can be considered as a benchmark or as the ground truth. They describe quantities that, though similar and correlated, may differ significantly among them. To perform well, the method requires sufficient information on the location of the bedding traces. Where bedding traces are not present in a study area, or are not visible in the remote imagery used to identify the traces, the quality of the result will suffer. It is also important that the traces are mapped accurately i.e., that they represent faithfully the geometrical relationships between the bedding and the terrain. The quality of the DEM used to capture the

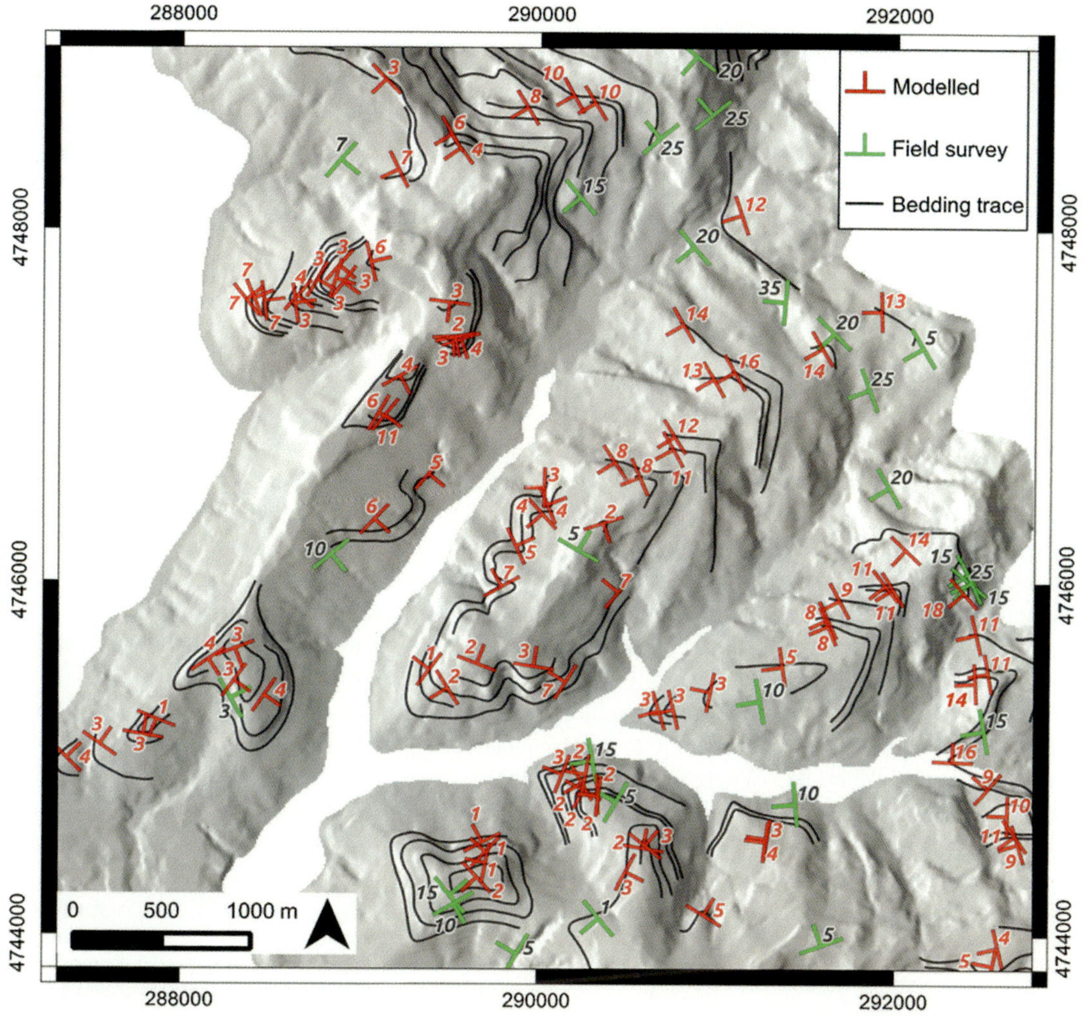

Fig. 7 Comparison of the field measured and modelled bedding attitude data for a portion of the study area. *Green symbols* indicate field measures, *red symbols* indicate modelled data, *black lines* show the bedding traces. From Santangelo et al. (2015)

elevation information along the bedding traces is also important. Not only the elevation values have to be sufficiently accurate, but they should also reflect the morphology of the terrain where the bedding traces are located. Mismatches between the DEM and the bedding traces affect negatively the result. For our experiment, we used a 10 m × 10 m DEM. Use of a very high resolution DEM (e.g., a 1 m × 1 m DEM captured by a Lidar survey) may not have improved the result significantly, and may have introduced unnecessary noise. Conversely, use of a coarse resolution DEM (e.g., the 3 arc-s elevation coverage provided by the Shuttle Radar Topography Mission) may not be sufficient to capture the relationships between bedding and terrain elevation.

The method assumes that the bedding plane is planar, and curvature of single bedding is not admitted. This is a limitation relevant in highly deformed (folded) terrains. However, since the GIS algorithm calculates separately the bedding attitude of each one of the available bedding traces in area, before of the GIS analysis, those traces can be split (manually or automatically) in short segments that reasonably approximate planar bedding planes. Circular statistics (Davis 1990; Butler 1992; Nichols 2009) can be used to identify variations in the interpolated plane that may indicate the presence of discontinuities, local folding, or interpolation errors. Where a sufficient number of bedding traces is available in an area, and the different traces capture geographical variations in the attitude of the bedding, the method is capable of reconstructing folded structures.

In general, the geometrical information obtained by thinly layered rocks is more accurate than the information obtained from single, very thick layers. This is because the terrain elevation of a thick layer (e.g., of sandstone or conglomerate) may vary significantly, depending on where the elevation is measured. Further, the method performs better in areas where bedding dips gently to moderately. In areas where bedding is very steep, or vertical, the bedding traces are linear, regardless of the local topographical settings, and linear traces are difficult to interpolate accurately (Fig. 3). Where bedding is nearly horizontal, the method is capable of determining the attitude (strike and dip) of the representative plane. However, the circular variance or the strike will be large.

7 Conclusions

In landscapes dominated by layered rocks, the geometrical relationships between the attitude of bedding planes and the attitude of the terrain condition the type, abundance, and distribution of landslides. For example, Santangelo et al. (2015) and Marchesini et al. (2015) demonstrated, also using the algorithm proposed in this paper, that bedding attitudes control the abundance and type of landslides for a given study area. We have proposed a method to obtain quantitative information on the 3-dimensional geometry of bedding planes from the analysis of bedding traces, where a bedding trace is the geometrical intersection of individual beddings, or groups of beddings, with the local terrain described by a Digital Elevation Model. The main advantages of the procedure include: (i) a quantitative estimation of the inclination and dip direction of multiple traces mapped through API, (ii) a significant increase in the number of bedding attitude features compared to those obtained during field surveys, (iii) the rapid execution of the procedure, compared to long and expensive field surveys, and (iv) the fact that the obtained bedding attitude features are not influenced by local anomalies. Bedding attitudes obtained exploiting API and the GRASS GIS script are quantitative, distributed geographically, and representative of the dominant geological structures in the study area. Using the proposed method it is possible to obtain a dense spatial distribution data, useful for bedding attitude spatial interpolation (Meentemeyer and Moody 2000; Günther 2003; Günther et al. 2004; Ghosh et al. 2010; Santangelo et al. 2015; Marchesini et al. 2015). The script is available for download at the following web address:

http://geomorphology.irpi.cnr.it/tools/gis-and-interpretation-of-aerial-photographs.

References

Ambrosetti P, Carboni M, Conti M, Esu D, Girotti O, La Monica G, Landini B, Parisi G (1987) Il Pliocene ed il Pleistocene inferiore del bacino del fiume Tevere nell'Umbria meridionale. Geografia Fisica e Dinamica Quaternaria 10:10–33

Antonini G, Ardizzone F, Cardinali M, Galli M, Guzzetti F, Reichenbach P (2002) Surface deposits and landslide inventory map of the area affected by the 1997 Umbria-Marche earthquakes. Boll Soc Geol It 121:843–853

Ardizzone F, Fiorucci F, Santangelo M, Cardinali M, Mondini AC, Rossi M, Reichenbach P, Guzzetti F (2013) Very-high resolution stereoscopic satellite images for landslide mapping. In: Margottini C, Canuti P, Sassa K (eds) Landslide science and practice, vol 1. Springer, Berlin, pp 95–101. Available from: http://link.springer.com/10.1007/978-3-642-31325-7_12

Barchi M, Brozzetti F, Lavecchia G (1991) Analisi strutturale e geometrica dei bacini della media valle del Tevere e della valle umbra. Boll Soc Geol It 110:65–76

Barchi M, Landuzzi A, Minelli G, Pialli G (2001) Outer northern Apeninnes. In: Vai GB, Martini IP (eds) Anatomy of an orogen: the Apennines and adjacent Mediterranean Basins. Kluwer Academic Publishers, Great Britain, pp 215–254

Bodien V, Tipper JC (2013) An image analysis procedure for recognising and measuring bedding in seemingly homogeneous rocks. Sed Geol 284–285:39–44. doi:10.1016/j.sedgeo.2012.11.002

Brardinoni F, Slaymaker O, Hassan MA (2003) Landslide inventory in a rugged forested watershed: a comparison between air-photo and field survey data. Geomorphology 54(3–4):179–196. doi:10.1016/S0169-555X(02)00355-0

Butler RF (1992) Paleomagnetism: Magnetic domains to geologic terranes. Blackwell Scientific Publications, Boston, Oxford. http://www.geo.arizona.edu/Paleomag/book/

Cardinali M, Antonini G, Reichenbach P, Guzzetti F (2001) Photo geological and landslide inventory map for the upper Tiber River basin. Publication CNR GNDCI n. 2116, Scale 1:1,200,000

Carrara A, Cardinali M, Guzzetti F (1992) Uncertainty in assessing landslide hazard and risk. ITC J 2:172–183

Clegg P, Bruciatelli L, Domingos F, Jones RR, De Donatis M, Wilson RW (2006) Digital geological mapping with tablet PC and PDA: a comparison. Comput Geosci 32(10):1682–1698. doi:10.1016/j.cageo.2006.03.007

Conti MA, Girotti O (1977) Il Villafranchiano nel "Lago Tiberino", ramo sud-occidentale: schema stratigrafico e tettonico. Geol Romana 16:67–80

Crozier MJ (1986) Landslides: causes, consequences and environment. Croom Helm Pub, London, p 252

Cruden DM (2003) The shapes of cold, high mountains in sedimentary rocks. Geomorphology 55:249–261

Davis JC (1990) Statistics and data analysis in geology, 2nd edn. John Wiley & Sons, Inc., New York

De Donatis M, Bruciatelli L (2006) MAP IT: the GIS software for field mapping with tablet pc. Comput Geosci 32(5):673–680. doi:10.1016/j.cageo.2005.09.003

Esu F (1977) Behaviour of slopes in structurally complex formations. In: Proceedings IntERl Symposium on the geotechnics of structurally complex formations, Capri, vol 2, pp 292–304

Fernández O (2005) Obtaining a best fitting plane through 3D georeferenced data. J Sruct Geol 27:855–858

Fookes PG, Wilson DD (1966) The geometry of discontinuities and slope failures in Siwalik Clay. Geotechnique 16(4):305–320

Frattini P, Crosta GB, Carrara A, Agliardi F (2008) Assessment of rockfall susceptibility by integrating statistical and physically-based approaches. Geomorphology 94:419–437. doi:10.1016/j.geomorph.2006.10.037

Ghosh S, Günther A, Carranza EJM, van Westen CJ, Jetten VG (2010) Rock slope instability assessment using spatially distributed structural orientation data in Darjeeling Himalaya (India). Earth Surf Process Landforms 35:1773–1792

Goodman RE, Bray JW (1976) Toppling of rock slopes. Proceedings special conference on rock engineering for foundations and slopes. ASCE, Boulder, pp 201–234

Goudie A (2004) Encyclopedia of geomorphology, vol 2. Routledge, p 1156

GRASS Development Team (2016) Geographic resources analysis support system (GRASS) Software, version 7.0. Open Source Geospatial Foundation. http://grass.osgeo.org

Grelle G, Revellino P, Donnarumma A, Guadagno FM (2011) Bedding control on landslides: a methodological approach for computer-aided mapping analysis. Nat Hazard Earth Syst Sci 11:1395–1409

Günther A (2003) SLOPEMAP: programs for automated mapping of geometrical and kinematical properties of hard rock hill slopes. Comput Geosci 29:865–875

Günther A, Carstensen A, Pohl W (2004) Automated sliding susceptibility mapping of rock slopes. Nat Hazards Earth Syst Sci 4:95–102

Guzzetti F, Cardinali M, Reichenbach P (1996) The influence of structural setting and lithology on landslide type and pattern. Environ Eng Geosci 2(4): 531–555

Guzzetti F, Reichenbach P, Ardizzone F, Cardinali M, Galli M (2006) Estimating the quality of landslide susceptibility models. Geomorphology 81:166–184

Guzzetti F, Ardizzone F, Cardinali M, Galli M, Reichenbach P, Rossi M (2008) Distribution of landslides in the Upper Tiber River basin, central Italy. Geomorphology 96:105–122. doi:10.1016/j.geomorph.2007.07.015

Guzzetti F, Ardizzone F, Cardinali M, Rossi M, Valigi D (2009) Landslide volumes and landslide mobilization rates in Umbria, central Italy. Earth Planet Sc Lett 279 (3–4):222–229. doi:10.1016/j.epsl.2009.01.005

Katz O, Aharonov E (2006) Landslides in vibrating sand box: what controls types of slope failure and frequency magnitude relations? Earth Planet Sc Lett 247 (3–4):280–294

Koukis G, Ziourkas C (1991) Slope instability phenomena in Greece: a statistical analysis. Int As Eng Geol Bull 43:47–60

Marchesini I, Santangelo M, Fiorucci F, Cardinali M, Rossi M, Guzzetti F (2013) A GIS method for obtaining geologic bedding attitude. In: Margottini C, Canuti P, Sassa K (eds) Landslide science and practice, vol 1. Springer, Berlin, pp 243–247. (online) Available from: http://link.springer.com/10.1007/978-3-642-31325-7_32

Marchesini I, Santangelo M, Guzzetti F, Cardinali M, Bucci F (2015) Assessing the influence of morpho-structural setting on landslide abundance. Georisk Assess Manag Risk Eng Syst Geohazards 9 (4):2015

Martini L, Sagri M (1993) Tectono-sedimentary characteristics of late Miocene-quaternary extensional basins of the Northern Apennines, Italy. Earth Sci Rev 34:197–233

Meentemeyer RK, Moody A (2000) Automated mapping of conformity between topographic and geological surfaces. Comput Geosci 26:815–829

Nichols G (2009) Sedimentology and stratigraphy. Wiley, Blackwell, Chichester, UK

Rib HT, Liang T (1978) Recognition and identification. In: Schuster RL, Krizek RJ (eds) Landslide analysis and control. Transportation research board special report, 176. National Academy of Sciences, Washington, pp 34–80

Rossi M, Guzzetti F, Reichenbach P, Mondini AC, Peruccacci S (2010) Optimal landslide susceptibility zonation based on multiple forecasts. Geomorphology 114(3):129–142. doi:10.1016/j.geomorph.2009.06.020

Ruff M, Czurda K (2008) Landslide susceptibility analysis with a heuristic approach in the Eastern Alps (Vorarlberg, Austria). Geomorphology 94(3–4):314–324. doi:10.1016/j.geomorph.2006.10.032

Sakov P (2012) nn-c: Natural neighbours interpolation library. Google Project Hosting. http://code.google.com/p/nn-c/. Date of access: 08 Aug 2013

Santangelo M, Gioia D, Cardinali M, Guzzetti F, Schiattarella M (2013) Interplay between mass movement and fluvial network organization: an example from southern Apennines, Italy. Geomorphology 188:54–67. doi:10.1016/j.geomorph.2012.12.008

Santangelo M, Marchesini I, Cardinali M, Fiorucci F, Rossi M, Bucci F, Guzzetti F (2015) A method for the assessment of the influence of bedding on landslide abundance and types. Landslides 12(2):295–309

Thiery Y, Malet J-PP, Sterlacchini S, Puissant A, Maquaire O (2007) Landslide susceptibility assessment by bivariate methods at large scales: application to a complex mountainous environment. Geomorphology 92(1–2):38–59

Varnes DJ (1978) Slope movement, types and processes. In: Schuster RL, Krizek RJ (eds) Landslides analysis and control. Transportation Research Board, National Academy of Sciences, Washington, DC, pp 11–33 (Special report 176)

Weirich F, Blesius L (2007) Comparison of satellite and air photo based landslide susceptibility maps. Geomorphology 87(4):352–364. doi:10.1016/j.geomorph.2006.10.003

Zaruba Q, Mencl V (1969) Landslides and their control. Elsevier-Academia, Prague, p 205

TXT-tool 1.086-1.1 Distribution of Island-like Permafrost in the Lesser Khingan Mountains of Northeast China Using Landsat7 ETM+ Imagery

Chunjiao Wang and Wei Shan

Abstract

In northeast China, at high-latitudes permafrost shows remarkable degradation in response to global changes, evidenced by the northward retreat of the southern boundary, and by shrinkage and fragmentation of the distribution of the permafrost. In these areas, the difficulty of establishing the spatial distribution of permafrost via conventional survey methods inhibits permafrost research, and has practical implications for activities such as highway engineering and construction. Permafrost thaw has serious implications and is associated with increased incidence of landslides and subsidence of many highway roadbeds constructed in the northern area of the Heilongjiang Province, China. To study the regional distribution of island-like permafrost, we selected the Bei'an to Heihe highway, from K153 to K183. We use three Landsat7 ETM+ images obtained in March, May, and September 2009. The land surface temperature (LST) in the research area was retrieved using the radiation transfer equation method. We thereby derived a surface temperature map corresponding to the image acquisition times. The reliability of the surface temperature map was verified by comparison with weather station data. According to the distribution characteristics of permafrost fragments in the study area, LST was selected as the main energy index of permafrost distribution. Permafrost distribution criteria were developed to obtain a permafrost distribution map of the study area in 2009, assisted by empirical interpretation. The estimated permafrost distribution shows a good agreement with field survey data. The findings can contribute to surveying, route selection, and maintenance of highways.

Keywords

Permafrost degradation • Landsat7 ETM+ • Thermal infrared land surface temperature retrieval • Permafrost distribution

W. Shan (✉)
Northeast Forestry University, Harbin, Heilongjiang, China
e-mail: shanwei456@163.com

C. Wang
Institute of Cold Region Science and Engineering, Northeast Forestry University, Harbin, Heilongjiang, China
e-mail: zlj_832008@126.com

K. Sassa et al. (eds.), *Landslide Dynamics: ISDR-ICL Landslide Interactive Teaching Tools*, https://doi.org/10.1007/978-3-319-57774-6_14

Contents

1 Introduction 188
2 Satellite Data 189
3 Calculation of Land Surface Emissivity 191
4 Land Surface Temperature Map 192
5 Permafrost Distribution Map 194
References 195

1 Introduction

Climate warming accelerates permafrost melting in the northeast, high-latitude zone of China. The geographical position of research area is between 126.94°E–127.63°E and 49.32°N–49.76°N, an area about 1600 km^2. Several sections of the Bei'an to Heihe highway in the research area have suffered settlement as a result of the melting of island-like permafrost, which represents a serious threat to roadbed stability and operational safety. The survey conducted in 2009 showed that within 10 m of the left roadbed of Bei'an to Heihe highway in the K177 + 400 to K179 + 200 section there were four landslides with a total volume that exceeded 20,000 m^3 (Fig. 1). In addition, there were numerous landslides within 3 km of this road section (Fig. 2) (Shan et al. 2016).

Russia was the first country to study permafrost distribution, using a series of Siberian soil data to produce the first simple permafrost regionalization map in 1882. According to the on-site permafrost survey data and climate data, Xin and Ren (1956) prepared the first preliminary permafrost distribution map of northeast China and concluded that the permafrost distribution limit was equivalent to the 0 °C line of annual temperature, and −24 °C temperature line in January. The rapid development of geographical information systems (GIS) facilitated the investigation of the distribution of permafrost and to establish a permafrost spatial distribution model. Remote sensing has also become an important tool, which has been gradually applied to the permafrost distribution model. Leverington et al. (1997) used a neural network to predict the presence or absence of the permafrost table within 1.5 m below the ground surface, as classified by land cover derived from Landsat TM imagery, equivalent latitude, aspect, and TM band 6 data. The approach provided a method to study the local distribution of permafrost. Gruber et al. (2001) proposed a GIS spatial model of the distribution of permafrost based on elevation and parameterization of solar radiation in Summer, which was refined by vegetation cover obtained from Landsat TM and albedo on snowless days. This model was used to prepare a global permafrost distribution map. The part of the map for China is shown in Fig. 3. Our study area is within the region of uncertain permafrost coverage (green in Fig. 3). Some of the discussed methods adopt a regional approach that cannot be used in other areas, while other methods operate at large spatial scales that lack local applicability. This calls for a new method to obtain the distribution of permafrost in our study area. According to the energy balance equation, for the same net solar radiation input, island-like permafrost need to use additional calories in order to increase the temperature of the surrounding land, and melt the permafrost, relative to that of a non-permafrost region. Therefore, the temperature in island-like permafrost zone is lower than that of anon-permafrost region. Zhang et al. (2001) investigated the characteristics of permafrost distribution along the Bei'an-Heihe highway, in the Lesser Khingan mountains, following the high-latitude permafrost distribution rules: (i) the land surface temperature (LST)decreased gradually from the slopes to the valleys, and (ii) the (LST) of valley is the lowest, while that of the mountains is the highest. The LST on sunny slopes is higher than on shaded slopes. From the mountaintops to the valley bottoms, sunny slopes to shaded slopes, permafrost distribution is from no permafrost distribution to permafrost distribution area. Generally, there is less permafrost in mountaintops and sunny slopes. LST is selected as the main energy index to describe the distribution of permafrost. Three Landsat7

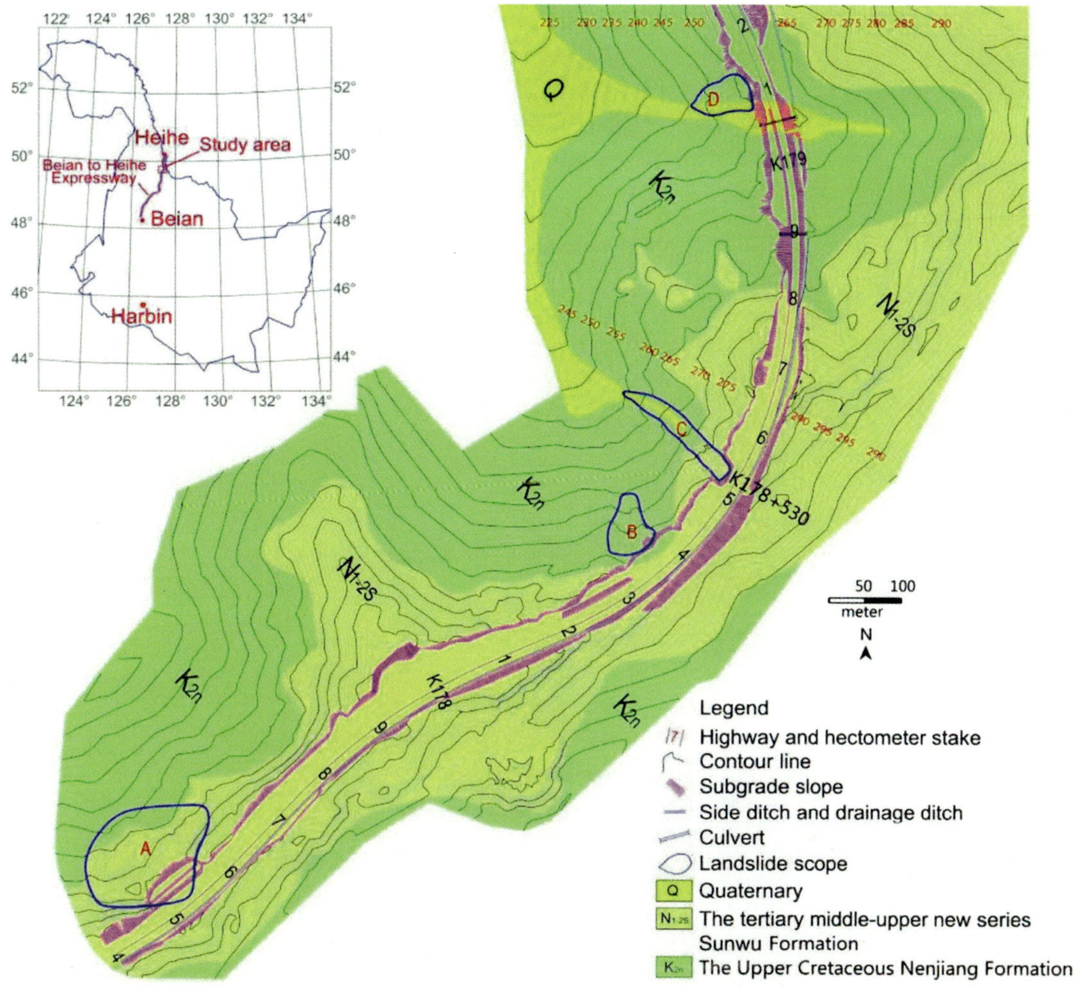

Fig. 1 Geological map of K177+400-K179+200 section of Bei'an-Heihe Highway

ETM + images covering the Bei'an to Heihe highway from 153 to 183 km are used for land surface temperature retrieval via the radiation transfer equation method (RTE). Permafrost division standards from LST map are produced using computational regional average LST as division thresholds, assisted by empirical interpretation. The low-temperature areas on the three LST maps are overlaid and analyzed according to the standards, their intersection section forming the regional permafrost distribution map.

2 Satellite Data

Landsat 7, launched on April 15, 1999, is the seventh satellite of the Landsat program. Landsat 7's primary goal is to refresh the global archive of satellite photos, providing up-to-date and cloud-free images. The Landsat Program is managed and operated by the USGS, and data from Landsat 7 is collected and distributed by the USGS. Landsat 7 ETM+ images consist of eight spectral bands with a thermal infrared band for bands 6, which has a resolution of 60 m. On May

Fig. 2 Geological map of K177+400-K179+200 section of Bei'an-Heihe Highway

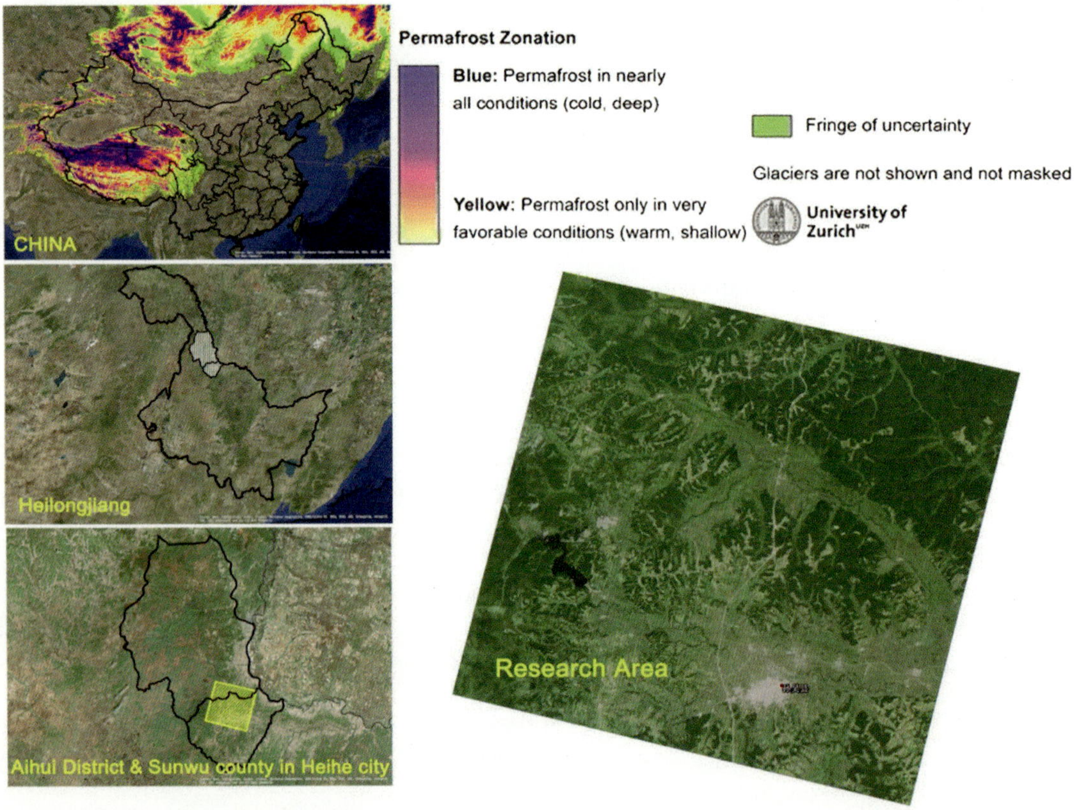

Fig. 3 Location of the study area

31, 2003 the Scan Line Corrector (SLC) in the ETM + instrument failed. The net effect is that approximately 22% of the data in a Landsat 7 scene is missing when acquired without a functional SLC.

Three Landsat7 ETM+ images were obtained on March 3, May 22, and September 11, 2009. The thermal infrared band is used for retrieving the land surface temperature (LST). The missing information is corrected by vertical strip-patch interpolation of single images using the ENVI software. This paper uses the Landsat7 L1 standard topographic correction data. We performed data-strip repair, geometric correction, radiometric calibration, atmospheric correction, cutting research area, and calculation of atmospheric parameters. After correction, the imagery gray value (DN) was transformed to radiant intensity received by the sensor. We executed the atmospheric correction of imagery following radiometric calibration, using the MODTRAN5 radiation transfer model for FLAASH atmospheric correction of the imagery.

3 Calculation of Land Surface Emissivity

Land Surface Emissivity (LSE) is a key parameter to retrieving LST. It is first necessary to calculate the land surface vegetation coverage to calculate the LSE, it can be expressed by the *NDVI* value as calculated from the reflectance of near-infrared and infrared light (Qin et al. 2001b). Higher *NDVI* values indicate greater surface vegetation coverage, whereas a lower *NDVI* values represent bare soil. Because most of the research area consists of natural surfaces, the dimidiate pixel model is adopted which means a combined pixel value for *NDVI* can be expressed by $NDVI_v$ contributed by the green vegetation part, and by $NDVI_s$ contributed by the bare soil part. Therefore, the vegetation coverage can be expressed as:

$$P_v = (NDVI - NDVI_s)/(NDVI_v - NDVI_s) \quad (1)$$

When calculating the vegetation coverage, the values of $NDVI_s$ and $NDVI_v$ cannot be fixed in the same image. (Qin et al. 2001b) proposed that where dense vegetation areas are present, the average *NDVI* of the areas is captured by $NDVI_v$, which is the same as $NDVI_s$. Li and Wu (2004) suggested that in the absence of field data, $NDVI_v$ and $NDVI_s$ should be decided by the maximum and minimum values of confidence degree in the confidence interval. According to the accumulative frequency of *NDVI*, we chose the 5% frequency of *NDVI* value as $NDVI_s$, and the 95% value as $NDVI_v$, due to the lack of testing data. Exploiting the method proposed by (Qin et al. 2001b), we calculate the land surface emissivity value for natural surfaces and buildings, as follows.

$$\varepsilon_{surface} = P_v R_v \varepsilon_v + (1 - Pv) R_s \varepsilon_s + \mathrm{d}\varepsilon \quad (2)$$

$$\varepsilon_{\text{building}} = P_v R_v \varepsilon_v + (1 - Pv) R_m \varepsilon_m + \mathrm{d}\varepsilon \quad (3)$$

Here, R_v is the vegetation temperature ratio, R_m is the building surface temperature ratio, R_s is the bare soil surface temperature ratio, ε surface is the land surface emissivity value of the natural surface, ε building is the building surface emissivity value, and $d\varepsilon$ shows the surface geometric distribution and the scattering effect. For typical ground temperature ratio estimation,

$$R_s = 0.9902 + 0.1068 P_v \quad (4)$$

$$R_v = 0.9332 + 0.0585 P_v \quad (5)$$

$$R_m = 0.9886 + 0.1287 P_v \quad (6)$$

With typical ground objects emissivity value, LSE can be calculated as:

$$\varepsilon_{surface} = 0.9626 + 0.0613 P_v - 0.0461 P_v^2 + \mathrm{d}\varepsilon \quad (7)$$

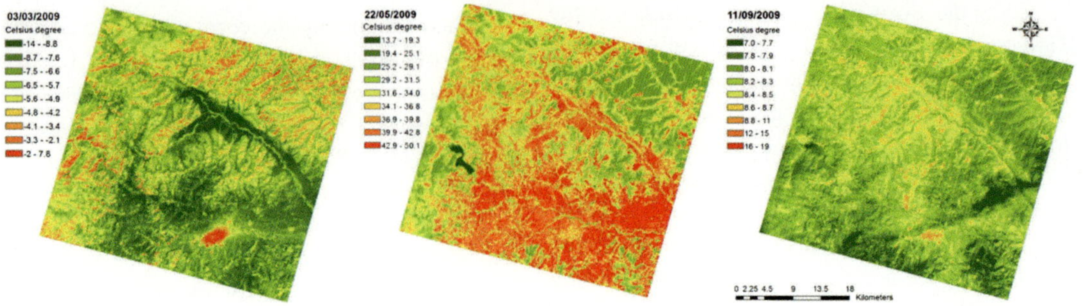

Fig. 4 Land Surface Temperature maps obtained from Landsat7 ETM+ imagery for March 3, May 22, and September 11, 2009

$$\varepsilon_{\text{building}} = 0.9589 + 0.086P_v - 0.0671P_v^2 + \mathrm{d}\varepsilon \tag{8}$$

where the topographic surface is horizontal, dε can be ignored. However, $d\varepsilon$ must be taken into account for heterogeneous and rough surfaces, and cannot be ignored in the present study. Where $P_v \leq 0.5, d\varepsilon = 0.0038P_v$, and where $P_v > 0.5, \mathrm{d}\varepsilon = 0.0038(1 - \mathrm{Pv})$.

4 Land Surface Temperature Map

The atmospheric correction method (Qin et al. 2001a) described as:

$$L_\lambda = [\varepsilon \cdot B(T_s) + (1 - \varepsilon)L\downarrow] \cdot \tau + L\uparrow \tag{9}$$

where, L_λ (W/m^2 sr μm) is thermal radiation intensity, which can be calculated by the DN value of the thermal infrared band (TM6), ε is the land surface emissivity, B(Ts) is the blackbody radiation intensity by Planck function, Ts is LST(K), τ is atmospheric transmittance, and L↑ and L↓ are atmospheric upward radiation and atmospheric downward radiation, respectively. Subsequently, blackbody radiation intensity with temperature Ts in the thermal infrared band can be express as:

$$B(T_s) = [L_\lambda - L\uparrow - \tau \cdot (1 - \varepsilon)L\downarrow]/(\tau\varepsilon) \tag{10}$$

According to the inverse function of the Planck formula, true LST can be calculated as:

$$T_s = K_2/\ln(K_1/B(T_s) + 1) \tag{11}$$

K1 and K2 are calibration parameters. K1 and K2 for Landsat7 ETM+ are severally 666.09 w/m^2 sr μm and 1282.71 K. According to the meteorological data from website (http://www.

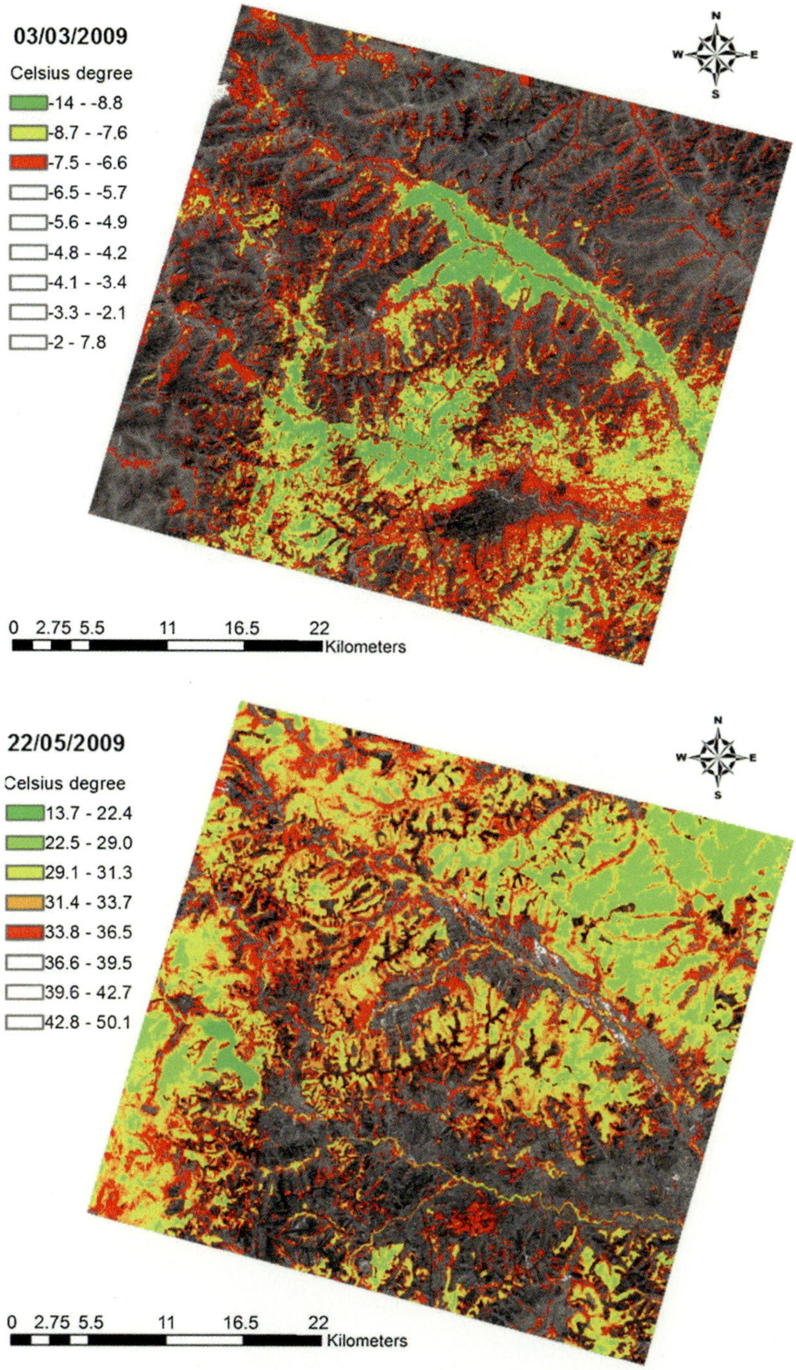

Fig. 5 Low LST results from *top* to *bottom* are 03/03/2009, 22/05/2009, and 11/09/2009

wunderground.com/), Surface meteorological data were obtained for Sunwu County for the satellite transit times. The estimated atmospheric profile parameters were obtained from the NASA website (http://www.atmcorr.gsfc.nasa.gov/). The LST map is generated by radiative transfer equation. Figure 4 shows the retrieval data for March 3, May 22, and September 11, 2009.

Fig. 5 (continued)

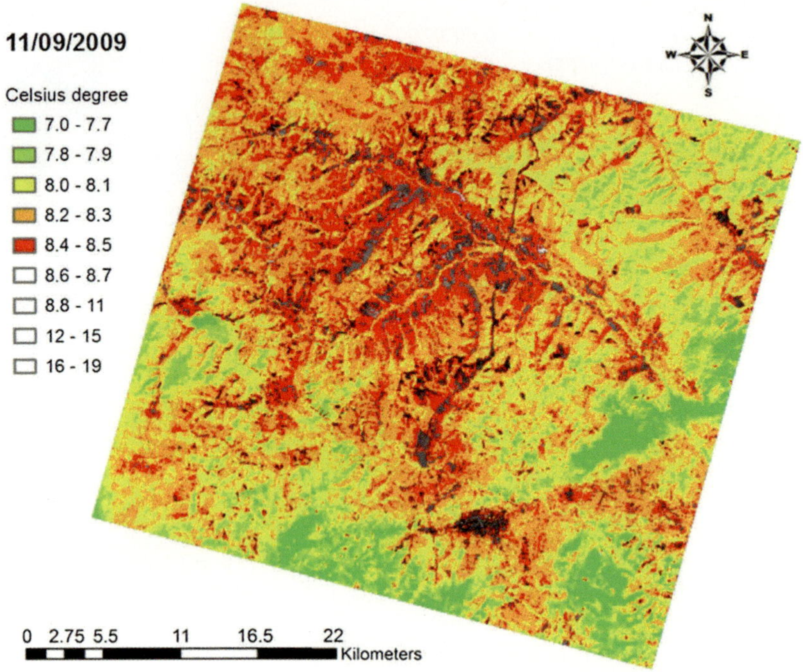

5 Permafrost Distribution Map

The LST maps on 3 March and 11 September, 2009 were classified by 9 grades with natural classification method in ArcGIS, for the LST map on on 22 May, the number of grade was confirmed through empirical interpretation and decided as 8. The low-temperature zone was the area whose temperature lower than average temperature. The low-temperature area coverage is shown in Fig. 5. The distribution of permafrost in the research area (shown in Fig. 6) is obtained from the intersection of the low-temperature zones.

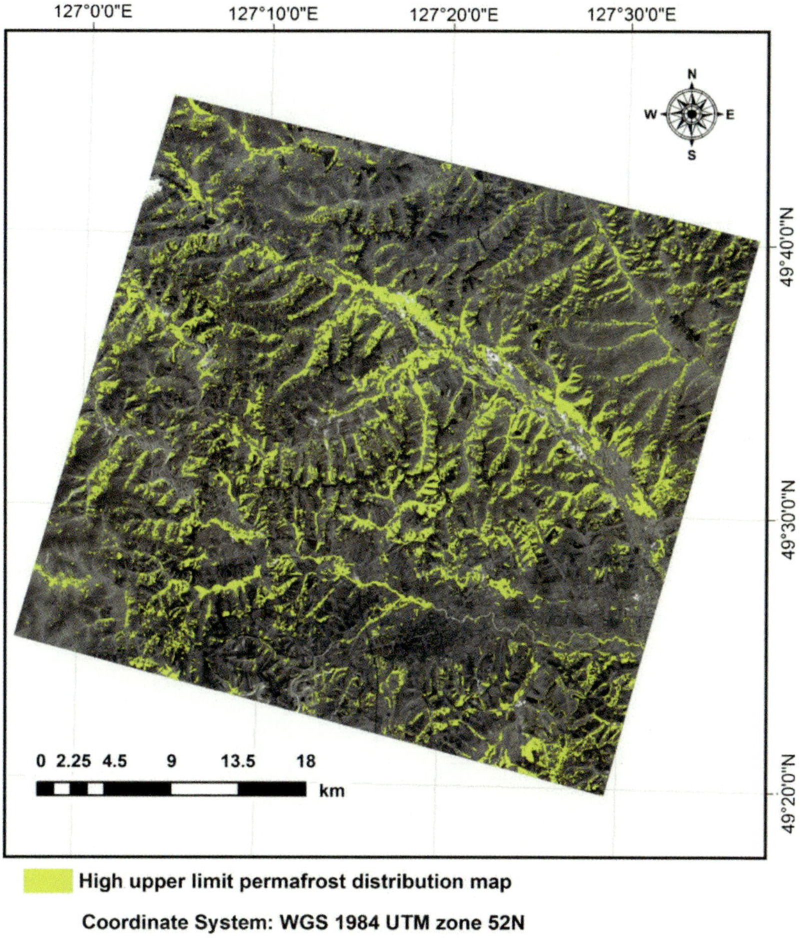

Fig. 6 Island-like permafrost distribution map in study area. Background is Landsat7 image taken on 24 September 2002

References

Gruber S, Hoelzle M (2001) Statistical modeling of mountain permafrost distribution: local calibration and incorporation of remotely sensed data. Permafr Periglac 12:69–77

Leverington DW, Duguay CR (1997) A neural network method to determine the presence or absence of permafrost near Mayo, Yukon Territory, Canada. Permafr Periglac 8:205–215

Li M, Wu B (2004) Vegetation coverage estimate of Miyun reservoir upstream by remote sensing. Recourses Sci 26(4):153–159

National Aeronautics and Space Administration (NASA). Atmospheric Correction Parameter Calculator. http://www.atmcorr.gsfc.nasa.gov/

Qin ZH, Zhang M,Karnieli A, Berliner P (2001a) Single window algorithm for land surface temperature based on land satellite TM6 [J]. J geogr 56(4):456–466

Qin Z, Li W, Xu b, Chen Z, Liu J (2001b) Land surface emissivity estimate of Landsat TM6 band range. Land Resour Remote Sens 3:28–41

Shan W, Guo Y, Hu Z, Wang C, Zhang C (2016) Landslides caused by climate change and groundwater movement in permafrost mountain. http://dx.doi.org/10.5772/63068

The Historical Meteorological Data Query Website, http://www.wunderground.com/. http://en.wikipedia.org/wiki/Weather_Underground_%28weather_service%29

Xin K, Ren Q (1956) Permafrost distribution in Northeast China. Geol Knowl 1(10):15–18

Zhang Y, Qinbai W, Liu J (2001) Heihe region of Lesser Khingan range—permafrost distribution characteristics on Bei'an section. J Glaciol Cryopedology 23 (3):312–317

TXT-tool 1.081-7.1 Investigating Landslides in the Field Using Google Earth and PowerPoint: A Case Study of Altos de Loarque in Honduras

Kiyoharu Hirota

Abstract

This teaching tool shows how to make a landslide map using Google Earth at the investigation site. The mapping technique was based on an empirical knowledge of the site Altos de Loarque in Tegucigalpa, Honduras. In this case, we used an image from Google Earth and a Brunton compass in the field. We used the compass to measure levels in the field, as there is no precise topographic map of the site, due to security concerns, and limited time to investigate.

Keywords

Landslide mapping · Google earth · PowerPoint · Field investigation

Contents

1 **Introduction** ... 197
2 **Landslide Location and Methodology** ... 197
3 **Results and Applications** ... 203
4 **Conclusion** ... 206
References ... 207

1 Introduction

This teaching tool introduces how to make a map for landslides during field surveys using Google Earth satellite imagery, even if precise topographic maps are not available. The following are necessary to prepare before making the map: (1) an internet connection, (2) installation of the Google Earth programme, (3) field research data.

2 Landslide Location and Methodology

The landslide being investigated is located at Altos de Loarque in Tegucigalpa, Honduras. The methods used involve a combination of the field data and Google Earth images, and using the results for landslide mapping. We show the study site in Fig. 1.

This landslide failed by the movement of blocks near the geological boundary between tuff units "Tep" and "Tpm" (Fig. 2). The landslide was triggered by heavy rainfalls in October 2011 (Hirota and Eveline 2013; Hirota and Kamiya 2014).

K. Hirota (✉)
Geosphere Engineering and Disaster Management Office, International Consulting Operations, NIPPON KOEI CO., LTD., 1-14-6 Kudan-Kita Chiyoda-Ku, Tokyo 102-8539, Japan
e-mail: HIROTA-KY@n-koei.jp

K. Sassa et al. (eds.), *Landslide Dynamics: ISDR-ICL Landslide Interactive Teaching Tools*, https://doi.org/10.1007/978-3-319-57774-6_15

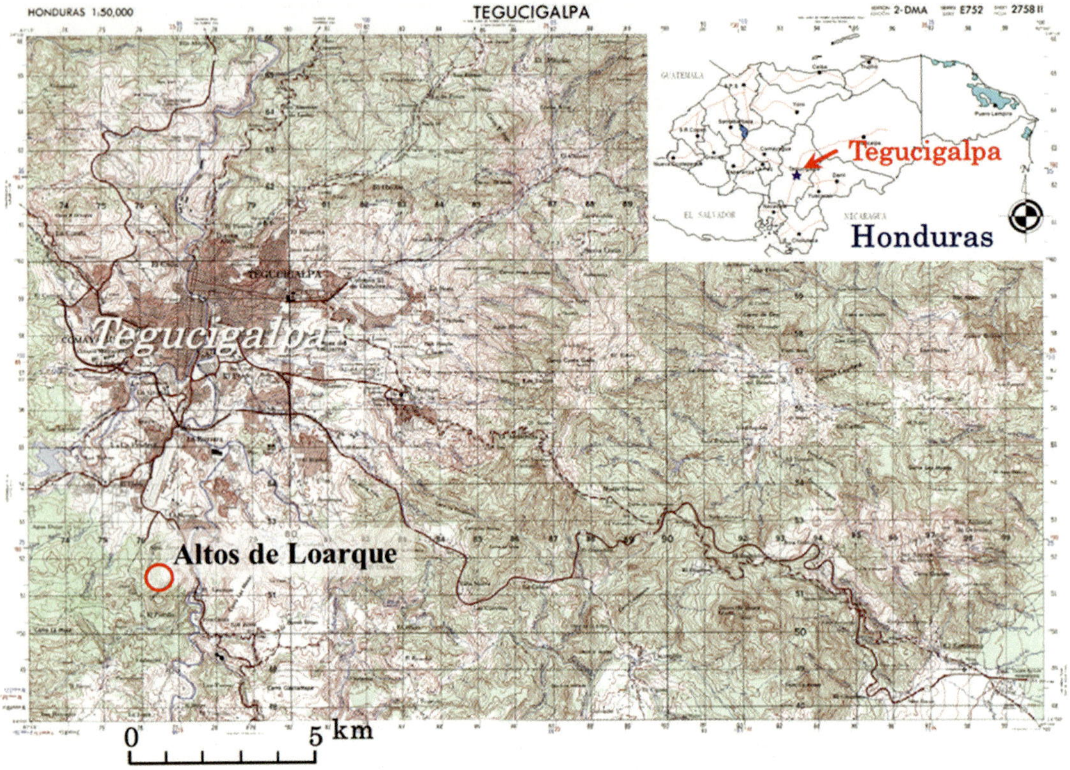

Fig. 1 The topographic map of "Tegucigalpa" (modified from Instituto Geográfico Nacional de Honduras: National Geographic Institure of Honduras, 1/50,000 scale)

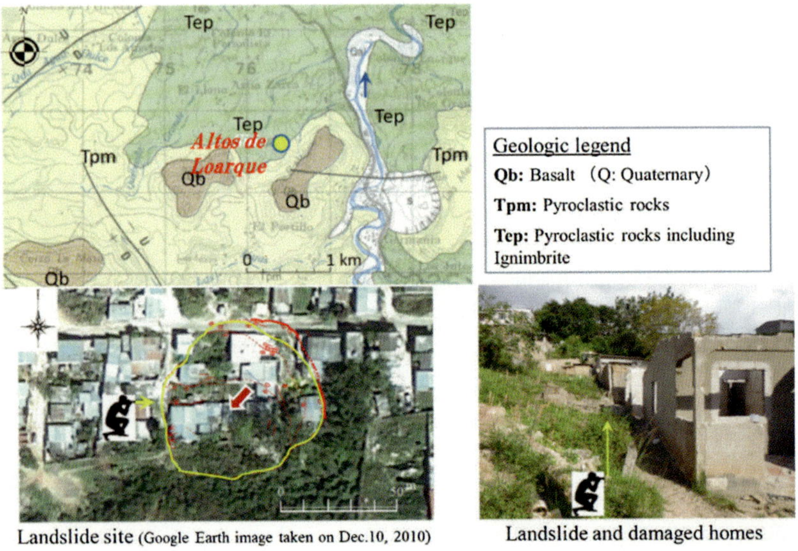

Fig. 2 The study site of the Altos de Loarque landslide

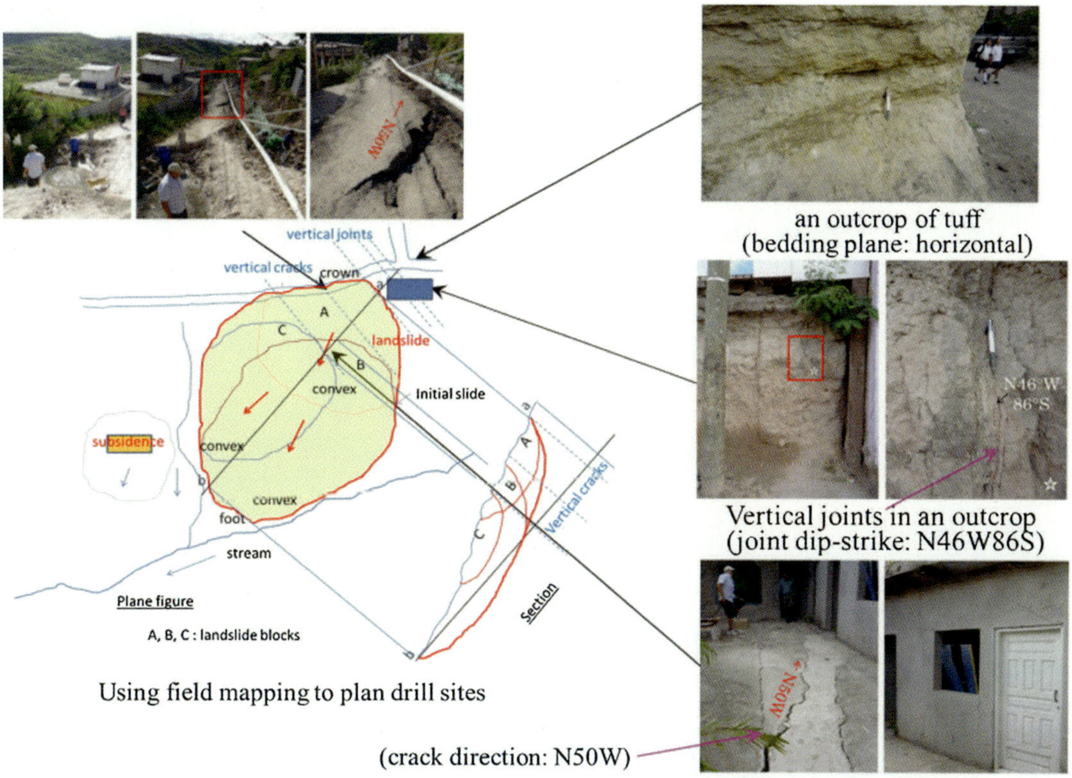

Fig. 3 Field mapping and selected drill sites

Fig. 4 The process of mapping of landslides using Google Earth and PowerPoint

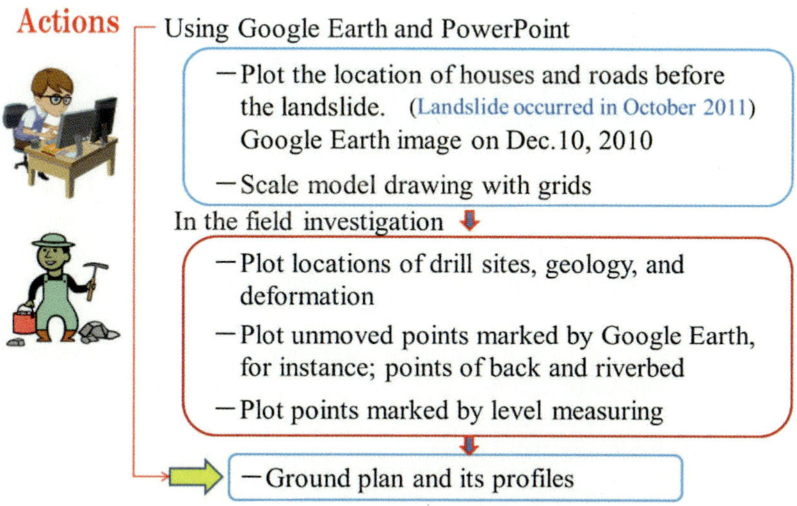

Fig. 5 The study site showing several displacements plotted in Google Earth and photos of the displacements

Fig. 6 The study site; *Left* drill sites, *Right* setting 10 m grid *squares* on a Google Earth image

Figure 3 shows a landslide map without the Google Earth image and the surrounding geology and cracks

The process of how to carry out landslide mapping is shown in Fig. 4.

During the field surveys, I recorded evidence of the landslide on the Google Earth image (Fig. 5).

For the investigation, it is useful to make a square grid. The square grid is 10 m × 10 m. Auxiliary lines of 10 and 50 bars are set to discern lines of distance in the field (Fig. 6).

The location of houses and road locations on Google Earth before the landslide were verified (Fig. 7).

Measuring of levels is done by hand leveling based on the height of my eyes from the soles of my feet, for instance, 1.6 m (using a Brunton compass), see Fig. 8.

To make the map, use PowerPoint and open **Insert** and choose **Shapes**. Then use **Lines** menu to choose **Scribble** (Fig. 9).

We plot the level measuring points on a Google Earth image (Fig. 10). After level

Fig. 7 Tracing the location of houses and roads before the landslide (*Red circles* show drill holes, *red-lined quadrangle* indicates houses before the landslide (Google Earth on Dec. 10, 2010)

Fig. 8 Measuring levels in the field

Fig. 9 Screen of Microsoft PowerPoint (the *red arrow* shows **Scribble**)

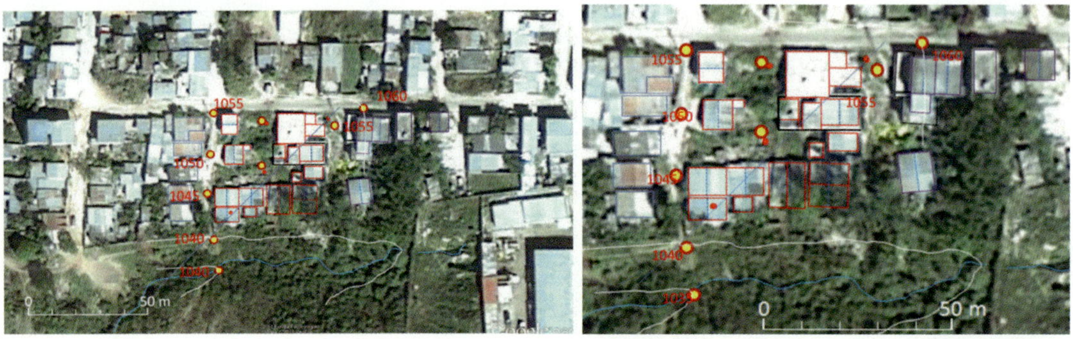

Fig. 10 Measuring: *Left* Level measuring points, *Right* an enlarged image of the *left*. Plot the level measuring points on the Google Earth image (*brown colored circles*: 5 m height, interval points set by level measuring)

Fig. 11 Drawing contour lines

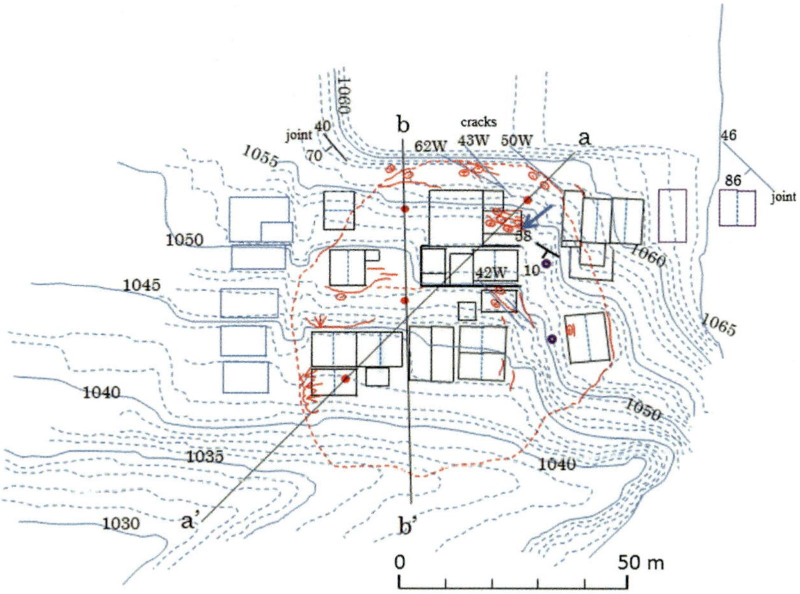

Fig. 12 A ground plan from the field investigation

surveying from the riverbed by 1.6 m proportional allocation, we marked the position of each 5 m, as shown by brown colored circles. Red numbers indicate their altitudes in meters at each point.

After plotting level measuring points, draw contour lines using "Scribble" in PowerPoint (Fig. 11).

First of all, draw the contour line of each 5 m (solid line in Fig. 11). Secondarily, draw the contour line for each 1 m by the interpolation technique (broken line in Fig. 11).

After drawing the contour lines, remove the Google Earth image. The contour lines do not show the foundation level of houses precisely, because these lines are drawn by the interpolation technique (Fig. 12).

3 Results and Applications

Results and applications are as follows;

- Plots of drill site data points can be obtained.

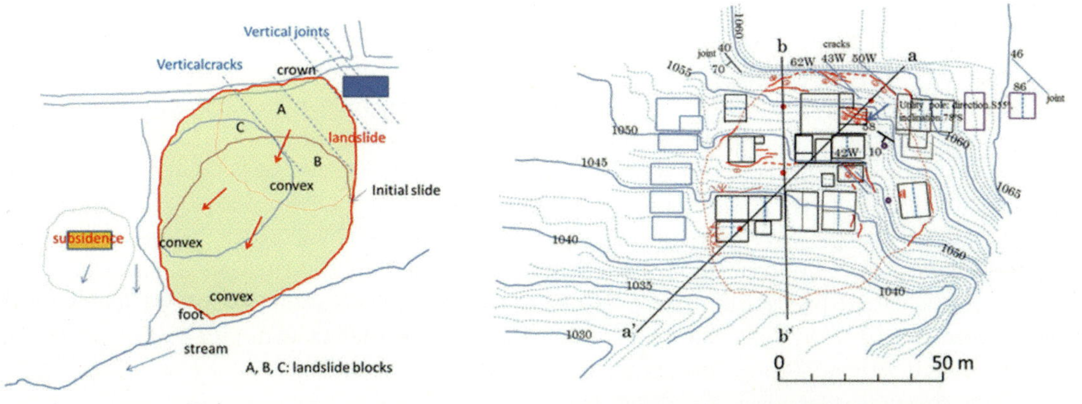

(rough mapping without Google Earth image) (mapping with Google Earth image)

Fig. 13 Comparisons of the mapping

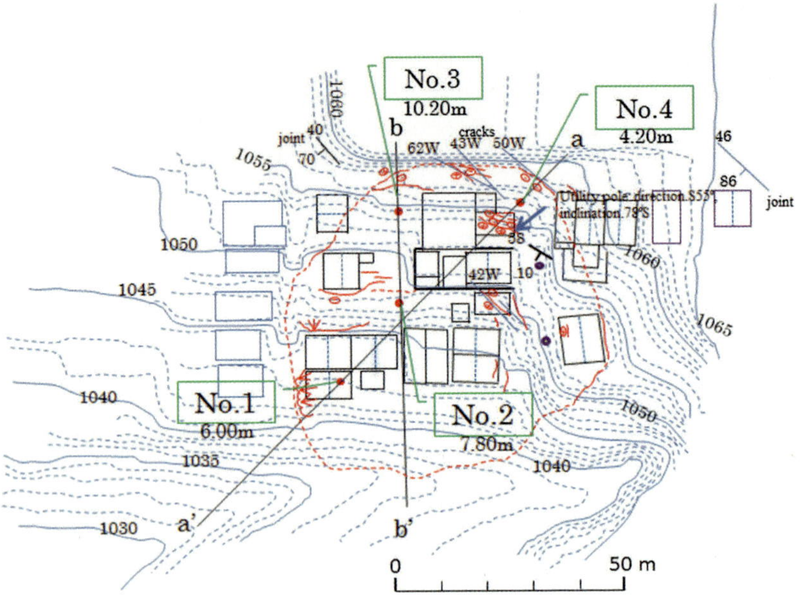

Fig. 14 The drilling sites map

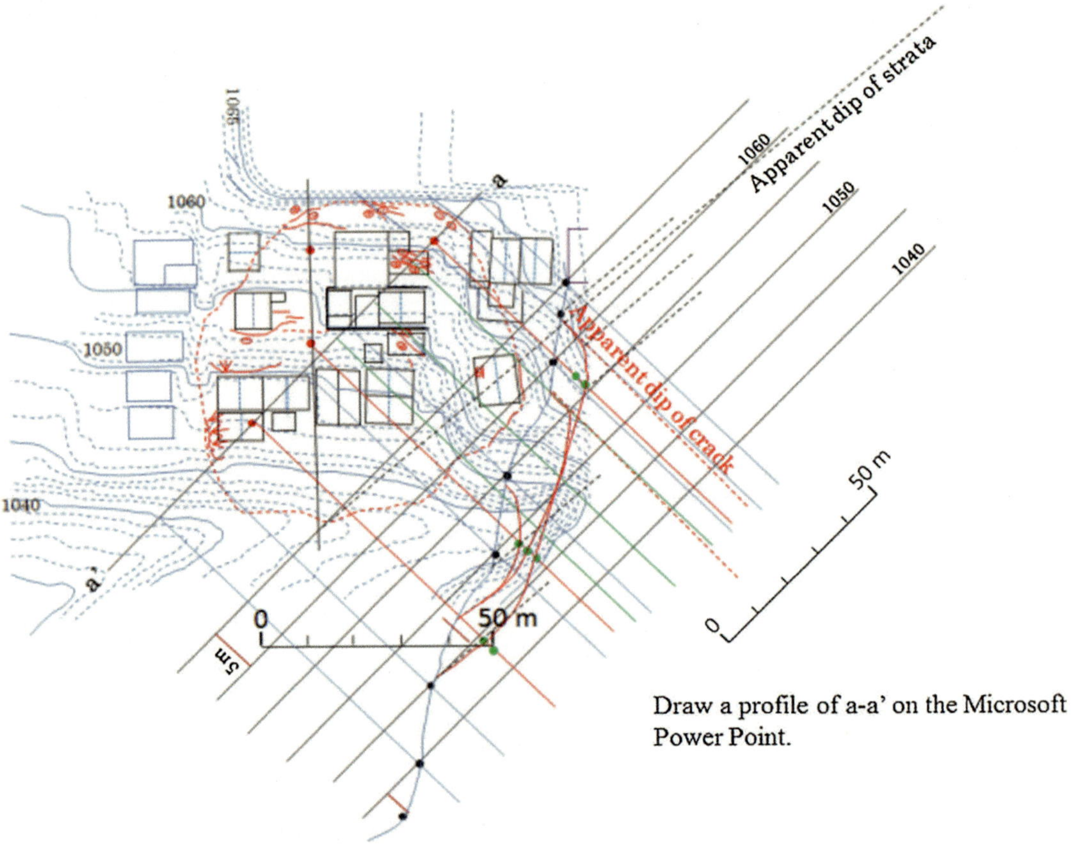

Fig. 15 Profile of a–a′

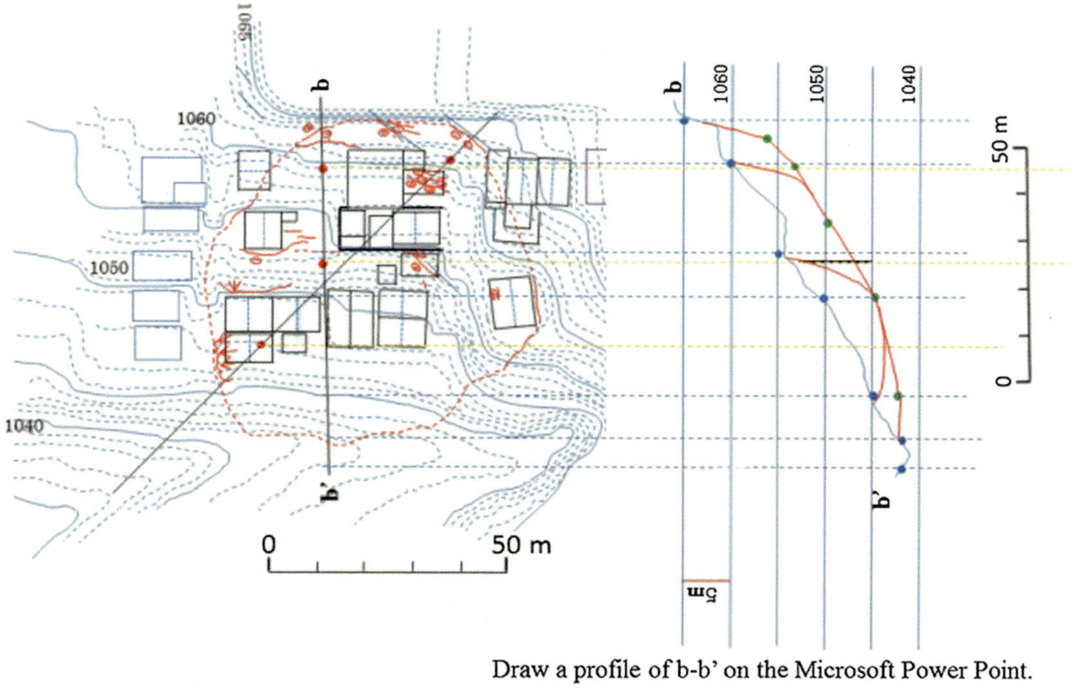

Fig. 16 Profile of b–b′ on the landslide

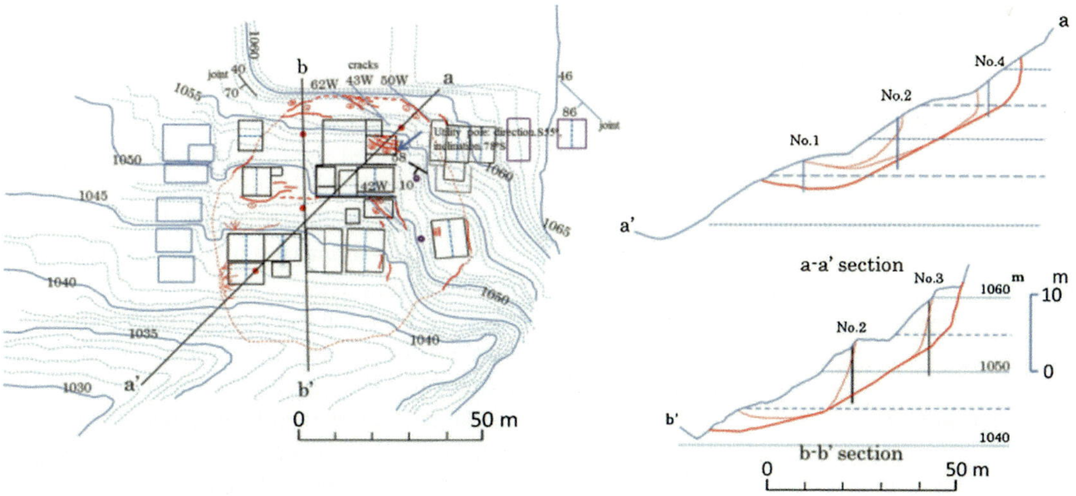

Fig. 17 A ground plan and its profiles

- Comparison can be made between ground plans and mapping by field research and Google Earth image.
- Cross sections of the landslide can be made easily.
- Planning of countermeasures based on mapping is possible.

Figure 13 shows comparisons between field research and using a Google Earth image.

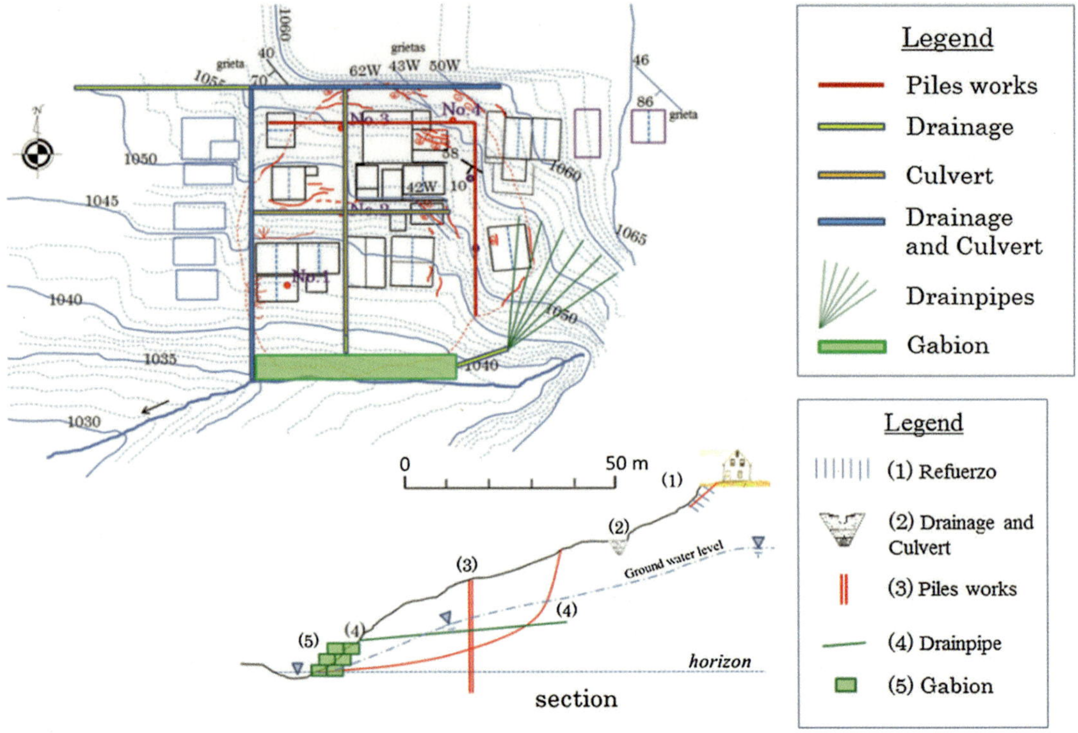

Fig. 18 An example of countermeasure planning

Drilling sites data can then be plotted (Fig. 14).

We can draw a profile of a–a′ of the landslide on the PowerPoint (Fig. 15).

Then, draw a profile of b–b′ of the landslide on the PowerPoint (Fig. 16).

Figure 17 shows a ground plan and its profiles.

Figure 18 shows countermeasure planning.

4 Conclusion

I introduce how to make a landslide maps can be made during field research and using Google Earth in order to better plan countermeasures. Maps obtained using Google Earth are not necessarily complete. These maps are useful when we analyze the underlying causes of landslides and to develop measures to prevent further movements of existing landslides.

This method will work well, because the visibility of the study site without vegetation was very good. I could tell the difference between stable parts and unstable parts at the site to find the landslide form.

If we have the opportunity in the future, we plan to increase the accuracy of our field data using LIDAR and GPS as measurement. It is shown another method for mapping with Google Earth on the web site as "Examples for mapping with Google Earth" in References.

Acknowledgements Many thanks to Mr. Luis Eveline as a director of Universidad Politécnica de Ingeniería (UPI) in Tegucigalpa, Honduras and his students who helped me during the field investigation in 2012.

References

Google Earth: Google Earth image taken on Dec. 10 (2010) 2011 Europa technologies image of 2011 Google Eye, Data SIO. NOAA, U.S. Navy, NGA, GEBCO, US Dept. of State Geographer

Hirota K, Eveline L (2013) A case study of landslide at Altos de Loarque, Tegucigalpa, Honduras. In: First Central American and Caribbean Landslide Congress (CD document). p 5

Hirota K, Kamiya S (2014) Re-evaluation of landslide caused by Hurricane Mitch 11998, Tegucigalpa Honduras. In: Sassa K (ed) Landslide science for a safer geo-environment, vol 3. pp 292–400

Instituto Geográfico Nacional de Honduras (National Geographic Institute of Honduras): an area map as "TEGUCIGALPA" 1/50,000 scale, Honduras (mid-level vertical data of the sea of 1951)

Examples for mapping with Google Earth

Google Earth Pro Map-Making Tool. https://www.youtube.com/watch?v=PjBDWkbwBYc. 6 Sept 2016)

Google Earth Field Mapping. https://www.youtube.com/watch?v=Jw2dNAB13eQ. 6 Sept 2016

Google Earth Reconnaissance: Robert Kayen-US Geological Survey. http://www.geerassociation.org/images/GEER_Activities/GEER%20Workshop%2009/Kayen%20GEER%20GE%20Mapping.pdf. 6 Sept 2016

TXT-tool 1.064-1.1 Field Guide for the Identification and Assessment of Landslide and Erosion Features and Related Hazards Affecting Pipelines

Chris Massey, Graham Hancox and Mike Page

Abstract

This Field Guide was originally prepared by GNS Science in 2007 as part of the Institute of Geological and Nuclear Science (GNS Science) Short Course on the Identification and Assessment of Landslide and Erosion Hazards aimed at pipeline overseers and technicians. The guide was originally designed as a reference for technicians to use in the field during pipeline inspections, but can also be applied to other engineered structures. It contains background information on landslides and other erosion processes, including: some definitions; a short glossary of relevant landslide terms; discussion on observation and recording of geological and geomorphic information in the field; a landslide and erosion classification scheme developed for pipeline technicians; and an example "Relative Hazard Exposure Matrix" that could be used as an initial tool to assess and rank the potential consequences to a pipeline should landslide and other erosion hazards occur.

Keywords

Landslides · Erosion · Pipelines · Hazard assessment

Contents

1 **1 Introduction** 210
1.1 Teaching Tool Structure 210
2 **2 Identifying and Recording Landslide and Erosion Hazards** 210
2.1 Some Definitions 210
2.2 Landslide Activity 211
2.3 Identification of Landslide and Erosion Features 212
2.4 Aerial Inspections 214
2.5 Ground Inspections 215
2.6 What to Record 217
3 **3 Landslide Hazard Models and Classification Schemes** 218
3.1 Landslides 218
3.2 Erosion 218
3.3 Material Types 218
4 **4 Landslides and Their Impacts** 220
4.1 Open Slope Flows 220
4.2 Channelised Flows 222

C. Massey (✉) · G. Hancox · M. Page
Active Landscapes Department, GNS Science, Avalon, Lower Hutt 5040, New Zealand
e-mail: c.massey@gns.cri.nz

G. Hancox
e-mail: g.hancox@gns.cri.nz

M. Page
e-mail: m.page@gns.cri.nz

K. Sassa et al. (eds.), *Landslide Dynamics: ISDR-ICL Landslide Interactive Teaching Tools*, https://doi.org/10.1007/978-3-319-57774-6_16

4.3 Slides 223
4.4 Rock Falls 224
4.5 Other Hazards and Indicators of Potential Hazards 224
4.6 Hazard Assessment 224
4.7 Risk Assessment 224
5 5 **Conclusions** 227
6 **Appendix 1. Example Relative Hazard Exposure Matrix** 228
References 232

1 Introduction

This information contained in this Landslide Teaching Tool has been taken from a field guide prepared by GNS Science as part of their Short Course on the Identification and Assessment of Landslide and Erosion Hazards aimed at pipeline overseers and technicians (Massey et al. 2007). The aim of this Landslide teaching Tool is to present an example of how to standardise the way in which landslide and other erosion hazards are identified, recorded, classified and reported. The original guide and this Landslide Teaching Tool are designed as a reference for pipeline technicians to use in the field.

1.1 Teaching Tool Structure

Initially, this Landslide Teaching Tool defines the landslides and erosion features typically found along pipelines in New Zealand, and next focuses on how to identify and record these features in the field from both aerial and ground reconnaissance (Sect. 2). The other parts of the guide deal with landslide classification schemes, and a simplified scheme for use by pipeline technicians is presented (Sect. 3). The impacts on the pipeline from landside and other erosion hazards are then discussed and an example relative hazard exposure matrix is presented (Sect. 4). Conclusions are then presented (Sect. 5).

2 Identifying and Recording Landslide and Erosion Hazards

2.1 Some Definitions

(a) Landslide:—Landslides are one type/form of erosion and for the purposes of this guide are being dealt with separately. A landslide is a gravitational movement of rock or soil down a slope as a mass along discrete shear surfaces, owing to failure of the material. As defined in the Cruden and Varnes Landslide Classification (Cruden and Varnes 1996), soil includes earth (material smaller than 2 mm) and debris (material larger than 2 mm). Rock is a hard or firm intact mass and in its natural place before movement. Landslides are most often triggered by heavy rainfall or strong earthquakes, but also occur 'spontaneously' without an obvious triggering event. Such failures are often caused by undercutting slopes by natural erosion, or slope modification by people, together with long-term weathering and weakening of slopes. Strong earthquake shaking of Modified Mercalli (MM) intensity MM7 can cause small failures ($<10^3$ m^3), but MM8 or greater is generally required for larger landslides ($\geq 10^3$–10^6 m^3) (Hancox et al. 2002, Dowrick et al. 2008).

Landslides are usually classified, or described in terms of: (a) the *type of material* involved (*rock, earth, debris*, or sometimes *sand, mud* etc.), and (b) the *type of movement* —*fall, topple, slide, flow, spread*, which are distinct modes of movement. Combining these two terms gives a range of landslide types such as: *rock fall, rock slide, rock topple, debris fall, debris slide, debris flow, earth flow* etc. Landslides involving soils and bedrock are often called *slips or landslips*, while *small failures with rotational slide surfaces are generally referred to as slumps*. Small landslides often do little damage, but very large failures of thousands or millions of cubic metres moving downslope can runout and bury buildings and roads, or cause foundation collapse on or at the tops of slopes. Effects of landslides can

Table 1 Definitions and characteristics of the main landslide types

Landslide type (*based on movement*)	General characteristics (*after Cruden and Varnes* 1996)
Falls	**Falls** are masses of rock, soil, or debris that move rapidly down very steep slopes (>40°) by free fall, bounding or rolling. Disrupted soil and debris falls most common
Slides	**Slides** are masses of rock, soil, or debris that slide down planes of weakness (bedding, joints, faults) and other surfaces. Rotational slides (or slumps) in soft rocks and soils move on curved failure surfaces. Disrupted soil and debris slides are most common. Landslides are also referred to (non-specifically) as slips, landslips, or slippages
Avalanches	**Rock and debris avalanches** are very rapid, long run-out failures on steep slopes (>35–40°) more than 150–200 m high. They may start as falls or slides, and transform into flows (wet or dry). Occur mainly on hill country and high mountain slopes
Debris floods and debris flows	**Debris floods** are rapid hyper-concentrated flows in streams, of water charged with sediment, often coarse gravel and sand. **Debris flows** are a type of landslide: they have much higher sediment concentrations (like wet concrete) than debris floods, and are potentially much more hazardous and destructive. Objects impacted by debris floods are surrounded or buried by gravel, but are often largely undamaged

range from minor deformation of foundations and structural failures to total destruction of sites and all buildings, lifelines and infrastructure above or below slopes. General definitions of the main landslide types are shown in Table 1 and illustrated in Fig. 1, and briefly discussed below, these landslide and erosion types will be discussed in detail in Sect. 3.

Debris flows and debris floods: These are both hydrological mass-transport phenomena but have different hazard and risk implications. *Debris floods* are rapid hyper-concentrated flows in stream channels of water charged with sediment. *Debris flows* are a type of landslide: they have higher sediment concentrations than debris floods, with a consistency rather like wet concrete. *Debris flows* have the ability to transport large boulders, and are therefore potentially much more hazardous and destructive. A *debris flood* is not a landslide and is less hazardous, with destructiveness similar to that of water, but less than *debris flows*; both generally occur on *alluvial fans*. Figure 2 shows the different landslide types that occur along pipeline alignments in New Zealand.

(b) *Erosion*:—*Erosion* is a term referring to those processes of *denudation* which wear away the land by the mechanical actions of erosional agents such as *rivers, streams*, *waves* on the shores of the sea and lakes, *glaciers*, and *wind*. The process of erosion thus involves *transportation* of soil and rock debris by the various agents, and must be distinguished from (chemical) weathering, in which no transportation is involved. *Gravity* is regarded as a prime factor in erosional processes, along with *abrasion* and *corrosion* (the mechanical erosion of a rock surface by material (debris) being transported across it by streams, rivers, glaciers, waves, and wind). *Landsliding* (gravitational downslope mass movements) is therefore a significant erosional process.

Typical erosional features and geomorphic landforms include: erosional terraces and banks in stream and river channels, and *water-eroded channels and gullies* formed on slopes and other geomorphic surfaces by rapid runoff during rainstorms. Steep *cliffs* along streams, rivers, glaciers, and the shores of lakes and coastal areas are also erosion features, *formed by progressive erosional under cutting and collapses* of these over-steepened slopes. Typical erosion features are illustrated in Fig. 3.

2.2 Landslide Activity

Landslides are often described by terms relating to their activity, or timing of the landslide movements, which generally reflects their hazard potential. **Active** landslides are those that are currently moving, while those that have moved within the last year, but are not moving at

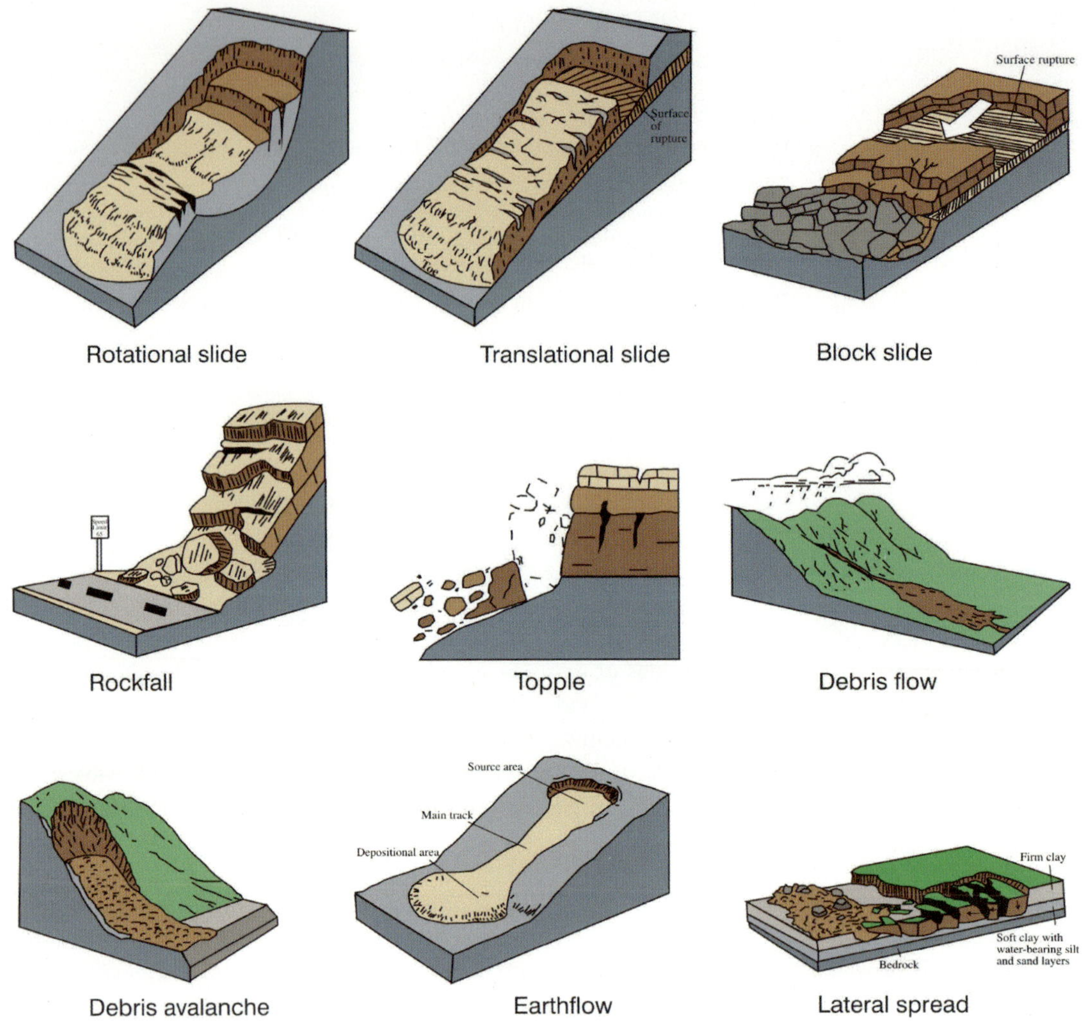

Fig. 1 Main types of landslides (*generalised after Cruden and Varnes* 1996)

present, are said to be suspended. **Inactive** landslides last moved more than one year ago, and are currently not moving. Inactive landslides are subdivided into: **Dormant**—the causes of (and potential for) movement remain. **Stabilised**—remedial measures have stopped landslide movement. **Relict**—landslides that developed under different geomorphological or climatic conditions, and may be several hundreds or thousands of years old (prehistoric, see Fig. 4). A landslide that has been inactive but has started to move again is called **reactivated**.

2.3 Identification of Landslide and Erosion Features

Landslides have geomorphic features that make them identifiable as mass movement landforms. Typical landslide features and their internationally-recognised names are shown in Fig. 5, which are essentially unchanged from Varne's original classification (1978), apart from the addition of some minor features (landslide pond, hummocky ground, springs, and seeps). These features are important as their recognition

Fig. 2 Multiple shallow earth slides and a large rotational slide (sl) and earth flows occurred in the Mangawhero valley (North Island, New Zealand) during the February 2004 rainstorm. The large (>c. 100 million m^3) prehistoric 'relict' landslide (rsl) opposite, with its vegetated, hummocky topography and landslide ponds (p) was apparently unaffected by the storm

Fig. 3 Erosion gully and debris flood gravel deposits (**a**), and erosive effects of flood water on road-edge fill (**b**) at Paekakariki, October 2003 (*from Hancox* 2003)

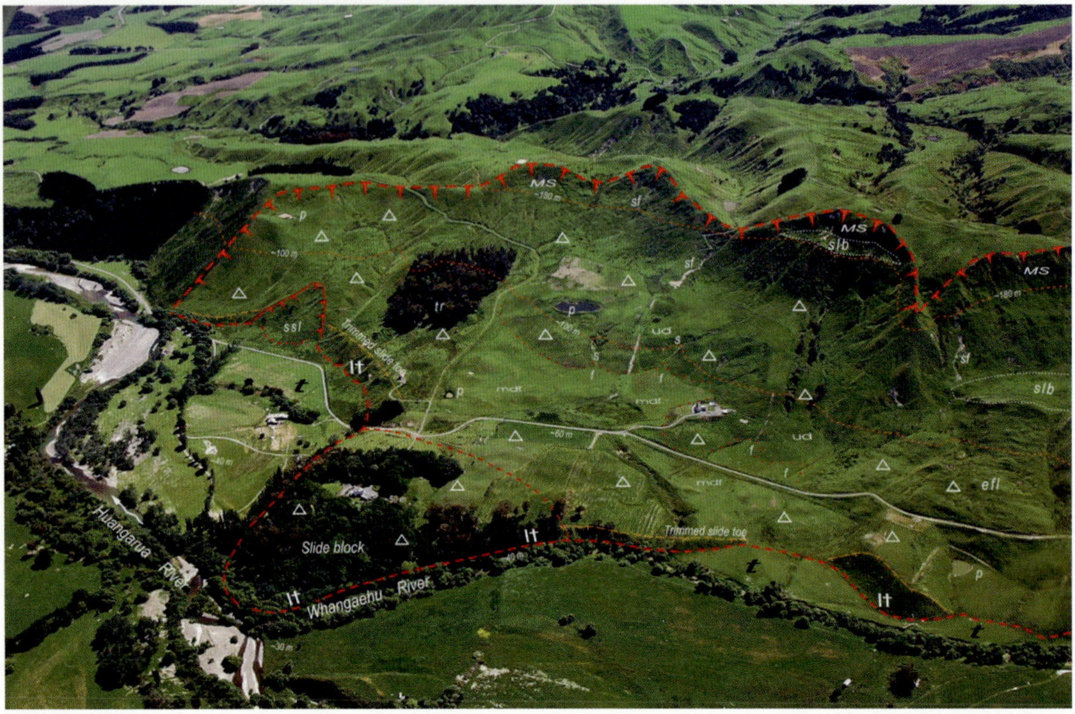

Fig. 4 Very large (~30 million m^3) complex of multiple landslides relict landslide near Martinborough (North Island, New Zealand). The larger relict landslide is thought to be several thousand years old, its main scarp (*MS*) is eroded, and its toe (*lt*) trimmed by the river when the terrace (*t*) was formed >1000 years ago. Field evidence suggests that the larger landslide periodically reactivates. Landslide ponds (*p*) are present on the debris, and more recent active earth flows (*efl*) with hummocky ground, and shallow soil slides (*sl*) have formed on the main scarp of the larger relict landslide in February 2004, other superficial landslides within the larger slide mass, are also apparent

makes it possible to identify landslides, from aerial and ground inspections. Important landslide features, their significance, and some simple ways to recognise both *active and inactive landslides,* and *erosion features* are summarised in Table 2; Fig. 5.

Typical erosion features and geomorphic landforms include: erosional river terraces, river/stream banks and bed, and *water-eroded rills and gullies* formed on slopes and other geomorphic surfaces by rapid runoff during rainstorms. Steep *cliffs* along streams, rivers, glaciers, and the shores of lakes and coastal areas are also erosion features, *formed by progressive erosional under cutting and collapses* of these over-steepened slopes.

Areas of bare ground including flat or slightly concave fields, unsealed roads and tracks, areas of heavy stock concentration, landslide scars and landslide debris trails are all prone to sheet erosion. If sheet erosion is left unchecked rills can develop, and once established, rills can develop into gullies and ultimately landslides and washouts. Different erosion processes and their impacts are discussed in more detail in Sects. 3 and 4. The typical landslide features described above can be used to identify the existence of landslides. The next two sections explain how this should be done on aerial and ground inspections, along with the benefits of both approaches.

2.4 Aerial Inspections

Aerial reconnaissance inspections are used routinely used to inspect long sections of pipelines, using both helicopters and fixed wing aircraft, for landslide and other erosion hazards. The latter

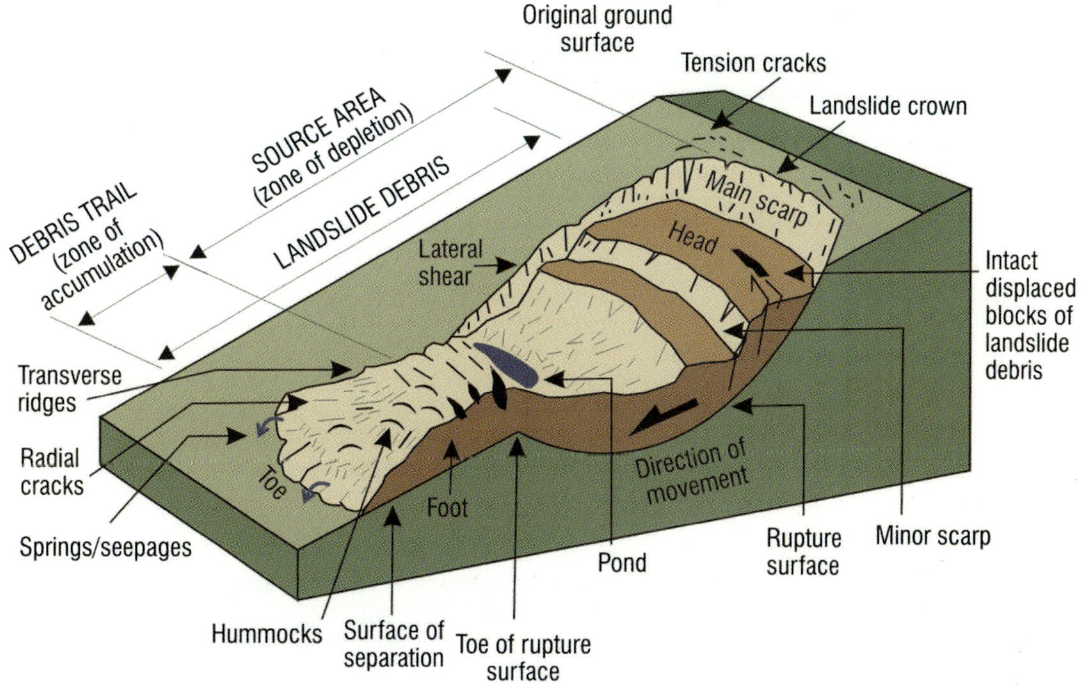

Fig. 5 Block diagram of an idealised complex rotational earth slide and earth flow showing typical landslide features (*after Cruden and Varnes* 1996)

are cheaper and can fly faster, but are not as good for accurate observations and photography. Fixed wing aircraft usually fly at higher altitude, which not only affects the quality of the observations, but also any photos and video shots taken on the flight. Helicopters are more expensive, but are much better for observing landslide and erosion effects as they can fly slowly along and a few hundred metres above pipelines, hovering over critical areas for close observation and photography. As well as routine inspections, aerial inspections should also be carried out following severe rainstorms/flooding events and large magnitude earthquakes (e.g., ~M 5 or >).

Accurate locations of pipeline observations and photographs are critical during aerial inspections. Topographic maps (1:50,000) or pipeline route maps allow good locations to be obtained, as do hand-held GPS units. Some digital cameras allow photographs to be linked directly to a GPS unit, with grid references added to metadata for photographs taken on a flight. Photographs taken using most digital cameras can also be linked to grid references using GPS-photo link software. This works by linking the date and time of the hand held GPS unit to the date and time of the camera (usually by photographing the screen of the GPS with the camera). The software then links the time stamp of the photos to the time stamp in the GPS track log. Any new landslide features that affect the pipeline, or have the potential to affect or damage the pipeline, should be photographed and noted for future ground inspection. On future flights photos should be taken from a similar distance and direction to monitor any changes in the size and activity of the feature. Is also worth taking flights at different times of day and times of the year in order to observe the ground under different lighting (shadow) conditions.

2.5 Ground Inspections

Ground inspections are also routinely used by technicians to inspect parts of a pipeline, especially in areas where erosion or landslide features

Table 2 Landslide features and criteria for field recognition

Landslide features	Description of features
Active landslides (and recently active or dormant landslides)	
Landslide scar	Includes the *source area and debris* trail
Source area	The area at the head of the landslide (zone of depletion) where the landslide mass (debris) is derived from
Landslide debris	Material (rock, soil, vegetation) displaced from the source area and transported down-slope by gravity
Main scarp	The *main scarp* is the steep slope in undisturbed ground at the head of the slide (head scarp)—the visible part of the failure surface. *Minor (secondary) scarps* may be present within the displaced material of the landslide mass
Tension cracks	Often located upslope of the landslide main scarp and tend to be aligned in an arc, and can be continuous or discontinuous, but are essentially linear. These indicate horizontal (pull-apart) movement, but may also show vertical and shear movement
Hummocky ground	Ground surface irregular, often formed of low amplitude hummocks, resulting from differential (compressional and shear) deformations within the displaced material—a feature of many landslides (active and inactive)
Ponds (un-drained)	Ponds formed in depressions, which are often un-drained, are present within the displaced material of many landslides, especially at the slide head; they may be filled by seepage from springs, or by rainfall
Springs, seepages	Give rise to areas of swampy or boggy ground; seepage water may accumulate in ponds
Trees with curved trunks or leaning backwards	Wind, steep topography and ground movement can all give rise to non-vertical tree trunks, so care is required in their interpretation so additional supporting evidence of landslide movement is required
Disruption of natural drainage	May be seen directly or inferred from seepages. Also, where landslide debris may have totally/partially blocked a drainage line, or where the drainage line has been forced to alter its course
Cracking to structures and paved surfaces and dislocation of drainage structures	These can also be related to local settlement of fill and foundations, so additional supporting evidence is required, e.g. presence of a source areas/landslide debris, tension cracks, trees leaning backwards

Notes Relative positions of features referred to in this table are shown in Fig. 5

were observed on aerial inspections. Ground inspections are often carried out during fine weather, or when a storm has passed. There is also, however, value in inspecting pipeline sites during bad weather. Although this may be difficult, and possibly hazardous, because of flooding and track wash-outs, it does allow direct observation of the erosional effects of rapid runoff on slopes, streams, and rivers in the vicinity of the pipeline. If sites of known landslides and erosion features cannot be inspected during a storm, they should be inspected as soon as the weather clears to determine the effects of the storm, and especially to observe and record any changes in the known features in relation to the pipeline. The main activities and data collected during a typical ground inspection may include:

- Precisely locate new landslide or erosion features, or changes in old features, on topographic or pipeline route maps, or by GPS.

- Inspect and take photos of any known features from various angles. Take both close-up and distant photos, and if possible from sites used to take previous photos. Photos should show the slopes above and below the pipeline, as well as the areas immediately adjacent to the pipeline.
- Map the extent of the features using GPS, tape measure, or by pacing out the features. Draw a rough sketch of the features and annotate with brief descriptive notes.
- Note the nature of the materials present in the area (e.g. silt, clay, sand, gravel, cobbles/boulders or rock) and strength (soft, strong, very strong etc.), refer to a later section for details.
- Note areas of water seepage and ponding and whether the materials are dry, moist or wet.
- Note any changes in stream flows in the area near the pipeline, and any changes in erosive effects.

2.6 What to Record

A variety of information should be recorded on any landslide and erosion features that are impacting, or could impact the pipeline. The main points that should be recorded are described briefly below.

(a) *Topographic location*: The positions of all landslide and erosion features in the immediate vicinity (within c. 10–20 m) of the pipeline should be recorded, especially landslides upslope of the pipeline, or gullies and drainage features that cross it. Locations of features should be marked on a sketch map and GPS locations used where possible.

(b) *Site details*: Site details that should be recorded include:
- Position on slope (upper, middle, lower).
- Slope steepness (measure or estimate slope angle).
- Distance from pipeline, streams, rivers, tracks etc.
- Nature of slope above features and pipeline.

(c) *Size of landslide and erosion features*:
- From the air the size of any landslide and erosion features can be estimated in relation to the pipeline dimensions, or the size of known features such as access tracks, roads, farm gates etc.
- On the ground the size of any features should be measured using GPS (also records location), tape measure, or by pacing (a last resort, but better than nothing). Record vertical limits (height difference) across feature (m), and also estimate the slope or gradient (i.e. the horizontal distance (m) for 1 m vertical rise) in two or three locations above and below the pipeline.

(d) *Material type*:
- The nature of the materials present in and around the feature should be recorded and their locations marked on the sketch map.
- The strength and moisture conditions of the materials should also be noted using the details contained in Sect. 3.3.

(e) *Activity*:
- The current activity and rates of movement of landslide features should be estimated (use terms outlined above). Note whether the feature is currently active and is still moving, or shows signs of recent movement, or signs of the potential for possible future movements or slope failures (such as ground cracking above the landslide scar or failure scarp, stream erosion at toe etc.).

(f) *Potential for Future Activity*:
- Consider (your best guess) what might happen to the landslide or erosion feature in the future, based on the visible features (and using the information contained in Appendix 4). Provide photographic or sketch record of the evidence.

(g) *Photographic Record*:
- It is always valuable to record and show the location of features. Use both aerial and ground photos.
- *Aerial photos*: Take close up (high resolution) photos of the entire feature, and also more distant shots showing the pipeline and terrain above and below the feature and pipeline. Record the date and time; repeat photographs from the same

height and angle on future inspections; take several photographs over several weeks or months to monitor feature.

- *Ground photos*: Use the same approach as for aerial photos. Take a series of photographs to show details of the feature at its most active location, and also photographs showing its position relative to the pipeline. Repeat photographs on subsequent inspections.

This information should be recorded using an agreed *pro forma* (procedure or format) in order to standardise the process. A sketch map showing the fey features identified on site should also be produced using an agreed set of map symbols.

3 Landslide Hazard Models and Classification Schemes

3.1 Landslides

When carrying out a landslide assessment for a pipeline or other key network it is usual to establish a series of landslide-hazard models, which best describe the landslides that either have occurred, or could occur along the alignment. Landslide-scenario models are landslides that have been grouped together on the basis of:

- Material type—nature of the displaced material—rock, debris or earth
- Type of movement—how the debris from the landslide is transported e.g. by falling, toppling, rolling and bouncing, sliding, flowing or as a combination, e.g. a slide that develops into a flow etc.

Additional descriptions can also be used regarding: topographical location—where the landslide is located on the slope e.g. open slope (where the landslide and debris remains totally on the open hillside and is not channelised along a stream course), or channelised (where the landslide debris is channelised along a stream course); and velocity of the debris—extremely rapid (typical velocity 5 m/s) to extremely slow (typically <16 mm/year). These different descriptions are combined to then classify the landslide. Many different types of landslide have been defined in the literature, a useful classification scheme is the one proposed by Hungr et al. (2014) in a recent revision of the "Varnes Landslide Classification". Hungr et al. (2014) discusses the many different types of landslides, which although important, are difficult to identify in the field and many are in fact variations on a theme. For the purpose of this tool the landslide hazards contained in Hungr et al. (2014) have been simplified into those landslides, which predominantly occur in New Zealand and more importantly, occur along pipeline alignments in New Zealand. It is recommended that when trying to classify landslides the following main hazard types are used, the details of which are presented in Table 3.

3.2 Erosion

Other main forms of erosion which affect pipelines and associated infrastructure (e.g. access roads to and from the pipeline) are described in the Table 4. The defining aspects of each of these processes are highlighted in bold italics.

3.3 Material Types

In most cases, it is the nature of the material, whether soil, rock or man-made fill, which usually controls the susceptibility of the ground to erosion. Recording material properties in the field is essential to understanding the nature of the problem. Although many variations in materials exist, there are two essential and relatively straightforward properties that should be recorded: material type (Table 5a); and material strength (Table 5b). Material properties should be described following a relevant guideline, e.g. NZGS (2005). In addition to material properties,

Table 3 Landslide hazard types typically encountered in New Zealand along the Maui and Kapuni gas pipeline alignments

Hazard type	Description
FLOWS	*These types of hazard usually originate as shallow slides, however, the failed mass tends to break down, becoming saturated and remoulded to move as a flow*
(1) Open slope—FLOW	These types of hazard are slow to rapid (typically 3 m/min–13 m/year), with the debris remaining wholly on the open hillside and not channelised along a stream course
(2) Channelised—FLOW	These types of hazard are rapid to extremely rapid (typically 5 m/s–3 m/min), with the debris becoming channelised along a stream course. Channelised debris flows generally have much greater mobility than open slope flows, and normally develop when debris from one or more landslides enters a stream course, and becomes mixed with stream water. Deposition of the debris tends to occur once it reaches low angle open slopes, forming debris fans
(3) SLIDES (deep-seated)	Movement of an intact mass by sliding along a basal rupture (failure) surface. These types of hazard are generally slower moving (1.6 m/year–16 mm/year) and have deeper rupture (slip) surfaces than the type of shallow slides that lead to flow-type landslides. The displaced material tends to move along the surface of rupture as a series of discrete intact blocks (these types of landslide are often referred to as translational block-slides)
(4) ROCK FALLS/TOPPLES	This type of hazard results from one or more rock fragments being transported initially by free-falling but may include sliding, rolling and bouncing

Table 4 Erosion hazards typically encountered along gas pipe alignments in New Zealand

Erosion type	*Description*
Sheet erosion	The removal of surface material ***by non-channelised overland flow of water***
Rills	Typically less than 50 cm deep and less than 100 cm wide, near linear features, which occur either on their own or as multiple features. ***They are features that can be smoothed out/removed by cultivation using normal farm equipment.*** Rills typically develop as sheet erosion becomes more established
Gullies	Unlike rills, these are large ***permanent features*** that cannot be removed using normal farm equipment. Like rills, gullies are formed by the channelised flow of water, including headword migration of the channel. Gullies tend to develop once rills become established
Stream bank and stream bed erosion	Stream bank and bed erosion refers to the removal of material from the banks and bed of a stream during periods of high water flows (important at pipeline river crossings)
Wind erosion	Refers to the removal and transportation of particles (soil, sediment etc.) by wind action
Coastal cliff erosion	Coastal cliffs and escarpments (not necessarily located on the coast) can retreat (erode backwards) due to removal of material from the slope toe (for coastal cliffs caused by wave action) and exposed (bare) slope surfaces (through rock falls, rills, gullies etc.)
Tunnel gullies (aka: pipe; shaft erosion, under runners)	Caused by the ***subsurface flow and concentration of water***, resulting in the removal of material by water, forming narrow conduits, tunnels, voids or pipes
Deposition of sediment/debris	Deposition refers to sediment (including vegetation) that has been eroded, transported and deposited by running water. This material may be deposited in channels, on terrace surfaces by overbanking of streams or rivers, or on fans (possibly deposited by landslide processes and not just water)

Table 5 (a) Material type descriptions, (b) Material strength descriptions

	Description
(a) Material type	
Organic (peat)	Contains much organic vegetable matter; often has noticeable smell and changes colour on oxidation
Clay	Grain size <0.002 mm, not visible to eye. Plastic (cohesive), sticks to the fingers and dries slowly; shrinks appreciably on drying, usually showing cracks
Silt	Grain size 0.06–0.002 mm, not visible to eye. Dilatant; slightly granular or silky to the touch; disintegrates in water, lumps dry quickly; can be plastic but can be powdered between fingers
Sand	2–0.06 mm, grains visible to the eye. Contains little or no cohesion; grading can be described, well-grade or poorly-graded
Gravel	60–2 mm. Shape and grading can be described
Cobbles/boulders	>60 mm. Shape and grading can be described
Rock (include type e.g. greywacke etc.)	Describe rock: strength; colour; texture; and name if possible; also include a description on weathering
(b) Material strength	
Soil—sands, gravels and boulders	
Loose	Obvious voids between grains
Dense	No obvious voids between grains
Soil—clays, silts and organics	
Very soft	Exudes between fingers
Soft	Moulded by light finger pressure
Firm	Moulded by strong finger pressure
Stiff	Cannot be moulded by finger pressure
Rock	
Weak	Easily crumbled by hand
Moderately weak	Broken with difficulty in two hands
Strong	Firm blows with point of pick causes only superficial damage
Very strong	Many hammer blows required to break specimen

the moisture content of a material should also be recorded (e.g. dry, moist and wet), as soft wet materials tend to be more prone to erosion than dry materials.

4 Landslides and Their Impacts

This section looks at the possible impact scenarios each of the main landslide hazard types discussed in Sect. 3 could have on the pipeline alignment. Impacts to the pipeline alignment (or access roads) from most types of erosion hazard will depend upon the topographic position of the feature with respect to the at-risk facility (Fig. 6).

4.1 Open Slope Flows

Figure 6 shows the main landslide types associated with open slope flows. If the landslide source area is located upslope of the pipeline it is possible that debris from the landslide may run out and impact the pipeline. Open slope flows do not tend to run out long distances and in most

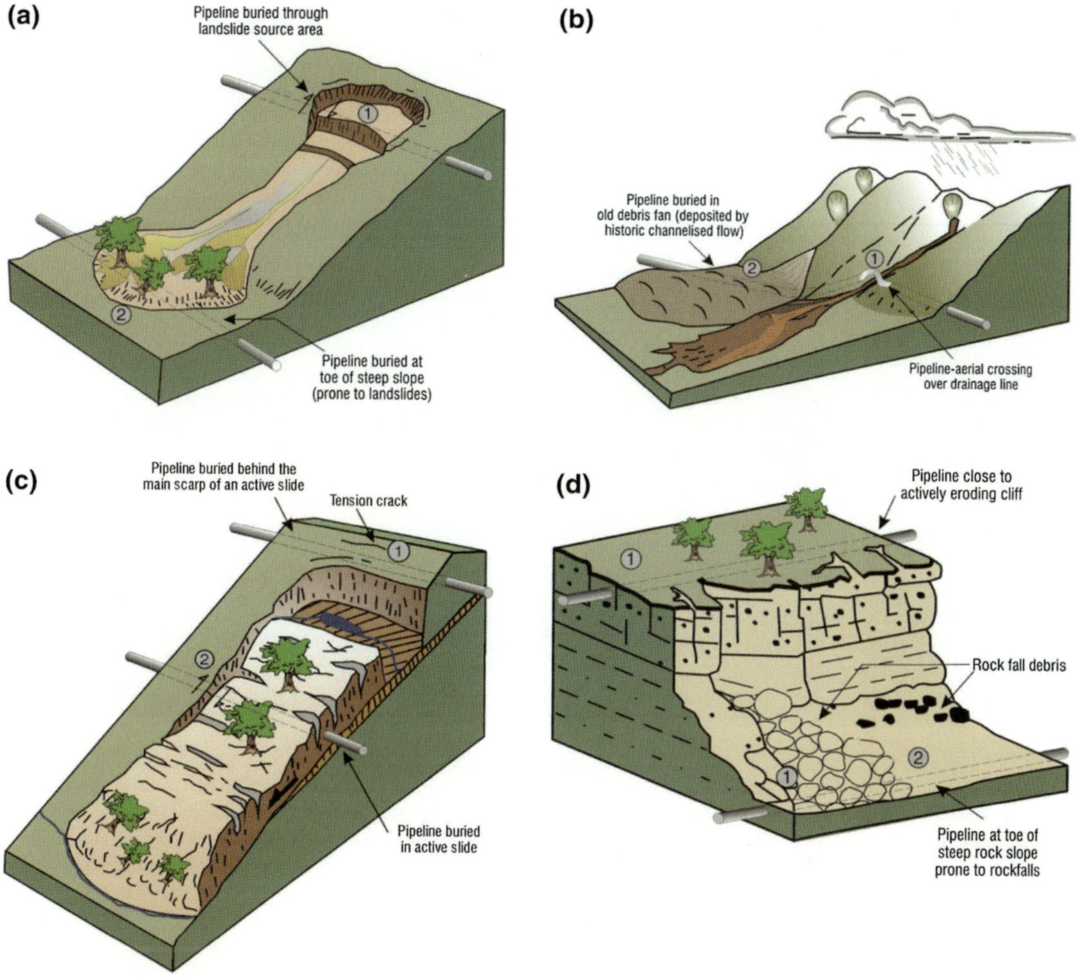

Fig. 6 Generalised summary diagrams showing the main landslide types (**a**–**d**) and their potential impact (*scenarios 1* and *2*) found along pipelines in New Zealand; **a** open slope flows; **b** channelised flows; **c** slides; and **d** rockfalls

cases (if the landslide is relatively small) may have little impact on the pipeline itself. However, debris from the flow may block access along the pipeline. If the landslide is of significant size and debris (several meters in thickness) were to be deposited on top of the pipeline, then this could place additional load on to the pipeline. In this situation the debris should be removed and if deemed necessary pipeline integrity should be checked. If the source area is located down slope of the pipeline it is unlikely that debris from the landslide will have any effect. However, the source area may undercut the pipeline and in the worst case this could lead to the pipeline being exposed. It is also possible that a source area located immediately down slope of the pipeline could retrogress (erode backwards) undercutting the pipeline.

Scenario 1—Slide movement in the landslide source area could cause the pipe to shear at the flanks of the landslide, or undercut the pipeline if located above the main scarp.

Scenario 2—Debris sourcing from open slope flows on the steep slope could deposit material on top of the pipeline, placing additional load on the pipe (Fig. 7).

Fig. 7 Oblique aerial photo of the large (c. 5000 m^3), rotational slide (*sl*) and earthflow in mudstone, which threatened several houses (*H1*, *H2*) in Hunterville during a rainstorm in July 2006. The landslide, which formed on the headscarp of a 'relict' (prehistoric) landslide, is c. 75 m wide and 3–12 m deep at its head (*left*). It transformed into a slow-moving earthflow, which extended c. 265 m down a gentle slope towards the houses. Other features seen here are: tension cracks around the slide head (*tc*); shallow soil (earth) slides formed during an earlier (February 2004) storm; older landslide debris *(osl);* and a bund *(b)* and channel *(c)* to direct further debris and protect the houses

4.2 Channelised Flows

Unlike open slope flows, debris from channelised flows can travel long distances. If the debris remains channelised along the stream course it can cause significant erosion. Some channelised flows may have only small source areas, but can bulk to many times their original size due to erosion and entrainment (inclusion) of additional material from along the stream course. Debris from these types of landslide tends to deposit if: the debris overtops the stream course (becoming non-channelised); or when the channel flattens out. If the latter occurs the debris typically forms distinctive fan-like features. A catchment prone to debris flows usually has a distinctive debris fan located at its mouth. When assessing the potential impact from channelised flows it is important to look further-a-field (than for open slope flows and slides) at entire drainage catchments rather than individual slopes. If the pipeline crosses a potential debris-flow path it is important to assess whether the crossing is located at a channelised or fan section of the stream course. For channelised sections (where the debris is channelised along a stream course) the pipeline is at greatest risk. Channelised flows are high in energy and can easily scour cover materials and expose and damage buried pipelines. At aerial pipeline crossings it is also possible that debris could impact the pipeline if the debris overtops the channel. Figure 6 shows the main impacts associated with channelised flows.

Scenario 1—Stream course prone to channelised flows. Potential exists for the pipeline to be scoured (buried crossing) or impacted (aerial crossing).

Scenario 2—Potential exists for future debris flows to either scour out the pipeline or place additional load on top of the pipeline.

4.3 Slides

Slide-type landslides tend to pose the greatest risk to pipelines as they are often misidentified or assumed to be inactive. In most cases the slip plane usually corresponds to a change in material type or zone of weakness. Many of the slides observed in the hill country around Taihape, Wanganui and New Plymouth (Lower North Island, New Zealand) have slip planes which correspond to the boundary between weathered soil and rock (mudstone/siltstone) and tend to be relatively shallow (slip plane <5 m below ground surface). In most cases open slope and channelised flows originate as slides. Other types of slide, e.g. the Waikorora slide (Fig. 8) have slip planes that are relatively deep (>10 m below ground surface), which correspond to weak zones (clay seams) within the rock. Figure 6 shows the main impacts associated with slides.

Scenario 1—Retrogressive failure (eroding backwards) of the landslide main scarp could undercut the pipeline. Evidence of potential retrogression is shown by the presence of tension cracks (located upslope from the main scarp).

Scenario 2—As the pipeline is within the sliding mass, shear forces can develop on the pipe at the flanks of the landslide, leading to deformation or even rupture of the pipe.

Fig. 8 Oblique aerial photograph of the Waikorora Bluff slide

4.4 Rock Falls

Rockfalls are triggered mainly by earthquakes and less frequently by periods of high rainfall and frost shattering (in areas of extreme cold). Areas particularly at risk from rockfalls are those located at the toe of steep slopes, e.g. coastal cliffs etc. Coastal cliffs are particularly vulnerable to rockfalls as they are constantly being eroded by wave action, leading to retrogression (eroding backwards) of the cliff top. Figure 6 shows the main impacts associated with rockfalls.

Scenario 1—Continued erosion of the steep cliff by either: sea erosion (if forming a sea cliff, e.g. Tongaporutu, North Taranaki); rainfall; and even earthquakes, could lead to undercutting of the pipe as a result of cliff-line regression.

Scenario 2—If the pipeline is located along the toe of an actively eroding cliff (evidence in the form of scree/talus or boulders at the toe of the slope), it could be susceptible to impact loading from rockfalls.

4.5 Other Hazards and Indicators of Potential Hazards

As well as landslide and erosion hazards, other potential hazards along the pipeline alignment that should be assessed are listed in Table 6.

4.6 Hazard Assessment

Figure 9 is a schematic diagram showing how the different hazards discussed in this section interact in the landscape. Typically an assessment of the hazards along the pipeline alignment is carried out prior to route selection and construction. The purpose of this assessment is to identify the locations of the different hazards so that mitigation measures can be considered and evaluated to safeguard the pipeline (e.g., Fig. 10).

When assessing what mitigation measure(s), are applicable to a particular hazard type the Engineering geologist/Geotechnical engineer usually carries out a detailed investigation and hazard assessment, which includes some or all of the following tasks:

- ***Desk study*** (review of all existing information, at both the regional and site-specific scales), including: geology and geomorphology maps, aerial photographs; results from any previous investigations; rainfall records etc.
- ***Field Mapping*** (to field verify the findings from the desk study), including: site-specific geology and geomorphology, material index testing etc.
- ***Ground investigation*** (to quantify the materials and conditions on site, e.g. depth to slip plane, groundwater conditions etc.; and to collect samples for laboratory testing), including: drilling of boreholes, excavation of inspection trenches, in situ testing, geophysics etc.
- ***Laboratory testing*** (to determine the strength characteristics of the materials derived from the ground investigation)
- ***Analyses*** (to analyse the information collated from the desk study, field mapping, ground investigation and laboratory testing stages), this may include: setting up critical cross sections through the slope/landslide; numerical stability analysis; identification of movement triggering factors (in the case of landslides)
- ***Design*** the findings from the hazard assessment are then used to determine the most appropriate mitigation strategy.

The hazards discussed in this section relate to those hazards more frequently encountered along the pipeline. This discussion does not include hazards such as earthquakes, Tsunami, flooding and volcanic eruptions, as well as any health and safety hazards.

4.7 Risk Assessment

The information collected from the hazard assessment can then be used to carry out a risk assessment for the pipeline. Such a risk

Table 6 Other potential hazards affecting the pipeline

No.	Potential hazard	Description
Ground vulnerable to landslides		
1	Steeply sloping ground in colluvium (typically >25° in slope angle or gradient)	Typically slope angles >25 in colluvium (old landslide debris) tend to be unstable and if possible these slopes should be identified and routinely inspected. Things to look for when identifying these features are: presence of relict landslide features; hummocky ground; ponding of water; and tension cracks (see Table 2 for detailed descriptions)
2	Slopes where water can be seen to collect and/or seeping	Slopes usually fail as a result of locally high groundwater levels. Therefore areas of sloping ground which appear to have seepages, springs or ponded water on them, could be either large relict slides, or ground susceptible to future failure
3	Slopes formed in adverse geological structures and materials (e.g. the Tertiary materials of North Island NZ)	Certain material types are more prone to landslides and erosion, due to the properties of the materials, mainly: strength; and composition (clay, silt, sand etc.). For example the Tertiary sediments, (through which a large section of the Maui pipeline traverses), are low strength materials formed of sand and silt, which are prone to wide-scale landslides
4	Slopes adjacent to active fault zones	Geological faults cause materials to become sheared and fragmented, usually leading to the material having a lower strength and therefore more prone to landslides
5	Slopes likely to be prone to river or stream scour at their base	Undercutting of slopes by rivers/stream courses can cause the slopes to become over-steep, and unstable. This process can also lead to reactivation of old relict landslides, as material from the landslide toe is removed
6	Sheet erosion (including areas of rilling/gullying)	Bare surfaces such as: paddocks; unsealed roads and tracks; areas of heavy stock concentration; landslide scars; and landslide debris trails. Once sheet erosion becomes established it can lead to the development of rills and gullies. Therefore once identified it should be addressed immediately
7	Areas devoid of vegetation	Including areas of recently felled forest (also refer to number 10)
8	Stream banks	As number 10
9	Landslide flow-paths	Landslides tend to leave scars on the hillside due to removal of vegetation. As a result, the bare ground is prone to erosion, which can lead to further landsliding
Human interference		
10	Slopes modified by man, e.g. slope over steepening, undercutting of toe and removal of vegetation	Steep man-made cuts are prone to landslides, as slope cutting either leads to the over-steepening or undercutting of slopes. If not done correctly this can lead to landslides. Removal of vegetation can also lead to shallow landslides, as vegetation intercepts rainwater and the roots tend to hold the near surface materials together, increasing their strength. It should be noted that vegetation has little stabilising effect on

(continued)

Table 6 (continued)

No.	Potential hazard	Description
Ground vulnerable to landslides		
		large deep-seated slides, as the slip planes tend to be below the depth of root penetration
11	Excavations along the pipeline alignment (unrelated to slope modifications)	Unauthorised excavations along the pipeline right of way can lead to exposure and damage to the pipeline. In New Zealand, current code requirements state that there must be a minimum of 800 mm of cover
12	Placement of fill along the pipeline alignment	Placement of fill material (or any other material) on top of the pipeline increases the vertical earth load acting on the pipeline. This is primarily a consideration for non-operating conditions of buried steel pipelines (when the pipeline is under no internal pressure). Under most operating conditions, the external earth pressures are insignificant in comparison to the internal pipe pressures. Vertical earth load is an important consideration when designing pipe casings used for rail and road crossings. Placement of additional load onto the pipeline can lead to settlement of the ground and deformation of the pipe. Areas susceptible to settlement tend to be soft (saturated) ground (e.g. river flood plains and old terraces, valley bottoms and areas of landslide debris)
13	Creation of unauthorised access routes across the pipeline alignment	In addition to supporting dead loads imposed by earth cover, buried pipelines can also be exposed to superimposed concentrated or distributed live loads. Large concentrated loads such as those caused by truck-wheel loads can place additional load onto the pipe and if the pipeline has not been designed to take these loads (in the case of unauthorised access routes over the pipeline) problems could occur
14	Changes in land use	Removal of vegetation (as numbers 6 and 8), changes in farming practices and water course deviations caused by man could all impact the pipeline by changing the local conditions making some areas more prone to landslides/erosion

assessment can be: (a) qualitative; (b) quantitative; or (c) semi quantitative, depending on the amount of information available and the nature of the impact should the hazard occur.

The risk assessment framework relating to landslide hazards is described in AGS (2007). An example of a qualitative risk assessment framework, called a "Relative Hazard Exposure Matrix" is presented in Appendix 1. The simple relative hazard exposure matrix (in Appendix 1) was developed by GNS Science to help pipeline overseers and technicians prioritise the landslide and other erosion hazards in terms of future investigations and the possible requirements needed to manage them, e.g. mitigation measures. The relative hazard exposure matrix is broadly based on the risk management framework contained in the original Risk Management Guidelines Companion to AS/NZS 4360:2004, which is now superseded by 31,000:2009. The risk relative assessment matrix contains several hazard classes (A–L) and three consequence classes—Low, Intermediate, and High (Appendix 1). The hazard classes are based on the assessed dominant type of landslide and other erosion hazards described in Tables 3 and 4.

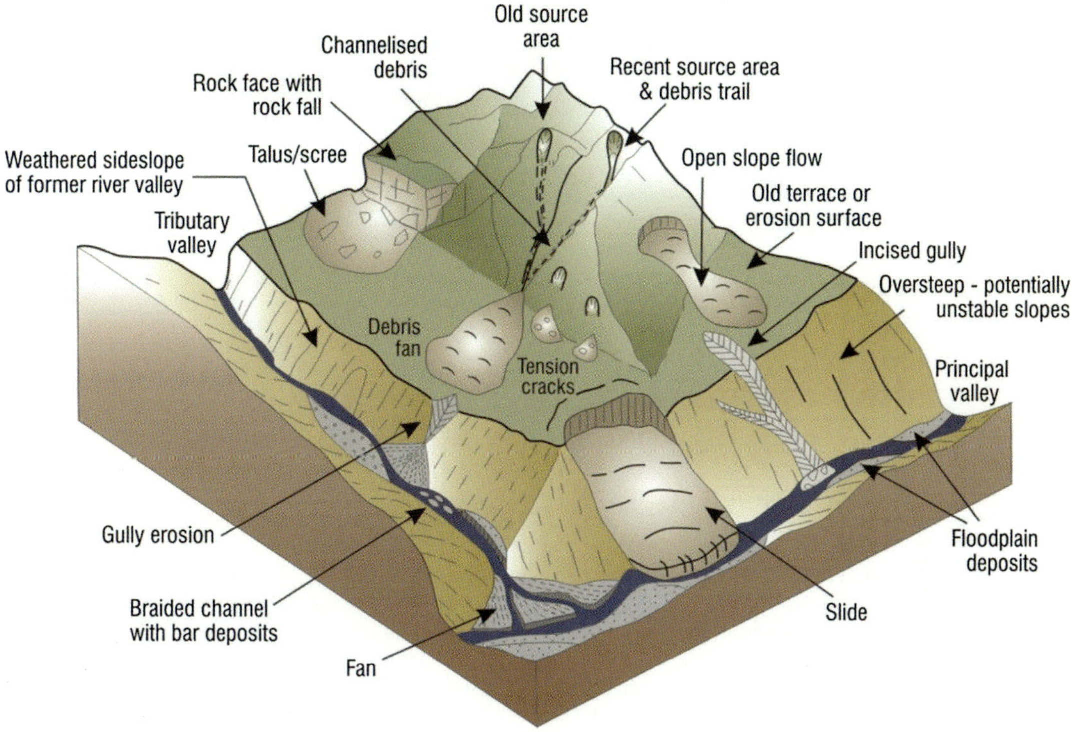

Fig. 9 Schematic diagram showing the different landslide and erosion hazards discussed in this section

A typical framework that can be used to manage the impacts to a pipeline from landslide and other erosion hazards is based on the following assessment steps:

1. Identify the hazard
2. Classify the hazard
3. Identify the cause of the hazard
4. Assess the impacts to the pipeline should the hazard occur
5. Evaluate the impacts with those from other hazards along the pipeline using a systematic and standardised approach
6. If required, design mitigation measures that target the cause of the hazard
7. Monitor for change, and reassess the hazard and impacts.

If necessary, the identified landslide hazards assessed in the Relative Hazard Exposure Matrix can be further evaluated by means of a Quantitative Risk Assessment, as described by the AGS (2007). This would likely involve estimating the frequency and annual probability of various landslide hazards (from geomorphic data and event dating), the size and impacts of the landslides should they occur, including an assessment of the vulnerability of a pipeline at specific sites.

5 Conclusions

Managing the impacts of landslide and other types of erosion hazards on existing pipelines usually begins with aerial reconnaissance surveys, coupled with field observations. Such surveys are typically used to identify, report and monitor such hazards. The aim of this Landslide teaching Tool is to present an example of how to standardise the way in which landslide and other erosion hazards are identified, recorded,

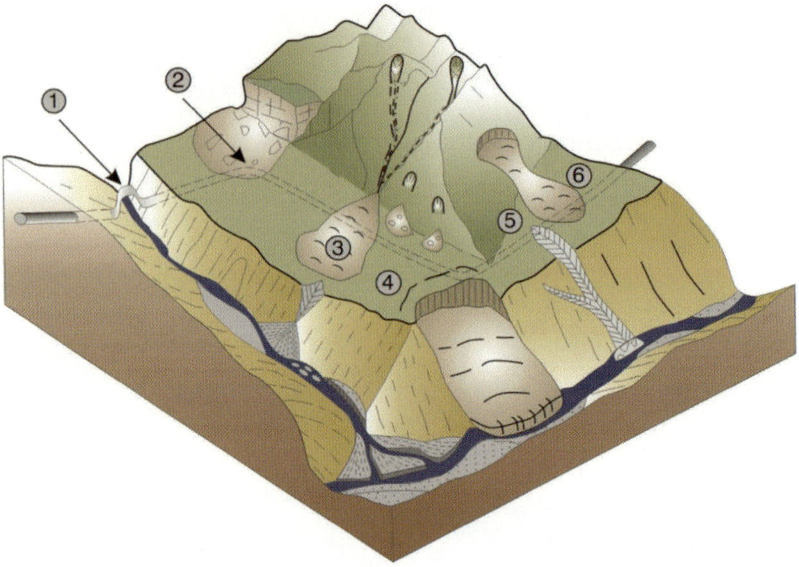

Fig. 10 Schematic diagram showing the potential impact scenarios on a pipeline from the different landslide hazards shown in Fig. 9. *1* Aerial crossing over river—potential for streambank erosion and impact from channelised flows; *2* Pipeline at toe of steep cliff, with rock fall debris on surface—potential for rockfalls; *3* Pipeline traverses debris fan—potential for burial or scour by channelised flows; *4* Pipeline traverses the main scarp of an active landslide with tension cracks visible on surface—potential for the landslide main scarp to retrogress, undercutting the pipeline; *5* Pipeline passes near the head of actively incising erosion gully—potential that the gully could retrogress, exposing the pipe; and *6* Pipeline traverses the toe of a steep slope, with evidence of historic open slope flows—potential for the debris from future open slope flows impacting and burying the pipe

classified and reported, therefore allowing the most appropriate course of action to be determined, whether this be: do nothing, monitor, install prescriptive measures; or refer the problem to an experienced consultant. Appendix 1 contains an example "Relative Hazard Exposure Matrix" that could be used as an initial tool to assess and rank the different landslide and other erosion hazards along a pipeline, and the impacts to the pipeline should these hazards occur.

Acknowledgements The authors would like to acknowledge S. Dellow and M. McSaveney of GNS Science for reviewing previous drafts of this document. We would also like to thank D. Coombe at First Gas Ltd (formerly Vector Gas Ltd) for his input at the early stage of this work.

Appendix 1. Example Relative Hazard Exposure Matrix

This is an example "Relative Hazard Exposure" matrix, established by GNS Science for pipeline engineers working in New Zealand, and reflects the type of landside and other erosion hazards present along a typical gas pipeline in New Zealand. It should be noted that the types of hazard and the impacts on a pipeline should the hazard occur, would change from site to site. This table is one example of a hazard exposure matrix that could be developed once the different landslide and other erosion hazards along a pipeline alignment have been assessed.

Hazard class	Impact scenario class		
	1. High High risk to pipeline, where the pipeline has been exposed at the ground surface, or where the hazard is highly active and could lead to failure of the pipe	2. Intermediate No immediate risk to the pipeline, however, ongoing development of the hazard could impact the pipeline in the future (months)—hazard assessed as active	3. Low No risk to the pipeline at present or in the future (years)—hazard assessed as inactive
Land movement			
(A) Open-slope flows (B) Channelised Flows (C) Slides	Retrogressive failure of existing landslide scarp or new landslide source area has undercut the pipeline. Pipeline may be exposed (even ruptured), or tension cracks and/or displaced blocks (of ground) can be observed across the pipeline alignment OR Landslide debris (from either open-slope or channelised flows), sourcing from up slope (above the pipeline) has impacted the pipeline; either eroding material (exposing the pipe), or depositing significant material (>3 m in height) over the pipeline alignment	Landslide main scarp is <5 m from pipeline alignment, with evidence of ongoing instability (retrogression) in the form of tension cracks, or fresh debris apparent OR Debris from a landslide (either open-slope or channelised flows), sourcing from up slope (above the pipeline) has impacted the pipeline; either eroding material (not exposing the pipe, but < 0.8 m between the pipe invert and ground surface, from probing), or depositing significant material (1–3 m in height) over the pipeline alignment. Or where potential exists for future landslide debris to impact the pipeline	Landslide main scarp is >5 m from pipeline alignment, no evidence of ongoing instability apparent upslope of landslide main scarp OR Debris from a landslide sourcing up slope (above the pipeline) has impacted the pipeline but no evidence of erosion and/or only minor deposition (<1 m) apparent
(D) Slides continued	For slides and where the pipeline alignment passes through the debris associated with an existing slide, reactivation (movement) of the slide block can cause significant damage to the buried pipeline. This movement may not always be visible on the surface, however evidence of fresh landslide scarps, hummocky/rolled ground, tension cracks, deformed vegetation and ponded water in and around the pipeline alignment are indicators that deep ground movements (associated with slides) may have occurred. If possible, the area of movement and movement rates (from knock out pegs, fence post measurements etc.) should be identified. Due to the nature and difficulties in assessing these types of hazard, they should be initially classified as HIGH, pending further investigation		
(E) Rock falls	Large, >10 m^3 volume of rock fall debris (either as one large block or as multiple smaller blocks), has fallen onto the pipeline alignment, with obvious signs of ground deformation (settlement) around the pipeline alignment	Intermediate, between 1 and 10 m^3 volume of rock fall debris (either as one large block or as multiple smaller blocks), has fallen onto the pipeline alignment, with no obvious signs of ground deformation (settlement) around the pipeline apparent. Evidence of future rock fall potential (dilated rock mass and loose boulders visible on slope) apparent	Small, <1 m^3 volume of rock fall debris (either as one large block or as multiple smaller blocks), has fallen onto the pipeline alignment

(continued)

Hazard class	Impact scenario class		
	1. High High risk to pipeline, where the pipeline has been exposed at the ground surface, or where the hazard is highly active and could lead to failure of the pipe	2. Intermediate No immediate risk to the pipeline, however, ongoing development of the hazard could impact the pipeline in the future (months)—hazard assessed as active	3. Low No risk to the pipeline at present or in the future (years)—hazard assessed as inactive
(F) Sheet, Rills and Gullies	Pipeline exposed due to severe rill and gully development in soil (gullies typically >0.5 m deep and >1 m wide), and where >20% of the area is affected	Areas of bare soil, where multiple rills have developed (typically <0.5 m deep and <1 m wide) in and around the pipeline alignment, with <0.8 m between the pipe invert and base of largest rill (from probing). Typically an area where 5–20% of bare soil is exposed	Areas of soil, either partially or totally devoid of vegetation (one or two rills may be apparent on the surface) either along, or immediately adjacent to the pipeline alignment. Typically an area where <5% of bare soil is exposed
(G) Tunnel gullies (sink holes)	Multiple tunnel gullies in and around the pipeline alignment. Diameters of tunnel gully/gullies (pipes) are >0.5 m, and where >20% of the area is affected. Evidence of tunnel gully roof collapse apparent	Multiple tunnel gullies in and around the pipeline alignment. Diameters of tunnel gullies (pipes) are >0.2 m but <0.5 m, and where 5–20% of the area is affected. No evidence of tunnel gully roof collapse apparent	One or two tunnel gullies identified in an area either along or immediately adjacent to the pipeline alignment. Diameters of tunnel gully/gullies (pipes) are <0.2 m, and where <5% of the area is affected
(H) Wind (removal or deposition of material by wind)	Pipeline exposed at surface. Typically where >20% or area affected by wind erosion	Area of active erosion with typically between 5 and 20% of area affected, and where there is <0.8 m of cover (from probing) between pipe invert and ground surface	Area of erosion with typically <5% of area affected, and where there is >0.8 m of cover (from probing) between pipe invert and ground surface
(I) Coastal erosion (cliff line retreat)	Coastal sea cliff actively retreating (evidence of recent landslides, rock falls from the cliff and related debris at toe of cliff) and where the crest of the sea cliff is < 0 m from the pipeline. Or, where the pipeline traverses across a zone/area of weakness (near an actively retreating crest of a sea cliff) and signs of subsidence (sink holes—disc like depressions) are apparent on the surface. Sea cliff erosion (in particular mapping extents of caves/sink holes) is difficult to assess without detailed investigations and therefore should be initially classified as HIGH, pending further investigation		Coastal sea cliff retreating but at relatively low rates (no evidence of recent landslides, rock falls from the cliff and related debris at toe of cliff) and where the crest of the sea cliff is >10 m from the pipeline, with no signs of subsidence apparent on the surface
River erosion			
(J) Bank erosion	Pipeline is exposed or < 1 m from an actively eroding stream bank (lateral bank erosion), in very soft to firm, silt, clay and organics; and/or loose sand and gravels (refer to field guide for material and strength definitions)	Where the pipeline alignment is between 1 and <10 m (horizontal distance) from an actively eroding stream bank, in very soft to firm, silt, clay and organics; and/or loose sand and gravels (refer to field guide for material and strength definitions). Or, where the horizontal distance	Where the pipeline alignment is: >10 m (horizontal distance) from a stream bank, in very soft to firm, silt, clay and organics; and/or loose sand and gravels (refer to field guide for material and strength definitions); or where the horizontal

(continued)

Hazard class	Impact scenario class		
	1. High High risk to pipeline, where the pipeline has been exposed at the ground surface, or where the hazard is highly active and could lead to failure of the pipe	2. Intermediate No immediate risk to the pipeline, however, ongoing development of the hazard could impact the pipeline in the future (months)—hazard assessed as active	3. Low No risk to the pipeline at present or in the future (years)—hazard assessed as inactive
		(from pipeline to stream bank) is <10 m and the height of the bank is >2 m	distance (from pipeline to stream bank) is <10 m and the height of the bank is <2 m
(K) Bed erosion	Erosion (along the stream/river course) has exposed the pipeline in the river bed	Stream/river course has either changed location (typically following storm events), with erosion now covering either a larger area, or different section of the pipeline crossing. Or where there is a constriction to the flow (new area of deposition, e.g. boulders, vegetation etc.) causing concentrated erosion in the area of the pipeline crossing. Due to the unpredictable nature of these hazards any changes to the stream course, which are likely to cause erosion to the pipeline (based on your assessment) should be classified as INTERMEDIATE	
Human activity			
(L) Placement of Fill (Settlement)	Where fill material >3 m in height has been placed over or adjacent to the pipeline alignment. Or where fill material has been stockpiled and potential exists for the material to form a landslide and impact the pipeline (refer to Hazard types A–C)	Where fill material >1 m in height has been placed over or adjacent to the pipeline alignment, and where soil types are soft to firm, silt, clay and organics; and/or loose sand and gravels (refer to field guide for material and strength definitions), and where there is ponded water apparent, or the area is known to water log (e.g. on river terraces). Or where there are signs of surface instability	Where fill material <3 m in height has been placed over or adjacent to the pipeline alignment, and where soil types are: stiff, silt and clays; dense, sands and gravels; and/or strong to very strong rock (refer to field guide for material and strength definitions). And where the area is not prone to water logging and where there are no signs of surface instability
(M) Slope modification	Where a steep cut slope (>3 m in height) has been created either immediately below or above the pipeline alignment in: soft to firm, silt, clay and organics; loose sand and gravels; or weak to moderately weak rock (refer to field guide for material and strength definitions), with evidence of severe erosion (gullies etc.), or landslide features (tension cracks, landslide scarps, fresh debris etc.) apparent on or around the slope	Where a steep cut slope (>3 m in height) has been created either immediately below or above the pipeline alignment, in: soft to firm, silt, clay and organics; loose sand and gravels; or weak to moderately weak rock (refer to field guide for material and strength definitions), with evidence of minor erosion (sheet and rills), but with no signs of surface water or instability on the surface	Where a steep cut slope (>3 m in height) has been created either immediately below or above the pipeline alignment, and where soil types are: stiff, silt and clays; dense, sands and gravels; and/or strong to very strong rock (refer to field guide for material and strength definitions) with evidence of only minor erosion (sheet and rills), but with no signs of surface water or instability

References

Australian Geomechanics Society (AGS) (2007) Practice note guidelines for landslide risk management. J News Aust Geomech Soc 42(1):63–114

Australia New Zealand Standard (AS/NZS) 4360:2004 companion guide

Cruden DM, Varnes DJ (1996) Landslide types and processes. In: Turner KA, Schuster RL (eds) Landslide: investigation and mitigation. Special report 247, Chapter 3. Transportation Research Board, National Research Council, Washington DC, pp 36–75

Dowrick DJ, Hancox GT, Perrin ND, Dellow GD (2008) The Modified Mercalli Intensity Scale—revisions arising from New Zealand experience. Bull New Zealand Soc Earthq Eng 41(3):193–205

Hancox GT (2003) Preliminary report on landslides, gully erosion, and debris flood effects in the Paekakariki area as a result of the 3 Oct 2003 flood. *Institute of Geological & Nuclear Sciences Client Report 2003/120, 19 p (now on open file as it was publically released by the Kapiti District Council)*

Hancox GT, Perrin, ND, Dellow GD (2002) Recent studies of historical earthquake-induced landsliding, ground damage and MM intensity in New Zealand. Bull New Zealand Soc Earthq Eng 35(2):59–95

Hungr O, Leroueil S, Picarelli L (2014) The Varnes classification of landslide types, an update. Review article Landslides 11(2):167–194

Massey C, Hancox G, Page M (2007) Field guide for the identification and assessment of landslide and erosion features and hazards. Part of the GNS Science short course on landslides and erosion hazards for Vector Ltd. Rev 2. October 2007

NZGS. 2005. Field description of soil and rock. Guideline for the field description of soil and rock for engineering purposes. New Zealand Geotechnical Society (NZGS) Inc. December 2005. http://nzgs.org/publications/guidelines/soil_and_rock.pdfVarnes (1978) "Landslide disaster database" Indonesian National Agency for Disaster Mitigation

Varnes DJ (1978) Slope movement types and processes Chapter 2 In: Schuster RL, Krizek RJ (eds) Landslides, analysis and control. National Academy of Sciences, Transport Research Board Special Report 176:11–33

Part II

Monitoring and Early Warning

TXT-tool 2.039-3.1 Satellite Remote Sensing Techniques for Landslides Detection and Mapping

Nicola Casagli, Veronica Tofani, Andrea Ciampalini, Federico Raspini, Ping Lu and Stefano Morelli

Abstract

In this work the application of remote sensing for landslide detection and mapping is described. Among Earth Observation (EO) techniques optical and radar images are very effective tools for these applications since very high spatial resolution obtained by optical systems (currently in the order of tens of centimeters) and by the launching of Synthetic Aperture Radar (SAR) sensors, purposely built for interferometric applications with revisiting times of few days. In this paper the potentiality of both satellite optical and radar data is explored in a selected case study, analyzing different slope instability processes at different scales. Thanks to them and to the support of existing bibliography, the main advantages and disadvantages are highlighted as well as some suggestions are proposed concerning the main fields of applications.

Keywords

Landslides · Remote sensing · Landslide mapping · SAR data
Optical VHR data · Messina province

N. Casagli (✉) · V. Tofani · A. Ciampalini · F. Raspini · S. Morelli
Department of Earth Sciences, University of Firenze, Via La Pira 4, 50121 Florence, Italy
e-mail: nicola.casagli@unifi.it

V. Tofani
e-mail: veronica.tofani@unifi.it

A. Ciampalini
e-mail: andrea.ciampalini@unifi.it

F. Raspini
e-mail: federico.raspini@unifi.it

S. Morelli
e-mail: stefano.morelli@unifi.it

P. Lu
College of Surveying and Geo-Informatics, Tongji University, Shanghai 200092, China
e-mail: luping@tongji.edu.cn

K. Sassa et al. (eds.), *Landslide Dynamics: ISDR-ICL Landslide Interactive Teaching Tools*, https://doi.org/10.1007/978-3-319-57774-6_17

Contents

1 Introduction 236
2 Remote Sensing Techniques 236
2.1 Optical Very High Resolution (VHR) Data 236
2.2 SAR Data 237
3 Remote Sensing Application for Landslide Detection and Mapping 240
3.1 Description of the Study Area and Landslide Event 240
3.2 Object-Oriented Analysis (OOA) for Mapping of Rapid-Moving Landslides 242
3.3 Detection and Mapping of Slow-Moving Landslides with SAR Data 244
4 Discussion and Conclusion 247
4.1 Satellite Optical 247
4.2 Satellite Radar 249
References 251

1 Introduction

Landslides are one of the most serious geological hazards, which threaten and influence the socio-economic conditions of many countries in Europe and worldwide, causing damages and casualties (Schuster 1996; Schuster and Highland 2001; Petley et al. 2005; Petley 2012).

The observation of the Earth from space has found many uses in the natural sciences, but it is only in the last decades that technological advances have also extended to landslides (Singhroy 1995; Mantovani et al. 1996; Massonnet and Feigl 1998; Ferretti et al. 2001; Canuti et al. 2004; van Westen et al. 2008; Martha et al. 2010; Ferretti et al. 2011; Tofani et al. 2013a; Lu et al. 2012).

Today, rapid advances are making Earth Observation (EO) techniques more effective for landslide detection, mapping, monitoring and hazard analysis. Applications are originating from nearly all types of sensors available today (Tofani et al. 2013b).

Rapid developments in this field are fostered by the very high spatial resolution obtained by optical systems (currently in the order of tens of centimeters) and by the launching of Synthetic Aperture Radar (SAR) sensors, purposely built for interferometric applications with revisiting times of few days such as TerraSAR X and COSMO-SkyMed (Tofani et al. 2013a). Landslide detection and mapping benefit from both optical (Hervás et al. 2003; Cheng et al. 2004; Nichol and Wong 2005; Marcelino et al. 2009; Martha et al. 2010; Lu et al. 2011) and radar imagery (Hilley et al. 2004; Catani et al. 2005; Strozzi et al. 2005; Colesanti and Wasowski 2006; Canuti et al. 2007; Casagli et al. 2009; Cascini et al. 2009; Lu et al. 2012; Righini et al. 2012; Ciampalini et al. 2014; Raspini et al. 2015a).

In this work we describe remote sensing techniques (optical and radar) and their applications to a selected case study for landslide detection and mapping. The selected study area is Messina, in Sicily, Southern Italy where a landslide event occurred on October 1, 2009. This contribution is organized as follows; in Chapter "Landslide Mapping Through the Interpretation of Aerial Photographs" we provide a description of the techniques for the analysis of optical very high resolution (VHR) data and SAR data; in Chapter "Landslide Mapping Through the Interpretation of Aerial Photographs and Topographic Maps" we report the application to the selected case study of both optical and radar remote sensing data for the detection and mapping of landslides (both shallow rapid moving flows and deep slow moving slides); in Chapter "Landslide Recognition and Mapping Using Aerial Photographs and Google Earth" we provide the main advantages and disadvantages of the two techniques and we propose some suggestions for their use in different stages of landslide analysis.

2 Remote Sensing Techniques

2.1 Optical Very High Resolution (VHR) Data

Space-borne multi- and hyper-spectral sensors represent important alternative data sources able to characterize the spectral properties of landscapes and provide useful information on landslide and neighboring areas. The most important active optical satellites are reported in Fig. 1. Optical data are usually used for landslide detection and mapping though visual inspection or analytical methods (Fiorucci et al. 2011; Parker et al. 2011; Guzzetti et al. 2012; Mondini et al. 2014). For example, several optical derivative products (panchromatic, pan sharpen, false colour composits, ratioing) can help in visual interpretation for landslide mapping (Casagli et al. 2005; Marcelino et al. 2009; Ma et al. 2016). In image fusion procedures, multispectral channels, characterized by a coarser spatial resolution than the panchromatic, are downscaled through analytical models based on the panchromatic-derived spatial information (Eyers et al. 1998; Chini et al. 2011; Martha and Kerle 2012; Kurtz et al. 2014). The false colour composites (FCCs) of the VHR images are often used to discriminate lithology or terrain having different characteristics (weathering, water content, vegetation cover). This basic technique uses three selected bands as the primary colours of red, green and blue (RGB) to emphasise the discrepancies between areas characterised by different lithology (Ciampalini et al. 2012; Lamri et al. 2016) or areas subjected to land cover changes (Mwaniki et al. 2015). Image ratioing consists of a ratio operation between the values of corresponding pixels of two registered images of different dates (Hervas et al. 2003) or the values resulting from at least two different bands of the same multi- or hyper-spectral image. Band ratioing using the red and near infrared bands is one of the most simple enhancement technique that can be very useful to reduce the problem related to the shadowing because it decreases the effects of variation in illumination by considering relative reflectance differences in two bands (Cheng et al. 2004). Another important advantages of the ratio Red/IR is represented by the maximizing of the contrast between vegetated and un-vegetated ground (Joyce et al. 2009). The Normalized Difference Vegetation Index (NDVI) derived from optical images, is another widely indices used map landslides which is used to evaluate the vegetation cover rate (Lin et al. 2004). Higher values of NDVI can be related to a wide vegetation cover, whereas lower values can represents areas affected by landslides. Furthermore multitemporal NDVI can lead information on temporal and spatial dynamics of vegetation (Birky 2001) and can be used for indirect measurement of soil water content (Senay and Elliott 2000) or rainfall patterns (Tucker et al. 1985). Furthermore, multispectral images can be enhanced to detect landslides using, more complicated, analytical methods based on the spectral characteristics of the land surface. Widely adopted automatic approaches focus on the classification of image pixels, used as reference spatial mapping unit. Based on its spectral properties, each pixel is labelled independently from the neighboring areas as a given land cover class or geomorphic typology. Such approach is usually termed "pixel-oriented". Pixel-oriented approaches are generally time-saving and represent useful tools for landslide mapping (Mondini et al. 2011). However, pixels are ill-suited to represent a hillslope processes, such as a landslide, strongly identified by spatial patterns especially in case of very high resolution images (Martha et al. 2010). Few studies have described the use of hyperspectral data for recognition and classification of landslides since most of the hyperspectral satellite sensors are still under development. Since their higher spectral resolution, hyperspectral data are commonly used in order to map specific Earth surface characteristics (i.e. lithology, mineralogy, terrain roughness, soil moisture) over wide areas (Scaioni et al. 2014).

2.2 SAR Data

SAR (Synthetic Aperture Radar) is an active object-detection system which operates in the microwave range of the electromagnetic spectrum. Figure 2 shows all the radar satellites. The great benefit derived by using such illuminating source (generated by a transmitting antenna), characterized by lower frequency and longer wavelength with respect to the visible right, is that radar sensors are capable of penetrating clouds and can operate independently of weather

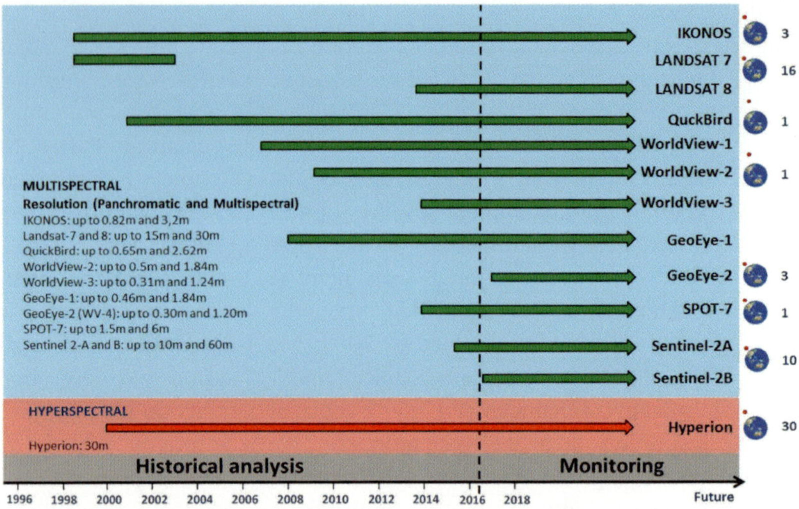

Fig. 1 Most important active optical satellites

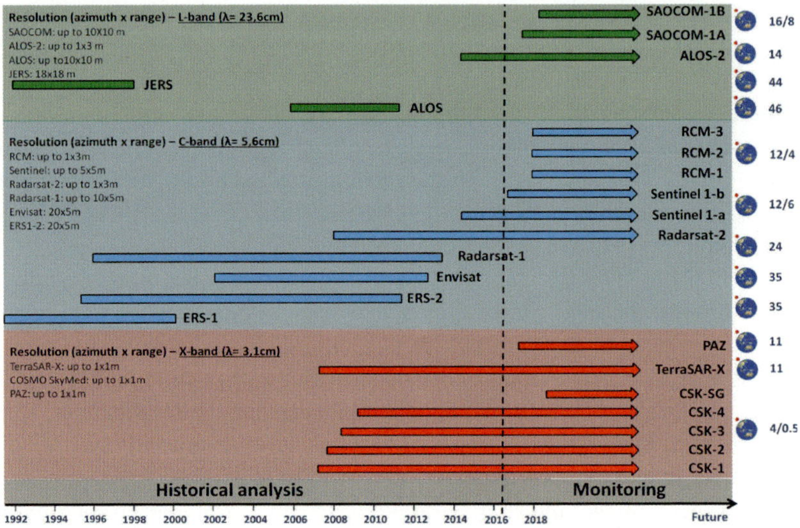

Fig. 2 Most important radar satellites used in interferometry

conditions and illumination. Transmitting their own energy, SAR sensors are designed to carry out their task 24 h per day, 365 days per year.

SAR sensors emit signals with a specific central frequency, the so-called operating frequency, which characterizes signal propagation and penetration features. The most commonly used bands in satellite radar applications are C-band (5–6 GHz, ~5.6 cm wavelength), X-band (8–12 GHz, ~3.1 cm wavelength) and L-band (1–2 GHz, ~23 cm wavelength). The family of satellites, carrying platforms hosting the SAR sensors, orbits the Earth at an altitude ranging from 500 to 800 km above the Earth's surface, following sun-synchronous, near-polar orbits, slightly inclined with respect of Earth meridians. The 'sensor to target' direction, inclined of an angle 'θ' with respect to the vertical, is referred to as Line Of Sight (LOS) and it varies accordingly to satellite employed (θ usually ranges from 23° to 34°). The combination between the Earth's rotation movement and the polar orbits of all SAR satellites, allows the sensor to scan along predetermined paths and to gather information of the same target from two opposite acquisition geometries: ascending and descending. Satellite SAR systems are designed to acquire images through different operating modes, with increasing spatial coverage but coarsening resolution. Each mode is specifically designed to acquire a target area with different swaths and resolutions, to fit the specific needs requested in different application fields, ranging from basin-scale mapping to localized emergency situations. Besides the spatial coverage, also temporal resolution of satellite SAR data depends on the SAR platform. SAR sensors can only acquire images at the frequency of the satellite repeat orbit. The time taken for a satellite to re-pass over the same area is called the 're-visiting time'.

A SAR image is composed of pixel corresponding to a ground area. Each pixel is defined by amplitude and phase values. The amplitude values depend on the intensity of electromagnetic field back-scattered by the illuminated objects towards the satellite. Phase values of a single SAR image is related to the sensor-target distance, to a contribution due to propagation of the microwave through the atmosphere plus a noise term inherent of the acquisition system. Recorded phase information is the key element to detect occurred displacement. SAR Interferometry is the technique specifically thought, designed and implemented to measure changes of signal phase over time through the analysis of two or more SAR images.

A single SAR image is of no practical use, since it is not possible to distinguish the different phase contributions related to atmosphere, topography and noise. A far more suitable approach to exploit SAR data is the Differential Interferometric SAR (DInSAR), which relies on the analysis of phase variations, or interference, between two (the master and the slave) different radar images, acquired over the same target area at different times. Once the topography contribution has been modelled and removed, the phase-difference image (called interferogram) maps and quantifies, along the LOS of the satellite, possible ground deformations that may have occurred between the first and the second acquisition. Since the pioneering studies performed during the nineties (Fruneau et al. 1996; Singhroy et al. 1998; Rott et al. 1999), in which SAR data were used to retrieve geohazard-related ground deformation, satellite SAR data have been fully exploited, becoming a standard operational tools applied the field of mapping and monitoring of landslide hazards (Berardino et al. 2003; Catani et al. 2005; Strozzi et al. 2005). Two main factors reduce (or even compromise) the quality and reliability of conventional, single-pairs DInSAR results, namely geometrical and temporal decorrelation caused by the variation of the phase reflectivity value of some categories of radar targets. A further limitation encountered is the phase distortions introduced by atmospheric effects, which cannot be estimated with conventional approach.

To overcome the limitations of the single-pair interferogram methods, InSAR-based information can be enhanced through multi-temporal interferometric approaches, based on analysis of long stacks of co-registered SAR imagery. Exploitation of long temporal series of SAR scenes leads to redundant measurements that can be utilized for more advanced time series methods reducing the atmospheric impact on the estimated displacement values. Two classes of multi-interferometric techniques (MIT), the Small Baseline (SB) techniques (Berardino et al. 2002) and the Persistent Scatterer Interferometry (PSI) method (Ferretti et al. 2000, 2001), are used for processing long series of SAR imagery. Within the last family, PSInSAR (Permanent Scatterers InSAR) (Ferretti et al. 2000, 2001) was the first technique specifically implemented for the processing of long archives of SAR images. Signal analysis of a network of coherent radar targets (Permanent Scatterers, PS), identified with amplitude-based algorithms and exhibiting high phase stability over the entire observation time period, allows estimating occurred displacement, acquisitions by acquisition. PSInSAR has been demonstrated to be a powerful tool for measuring ground movements in urban areas. Over natural, less urbanized and agricultural environments, the lack of continuous urban fabric led to less dense network of stable targets and MIT approaches fail due to complete decorrelation of the great majority of scatterers. Different algorithms have been developed to extend the ability of PSI to natural terrain, among which the StaMPS (Hooper et al. 2004) and SqueeSAR (Ferretti et al. 2011) techniques. In order to overcome the limitation related to the lack of coherent radar targets, the usage of artificial reflectors is a very promising alternative. Outputs of MIT processing are represented by ground displacement maps and time series of deformation. LOS deformation rate can be estimated with an accuracy theoretically better than 0.1 mm/yr. Each measurement is referred temporally and spatially to a unique reference image and to a stable reference point.

Over the last two decade geoscientists have widely exploited existing SAR archives chiefly to resolve the spatial distribution and to analyze the temporal evolution of displacements in areas affected by slow moving landslides (Hilley et al. 2004). The ability to make numerous point measurements of displacement over the landslide body allows one the detection and mapping of the actively deforming slopes (e.g. Righini et al. 2012), the characterization and monitoring of landslide mechanism (Tofani et al. 2013b) and, through the analysis of time series of deformation, the identification of velocity changes in the landslide evolution (Berti et al. 2013) and the modelling of large slope instability (Berardino et al. 2003).

The potential of SAR data has been exploited for landslide investigations at different scales, from nation (Adam et al. 2011) to regional (Meisina et al. 2008) to basin (Lu et al. 2012) to local (Ciampalini et al. 2014) to building scale (Bianchini et al. 2015) and in different phases of landslide response (Canuti et al. 2007) and Civil Protection practice (Farina et al. 2008).

3 Remote Sensing Application for Landslide Detection and Mapping

In this section we show the application of both optical and radar remote sensing data for the detection and mapping of landslides (both shallow rapid moving flows and deep slow moving slides). The selected study area is Messina, in Sicily, Southern Italy where a landslide event occurred on October 1, 2009.

3.1 Description of the Study Area and Landslide Event

The study area (Messina Province, Fig. 3) represents the north eastern sector of Sicily Island. From a geologic point of view Messina Province is part of the Sicilian fold-and-thrust orogenic belt, formed during the Tertiary, due to the convergence between the European and African plates (Dewey et al. 1989; De Guidi and Scudero 2013). The area comprises two different mountain chains: The Peloritani, located along the Ionian coast and the Nebrodi along the Thyrrenian coast. The geomorphology of both mountain chains is characterized by the typical features of recently uplifted areas, developed on a crystalline basement with steep slopes eroded by torrent-like straight watercourses. The stream network is formed by regular and parallel paths, perpendicular to the ridges, leading to the coast. River catchments are affected by a reduced concentration time with a significant solid transport. Their trend seasonally varies and regime is torrential with catastrophic transport of solid materials due to heavy rain falls. Landslides, mainly triggered by rainfalls, are frequent and include shallow soil slides and debris flows, deep-seated rotational and translational slides, as well complex and compound failures.

In the late afternoon of October 1st 2009, an intense storm affected the area between the Peloritani Mountains ridge and the Ionian coastline (Ciampalini et al. 2015a; Del Ventisette et al. 2012), where the main villages are located. The rain gauge of Santo Stefano di Briga recorded a rainfall of 225 mm in 8 h, with a peak of 115 mm in three hours. During the night, the persisting rainfall triggered more than 600 hillslope processes (shallow soil slides, debris flows and debris avalanches) on an area of about 50 km^2. The worst damages were reported in the villages of Giampilieri (Fig. 3). The assessed

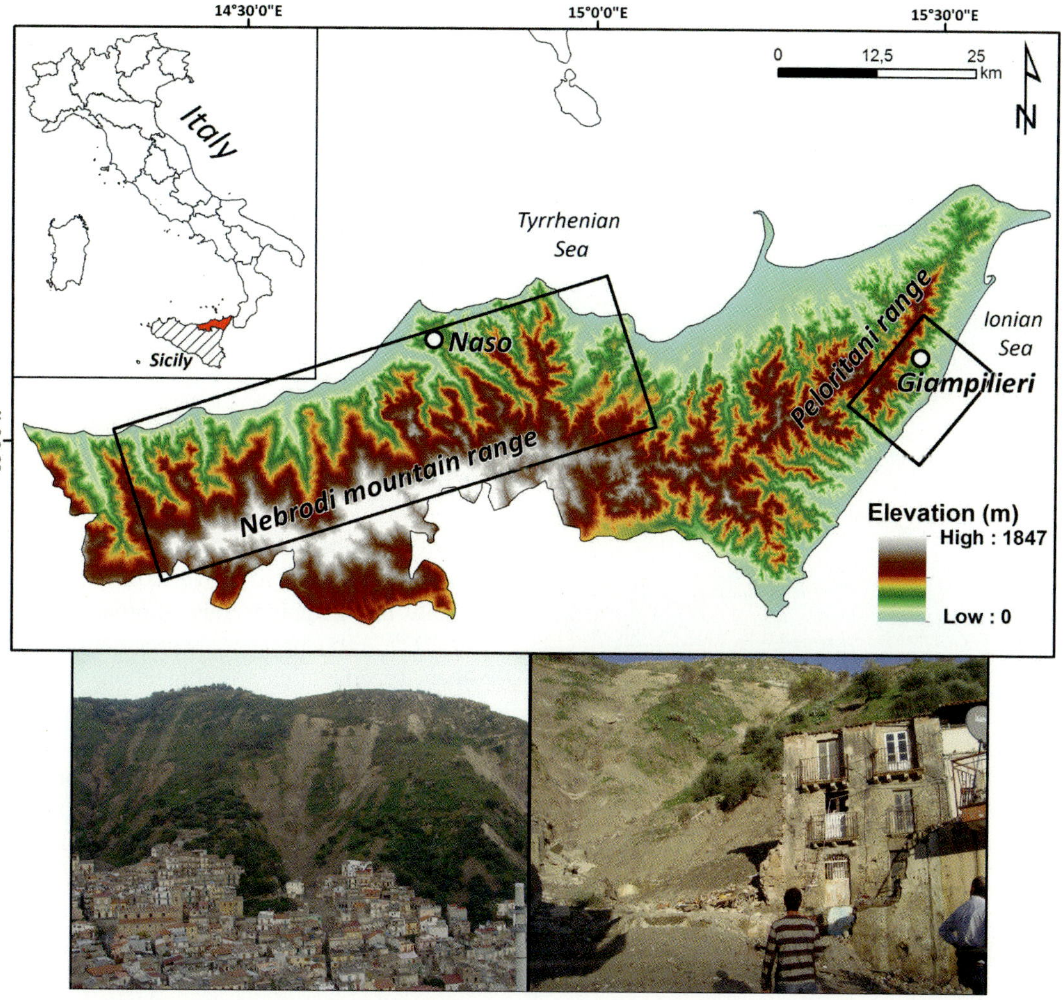

Fig. 3 The area of the Messina province affected by the 2009–2010 landslides (in the *black square*). In the *inset below* effects on the landscape and in the village of Giampilieri are reported

number of fatalities caused by landslides and inundation amounts to 37 (including 31 deaths and 6 missing persons), with 122 injured people and 2019 evacuated people. The area is located along the eastern-facing slope of the Peloritani Mountains. It is located in the southern periphery of the city of Messina with a total extension of about 75 km^2. The main inhabited centers are concentrated along the strongly urbanized coastal area. The middle hill area is characterized by the presence of smaller settlements, connected with the coastline by a single road axis climbing upward along very steep slopes. Between the 2009 and the 2010, Nebrodi Mountains several municipalities (San Fratello, Caronia, Naso, Ucria, and many others) located within the Nebrodi Mountains were strongly affected by several complex, rotational and deep-seated landslides which damaged buildings and infrastructures (Ciampalini et al. 2014; Bardi et al. 2014).

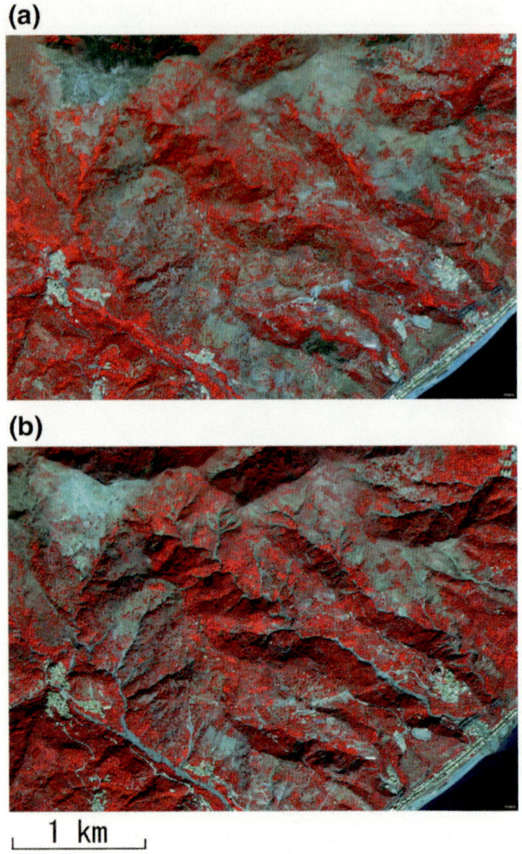

Fig. 4 The used Quickbird imagery: **a** pre-event QuickBird image, **b** post-event QuickBird image false color 4-3-2)

3.2 Object-Oriented Analysis (OOA) for Mapping of Rapid-Moving Landslides

The application with the optical data (Lu et al. 2011) is based upon the OOA (object oriented analysis). Two Quickbird images were used in the study (Fig. 4), acquired on September 6, 2006 and October 8, 2009, with 0%, 3% and zero cloud cover respectively. For each image, only four multispectral bands (Blue: 450–520 nm, Green: 520–600 nm, Red: 630–690 nm, NIR: 760–900 nm) with the spatial resolution of 2.4 m were used.

OOA is mainly dealing with the measuring unit of 'object'. The term 'object' inside OOA can be defined as 'individually resolvable entities located within a digital image which are perceptually generated from high-resolution pixel groups' (Hay et al. 2003). In detail, OOA initiates with an image segmentation approach that spatially divides the digital image (including remote sensing imagery) into several homogeneous segments which contain high spectral autocorrelation, so as to form these 'objects', and the following analysis can be then performed on the unit of these segmented objects instead of original pixels (Hay et al. 2003).

The purpose of this work is to introduce a new approach for a rapid mapping of newly-triggered landslides using an objected-oriented change detection technique. The methodology aims at a semi-automatic and rapid analysis with a minimum of operator involvement and manual analysis steps. Compared to conventional approaches for landslide mapping, this approach benefits from (1) an image segmentation with problem-specified scale optimization, and (2) a multi-temporal analysis at object level with several systemized spectral and textural metrics.

The multi-resolution segmentation based on Fractal Net Evolution Approach (FNEA) described by Baatz and Schaepe (2000) and implemented in Definiens eCognition software (Benz et al. 2004), is employed for the initial segmentation, parameterized according to the specific needs of event-based rapid mapping of landslides, and incorporated in a multi-scale optimization routine.

The landslide detection approach presented here makes use of additional spectral and textural measurements: change detection using temporal Principal Component Analysis (PCA), image matching through Spectral Angle Mapper (SAM), anomaly detection by Reed-Xiaoli detector (RXD), and textural analysis with grey level co-occurrence matrix (GLCM).

In the area of Giampilieri the event caused severe damages to several towns, which were isolated due to the destruction of roads and railways. Two of the most damaged areas were studied, including a training area (ca. 1.8 km^2) for algorithm development, and a larger independent testing area (ca. 8.1 km^2). The latter allows the robustness and transferability of the

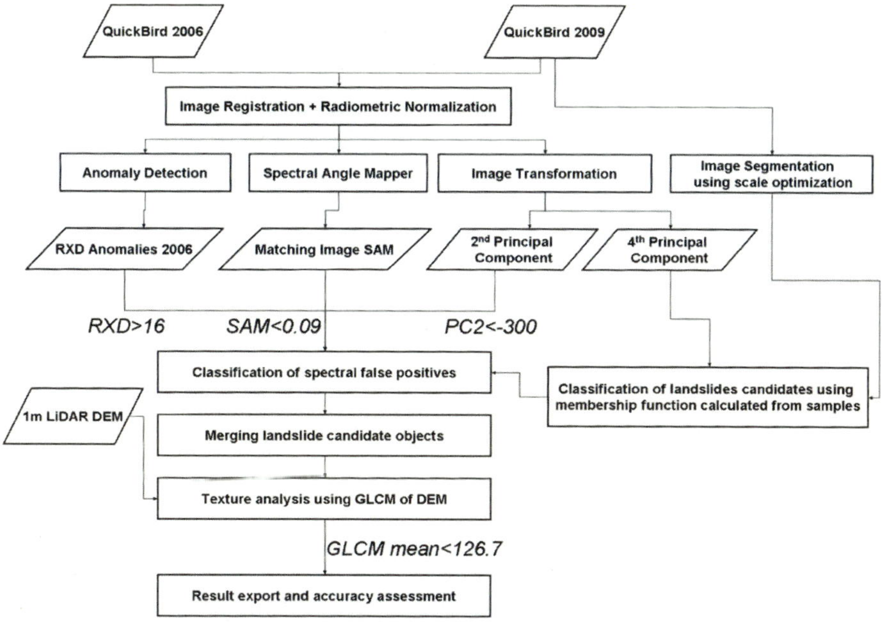

Fig. 5 General flowchart of landslide mapping by OOA change detection. *RXD* Reed-Xiaoli Detector; *SAM* Spectral Angle Mapper; *PC* Principal Component; *GLCM* grey level co-occurrence matrix (Lu et al. 2011)

Fig. 6 The result of OOA landslide mapping in the independent testing area

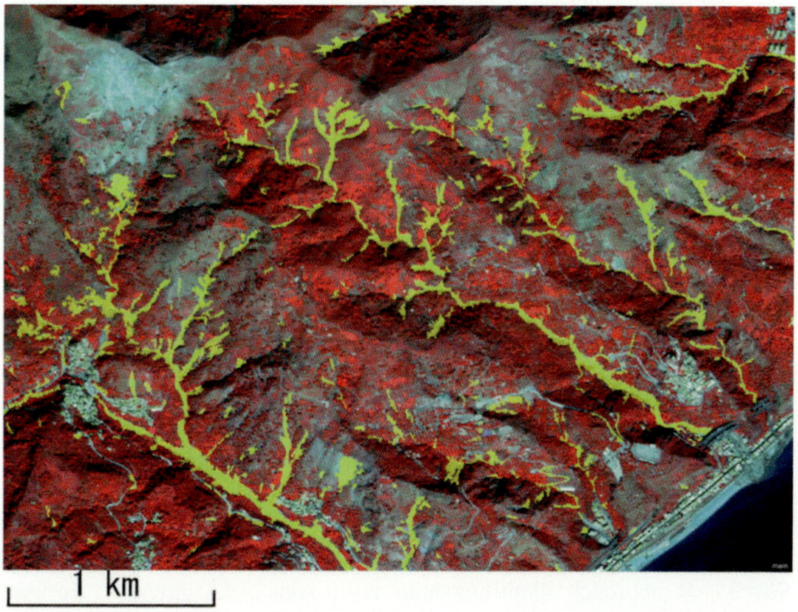

algorithm (without any change of ruleset and threshold) and the corresponding accuracy to be assessed by comparison with a manually mapped landslide inventory prepared from field works and subsequent modifications from image interpretation. The general methodology is shown in Fig. 5.

The algorithm developed based on the training area was directly applied in the testing area. The final outputs for the testing area are shown in

Fig. 6. To evaluate the accuracy of this approach, OOA-derived landslides were compared with a manually-mapped landslide inventory. The accuracy assessment was carried out for the number and the spatial extent of mapped landslides. The accuracy assessment calculates the commission and omission errors, which are measures of the user's and producer's accuracies (Congalton 1991; Story and Congalton 1986) of the mapped landslides, respectively.

For the spatial extent of landslides a user's accuracy of 75.9% and a producer's accuracy of 69.9% were achieved. In terms of the number of landslides, user's and producer's accuracies of 81.8 and 69.5%, respectively, were reached. For both number and spatial extent of landslides the results show a lower producer's accuracy than user's accuracy: specifically, ca. 31% of all manually mapped landslides were omitted in the OOA- based detection. This indicates an overestimation of false positives during their classification, accompanied with an underestimation of true positives obtained from the membership function of the selected samples. Further improvements should include a more accurate definition of these thresholds for classifying false positives and a more careful selection of representative samples.

3.3 Detection and Mapping of Slow-Moving Landslides with SAR Data

Satellite analysis of ground deformation has been performed using the SqueeSAR approach (Ferretti et al. 2011), a second generation PSInSAR algorithm (Ferretti et al. 2001), one of the several multi-interferometric techniques available today to process long series of SAR images. The SqueeSAR approach identifies both point-wise Permanent Scatterers (as previous PSInSAR technique) and spatially Distributed Scatterers (DS), providing a significantly increased coverage of ground deformation measurement points, especially over non-urban areas. DS are identified in low-reflectivity homogeneous areas, scattered outcropping rocks, debris flows, non-cultivated lands and desert areas. The SqueeSAR algorithm has been applied to C-band SAR dataset acquired by ERS (08/09/1992–24/11/2000) and Envisat (22/01/2003–20/05/2009) missions along ascending orbits.

Following the approaches proposed by Farina et al. (2008) and Bianchini et al. (2012), deformation measurements extracted by means of SqueeSAR technique have been coupled and integrated with thematic maps (topographic and geological maps), optical data (ortophoto, VHR optical satellite images and multi-temporal aerial photos) and available landslides inventory maps to identify the areas characterized by high hydro-geological hazard (hotspot mapping) related to the occurrence of extremely and very slow moving landslide (according to the classification of Cruden and Varnes 1996). Twenty-six sites have been identified. For these areas landslides have been detected and mapped (Fig. 7).

On the basis of available multi–interferometric data these sites were assessed as the most critical in terms of hydro-geological hazard, both for the type of instability detected and/or the extent of the mapped phenomena and/or the measured deformation velocities and/or the presence of elements at risk.

In the inset of Fig. 7 the hotspot analysis for the area of the village of Croce (identified with number 9) is reported. The area is located on the right bank of Guidomandri creek and is characterized by the presence of several elements at risk, including, beside the village itself, isolated buildings and minor settlements connected by stretches of paved roads and a cemetery. Pre-existing landslide inventory maps do not report slope instability in the study area. On the contrary, the SqueeSAR results show a large sector of the slope characterized by a displacement with velocity ranging from 1.6 mm/yr to 4.8 mm/yr Envisat dataset (2003–2009). Photo-interpretation of stereoscopic colour images (1:3500 scale) and analysis of information provided by SqueeSAR results allow to detect and map a large complex system of active continuous slides affecting the area. Such deformation rates don't pose threat to population, but can cause, persisting for many years, damages to

Fig. 7 Location of the twenty-six sites characterized by high hydro-geological risk (*hotspot mapping*). In the *inset* the landslide mapping of the village of Croce is reported

buildings and manmade infrastructures. Cracks and damages have been surveyed during field validation, which also helped to confirm the presence and the extension of the active movements through the identification of tension cracks, scarps and counterscarps.

After the 2009 and 2010 events that affected the Nebrodi Mountains, the SqueeSAR technique has been applied to characterize the triggered hillslope phenomena both at the basin and at the local scale.

At the basin scale, SqueeSAR PSI data have been used to update the available Landslide Inventory Map (LIM) including information on typology and state of activity of each identified landslide. The updating procedure has been performed using: (i) radar interpretation of four different available SAR datasets; (ii) photo-interpretation of 1:33,000 scale aerial photographs flown in 1954, 1955 and 2005; and (iii) field surveys.

InSAR displacement measurements were acquired in different periods (2006–2009, RADARSAT-1 scenes and 2011–2012, COSMO-SkyMed images). Considering the limitation of the adopted technique, the updating of the per-existing LIM has been limited to the extremely slow and very slow moving landslides.

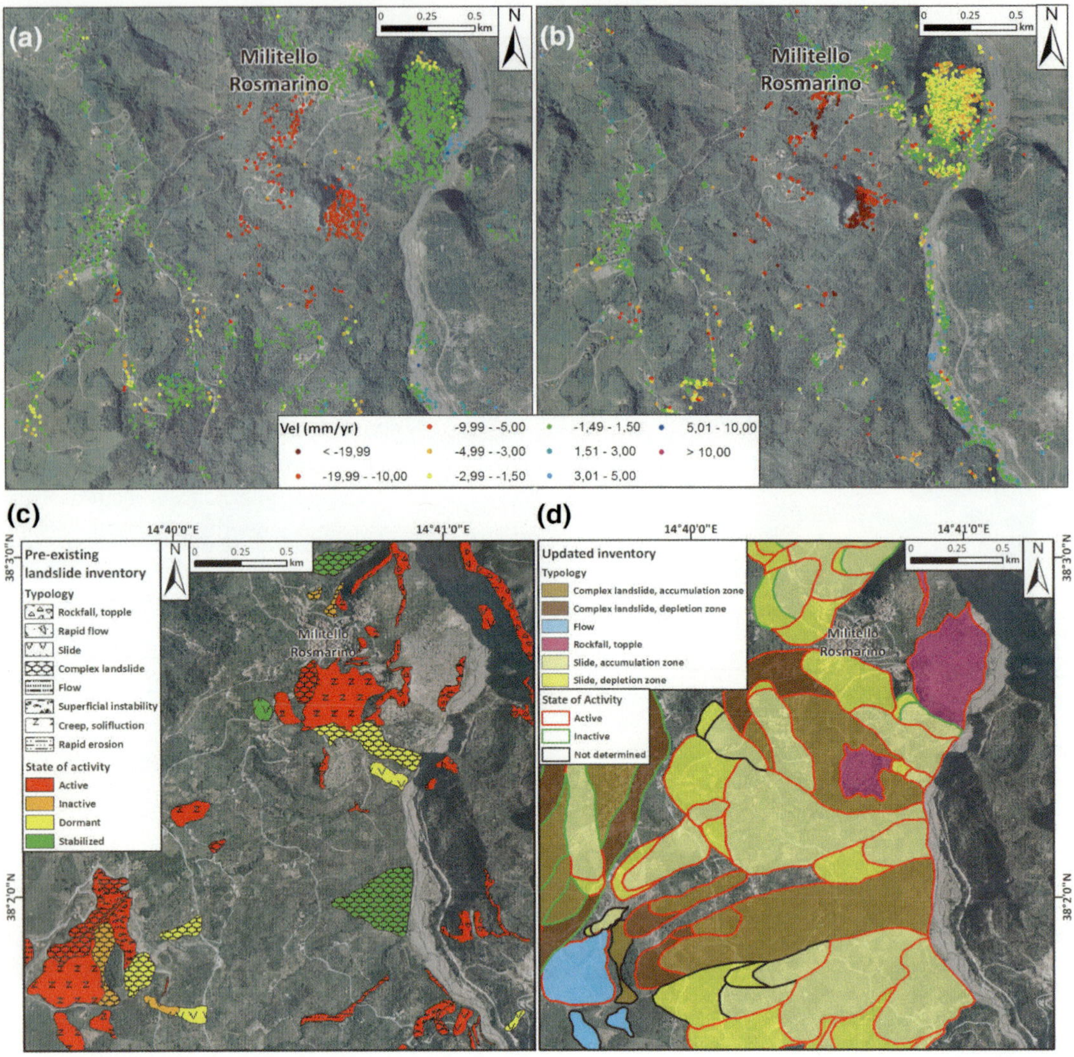

Fig. 8 Ground deformation velocity maps obtained using Radarsat-1 (**a**) and COSMO-SkyMed (**b**) PSI data. Pre-existing LIM (**c**) and updated LIM (**d**)

Faster phenomena have been excluded due to their rapid kinematics. The new LIM (Fig. 8) includes 566 events: 15 (2.7%) rockfalls and topples, 136 (24.0%) complex landslides, 188 (33.2%) flows and 227 (40.1%) slides, covering an area of 74.1 km^2. The comparison between the pre-existing and the new LIMs has been performed using three classes: (i) confirmed, (ii) enlarged with respect to the pre-existing LIM and (iii) new (landslide not included in the pre-existing LIM). This approach led to the enlargement of 120 events (21.2%) of the pre-existing LIM, to the confirmation of 155 events (27.4%), and to the recognition of 291 (51.4%) new phenomena.

At the local scale, PSI data can be also used for faster landslides such as rockfalls and rock-slides because they can be preceded by a period of slow movements prior to slope failure. PSI analysis is designed to generate time-series of ground deformations for individual elementary reflectors (i.e. the Persistent Scatterers, PS), assuming different types of deformation models (e.g. linear, nonlinear or hybrid). Each measurement on a time series corresponds to a single satellite acquisition. The accuracy of the single measurement in correspondence of each SAR acquisition ranges from 1 to 3 mm. Time series of deformation represent the most advanced PSI product, providing the deformation history over the observed period. Time series shows the temporal pattern of the deformation, highlighting non-linear movements, seasonal trends, ground acceleration and any potential changes occurred during the monitoring period, supporting the identification of precursor movements prior a major events (Frodella et al. 2016).

PSI data acquired between 2006 and 2010 (Radarsat-1 in ascending geometry) were used to detect the presence of precursory phenomena in the surrounding of the village of Naso (Ciampalini et al. 2015b). Naso is historically affected by slope instability phenomena. The most common hillslope processes were represented by complex, rotational and translational landslides. Naso, for its peculiar geological framework, has been also affected by rockfalls phenomena.

The slope back-monitoring has been performed following the approach described in Cigna et al. (2011), under-sampling the whole time interval of observation into different sub-samples in which the deformation rate has been evaluated using a simple linear regression.

Radarsat-1 PSI data clearly highlight the ground deformation which affecting the southern slope of Naso. In particular the area close to the municipality (point 3 in Fig. 9a) is characterized by an average ground deformation velocity of −3.05 mm/yr. The highest velocities (−5.88 mm/yr) were detected along the SS116 (Points 1 and 2 in Fig.). Time series analysis (Fig. 9b), shows the presence of two periods of acceleration: the first one between August and October 2008 and the second one between April and July 2009.

To understand the triggering factors of the slope instability processes that affect the village of Naso, the time series of moving PS have been compared to the rainfall data from the Naso measuring station. Results suggest that each time that the monthly rainfall exceed 60 mm the time series show an increase of the displacement. Time series interpretation can be successfully used at the local scale to monitor and forecast hillslope processes for specific triggering factor establishing, as in this case, a rainfall threshold over which new events or reactivation of pre-existing landslides can occur.

4 Discussion and Conclusion

In recent years spaceborne techniques have been interested by several technical and scientific improvements. That has not led to clearly identify a technology which can be undoubtedly and universally considered better than others. Conversely, the reliability and effectiveness of every spaceborne technique have been enhanced, providing for each geological problem a wide range of equally good solutions. As a consequence, at present end-users can select the most proper methodology with respect to their specific needs, which can be related to technical issues, economic budget, environmental factors and specific features of the landslide(s) to be monitored.

It should be also stressed that the various proposed methodologies described in Sect. 3 should not be seen one in competition with the other ones; in contrast many of them could conveniently be used in close cooperation to obtain a very effective monitoring system.

Here below some principal advantages and limitations for satellite optical and satellite radar.

4.1 Satellite Optical

With the increasing large fleet of VHR satellites, imagery can be acquired timely after major events and with daily temporal resolution at nearly global coverage. The main advantage of VHR imagery is the great density of spatial information, whereas, with more competing satellite operators entering the market, prices constantly decrease. A main advantage of optical datasets is their synergetic values for several other applications such as post-disaster damage assessment and updating of land cover and

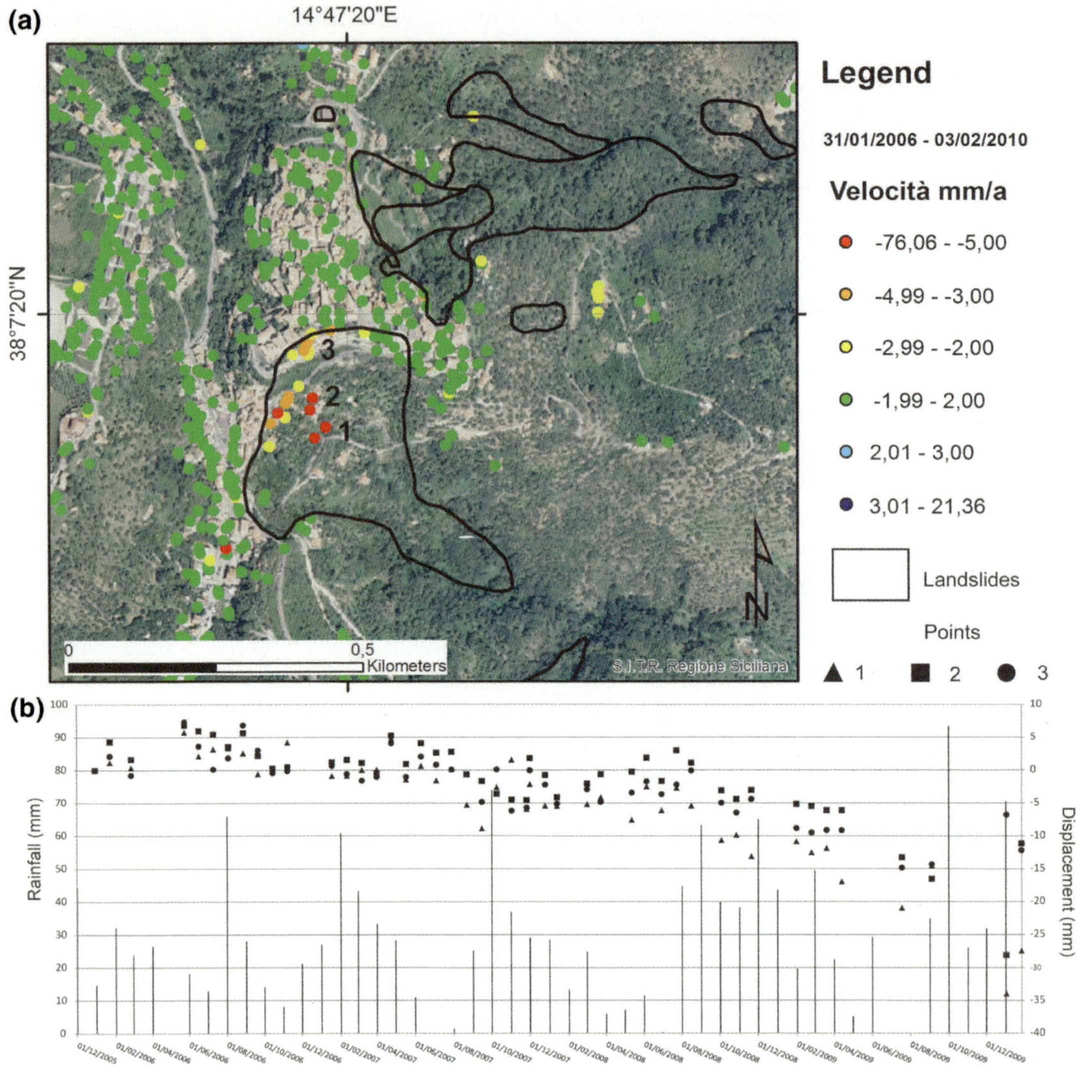

Fig. 9 **a** Ground deformation velocity map of Naso using Radarsat-1 PSI data and **b** comparison between time series of cumulative displacement and the monthly rainfall data

landslide inventory maps and corresponding archives are often available over a given area.

A greater diversity of platforms increases the chance to acquire cloud free imagery of a given area with a specified time frame but atmospheric conditions remain an important factor that, depending on the climate zone and the season, may delay the acquisition of suitable images considerable. For this and other reasons, satellite tasked for images with sub-meter resolution can still be associated with considerable costs.

Higher spatial, spectral (e.g. WorldView-2) and temporal resolutions strongly increase the computational load for the storage and analysis of the datasets. Especially for mapping over wide areas. This can considerably slow down the analysis and may need for further investments in hardware and software. This is closely related to the desirable exploitation of spatial context which is typically computational intensive.

Pixel-based change detection (typically image differencing) is relatively easy to apply and can be accurate when most of surface changes are caused by landslides. In many cases, it might also be possible to account for sensor and illumination differences by cross calibration and image transformation. However, only limited accuracy can be expected from such approaches in situation where other similar surface changes such as deforestation or barren fields are present in the same scene. The selection of an appropriate threshold to distinguish between changed and unchanged areas remains as a general difficulty for the application of pixel-based methods. Further problems are usually encountered when pixel-base change detection is applied on VHR imagery because of the higher spectral variance and stronger impacts of small co-registration errors.

Due to a better exploitation of the spatial context within remote sensing images OOA approaches generally yield better results than could be achieved with per pixel analyses (Blaschke 2010). It has been demonstrated that OOA rule sets are not only capable to accurately delineate areas affected by landslides but can also be used to distinguish among different landslide types (Martha et al. 2010).

In the study area of Messina (Lu et al. 2011) the OOA approach has proved an effective tool to map rapid landslides.

4.2 Satellite Radar

As discussed by Colesanti and Wasowsky (2006), due to the inherent limitations of current space observation systems and relevant data processing techniques, the practical applicability of PSI approaches is usually limited to two classes of the Cruden and Varnes (1996) classification: extremely slow and very slow movements (vel < 16 mm/year and 16 mm/year $\leq$ vel < 1.6 m/year, respectively). These movements are suitable for analysis based on PSI techniques, as long as they evolve with very low displacement rates (few tens of centimeters per year) and their velocities do not exceed the intrinsic limits of the techniques, related to the radar wavelength, revisiting time of the platform and the spatial density of measurement points. Moreover, landslide-induced displacements, detectable through PSI techniques, are restricted to "coherent" landslides with very slow dynamics, i.e. with little internal deformation such as deep-seated deformations (García-Davalillo et al. 2014), creep (Cascini et al. 2010), and, in some cases, slides (Raspini et al. 2015b) roto-translational slides (Tofani et al. 2013b), rockslide (Lauknes et al. 2010), complex landslides (Bardi et al. 2014), slow earth flows (Herrera et al. 2011) and badlands (Herrera et al. 2009). InSAR data can provide useful information about pre-event movements, often characterized by low displacement rates (few cm/year) persisting over long time periods (Bardi et al. 2016; Frodella et al. 2016). This deformation regime is quite different to failure events, which occur suddenly and may produce ground displacements of several meters (Raspini et al. 2015a).

With reference to landslide detection/mapping the most advantageous aspects of the multipass DInSAR approach are (Colesanti and Wasowski 2006):

- The cost-effectiveness for wide-area (hundreds and thousands of km^2) applications, typical of spaceborne remotely sensed data.
- The high density of benchmarks (up to several hundred per km^2).
- The use of "natural" benchmarks not requiring deployment and maintenance.
- The possibility of geo-locating the benchmarks with a precision in the order of 1–5 meters.
- The availability of the extremely valuable ESA (European Space Agency) ERS archive spanning about 20 years, which enables to carry out retrospective studies.
- Recent studies proved the feasibility of combining (stitching) SAR data acquired by different sensors (e.g. ENVISAT with ERS or RADARSAT-1 with RADARSAT-2), despite slight differences in critical image acquisition parameters.

- Regular revisiting time in the order of 20–40 days (up to 6 days with the new Sentinel-1 ESA mission).

On the whole, the case studies described in the scientific literature highlight that with reference to the detection/mapping of slow-moving landslide phenomena the main benefits regard:

- the definition of the boundaries of already detected mass movements;
- the definition of the states of activity;
- the detection of previously unmapped unstable areas.

However, several limiting factors need to be properly taken into account (Colesanti and Wasowski, 2006):

1. Displacement data represent the one dimensional projection along the Line Of Sight (1D LOS projection) of a deformation that can actually occur in all three dimensions.
2. The ambiguity of phase measurements implies the impossibility to track correctly (i.e. unambiguously) the relative LOS displacement between two scatterers exceeding $\lambda/4$ (=1.4 cm for ERS) within one revisiting time interval (35 days for ERS), i.e. approximately 14.5 cm/yr. In practice it is extremely difficult to detect LOS displacement rates exceeding 8–10 cm/yr in the presence of low density of stable scatterers, such as in the case of landslides where topography and vegetation introduce a limitation in the number of detected scatterers. This limits the use of multi-interferometric approaches only to landslides ranging from extremely to very slow phenomena according to the velocity classification of Cruden and Varnes (1996).
3. Limited versatility in terms of (a) positioning of the measurement points and (b) revisiting time. Both factors (a) and (b) cannot be optimized as degrees of freedom while planning an analysis.
4. Finally, it is still difficult to forecast the coherent pixel density in rural areas without carrying out at least several processing steps on a significant number (15–20) of SAR images.

Based on advantages and limitations of the two techniques, the applications shown in Sect. 3 and relevant bibliography, it is possible to define for which applications (Table 1) these techniques can provide more sound results. The applications taken into account are:

- Detection: new landslides recognition from spaceborne imagery
- Mapping: identification of landslide perimeter and extension
- Characterization: retrieving information on failure mechanism, velocity and volume involved
- Hotspot mapping: detecting and mapping extremely slow and very slow landslides and consequently recognizes the areas on which field surveys, in situ monitoring and validation campaigns have to be successively planned with higher priority (Bianchini et al. 2013)
- Back monitoring: processing past data for retrieving past deformation patterns and time series

Table 1 Main applications of satellite optical and satellite radar for landslide analysis

	Landslide detection	Landslide mapping/landslide inventory	Characterization	Hotspot mapping	Back-monitoring	Monitoring/early warning	Time series analysis
Satellite optical	Yes	Yes	No	No	No	No	No
Satellite radar	Yes	Yes	Yes	Yes	Yes	No	Yes

- Early warning: the set of capacities needed to generate and disseminate timely and meaningful warning information to enable individuals, communities and organizations threatened by a hazard to prepare and to act appropriately and in sufficient time to reduce the possibility of harm or loss.

References

Adam N, Rodriguez-Gonzalez F, Parizzi A, Liebhart W (2011) Wide area persistent scatterer interferometry. In: Proceedings of IGARSS 2011, Vancouver, Canada

Baatz M, Schäpe A (2000) Multiresolution Segmentation—an optimization approach for high quality multi-scale image segmentation. In: Strobl, Blaschke, Griesebner (eds) Angewandte Geographische In formationsverarbeitung XII. Wichmann-Verlag, Heidelberg, pp12–23

Bardi F, Frodella W, Ciampalini A, Bianchini S, Del Ventisette C, Gigli G, Casagli N (2014) Integration between ground based and satellite SAR data in landslide mapping: the san fratello case study. Geomorphology 223:45–60

Bardi F, Raspini F, Ciampalini A, Kristensen L, Rouyet L, Lauknes TR, Casagli N (2016) Space-borne and ground-based InSAR data integration: the åknes test site. Remote Sens 8(3):237

Benz UC, Hofmann P, Willhauck G, Lingenfelder I, Heynen M (2004) Multi-Resolution, object-oriented fuzzy analysis of remote sensing data for GIS-ready information. ISPRS J Photogram Remote Sens 58:239–258

Berardino P, Fornaro G, Lanari R, Sansosti E (2002) A new algorithm for surface deformation monitoring based on small baseline differential SAR interferograms. IEEE Trans Geosci Remote Sens 40(11):2375–2383

Berardino P, Costantini M, Franceschetti G, Iodice A, Pietranera L, Rizzo V (2003) Use of differential SAR interferometry in monitoring and modelling large slope instability at Maratea (Basilicata, Italy). Eng Geol 68(1–2):31–51

Berti M, Corsini A, Franceschini S, Iannacone JP (2013) Automated classification of persistent scatterers interferometry time-series. Nat Hazards Earth Syst Sci Discuss 1(1):207–246

Bianchini S, Cigna F, Righini G, Proietti C, Casagli N (2012) Landslide hotspot mapping by means of persistent scatterer interferometry. Environ Earth Sci 67(4):1155–1172

Bianchini S, Pratesi F, Nolesini T, Casagli N (2015) Building deformation assessment by means of persistent scatterer interferometry analysis on a landslide-affected area: the Volterra (Italy) case study. Remote Sens 7(4):4678–4701

Bianchini S, Herrera G, Mateos RM, Notti D, Garcia I, Mora O, Moretti S. (2013) Landslide activity maps generation by means of persistent scatterer interferometry. Remote Sens 5:6198–6222

Birky AK (2001) NDVI and a simple model of deciduous forest seasonal dynamics. Ecol Model 143:43–58

Blaschke T (2010) Object based image analysis for remote sensing. ISPRS J Photogram Remote Sens 65:2–16

Canuti P, Casagli N, Ermini L, Fanti R, Farina P (2004) Land-slide activity as a geoindicator in Italy: significance and new perspectives from remote sensing. Environ Geol 45:907–919

Canuti P, Casagli N, Catani F, Falorni G, Farina P (2007) Integration of remote sensing techniques in different stages of landslide response. In: Sassa K, Fukuoka H, Wang F, Wang G (eds) Progress in landslide sciences. Springer-Verlag, Berlin, pp 251–259

Casagli N, Fanti R, Nocetini M, Righini G (2005) Assessing the capabilities of VHR satellite data for debris flow mapping in the Machu Picchu area. In: Sassa K, Fukuoka H, Wang F, Wang G (eds) Landslides: risk analysis and sustainable disaster management, vol 1539. Springer, pp 61–70

Cascini L, Fornaro G, Peduto D (2009) Analysis at medium scale of low-resolution DInSAR data in slow-moving landslide- affected areas. ISPRS J Photogram Remote Sens 64(598–611):2009. doi:10.1016/j.isprsjprs.2009.05.003

Cascini L, Fornaro G, Peduto D (2010) Advanced low-and full-resolution DInSAR map generation for slow-moving landslide analysis at different scales. Eng Geol 112(1):29–42

Catani F, Farina P, Moretti S, Nico G, Strozzi T (2005) On the application of SAR interferometry to geomorphological studies: estimation of landform attributes and mass movements. Geomorphology 66:119–131

Cheng KS, Wei C, Chang SC (2004) Locating landslides using multi-temporal satellite images. Adv Space Res 33:296–301

Chini M, Cinti FR, Stramondo S (2011) Co-seismic surface effects from very high resolution panchromatic images: the case of the 2005 Kashmir (Pakistan) earthquake. Nat Hazards Earth Syst Sci 11:931–943

Ciampalini A, Garfagnoli F, Antonielli B, Del Ventisette C, Moretti S (2012) Photo-lithological map of the southern flank of the Tindouf Basin (Western Sahara). J Maps 8:453–464

Ciampalini A, Bardi F, Bianchini S, Frodella W, Del Ventisette C, Moretti S, Casagli N (2014) Analysis of building deformation in landslide area using multisensor PSInSAR™ technique. Int J Appl Earth Obs Geoinf 33:166–180

Ciampalini A, Raspini F, Bianchini S, Frodella W, Bardi F, Lagomarsino D, Di Traglia F, Moretti S, Proietti C, Pagliara P, Onori R, Corazza A, Duro A, Basile G, Casagli N (2015a) Remote sensing as tool for development of landslide databases: the case of the Messina Province (Italy) geodatabase. Geomorphology 249:103–118

Ciampalini A, Raspini F, Moretti S (2015b) Landslide back monitoring and forecasting by using PSInSAR technique: the case of Naso (Sicily, Southern Italy). Atti Soc Tosc Sci Nat Mem Serie A 122. doi:10.2424/ASTSN.M.2015.16

Cigna F, Del Ventisette C, Liguori V, Casagli N (2011) Advanced radar-interpretation of InSAR time series for mapping and characterization of geological processes. Nat Haz Earth Syst Sci 11:865–881

Colesanti C, Wasowski J (2006) Investigating landslides with space-borne Synthetic Aperture Radar (SAR) interferometry. Eng Geol 88:173–199

Congalton RG (1991) A review of assessing the accuracy of classifications of remotely sensed data. Remote Sens Environ 37:35–46

Cruden DM, Varnes DJ (1996) Landslide types and processes. In: Turner AK, Schuster RL (eds) Landslides: investigation and mitigation, Sp. Rep. 247, Transportation Research Board, National Research Council, National Academy Press, Washington DC, USA, pp 36–75

De Guidi G, Scudero S (2013) Landslide susceptibility assessment in the Peloritani Mts. (Sicily, Italy) and clues for tectonic control of relief processes. Nat Hazards Earth Syst Sci 13:949–963

Dewey JF, Helman ML, Knott SD, Turco E, Hutton DHW (1989) Geol Soc London Spec Publ 45(1 Jan): 265–283. https://doi.org/10.1144/GSL.SP.1989.045.01.15

Del Ventisette C, Garfagnoli F, Ciampalini A, Battistini A, Gigli G, Moretti S, Casagli N (2012) An integrated approach to the study of catastrophic debris-flows: geological hazards and human influence. Nat Haz Earth Syst Sci 12:2907–2922

Eyers R, Moore JMM, Hervás J, Liu JG (1998) Integrated use of landsat TM and SPOT panchromatic imagery for landslide mapping: case histories from southeast Spain. Geol Soc Lond Eng Geol Spec Publ 1 Jan 133–140

Farina P, Casagli N, Ferretti A (2008) Radar-interpretation of InSAR measurements for landslide investigations in civil protection practices. Proceedings of 1st North American landslide conference, Vail, Colorado, pp 272–283

Ferretti A, Prati C, Rocca F (2000) Nonlinear subsidence rate estimation using permanent scatterers in differential SAR interferometry. IEEE Trans Geosci Remote Sens 38(5):2202–2212

Ferretti A, Prati C, Rocca F (2001) Permanent scatterers in SAR interferometry. IEEE Trans Geosci Remote Sens 39(1):8–20

Ferretti A, Fumagalli A, Novali F, Prati C, Rocca F, Rucci A (2011) A new algorithm for processing interferometric data-stacks: SqueeSAR™. IEEE Trans Geosci Remote Sens 99:1–11

Fiorucci F, Cardinali M, Carlà R, Rossi R, Mondini AC, Santurri L, Ardizzone F, Guzzetti F (2011) Seasonal landslide mapping and estimation of landslide mobilization rates using aerial and satellite images. Geomorphology 129:59–70

Frodella W, Ciampalini A, Gigli G, Lombardi L, Raspini F, Nocentini M, Scardigli C, Casagli N (2016) Synergic use of satellite and ground based remote sensing methods for monitoring the San Leo rock cliff (Northern Italy). Geomorphology 264:80–94

Fruneau B, Achache J, Delacourt C (1996) Observation and modeling of the Saint-Etienne-de-Tine´e Landslide using SAR interferometry. Tectonophysics 265

García-Davalillo JC, Herrera G, Notti D, Strozzi T, Álvarez-Fernández I (2014) DInSAR analysis of ALOS PALSAR images for the assessment of very slow landslides: the Tena Valley case study. Landslides 11(2):225–246

Guzzetti F, Mondini A, Cardinali M, Fiorucci F, Santangelo M, Chang K-T (2012) Landslide inventory maps: new tools for an old problem. Earth Sci Rev 112:42–66

Hay GJ, Blaschke T, Marceau DJ, Bouchard A (2003) A comparison of three image-object methods for the multiscale analysis of landscape structure. ISPRS

Herrera G, Davalillo JC, Mulas J, Cooksley G, Monserrat O, Pancioli V (2009) Mapping and monitoring geomorphological processes in mountainous areas using PSI data: central pyrenees case study. Nat Hazards Earth Syst Sci 9:1587–1598

Herrera G, Notti D, Garcıa-Davalillo JC, Mora O, Cooksley G, Sanchez M, Arnaud A, Crosetto M (2011) Landslides analysis with C- and X-band satellite SAR data: the portalet landslide area. Landslides 8:195–206

Hervas J, Barredo JI, Rosin PL, Pasuto A, Mantovani F, Silvano S (2003) Monitoring landslides from optical remotely sensed imagery: the case history of Tessina landslide, Italy. Geomorphology 54:63–75

Hilley GE, Bürgmann R, Ferretti A, Novali F, Rocca F (2004) Dynamics of slow-moving landslides from permanent scatterer analysis. Science 304 (5679):1952–1955

Hooper A, Zebker HA, Segall P, Kampes B (2004) A new method for measuring deformation on volcanoes and other natural terrains using InSAR persistent scatterers. Geophys Res Lett 31

Joyce KE, Belliss SE, Samsonov SV, McNeill SJ, Glassey PJ (2009) A review of the status of satellite remote sensing and image processing techniques for mapping natural hazards and disasters. Prog Phys Geogr 33(2):183–207

Kurtz C, Stumpf A, Malet J-P, Gançarski P, Puissant A, Passat N (2014) Hierarchical extraction of landslides from multiresolution remotely sensed optical images. ISPRS J Photog Remote Sens 87:122–136

Lamri T, Djemaï S, Hamoudi M, Zoheir B, Bendaoud A, Ouzegane K, Amara M (2016) Satellite imagery and airborne geophysics for geologic mapping of the Edembo area, Eastern Hoggar (Algerian Sahara). J Afr Earth Sci 115:143–158

Lauknes TR, Piyush Shanker A, Dehls JF, Zebker HA, Henderson IHC, Larsen Y (2010) Detailed rockslide mapping in northern Norway with small baseline and persistent scatterer interferometric SAR time series methods. Remote Sens Environ 114:2097–2109

Lin CY, Lo HM, Chou WC, Lin WT (2004) Vegetation recovery assessment on the Jou-Jou Mountain landslide area caused by the 921 earthquake in the Central Taiwan. Ecol Model 176:75–81

Lu P, Stumpf A, Kerle N, Casagli N (2011) Object-oriented change detection for landslide rapid mapping. Geosci Remote Sens Lett 8(701–705):2011

Lu P, Casagli N, Catani F, Tofani V (2012) Persistent Scatterers Interferometry Hotspot and Cluster Analysis (PSI-HCA) for detection of extremely slow-moving landslides. Int J Remote Sens 33(2):466–489

Ma H-R, Cheng X, Chen L, Zhang H, Xiong H (2016) Automatic identification of shallow landslides based on worldview2 remote sensing images. J Appl Remote Sens 10. doi:10.1117/1.JRS.10.016008

Mantovani F, Soeters R, Van Westen CJ (1996) Remote sensing techniques for landslide studies and hazard zonation in Europe. Geomorphology 15:213–225

Marcelino EV, Formaggio AR, Maeda EE (2009) Landslide inventory using image fusion techniques in Brazil. Int J Appl Earth Obs Geoinf 11:181–191

Martha TR, Kerle N (2012) Creation of event-based landslide inventory from panchromatic images by object oriented analysis. In: Proceedings of the 4th GEOBIA, 7–9 May 2012, Rio de Janeiro, Brazil, p 053

Martha TR, Kerle N, Jetten V, van Westen CJ, Kumar KV (2010) Characterizing spectral, spatial and morphometric properties of landslides for semi-automatic detection using object-oriented methods. Geomorphology 116:24–36

Massonnet D, Feigl KL (1998) Radar interferometry and its ap- plication to changes in the earth's surface. Rev Geophys 36:441–500

Meisina C, Zucca F, Notti D, Colombo A, Cucchi A, Savio G, Bianchi M (2008) Geological interpretation of PSInSAR data at regional scale. Sensors 8 (11):7469–7492

Mondini AC, Guzzetti F, Reichenbach P, Rossi M, Cardinali M, Ardizzone F (2011) Semi-automatic recognition and mapping of rainfall induced shallow landslides using optical satellite images. Remote Sens Environ 115:1743–1757

Mondini AC, Viero A, Cavalli M, Marchi L, Herrera G, Guzzetti F (2014) Comparison of event landslide inventories: the Pogliaschina catchment test case. Italy Nat Hazards Earth Syst Discuss 2:1093–1125

Mwaniki MV, Agutu NO, Mbaka JG, Ngigi TG, Waithaka EH (2015) Landslide scar/soil erodibility mapping using Landsat TM/ETM + bands 7 and 3 normalised difference index: a case study of central region of Kenya. App Geography 64:108–120

Nichol J, Wong MS (2005) Satellite remote-sensing for detailed landslide inventories using change detection and image fusion. Int J Remote Sens 26:1913–1926

Parker RN, Densmore AL, Rosser NJ, de Michele M, Li Y, Huang R, Whadcoat S, Petley DN (2011) Mass wasting triggered by the 2008 Wenchuan earthquake is greater than orogenic growth. Nat Geosci 4(7):449–452

Petley D (2012) Global patterns of loss of life from landslidesm. Geology 40:927–930

Petley DN, Dunning SA, Rosser NJ (2005) The analysis of global landslide risk through the creation of a database of world-wide landslide fatalities. In: Hungr O, Fell R, Couture R, Eberhardt E (eds) Landslide risk management, Taylor & Francis Group, London, ISBN 04 1538 043 X

Raspini F, Ciampalini A, Del Conte S, Lombardi L, Nocentini M, Gigli G, Casagli N (2015a) Exploitation of amplitude and phase of satellite sar images for landslide mapping: the case of montescaglioso (South Italy). Remote Sensing 7(11):14576–14596

Raspini F, Ciampalini A, Bianchini S, Bardi F, Di Traglia F, Basile G, Moretti S (2015b) Updated landslide inventory of the area between the Furiano and Rosmarino creeks (Sicily, Italy). J Maps 1–10

Righini G, Pancioli V, Casagli N (2012) Updating landslide inventory maps using Persistent Scatterer Interferometry (PSI). Int J Remote Sens 33(7):2068–2096

Rott H, Scheuchl B, Siegel A, Grasemann B (1999) Monitoring very slow slope motion by means of SAR interferometry: a case study from a mass waste above a reservoir in the Ötztal Alps. Austria Geophys Res Lett 26:1629–1632

Scaioni M, Longoni L, Melillo V, Papini M (2014) Remote sensing for landslide investigations: an overview of recent achievements and perspectives. Remote Sens 6. doi:10.3390/rs60x000x

Schuster, RL (1996) Socioeconomic significance of landslides. In: Turner AK, Schuster RL (eds) Landslides: investigation and mitigation. Transportation Research Board, Special Report 247, National Academy Press, WA (1995), 12–35

Schuster RL, Highland LM (2001) Socioeconomic and environmental impacts of landslides in the western hemisphere. US geological survey open-file report 01-0276, 47 pp. available at: http://pubs.usgs.gov/of/2001/ofr-01-0276/

Senay GB, Elliott RL (2000) Combining AVHRR-NDVI and land use data to describe temporal and spatial dynamics of vegetation. For Ecol Manage 128:83–91

Singhroy V (1995) SAR integrated techniques for geohazard assessment. Adv Space Res 15:67–78

Singhroy V, Mattar KE, Gray AL (1998) Landslide characterisation in Canada using interferometric SAR and combined SAR and TM images. Adv Space Res 21:465–476

Story M, Congalton RG (1986) Accuracy assessment: a user's perspective. Photogram Engand Remote Sens 52:397–399

Strozzi T, Farina P, Corsini A, Ambrosi C, Thüring M, Zilger J, Wiesmann A, Wegmüller U, Werner C (2005) Survey and monitoring of landslide displacements by means of L-band satellite SAR interferometry. Landslides 193–201

Tofani V, Segoni S, Agostini A, Catani F, Casagli N (2013a) Technical note: use of remote sensing for landslide studies in Europe, Nat Hazards Earth Syst Sci 13:1–12

Tofani V, Raspini F, Catani F, Casagli N (2013b) Persistent Scatterer Interferometry (PSI) technique for landslide characterization and monitoring. Remote Sens 5(3):1045–1065

Tucker CJ, Townshend JRG, Goff TE (1985) African land-cover classification using satellite data. Science 227:369–375

van Westen CJ, Castellanos E, Kuriakose SL (2008) Spatial data for landslide susceptibility, hazard, and vulnerability assessment: An overview. Eng Geol 102:112–131

TXT-tool 2.039-3.2
Ground-Based Remote Sensing Techniques for Landslides Mapping, Monitoring and Early Warning

Nicola Casagli, Stefano Morelli, William Frodella, Emanuele Intrieri and Veronica Tofani

Abstract

The current availability of advanced remote sensing technologies in the field of landslide analysis allows rapid and easily updatable data acquisitions, improving the traditional capabilities of detection, mapping and monitoring, optimizing field work, and allowing to investigate hazardous and inaccessible areas while granting at the same time the safety of the operators. In the recent years in particular, ground-based remote sensing techniques have undergone a significant increase of usage, thanks to their technological development and quality data improvement, offering advantages with respect to air- or spaceborne remote sensing techniques, in terms of data spatial resolution and accuracy, fast measurement and processing times, and portability and cost-effectiveness of the acquiring instruments. These advantages can be highlighted in the framework of landslide emergency management, when it is often urgently necessary to minimize survey time when operating in dangerous environments and gather all the required information as fast as possible. In this paper, the potential of some ground-based remote sensing techniques and the effectiveness of their synergic use is explored in several case studies, analyzing different slope instability processes at different scales of

N. Casagli (✉) · S. Morelli · W. Frodella · E. Intrieri · V. Tofani
Department of Earth Sciences, University of Firenze, via G. La Pira n.4, 50121 Firenze, Italy
e-mail: nicola.casagli@unifi.it

S. Morelli
e-mail: stefano.morelli@unifi.it

W. Frodella
e-mail: william.frodella@unifi.it

E. Intrieri
e-mail: emanuele.intrieri@unifi.it

V. Tofani
e-mail: veronica.tofani@unifi.it

K. Sassa et al. (eds.), *Landslide Dynamics: ISDR-ICL Landslide Interactive Teaching Tools*, https://doi.org/10.1007/978-3-319-57774-6_18

emergency or post-emergency management. Thanks to them and to the support of existing bibliography, the most common fields of application are suggested for all the considered ground-based sensor technologies and their level of effectiveness is evaluated in relation to the dynamics of landslide types.

Keywords
Landslide mapping and monitoring · Early warning
Terrestrial laser scanning · Infrared thermography · GB-InSAR

Contents

1 Introduction 256
2 Theoretical Principles of Techniques 257
2.1 Terrestrial Laser Scanner (TLS) 257
2.2 Ground-Based Interferometric Synthetic Aperture Radar (GB-InSAR) 257
2.3 InfraRed Thermography (IRT) 259
2.4 Robotic Total Station (RTS) 259
2.5 Digital Photogrammetry Workstation (DPW) 260
3 Applications in Case Studies 261
3.1 Western Elba Island Coastline 262
3.2 San Leo Rock Cliff 264
3.3 Santa Trada Landslide 265
4 Discussion and Conclusion 269
References 271

1 Introduction

Landslides play an important role in the evolution and shaping of subaerial landscapes (Brunetti et al. 2015) and are a major cause of loss of life, injuries, property damage, socio-economic disruption and environmental degradation in many areas of the emerged continents (Alexander 1993), especially if they are associated with other natural hazards (like earthquakes, volcanic eruptions, meteorological events and wildfires). Because of such habitual combinations, reliable numbers for the social impact only due to landslides are difficult to obtain on a global scale and the economic losses are certainly underestimated (or not quoted at all). This general condition often contributes to reducing the concern of individuals and authorities have about landslide risk (Kjekstad and Highland 2009). However, some remarkable statistics are nowadays available for the worldwide exposure to landslide hazard issues. For example, a recent World Bank report (Dilley et al. 2005) presents the following profile:

- Land area of the globe exposed to landslides: 3.7 million km^2
- Population exposed: 300 million, or 5% of the world population
- Land area identified as high risk zones: 820,000 km^2
- Population living in high risk areas: 66 million people

Although in most of the disaster-prone areas the consideration of the social-cultural and socio-economic conditions in relation to their physical safety is still very confused, the application of appropriate technologies for landslide detection, monitoring and early warning systems are increasingly considered crucial by local authorities in reducing the risk of landslide disasters.

In the field of slope instability detection, mapping and monitoring, during a short/long term landslide management (real time and near real time and deferred time), advanced ground-based remote sensing technologies are nowadays available (Lillesand et al. 2014). They are characterized by operational efficiency and accuracy of data not reached by traditional methods: high-resolution acquisition, multifunction versatility, device portability, low cost sensors, easy and fast data processing. Such equipment allows for systematic and easily updatable acquisitions of data that may also enhance the implementation of effective early warning systems at slope scale. In this paper, the potential of the most used terrestrial remote sensing techniques is introduced through a description of their main technical features and applicability in different observed scenarios (typology of landslide and geomorphological setting). Some case studies affected by criticalities in the framework of Civil Protection, are also shown and discussed in order to exhibit good practices in landslide characterization and prediction by means of different techniques and sensors in synergic action.

2 Theoretical Principles of Techniques

2.1 Terrestrial Laser Scanner (TLS)

A Terrestrial Laser Scanner (TLS) device consists of a directional, coherent and in-phase electromagnetic radiation beam (or discrete pulses) transmitter, a back-scattered signals receiver and a scanning mechanism (e.g. a rotating mirror) (Fig. 1a). The TLS device, by measuring with high accuracy (millimeter or centimeter) the laser time-of-flight between the sensor and the reflecting targets, is capable of obtaining the exact position of a mesh of points (point cloud), characterized by (x, y, z) Cartesian coordinates and reflectance value R (x, y, z), providing in a short time a high resolution 3D surface digital model of the scanned object (Slob et al. 2007; Frohlich and Mettenleiter 2004) (Fig. 1b). The high acquisition rate (up to hundreds of thousands points per second) makes the detailed 3-D shape of the object immediately available. TLS is more and more used for the analysis of slopes characterized by instability processes, as it safely allows in a short time a high detailed and accurate 3D representation of the investigated rock mass plano-altimetric morphological and geostructural setting, and is therefore particularly suitable for geostructural characterization and unstable rock cliffs monitoring applications (Abellán et al. 2006, 2011; Ferrero et al. 2009; Gigli et al. 2014a, b; Oppikofer et al. 2009). Thanks to the high resolution of the laser scanning survey it is also possible to extract even the smallest features, such as the structural crack pattern, the crack opening direction (Gigli et al. 2009), and the orientation of critical discontinuities within the rock mass (Gigli and Casagli 2011; Gigli et al. 2014b). Furthermore this technique is capable of measuring ground 3D temporal displacements by comparing sequential datasets of the same scenario (Rosser et al. 2005; Kasperski et al. 2010; Abellán et al. 2011). Furthermore, for each point the intensity of the reflected signal is acquired. The intensity data can provide some information about the type of material and the soil moisture content of the targets (Pesci and Teza 2008; Voegtle et al. 2008; Franceschi et al. 2009). The point cloud high resolution and the variable laser return intensity values can also add information regarding the landslide main geomorphologic features. Finally, by placing several laser reflectors within the surveyed area, and defining their coordinates by means of a differential RTK-GPS survey (Tapete et al. 2015; Morelli et al. 2012), it is possible to link the different acquired point clouds to a global reference system.

2.2 Ground-Based Interferometric Synthetic Aperture Radar (GB-InSAR)

Ground-Based Interferometric Synthetic Aperture Radar (GB-InSAR) is an apparatus adopting a technique derived from satellite interferometry, with which shares the same base working principles but on a ground-based platform (Fig. 1c). Basically, the system emits two microwave signals in two different times; the waves reach the target (e.g. the landslide) and are backscattered to the radar, where their amplitude and phase are

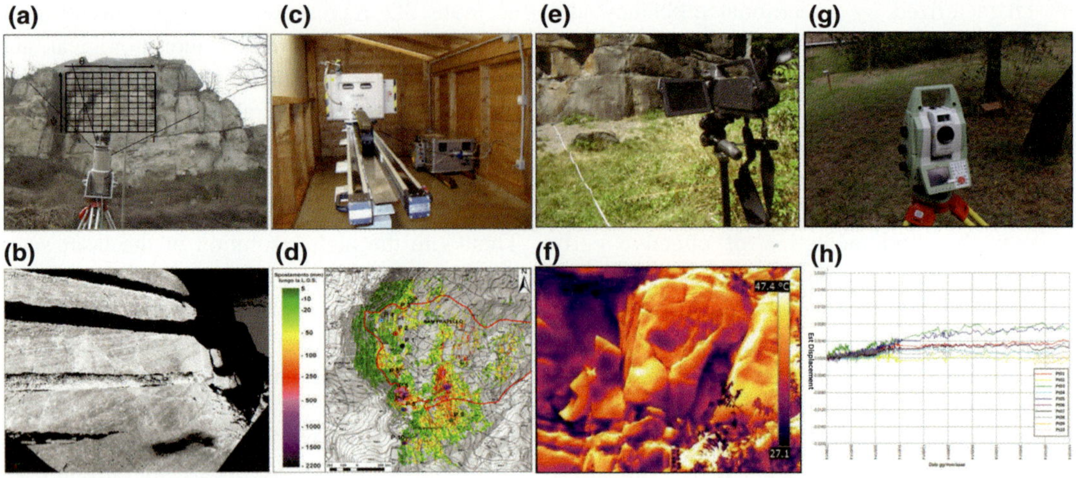

Fig. 1 The employed ground-based landslide surveying devices and related product: **a**, **b** terrestrial laser scanner; **c**, **d** GB-InSAR system and displacement map; **e**, **f** infrared thermal camera and surface temperature map; **g**, **h** robotized total station and target time series

measured. Should any movement occur between the two acquisitions, a phase difference (called interferometric phase) is measured (Fig. 2). Then from the phase difference it is possible to calculate the actual surface displacement along the radar line of sight (LOS) of the with millimeter accuracy (Figs. 1d and 2b). This means that interferometry provides only the displacement component along the radar LOS. From the amplitude, power images relative to the radar reflectivity of the scenario can be obtained; this is useful to understand the main morphological features on the radar image. On the other hand, from the phase it is possible to compute interferograms, i.e. images in which every pixel corresponds to a phase difference, from which it is possible to obtain 2D displacement maps (Figs. 1d and 2b).

In an interferogram the chromatic scale that indicates the displacement (directly related to the phase difference) covers a total value corresponding to half of the wavelength used. However, since the phase is periodic, it cyclically assumes the same values. This may cause problems in interpreting the image as the same colour can indicate any displacement equal to the value represented on the scale plus a multiple of $\lambda/2$ (Fig. 2b). This issue, known as phase ambiguity, is the same phenomenon that makes it possible to confuse 12 P.M. with 12 A.M. on an analogical watch. It can be solved through interpretation based on geological knowledge and through adopting apposite phase unwrapping algorithms that count the number of cycles performed by the wave (Ghiglia and Romero 1994). Unwrapping algorithms permit to represent the displacement as a linear colour bar in which the correspondence between colour and displacement is one-to-one. The resulting images are usually called cumulated displacement maps. In order to calculate the displacement from the phase difference, it must be assumed that a series of other contributions influencing the phase (such as air humidity and temperature, the dielectric constant of the target, relative position of the main reflectors within a pixel cell) keep constant and therefore are elided during the computation of the interferometric phase. A measure of how much the scenario is "stable" is given by the coherence parameter, ranging from 0 (total change) to 1 (maximum stability). The pixel of a radar image are not square and are instead characterized by two different resolutions: range resolution (along the LOS direction) and azimuth resolution (perpendicular to the LOS direction). These vary with certain parameters as indicated in Fig. 2. The main research applications of GB-InSAR

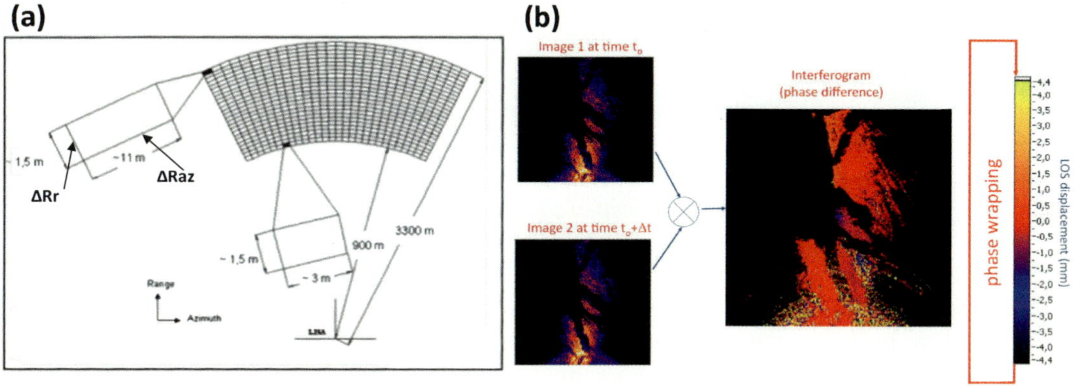

Fig. 2 **a** Resolution grid size and parameters; **b** simplified interpherometric principles

soon became focused on slope monitoring (Tarchi et al. 2003; Pieraccini et al. 2003), for civil protection purposes (Del Ventisette et al. 2011; Intrieri et al. 2012; Bardi et al. 2014) and, more recently, for mining safety (Farina et al. 2011; Severin et al. 2014). Other successful or promising fields include monitoring volcanoes (Casagli et al. 2003; Di Traglia et al. 2013, 2014a, b; Intrieri et al. 2013; Calvari et al. 2016), cultural heritage (Tapete et al. 2013; Pratesi et al. 2015), glaciers and snowpack (Caduff et al. 2015a), sinkholes (Intrieri et al. 2015).

2.3 InfraRed Thermography (IRT)

InfraRed Thermography (IRT), or thermal imaging, is a remote sensing technique capable of mapping the surface temperature pattern evolution, leading to the detection of thermal anomalies within the investigated scenario (Spampinato et al. 2011). In recent years, IRT has undergone a significant increase of applications in the geosciences, thanks to the technological development of portable and cost-effective thermal imaging cameras, as well as the fast measurement and processing times of thermographic data. Nevertheless in the study of slope instability processes IRT is still not widely applied, apart from a few interesting experimental studies (e.g., Wu et al. 2005; Squarzoni et al. 2008; Baroň et al. 2012; Teza et al. 2012; Gigli et al. 2014a; Frodella et al. 2014a, b, 2015). The product of an infrared thermographic survey is a pixel matrix (thermogram), collected through the thermal camera array detector (Fig. 1e, f). After the correction of the sensitive parameters (object emissivity, path length, air temperature and humidity), the thermogram represents a radiant temperature map of the investigated object, which is a function of heat flow crossing the analyzed surface and of local boundary conditions (Maldague 2001). The presence of any inhomogeneity within the surface (i.e. cracks and fractures, moisture and seepage zones, subsurface voids), will influence the material thermal parameters (density, thermal capacity and conductivity), modifying its heat transfer (Teza et al. 2012). Therefore, the inhomogeneity will be displayed in the radiant temperature map of the analyzed scenario as a “thermal anomaly” (an irregular thermal pattern with respect to the surroundings). As regards the analysis of instability phenomena, thermal anomalies can reveal the presence of potential criticalities such as open fractures, scarps, moisture and seepage zones.

2.4 Robotic Total Station (RTS)

A RTS is an advanced electronic instrument used in modern field surveying for topographic measurements (macro- and micro-scale application), as it allows the operator to control the instrument from a distance via remote control (Kvamme et al. 2006). Its main operational advantage is the ability to collect, store and process data at very high accuracy (10^{-3} to 10^{-2} m), for multiple benchmarks on the landslide surface with a high measurement frequency (Giordan et al. 2013). This device is based on an electronic theodolite, which exactly measures angles, and it is integrated with an electronic distance meter capable of reading distances from the instrument to points on the landslide surface (Fig. 1g). After a first processing within its own polar coordinate system x, y, z, having a common origin point coincident with the optical center of the instrument, it provides three-dimensional topographic measurements in reference to a standard grid system (e.g. UTM-WGS84) (Fig. 1h). RTSs require the installation of optical benchmarks on the point to be monitored. However, installation of benchmarks for TSs is less onerous (in terms of cost, time and technical skills deployment) than the installation of GPS antennas in the same observed points. Similarly to GPS technique, TSs can be used to perform repeated surveys at defined time intervals on medium-long periods, or high frequency measurements (<20 min, depending on the network characteristics). Another advantage of RTSs consists in the ability to measure a large spectrum of topographic change rates, from mm day^{-1} to m day^{-1}, without changing the instrumentation setting. All these properties are significant for a landslide early warning system which exploits topographical measurements, especially when monitoring landslide types characterised by large areas and variable temporal modification rates. On the other hand, the effectiveness of RTSs may be reduced by: (i) adverse weather conditions (e.g. mist and intense rainfall) influencing the accuracy measurement; (ii) high distance (measurements are performed along the free optical LOS between the instrument and the individual optical benchmarks) and (iii) great plano-altimetric changes (the operational lifetime of individual benchmarks is limited where the rates of plano-altimetric change are >2 m day^{-1}). State-of-the-art image processing technology delivers live video streaming with images of very high quality. By means of the installed remote control system, it is possible to steer the RTS, transfer and process all the collected data according to a planned schedule that may vary according to the landslide speed and the total amount of monitored points.

2.5 Digital Photogrammetry Workstation (DPW)

Terrestrial digital photogrammetry is a well-established technique for acquiring dense 3D geometric information in slopes from stereoscopic overlaps of photo sequences captured by a calibrated digital camera set in a fixed position on the ground in front of scenario and without any physical contact (Zhang et al. 2004). During past few years, with the rapid development of digital photogrammetry techniques and the availability of ease-using, focusable and relatively cheap digital cameras, the method and device used in terrestrial photogrammetry changed greatly. Therefore, thanks to its increasing efficiency through time it gained wide applications in many fields such as 3D building reconstruction, heritage protection and landslides studies (Fan et al. 2016; Grussenmeyer et al. 2008; Scaioni et al. 2015). However, in spite of great development of traditional terrestrial photogrammetry over years, it is still necessary to set some control points on the slopes to be measured (Stavroulaki et al. 2016) if not supported by other technologies (Forlani et al. 2014). For the complex nature of the observed scenarios the setting control points around the objects is sometimes a very time-consuming and labour-intensive job. Principally, photogrammetry can be divided into two fields of activity depending on the lens-setting: Far range (with camera distance setting to indefinite) and Close range (with camera settings to finite values; i.e. up to about 300 m) (Gopi 2007). In the field of landslides studies and disaster management the first type is usually more exploited for landslide characterization and general mapping (Wolter et al. 2014) while the second type finds a wide applicability in high precision metrological and deformation monitoring applications (Cardenal et al. 2008; Scaioni et al. 2015). The processed outputs that address the needs of these objectives can be summarized as follows: (i) coordinate of required object points; (ii) topographic and thematic maps; (iii) rectified photographs or orthophotos (Gopi 2007).

3 Applications in Case Studies

In this section, we present three exemplifying case studies, in which some of the above-mentioned ground-based techniques were profitably exploited for the landslide emergency management in Italy (Fig. 3): TLS, GB-InSAR, IRT.

The employed TLS is a Long Range and High Accuracy 3-D time-of-flight (model Riegl LMS-Z420i; RIEGL 2010). This device is able to acquire the position of up to 12,000 points/s calculating the round-trip travelling time of the pulsed laser beam (near-infrared wavelength) from the instrument to the scanned object, with a maximum angular resolution of 0.008°, an accuracy of ±10 mm (one s at 50 m range under RIEGL test conditions) and from a maximum distance of 800 m.

The applied GB-InSAR, is a system designed and implemented by the Joint Research Centre (JRC) of the European Commission and its spin-off company Ellegi-LiSALab. This device is able to acquire high resolution 2D displacement data every few tens of seconds, making it an invaluable tool for both early warning and characterization purposes. It operates automatically round-the-clock and in all-weather conditions, but the main constraint is probably the need for a

Fig. 3 **a** Case studies location; **b** Western Elba Island (unstable rock masses); **c** San Leo (collapsed rock cliff); **d** Santa Trada (planar slide in metamorphic weathered rocks)

connection to the electricity network since the system cannot practically operate with batteries, generators or solar panels.

The used thermal camera is a hand-held or tripod-mounted FLIR SC620 model, characterized by a focal plane array microbolometer sensor with a 640-by-480 pixel matrix (leading to a 13 cm spatial resolution considering the specific acquiring distance), capable of measuring electromagnetic radiation in the thermal infrared band between 7.5 and 13 μm (Long Wavelength IR band), with a thermal accuracy of ±2 °C and a 0.65-mrad angular resolution. A built-in 3.2 Mpixel digital camera allowed the comparison between IR and photos taken in the visible range in order to improve the acquired thermograms interpretation.

3.1 Western Elba Island Coastline

The investigated area is located on the western hillside of Mount Capanne, overlooking the westernmost portion of the Elba Island coastline (Central Italy), along a 250 m stretch of the provincial roadway n° 25 (Fig. 3a, b). This roadway plays a key role in the Elba Island transit conditions, representing the only linear infrastructure connecting the villages located on the island western coastline, furthermore, being part of the Tuscan Archipelago National Park it has a panoramic relevance for tourism, the local population and authorities. This area is characterized by very steep rock slopes overlooking the roadway, which, due to its geostructural setting and degree of fracturing, in the recent years underwent the detachment of rock mass portions and rock debris, invading the roadway and in some occasions severely damaging its protection system (catchment nets and barriers). In order to define the risk scenarios for the roadway transportation security conditions, these instability occurrences were investigated through a methodology based on the integration of accurate geological and geomechanical field surveys and terrestrial remote sensing techniques, such as TLS and IRT (Frodella and Morelli 2013; Gigli et al. 2014a). The final aim was to define the area 3D geo-structural and geomechanical rock mass characterization, detect the potentially more hazardous rock mass portions, calculate their volume and perform a detailed stability analysis. IRT surveys in particular were carried out in correspondence of rock mass most critical sectors, in order to detect thermal anomalies connected to open fractures, water seepage and moisture zones, validate the unstable block volume calculation, rapidly assess the hydraulic conditions along the more critical rock mass discontinuities, contributing together with the TLS semiautomatic analysis to a more detailed remote 3D geometrical and geomechanical characterization.

From a geological point of view, the study area is characterized by complex structural setting (Gigli et al. 2014a). Figure 4c shows the stereographic projection of the collected data: five main discontinuity sets were identified, JN3 set in particular includes high persistent decimetric-spaced discontinuity planes (=exfoliation joints, EJ), which represent slipping planes that isolate large rock mass portions (Fig. 4c). The obtained TLS 3-D surface model, obtained by merging the different point clouds, contributed to characterize the morphological variability of the investigated area: a rough morphology, characterized by creek erosion gullies isolating jutting rock mass portions, was revealed (Fig. 4a). Furthermore a semi-automatic geo-structural survey was performed by means of a Matlab tool (DiAna = Discontinuity Analysis; Gigli and Casagli 2011), on a limited sector of the rock mass not covered by nets, rock bolts, and fences. Figure 4d reports the poles of the semi-automatically extracted discontinuities (labeled from D1 to D7, and represented in 3D in Fig. 4b), which improved the rock mass structural characterization, adding two more discontinuity sets to the five detected by means of the traditional field survey. Given the geological setting of the investigated area, and the most probable detected failure mechanism occurring (planar failure along JN3 discontinuity set), an iterative procedure was applied with the aim of identifying the maximum credible scenario. A Matlab routine was built for this purpose by moving on the 3-D surface a plane with the same orientation of JN3 set.

By selecting a volume threshold value of 1000 m^3, three protruding rock masses were detected and labeled from north to south as M1, M2, and M3 (Fig. 4a). The latter masses for their considerable extension, overhanging position, and shape (mainly elongated along the direction of maximum slope) were identified as potential critical rock mass sectors with regards to instability mechanisms. M3 rock mass, in addition to the basal slipping plane, is also delimited southeastward from the stable portion of the rock slope by a second sub-vertical plane (belonging to JN2 set in Fig. 4c and D3 in Fig. 4d). The resulting rock masses volumes (expressed in cubic meter) are 3706 (M1), 4359 (M2), and 1293 (M3). The obtained IRT surface temperature maps (thermograms), are represented by means of a color scale, in which the higher temperatures are displayed by the lighter colors, whereas the colder temperatures by the darker ones (Fig. 5). Warm thermal anomalies connected to air circulation were detected in correspondence of the open portions of the JN3 discontinuities delimiting the detected M1, M2, and M3 critical masses. The abovementioned discontinuities detected on the thermograms follow closely the EJ basal planes; this interpretation was strengthened by the comparison of the thermograms with the optical images that confirmed no evidence of water flow along the detected discontinuities. For these reasons, dry conditions were diagnosed for all M1, M2, and M3 basal slipping planes, and the absence of water pressure was considered in the stability analysis.

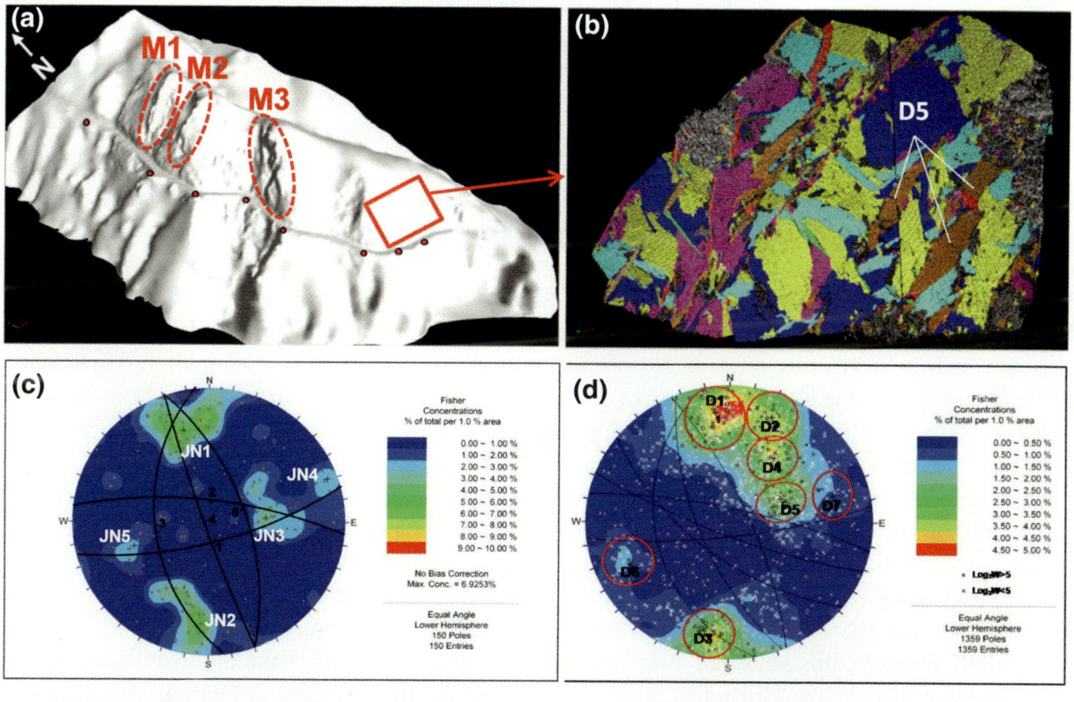

Fig. 4 **a** High-definition 3-D surface and maps of the investigated rock slope (**a**) (*dots* mark the different scan positions labeled from 1 to 8), while the *square* delimitates the sector where the semiautomatic geomechanical survey was carried out. **b** 3-D representation of all the joint sets extracted from the point cloud selected sector. Stereographic projection of discontinuity poles and modal planes of the main sets collected in the investigated area through the traditional field surveys (**c**) and the semi-automatic analysis (**d**)

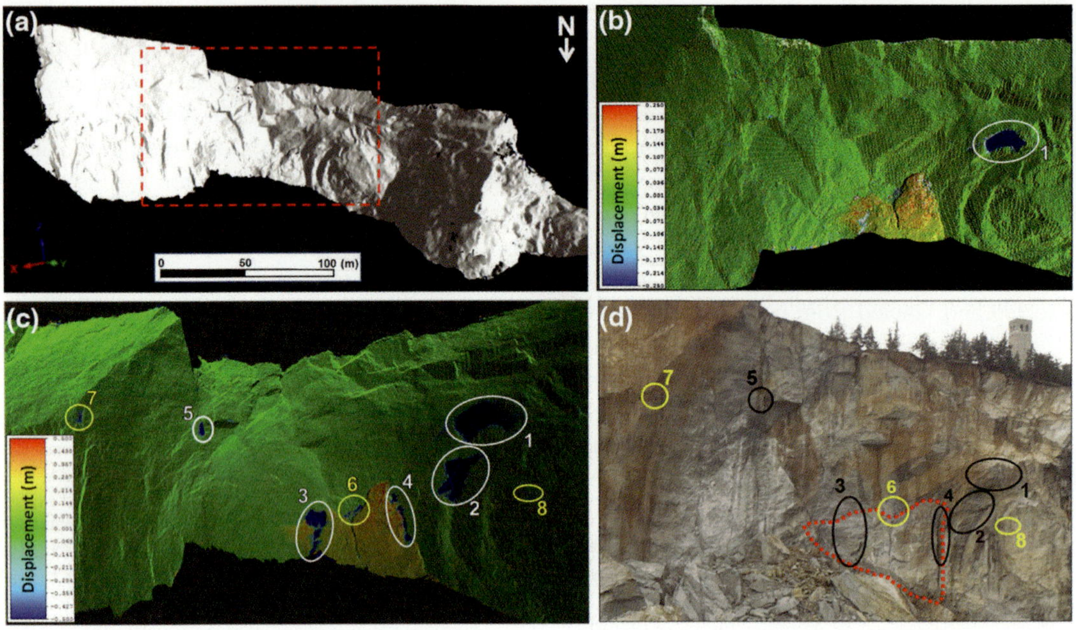

Fig. 5 Optical images of the unstable rock masses taken from the digital camera integrated in the laser scanner device from different scanning positions (**a** = M1; **c** = M2; **e** = M3); related 3-D digital model with the detected basal and lateral slipping planes (**b**, **d**, **f**), and mosaicked thermograms (**g** = M1, **h** = M2, and **i** = M3) acquired around 1 p.m., November 2011. *Dotted lines* mark the basal slipping planes; *white squares* on the thermogram allow a comparison with the correspondent sectors in the optical images, acquired by the built-in digital camera

3.2 San Leo Rock Cliff

The town of San Leo, renowned for its medieval Fortress and Romanesque churches, is located in the southwestern sector of the Emilia Romagna Region (northern Italy) (Fig. 3a, c), on top of a limestone isolated rock massif overlying clayey slopes, which is historically affected by instability phenomena (Benedetti et al. 2013). On February 27th, 2014 an entire portion of the rock plate north-eastern sector collapsed, causing a huge rock fall of about 0.30 Mm^3 (Borgatti et al. 2015). Minor structural damages were reported in the town and a few buildings were evacuated as a precautionary measure. Although no fatalities occurred and the San Leo cultural heritage suffered no damages, this landslide event caused a consistent retreat of the cliff edge, and minor rock falls kept taking place on the newly formed rock wall, worsening this hazardous situation. For these reasons a GB-InSAR monitoring activity coupled with TLS surveys were carried out, in order to manage the post-event emergency phase and evaluate the residual risk (Frodella et al. 2016). Furthermore IRT surveys were performed in order to integrate the TLS and GB-InSAR data in the analysis of the rock wall morphological and structural features.

In order to completely cover the intervention areas, four different laser scanning surveys (March 7th 2014, April 9th 2014, June 11th 2014 and December 8th 2014) were performed from different viewpoints. The 3D terrain model of the analysed rock wall sector revealed a rock wall surface, characterized by criticalities such as overhanging sectors, ledges and niches (Fig. 6a). 3-D temporal variations of the terrain model were detected by comparing sequential datasets acquired in the carried out different laser scanning surveys; the resulting 3-D rock wall temporal variations from March 7th 2014 to December 18th 2014 is shown in Fig. 6b, c. For each scanned point, the displacement vector component was determined perpendicularly to the local surface of the 3D model of the former scan. The deformational field analysis provided evidence of an ongoing rock block toppling (with an estimated volume of 450 m^3), which is almost completely detached from the rock wall central sector (Fig. 6b, c). The latter displacement evolution reached values ranging from 12 cm (March 7th–April 9th comparison), to 25 cm (March 7th–June 11th comparison; Fig. 6b) and finally to about 50 cm (March 7th–December 18th comparison; Fig. 6c). The scan comparison also provided the detection of block fall (areas colored in blue in Fig. 6b, c) that took place during the laser scanning monitoring period. The detachment of rock blocks is in fact the mass movement of highest frequency and lowest magnitude in the rock wall monitored area. The major rock block detachment occurred in between the April 9th–June 11th TLS monitoring period, involving a 94 m^3 rock volume (oval 1 in Fig. 6c, d), whereas the rest took place in between June 11th and December 18th, involving consistent detachments (ranging from 66 to 10 m^3 in volume) and minor block falls (ranging from 2 to 1 m^3 in volume; yellow ovals in Fig. 6c, d).

On March 7th 2014, a GB-InSAR system was installed opposite to the north-eastern sector of the analyzed rock wall, for the real time monitoring of the residual movements of both the rock wall collapse-affected sector and the rock fall deposits. The employed radar system operates in the k_u-band (wavelength range of 2.5–1.67 cm; central frequency of 17.2 GHz and 200 MHz band width), and is capable of achieving a 3 m synthetic aperture in a 2 min scanning time. The TLS 3D model was merged with the GB-InSAR data obtaining a 3D GB-InSAR cumulative displacement map, which allowed to read detected LOS displacements directly on the observed scenario 3D representation, and therefore both to better localize the most critical areas, and compare the different techniques displacement data (Fig. 7a).

The GB-InSAR data acquired during the first monitoring year allowed us to assess a relative stability of the rock cliff and the town structures observed, and to detect critical areas, corresponding to: (i) a detensioned rock block located at the foot of the monitored rock wall central sector (confirming the TLS analysis of an ongoing rock block toppling); and (ii) the rock fall deposits (metric and decametric size boulders and blocks in a coarse sandy-clayey matrix (maximum recorded cumulative displacement in the investigated area).

Infrared surface temperature maps collected on April 9th 2014 (following a period characterized by local intense rainfall) allowed to detect widespread seepage sectors in correspondence of a rock mass key discontinuity, corresponding to a high persistent normal fault dissecting the whole rock massif (oval 1 in Fig. 7b, c). In this geological, morphological and structural context discontinuities affected by seepage represent potential criticalities with respect to instability phenomena, as confirmed by minor seepage sectors (ovals 4 in Fig. 7b, c), which are located in correspondence of rock wall sectors affected by a widespread fracture network and block detachments (ovals 1–2 in Fig. 6c, d).

3.3 Santa Trada Landslide

The Santa Trada landslide (Calabria Region, Southern Italy; Del Ventisette et al. 2011) offers a good example of emergency management through the application of GB-InSAR in emergency conditions to assess the risk impending on a critical infrastructure. The landslide is a 100 m high, 90 m wide translational slide, developed in sand and conglomerates originating from metamorphic weathered rocks (Fig. 8); its thickness is estimated to be between 3 and 5 m. It occurred on 30th January 2009 after a period characterized by heavy rainfall (146.0 mm in December 2008 and 188.6 mm in January 2009) and its sudden movement threatened a viaduct sector along the A3 motorway, an important communication route connecting Salerno and Reggio Calabria. Two possible risk scenarios were consequently hypothesized:

- Progressive mobilization of small shallow landslides: the landslides could obstruct a minor road present at the foot of the slope. In this case probably the debris would not have concerned the motorway or its pillars;

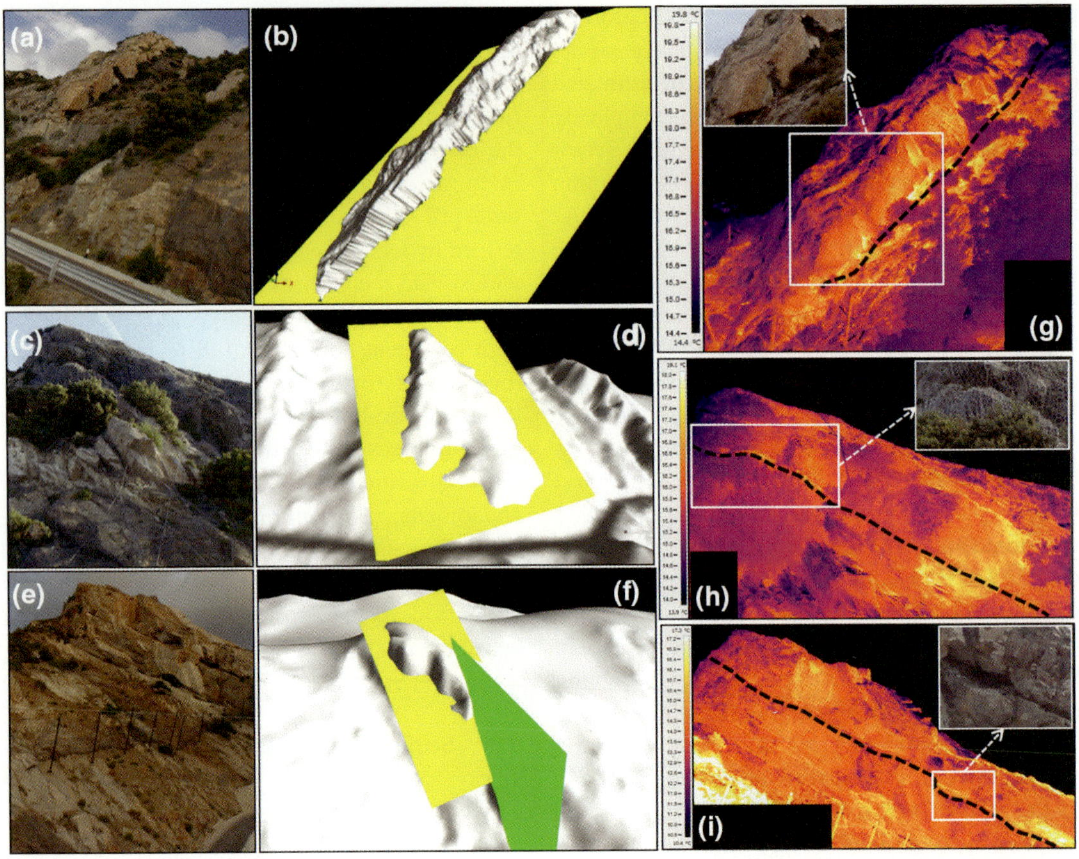

Fig. 6 High resolution 3D surface of the surveyed rock wall **a** (*red dashed square* corresponds to the area affected by rock block detachments). **b** Comparison between March 7th 2014 and April 9th scans: *white oval* showing the first monitored rock block detachment; *orange-yellow areas* enhance rock wall sector characterized by displacements. **c** Comparison between March 7th 2014 and December 18th scans, enhancing the occurred detached rock block sectors (*in blue*); *yellow ovals* enhance the minor block detachments. **d** Correspondent sectors in optical image (*black ovals* enhance major detachments); *dashed line* delimits the rock wall displaced sector

- Mobilization of the whole landslide: in this case, considering the estimated volume of several tens of thousands of cubic meters, the motorway pillars could have been affected by the failure. In addition, given the characteristics of the material involved and the geomorphology of the area, in case of collapse of the whole landslide an unstable dam on the Santa Trada seasonal stream (at the foot of the landslide) could have developed.

For safety reasons this tract of the motorway was closed to traffic. The GB-InSAR was a suitable and versatile tool to assess the actual hazard of the landslide in order to enable the re-opening of the motorway.

On 31st January a ground-based had been installed. Already on 2nd February the motorway was partially reopened thanks to the first monitoring data. During the mid-term monitoring campaign (lasted until 24th April) two approaches for calculating the interferograms and displacement maps were adopted:

- Differential: the time span between the first and last image composing the interferogram is kept constant (e.g. 1 day) and the interferograms represent sequential moments (e.g.

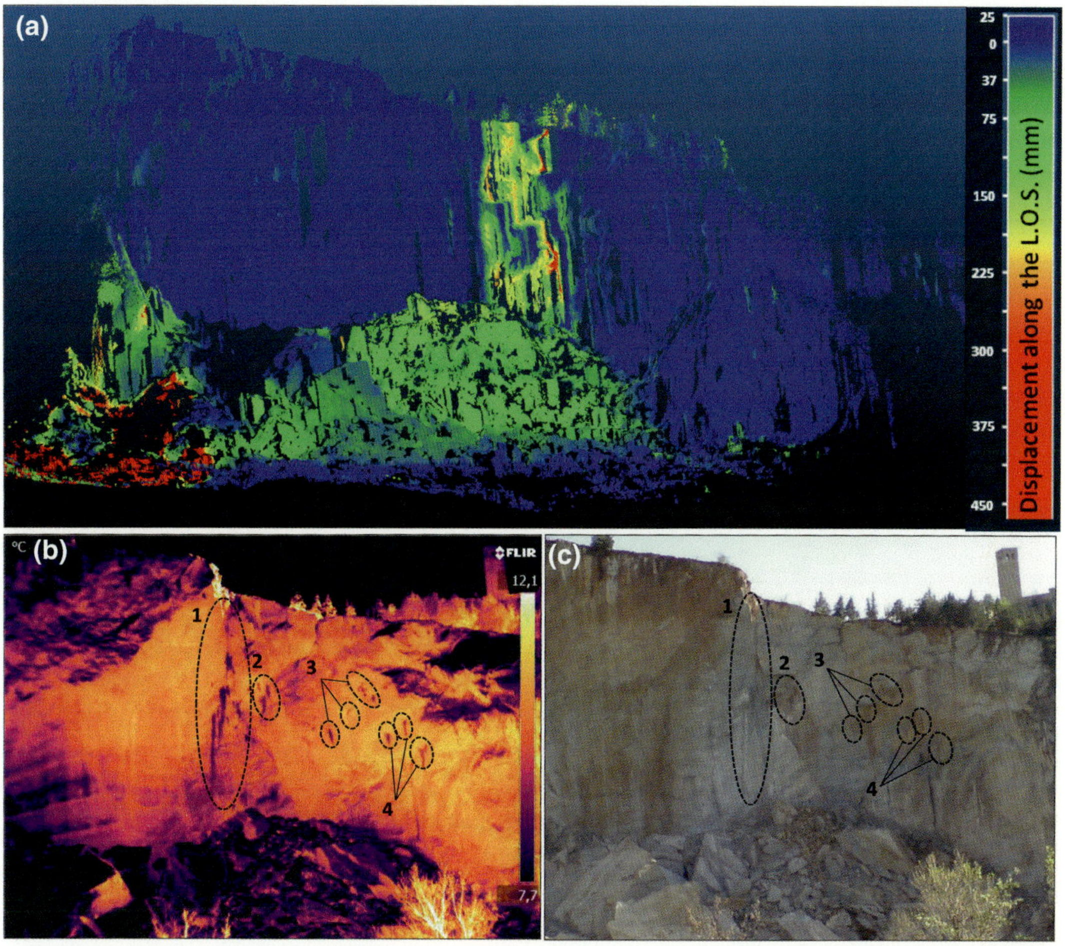

Fig. 7 GBInSAR and TLS data integration. **a** San Leo rock wall March 7th 2014–2015 3D GB-InSAR cumulative displacement map (after Frodella et al. 2016). **b** Thermogram acquired during April 9th 2014 and **c** correspondent photo (seepage sectors 1–4 are characterized by lower temperatures, due to local rock wall surface cooling caused by water evaporation)

day by day displacement). This approach is used in particular for kinematic analyses as it permits to identify acceleration phases because it enables to evaluate two comparable time periods.

- Incremental: the first image is taken as a reference and the interferograms are all calculated between the reference and the last image; in this way the time span is increasing with time. This approach is useful to evaluate the total cumulative displacement and to measure the displacement also in the slower portions of the landslide. Furthermore, it is suitable for spatial analyses as the total extension of the unstable area can be assessed. On the other hand, long time intervals can cause phase ambiguity and loss of coherence. In fact, for Santa Trada landslide, thanks to the displacement maps provided by the GB-InSAR it has been possible to delimitate the area affected by the movement (Fig. 9) and to identify some temporal phases characterized by different activity levels and to assess the risk scenarios' temporal evolution.

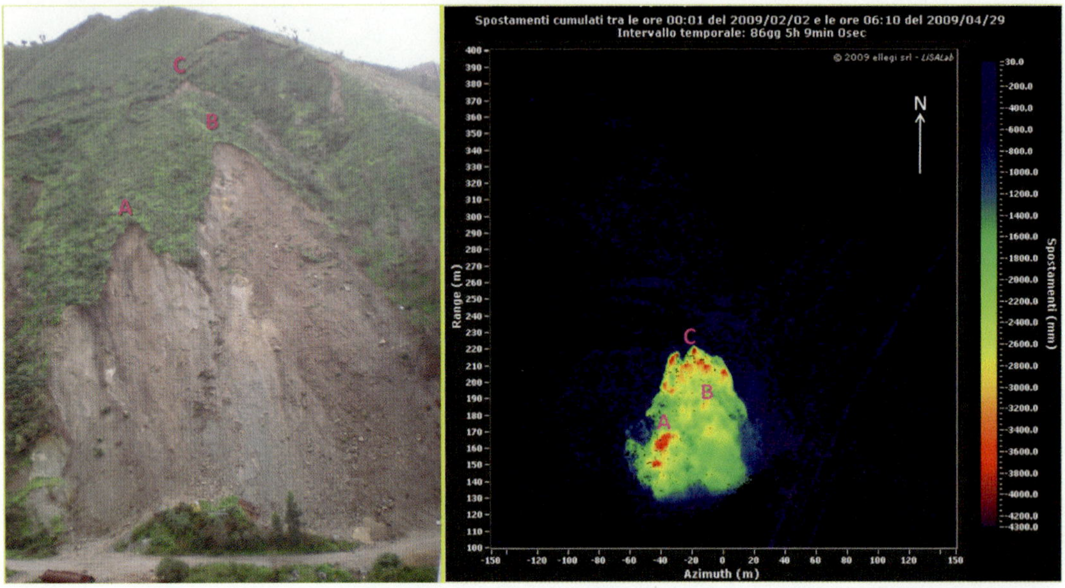

Fig. 8 *Left* Photo of the Santa Trada landslide from 1st February 2009. *Right* Cumulated displacement map calculated with incremental approach spanning from 2nd February to 29th April 2009. The *letters* indicate the corresponding points between the two images

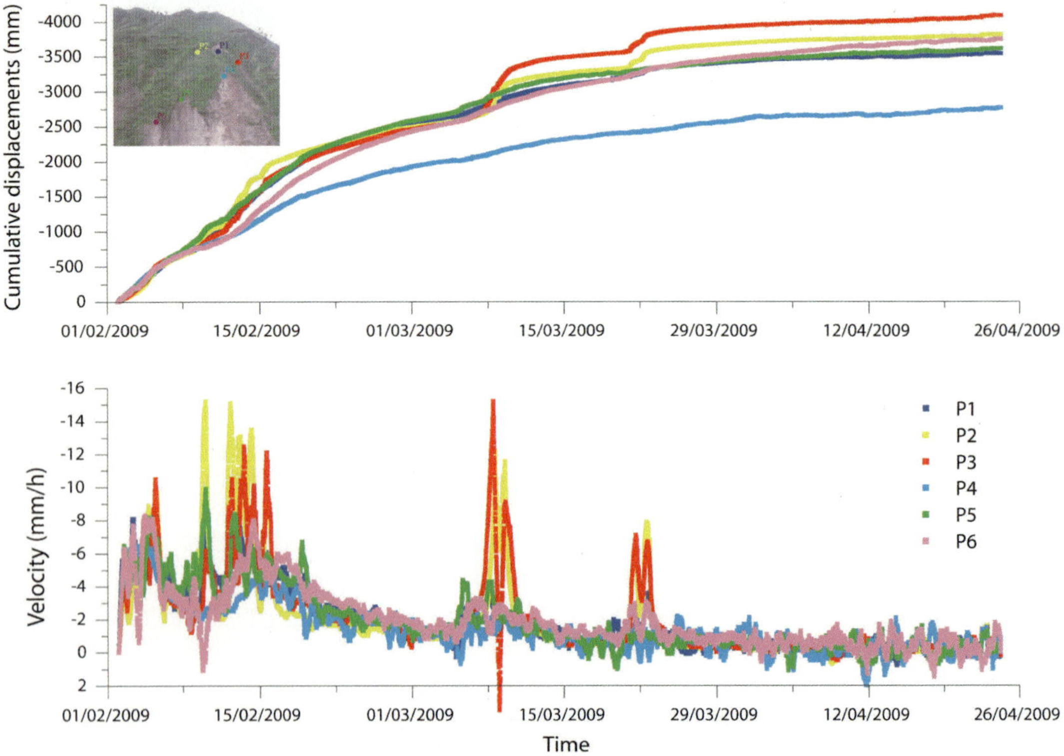

Fig. 9 Displacement and velocity time series of Santa Trada landslide obtained from a GB-InSAR (from Del Ventisette et al. 2011)

4 Discussion and Conclusion

Ground-based remote sensing refers to the science aimed at collecting Earth Observations by using non-contact techniques from terrestrial platforms (Lillesand et al. 2014). According to this general definition, a number of advanced sensors are available today, which may contribute differently to analyse and reduce the landslide risk (Qiao et al. 2013). Their application are usually categorized into three main groups (Mantovani et al. 1996; Scaioni et al. 2014):

(i) recognition, mapping, classification and analysis;
(ii) monitoring (continuous and periodical);
(iii) hazard assessment, forecasting and early warning.

However, each domain could be further subdivided on the basis of the adopted sensor technology and the same sensors may be employed for more different purposes and activities (sometimes simultaneously and sometimes changing slightly their settings). In this paper, the main ground-based instruments used in the field of civil protection have been shown and in this framework some paradigmatic case studies have been presented in order to show the synergic use of most of these analytic techniques (Fig. 10).

In western Elba Island, an integrated methodology based on traditional field and remote surveys such as terrestrial laser scanning and terrestrial infrared thermography was tested on a rock slope threatening a coastal panoramic roadway. The aim was to define susceptibility scenarios connected to rock slopes affected by periodical instability processes. In particular, the methodology was oriented to obtain an accurate three-dimensional rock mass characterization in order to detect the potentially more hazardous rock mass portions, calculate their volume, and collect all the required geomechanical and geometrical parameters to perform a detailed stability analysis. The proposed approach proved to be an effective tool in the field of emergency management, when it is often urgently necessary and gather all the required information (characterization and mapping) as fast as possible in dangerous environments.

In the San Leo rock cliff a monitoring system based on remote sensing techniques, such as radar interferometry, infrared thermography and terrestrial laser scanning, was planned in order to monitor the ground deformation and the residual risk of a rock wall affected by a huge sector

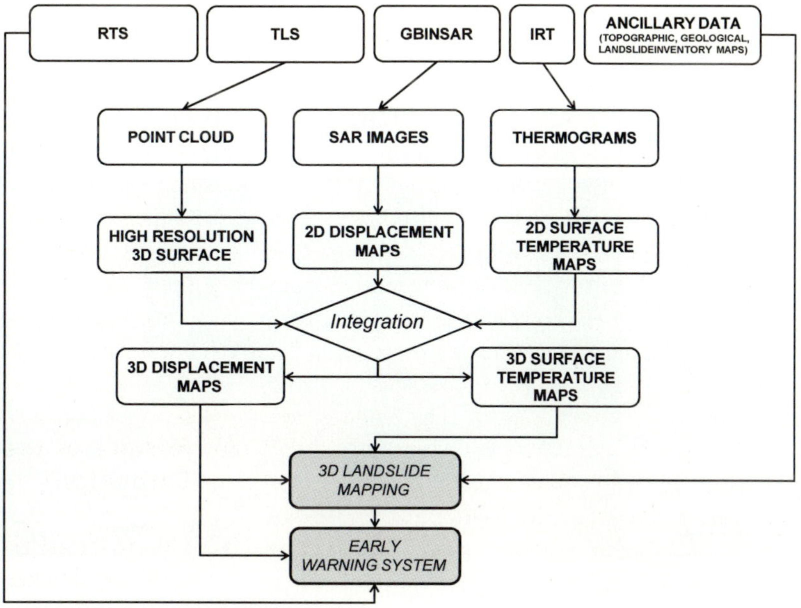

Fig. 10 Logic scheme of the presented case study methodologies

collapse. The analysed area is characterized by complex geomorphological and geometric features, different ongoing landslide processes with various state of activity. Each single employed monitoring technique can be considered not adequate for their intrinsic limitations. The combined use of the abovementioned techniques provided an effective monitoring system for landslide characterization and state of activity monitoring, thanks to the different instrument characteristics (LOS, range of detectable velocity, repetition time), which allowed to overcome the limitations of each single employed technique.

In the Santa Trada area, a GB-InSAR device was promptly installed in order to understand the temporal evolution of a landslide that seriously threatened the functionality and the safety of a strategic road infrastructure. This technique worked with all weather conditions and with a continuous surveillance for all the time of emergency, allowing the rapid assessment of the overall dynamics of the instable slope and related risks scenarios. This application was among the first to demonstrate the full effectiveness of this system in managing landslides emergencies since it greatly facilitated the intervention operations by designated authorities that aimed to restore a normal service in the shortest possible time after a precautionary closure of the motorway. It also granted the possibility to acquire data during precipitations, which represent the most critical moment in stability terms; indeed, usually traditional monitoring instruments are not able to work in such conditions and to provide such useful information real-time, since they normally require a longer time for the installation and a direct access to the unstable and unsafe sites.

With all these cases study we show a brief overview of applications (individual or combined) which include all the above-mentioned categorized groups, from the identification to the early warning system. In view of these, their supporting literature and existing best practices at national and international level in the framework of Civil Protection (e.g., Caduff et al. 2015b; Calvello et al. 2015; Fekete et al. 2015), the most common fields of application can be extrapolated (Table 1).

This synthesis suggests that each instrument is usable in more than one application area and their effectiveness is certainly greater if coupled with other devices since usually proper settings does not allow having redundant information but rather a more complete evaluation. The absence of applications (red areas in the table) indicates that existing analytical techniques based on these instruments are still at a research level or, at most, in a testing phase on few cases of general

Table 1 Ground-based devices in the application fields of civil protection: green = convenience of use, red = no convenience of use

	CHARACTERIZATION AND MAPPING	PERIODICAL CHECK	CONTINUOS MONITORING	EARLY WARNING
TLS	green	green	red	red
DPW	green	green	red	red
IRT	green	green	red	red
GB-INSAR	green	red	green	green
RTS	red	red	green	green

Table 2 Evaluation of the effectiveness of existing ground-based techniques for different landslide types

Technique	Effectiveness				
	Slide	Topples/Falls	Flows	Lateral spreads	Complex mechanisms
TLS	High	High	High	High	High
DPW	High	High	Medium	Low	Medium
IRT	High	High	Medium	Low	Medium
GB-INSAR	High	Medium	Medium	High	High
RTS	High	High	High	High	High

interest. Because of a growing demand for effective Civil Protection procedures in pre- and post-disaster initiatives in landslide-prone areas, the purpose of researchers in the near future is to improve the investigative capacity of the such instruments and consequently to extend their fields of application. In fact, landslides managing in order to reduce vulnerability is currently considered more feasible (in terms of faster and extensive results) than governing all the natural conditions leading to instability, such as the spatial distribution of geology and geomorphology and the climatic influence. In such new global awareness we have extrapolated also the effectiveness of the performance of each ground-based technique with respect to the dynamics of landslide types (as defined by Cruden and Varnes 1996) (Table 2).

This evaluation can be considered a first crucial step towards the choice of the methodology to adopt which grants the best-quality results. These levels of effectiveness are complied if the effects of the techniques limitations (SafeLand 2012) are reduced to a minimum and if the movement speed of the observed scenario does not exceed the instrumental acquisition and processing times.

In conclusion, ground-based techniques have been recently interested by several technical and scientific improvements, providing for each instability process a wide range of equally effective solutions. Currently landslide analysis operators can select the most proper methodology with respect to their specific needs, which can be related to the different civil protection phases, technical issues, financial budget, environmental factors and specific features of the studied unstable slope. It should be also stressed that all the listed techniques provide different kind of measurements (e.g. regarding single targets or widespread areas), that could be conveniently used in close cooperation in order to overcome each single intrinsic limitations and arrange effective action plans and management strategies.

Acknowledgements The Santa Trada and San Leo landslides are case studies described in this paper are relative to monitoring campaigns financed and supported by the Italian National Civil Protection Department.

References

Abellán A, Vilaplana JM, Martínez J (2006) Application of a long-range terrestrial laser scanner to a detailed rockfall study at Vall de Núria (Eastern pyrenees, Spain). Eng Geol 88:136–148

Abellán A, Vilaplana JM, Calvet J, Garcıa-Selles D, Asensio E (2011) Rockfall monitoring by Terrestrial Laser Scanning—case study of the basaltic rock face at Castellfollit de la Roca (Catalonia, Spain). Nat Hazards Earth Syst Sci 11:829–841

Alexander DE (1993) Natural disasters. Springer Science & Business Media

Bardi F, Frodella W, Ciampalini A, Bianchini S, Del Ventisette C, Gigli G, Fanti R, Moretti S, Basile G, Casagli N (2014) Integration between ground based and satellite SAR data in landslide mapping: The San Fratello case study. Geomorphology 223:45–60

Baroň I, Bečkovský D, Míča L (2012) Application of infrared thermography for mapping open fractures in deep-seated rockslides and unstable cliffs. Landslides 1–13

Benedetti G, Bernardi M, Borgatti L, Continelli F, Ghirotti M, Guerra C, Landuzzi A, Lucente CC, Marchi G (2013) San Leo: centuries of coexistence with landslides. In: Margottini C, Canuti P, Sassa K (eds) Landslide Science and Practice. Springer, Heidelberg, Germany 529–537

Brunetti MT, Xiao Z, Komatsu G, Peruccacci S, Guzzetti F (2015) Terrestrial and extraterrestrial landslide size statistics. In: European planetary science congress 2015, 27 Sept–2 Oct 2015, Nantes, France. Copernicus. org/EPSC2015, id. EPSC2015-776 (10: 776)

Caduff R, Schlunegger F, Kos A, Wiesmann A (2015a) A review of terrestrial radar interferometry for measuring surface change in the geosciences. Earth Surf Proc Land 40(2):208–228

Caduff R, Wiesmann A, Bühler Y, Pielmeier C (2015b) Continuous monitoring of snowpack displacement at high spatial and temporal resolution with terrestrial radar interferometry. Geophys Res Lett 42(3):813–820

Calvari S, Intrieri E, Di Traglia F, Bonaccorso A, Casagli N, Cristaldi A (2016) Monitoring crater-wall collapse at open-conduit volcanoes: the case study of the 12 January 2013 event at Stromboli. Bull Volc 78 (39):1–16

Calvello M, d'Orsi RN, Piciullo L, Paes N, Magalhaes M, Lacerda WA (2015) The Rio de Janeiro early warning system for rainfall-induced landslides: analysis of performance for the years 2010–2013. Int J Disaster Risk Reduction 12:3–15

Cardenal J, Mata E, Perez-Garcia JL, Delgado J, Andez M, Gonzalez A, Diaz-de-Teran JR (2008) Close range digital photogrammetry techniques applied to landslide monitoring. Int Arch Photogrammetry, Remote Sens Spat Inf Sci 37

Casagli N, Farina P, Guerri L, Tarchi D, Fortuny J, Leva D, Nico G (2003) Preliminary results of SAR monitoring of the Sciara del Fuoco on the Stromboli volcano. In: International workshop "Occurrence and mechanisms of flow-like landslides in natural slopes and earthfills", Sorrento, Italy, 14–16 May 2003, vol 2, pp 291–295

Cruden DM, Varnes DJ (1996) Landslide types and processes. In Landslides: investigation and Mitigation, Sp. Rep. 247, Transportation Research Board, National Research Council, Turner AK, Schuster RL (ed) Washington DC: National Academy Press. 36–75

Del Ventisette C, Intrieri E, Luzi G, Casagli N, Fanti R, Leva D (2011) Using ground based radar interferometry during emergency: the case of the A3 motorway (Calabria Region, Italy) threatened by a landslide. Nat Hazards Earth Syst Sci 11(9):2483–2495

Di Traglia F, Del Ventisette C, Rosi M, Mugnai F, Intrieri E, Moretti S, Casagli N (2013) Ground-based InSAR reveals conduit pressurization pulses at Stromboli volcano. Terra Nova 25(3):192–198

Di Traglia F, Intrieri E, Nolesini T, Bardi F, Del Ventisette C, Ferrigno F, Frangioni S, Frodella W, Gigli G, Lotti A, Tacconi Stefanelli C, Tanteri L, Leva D, Casagli N (2014a) The ground-based InSAR monitoring system at Stromboli volcano: linking changes in displacement rate and intensity of persistent volcanic activity. Bull Volc 76(2):1–18

Di Traglia F, Nolesini T, Intrieri E, Mugnai F, Leva D, Rosi M, Casagli N (2014b) Review of ten years of volcano deformations recorded by the ground-based InSAR monitoring system at Stromboli volcano: a tool to mitigate volcano flank dynamics and intense volcanic activity. Earth Sci Rev 139:317–335

Dilley M, Chen RS, Deichmann U (2005) Natural disaster hotspots: a global risk analysis. World Bank

Fan YB, Yang SW, Xu LK, Li SH, Feng C, Liang BF (2016) Real-time monitoring instrument designed for the deformation and sliding period of colluvial landslides. Bull Eng Geol Environ 1–10

Farina P, Leoni L, Babboni F, Coppi F, Mayer L, Ricci P (2011) IBIS-M, an innovative radar for monitoring slopes in open-pit mines. In: Proceedings, slope stability 2011: international symposium on rock slope stability in open pit mining and civil engineering, Vancouver (Canada), 18–21 Sept 2011

Fekete A, Tzavella K, Armas I, Binner J, Garschagen M, Giupponi C et al (2015) Critical data source; tool or even infrastructure? Challenges of geographic information systems and remote sensing for disaster risk governance. ISPRS Int J Geo-Inf 4(4):1848–1869

Ferrero AM, Forlani G, Roncella R, Voyat HI (2009) Advanced geostructural survey methods applied to rock mass characterization. Rock Mech Rock Eng 42:65–631

Forlani G, Pinto L, Roncella R, Pagliari D (2014) Terrestrial photogrammetry without ground control points. Earth Sci Inf 7(2):71–81

Franceschi M, Teza G, Preto N, Pesci A, Galgaro A, Girardi S (2009) Discrimination between marls and limestones using intensity data from terrestrial laser scanner. ISPRS J Photogramm 64:522–528

Frodella W, Morelli S (2013) High-resolution 3D geomechanical characterization for the evaluation of rockslide susceptibility scenarios. Rendiconti Online della Società Geologica Italiana 24:143–145. ISSN 2035-8008

Frodella W, Morelli S, Fidolini F, Pazzi V, Fanti R (2014a) Geomorphology of the Rotolon landslide (Veneto Region, Italy). J Maps 10(3):394–401

Frodella W, Morelli S, Gigli G, Casagli N (2014b) Contribution of infrared thermography to the slope instability characterization. In: Proceedings of world landslide forum 3, vol 4, 2–6 June 2014, Beijing, China, pp 144–147

Frodella W, Fidolini F, Morelli S, Pazzi F (2015) Application of infrared thermography for landslide mapping: the Rotolon DSGDS case study. Rend Online Soc Geol It 35:144–147. Società Geologica Italiana, Roma 2015

Frodella W, Ciampalini A, Gigli G, Lombardi L, Raspini F, Nocentini M, Scardigli C, Casagli N (2016) Synergic use of satellite and ground based remote sensing methods for monitoring the San Leo rock cliff (Northern Italy). Geomorphology 264:80–94

Frohlich C, Mettenleiter M (2004) Terrestrial laser scanning: new perspectives in 3D surveying. In: Thies M, Koch B, Spiecker H, Weinacker H (eds) Laser scanners for forest and landscape assessment. Int Arch Photogrammetry, Remote Sens Spatial Inf Sci 36:8/W2

Gigli G, Casagli N (2011) Semi-automatic extraction of rock mass structural data from high resolution LIDAR point clouds. Int J Rock Mech Min Sci 48:187–198

Gigli G, Mugnai F, Leoni L, Casagli N (2009) Analysis of deformations in historic urban areas using terrestrial laser scanning. Nat Hazards Earth Syst Sci 9: 1759–1761

Gigli G, Frodella W, Garfagnoli F, Mugnai F, Morelli S, Menna F, Casagli N (2014a) 3-D geomechanical rock mass characterization for the evaluation of rockslide susceptibility scenarios. Landslides 11(1):131–140

Gigli G, Morelli S, Fornera S, Casagli N (2014b) Terrestrial laser scanner and geomechanical surveys for the rapid evaluation of rockfall susceptibility scenarios. Landslides 11(1):1–14

Ghiglia DC, Romero LA (1994) Robust two-dimensional weighted and un-weighted phase unwrapping that uses fast transforms and iterative methods. J Opt Soc Am 11(1):107–117

Giordan D, Allasia P, Manconi A, Baldo M, Santangelo M, Cardinali M, Corazza A, Albanese V, Lollino G, Guzzetti F (2013) Morphological and kinematic evolution of a large earthflow: The Montaguto landslide, southern Italy. Geomorphology 187:61–79

Gopi S (2007) Advanced surveying: total station, GIS and remote sensing. Pearson Education India

Grussenmeyer P, Landes T, Voegtle T, Ringle K (2008) Comparison methods of terrestrial laser scanning, photogrammetry and tacheometry data for recording of cultural heritage buildings. ISPRS Arch Photogramm Remote Sens 37:W5

Kasperski J, Delacourt C, Allemand P, Potherat P, Jaud M, Varrel E (2010) Application of a terrestrial laser scanner (TLS) to the study of the Séchilienne landslide (Isère, France). Remote Sens 2:2785–2802

Kjekstad O, Highland L (2009) Economic and social impacts of landslides. In: Landslides–disaster risk reduction. Springer, Berlin, pp 573–587

Kvamme KL, Ernenwein EG, Markussen CJ (2006) Robotic total station for microtopographic mapping: an example from the Northern Great Plains. Archaeol Prospection 13(2):91–102

Intrieri E, Gigli G, Mugnai F, Fanti R, Casagli N (2012) Design and implementation of a landslide early warning system. Eng Geol 147–148:124–136

Intrieri E, Di Traglia F, Del Ventisette C, Gigli G, Mugnai F, Luzi G, Casagli N (2013) Flank instability of Stromboli volcano (Aeolian Islands, Southern Italy): integration of GB-InSAR and geomorphological observations. Geomorphology 201:60–69

Intrieri E, Gigli G, Nocentini M, Lombardi L, Mugnai F, Casagli N (2015) Sinkhole monitoring and early warning: an experimental and successful GB-InSAR application. Geomorphology 241:304–314

Lillesand T, Kiefer RW, Chipman J (2014) Remote sensing and image interpretation. Wiley, New York

Maldague X (2001) Theory and practice of infrared technology for non destructive testing. Wiley, New York, 684 p

Mantovani F, Soeters R, van Westen CJ (1996) Remote sensing techniques for landslide studies and hazard zonation in Europe. Geomorphology 15:213–225

Morelli S, Segoni S, Manzo G, Ermini L, Catani F (2012) Urban planning, flood risk and public policy: the case of the Arno River, Firenze, Italy. Appl Geogr 34: 205–218

Oppikofer T, Jaboyedoff M, Blikra L, Derron MH, Metzer R (2009) Characterization and monitoring of the Åknes rockslide using terrestrial laser scanning. Nat Hazards Earth Syst Sci 9:1003–1019

Pesci A, Teza G (2008) Effects of surface irregularities on intensity data from laser scanning an experimental approach. Ann Geophys-Italy 51:839–848

Pieraccini M, Casagli N, Luzi G, Tarchi D, Mecatti D, Noferini L, Atzeni C (2003) Landslide monitoring by ground-based radar interferometry: a field test in Valdarno (Italy). Int J Remote Sens 24(6):1385–1391

Pratesi F, Nolesini T, Bianchini S, Leva D, Lombardi L, Fanti R, Casagli N (2015) Early warning GBInSAR-based method for monitoring Volterra (Tuscany, Italy) city walls. IEEE J Sel Top Appl Earth Obs Remote Sens 8(4):1753–1762

Qiao G, Lu P, Scaioni M, Xu S, Tong S, Feng T, Wu H, Chen W, Tian Y, Wang W, Li R (2013) Landslide investigation with remote sensing and sensor network: from susceptibility mapping and scaled-down simulation towards in situ sensor network design. Remote Sens 5:4319–4346

RIEGL (2010) Data sheet of long range & high accuracy 3D terrestrial laser scanner LMS- Z420i. http://www.riegl.com/uploads/tx_pxpriegldownloads/10_DataSheet_Z420i_03-05-2010.pdf

Rosser NJ, Petley DN, Lim M, Dunning SA, Allison RJ (2005) Terrestrial laser scanning for monitoring the process of hard rock coastal cliff erosion. Q J Eng Geol Hydrogeol 38:363–375

SafeLand (2012) SafeLand–FP7, Deliverable 4.5, Evaluation report on innovative monitoring and remote sensing methods and future technology, 280 p http://cordis.europa.eu/result/rcn/54948_en.html. Last accessed 10 May 2016

Scaioni M, Longoni L, Melillo V, Papini M (2014) Remote sensing for landslide investigations: an overview of recent achievements and perspectives. Remote Sens 6(10):9600–9652

Scaioni M, Feng T, Lu P, Qiao G, Tong X, Li R, Barazzetti L, Previtali M, Roncella R (2015) Close-range photogrammetric techniques for deformation measurement: applications to landslides. In: Modern technologies for landslide monitoring and prediction. Springer, Berlin, pp 13–41

Severin J, Eberhardt E, Leoni L, Fortin S (2014) Development and application of a pseudo-3D pit slope displacement map derived from ground-based radar. Eng Geol 181:202–211

Slob S, Hack HRGK, Feng Q, Röshoff K, Turner AK (2007) Fracture mapping using 3D laser scanning techniques. In: Proceedings of the 11th congress of the

International Society for Rock Mechanics, Lisbon, Portugal, vol 1, pp 299–302
Spampinato L, Calvari S, Oppenheimer C, Boschi E (2011) Volcano surveillance using infrared cameras. Earth Sci Rev 106:63–91
Squarzoni C, Galgaro A, Teza G, Acosta CAT, Pernito MA, Bucceri N (2008) Terrestrial laser scanner and infrared thermography in rock fall prone slope analysis. Geophysical research abstracts 10, EGU2008-A-09254, EGU General Assembly 2008
Stavroulaki ME, Riveiro B, Drosopoulos GA, Solla M, Koutsianitis P, Stavroulakis GE (2016) Modelling and strength evaluation of masonry bridges using terrestrial photogrammetry and finite elements. Adv Eng Softw
Tapete D, Casagli N, Luzi G, Fanti R, Gigli G, Leva D (2013) Integrating radar and laser-based remote sensing techniques for monitoring structural deformation of archaeological monuments. J Archaeol Sci 40 (1):176–189
Tapete D, Morelli S, Fanti R, Casagli N (2015) Localising deformation along the elevation of linear structures: an experiment with space-borne InSAR and RTK GPS on the Roman Aqueducts in Rome, Italy. Appl Geogr 58:65–83
Tarchi D, Casagli N, Fanti R, Leva D, Luzi G, Pasuto A, Pieraccini M, Silvano S (2003) Landslide monitoring by using ground-based SAR interferometry: an example of application to the Tessina landslide in Italy. Eng Geol 1(68):15–30
Teza G, Marcato G, Castelli E, Galgaro A (2012) IRTROCK: a matlab toolbox for contactless recognition of surface and shallow weakness traces of a rock mass by infrared thermography. Comput Geosci 45:109–118
Voegtle T, Schwab I, Landes T (2008) Influences of different materials on the measurements of a terrestrial laser scanner (TLS). In: Proceedings of the XXI congress, The International Society for Photogrammetry and Remote Sensing, ISPRS2008, vol 37, pp 1061–1066
Wolter A, Stead D, Clague JJ (2014) A morphologic characterisation of the 1963 Vajont Slide, Italy, using long-range terrestrial photogrammetry. Geomorphology 206:147–164
Wu JH, Lin HM, Lee DH, Fang SC (2005) Integrity assessment of rock mass behind the shotcreted slope using thermography. Eng Geol 80:164–173
Zhang Z, Zheng S, Zhan Z (2004) Digital terrestrial photogrammetry with photo total station. In: International archives of photogrammetry and remote sensing, Istanbul, Turkey, pp 232–236

TXT-tool 2.386-2.1 SAR Interferometry as a Tool for Detection of Landslides in Early Phases

Marko Komac, Tina Peternel and Mateja Jemec Auflič

Abstract

This teaching tool describes applications of Persistent Scatterer Interferometry (PSI), which represents an effective method for monitoring landslides and soil creeping. In this paper, the usefulness, main advantages and disadvantages of the abovementioned radar method are shown on the basis of different case studies in Slovenia. Furthermore, displacement rates that were observed have been explained and compared with other available in situ data.

Keywords

SAR interferometry · Landslide · Soil creeping · Monitoring · Slovenia

Contents

1 Introduction ... 275
2 Methodology ... 276
2.1 Persistent Scatterer Interferometry ... 276
2.2 InSAR Campaigns in Slovenia ... 279
3 Results of the PSInSAR Campaigns ... 280
3.1 Results of the Central Slovenia (Škofjeloško-Cerkljansko Area) Campaign ... 280
3.2 The Results of the Potoška Planina Location Campaign ... 281
4 Conclusions ... 283
References ... 284

M. Komac (✉)
Faculty of Civil and Geodetic Engineering, University of Ljubljana, Ljubljana, Slovenia
e-mail: m.komac@telemach.net

T. Peternel · M.J. Auflič
Geological Survey of Slovenia, Ljubljana, Slovenia
e-mail: tina.peternel@geo-zs.si

M.J. Auflič
e-mail: mateja.jemec-auflic@geo-zs.si

1 Introduction

In Slovenian territory predominantly shallow landslides and soil slips occur after short or long lasting rainfall. Another very common slope process is slope creeping, which represents a pre-failure phenomenon in slopes. Soil creep is extremely slow movement of surficial soil layers due to climate-driven cyclical volume changes (the shrink swell and the freeze thaw processes) (Hungr et al. 2014). Here shear deformations cause dilation and development of negative pore pressures, which do not develop uniformly throughout the sample, but concentrate along planes where the greatest shearing stresses develop (Fell et al. 2000). In general, creeping slopes can be categorized either as shallow or as deep seated. The formation of shallow landslides mostly occurs in response to extreme rainfall

K. Sassa et al. (eds.), *Landslide Dynamics: ISDR-ICL Landslide Interactive Teaching Tools*, https://doi.org/10.1007/978-3-319-57774-6_19

events (Terlien 1998) and depends on near-surface structures and processes. Deep seated landslides on the other hand have a vertical extent up to several tens of meters and spread horizontally from a hundred to a few thousand square meters. They are defined as fast moving or creeping landslides. Creeping landsides move several centimetres or decimetres per year (Vulliet 1999). The velocity of these slides varies between 1 and 10 cm/year, thus the movements do not create visible deformations to buildings and roads on the surface, at least not in the phase when the process could be substantially slowed and consequences mitigated. Short-term landslide triggers are heavy rainfall events, vadose zone and groundwater flows, and pressure dynamics, but they can also include river bank or hill foot erosion during flood events, depending on site conditions. Long-term landslide triggers include seasonal changes to self-loading, due to seasonal soil moisture variations and snow cover, and the contribution of trees and infrastructure to the self-load. Although creeping and deep seated landslides appear less spectacular than active landslides, in the long term they cause steadily increasing damage to buildings and infrastructure. Various upgrades of Synthetic Aperture Radar interferometry technique (InSAR), including Persistent Scatterer Interferometry (PSI), enable monitoring and identification of slope instability. Often PSI is complementary with other techniques, e.g. GPS measurements or ground control points (Ciampalini et al. 2015). Many researchers (Ferretti et al. 2000; Colesanti et al. 2003; Canuti et al. 2004; Colesanti and Wasowski 2006; Hilley et al. 2004; Žibret et al. 2012) have proved usefulness of PSI data in monitoring slow slope mass movements (Tofani et al. 2013). Radar interferometry can be also applied with a ground-based microwave interferometer (GBInSAR), which has become increasingly widely used for monitoring landslides and unstable slopes in recent years. In comparison with space-born radar methods GBInSAR methods provides a higher image and consequently enables real-time monitoring of landslide displacements (Casagli et al. 2010; Del Ventisette et al. 2011; Monserrat et al. 2013; Nolesini et al. 2013; Bardi et al. 2014). However, the GBInSAR is useful in cases where good close-range visibility of the landslide site is available. Yet, the cost of GBInSAR technique is comparatively high (Monserrat et al. 2013).

In Slovenia, PSI campaigns have been used to detect slope mass movements in the Škofjeloško Cerkljansko area (Žibret et al. 2012; Jemec et al. 2013; Čarman et al. 2014), Maribor and the area surrounding Ljubljana (Capes and Komac 2011; Komac et al. 2012b, 2013) and in Potoška planina (Komac et al. 2012a, 2014). Displacement rates at the first three territories have been obtained by natural point targets, referred to as persistent scatterer (PSs), such as buildings, antennas, bridges, urban structures or stable natural outcrops, while the deformation rates in Potoška planina have been monitored by artificial scatterers in the form of compact active transponders (CATs). In this article authors will describe results of PSI data sets for the Škofjeloško Cerkljansko area and Potoška planina where slope mass movements have been detected, meanwhile in Ljubljana and Maribor area PSI data are indicated mostly by the subsidence of the terrain.

2 Methodology

2.1 Persistent Scatterer Interferometry

PSI is an interferometric technique based on the phase comparison of synthetic aperture radar (SAR) images, gathered at different times with slightly different looking angles (Gabriel et al. 1989; Bamler and Hartl 1998; Rosen et al. 2000). The PSI technique represents an upgraded version of the differential radar interferometry technique in which the main limitations are linked to the loss of coherence with time due to the influence of atmospheric artefacts and the presence of uncompensated topography. The PSI technique overcomes some of these drawbacks and is capable of detecting isolated coherent pixels and tackling the problem of atmospheric delay errors at the expense of a large number of required images (>20) and a sparse, pixel-by-pixel-based evaluation (Ferretti et al. 2000). Point targets, referred to as persistent scatterers (PSs), are not affected by temporal decorrelation and are recognised by a statistical analysis of their amplitude in all available SAR images. The contributions of topography, deformation and the atmosphere can be estimated by carefully exploiting their different behaviours in time and space. A large archive of SAR satellite imagery allows analyses of the PS displacements since 1992, which is not possible with traditional methods such as GPS/GNSS surveys. As deformations can be measured with millimetre accuracy, the PSI method is suitable for detecting and investigating slow slope deformation induced by natural and man-induced phenomena such as land subsidence, landslides, seismic faults, and tectonic deformation, with a very high spatial resolution, both at a local and at a regional scale (Colesanti et al. 2003). In addition, the microwave radar is only sensitive to displacement changes in the radar line-of-sight (LOS) direction in relation to a stable reference point, which is assumed to be motionless.

One of the most obvious SAR advantage is the weather and daylight independence of radar systems, which ensure a guaranteed acquisition of the area of interest. This also enables consistent monitoring independent of lighting, weather or cloud-cover conditions. Most radar satellites operate at wavelengths between 0.5 and 75 cm. Table 1 shows the frequency and wavelength of commonly used radar remote sensing bands.

The processing technique is based on the identification of the radar benchmarks, named Permanent Scatterers, which are stable natural reflectors (Fig. 1) (rock outcrops, buildings and urban structures), characterized by stable individual radar amplitude and radar phase over long temporal series of interferometric SAR images. The final results are (Ferretti et al. 2000):

1. A map of the PS identified in the image and their coordinates: latitude and precise elevation (accuracy of elevation better than 1 m); each PS is labelled by a code;
2. Average LOS deformation rate of every PS (expressed as average velocity with accuracy usually between 0.1 and 1 mm $year^{-1}$, depending on the number of available interferograms and on the phase stability of each single PS)
3. Displacement time series showing the relative (i.e., with respect to a unique reference image) LOS position of PS in correspondence in each SAR acquisition. The time series represent, therefore the motion component of the PS in the direction of the Line of Sight (LOS) as a function of the time (accuracy on single measurements usually ranging from 1 to 3 mm).

As in all differential interferometry applications, the results are not absolute both in time and space, but the deformation data are referred to the master image (in time) and the results are computed with respect to a reference point of known elevation and presumed motionless (in space). Traditional radar interferometry can be affected

Table 1 Frequency and wavelength of commonly used radar remote sensing bands

Band	Frequency (GHz)	Wavelength (cm)	Radar system
X	12.5–8	2.4–3.75	TerraSAR-X, TanDEM-X, COSMO-SkyMed
C	8.0–4.0	3.75–7.5	RADARSAT, ERS-1, ERS-2, sentinel
S	4.0–2.0	7.5–15.0	Sirius XM holdings
L	2.0–1.0	15.0–30.0	ALOS PALSAR
P	1.0–0.3	30.0–100.0	AirMOSS

Fig. 1 Examples of natural reflectors

by atmospheric noise, but the PSI technique can greatly reduce the effects of time decorrelation and atmospheric phase (Ferretti et al. 2000).

In some cases an upgrade of the InSAR technique with other techniques is required to assess more detailed and more complex landslide movements, i.e. very precise measurements of horizontal displacements as well as vertical displacements of monitored area. In order to improve survey resolution of horizontal and vertical landslide movement components a novel device (Fig. 2), consisting of a Compact Active Transponder (CAT) and a Global Navigation Satellite System (GNSS) antenna, was developed in the frame of the European Union funded project "Integrated Interferometry and GNSS for Precision Survey (I2GPS)" found by Seventh Framework programme (FP7-GALILEO-2008-GSA-1) (02/2010–09/2011).

Fig. 2 The I2GPS unit which record InSAR and GNSS data

The integrated approach integrates radar interferometry in order to provide accurate vertical displacements and GNSS measurements, which ensure temporal interpolation and accurate horizontal displacements. Integrating the GNSS signals with InSAR technique ensure millimetric co-registration and a coherent cross-reference. The benefits are (1) all advantages of CATs, such as a small size and a light-weight device with unobtrusiveness and usability with multiple satellites and tracks; (2) absolute calibration for PSI data; (3) high sampling rate of GNSS enables detection of abrupt ground motion in 3D and (4) vertical components of the local velocity field can be derived from single-track InSAR line-of-sight displacements (Komac et al. 2014). This technique was tested on a specific landslide site in Slovenia that is described in this paper.

2.2 InSAR Campaigns in Slovenia

In the period from 2004 to 2014 four major PSInSAR campaigns were performed and additionally one smaller, very focused one. The PSI data that covered the Julian Alps and their foreland (NW Slovenia) were acquired in 2005 during the ClimChAlp project under the Alpine Space project framework, the project that covered the Ljubljana Basin area was performed within the ESA TerraFirma programme in 2008, the PSI data for the Central Slovenia (Škofjeloško Cerkljansko area) were acquired in 2010 within the project AdaptAlp, and the PSI data for wider Maribor area were acquired within the PanGeo EU FP7 project. The focused analysis that covered only one landslide site was performed without the PSI data from the archives, but only with the pre-programmed PSI data acquisition in the six month period in 2011. The latter was done within the I2GPS EU 7FP project. Here we focus only on one larger the smaller campaign due to their direct link to landslides and more in-depth analyses related to these phenomena. Figure 3 shows the locations of the four larger campaigns' areas and two landslide

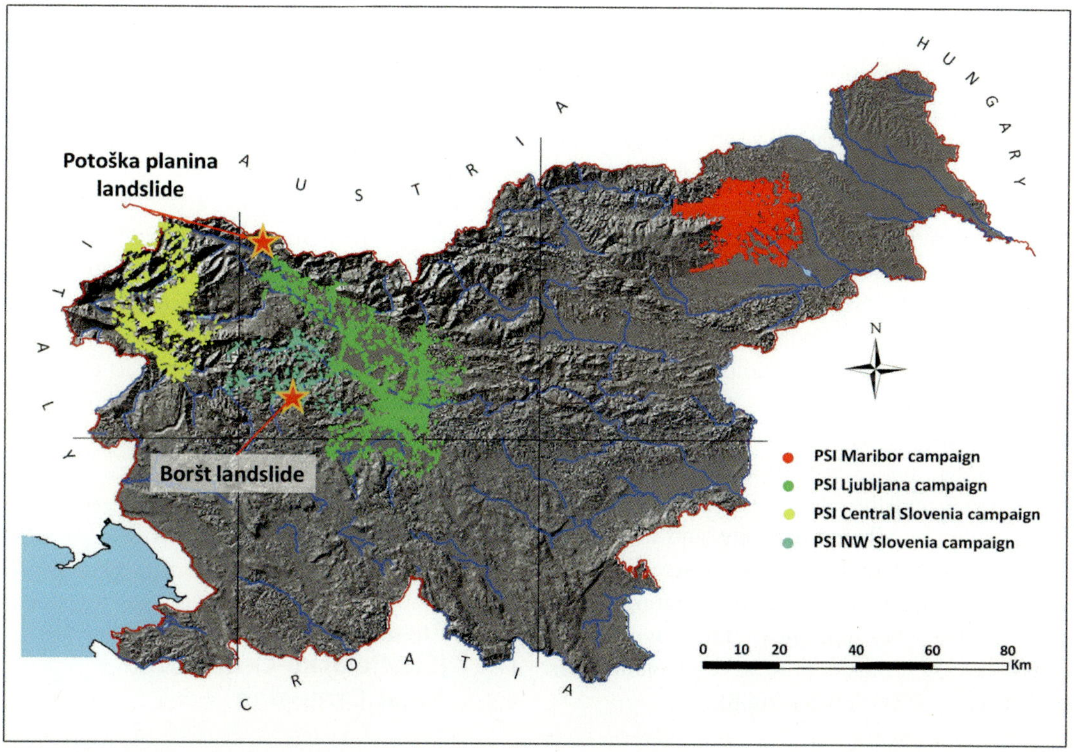

Fig. 3 Map showing the locations of the PSI campaigns in Slovenia. The two landslide locations described in this paper are marked with *stars*. The northern location—Potoška planina also represents a separate and focused PSI campaign in addition to the four larger ones. PSI data for the southern location was acquired during the Central Slovenia campaign

locations that are in more detail presented in this paper marked with stars. Potoška planina landslide represents the abovementioned smaller, specific site oriented PSI campaign, while the Boršt landslide PSI data is a part of the Central Slovenia PSI campaign.

PSI campaign in the Central Slovenia (Škofjeloško Cerkljansko area)

In the Central Slovenia, the Škofjeloško Cerkljansko area, 67 ERS scenes were acquired between June 1992 and December 2000 along descending orbits and were used for the interferometric analyses of the Škofja Loka and Cerkljansko areas. A total of 2787 PSs were identified, mainly corresponding to outcrops and man-made structures such as building, poles and antennas. The geocoding accuracy of the PS locations was approximately ±15 m in the east–west and ±4 m in the north–south directions, while the estimated accuracy of the elevation values was usually better than 2 m. The LOS direction displacement rates varied from +7.26 to −18.75 mm/year, relative to the reference point located in the town Železniki (46° 13′ 29.13″N, 14° 10′ 31.77″E). This means that for each PS, every measurement of displacement value in the LOS is expressed in relation to the 'stable' (presumed motionless) reference point.

In general, the Škofjeloško and Cerkljansko hill area is constituted mainly of Mesozoic carbonate rocks (limestone and dolomite), Mesozoic clastic rocks (shale, siltstone, marlstone, greywacke, sandstone, conglomerate, breccias, tuff), and Permian and Carboniferous clastic rocks (claystone, sandstone, shale). The valleys are mainly filled by alluvial deposits, and the feet of the slopes are regularly covered by scree, both of Quaternary age. Also, the tectonic setting of the area is also very complex. The investigated region lies on contacts with several fault zones (Poljak et al. 2000; Placer 2008).

The Potoška planina landslide campaign

The developed methodology of coupling the PSI and GNSS measurements that is briefly described in Sect. 2.1 was tested on the Potoška planina landslide in order to provide the 3D displacement assessments of monitored locations on the landslide and in its vicinity and to evaluate the applicability for operational monitoring of natural hazards.

The Potoška planina landslide is situated on the Karavanke mountain ridge in north-western Slovenia. The landslide lies above the densely populated village Koroška Bela (approximately 2200 inhabitants), which occupies an area of 1.02 km^2. The area has experienced severe debris-flow events in the recent geological past, the latest dating to the end of 18th century, when more than 40 houses were damaged (Komac et al. 2014). The landslide Potoška planina covers an area of 300 m × 500 m, with an estimated average depth of 10 m and an estimated volume of the sliding mass of 1.8 million m^3.

The monitoring campaign was performed in the period of February to August 2011 during which six SAR (Side Aperture Radar) images were acquired. These were compared with the reference (first) one to assess the surface displacements in the given period. As the campaign was focused on testing the newly developed equipment several of units were deployed at the site. Four CAT (compact active transponder) units, two of them also equipped with the GNSS antenna were installed on the landslide body to assess the landslide movement, three CAT units outside the mapped landslide area (one of them equipped with the GNSS antenna) to assess the presumed stable area movement. In addition, one CAT was installed in the valley below.

3 Results of the PSInSAR Campaigns

3.1 Results of the Central Slovenia (Škofjeloško-Cerkljansko Area) Campaign

The results of the PSI survey in the Škofjeloško-Cerkljansko area show that the PSI method is suitable for assessing the temporal evolution of slow and extremely slow landslides with constant deformation velocities, hereafter called the soil creep (Jemec Auflič 2012; Žibret et al. 2012; Jemec Auflič et al. 2013). Due to the high PS density in some smaller areas within the broader study area, it was possible to identify three areas with a constant deformation—areas above Cerkno, Planina and Čeplez (Komac and Jemec 2010), indicating the presence of creeping. In these areas, the analysis of average annual creeping rates revealed that lithology and slope inclination are among the key precondition factors for the occurrence of the slope creeping process. Due to the ability of PSI methods to monitor the deformations in time, an assessment was also made of possible climatic triggering factors, especially rainfall events (Žibret et al. 2012). The results show that soil creeping is induced not only by heavy precipitation events (>100 mm/day), but also by less intense precipitation (>20 mm/day), or cumulative totals of 50 mm of rain in three days. The first event produced 0.4–0.76 mm of elevation change, and events with higher precipitation showed average elevation decreases of −1.2 mm.

The Boršt location

More focused analysis of the PSI measurements were conducted in the location of a uranium mill tailing deposit at Boršt location (Fig. 3), where the favourable positions of the PSs that lie close to the deposit site—approximately 250 m away enabled such an exercise. At this location, a landslide occurred in November 1990 after heavy rainfall (Čarman et al. 2014) and due to the delicate content of the deposit the site was under periodical detailed monitoring. Consequentially, PSI results were compared with the in situ observations results. Field surveys confirmed the interpretation from the ortho-photo that two PSs coincide with the outcrop, while one was in the rubble deposit and consequently, the values measured at this location are most probably subject to high uncertainty and may not represent the real natural ground movements. The results of the detailed geological mapping of the area indicate that, aside from the mill tailing site, the whole slope might also be subject to sliding, approximately in a north-west direction. To simplify comparison between PSI and in situ data, we assumed that the part of the slope where the PSs are located is also moving as one mass towards the valley bottom. Still, when considering only the till deposit location the surroundings were regarded as a background, independently moving from the landslide, which enabled to assess the success of the remediation works performed on the landslide body in the mid-1990s. The results show that the remediation works, in the form of a drainage tunnel, were very effective, as post-remediation velocities at the landslide closer to the PSs resembled the velocities of the PSs, while the velocities of the landslide mass above the drainage tunnel slowed down, even to below the background velocities. The high correlation values between the movements of the benchmarks and the PSs also confirm that the remediation works were effective, as the fluctuations in the displacement values of the landslide were very similar to those of the PSs. In addition, the results also contribute to understanding the dynamics and triggering mechanisms of landslides and their elaboration on hazard maps. Figure 4 shows the Boršt mill tailing deposit site after remediation works in 2007, the position of PSs and the landslide area.

3.2 The Results of the Potoška Planina Location Campaign

In comparison to the Boršt landslide, the Potoška planina landslide and its surroundings were not analysed in such a detailed manner considering the size and dynamics through time, as the later location is not a radioactive tiling deposit and it was only discovered ten years ago. Nevertheless, due to its estimated size and potential hazard to the people (Mikoš et al. 2008; Jež et al. 2008) the Potoška planina site was equipped with a special system for the monitoring of the landslide movement using a combination of radar interferometry and GNSS technologies. Figure 5 represents the overview of the Potoška planina landslide in the direction towards south with the positions of all units that were installed at the

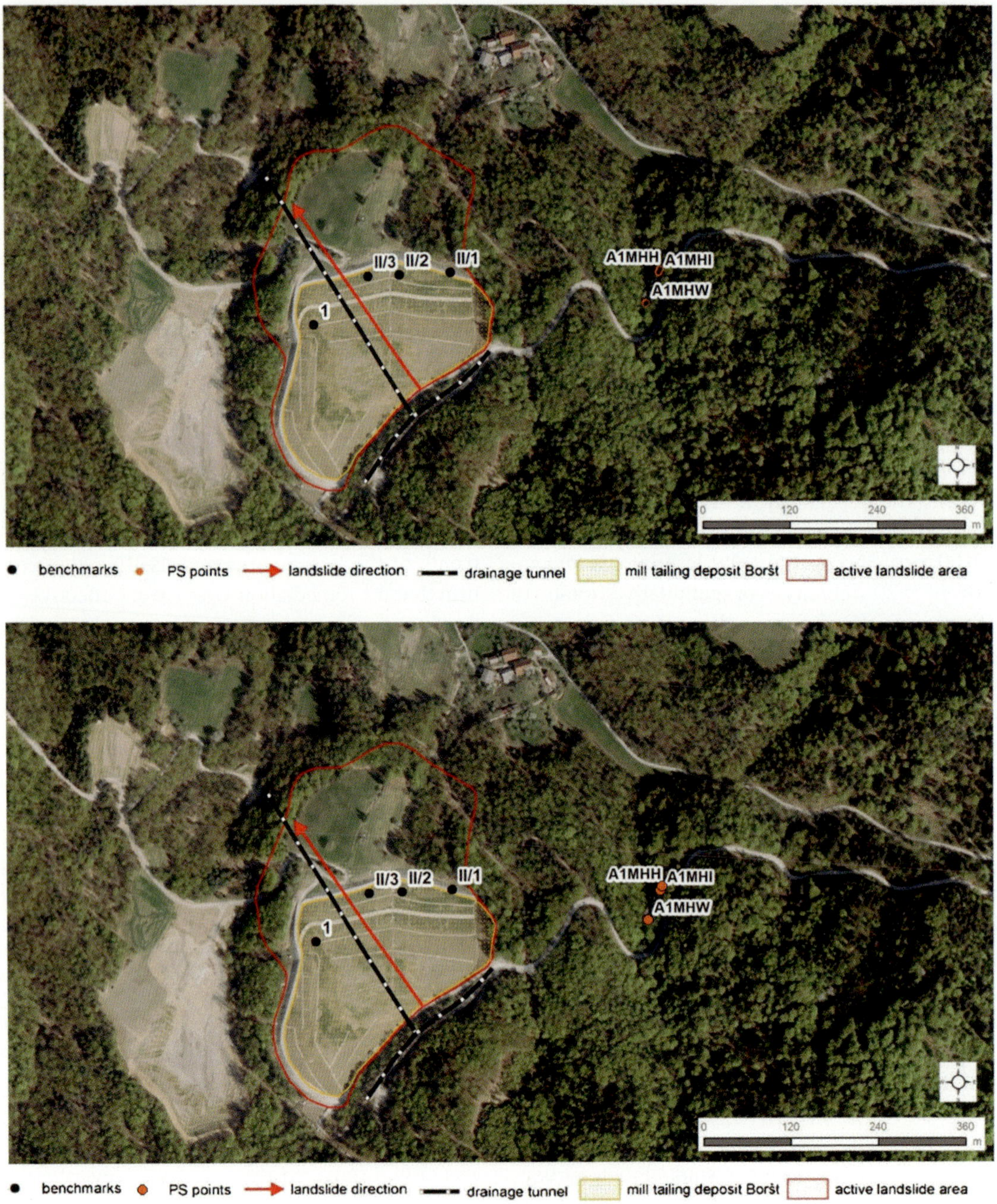

Fig. 4 Boršt uranium mill tailing deposit site after remediation works in 2007. *Black dots* represent the position of the benchmarks (1, II/1, II/2, II/3); *orange dots* represent PSs (A1MHH, A1MHI, A1MHW); *red arrow* represents landslide direction; *red polygon* indicates active landslide area and *black–white line* shows the position of the drainage tunnel

site. Units CAT1–CAT4 were installed onto the landslide body, units Ref1–Ref3 were positioned in the supposedly stable landslide surroundings and the unit CAT5 was positioned in the valley below. SAR and GNSS results ranged in relation to the Ref2 point from −0.41 mm (Ref1) to

Fig. 5 Position of CAT units on the landslide body (*encircled with red line*), in its vicinity and in the valley below

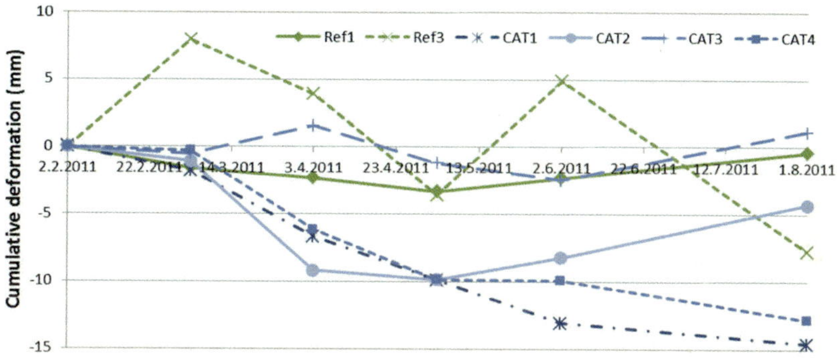

Fig. 6 Cummulative displacements at observation locations located in different parts of the landslide. All displacements are relative to the reference point Ref2 (Komac et al. 2014)

−7.7 mm (Ref3) for the presumed stable reference points, and from 1.2 mm (CAT3) and −0.05 mm (CAT5) to −12.8 (CAT4) to −14.6 (CAT1), all obtained from six observations during the monitoring period from February to August 2011. These results indicated displacements of the central-upper and south-eastern parts of the landslide body (Komac et al. 2014) (Fig. 6), still for the more reliable conclusions longer observation periods are needed. Newer monitoring results on the described site are available in Peternel et al. (2016; online first).

4 Conclusions

We've shown that PSI technology could be used as a good supporting tool in the area of landslide monitoring. For now the PSI technology offers an insight into slow movement monitoring that could especially be utilised in the pre-sliding phase of faster and more notable landslides movements. The main advantage of this movement monitoring and detecting methodology is the coverage of a large areas where buildings and other man-made structures often act as PS due to their corner reflector-like scattering behaviour and high radar reflectivity, and the possibility of data acquisition regardless of the weather conditions or sunlight availability. The main drawbacks (or disadvantages) of the technique are the inability to acquire data in natural terrain (vegetated areas and mountainous area with steep slopes) oriented away from the radar signal LOS, the (still scarce) frequency of the radar data that governs the ability of the method to detect faster movements, and still relatively costly processing of the radar interferometric data. In natural terrain the lack of man-made objects makes reliable estimation of deformation using PSI technology a challenging task.

Acknowledgements Authors would like to thank Geological Survey of Slovenia, the Slovenian Research Agency, and the European Commission (Interreg IIIB, Alpine Space, Galileo programme and 7FP) for the funding the above-mentioned research projects in the period from 2006 to 2016.

References

Bamler R, Hartl P (1998) Synthetic aperture radar interferometry. Inverse Probl. doi:10.1088/0266-5611/14/4/001

Bardi F, Frodella W, Ciampalini A, Bianchini S, Del Ventisette C, Gigli G, Fanti R, Moretti S, Basile G, Casagli N (2014) Integration between ground based and satellite SAR data in landslide mapping: The San Fratello case study. Geomorphology 223:45–60

Canuti P, Casagli N, Ermini L, Fanti R, Farina P (2004) Landslide activity as a geoindicator in Italy: significance and new perspectives from remote sensing. Environ Geol 45:907–919

Capes R, Komac M (2011) Pangeo enabling access to geological information in support of GMES. In: Banjac N (ed) The geology in digital age. Proceedings of the 17th meeting of the Association of European Geological Societies. The Serbian Geological Society, Belgrade, p 50

Čarman M, Jemec Auflič M, Komac M (2014) Landslides at a uranium mill tailing deposit site Boršt (Slovenia) detected by radar interferometry. Landslides 11 (3):527–536

Casagli N, Catani F, Del Ventisette C, Luzi G (2010) Monitoring, prediction, and early warning using ground-based radar interferometry. Landslides 7:291–301

Ciampalini A, Raspini F, Bianchini S et al (2015) Remote sensing as tool for development of landslide databases: the case of the Messina Province (Italy) geodatabase. Geomorphology 249:103–118

Colesanti C, Wasowski J (2006) Investigating landslides with space-borne Synthetic Aperture Radar (SAR) interferometry. Eng Geol 88:173–199

Colesanti C, Ferretti A, Prati C, Rocca F (2003) Monitoring landslides and tectonic motion with the Permanent Scatterers technique. Eng Geol 68(1–2):3–14

Del Ventisette C, Intrier E, Luzi G, Casagli N, Fanti R, Leva D (2011) Using ground based radar interferometry during emergency: the case of the A3 motorway (Calabria Region, Italy) threatened by a landslide. Nat Hazards Earth Syst Sci 11:2483–2495

Fell R, Hungr O, Leoroueil S, Riemer W (2000) Keynote Lecture—geotechnical engineering of the stability of natural slopes and cuts and fills, GeoEng 2000. An international conference on geotechnical & geological engineering. Techomic Publishing Company, Melbourne, pp 21–120

Ferretti A, Prati C, Rocca F (2000) Nonlinear subsidence rate estimation using permanent scatterers in differential SAR interferometry. IEEE Trans Geosci Remote Sens 38(5):2202–2212

Gabriel AK, Goldstein RM, Zebker HA (1989) Mapping small elevation changes over large areas—differential radar interferometry. J Geophys Res 94(B7):9183–9191

Hilley GE, Burgbman R, Ferretti A, Novali F, Rocca F (2004) Dynamics of slow moving landslides from Permanent Scatterer Analysis. Science 304:1952–1955

Hungr O, Leroueil S, Picarelli L (2014) The Varnes classification of landslide types, an update. Landslides 11(2):167–194

Jemec Auflič M (2012) Application of the permanent scatterers technique in geology as a support for the landslide prone areas investigation. PhD thesis, University of Ljubljana, Ljubljana, Slovenia

Jemec Auflič M, Čarman M, Komac M (2013) Assessing of spatio-temporal factors influencing landslides using PSInSAR. In: Margottini C, Canuti P, Sassa K (eds) Landslide science and practice, vol. 2: Early warning, instrumentation and monitoring. Springer, Heidelberg, pp 371–378

Jež J, Mikoš M, Trajanova M, Kumelj Š, Bavec M (2008) Vršaj Koroška Bela-Rezultat katastrofičnih pobočnih dogodkov=Koroška Bela alluvial fan—the result of the catastrophic slope events (Karavanke Mountains, NW Slovenia). Geologija 51(2):219–227

Komac M, Jemec Auflič M (2010) Ocena premikov počasnega plazenja z interferometrično metodo permanentnih sipalcev in GIS. In: Perko D, Zorn M (eds) Geografski informacijski sistemi v Sloveniji 2009–2010, GIS v Sloveniji. ZRC, Ljubljana, pp 121–129

Komac M, Jemec Auflič M, Milanič B, Bavec M, Jež J, Peternel T, Krivic J (2012a) Enabling access to geological information in support of GMES.

Deliverable D7.1.35 of Pangeo Project: Geohazard description for Ljubljana, Slovenia, 209 p
Komac M, Jež J, Celarc B, Milanič B, Bavec M (2012b) Prvi rezultati merjenja premikov površja na območju Jesenic in Potoške planine s kombinacijo InSAR in GPS meritev. In: Ciglič R, Perko D, Zorn M (eds) Geografski informacijski sistemi v Sloveniji 2011–2012, GIS v Sloveniji. ZRC, Ljubljana, pp 25–31
Komac M, Jemec Auflič M, Milanič B, Jež J, Peternel T, Rižnar I (2013) Enabling access to geological information in support of GMES. Deliverable D7.1.35 of Pangeo Project: Geohazard description for Maribor, Slovenia, 165 p
Komac M, Holly R, Mahapatra P, Van der Marel H, Bavec M (2014) Coupling of GPS/GNSS and radar interferometric data for a 3D surface displacement monitoring of landslides. Landslides. doi:10.1007/s10346-014-0482-0
Mikoš M, Bavec M, Budkovič T, Durjava D, Hribernik K, Jež J, Klabus A, Komac M, Krivic M, Kumelj Š, Maček M, Mahne M, Novak M, Otrin J, Petje U, Petkovšek A, Ribičič M, Sodnik J, Šinigoj J, Trajanova M (2008) Ocena ogroženosti zaradi delovanja drobirskih tokov (final report; in Slovenian), 224 pp
Monserrat O, Moya J, Luzi G, Crosetto M, Gili JA, Corominas J (2013) Non-interferometric GB-SAR measurement: application to the Vallcebre landslide (eastern Pyrenees, Spain). Nat Hazards Earth Syst Sci 13:1873–1887
Nolesini T, Di Traglia F, Del Ventisette C, Moretti S (2013) Deformations and slope instability on Stromboli volcano: integration of GBInSAR data and analog modeling. Geomorphology 180–181:242–254
Peternel T, Kumelj Š, Oštir K, Komac M (2016) Monitoring the Potoška planina landslide (NW Slovenia) using UAV photogrammetry and tachymetric measurements. Landslides. doi:10.1007/s10346-016-0759-6
Placer L (2008) Principles of the tectonic subdivision of Slovenia. Geologija 51(2):205–217
Poljak M, Živčič M, Zupančič P (2000) The seismotectonic characteristics of Slovenia. Pure Appl Geophys 157:37–55
Rosen PA, Hensley S, Joughin IR, Li FK, Madsen SN, Rodríguez E, Goldstein RM (2000) Synthetic aperture radar interferometry. Proc IEEE 88(3):333–382
Terlien MTJ (1998) The determination of statistical and deterministic hydrological landslide-triggering thresholds. Environ Geol 35(2–3):124–130
Tofani V, Raspini F, Catani F, Casagli N (2013) Persistent Scatterer Interferometry (PSI) technique for landslide characterization and monitoring. Remote Sens 5 (3):1045–1065
Vulliet L (1999) Modelling creeping slopes. Rivista Italiana di Geotecnica 23(1):71–76
Žibret G, Komac M, Jemec Auflič M (2012) PSInSAR displacements related to soil creep and rainfall intensities in the Alpine foreland of western Slovenia. Geomorphology 175–176:107–114

TXT-tool 2.039-3.3 Ground-Based Radar Interferometry for Landslide Monitoring

Nicola Casagli, Filippo Catani, Chiara Del Ventisette and Guido Luzi

Abstract

Landslide detection and monitoring represent a starting point to produce hazard and risk maps useful to define adequate prevention measures and to manage landslide emergencies. This paper reports the experience of our research group on the ground-based synthetic aperture radar interferometry. In particular, we discuss the use of a ground-based synthetic aperture radar interferometer (GB-InSAR) not only as a monitoring system but also as a tool to obtain spatial information on the landslide displacements to be integrated with rainfall data. This technique can be incorporated in an early warning system for the detection of slope acceleration patterns indicating the upcoming occurrence of a slope failure. The case studies reported demonstrate the capability of this technique to operate in different operative settings (i.e., different phenomena and geological framework) and for different aims (monitoring for prevention, early warning, and emergency assessment). This methodology has also been proved by national and regional authorities of civil protection in order to provide a real-time monitoring for emergency management.

Keywords

Landslides · Interferometry · Ground-based SAR
Ruinon (Valfurva, Italy) · Stromboli (Italy)

N. Casagli · F. Catani · C. Del Ventisette (✉)
Department of Earth Sciences, University of Firenze, Via La Pira 4, Florence, Italy
e-mail: chiara.delventisette@unifi.it

N. Casagli
e-mail: nicola.casagli@unifi.it

F. Catani
e-mail: filippo.catani@unifi.it

G. Luzi
Division of Geomatics, Centre Tecnològic de Telecomunicacions de Catalunya (CTTC), Av. Gauss 11, 08860 Castelldefels, Barcelona, Spain
e-mail: guido.luzi@cttc.cat

K. Sassa et al. (eds.), *Landslide Dynamics: ISDR-ICL Landslide Interactive Teaching Tools*, https://doi.org/10.1007/978-3-319-57774-6_20

Contents

1 Introduction ... 288
2 A Monitoring Case: Ruinon Landslide (Valfurva, Italy) ... 288
2.1 Data Analysis ... 289
3 Monitoring and Early Warning at Stromboli 290
3.1 Data Analysis ... 292
4 Discussion and Conclusions ... 293
References ... 294

1 Introduction

Landslide represents one of the major natural hazards causing fatalities and economic damage worldwide. Their detection and real-time monitoring, that considers the triggering factors and landslide kinematics, represent a starting point to produce hazard and risk maps useful for a proper urban planning and to define adequate prevention measures for the mitigation of loss of human life and of assets. The remote sensing approach allows obtaining the two-dimensional deformation maps can overcome some limitations of early warning system based on in situ instrumentation (i.e. provides measurements obtained in a few points of the monitored area, which are often not adequately representative of the whole unstable area).

The GB-InSAR is a remote sensing technique based on microwaves interferometry which permits, starting from two or more images, the production of 2D displacement maps of a region (also called interferograms) with millimetre precision.

The used GB-InSAR has metre-scale spatial resolution and an acquisition frequency variable between 5 and 11 min. Interferograms, that does not contain topographic information, since the position of the antennas remains the same during different scans (zero baseline condition), are obtained using pairs of averaged sequential images. Through the phase difference of the backscattered signal in different times, it is possible to estimate the displacement (it is possible to assess only the component of the displacement vector along the line-of-sight—LoS); minus and positive signs indicate movement towards and away from the sensor, respectively.

This technique has been used in a variety of applications from landslide early warning and investigation (Tarchi et al. 2003; Del Ventisette et al. 2011, 2012; Intrieri et al. 2013; Bardi et al. 2014; Tofani et al. 2014) to civil engineering applications (Tarchi et al. 1997), glacier monitoring (Strozzi et al. 2012) and snow characterization (Luzi et al. 2009).

2 A Monitoring Case: Ruinon Landslide (Valfurva, Italy)

The example chosen to show the application of GB-InSAR technique to monitor active landslide is the Ruinon landslide, which is a 30 × 106 m^3 active rock slide located on the North-East sector of the Valfurva valley (Fig. 1). A potential risk from this landslide, which is suspended above the valley, is the possibility of a fast-moving rock avalanche affecting the important tourist road connecting Bormio to Santa Caterina Valfurva and the consequent development of a strongly unstable landslide dam on the Frodolfo River.

The landslide that affects a steep slope made by schistose metapelites locally covered by slope debris (Agliardi et al. 2001; Tarchi et al. 2003). From a geomorphological point of view the landslide is characterized by two main scarps oriented northwest–southeast, parallel to the main fracture system recorded on the slope.

Geological aspects of Ruinon landslide were thoroughly analyzed by several authors (Agliardi et al. 2001; Crosta and Agliardi 2003; Tarchi et al. 2003 cum biblio).

Since 1997 a permanent monitoring network (total station, GPS receivers, inclinometer tubes and extensometers) has been installed by the ARPA.

The first ground-based SAR interferometric measurement campaign about this landslide dates back to summer 2000 (Tarchi et al. 2003; Antonello et al. 2004). In June 2006, a permanent

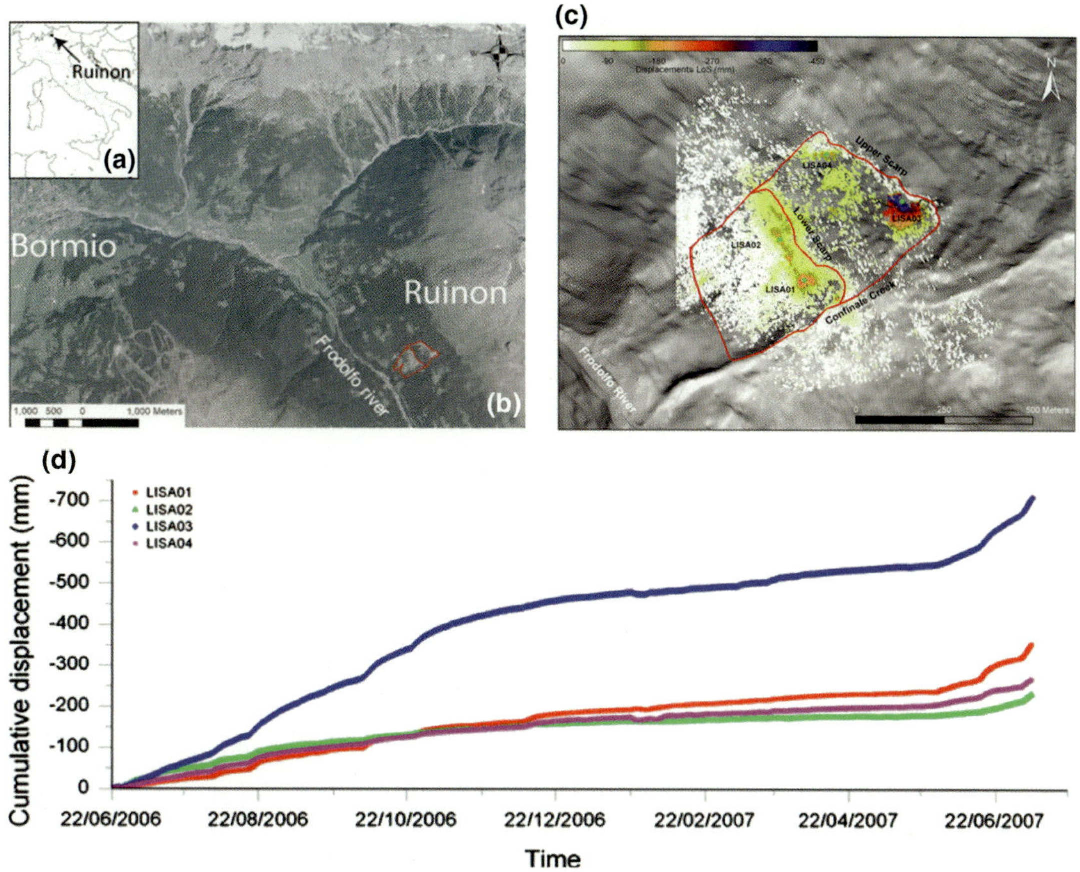

Fig. 1 **a**, **b** Location of the Ruinon landslide. The Ruinon landslide developing in the Valfurva Valley on the hydrographic right of the Frodolfo river. **c** Cumulative displacement map spanning a period of 364 days, 14 h and 50 min between 22/06/2006 and 22/06/2007. LISA01, LISA02, LISA03, and LISA04 identify the points whose displacement-time diagram are plotted in **d**

GB-InSAR monitoring system was installed for early warning purposes.

The acquired data have permitted evaluation of the capability of the apparatus to detect the movement of the landslides' surface and to integrate them with data coming from in situ instrumentations, allowing arrangement of early warning systems.

2.1 Data Analysis

In this paper results of the GB-InSAR monitoring from June 2006 to June 2007, are briefly presented. Thanks to the high sampling rate of the GB-InSAR system, multitemporal maps showing the complete displacement field of the landslide have been generated and used to derive velocity–time diagrams for selected pixels.

The analysis of cumulative displacement maps in a quasi-continuous surface allows identification of landslide sectors characterized by different displacement patterns, indicating partial activations or site-specific surface processes.

The deformations are mainly focused on two sectors (Fig. 1b); (i) the eastern part of the Upper Scarp (total estimated displacement along LoS for the entire period of monitoring is more than

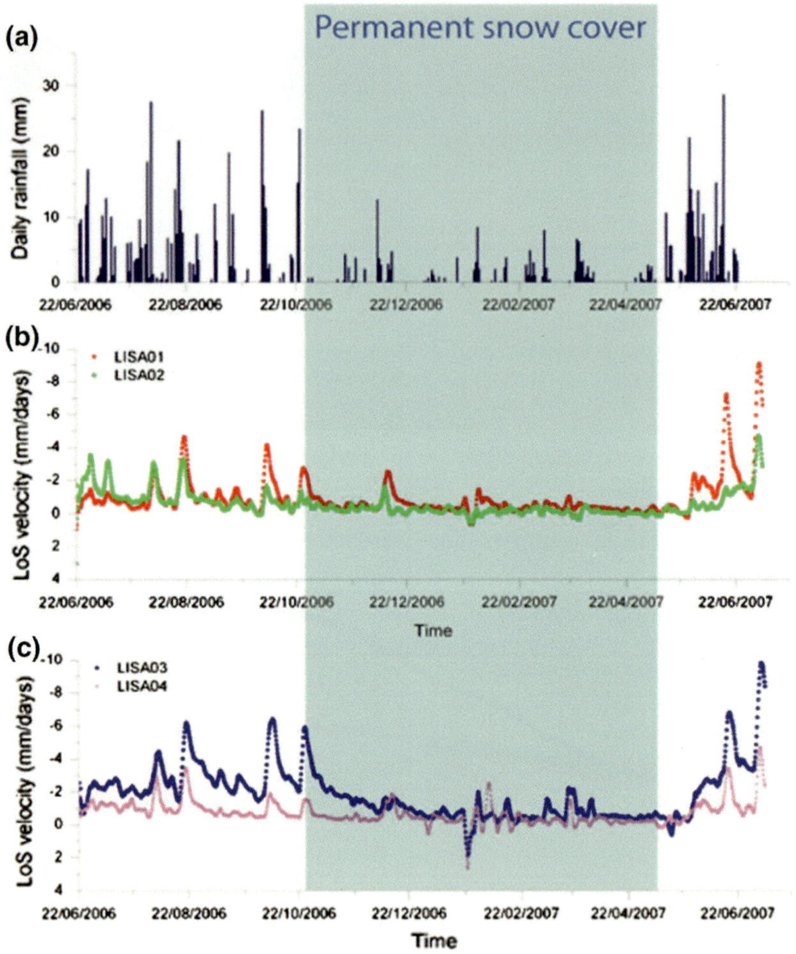

Fig. 2 Comparison among rainfall events (daily rainfall and cumulative rainfall) and velocity displacement as a function of time. **a** Daily rainfall; **b** LOS velocity versus time diagram of the point located on the *Lower Scarp* (LISA01 and LISA02); **c** LOS velocity versus time diagram of the points located on the *Upper Scarp* (LISA03 and LISA04)

630 mm, Fig. 1c), and (ii) the upper sector of the Lower Scarp (total displacement along the LoS more than 300 mm, Fig. 1c).

The cumulative displacement versus time shown a clear seasonality with the maximum speed recorded in spring and the minimum recorded during the winter. These plots demonstrate that the seasonal accelerating phase is generally rapid and followed by a slower deceleration phase.

The area is characterized by a typical continental-alpine rainfall regime (with rainy summer and autumn).

The data recorded by the GB-InSAR show a good correlation between landslide acceleration peaks and rainfalls (Fig. 2).

The recorded acceleration peaks are mainly located after major rainfall events (defined as the occurrence of two o more rainy days separated by at least 2 days with rainfall <4 mm). Although the water amount resulting from snow melting, which of course occurs later than precipitation, has not been taken into account in this analysis, is possible to clearly distinguish the effects of rainfall on the landslide.

3 Monitoring and Early Warning at Stromboli

As example of the capability to use the GB-InSAR system as continuous and long-tome monitoring system we present the Stromboli case study, a composite volcano forming the northernmost island of the Aeolian Archipelago (Tyrrhenian Sea; Fig. 3).

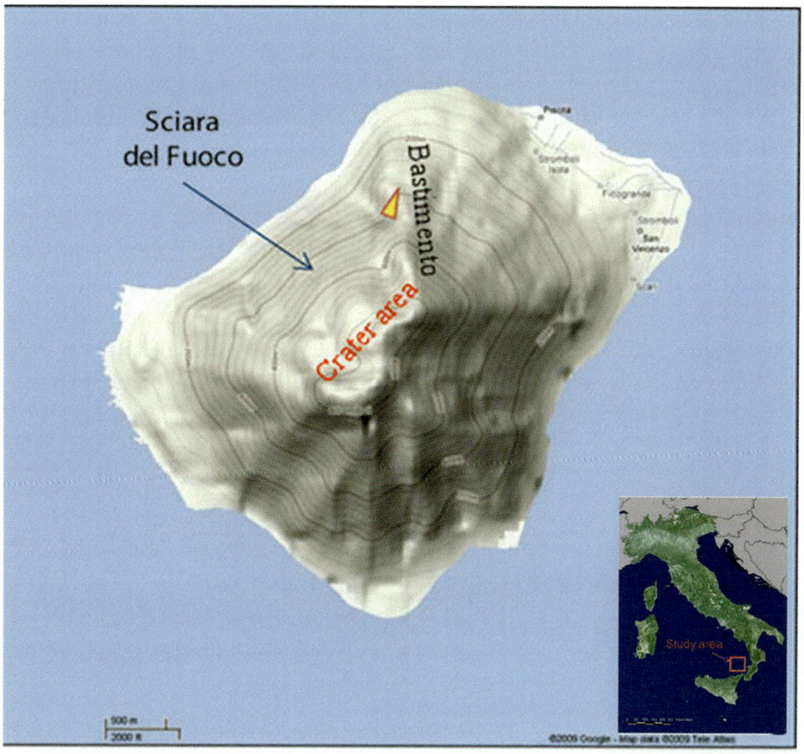

Fig. 3 Localization of: **a** Stromboli volcano and **b** Sciara del Fuoco. The *yellow triangle* in **b** marks the radar site

In this site, during the last 13 years, the GB-InSAR technique has disclosed its reliability and robustness in a real operational case (Barbieri et al. 2004; Bertolaso et al. 2009; Casagli et al. 2009; Tarchi et al. 2009; Di Traglia et al. 2013, 2014a, b).

Although the activity of Stromboli volcano is generally characterized by a persistent mild explosive activity at the summit craters, in the last 13 years, the usually mild "Strombolian" activity of the volcano has been interrupted by instability periods that influenced the stability of the Sciara del Fuoco landslide (SdF). From February 2003 GB-InSAR system was operative to monitor the crater area and the SdF (Bertolaso et al. 2009; Tarchi et al. 2009 cum biblio; Nolesini et al. 2013; Di Traglia et al. 2014b) with the aim to understand triggering mechanisms and dynamics of the SdF slope and, applying forecast methods as those proposed by Fukuzono (1985a, b, 1990) and Voight (1988, 1989, 2000), to built a early warning system.

The static loading induced by the continuous deposition of lava at the top of the SdF, the increased magma pressure inside dykes, and the dynamic loading caused by explosions and eruptive processes appear to be the main causes of the landslide initiation.

The presence of lava flows along the unstable slope inhibited the use of traditional topographic surveys, such as total stations and GPS receivers which require the access to the unstable slope for sensor placement. In this occasion, interferometric processing of spaceborne radar imagery, even if frequently applied to volcanoes monitoring (Amelung et al. 2000; Pritchard and Simons 2002), was not suitable for its satellites inadequate revisiting time, with respect to the velocity of the observed phenomena and for the geometric distortions induced by the SdF steepness.

The GB-InSAR data have been used both for long-term monitoring on areas with low displacement rates as the SdF (using interferograms calculated on long time intervals spanning from weeks to months) and for early warning studying explosions and rapid movements using interferograms calculated on relatively brief time intervals (spanning from a few minutes to a few days). The use of interferograms spanning a long time interval permits us to distinguish areas affected by different kinds of movement probably due to different triggering mechanism.

The capability of GB-InSAR instrumentation of reproducing the ground field of motion in a quasi-real time and in a detailed scale, allows the continuous monitoring of the deformational evolution of the volcano flank and consequently, the use of the system for early warning.

3.1 Data Analysis

In this paper we report some GB-InSAR data used as long-time deformation monitoring of SdF (2007–2012) and as early warning system (during the 2007 "crisis").

Interferograms calculated over a time interval of 28 days over the entire period of monitoring (2003–2016) show the presence of two sectors on the SdF characterized by different deformation rate and geometry. For example, during December 2008, the upper sector of the SdF is characterized by two separate elongated, tongue-like movement patterns (Fig. 4). The lower sector displays higher displacement rates (0.23 mm/day); here the moving area is narrower and nail shaped. Moreover during this time a major explosion and a high intensity Strombolian explosion occurred, respectively on the 6th and the 17th of December, inducing a total decorrelation of the crater area.

From January 15th 2007, deformations on the crater area began to increase with respect to the low usual value measured in the "normal Strombolian activity". On February 27th, the effusive phase occurred; it was characterized by an explosion in the lower sector of the crater

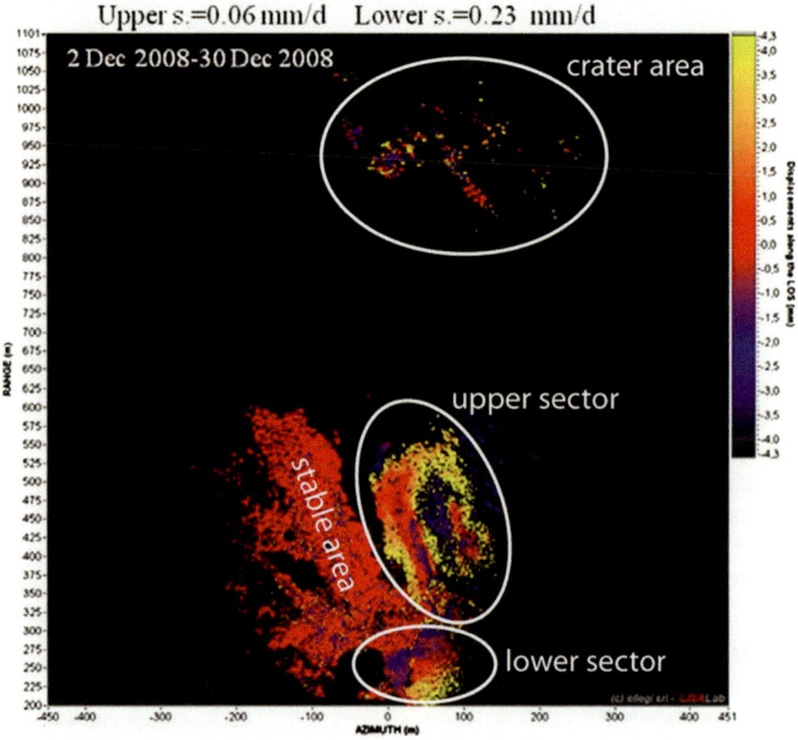

Fig. 4 Example of 28-day interferograms representative of the period 2–30 December 2008

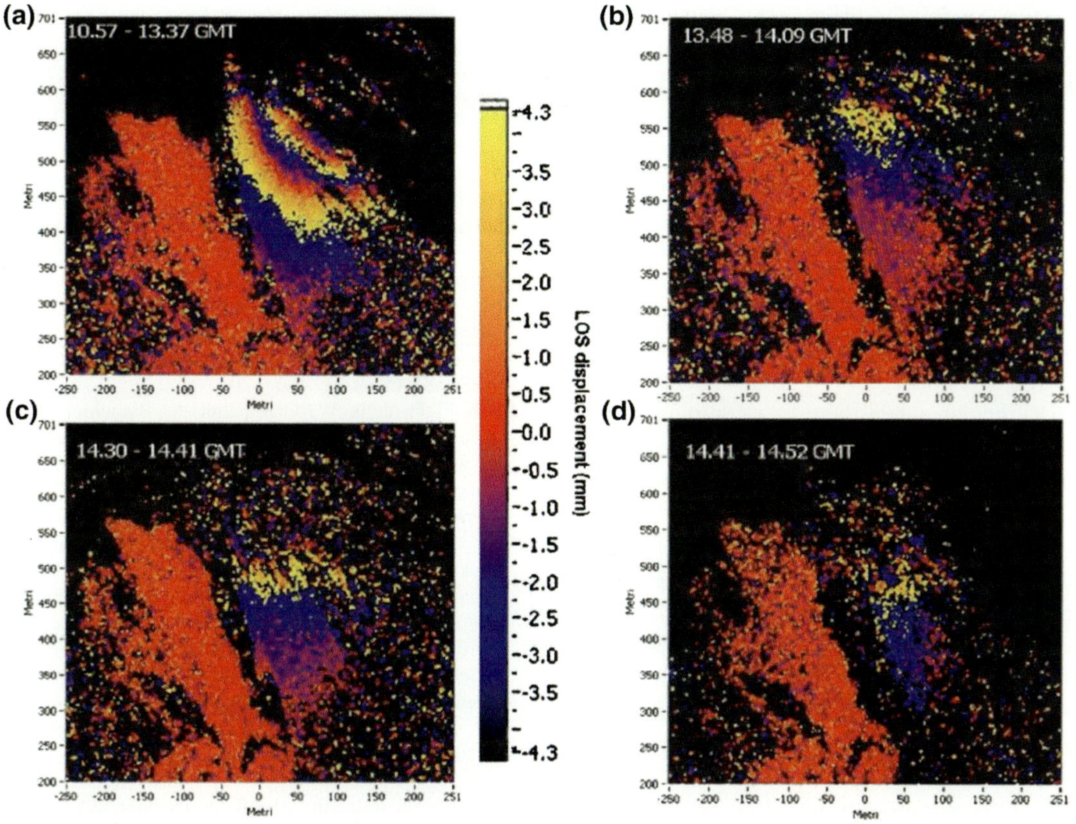

Fig. 5 Interferograms expressed through displacement map obtained on 27 February 2007, spanning different time intervals: **a** 10:57 GMT to 13:37 UT; **b** 13:18 UT to 14:09 UT; **c** 14:30 UT to 14:41 UT; and **d** 14:41 UT to 14:52 UT

area, which induced a partial collapse of the crater flank and was followed by the opening of a new vent at 600 mslm. This event induced inducing deformations also on the SdF.

The effect of these events can be clearly observed on the interferogram where the region corresponding to the craters appears totally decorrelated (see the "salt and pepper" texture) due to the rapidly occurred morphological modifications. Moreover, in the SdF portion of the map, a fringe is observable due to surface deformation related to the vent opening. An outline of the deformations occurring on the SdF on February 27th can be formulated in detail observing the sequence of interferograms in Fig. 5. In the first phase (10:57–13.37 UT), the interferometric fringes are parallel to the slope direction (Fig. 5a), suggesting a volcanic deformation on the flank. On the other hand, the direction of the fringes in the residual figures (Fig. 5b–d) is predominantly perpendicular to this direction, which can be interpreted as the signature of a gravitational movement towards the sea.

4 Discussion and Conclusions

This paper highlights the utility to use a ground-based interferometric SAR to monitor fast landslides and unstable slopes. The main characteristics of a GB-InSAR monitoring is to acquire information on the displacements over the entire observed area, permitting to gain information of dangerous and fast-moving areas, where no in situ instrumentation can been installed. As demonstrated in the Ruinon test site and in long-term monitoring of Stromboli, this features permits to identify the landslide sectors characterized by a different behaviour and also related to possible triggering mechanism.

The analysis of the Ruinon landslide kinematics, in particular, the abrupt acceleration phases, suggests a relation between acceleration peaks and rain.

The Stromboli test site has also demonstrated the system capability of being employed over long time periods, as a permanent monitoring system and also in presence of fast-evolution processes in unfavourable environmental settings. In particular, when there is a complex displacement pattern derived from the superimposition and interference of different processes (i.e., the deformation due to the lava flow and the bulging related to the opening of new vent superposed on that due to the slope instability), it is possible to understand the behaviour of the monitored portion of the SdF during particular events.

The accuracy of GBInSAR data in mapping and in monitoring landslides is fundamental in risk management and in preparation of emergency plans.

Acknowledgements This work has been partly sponsored by the National Civil Protection Department (DPC), in the framework of the SAR.net Project, partially by PRIN Project 2007 (Development and validation of hydraulic and geological tools for supporting a Tsunami Early Warning System: implementation to the Stromboli landslide case) and partly from the Valfurva Municipality. The DPC is acknowledged for the support of the project and for the permission given for the publication. The authors are grateful to Dario Tarchi and Joaquim Fortuny-Guasch for their support to data processing and interpretation, to the Ellegi-Lisalab for providing the systems used for data acquisition and to Dr. Mannucci (ARPA-Lombardia) for making available rainfall data. Thanks are due to the staff of the Engineering Geology group of the Earth Science Department of the University of Florence, for their valuable and constant work.

References

Agliardi F, Crosta G, Zanchi A (2001) Structural constraints on deep-seated slope deformation kinematics. Eng Geol 59(1–2):83–102

Amelung F, Jonsson S, Zebker H, Segall P (2000) Widespread uplift and "trapdoor" faulting on Galapagos volcanoes observed with radar interferometry. Nature 407:993–996

Antonello G, Tarchi D, Casagli N, Farina P, Guerri L, Leva D (2004) SAR interferometry from satellite and ground-based system for monitoring deformations on the stromboli volcano. In: Proceedings IGARSS 2004—international geoscience and remote sensing symposium, Anchorage, Alaska (USA)

Barbieri M, Corsini A, Casagli N, Farina P, Coren F, Sterzai P, Leva D, Tarchi D (2004) Spaceborne and ground-based SAR interferometry for landslide activity analysis and monitoring in the apennines of Emilia Romagna (Italy): review of methods and preliminary results European Space Agency, (Special Publication) ESA SP (550), pp 463–470

Bardi F, Frodella W, Ciampalini A, Bianchini S, Del Ventisette C, Gigli G, Fanti R, Moretti S, Basile G, Casagli N (2014) Integration between ground based and satellite SAR data in landslide mapping: The San Fratello case study. Geomorphology 223:45–60

Bertolaso G, De Bernardinis B, Bosi V, Cardaci C, Ciolli S, Colozza R, Cristiani C, Mangione D, Ricciardi A, Rosi M, Scalzo A, Soddu P (2009) Civil protection preparedness and response to the 2007 eruptive crisis of Stromboli volcano, Italy. J Volcanol Geotherm Res 182:269–277

Casagli N, Tibaldi A, Merri A, Del Ventisette C, Apuani T, Guerri L, Fortuny-Guasch J, Tarchi D (2009) Deformation of Stromboli Volcano (Italy) during the 2007 crisis by radar interferometry, numerical modeling and field structural data. J Volcanol Geoth Res 182:182–200

Crosta G, Agliardi F (2003) Failure forecast for large rock slides by surface displacement measurements. Can Geotech J 40:176–191

Del Ventisette C, Intrieri E, Luzi G, Casagli N, Fanti R, Leva D (2011) Using ground based radar interferometry during emergency: the case of the A3 motorway (Calabria Region, Italy) threatened by a landslide. Nat Hazards Earth Syst Sci 11:2483–2495

Del Ventisette C, Casagli N, Fortuny-Guasch J, Tarchi D (2012) Ruinon landslide (Valfurva, Italy) activity in relation to rainfall by means of GBInSAR monitoring. Landslides 9:497–509

Di Traglia F, Del Ventisette C, Rosi M, Mugnai F, Intrieri E, Moretti s, Casagli N (2013) Ground-based InSAR reveals conduit pressurization pulses at Stromboli volcano. Terra Nova (25):192–198

Di Traglia F, Intrieri E, Nolesini T, Bardi F, Del Ventisette C, Ferrigno F, Frangioni S, Frodella W, Gigli G, Lotti A, Tacconi Stefanelli C, Tanteri L, Leva D, Casagli N (2014a) The ground-based InSAR monitoring system at Stromboli volcano: linking changes in displacement rate and intensity of persistent volcanic activity. Bull Volcanol 76:786

Di Traglia F, Nolesini T, Intrieri E, Mignai F, Leva D, Rosi M, Casagli N (2014b) Review of ten years of volcano deformations recorded by the ground-based InSAR monitoring system at Stromboli volcano: a tool to mitigate volcano flank dynamics and intense volcanic activity. Earth Sci Rev 139:317–335

Fukuzono T (1985a) A new method for predicting the failure time of a slope failure. In: Proceedings of the 4th international conference and field workshop on landslides, Tokyo (Japan), pp 145–150

Fukuzono T (1985b) A method to predict the time of slope failure caused by rainfall using the inverse number of velocity of surface displacement. J Jpn Landslide Soc 22:8–13

Fukuzono T (1990) Recent studies on time prediction of slope failure. Landslide News 4:9–12

Intrieri E, Di Traglia F, Del Ventisette C, Gigli G, Mugnai F, Luzi G, Casagli N (2013) Flank instability of Stromboli volcano (Aeolian Islands, Southern Italy): integration of GB-InSAR and geomorphological observations. Geomorphology 201:60–69

Luzi G, Noferini L, Mecatti D, Macaluso G, Pieraccini M, Atzeni C, Schaffhauser A, Fromm R, Nagler T (2009) Using a ground-based SAR interferometer and a terrestrial laser scanner to monitor a snow-covered slope: results from an experimental data collection in Tyrol (Austria). IEEE Trans Geosci Remote Sens 47:382–393

Nolesini T, Di Traglia F, Del Ventisette C, Moretti S, Casagli N (2013) Deformations and slope instability on Stromboli volcano: Integration of GBInSAR data and analog modeling. Geomorphology 180–181:242–254

Pritchard ME, Simons MA (2002) A satellite geodetic survey of large-scale deformation of volcanic centres in the central Andes. Nature 418:167–171

Strozzi T, Werner C, Wiesmann A, Wegmuller U (2012) Topography mapping with a portable real-aperture radar interferometer. IEEE Geosci Remote Sens Lett 9:277–281

Tarchi D, Ohlmer H, Sieber AJ (1997) Monitoring of structural changes by radar interferometry. Res Nondestr Eval 9:213–225

Tarchi D, Casagli N, Moretti S, Leva D, Sieber AJ (2003) Monitoring landslide displacements by using ground-based radar interferometry: application to the Ruinon landslide in the Italian Alps. J Geophys Res 108(B8–2387):101–114

Tarchi D, Casagli N, Fortuny-Guash F, Guerri L, Antonello G, Leva D (2009) Ground deformation from ground-based SAR interferometry. Learning from Stromboli and its 2002–03 eruptive crisis, Geophysical Monograph

Tofani V, Del Ventisette C, Moretti S, Casagli N (2014) Integration of remote sensing techniques for intensity zonation within a landslide area: a case study in the Northern Apennines, Italy. Remote Sens 6:907–924

Voight B (1988) Material science law applies to time forecast of slope failure. Landslide News 3:8–11

Voight B (1989) A relation to describe rate-dependent material failure. Science 243(10):125–130

Voight B (2000) Structural stability of andesite volcanoes and lava domes. Philos Trans R Soc Lond 358: 1663–1703

TXT-tool 2.062-1.1 A Landslide Monitoring and Early Warning System

Teuku Faisal Fathani and Dwikorita Karnawati

Abstract

Landslides are one of most common major disasters in Indonesia due to the physical susceptibility of the region and socio-economical conditions within the country. Efforts are urgently needed to avoid or reduce the risk of landslides. Unfortunately, most landslide-susceptible areas have very fertile soils and abundant very good quality water. The susceptible areas are thus usually densely populated, with serious consequences with respect to slope instability. Despite efforts to establish slope protection zones that are restricted for any development and settlement, relocation programs cannot easy to be carried out due to socio-economic constraints. Therefore, landslide monitoring, prediction and early warning systems are urgently required to guarantee the safety of communities in such areas. A long running and sustainable community-based landslide monitoring and early warning system has been developed in Indonesia that includes collaboration among the local government, universities, private sectors, NGOs, and the disaster management community. The program has three stages: the development of a simple and low-cost landslide early warning system, the design and implementation of the warning system in real time, and the establishment of a socio-technical strategic approach to disaster risk reduction. These activities have already met community needs and helped save lives, and they continue to obtain solid community support. Now the further challenges are to expand the project coverage, and propose more effective landslide monitoring, early warning, analysis, and visualization. In addition, the capabilities in socio-economic risk assessment need to be expedited to help identify those most at-risk within the community. This chapter describes the achievements and the current activities of the IPL-158 Project "Development of Community-based Landslide Early Warning System".

T.F. Fathani (✉)
Center for Disaster Mitigation and Technological Innovation (GAMA-InaTEK), Department of Civil and Environmental Engineering, Universitas Gadjah Mada, Jl. Grafika No. 2, Yogyakarta 55281, Indonesia
e-mail: tfathani@ugm.ac.id

D. Karnawati
Center for Disaster Mitigation and Technological Innovation (GAMA-InaTEK), Department of Geological Engineering, Universitas Gadjah Mada, Jl. Grafika No. 2, Yogyakarta 55281, Indonesia
e-mail: dwiko@ugm.ac.id

K. Sassa et al. (eds.), *Landslide Dynamics: ISDR-ICL Landslide Interactive Teaching Tools*, https://doi.org/10.1007/978-3-319-57774-6_21

Keywords
Landslide monitoring • Early warning • Socio-technical strategic approach
Community participation • Risk assessment

Contents

1 **Introduction** ... 298
2 **Stage of Development of the Landslide Monitoring and Early Warning System** ... 298
3 **Simple and Low-Cost Technology for Landslide Monitoring** ... 299
4 **Real-Time Landslide Monitoring** ... 301
4.1 First Stage of Development (2007–2009) ... 301
4.2 Second Stage of Development (2010–2012) ... 304
5 **Landslide Monitoring Devices** ... 305
6 **Strategic Approach: Community-Based with Linkages to Local Government and the University** ... 306
References ... 308

1 Introduction

Areas prone to mass movement are widespread in Indonesia. The potential for landslides in several regions is controlled by geotechnical and geological conditions, with landslides often being triggered by earthquake activity and high rainfall intensity. The vulnerability to landslides is made worse by intensive land-use development. The urgent issues to be addressed are the dense populations residing in areas prone to mass movement and the difficulty, for socio-economic reasons, of relocating local people to safer areas. Meanwhile, the potential for mass movement occurrence has been rising due to increases in rainfall intensity, earthquakes, and the effects of human interference on slope stability (Karnawati et al. 2005; Karnawati and Fathani 2008). As one of the efforts to protect the people living in hazardous areas, a landslide monitoring and early warning system is highly needed, while simultaneously efforts to improve the community's resilience are carried out.

The early warning system is a set of instruments working simultaneously and coordinated for natural disaster mitigation. To set up a landslide monitoring and early warning system, a preliminary survey and investigation should be conducted to understand the physical characteristics and social conditions of the disaster area. The next phase is the design and the installation of the most adaptive landslide monitoring and early warning system, which must be conducted simultaneously with socialization and training so that the local people can operate and maintain the system independently (Fathani et al. 2008, 2010).

This paper presents a long-running and sustainable community-based landslide monitoring and early warning system developed in Indonesia. The first challenge was the development of a simple and low-cost system that is easily operated and can be maintained by the local community. A further advance was the design and implementation of real-time landslide monitoring, involving several trials of telemetry systems. In order to support this technical program, a socio-technical strategic approach was introduced by establishing collaborations between the local government, universities, private sectors, NGOs, and the disaster management community.

2 Stage of Development of the Landslide Monitoring and Early Warning System

One method for mitigating landslides disasters is the development of a highly adaptive monitoring and early warning system, so that the local people and governmental officers are capable of preparing and operating the system. Moreover, the local people, supported by the local government, are expected to be able to reproduce the system independently so it can be applied in other vulnerable areas. The following are steps in the development and application of landslide monitoring and early warning.

(a) Carry out surveys and field investigations of the geological and geotechnical conditions, the factors controlling the landslides, and social condition of the people in the vulnerable areas, to determine the most appropriate monitoring equipment and the best location for installing an early warning system.
(b) Design the most adaptive early warning system by taking into account the results of the geological, geotechnical and social surveys in several pilot locations in the study area.
(c) Carry out socialization and training to improve the capacity of the local people living in the pilot locations, so that they can operate and maintain the system.
(d) Install the early warning system and provide consultation on the maintenance of the instruments in the pilot locations in the landslide prone areas, followed by testing and calibration of the instruments to ensure the continuity and accuracy of the system.

The proposed landslide monitoring and early warning system is an integrated part of the technical and social network, and is based on community participation in the application of the most adaptive and appropriate technology (Fathani and Karnawati 2010, 2012). Through this system, the community is empowered to actively participate in the preparation of the technical network (consisting of the hardware for landslide monitoring and early warning), and the social networks are strengthened by Village Disaster Management Forums supported by local government (Karnawati et al. 2011).

Through the application of these systems, the technical network of landslide monitoring and early warning can be developed and the capacity building of the local people on disaster preparedness can be enhanced (Andayani et al. 2008). In the end, the local people are expected to be more motivated, more capable, and more empowered in participating in landslide disaster mitigation in their village.

The stages of development of a landslide early warning system, starting from the simple and low-cost manual monitoring system to the real-time monitoring system, are discussed in the following section. In addition, newly developed monitoring devices such as extensometers, tiltmeters, and raingauge, and the proposed network diagram of the telemetry system are described.

3 Simple and Low-Cost Technology for Landslide Monitoring

Since 2007, Universitas Gadjah Mada Indonesia has developed simple and low-cost equipment for landslide monitoring and early warning. At the initiation of quantitative investigations, two types of simple extensometers and raingauges were installed at several pilot areas in Central Java and East Java Provinces. The first type of extensometer is a handmade manual reading extensometer with an accuracy of 1 mm (Fig. 1). Another type is an automatic extensometer to monitor ground surface movement with an accuracy of 0.2 mm, in which the relative movement between two points is mechanically enlarged by 5 times and recorded continuously on paper (Fig. 2). Both types of extensometers, coupled with an automatic rainfall recorder, are connected to an alarm system in order to directly warn the local community to take necessary actions in dealing with a potential landslide disaster.

The implementation of this landslide early warning system, initially started by the Faculty of Engineering Universitas Gadjah Mada, had several failures before it finally succeeded through appropriate technology, the community's understanding of disaster preparedness, and the effectiveness of a socio-technical strategy approach to implement the system. In one example, this simple and low-cost monitoring equipment successfully saved 35 families in Kalitelaga village at Banjarnegera Regency, Central Java from a landslide on November 7, 2007 (Fathani and Karnawati 2010).

The installation of a simple and low-cost landslide monitoring and early warning system in

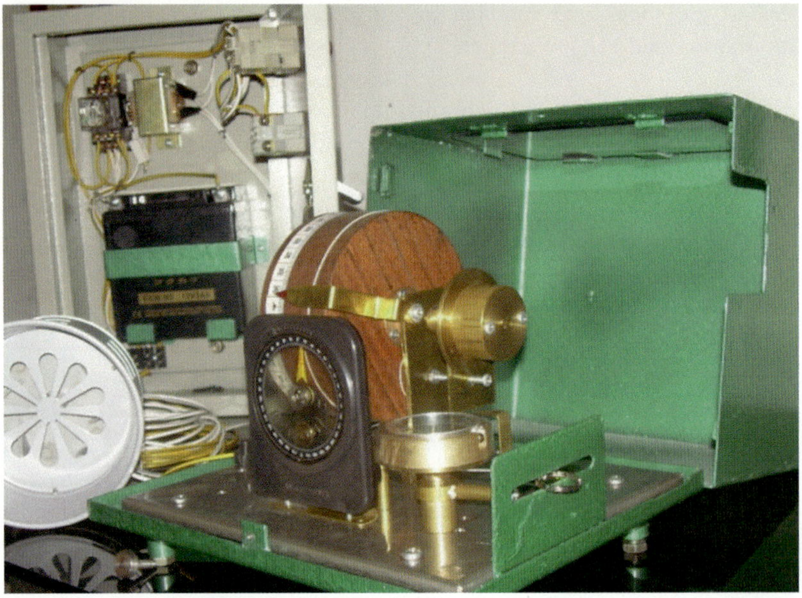

Fig. 1 Manual reading extensometer

Fig. 2 Paper recorded, automatic extensometer

Kalitelaga Village is shown in Figs. 3 and 4. Figure 3 shows an automatic extensometer installed in the upper part of an unstable slope in order to monitor the ground movement on a crack, in a location where houses are located on the lower slope. Meanwhile, Fig. 4 shows a manual reading extensometer at the upper part of a collapsed slope. This extensometer warned the local community about four hours before the occurrence of a landslide and saved 35 families living downslope.

In order to ensure that the simple landslide early warning system can work properly, the local authority (local government at a village level) should establish the warning information flow and command flow as shown in Fig. 5. Since this simple early warning system equipment (extensometer and automatic rainfall

Fig. 3 Placement of an automatic extensometer

Fig. 4 Manual reading extensometer on the upper part of a collapsed slope

recorder) directly sends the warning alert to local community, hence the community can immediately prepare for evacuation. The head of village then gives an order to the selected local operators to lead the evacuation process (if necessary). This command flow worked well during the occurrence of a landslide on November 7, 2007. The selected local operators (under the supervision of the head of village) can decide when and where to evacuate, and convince the evacuees to stay in a safer area while the local operator keeps monitoring the slope condition.

4 Real-Time Landslide Monitoring

4.1 First Stage of Development (2007–2009)

In line with the installation of simple monitoring equipment, on September 2007, the Asian Joint Research Project for Early Warning of Landslides has conducted a field survey to support the installation of real-time landslide monitoring equipment. A pilot area has been established in

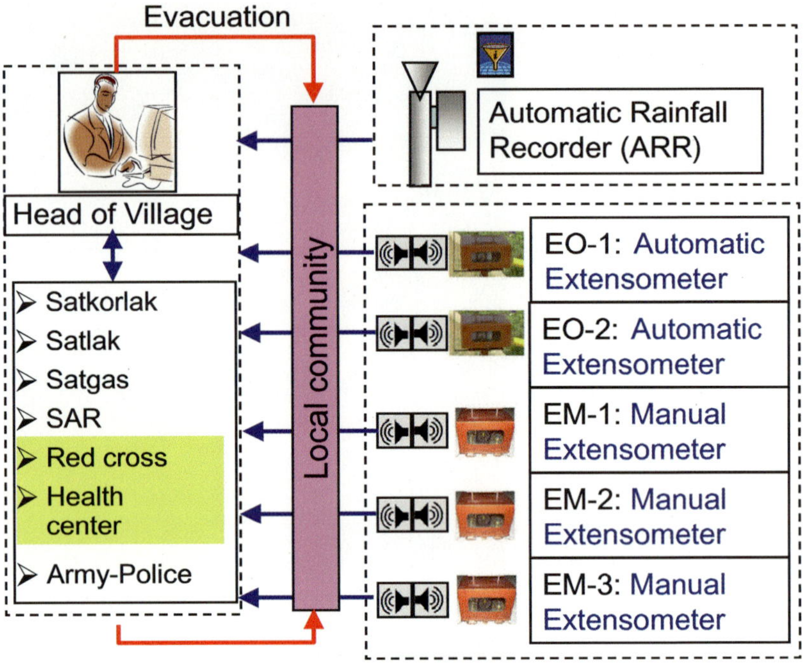

Fig. 5 Flow of warning information and evacuation command for a simple landslide early warning system at village level

Banjarnegara Regency, Central Java Province. Figure 6 shows the aerial photo and topography map of a landslide area mapped by using a photogrammetry system.

Based on the site investigation, it is clear that not only the rain intensity but also the morphology and geological conditions of the study area significantly control the occurrence of landslides. The unstable zone in the study area is situated on the lower slopes of the mountains, with slope inclinations of 20°–60°. The moving materials consist of colluvial deposits of silty clay overlying an inclined impermeable layer of clay, which is situated on the lower part of the andesitic breccia of the mountain. The moving zone is saturated for most of the rainy season due to its low position relative to the surrounding mountain slopes.

The existence of an impermeable clay layer underneath the colluvial deposits creates a saturated condition within the colluvial deposits that gradually increases and is maintained during the rainy season, until the rise of pore water pressure within the soil induces movement. Therefore, monitoring the pore water pressure in response to rain infiltration should be the main concern in establishing an early warning.

This monitoring system presents the results of real-time measurements, using long-span extensometers, raingauges, pore pressure sensors, and monitoring of the scene by an Internet Protocol (IP) camera. The real-time monitoring equipment consists of an outdoor unit and an indoor unit. The outdoor unit is fixed on a center pole and consists of a field server, two extensometers, a raingauge, an IP camera and a water pressure sensor (Fig. 7). The field server is a sensing device with real-time online data display system which gathers the data from multiple sensors and shows them in a webserver.

The extensometer is placed at two positions connected by a pulley and a super invar wire which can measure both extension (+) and compression (−). The indoor unit has two crucial components, i.e. a processing unit and a GPRS modem. The indoor unit collects the data, then processes and stores it on the monitor, and sends the data to the server every hour. This unit also can implement an early warning that can be adjusted depending on the site conditions.

Fig. 6 Aerial photo with a topographic map and position of real-time monitoring equipment in the pilot study area in Banjarnegara Regency, Central Java Province. Landslide damage and fatalities are shown *on the right*

Fig. 7 Outdoor real-time monitoring devices in the pilot area

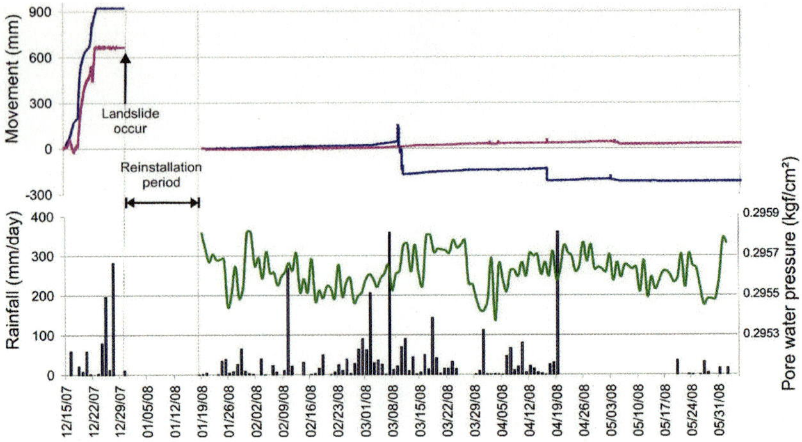

Fig. 8 Pilot area measurements from extensometers, a raingauge and pore water pressure sensor

The position of the long-span extensometer poles (P1–P6), raingauge, pore water pressure sensor and indoor unit are shown in Fig. 6. Three poles (P1–P3) were installed on December 15th, 2007. The installation process had faced some problems, as the slide occurred on day the system was set up. As shown in Fig. 8, starting from December 23rd, 2007, the extensometer had been saturated (up to 660–920 mm of displacement), therefore it could not measure the movement when the landslide occurred on December 30th, 2007. The landslide destroyed the center pole (P2), buried the lowest pole (P3) and also struck several houses, farm land and a district road.

On January 19th, 2008, the monitoring system had been reinstalled at a new location about 150 m from the previous destroyed place (Fig. 6). Three new poles (P4–P6) were erected with two long-span extensometers, a raingauge and an IP camera connected to the center pole (P5). A pore water pressure sensor was placed inside a well near P5, while the indoor unit was located in a house belonging to a volunteer resident near P4.

The result of measurement of two extensometers, daily rainfall and pore water pressure fluctuation are shown in Fig. 8. The accumulated movement of the extensometer starting from January 19th until May 31st 2008 reached about 30 and 220 mm for Extensometers P4–P5 and P5–P6, respectively. Meanwhile the maximum rate of rainfall had reached 200–360 mm/day. It can be seen that the extensometer movements on March 7th and April 13th, 2008 were strongly related to the occurrence of rainfall.

4.2 Second Stage of Development (2010–2012)

A field survey to support the establishment of the new real-time landslide monitoring and warning system was conducted. This system presents the results of real-time measurements from an upper ground and underground extensometer, a tiltmeter, groundwater measurements and a raingauge. The monitoring equipment is connected with a data logger and integrated in a field server. This sensing device provides a real-time online data display; the system collects the data from multiple sensors and shows them in a webserver. This unit also implements early warnings that can be adjusted depending on the site conditions.

Fathani et al. (2011) have determined the warning criteria for rainfall-induced landslides in Central Java by deploying this real-time monitoring system and assessing the effects of rainfall intensity on landslide activity using various setting methods. Rainfall thresholds and slope movement acceleration are used to determine the early warning criteria and time for evacuation.

Figure 9 shows a network diagram of a telemetry system for real-time monitoring of landslides. The system comprises three sensors—an extensometer, a tiltmeter and a raingauge, and other sensors may be added, such as pore water

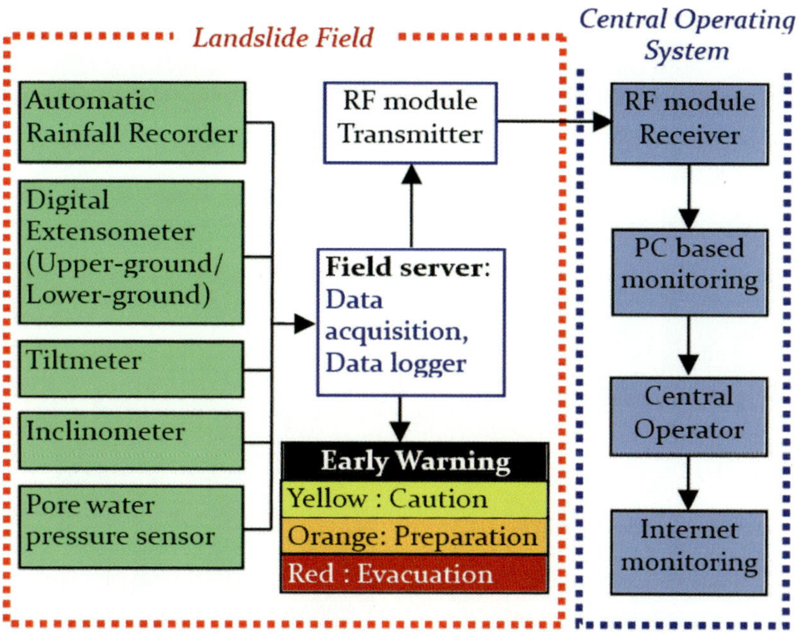

Fig. 9 Network diagram of a telemetry system for real-time monitoring and early warning of landslides

pressure and inclinometer sensors. The collected data, processed by a microcontroller, is sent point to point in a wireless network. The data is received by a field server; its functions are to receive, store, analyze and resend data to the central server, and decide when to send an early warning to the local residents. The maximum distance of the radio frequency is 2 km Line of Sight (LOS) between the sensor and local server, while the distance from the local server to the central server is from 15 to 20 km (LOS).

The data received is then stored in a digital storing media (memory card). The central server consists of receiver equipment and a specific Personal Computer (PC) to monitor the data visually. The PC, which is connected to the internet, will upload all of the data to a web server; therefore, the data can be monitored from any place with internet access.

5 Landslide Monitoring Devices

The landslide monitoring devices (left side of Fig. 9) were developed by Universitas Gadjah Mada, in cooperation with the International Consortium on Landslides (ICL), DPRI Kyoto University, National Board for Disaster Management (BNPB) and Ministry for the Development of Disadvantaged Regions (KPDT). Fathani and Karnawati (2012) proposed that the newly developed monitoring devices consist of an extensometers, raingauges, tiltmeters, inclinometers and pore water pressure sensors (Fig. 10).

An extensometer is a device to monitor ground surface movement in landslide-prone areas. It measures relative surface movement between two points at a ground crack on moving land. The digital extensometer can measure lateral ground movement and slope along X and Y axes. The working principle of the extensometer involves installing this device between two points, one of which is installed on the moving ground. An invar wire is connected to the ground crack on one end and to the potentially unstable ground on the other end.

When the ground moves, the invar wire will pull a spring mechanism. The sensor will then read this movement. The newly developed upper ground extensometer has been installed to monitor ground movement in a mining site in South Kalimantan. The international mining company prefers to utilize a rotary light signal instead of a

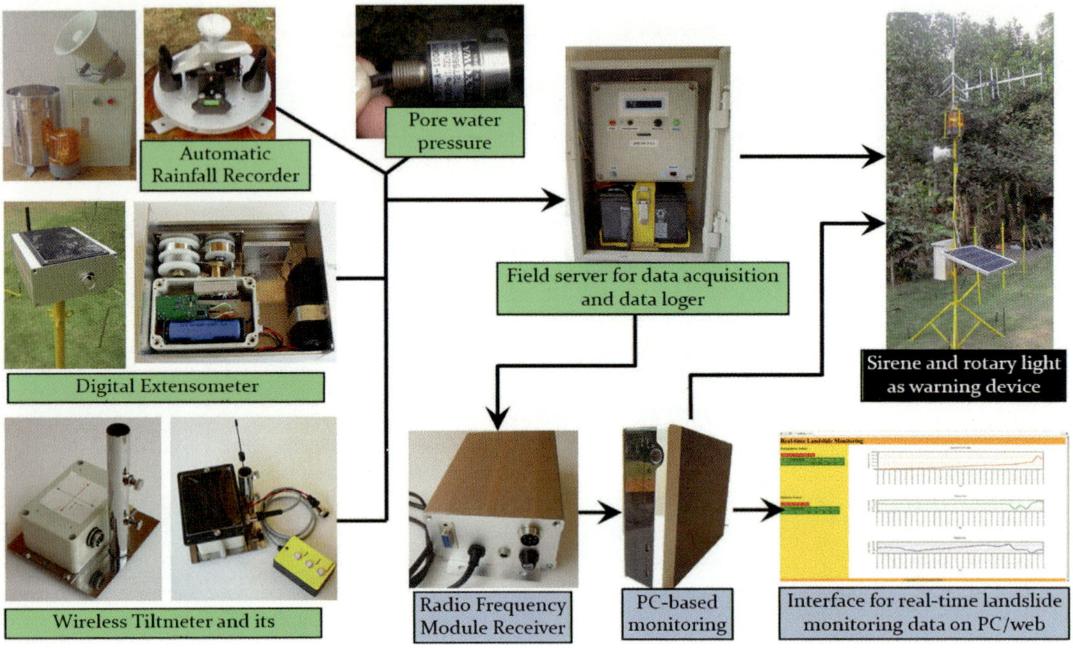

Fig. 10 Newly developed devices for landslide monitoring on a network diagram of a telemetry system

sound alarm system, due to the noise from heavy equipment at the mining site.

This extensometer has a mechanical system and an electrical system. The mechanical system consists of a spring and the spring housing with an invar wire roller. The system is used to measure the ground movement. The extensometer comprises two sensors: (1) an absolute rotary encoder sensor and (2) a tilt sensor. The absolute rotary encoder sensor, combined with the pulley and spring system, can measure a ground crack with an accuracy of 0.1 mm. A tilt sensor with a Micro-electromechanical System (MEMS) is used to measure a change in the slope inclination. The computed data, i.e., ground movement data (mm) and slope inclination data (degree) are then sent to a field server to be stored and processed.

A tiltmeter is a device to monitor the change in the slope in a landslide-prone area. The new developed tiltmeter may measure a slope along X and Y axes with an accuracy up to 0.1°. Other than measuring a slope, the tiltmeter can measure the diagonal angle of the ground movement direction. The preliminary model of the developed wireless tiltmeter is shown in Fig. 10. The tiltmeter will be further developed using a sensor with higher resolution i.e., 0.01°.

The raingauge that has been developed uses the common tipping-bucket system (Fig. 10). This raingauge uses a sensor that calculates the number of times the buckets tip and converts it to rainfall units (mm/h). This data is transmitted to the data logger, with a memory card as the storage media. Aside from that, the rainfall intensity can be seen on the LCD monitor. The raingauge is fitted with an alarm that gives an audio and visual warning when the rainfall intensity exceeds a certain threshold. The alarm can be adjusted with a keypad in the data logger.

6 Strategic Approach: Community-Based with Linkages to Local Government and the University

Universitas Gadjah Mada has been developing and implementing a community-based landslide risk assessment and early warning project over the last several years in collaboration with

International Consortium on Landslides (ICL), Disaster Prevention Research Institute (DPRI) Kyoto University, National Board for Disaster Management (BNPB), Ministry for the development of Disadvantaged Regions (KPDT), local governments and several private companies and NGOs. This effort has been sustained by the local government, university, the community and related stakeholders for years, and it has resulted in both increased capability and capacity at the local/village level to reduce disaster risk and increase community resilience. This landslide early warning system has been implemented at Java, Kalimantan and Sulawesi Islands in Indonesia and will also be installed in a mining area in Myanmar (Fig. 11).

Based on this extensive experience, landslide early warning systems should be based on the most appropriate and adaptive technology, with the participation and involvement of communities. Therefore, both technical skills and communication skills are the main requirements to foster the success of the early warning system program. The system should include some technical aspects such as geological surveys and site selection, the design of monitoring equipment that is simple (low-cost) but effective, the determination of early warning levels (warning criteria), equipment installation, and operation and maintenance, at the field site, and social aspects such as social mapping and evaluation, public consultation and dissemination of the program, and community empowerment, including technical training and evacuation drills for landslide hazard preparedness.

Over the years, these efforts have resulted in the successful implementation of a landslide early warning system at the community level. The pilot projects are now supported by the community, local government and local universities. Further challenges are to expand the project coverage within and outside of the pilot areas, and to store and process the data for a more effective monitoring, early warning, analysis, and visualization.

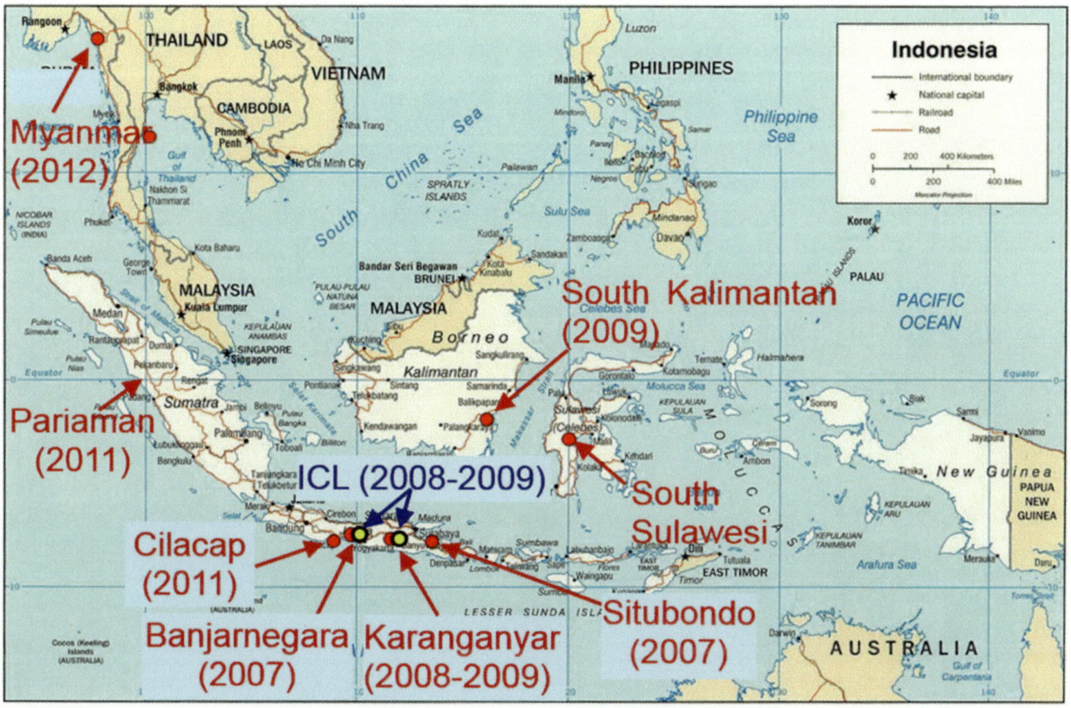

Fig. 11 Locations with the landslide early warning system installed (2007–2012)

Acknowledgements This research was conducted as a part of the Asian Joint Research Project for Early Warning of Landslides supported by the International Consortium on Landslides (ICL), Universitas Gadjah Mada (UGM), and the Disaster Prevention Research Institute of Kyoto University (DPRI/KU). The projects were financially supported by the Ministry of National Education, National Board for Disaster Management (BNPB), and Ministry for the Development of Disadvantaged Regions (KPDT). This research is recognized by ICL as IPL-158 Project.

References

Andayani B, Karnawati D, Pramumijoyo S (2008) Institutional frame work for community empowerment towards landslide mitigation and risk reduction in Indonesia. In: Proceedings of the 1st world landslide forum, IPL–ISDR, Tokyo, pp 57–59

Fathani TF, Karnawati D (2010) Early warning of landslide for disaster risk reduction in Central Java Indonesia. In: Sassa K, Yueping Y (eds) Early warning of landslides. Geological Publishing House, Beijing, pp 159–166

Fathani TF, Karnawati D (2012) Lessons learned in the development of landslide early warning system in Indonesia. In: Proceeding of IPL symposium 2012, UNESCO HQ, Paris, pp 30–34

Fathani TF, Karnawati D, Sassa K, Fukuoka H, Honda K (2008) Development of landslide monitoring and early warning system in Indonesia. In: Proceedings of the 1st world landslide forum, IPL–ISDR, Tokyo, pp 195–198

Fathani TF, Karnawati K, Sassa K, Fukuoka H (2010) Low-cost and adaptive technology to support a community-based landslide early warning system in developing countries. In: Kameda H et al (eds) Disaster reduction hyperbase-Asian Application (DRH-Asia) (EDM-NIED Editorial Team), pp 89–97

Fathani TF, Karnawati D, Legono D, Faris F (2011) Development of early warning system for rainfall-induced landslide in Indonesia. In: Proceeding of the 2nd international workshop on multimodal sediment disaster: Asian Cloud Network on Disaster Research, Tainan, Taiwan, pp 103–113

Karnawati D, Fathani TF (2008) Mechanism of earthquake induced landslides in Yogyakarta Province, Indonesia. The Yogyakarta Earthquake of 27 May 2006. Star Publishing Company Inc., Belmont, pp 8-1–8-8

Karnawati D, Ibriam I, Anderson MG, Holcombe EA, Mummery GT, Renaud JP, Wang Y (2005) An initial approach to identifying slope stability controls in Southern Java and to providing community-based landslide warning information. Glade T, Anderson MG, Crozier MJ (eds) Landslide hazard and risk. Wiley, New York, pp 733–763

Karnawati D, Fathani TF, Wilopo W, Setianto A, Andayani B (2011) Promoting the hybrid socio-technical approach for effective disaster risk reduction in developing countries. In: Brebbia CA, Kassab AJ, Divo EA (eds) Disaster management and human health risk II. WIT Press, Southampton, pp 175–182

TXT-tool 2.007-1.1 Monitoring Alarm System of Landslide and Seismic Safety for Potentially Hazardous Objects

Alexander Ginzburg, Alexey Nikolaev, Valentina Svalova, Anatoliy Manukin and Vladimir Savosin

Abstract

Analysis of seismological phone changes will give possibility to elaborate early warning system. Sergeev Institute of Environmental Geoscience RAS (IEG RAS) has developed and delivered the "System monitoring acceleration induced on the upper part of the offshore oil and gas platforms" for deposits Lunskoe-A (LUN-A) and Piltun-Astokhskoye (PA-B) for Sakhalin-2 project. The System is intended to ensure the safety of the operation of these facilities. Experience of the System creation was used for real time early-warning landslide monitoring system construction. This system was successfully used for landslides monitoring of coastal slope of the river Yenissei.

Keywords

Landslides · Oil and gas platforms · Early-warning monitoring Sakhalin-2 · Yenissei · Geotechnical instrumentation

Contents

1 **Introduction** ... 309
2 **Topsides Induced Acceleration Monitoring System for Oil and Gas Offshore Platforms—TIAMS** ... 310
3 **Description of the Topsides Induced Acceleration Monitoring System (TIAMS)** ... 312
4 **Landslide Monitoring System for Coastal Slope of the River Yenissei** ... 318
5 **Conclusion** ... 324
References ... 324

1 Introduction

Landslides, mudflows, debris flow and avalanches are the most dangerous and destructive processes that cause significant damage to economic objects and often resulting in death.

There are numerous examples of negative consequences of influence of debris flows, landslides, mudflows and avalanches in various regions of Russia. One of the main causes of such emergencies and consequences of the

A. Ginzburg · A. Nikolaev · V. Svalova (✉) ·
A. Manukin · V. Savosin
Sergeev Institute of Environmental Geoscience RAS, Ulansky per., 13, PB 145, Moscow 101000, Russia
e-mail: inter@geoenv.ru

K. Sassa et al. (eds.), *Landslide Dynamics: ISDR-ICL Landslide Interactive Teaching Tools*, https://doi.org/10.1007/978-3-319-57774-6_22

negative effect is the absence of a unified monitoring system for early warning of the dangerous state of controlled processes and the appearance of threats to economic objects (Nikolaev et al. 2011; Postoev and Svalova 2005; Svalova and Postoev 2008; Svalova 2009, 2011a, b, c, 2012, 2013, 2014a, b, c, 2015).

Development of a method for automated monitoring of exogenous geological processes for the purpose of early warning about dangerous development, with identification of the main informative parameters of the processes and for their control, with development monitoring schemes (placement of equipment), with development and testing of elements of systems for automated monitoring, the criteria for a hazardous condition are the main purposes of research (Ginzburg et al. 2008, 2009; Postoev and Svalova 2005; Svalova 2009, 2011a, b, c, 2012, 2013, 2014a, b, c, 2015; Svalova and Postoev 2008).

The development of modern socio-economic system provides for the establishment and operation of such grand and environmentally hazardous facilities like pipelines, nuclear power plants, chemical industry, etc.

Most ecologically dangerous objects or grandiose constructions are in seismic areas and tectonically active zone, in which there may be strong earthquakes, as well as landslides and mudflows. During operation it is necessary to ensure the safety of both the expensive facilities and safety of the environment.

Under these conditions, the timely detection of dangerous earthquakes and giving alarms and automatic shutdown of environmentally hazardous facilities is a paramount task.

IEG RAS for many years carries out all the work necessary to create the monitoring of environmental and seismic safety,

In 2005–2006. IEG RAS has developed and delivered the "System monitoring acceleration induced on the upper part of the offshore oil and gas platforms" for deposits Lunskoe-A (LUN-A) and Piltun-Astokhskoye (PA-B), located near about Sakhalin for Sakhalin-2 project. "System ..." is intended to ensure the safety of the operation of these facilities.

A second example of environmental safety monitoring system is a system of monitoring of dangerous geological processes. The system is designed for monitoring of landslides coastal slope of the river Yenissei in real time. The system provides rapid collection of measurement data on the state of the observed landslides, processing and analysis of the distribution of monitoring results between users and controls the security of shopping and entertainment complex, located on the coastal slope.

2 Topsides Induced Acceleration Monitoring System for Oil and Gas Offshore Platforms—TIAMS

According to the Sakhalin II Project Sakhalin Energy Investment Company is building offshore oil and gas platforms PA-B and LUN-A at the Sakhalin Island shelf. The platforms are situated within the seismically dangerous area where destructive earthquakes are likely to occur.

To reduce the risk of environmental accidents that can appear during oil and gas production the as a result of destructive earthquake Client took a decision to provide platforms with Topsides Induced Acceleration Monitoring System (further referred as TIAMS).

Institute of Environmental Geoscience RAS has won the Tender for design, development and manufacturing of the TIAMS arranged by Sakhalin Energy Investment Company (further referred as SEIC). Basing on the technical assignment Information and Measuring Systems Department, IEG RAS has designed and manufactured in 2005–2006 TIAMS packages for two offshore oil and gas platforms near Piltun-Astokhskoe (PA_B) and Lunskoe (LUN-A) fields (Ginzburg et al. 2008, 2009).

LUN-A and PA-B platforms are very complicated constructions. Each platform has three decks of the football ground size. The platforms are supported by four legs. Their diameters are from 16 to 24 m, height is approximately 60 m; depth of the sea at the site is 30–35 m. The lower decks are placed at the height of ~27 m, the upper decks are at the height of 50–60 m above the sea surface.

Friction pendulum bearings are placed at the tops of the legs to damp horizontal oscillations under seismic and load impacts to the platform supports.

The main function of the TIAMS is to determine dangerous earthquakes from other impacts induced to the platform (ice impacts, ship impacts, wave impacts, drill snatch, etc.) that can cause accelerations same to the dangerous earthquakes accelerations at the topsides of the platforms. In case the destructive earthquake has been detected and its acceleration level exceed the threshold of 0.5 g (assigned by the Client) in any key point of the platform the TIAMS shall initiate the Emergency Shutdown signal (ESD). In such away TIAMS shall provide safety of the oil and gas offshore platforms.

It is necessary to mention that there were no such systems in the world practice between earthquake detection systems that can detect earthquakes from other impacts that can cause the same accelerations as dangerous earthquakes.

IEG RAS has to solve the following tasks during the development and manufacturing that are due to the Client's specifications and platform construction:

TIAMS shall work within severe ambient conditions, i.e. −40 °C to +50 °C in aggressive medium (salt fogs);
TIAMS shall work in the gas dangerous area, i.e. it shall comply with the requirements of intrinsically and explosion proof safety;
TIAMS shall comply with the requirements if the IEC 61508 «Electrical, electronic, programmable electronic systems connected with functional safety»;
TIAMS shall pass certification to obtain certificates for its elements and overall system;
IEG RAS shall simulate all input loads during Factory Acceptance test;
IEG RAS shall determine requirements to the frequency and dynamic band of the measured accelerations;
IEG RAS shall determine platform responses to the dangerous loads;
IEG RAS shall select key points for the placement of sensors to minimize the configuration of the TIAMS.

The solution was based on the Unified automated equipment elements (UAE) designed in the Seismological center by IEG RAS for forecasting polygons of seismic dangerous regions of Russia from 1996 to 1998 in the frames of the Federal Program "Development of the federal seismic monitoring system and forecasts of earthquakes in 1996–2000" (for detailed description of UAE see IEG RAS web-site: www.geoenv.ru).

AMEC (affiliate of SEIC) has developed mathematical model of the platform in ABAQUS software and modeled seismic impact to the platform. IEG RAS specialists then formalized all external loads and simulated their impacts for ABAQUS platform models. Then IEG RAS specialists analyzed platform responses to the impact loads and earthquakes (more than 60 thousand diagrams and schemes).

During the first stage IEG RAS has done the following:

- Theoretical justification of the external non-seismic impacts to the platform, detection of their features: value and direction of the affecting forces and time dependences.
- Modeled of 17 variant of impacts using ABAQUS platform model developed by AMEC.
- Qualitative physical analyses of the topsides responses to the earthquakes and other impacts, detection of the main directions and methods of the mathematical processing of the modeling results.
- Developed the software program to process modeling results.
- Analyses of modeling results from seismic and non-seismic impacts.

- Determined the key point for the sensors and their numbers as 6.
- Developed requirements for sensors installation.
- Developed earthquake detection algorithm and algorithm of ESD signal initiation.

An experience in development of such systems and its operation in the severe environment are very important for solving the same problems at other hazardous ecological objects, such as atomic power plants, chemical plats, high dams and barrages. Such systems can also be used to provide safety of the mega polices.

Fig. 2 RMU form

3 Description of the Topsides Induced Acceleration Monitoring System (TIAMS)

TIAMS structural scheme is given at Fig. 1.

TIAMS includes 6 Remote Measuring Units (RMU) and equipments placed in the Unit Control Panel (UCP).

RMU layout is given at Figs. 2, 3 and 4. UCP layout is given at Figs. 5 and 6.

Three-component accelerometers (TA), RMS, CCT-R are constructively united into the Remote Measuring Unit (RMU), which is designed as the field device permitted to be used in the hazardous environment and severe severity conditions. RMU is located directly in the measuring point (Fig. 7).

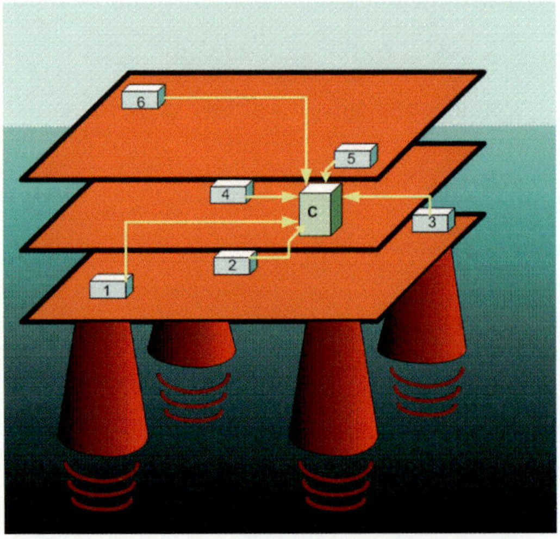

Fig. 1 TIAMS structural scheme

Fig. 3 RMU with open cover

The rest elements of the MCC are placed in the Unit Control Panel (UCP) placed in the Main Equipment Room (MER) of the oil and gas offshore platform.

Signals, coming from the RMU to the control panel are transferred by the standard interface RS-485. Date rate through the COM port is

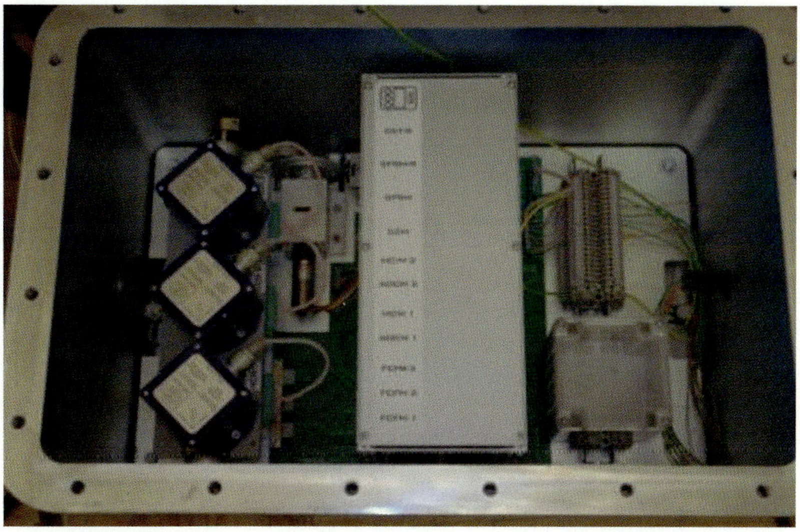

Fig. 4 RMU, view from above, open cover

115,200 bit/s. Acquisition and preprocessing of the data received from the RMU is effected in the DSCPS. DSCPS-M carries out the temporary storage of data received from the DSCPS, forms the data blocks in the defined format, processes the information according to the defined algorithm and transfers it through the standard serial interface RS-485 to the controlling computer for registration, viewing and current control of the equipment status. In case the DSCPS-M finds the hazardous seismic event it initiates the ESD signal. Storage and processing of the received information and the viewing are done on the processing computer.

TIAMS is designed for the automatic mode of operation and it provides control and operation of all system elements automatically. TIAMS has the embed means of diagnostic of the hardware and provides the automatic finding of the emergency situations (at that TIAMS provides the "FAILURE" signal).

The controlling computer and the processing computer provide the human–machine interface (HMI) within the operating system. The processing computer sets the operating modes for the TIAMS and makes changes the configuration. The current status of the system elements, results of the diagnostic, emergency reports, etc. are displayed at the monitor.

TIAMS modes

TIAMS has the following operation modes: configuration mode (job assignment); operation mode; testing mode.

Configuring mode allows determining the logic addresses of the RMU, calibration coefficients of the installed accelerometers, thresholds to track the dangerous seismic events, location and length of files with the recorded received data, etc.

After the power supply is switched on the TIAMS proceeds to the operation mode, and the work begins from the selection of the job entered in the configuring mode. After the self-diagnostic and evaluation of the real configuration job adequacy the TIAMS begins to monitor the accelerations in the selected points of the platform, then to analyze the events, and provide the ESD signal initiation when it finds the dangerous

Fig. 5 UCP

Fig. 6 UCP, view without back panel

Fig. 7 Three-component accelerometer (primary converters module and frequency characteristic)

earthquake according to the defined algorithm. It also provides the recurring calibration of the accelerometers.

The testing mode allows effecting the detailed examination of the modules capacity for work.

Switching of the system

The system switching is done in three stages as follows: self-testing; job of the configuration parameters; system starting.

Self-testing

On switching all programmable devices of the system (RMS, CCM.DSCPS, DSCPS-M, and Controlling Computer) are running the self-testing and inspection of the connection of the subordinate devices. The test functions of each device are described in the section "Efficiency monitoring".

After self-testing and detection of the efficiency of the connected programmable devices, the DCSPS is doing the initial calibration of the accelerometers. The results of self-testing are transmitted to the seignior devices upon requests. After self-testing of all modules have been finished, the results are displayed at the screen of the controlling computer and registered in the protocol. The results are as follows: devices' addresses tree; self-testing results of each device: "no-fault" or list of the malfunctions; calibration results from each accelerometer.

The DSCPS-M passes a "Failure" signal to the PCS in case a malfunction is found. If malfunctions are not resulted in a complete loss of system efficiency, then the system will go on operating with the following restrains: in case of failure of the calibration platform, the calibration shall not be running until it will be replaced; in case of failure of one of the accelerometers, the averaging-out shall be done with two remained; in case of failure of the RMU or DSCPS, the system will go on operating but the probability of false response will be increased; in case of failure of the active DSCPS-M, its the reserved one takes up its functions; in case of failure of the port or connection line RS-485-1 of the DSCPS-M the connection shall be through the reserved port RS-485-2; in case of failure of the Controlling computer, the doubling takes up its functions; In case there are no fatal malfunctions, the system proceeds to the assigning of the parameters of configuration, otherwise the system is considered as inoperative until fault removal.

Assigning of the configuration parameters

After the self-testing has been completed, the system checks up the availability of the job assignment with system parameters and its configuration. Parameters are stored in the file with a fixed name at the Controlling computer. If system configuration defined during self-testing concurs with the configuration from the job assignment file, then the automatic start-up of the system shall be done (if there is defined parameter of auto start).

In case there is no file with job assignment, or it is not correct, or it doesn't conform to the system configuration defined during the self-testing then, the system shall wait for the entering of the correct parameters which include the following: Tree of the serial numbers and addresses of the units; Acceleration numeralization frequency; RMS channels in use; Periodicity of testing and accelerometers calibration; Factors of the analogue digital readout conversion into acceleration; Threshold value of the acceleration; Earthquake detection algorithm parameters;

After the correct parameters have been entered, the specialist starts up the system.

System start up

On start up command the following is carried out: time synchronization to the Platform time system; transmission of the job assignment to the subordinate devices; starting of all units.

Efficiency monitoring

Each unit periodically does self-testing and informs the seignior device about results. Each unit monitors the subordinate devices. Each unit controls the connection with subordinate units. Two DSCPS-Ms check the operation of each other and in case of failure of the main DSCPS-M, the other one takes up its functions. In case any of the DSCPS-M fails then the «FAILURE» signal is initiated. The information of found malfunctions is displayed at the screen of the controlling computer and registered in the protocol. All diagnostic can be done in the automatic mode according to the program and under the commands from the operator. All processor units (RMS, CCM, DSCPS, DSCPS-M and Controlling Computer) have WatchDog timers to prevent the suspension of the program.

RMS diagnostic functions

RMS controller carries out the diagnostic of the following: RMS programs memory; RMS data memory; RMS EEPROM memory; efficiency of the analogue digital converter (ADC) of the RMS; conversion frequency error of the RMS ADC; correctness of the job assignment for the RMS ADC; keeps statistic of the faults in the connection line to the DSC PS.

The diagnostic is done in the automatic mode and under the commands from the DSCPS. The results of diagnostic am transmitted to the DSCPS on demands.

CCM diagnostic functions

CCM micro controller carries out the diagnostic of the following: CCM programs memory; CCM EEPROM memory; CCM non-volatile backing memory; availability of the job for calibration of the accelerometers; efficiency of the engine; engine rotation speed; oscillation frequency of the of the calibration platform; position of the platform under operation mode; yields a "SELF TEST" signal for electronic test of the accelerometers; keeps statistic of faults in the connection line to the DSCPS. The diagnostic is done in the automatic mode and under the commands from the DSCPS. The results of diagnostic are transmitted to the DSCPS on demands

DSCPS diagnostic functions

DSCPS micro controller carries out the diagnostic of the following: DSCPS programs memory; DSCPS data memory; sends the commands for testing and calibration of the RMU (RMS and CCM); requests the results of the diagnostic from RMU; analyses the data of accelerometers and estimates their efficiency and calibration parameters; analyses the efficiency of connection line with RMU; keeps the statistic of the faults in the connection line to the DSCPS-M. The diagnostic is done in the automatic mode and under the commands from the DSCPS-M. The results of diagnostic are transmitted to the DSCPS-M on demands

DSCPS-M diagnostic functions

DSCPS-M micro controller carries out the diagnostic of the following: DSCPS-M programs memory; DSCPS-M data memory; sends the commands for testing and calibration to all DSCPS; requests the results of the diagnostic from DSCPS; analyses the efficiency of connection line with all DSCPS; analyses the efficiency of the redundant DSCPS-M; keeps the statistic of the faults in the connection line to the redundant DSCPS-M; analyses the efficiency of the main and redundant controlling computer; keeps the statistic of the faults in the connection line to the main and redundant controlling computer; analyses the efficiency of connection line to the main and redundant controlling computer; controls the regularity of the secondary power supply sources. The diagnostic is done in the automatic mode and under the commands of the controlling computer. The results of diagnostic are transmitted to the controlling computer on demand. In case of failure the DSCPS-M initiates the "Failure" signal to the PCS.

Controlling computer (CC) diagnostic functions

The CC carries out the diagnostic as following: passes the job for caring out of the automatic diagnostic and calibration to the subordinate devices; passes the job for caring out of the automatic diagnostic and calibration to the subordinate device on command from operator, collects the results of diagnostic of all devices from DSCPS-M; displays at the screen the status of all units; registers in protocol the reports of all malfunctions and reestablishment of operation; controls 3 levels of access to work with TIAMS by the systems of passwords.

All functions of diagnostic and data storage are carried out at the same time at the main and redundant controlling computer. The commands can be given from any of the controlling computers, Laptop diagnostic functions. Laptop provides the possibility to do the autonomic diagnostic of the system modules (RMU, DSCPS, DSCPS-M) that are under repair or spared. Principle of the RMU operation is based on the RMU three-component accelerometers conversion of the induced linear accelerations into analog electrical signals with further analog-digital conversion. Three-component accelerometers do the conversion of the current values of the accelerations into output voltage through three mutually perpendicular axes. Each accelerometer has the self-testing and calibration devices. The self-testing is based on the sensing device inertial mass position changing upon the command "Self-test" and under the impact of the electrostatic force that approximately equal to the 20% of the calibration measuring range (full scale). This is the way to check the efficiency of the overall mechanical structure and electric circuit of the accelerometer.

Electric motor with reducer, driving gear, calibration platform, platform position transducer and calibration control module form the calibration unit that provides the calibration mode under which the position of the calibration platform should be changed according to the assigned principle. Control on the calibration and self-testing of the three-component accelerometer is done by the Calibration Control Module that forms the "Self-Test" signal and drives the electric motor (sets in motion the calibration platform). The CP is intended for the placement, fastening and mutual orientation of tree Measuring Converter Modules. It is fixed at the bearings that installed at the platform supports. CP could change its position relatively to the long axis within the range of 0°–45°. The moment that provides the oscillation movements of the platform revolving on its axis is passed by the crank- link mechanism that connects the drive shaft of the electric motor and CP. The oscillation frequency of the CP is equal to the value of the central frequency of the working range, which is 1, 0 Hz. The Platform position transducer is used to provide the return of the platform to its starting (measuring) position. The signal of this transducer is used by CCM to control the operation of the electric motor.

The calibration of the accelerometer-measuring channel is based on the periodical (with the frequency of 1 Hz) changing of the initial accelerometer converters position. These initial accelerometer converters are included into the Measuring Converter Module that is placed at the CP. This method shows the changing of the gravitational acceleration projection to the sensor sensitive axis. Variations amplitude of this projection is about 0.15 g (1.5 m/s^2) and serves as the calibration constant. The value of the calibration constant is a reference value for the periodical calibration of the accelerometer measuring channels.

Two Remote Measuring Systems do the conversion of the output analog signals from the three-component accelerometers into the digital code. Each RMS contains of Analog Digital Conversion Module and Micro-Controlling Module, which control the analog-digital conversion process, self-testing, receiving and transmission of the data.

The changing of the acceleration amplitude, testing and calibration are effected upon the commands from the UCP according to the operation software of TIAMS.

During the implementation of the contract TIAMS has received the following certificates: (1) Permission for use from Rostekhnadzor; (2) EMC certificate; (3) Explosion proof certificate; (4) Type approval certificate for system elements (RMS, DSCPS, CCT-C, CCT-R); (5) Type approval certificate for overall system is in process of issuing; (6) Type approval certificate for three-component accelerometer is in process of issuing; (7) IEG RAS has obtained Quality Management certificate ISO 9001.

Oil and gas platforms for deposits Lunskoe-A (LUN-A) and Piltun-Astokhskoye (PA-B) for Sakhalin-2 project are represented at Figs. 8, 9, 10, 11 and 12.

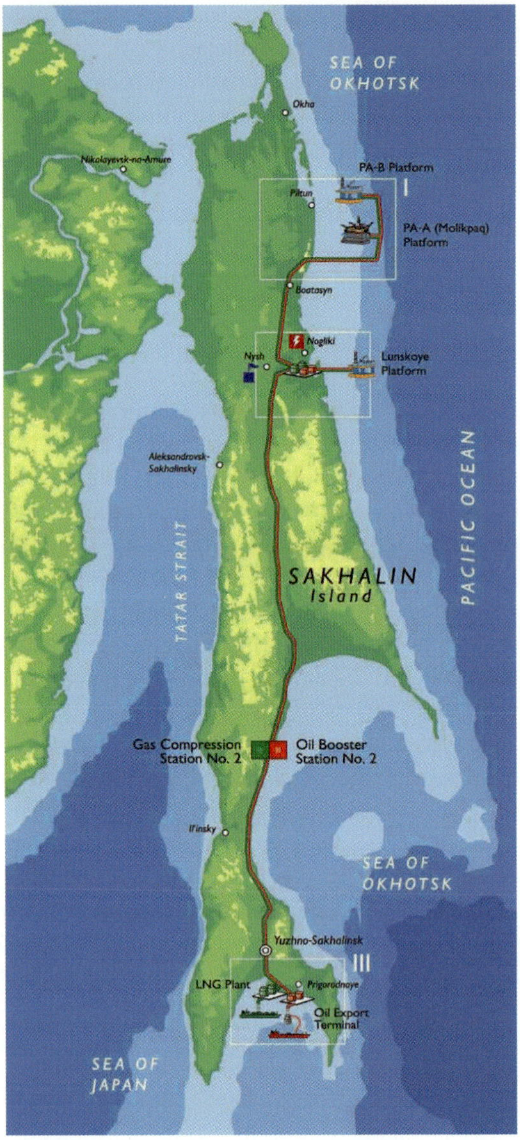

Fig. 8 Sakhalin Island and oil-gas platforms

4 Landslide Monitoring System for Coastal Slope of the River Yenissei

Geohazards monitoring system designed to monitor landslides coastal slope of the river Yenissei in real time. The system provides rapid collection of measurement data on the state of the observed landslides, processing and analysis of

Fig. 9 Platform for PA-B deposit

Fig. 10 Commissioning works in South Korea

the distribution of monitoring results between users and controls the security of "June" shopping and entertainment center, located on the coastal slope (Fig. 13).

The monitoring system includes eight mass displacements of ground points, two points of monitoring changes in the level of groundwater and automatic workplace of geologist, provides organizations with the process of collecting, processing and distributing data and carrying out management of all its elements (Fig. 14).

In order to collect and exchange information using cellular communications channels, which included two GSM-modems (primary and backup), and each control point it is a part of a GSM-modem.

The system has two modes of user access to information. Access to each of the modes of operation is carried out by a password. The first mode of access—"Operator", which is only possible to view all parameters of the control points and the region, measurement data and

Fig. 11 Platform for Lunskoe-A (LUN-A) deposit

Fig. 12 Platform for Lunskoe-A (LUN-A) deposit

alarms, and change the system configuration or structure of any device in this mode can not be. The second access mode—"Administrator", in which it is possible to make a change of settings and system structure. To transmit alarm signals provided connectivity with unified duty-dispatch service of Krasnoyarsk by the Internet.

Equipment set deep frame is designed to measure linear displacement of soil that occurs when the landslide processes caused by natural and man-made causes, by its (linear movement) transformation into a digital code (Fig. 15). The kit has controls equipment malfunctions and unauthorized access to and allows you to quickly transfer the alarm information. Complete registration of groundwater level is designed for continuous automated measurement level, water temperature and atmospheric pressure well and

Fig. 13 Trade and amusement complex "JUNE", located on the monitored coastal slope of the river Yenissei

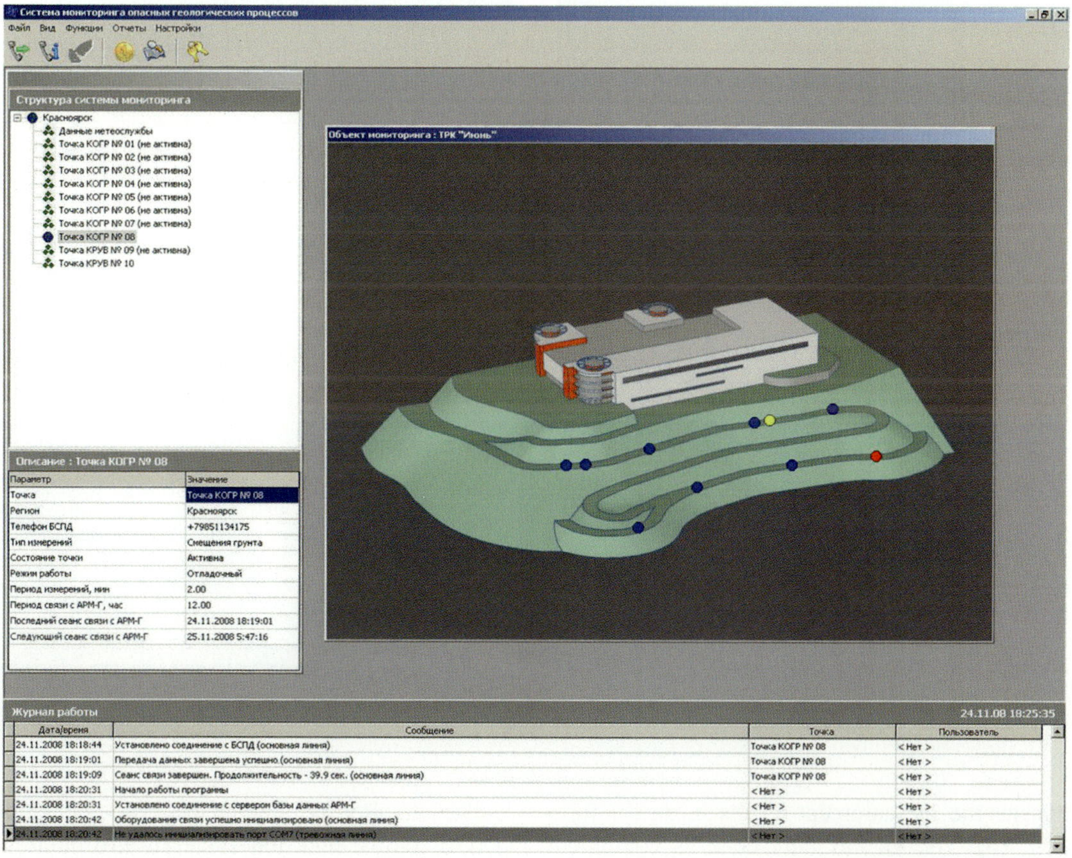

Fig. 14 The main window of the work program

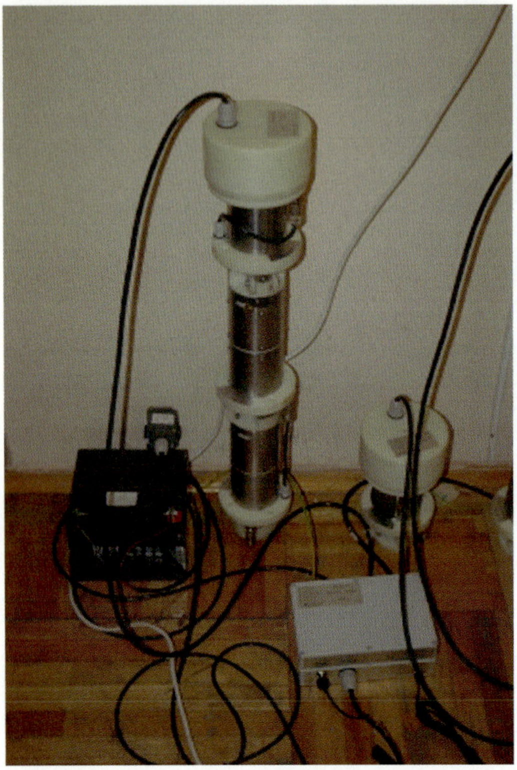

Fig. 15 Equipment set deep frame

Fig. 16 Complete registration of groundwater levels

Fig. 17 Installation and commissioning of geohazards monitoring system

transfer the measurement results in digital form (Fig. 16). The kit has fault controls equipment and unauthorized access to it.

The monitoring system has two operating modes: normal and abnormal. In any mode of functioning of the data from the hardware coastal

Fig. 18 Installation and commissioning of geohazards monitoring system

Fig. 19 Installation and commissioning of geohazards monitoring system

Fig. 20 Installation and commissioning of geohazards monitoring system

slope control points are processed in real time. If the ground speed displacement mass or velocity of groundwater level changes less than a predetermined threshold, the information is recorded and subsequently subjected to analysis and comparison with data obtained previously. In another case, when the speed of the displacement of soil mass or rate of change of groundwater level with some—any control point exceeds a predetermined threshold, the equipment together

Fig. 21 After installation and commissioning of geohazards monitoring system

with the data sends an alarm. Alarm is the basis for the transition to a freelance mode, in which decisions are made on a more detailed examination of the coastal slope and, if necessary, the evacuation of people from the building trade and entertainment complex "June" and further strengthening of the coastal slope (Figs. 17, 18, 19, 20 and 21).

5 Conclusion

The "System monitoring acceleration induced on the upper part of the offshore oil and gas platforms" for deposits Lunskoe-A (LUN-A) and Piltun-Astokhskoye (PA-B) for Sakhalin-2 project have been developed and constructed by Sergeev Institute of Environmental Geoscience RAS (IEG RAS). The system was successfully modified for landslides alarm monitoring and used for coastal slope of the river Yenissei and for a number of large industrial objects and urban areas.

References

Ginzburg AA, Itzko AS, Kondratenko AI, Manukin AB, Mironov OK, Novikova AV (2008) Simulation of different impacts on offshore oil and gas platforms for the detection of dangerous earthquakes systems. Questions Eng Seismol N2:v. 35

Ginzburg AA, Manukin AB, Mironov OK, Novikova AV (2009) The spatial and spectral characteristics caused by earthquakes and other effects of fluctuations in offshore oil and gas platforms. Geoecology N5

Nikolaev AV, Bashilov IP, Keh-Jian Shou SVB, Manukin AB, Zubko YN, Behterev SV, Kazantseva OS, Rebrov VI (2011) Some directions of works on maintenance of geological safety of engineering constructions. In: Proceedings of ENGEOPRO, Moscow, 7 pp

Postoev GP, Svalova VB (2005) Landslides risk reduction and monitoring for urban territories in Russia. In: Proceedings of the First General Assembly of ICL (International Consortium on Landslides), "Landslides: risk analysis and sustainable disaster management", Washington, USA. Springer, Berlin, pp 297–303

Svalova VB (2009) Mechanical-mathematical modeling for sedimentary movement and landslide processes. In: CD Proceedings of the international Association for Mathematical Geosciences Meeting (IAMG 2009), Stanford, California, USA, 15 pp

Svalova VB (2011a) Mechanical-mathematical modeling and monitoring for landslide processes. J Environ Sci Eng 5(10):1282–1287

Svalova V (2011b) Monitoring and modeling of landslide processes. Monit Sci Technol 2(7):19–27 (in Russian)

Svalova VB (2011c) Landslide process simulation and monitoring. In: Proceedings of ENGEOPRO, Moscow, 7 pp

Svalova VB (2012) Mechanical-mathematical modeling and monitoring for landslides. In: Proceedings of IPL (International Program on Landslides) Symposium, UNESCO, Paris, pp 63–68

Svalova VB (2013) Risk reduction for landslide hazards. In: Modeling and monitoring, Proceedings of the international conference natural risks: analysis, estimation, mapping, Moscow, MSU, pp 157–163 (in Russian)

Svalova VB (2014a) Modeling and monitoring for landslide processes. In: Linwood K (ed) Natural disasters—typhoons and landslides—risk prediction, crisis management and environmental impacts. Nova Science Publishers, New York, pp 177–198

Svalova VB (2014b) Mechanical-mathematical modeling and monitoring for landslide processes. IPL 163 Project. In: Proceedings of the world landslide forum 3, vol 4, Beijing, China, pp 24–27

Svalova VB (2014c) Modeling and monitoring for landslide processes: case study of Moscow and Taiwan. In: Proceedings of the world landslide forum 3, vol 4, Beijing, China, pp 628–632

Svalova VB (2015) Mechanical modeling and geophysical monitoring for landslide processes. In: Proceedings of IAEG XII congress "Engineering geology for society and territory", vol 2, Torino-2014, Italy. Springer, Berlin, 345–348

Svalova V, Postoev G (2008) Landslide process activization on sites of cultural heritage in Moscow, Russia. In: Proceedings of the first world landslide forum 2008, Tokyo, Japan, 4 p

TXT-tool 2.007-1.2 Landslide and Seismic Monitoring System on the Base of Unified Automatic Equipment

Alexander Ginzburg, Alexey Nikolaev, Valentina Svalova, German Postoev and Andrey Kazeev

Abstract

The operative monitoring systems of landslides have been developed by Sergeev Institute of Environmental Geoscience RAS (IEG RAS) for a number of large industrial objects and urban areas. The designed monitoring system of geological hazards for the "Blue Stream" gas pipeline and for Sochi territory are the examples.

Keywords

Landslide hazard · Early-warning monitoring · Geotechnical instrumentation · Unified automatic equipment · "Blue Stream" 2014 Olympics · Sochi

Contents

1 **Introduction** ... 327
2 **Unified Automatic Equipment** ... 328
3 **Russia–Turkey Gas Pipeline "Blue Stream" Monitoring System** ... 334
4 **Landslide Monitoring System for Objects of the 2014 Olympics in Sochi** ... 335
5 **Conclusion** ... 339
References ... 339

A. Ginzburg · A. Nikolaev · V. Svalova (✉) ·
G. Postoev · A. Kazeev
Sergeev Institute of Environmental Geoscience RAS,
Ulansky Per. 13, PB 145, Moscow 101000, Russia
e-mail: inter@geoenv.ru

K. Sassa et al. (eds.), *Landslide Dynamics: ISDR-ICL Landslide Interactive Teaching Tools*, https://doi.org/10.1007/978-3-319-57774-6_23

1 Introduction

IEG RAS has developed Unified Automatic Equipment (UAE) working for Federal target program "Development of the Federal Seismic Monitoring and Earthquake Forecast System" to use it at the forecast landslides in the seismically dangerous regions of Russia. The problem of landslides risk reduction on one hand concerns building in landslide-prone areas and on the other hand concerns providing an advanced warning about an activation of landslides which threaten the buildings. The complexity of this problem is compounded by a large number of different types of landslides, their occurrence along the slopes and the developmental dynamics during the landslide progression from the formation or activation phase and to the catastrophic destruction phase (Kazeev et al. 2014; Nikolaev et al. 2011; Postoev et al. 1989; Postoev and Svalova

2005; Svalova 2009, 2011a, b, c, 2012, 2013, 2014a, b, c, 2015; Svalova and Postoev 2008).

In preparation for the 2014 Winter Olympics in Sochi, a new road that combine highway with railway was constructed between Adler and Krasnaya Polyana. The combined road further goes on to the Alpika-Servis train station. It passes the valley of the river Mzymta. It is one of the most important objects in transportation infrastructure for the 2014 Olympics in Sochi. Olympic structures such as ski jumps, alpine skiing tracks and etc. are situated in the mountainous part of Sochi.

The goal of research was to identify monitoring technologies and criteria for landslide hazard assessment for effective early-warning monitoring of the combined road. The aim of this monitoring system is to support the transportation safety for the 2014 Olympics in Sochi.

The results of the research have shown that as primary monitoring instrumentations should be used horizontal extensometers and inclinometers. They provide the highest accuracy of landslide deformation measurements. The criteria for landslide hazard assessment were developed on the basis of 2010–2013 monitoring results. Mechanisms of landslides in the region have also been taken into account.

2 Unified Automatic Equipment

Unified Automatic Equipment (UAE) is assigned to develop wide application diffused informational-measuring and controlling complexes. If such complexes are connected to radio telemetry channels they provide the possibility to create area-monitoring systems of different application including seismic, geophysical regional monitoring systems, systems of ecological monitoring, etc.

The following devices are included into UAE: Remote measuring system (RMS); System for data storage, control and processing (DSCPS); Cable Connection Terminal for filed conditions, central (CCT-C); Cable Connection Terminal for filed conditions, remote (CCT-R). Remote measuring system (RMS) is assigned to convert voltage signals coming from the geophysical channels' outputs into analog-digital signals.

RMS does the synchronization of the received information with the Integrated Time System, calibration of the geophysical channels, calibration of geophysical channels control on the operation of the own hardware and software, forms data arrays, stores data and transmits the registered and preprocessed data through the serial connection channel of the RS-485 type through CCT-C directly to the DSCPS. General view of RMS is given at the Fig. 1. RMS operates under temperature range from −30 °C to +50 °C. Humidity is not more than 98% under the temperature of + 25 °C. RMS technical features are given in the Table 1.

System for data storage, control and processing (DSCPS) (Fig. 2) does the following: Reception of information directly from RMS or from RMS through CCT-C; Processing of received information; Injection of commands through CCT-C or directly to RMS; Synchronization of data with Common Time System; Transmission of the processed information through radio communication channel (or through cable communication line) to Information processing Center (IPC).

DSCPS can operate under the following conditions: Ambient temperature up to +50 °C; Ambient temperature up to −30 °C; Relative humidity—98% under temperature +25 °C.

Fig. 1 RMS general view

Table 1 RMS technical features

#	Parameters	Nominal value
1.	Number of differential analog inputs	8
2.	Input range, for each channel	−10 ÷ +10
3.	Input differential resistance (kΩ)	Not less than 20
4.	Reduction factor of the in-phase component, decibel, for each channel	Not less than 60
5.	Interference, decibel, one channel into other (for all)	Not more than −100
6.	Number of the effective bytes of the ADC under sampling frequency of 200 Hz for each channel	18
7.	ADC Least significant bit weight range, μV, for each channel[a]	2.60 ± 0.10
8.	Absolute error of the whole range, mV, not more than	±1
9.	Temperature coefficient of ADC (%, °C)[b]	±0.006
10.	Sampling frequency of the ADC (Hz)	
	Minimum	Not more than 0.001
	Maximum	Not less than 200
11	Calibration signal source (DAC) output signal range	
	True output	0–4
	Complementary output	−4 to 0
12	DAC bit number	12
13	DAC Least significant bit weight range (μV)	1
14	DAC limits of basic absolute accuracy (μV)	±25
15	Temperature coefficient of DAC (%, °C)	±0.16
16	DAC output resistance (Ω)	52 ± 3
17	Load resistance, Ω, not less	600
18	Power consumption (W)	Not more than 1.50

Note
[a]For each RMS parameter is be determined during manufacturing and stated in RMS passport
[b]% of the measured value

Fig. 2 DSCPS general view

DSCPS technical features are given in the Table 2.

Cable Connection Terminal for filed conditions, central (CCT-C) (Fig. 3) is assigned to the following: To be used in informational-measuring and controlling complexes as a part of area-spread cable monitoring system of several kilometers square for provision of communication through serial duplex channel between DSCPS and CCT-R; To provide galvanic isolation of informational-measuring and controlling complexes communication lines components from the connection line.

Table 2 DSCPS technical features

Parameter	Value
Supply voltage	+24
Power consumption from 24 V dc source, not more than	3 W
Number of RS-485 comports	2
Time synchronization accuracy (ms)	±1
RAM, not less than	6 Mbite
Data rate through serial ports, not more than RS-485	115,200 bit/s
Overall dimensions (mm)	
Length (with connectors)	264 ± 2
Width	68 ± 2
Height	136 ± 2

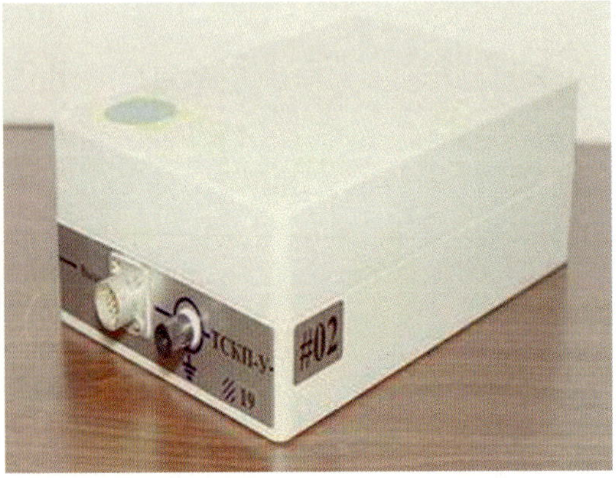

Fig. 3 CCT-C and CCT-R general view

CCT-C also can be used as the central monitor in an informational-measuring and controlling complex that provides tracing of the current time and initiation of DSCPS and other units therein according to set schedule. When CCT-C is connected to DSCPS in computer or communication line it can turn on DSCPS and turn on/off any lines out of schedule under DSCPS commands.

CCT-C can operate under the following conditions: Ambient temperature up to +50 °C; Ambient temperature up to −30 °C; Relative humidity—98% under temperature +25 °C. Acid or alkali vapors and any other aggressive dirt in the air are inadmissible. Interface type-optically isolated RS-485. Information rate—up to 115,200 bit/s. Number of input lines depends on modification. Line length—up to 1200 m. Primary power supply source voltage—8 to 30 V. Power consumption, Watt, not more than 1.0. IP rate—IP65.

Cable Connection Terminal for filed conditions, remote (CCT-R) (Fig. 3) is assigned to the following: To be used in informational-measuring and controlling complexes as a part of area-spread cable monitoring system with square of several kilometers for provision of communication through serial duplex channel between DSCPS and CCT-C; To provide galvanic isolation of informational-measuring and controlling complexes communication lines components from the connection line.

CCT-C can operate under the following conditions: Ambient temperature up to +50 °C; Ambient temperature up to −30 °C; Relative humidity—98% under temperature +25 °C. Acid or alkali vapors and any other aggressive dirt in the air are inadmissible. Interface type: optically isolated RS-485. Information rate—up to 115,200 bit/s. Number of input lines depends on modification. Line length—up to 1200 m. Primary power supply source voltage—8 to 30 V. Power consumption, Watt, not more than 1.0. IP rate—IP65.

Informational-measuring and controlling complex (IMCC) based on UAE. It includes the following units: UAE, DSCPS, CCT-C, CCT-R and RMS. Number of CCT-R and RMS depend on the objective of the IMCC. Types of measuring channels, their number and placement relatively to DSCPS, schedule of data acquisition are determined by main objective of the complex.

9 cables lines are connected to CCT-C with CCT-R at their ends. Maximum length of cable is 1200 m. Power goes through these cables from CCT-C to CCT-R. The cable line is protected with lightning protection devices that are installed in CCT-C and CCT-R. Information interchange rate for CCT-C and CCT-R is 115 kbit/s according to RS-485 protocol. 9 RMS can be connected to each of CCT-R and up to 8 analog-digital converters can be connected to each RMS. Power to RMS is transmitted through CCT-R.

The following restrictions shall be considered in design of IMCC based on UAE: Number of analog-digital channels in one system shall not be more than 72; Summary digital information recorded from all analog-digital converters of all RMSes included into one system shall not be more than 64 kbit/s.

There are two operating modes of IMCC that are determined by recording and data acquisition modes: Information accumulation mode (Information is collected only during visits of the IMCC); Real-time recording mode (IMCCes are combined into regional or local monitoring system by means of radiotelemetering or cable communication lines. In this case information will be recorded in real time mode).

IMCC operation under Information accumulation mode. In that case information will be stored in DSCPS main memory, which is 6 Mbytes. Laptop shall be connected to DSCPS before measuring. A program to configure IMCC shall be run at this laptop to set schedule of inquires for measuring channels connected to RMS ADCes. CCT-C is also coded with this program (measuring channels turn on/off schedule). Then special recording program shall be run. First it will test IMCC and then synchronize DSCPS time service and control module according to GPS signals. GPS is connected to the laptop. Then the laptop will be disconnected from DSCPS and IMCC will start autonomous recording of information according to assigned schedule.

According to that schedule CCT-C will send voltage to DSCPS and to measuring channels indicated by the program. DSCPS inquires all RMSes by turn (with time gap between each inquiry) then compact information and forms packets of set format. These packets are stored in RAM. DSCPS synchronizes all RMSes under special commands. DSCPS Operation software is coded in ROM and it is loaded after DSCPS is turned on or cleared. The software provides DSCPS operation as a main part of IMCC. It controls all connected RMS, stores and transmits measuring data to the laptop.

The software provides the following: Initial self-testing of DSCPS and RMS status inquiry; Receiving from the laptop and transmission to RMS test frequency for each ADC channel; Time control and RMS time synchronization; Simultaneous start of measuring in all RMS; RMS inquiry, storing of measuring results; Compacting and buffering of data; Re-coding and start of RMS after power supply failure that resulted in RMS re-initialization; Receiving from a laptop and transmission to RMS any random command in RMS-laptop interchange format and transmission of RMS response to a laptop; Transmission of DSCPS memory data to a laptop, transmission from laptop data packets to DSCPS memory (including program loading) and transmission of the control command to assigned address that provides remote loading and running of tests or running of new version of DSCPS software.

RMS software consists of RMS microprocessor program (DS80C320) and programs for personal computer like IBM-PC that allow autonomous operation of RMS outside RMS-DSCPS complex.

Microprocessor program is placed in microprocessor's ROM or external ROM and provides the following: Initial self-testing of RMS and its all ADCes; Receiving of commands and their transmission through communication line (RS-485); Transmission of system status and failures reports (system stays operative if some of ADC failed or disconnected); Self coding of test frequency for each ADC channel (from 5 ms to 1 month frequency of measuring or turning off); Measuring, storing and transmission of recorded data; ADC efficiency control, disconnection of channel in case of ADC failure (the channel can be disconnected under command of external computer); Operation from DAC: DAC will be connected to one of 8 channels under the command and output voltage will be changed according to assigned program; Time service control: RMS monitors time with accuracy of 1 ms and transmits milliseconds counter together with ADC values from the moment of measuring.

RMS time service can be controlled and adjusted under the commands from external computer or DSCPS. It allows keeping time accuracy as ± 1 ms during all time. RMS-DSCPS or RMS-external computer interchange rate is 115.2 or 19.6 k baud (switch is upon a command) and is always performed upon external computer initiative. Each RMS has its own address and responds only to the referenced commands. It is possible to connect up to 254 RMS to one line (some commands can be sent to all RMS at one time. RMS respond is not required in that case). The system provides effective operation even if communication cable line has a malfunctioning. An interchange is performed with packets and each packet is accompanied with check sum. In case of interchange error the packet will be ignored. Measuring results are stored in RMS and are transmitted asynchronously.

An information form DSCPS can be read as follows: A laptop is connected to DSCPS and information is copied from main memory to hard disk. Testing and synchronization of IMCC will be performed at the same time. IMCC operation in Data real-time recording mode. To carry out recording in real time mode IMCC are combined into regional or local monitoring systems by means of high speed telemetric connection or cable communication line. In that case information is recorded by controlling computer in the Information-processing Center—IPC that controls operation of all IMCCes. The number of IMCC included into monitoring system is restricted by hardware and software and can be not more than 16. Maximum interchange rate between IPC and all IMCC is restricted by throughput of communication system.

IPC controlling computer software is assigned to control IMCC equipment, finding out its failures, receiving data and information storing. It can be installed at IBM PC (PENTIUM based) with clock speed not less than 166 MHz.

Operation of regional or local monitoring system based on IMCC is determined by configuration file that will be created by special program. This file considers availability of strategic monitoring points, types of geophysical channels, their parameters and availability of time synchronization facilities (GPS).

After start of the program it reads configuration of the overall local seismic (geophysical) monitoring system and synchronizes time according to Greenwich with accuracy up to decimals of ms and calculates corrections of computer quartz frequency to provide system functioning in case of short time failure of GPS. Time synchronization is done from GPS every second.

After synchronization it inquires, codes, synchronizes and starts equipment of some IMCC. The program inquires status of IMCCes equipment (operation status, availability of channels, channel operation status, etc.). In case of any problem report will be displayed at the screen (this report is also recorded in operation protocol file). Operation program for each measuring channel, frequency of inquiry and gain factor plus one program of intense movements channel operation including number of channel are transmitted to central controller of monitoring point. Before start each IMCC is synchronized with IPC computer time (then synchronization is performed each 10 s). Time of signal transit from IPC to IMCC is also considered in synchronization procedure. After that DSCPS in IMCC starts collecting data from RMS.

IPC controlling computer cyclically inquires data from corresponding IMCC and records them into main memory buffers. After necessary data have been collected they are recorded into file that corresponds to channel type. Normally these files are stored at other computer and can be read through local network. Nevertheless for short-time experiments the system can be configured to use local disk of controlling computer. During short breaks of communication (not longer than 10 s) data will not be lost due to DSCPS of each IMCC has large memory buffer. Operator can carry out calibration of any sensor of the system during its operation. In case of contingency (disconnection of communication with IMCC, its failure, failure of some channels or all RMS channels or restoration of all above) and under some normal modes like recording of events or sensor calibration the system will display report at the screen stating Greenwich time of event start, name of IMCC and channel name plus event description. The report is recorded into buffer of 100 lines size and into protocol file. Operator can check the buffer at any time.

IMCC real-time recording mode (2). In that case IMCC is included into regional or local monitoring system that operates in real-time mode. Operating software of DSCPS is coded in ROM and will be started after turning on or clearing of DSCPS. This software is assigned to provide DSCPS functioning as a main part of IMCC control of RMS, storing and transmission of data (reading of data in attendance mode) to IPC after processing.

Its functions are as follows: Initial self-test of DSCPS and RMS status; Receiving from IPC or laptop test frequency for each ADC channel and transmission of these data to RMS; Time monitoring and synchronization of RMS time; Start of simultaneous measuring in all RMSes; RMS inquiry, storing of measuring results; Data compression and buffering.

If IMCC is used for complex geophysical monitoring systems with sensitive seismic sensors and sensors of intensive movements DSCPS operational software provides the following: Selection of intensive movement channels and storing their data for further transmission of event prehistory to IPC after detection of event; Transmission of other channels' measuring results upon IPC request; Calculation (in sliding window) of number of exceeding of the set threshold recorded by event identification channel; Event processing: transmission of event report, its prehistory and report on time that has passed after event finish to IPC; Receiving of program to change output voltage at DAC to perform sensors calibration from IPC and its transmission to RMS upon Operator's command; RMS and ADC efficiency monitoring, reporting to IPC on equipment and communication failures, recoding and restart of RMS after power supply failure that results in re-initialization of RMS; Receiving from IPC of random command in RMS—external computer interchange format, its transmission to RMS and transmission of RMS response to IPC; Transmission of DSCPS memory to IPC, receiving of the packet for DSCPS memory form IPC (including loading of the program) and transmission of control to the assigned address that allows performing remote loading and running of tests or new version of DSCPS main software. DSCPS time service can be control and adjusted upon commands from IPC, that includes GPS time signal receiver. It allows keeping time accuracy as ±1 ms during all time. RMS provides the same functioning as a part of IMCC operating either in autonomous mode and controlled by DSCPS or as a part of IMCC included into regional or local monitoring system UAE is certificated by GOSTANDARD of RF.

Type approval certificate RU.C.34.004.A. No 13659 for information-measuring and controlling complexes based on Unified automatic equipment has been issued according to the decision of the Russian Federation State Committee on standards and metrology on 25.12.2002. The complexes were included into State register of measuring equipment under the No 23981-02B.

Basing on this design IEG RAS has developed and delivered hazardous geological processes monitoring system to such big industrial enterprises like Russian–Turkey Gas pipeline "Blue Stream", Soligorsk Ore Mining and Processing Enterprise (Republic of Belarus), Topsides Induced Monitoring System For Oil and Gas Offshore Platforms and 2014 Olympics in Sochi objects.

Fig. 4 "Blue Stream" scheme

Fig. 5 "Blue Stream" during construction

Fig. 6 The compressor station "Beregovaya", view from the sea. Visible glade in which the buried pipeline

Fig. 7 The compressor station "Beregovaya", view from the sea. Visible glade in which the buried pipeline

3 Russia–Turkey Gas Pipeline "Blue Stream" Monitoring System

The construction of the Russia–Turkey gas pipeline "Blue Stream" was accomplished in 2002 (Figs. 4, 5, 6, 7 and 8).

The pipeline route of a total length of 1226 km crosses the Black Sea. The pipeline consists of two pipes each 610 m in diameter. The project volume of transported gas is equal to 16 billion cubic m per year. When approaching the Black Sea, the pipeline crosses the north-western slopes of the Big Caucasus Ridge, where landslides are widespread. Thirty-five landslides are registered at this section of the pipeline route, 7 of them being the most hazardous. The on-line operating automatic control system of landslide processes was projected for these sites. The following registering devices were installed at each of the 7 sites: the seismic acoustic control unit; the inclinometric control unit; the groundwater level control unit.

Each unit was disposed in a separate borehole drilled in the landslide-prone slope. The equipment of each borehole provides the control of only one parameter. The measurement complex included the gauge of seismic acoustic emission and two units of data registration and collection.

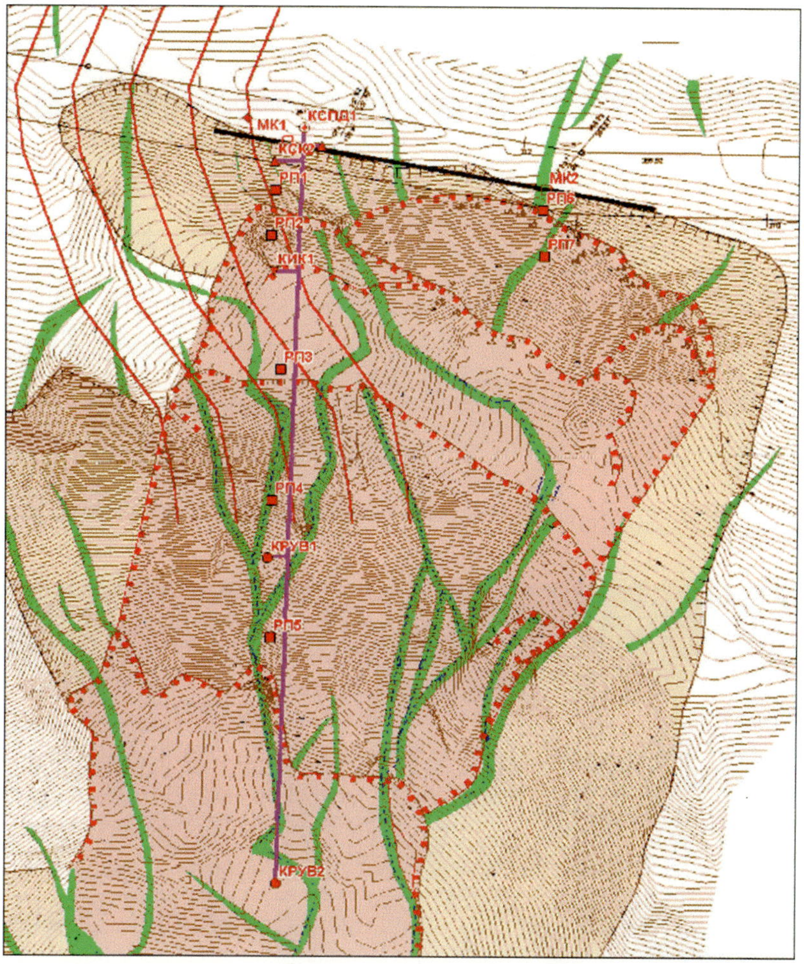

Fig. 8 The system of landslide-prone slope monitoring at the Russia–Turkey gas pipeline route with automatic unit for seismic acoustic measurements, unit for deviation control (inclinometer), unit for groundwater level control

A special hydrogeodynamic gauge was used for measuring the groundwater level, and the tree-point extensometer was applied for rock mass displacement measurement. The measured data are communicated to the monitoring center, where they are processed using the special software.

In addition to the surface automatic control, the remote sensing control based on the high-resolution space and aerial survey is used. The remote survey data are also processed using the special software. The developed monitoring system permits to control the conditions of the landslide-prone slopes and thus ensure the safety of pipeline operation at the site of high geological risk.

4 Landslide Monitoring System for Objects of the 2014 Olympics in Sochi

The landslide hazard is the main geologic hazard along the combined road from Adler to Krasnaya Polyana. The largest part of the area along the road is landslide-prone area and numerous landslides historically occurred on the bank slopes. At present, monitoring systems are installed at a number of Olympic structures (Figs. 9, 10, 11, 12 and 13).

Also monitoring systems were used during constructing the roads. They include an automatic monitoring of main parameters which

Fig. 9 Sea cluster. Imeretinskaya depression. Olympic objects during construction

Fig. 10 Ski slope

Fig. 12 Bobsleigh track

Fig. 11 Bobsleigh track

Fig. 13 Ski jump

Fig. 14 The location of landslide sites along the combined road from Adler to Alpika-service (Krasnaya Polyana)

characterize the state of the landslide area at each moment. Numerous factors contribute to such an active development of landslides in the region, such as a high degree of bedrock weathering on slopes. Abundant rainfalls and saturation of cover sediments also lead to the formation of numerous cracks and sliding. Nine landslide sites have been detected along the route of the combined road during geological investigations (Fig. 14). The stages of works are shown in Fig. 15. The results of monitoring are in Fig. 16.

It was determined that the most widespread type of landslides within the study area is a debris slide. Several sites, especially ones at the beginning of the route, exhibited block-type landslides of compression-extrusion. The investigations have shown that on landslide slopes with relatively long-time displacements and with periodic changes in displacement conditions, the acceleration of a displacement velocity up to hazardous levels may lead to the initiation of the sliding in new areas near upper and side borders of an active landslide. This increase of the size of a landslide can result in a catastrophic destruction.

Two general methods of observations were accepted in the automatic monitoring system, set along the combined highway and railway: (1) extensometric arrays (providing automatic measurements of displacements over the surface of a landslide prone slopes), and (2) inclinometric measurement in drills (supplying measurements of sliding parameters vs. depth, some in manual and automatic modes). The results of research have shown that the most useful parameters related to the characterization of an active landslide state and sliding dynamics, including the progressing development of a landslide during activation, are: landslide displacement velocity, depths of slip surfaces and

Fig. 15 The stages of works at railway Adler-Krasnaya Polyana construction

propagation of active displacements within the territory.

Landslide hazard criteria were proposed for the constructions of the road based on the monitoring data of an active landslide at one of the study sites along the railway from Adler to Krasnaya Polyana. These criteria are based on measurements of displacement velocities and distribution of landslide deformations (including new volumes of ground masses involved along the margins of active landslides) with area and dept (Kazeev et al. 2014; Postoev et al. 1989). As a result, several monitoring methods as related to the landslide hazard were recommended along the Adler-Krasnaya Polyana railway: automatic observations of displacements over the slope surface using extensometers; inclinometers (during site visits and in partly automatic mode).

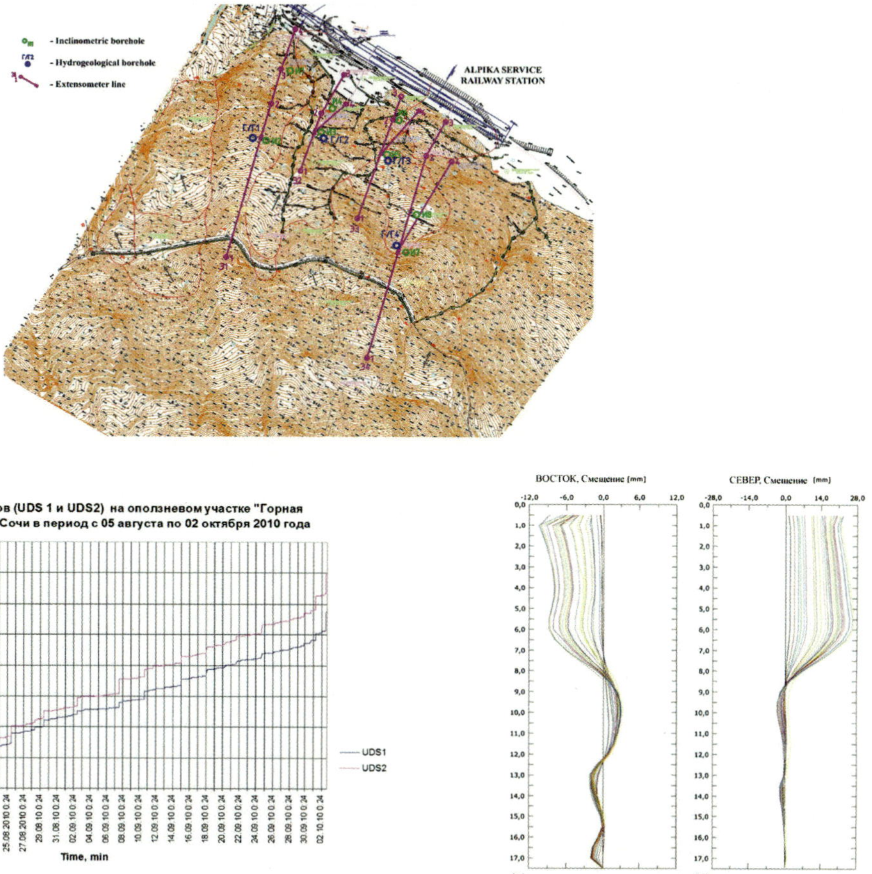

Fig. 16 Results of monitoring at railway Adler-Krasnaya Polyana. **a** Scheme of monitoring; **b** extensometer data; **c** inclinometer data

5 Conclusion

Unified Automatic Equipment (UAE) and landslide monitoring systems have been developed and constructed by Sergeev Institute of Environmental Geoscience RAS (IEG RAS). The systems were successfully used for a number of large industrial objects and urban areas such as the "Blue Stream" gas pipeline and for Sochi territory in connection with Winter 2014 Olympics.

References

Kazeev AI, Postoev GP, Fedotova KY (2014) Landslide hazard criteria for transportation safety of the 2014 Olympics in Sochi. In: 14th geoconference on science and technologies in geology, exploration and mining. conference proceedings,vol II (international multidisciplinary scientific geoconference; 17–26 June 2014, Bulgaria), pp 567–572

Nikolaev AV, Bashilov IP, Keh-Jian Shou, Svalova VB Manukin A B, Zubko YN, Behterev SV, Kazantseva OS, Rebrov VI (2011) Some directions of works on maintenance of geological safety of engineering constructions. In: Proceedings of ENGEOPRO, Moscow, 7 pp

Postoev GP, Svalova VB (2005) Landslides risk reduction and monitoring for urban territories in Russia. In: Landslides: risk analysis and sustainable disaster management. Springer, Washington, USA, pp 297–303 (Proceedings of the first general assembly of ICL (international consortium on landslides))

Postoev GP, Erysh IF, Salomatin VN et al (1989) Artificial activation of landslides. M: Nedra, Russia, p 134

Svalova VB (2009) Mechanical-mathematical modeling for sedimentary movement and landslide processes. In: CD Proceedings of the international association for

mathematical geosciences meeting (IAMG 2009), Stanford, California, USA, 15 pp
Svalova VB (2011a) Mechanical-mathematical modeling and monitoring for landslide processes. J Environ Sci Eng 5(10):1282–1287
Svalova VB (2011b) Monitoring and modeling of landslide processes. Monit Sci Technol 2(7):19–27 (in Russian)
Svalova VB (2011c) Landslide process simulation and monitoring. In: Proceedings of ENGEOPRO, Moscow, 7 pp
Svalova VB (2012) Mechanical-mathematical modeling and monitoring for landslides. In: Proceedings of IPL (international program on landslides) symposium, UNESCO, Paris, pp 63–68
Svalova VB (2013) Risk reduction for landslide hazards. Modeling and monitoring. In: Proceedings of the international conference natural risks: analysis, estimation, mapping. MSU, Moscow, pp 157–163 (in Russian)
Svalova VB (2014a) Modeling and monitoring for landslide processes. In: Linwood K (ed) Natural disasters—typhoons and landslides—risk prediction, crisis management and environmental impacts. Nova Science Publishers, NY USA, pp 177–198
Svalova VB (2014b) Mechanical-mathematical modeling and monitoring for landslide processes. IPL 163 project. In: Proceedings of the world landslide forum 3, vol 4. Beijing, China, pp 24–27
Svalova VB (2014c) Modeling and monitoring for landslide processes: case study of Moscow and Taiwan. In: Proceedings of the World landslide forum 3, vol 4. Beijing, China, pp 628–632
Svalova VB (2015) Mechanical modeling and geophysical monitoring for landslide processes. In: Proceedings of IAEG XII congress "engineering geology for society and territory", v. 2, Torino-2014. Springer, Italy, pp 345–348
Svalova V, Postoev G (2008) Landslide process activization on sites of cultural heritage in Moscow, Russia. In: Proceedings of the first world landslide forum 2008, Tokyo, Japan, 4p

TXT-tool 2.039-1.1 Italian National Early Warning System

Mauro Rossi, Ivan Marchesini, Gabriele Tonelli, Silvia Peruccacci, Maria Teresa Brunetti, Silvia Luciani, Francesca Ardizzone, Vinicio Balducci, Cinzia Bianchi, Mauro Cardinali, Federica Fiorucci, Alessandro Cesare Mondini, Paola Reichenbach, Paola Salvati, Michele Santangelo and Fausto Guzzetti

Abstract

In Italy rainfall-induced slope failures occur every year, claiming lives and causing severe economic disruptions. We have designed and implemented a warning system, named SANF (an acronym for national early warning system for rainfall-induced landslides), to forecast the possible occurrence of rainfall-induced landslides. The system is based on: (i) rainfall thresholds for possible landslide occurrence, (ii) sub-hourly rainfall measurements obtained by a nationwide network of 1950 rain gauges, and (iii) quantitative rainfall forecasts. All system components exploit Open Source software. Twice a day the system compares the measured and the forecasted rainfall amounts against pre-defined thresholds, and assigns to

M. Rossi (✉) · I. Marchesini · S. Peruccacci · M.T. Brunetti · S. Luciani · F. Ardizzone · V. Balducci · C. Bianchi · M. Cardinali · F. Fiorucci · A.C. Mondini · P. Reichenbach · P. Salvati · M. Santangelo · F. Guzzetti
Istituto di Ricerca per la Protezione Idrogeologica, Consiglio Nazionale delle Ricerche (CNR IRPI), Perugia, Italy
e-mail: mauro.rossi@irpi.cnr.it

I. Marchesini
e-mail: geomorfologia@irpi.cnr.it

S. Peruccacci
e-mail: geomorfologia@irpi.cnr.it

M.T. Brunetti
e-mail: geomorfologia@irpi.cnr.it

S. Luciani
e-mail: geomorfologia@irpi.cnr.it

F. Ardizzone
e-mail: geomorfologia@irpi.cnr.it

V. Balducci
e-mail: geomorfologia@irpi.cnr.it

C. Bianchi
e-mail: geomorfologia@irpi.cnr.it

M. Cardinali
e-mail: geomorfologia@irpi.cnr.it

F. Fiorucci
e-mail: geomorfologia@irpi.cnr.it

A.C. Mondini
e-mail: geomorfologia@irpi.cnr.it

P. Reichenbach
e-mail: geomorfologia@irpi.cnr.it

P. Salvati
e-mail: geomorfologia@irpi.cnr.it

M. Santangelo
e-mail: geomorfologia@irpi.cnr.it

F. Guzzetti
e-mail: geomorfologia@irpi.cnr.it

G. Tonelli
Independent Contractor, Via Emilia 231/a, San Lazzaro di Savena, Bologna, Italy
e-mail: g.tonelli@g-tonelli.it

K. Sassa et al. (eds.), *Landslide Dynamics: ISDR-ICL Landslide Interactive Teaching Tools*, https://doi.org/10.1007/978-3-319-57774-6_24

each rain gauge a probability of landslide occurrence. This information is used to prepare synoptic-scale maps showing where rainfall-induced landslides are expected. The system outputs are delivered to the National Civil Protection Authorities in different formats. Spatial outputs are published as standard OGC (Open Geospatial Consortium) web services (WMS, WFS, WCS) by the IRPI Spatial Data Infrastructure (IRPI SDI). A password protected WebGIS interface facilitates the use of the system by the Civil Protection personnel and gives access to current and past forecasts. In addition, bulletins containing the system information can be generated automatically and sent via e-mail to the Civil Protection personnel. In a more recent implementation, the system calculate hourly-based forecast using new regional rainfall thresholds and combine landslide forecasts with landslide susceptibility information available at synoptic scale in the national territory. Improvements of the validation procedures and of the landslide susceptibility layer are currently underway.

Keywords
Landslide • Rainfall • Early-warning • Rainfall threshold

Contents

1 **Introduction** 342
2 **System Structure** 343
3 **Data Input and Storage** 344
4 **Computation and Data Analysis** 344
5 **Early Warning System Outputs** 348
6 **System Validation and Future Improvements** 348
References 349

1 Introduction

Landslides are widespread phenomena that cause every year casualties and extensive damages. Landslides can be triggered by a variety of factors, like rainfall events, earthquakes, rapid snow melt, human activities.

In Italy, rainfall is the primary cause of landslides. Damaging failures occur every year in this country, and since 1950 landslides have caused more than 6300 casualties (Salvati et al. 2010). Hence, research activities aiming at the determination of the amount of rainfall needed to trigger landslides is of both scientific and social interest. In this context, the Italian National Department for Civil Protection, an Office of the Prime Minister, asked to the Research Institute for Geo-Hydrological Hazards Assessment of the National Research Council (IRPI CNR) to design and implement an early-warning system to forecast the possible occurrence of rainfall-induced landslides in the entire Italian territory. Following this request, in 2009 IRPI CNR conceived and built SANF (an acronym for national early warning system for rainfall-induced landslides). The SANF early-warning system is based on the comparison between rainfall measurements and forecasts and empirical rainfall thresholds for the possible occurrence of landslides. The SANF warning system uses rainfall measurements provided by the rain gauge network of the Italian Civil Protection and rainfall forecasts obtained from numerical weather forecast models, to generate synoptic-scale maps showing the occurrence probability of rainfall-induced landslides. The probabilities delimitate five critical levels for the occurrence of landslides, from “well below the threshold” to “well above the threshold”.

2 System Structure

The SANF early-warning system is based on the comparison between rainfall measurements and forecasts (rainfall data) and empirical rainfall thresholds.

Figure 1 shows the logical scheme of the SANF early-warning system. Specifically, the system consists of three main components:

- data input and storage,
- computation and data analysis, and
- data output.

The first component includes the importing procedures for the input data. The system manages rainfall data and information related to the occurrence of rainfall-induced landslides. Several procedures are implemented to import different data types in the system, both manually and automatically. Data are stored in a database managed by a Relational Database Management System (RDBMS).

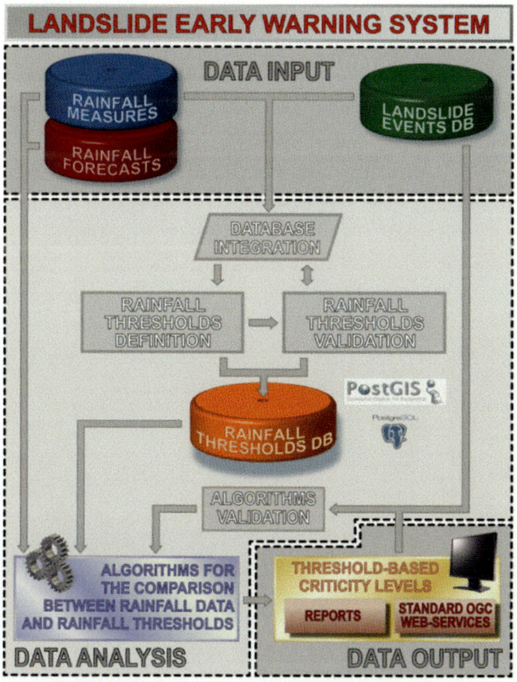

Fig. 1 Scheme of the Italian National Early Warning System for rainfall-induced landslide forecast

The second component, exploiting empirical thresholds for the possible landslide occurrence and the amount of rainfall measured and/or forecasted, calculates the corresponding probabilities of triggering of slope movements.

The third component generates and delivers the early-warning system outputs, which consist of a set of synoptic-scale maps of critical levels for the occurrence of rainfall-induced landslides. The system outputs are accessible via web as standard OGC (http://www.opengeospatial.org, Open Geospatial Consortium) services. Moreover, three specific WebGIS interfaces, showing maps of probabilities estimation based on measured data, forecasted data and archive data were set up for viewing, querying and filtering the system outputs. Daily reports are delivered via e-mail to Civil Protection authorities.

Almost all the system components are developed using open source software: (i) PostgreSQL (http://www.postgresql.org) with the spatial extension PostGIS (http://postgis.refractions.net), which provides database facilities for the data storage; (ii) Mono (http://www.php.net), PHP (http://www.php.net/) and R programming environments (R Development Core Team 2011), which are used to write system procedures; (iii) wgrib (http://www.cpc.ncep.noaa.gov/products/wesley/wgrib.html), GrADS (http://www.iges.org/grads/), libsim and DB.All.e (experimental libraries developed by the Hydro meteorological service of Emilia Romagna, http://www.arpa.emr.it/sim/), which are used to import rainfall forecasts; (iv) GeoServer (http://geoserver.org/), which is used to generate the output maps as standard OGC services; (v) ExtJS (http://www.sencha.com/products/extjs/), GeoExt (http://geoext.org/) and OpenLayers (http://openlayers.org/) software libraries, which are used to create the WebGIS system interfaces; (vi) Apache (http://www.apache.org/) and ApacheTomcat (http://tomcat.apache.org/), which provides HTTP webserver facilities;

(vii) two GNU/Linux distributions: Fedora (http://fedoraproject.org/) and Ubuntu (http://www.ubuntu.com/) are used to host the different system components.

3 Data Input and Storage

The input data of the SANF system are rainfall measurements, rainfall forecasts, and information on rainfall-induced landslides in Italy (Fig. 1). A network of 1950 rain gauges provides rainfall measurements for the entire Italian territory (Fig. 2). Rainfall measures are imported, checked, and stored automatically in the system every 6 h.

A specific procedure was implemented in the system to evaluate the reliability of the rain gauge. Daily, the procedures compares the total annual precipitation of each gauge to the mean annual precipitation (in the 30-year period 1961–1990). Empirical rules were set and used to detect anomalous gauges.

Three-day rainfall forecasts (Fig. 3) are generated twice a day by the Italian Local Area Model, LAMI and are automatically stored in the system.

Information on rainfall-induced landslides is aimed at the definition of empirical rainfall thresholds for the possible initiation of slope failures. In the SANF early warning system this information is manually entered in a database. For each rainfall event that triggered one or more slope failures, the information includes the location and the date of the landslide, and the rainfall conditions responsible of the failure, including the total event rainfall, the rainfall duration (D), and the mean rainfall intensity (I). A group of trained operators composed mainly of geologist and engineers compile the database seeking periodically landslide information in newspapers, technical reports and blogs. Operators pertain to different IRPI CNR operative units, and are coordinated by a multi-disciplinary team. Standard procedures are established to homogenize data collection. Double check tests are predisposed to ensure the quality of the database.

4 Computation and Data Analysis

The computation and data analysis component (i) defines and validates empirical rainfall thresholds for the possible initiation of landslides (task 1), and (ii) compare the rainfall measurements and forecasts with the thresholds (task 2). The two tasks are accomplished using different approaches: the first is a manual procedure, while the second is automatic and integrates different computational algorithms.

For rainfall-induced landslides, empirical thresholds for the possible initiation of failures may define the rainfall, soil moisture, or

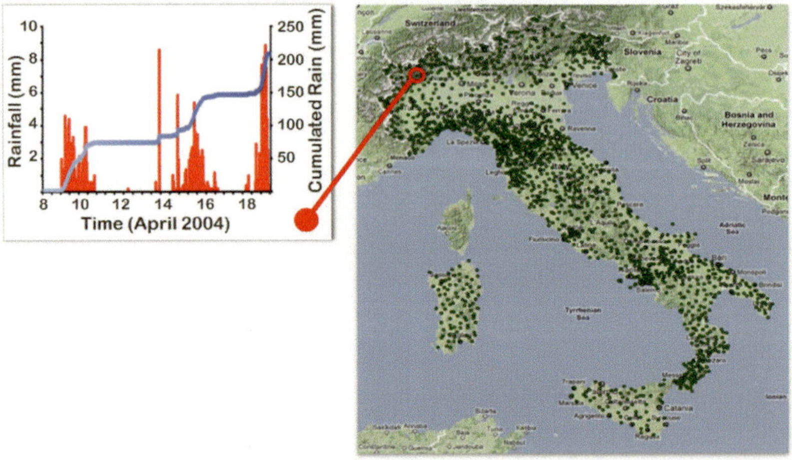

Fig. 2 Rain gauge network of the Italian Department for Civil Protection (1950 stations). An example of the typical rainfall series imported in the SANF system is shown in red in the inset

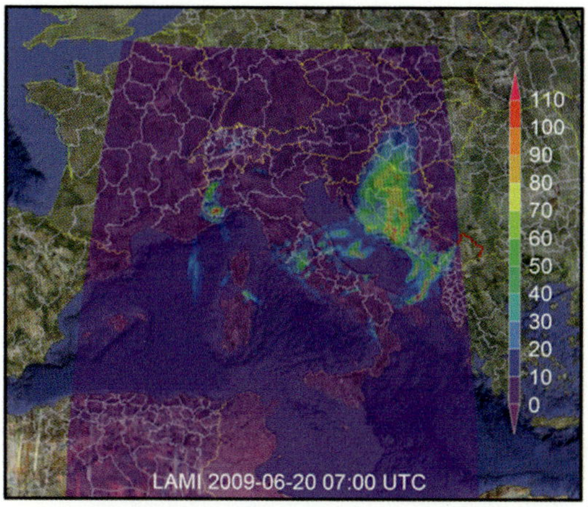

Fig. 3 Example of rainfall forecasts, produced by the Italian LAMI model, imported in the system

hydrological conditions that, when reached or exceeded, are likely to trigger landslides (Guzzetti et al. 2007, 2008; Brunetti et al. 2010). Empirical thresholds are obtained statistically, studying past rainfall events that have resulted in slope failures. The SANF early warning system uses mean intensity-duration (*ID*) rainfall thresholds, defined using an innovative and objective statistical technique. The system assumes that threshold curves are power laws: $I = \alpha \cdot D^{-\beta}$, where I is the mean rainfall intensity (in mm/h), D is the rainfall duration (in h), and α and β are positive coefficients. The rainfall thresholds are obtained using a frequentist statistical approach. The method allows deriving thresholds for different exceedance probability levels (Brunetti et al. 2010). Figure 4 shows the frequentist threshold at 1% exceedance probability defined for the entire Italian territory. Before the use in the early warning system the rainfall threshold was successfully validated against independent landslide rainfall conditions.

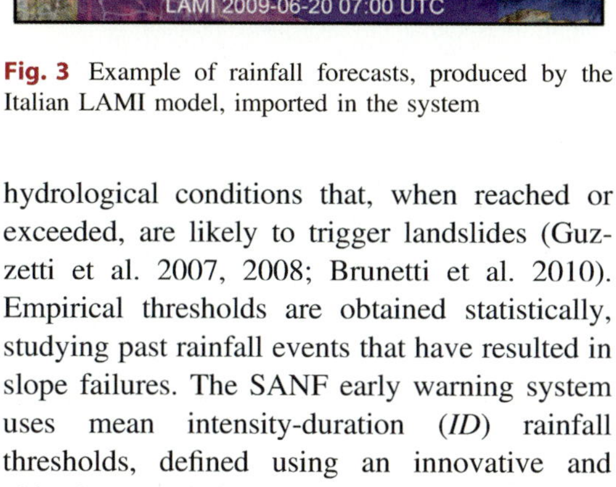

Fig. 4 Intensity-duration conditions (*blue dots*) that triggered landslides in Italy. *Red line* is the rainfall threshold at 1% exceedance probability defined using the frequentist approach implemented in the early warning system

In the SANF system the landslide occurrence probability is assessed for the entire Italian territory. For this purpose, five critical levels for the occurrence of landslides were defined using multiple thresholds at different exceedance probabilities.

The probabilistic scheme (Fig. 5) is based on four Frequentist thresholds at 0.005, 0.5, 1.5 and 5% of exceedance probabilities. In the scheme, the four thresholds, shown by different colours, separate five ID fields. For any given rainfall duration D, when the (measured or predicted) rainfall mean intensity I is lower than the 0.005% threshold, the rainfall condition is "well below the threshold" (level 1). Similarly, when the rainfall mean intensity I is in the range defined by the 0.005% and the 0.5% thresholds, the rainfall condition is "below the threshold" (level 2). When the rainfall mean intensity I is in the range between the 0.5% and the 1.5% thresholds or in the range between the 1.5% and the 5% thresholds, the rainfall condition is "on the threshold" (level 3) or "above the threshold" (level 4), respectively. Lastly, when the rainfall mean intensity I is equal to, or larger than, the upper 5% threshold, the rainfall condition is "well above the threshold" (level 5). In this area, landslides are typically expected, with a chance of false negatives of 5.0% or more.

Every 12 h, for each rain gauge, levels for rainfall measurements and forecasts are calculated. The critical level for rainfall measurements is calculated using the antecedent rainfall measured at duration of 24, 48, 72, and 96 h (Fig. 6a). Levels obtained from the comparison of the measured rainfall at each duration and the corresponding values on the rainfall thresholds (L_{-24}, L_{-48}, L_{-72}, and L_{-96}) are combined to

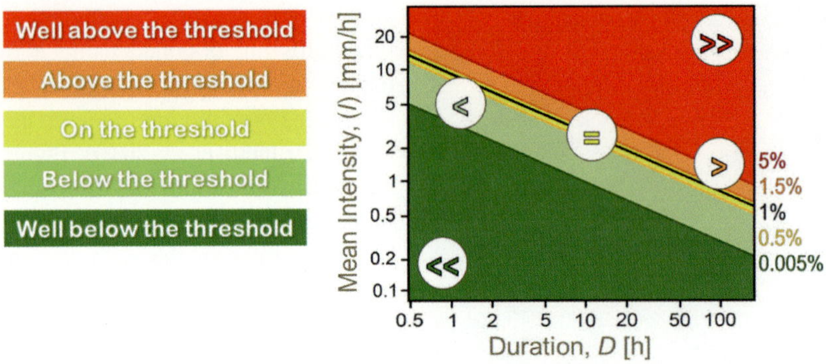

Fig. 5 Conditions of rainfall intensity and duration (*ID*) corresponding to different critical levels, delimited by rainfall thresholds defined using the frequentist approach considering different exceedance probabilities: 0.005, 0.5, 1.5 and 5% (*colored solid lines*). *Black line* shows the rainfall threshold at 1% exceedance probability

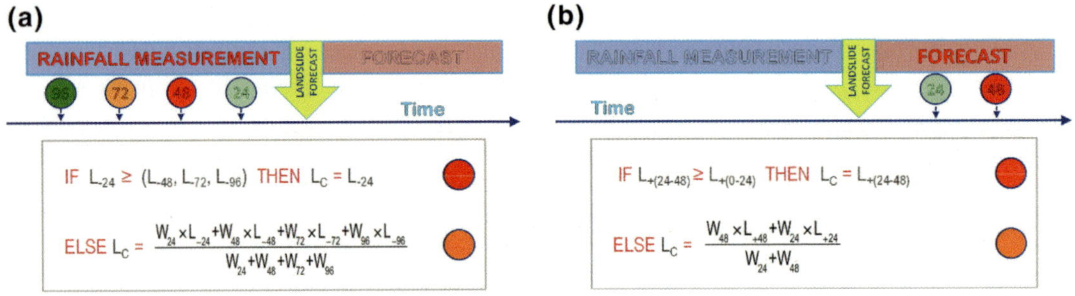

Fig. 6 Calculation of the critical level for: **a** rainfall measurements at duration of 24, 48, 72, and 96 h antecedent the time of landslide forecast, **b** rainfall forecasts estimated at duration of 24, 48 h successive the time of landslide forecast. The equations show the two possible criteria for the calculation of the critical level (L_C) based on rainfall measurements and forecasts

calculate the resulting critical level (L_C) following the criteria shown in Fig. 6a. If the level obtained for the antecedent 24 h rainfall (L_{-24}) is bigger than the other three levels, then it becomes the resulting critical level. Otherwise the rain gauge critical level is calculated with a weighted mean of the four levels, with weights (W_i) decreasing linearly with the duration of the considered antecedent periods. This algorithm for the calculation of the critical level for rainfall measurements is shown in Fig. 6a.

The rain gauge critical level for rainfall forecasts is calculated in a similar way, using the forecasted rainfall at 24 and 48 h (Fig. 6b).

Critical levels do not consider directly the real instability predisposition of the territory. To account that the system calculates critical/susceptibility combined levels. These are obtained multiplying the landslide susceptibility (*S*) for the exceedance probability (P_{sup}) of the rainfall Intensity-Duration conditions associated to rainfall events (Fig. 7). They are calculated similarly to critical levels by the weighted mean

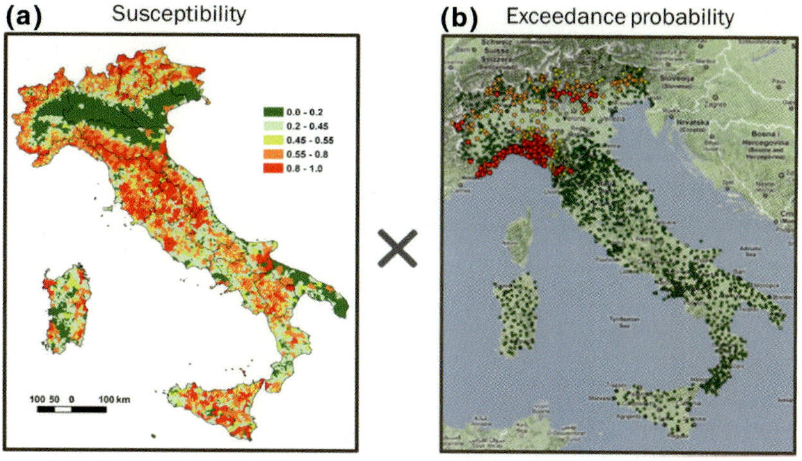

Fig. 7 Probabilistic scheme used for the calculation of the combined critical levels (L_{CS})

of the product of susceptibility (S) and the exceedance probability (P_{sup}) associated to the rainfall for different antecedent or forecast periods. The following equations are used to calculate the combined critical levels (L_{CS}):

- for the rainfall measures

$$L_{cs} = \frac{96 \cdot (P_{SUP-24} \cdot S) + 72 \cdot (P_{SUP-48}) + 48 \cdot (P_{SUP-72} \cdot S) + 24 \cdot (P_{SUP-96} \cdot S)}{96 + 72 + 48 + 24}$$

- for the 24 h rainfall forecast

$$L_{CS} = P_{SUP+24} \cdot S$$

- for the 48 h rainfall forecast

$$L_{CS} \frac{24 \cdot (P_{SUP+24} \cdot S) + 48 \cdot (P_{SUP+48} \cdot S)}{24 + 48}$$

The landslide susceptibility map is derived at municipal scale using a Linear Discriminant Analysis (Rossi et al. 2010) and landslide data from the AVI archive (Archivio Aree Vulnerate Italiane) and geological, geo-morphometrical, soil and climatic data available at synoptic scale for the entire Italian territory (Fig. 7a).

The Italian National Department for Civil Protection subdivided the national territory in 129 alert zones devised for hydrological purposes. Critical levels of rain gauges belonging to the same alert zone are aggregated to define alert zone critical levels. The system uses two different aggregation criteria based on the maximum and the modal value, respectively. Following the first criteria, the system assigns to the alert zone the highest level among those reached by rain gauges

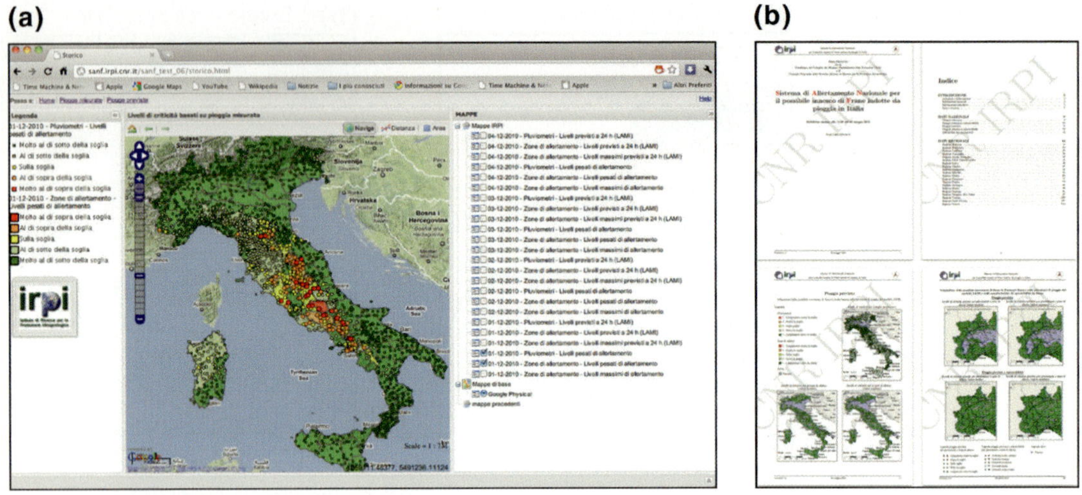

Fig. 8 Early warning system outputs: **a** WebGIS interface or **b** daily bulletins in PDF format

in that alert zone. Using the second criteria, the system assigns to the alert zone the most frequent (modal) level among those reached by rain gauges in that alert zone.

5 Early Warning System Outputs

The early warning system computes critical levels for each rain gauge and for each alert zone every 12 h. Once levels are calculated, the SANF system generates system outputs, consisting of synoptic-scale maps of critical levels for the occurrence of rainfall-induced landslides. Output maps are generated by the IRPI CNR spatial data infrastructure (SDI), as standard OGC web services (Open Geospatial Consortium, http://www.opengeospatial.org). In particular these services consist in vector (WFS) or raster (WMS) maps. Maps can be consulted using both open source and proprietary software (e.g. QGIS, ESRI ArcMap, etc.) and they can be easily integrated in existing web interfaces. All the maps generated by the system have been integrated in a WebGIS infrastructure (Fig. 8a) including tools for viewing, querying and filtering data. The WebGIS interface allows also the visualization and query of landslide synoptic maps including landslide inventory maps, landslide hazard maps and landslide and flood risk maps. Once a day, a specific procedure generates a summary report in PDF format (Fig. 8b) containing, for both rainfall measurements and rainfall forecasts, a map showing the rain gauge critical levels and two maps showing the alert zone critical levels (maximum and modal). The report is delivered via e-mail to the Italian National Department for Civil Protection.

6 System Validation and Future Improvements

Validation of the system outputs is a key part of the system, but is difficult to perform and results are uncertain. The validation of the thresholds, as well as the validation of the system algorithms is needed. Just a partial system of validation has been realized. The rainfall thresholds were validated against independent landslide information. Algorithms for the calculation of critical levels were validated using landslide information reported in a 6 month period in the Abruzzo region. Collection of data on landslide spatial and temporal occurrence is in progress for the validation of the system over large areas and for the identification of regional rainfall thresholds for the possible landslide occurrence.

Multifold system developments are possible and, in some cases, have been already preliminarily deployed: (i) increase of system forecasts at hourly scale; (ii) implementation of new regional thresholds; (iii) integration of different rainfall forecasts; (iv) improvement of the validation procedures; (v) improvement of landslide susceptibility; (vi) integration with hazard/vulnerability/risk information at synoptic scale.

References

Brunetti MT, Peruccacci S, Rossi M, Luciani S, Valigi D, Guzzetti F (2010) Rainfall thresholds for the possible occurrence of landslides in Italy. Nat Hazards Earth Syst Sci 10:447–458

Guzzetti F, Peruccacci S, Rossi M, Stark CP (2007) Rainfall thresholds for the initiation of landslides in central and southern Europe. Meteorol Atmos Phys 98:239–267

Guzzetti F, Peruccacci S, Rossi M, Stark CP (2008) The rainfall intensity-duration control of shallow landslides and debris flows: an update. Landslides 5(1):3–17

Rossi M, Guzzetti F, Reichenbach P, Mondini AC, Peruccacci S (2010) Optimal landslide susceptibility zonation based on multiple forecasts. Geomorphology 114(3):129–142

Salvati P, Bianchi C, Rossi M, Guzzetti F (2010) Societal landslide and flood risk in Italy. Nat Hazards Earth Syst Sci 10:465–483

TXT-tool 2.084-3.1 Rainfall Thresholds for Triggering Geohazards in Bac Kan Province (Vietnam)

Duc Ha Nguyen, Thi Hai Van Nguyen, Quoc Hung Le, Van Son Pham and Ho Khanh Nguyen

Abstract

This paper presents the integration of several developed methods to establish rainfall thresholds for triggering any type of geohazards in data-scarce areas. In the developed procedure, it is important to obtain the exact dates of the past disastrous events such as landslides, floods, river bank erosion, debris flows, etc. For the case study area of Bac Kan province, the available data comprises of historic information of 14 geohazard events that occurred in some years from 1986 to 2010 and in 2013, and daily rainfall values during this period. The results show that the relationships between the daily rainfalls and the 7 days-antecedent rainfalls are of very important to trigger series of disastrous events in Bac Kan province. It is a simple and useful approach to applying to the contexts that lack detailed data of a specific geohazard type. This approach is going to be further developed to obtain more accurate thresholds not only for Bac Kan province but also for other areas with similar setting by updating more information on historic geohazard events.

Keywords

Rainfall threshold • Geohazard • Landslide • Flood • Heavy rain

Contents

1 **Introduction** ... 351
2 **Methodology** ... 352
2.1 Study Area ... 352
2.2 Data and Methods Used ... 352
3 **Results and Discussion** ... 354
4 **Conclusion** ... 359
References ... 359

D.H. Nguyen · T.H.V. Nguyen · V.S. Pham · H.K. Nguyen
Department of Remote Sensing and Geomatics, Vietnam Institute of Geosciences and Mineral Resources, No. 67 Chien Thang Street, Van Quan Ward, Ha Dong District, Hanoi City, Vietnam

Q.H. Le (✉)
Vietnam Institute of Geosciences and Mineral Resources, No. 67 Chien Thang Street, Van Quan Ward, Ha Dong District, Hanoi City, Vietnam
e-mail: le.quoc.hung@vigmr.vn; hunglan@gmail.com

1 Introduction

Vietnam is one of the countries that have been deeply affected by global climate change. The abnormal weather phenomena often cause heavy rains, combined with human activities such as deforestation, mining, hill levelling, etc. induce

K. Sassa et al. (eds.), *Landslide Dynamics: ISDR-ICL Landslide Interactive Teaching Tools*, https://doi.org/10.1007/978-3-319-57774-6_25

lots of negative impacts to the environment, especially more serious for the mountainous areas. Various types of disasters related to geological processes, or so-called geohazards, such as landslide, river bank erosion, debris flow, flash flood, etc. have occurred in most of the mountainous region of Vietnam, causing much more damages and losses to people, infrastructure and the environment. It is urgently required more effort to prevent and mitigate those disasters by implementing one or more approaches to mapping, monitoring, forecasting and warning geohazards.

Studies on the investigation, assessment, zonation, mapping and warning in Vietnam have shown that the occurrences of most geohazard events are closely related to the extreme weather conditions that cause heavy rains or prolonged rains (Tran Van et al. 2002; Nguyen et al. 2011; Le et al. 2014). Establishment of rainfall thresholds is one of the effective measures to give a timely warning of landslides, (flash) floods, debris flows, etc., (Phan Van 2005; Zezere et al. 2005; Guzzetti et al. 2007; Vu et al. 2007a; Jaiswal et al. 2011; Chen and La 2012; Bui et al. 2013; Le et al. 2014). However, it is very difficult to obtain appropriate thresholds due to the lack of necessary data related to each type of the above-mentioned geohazards.

In Vietnam, few researches have studied on the determination of rainfall thresholds that trigger flash floods or landslides (Nguyen et al. 1995; La and Tran 2012; Bui et al. 2013), in which the historic events were identified for occurrences of only landsliding or flooding. The fact shows that disastrous events related to extreme rainfall phenomena are often not separately recorded for only landslides or flash floods or debris flows. The historic information on the occurrences of one type of geohazards (such as flash floods) is often mixed with that of other types (such as landslides, debris flows, river bank erosion, etc.). The reasons were due to the fact that one rain events can trigger several types of geohazards, and they can together happen at the same time or they give domino effects (one disaster can trigger another disaster). Therefore, it would be useful to make use of all collected data of all related geohazard types to establish rainfall thresholds for triggering geohazards in order to prevent a series of disastrous events.

2 Methodology

2.1 Study Area

The method is applied to the case study area of Bac Kan province, which is a mountainous province in the northeast (Fig. 1). This province has a diverse and complicated terrain with a dense river system, which covers an area of 4857.21 km^2, and comprises of hills, mountains and midland. The elevations range from 100 m to nearly 1700 m. The province has a humid monsoon tropical climate that differentiates with the altitudes of the terrain and aspects of the mountain. In Bac Kan, the rainy season starts from May to October that accounts for 70–80% of annual rainfall. Average annual rainfall in the 1400–1600 mm. Bac Kan is one of the mountainous provinces that is threatened by many types of geohazards (Vu et al. 2007b).

2.2 Data and Methods Used

The rainfall thresholds for triggering geohazards in Bac Kan province were established by adjusting several methods that had been developed and applied in many areas in Vietnam and in the world (Phan Van 2005; Zezere et al. 2005; Jaiswal et al. 2011; La and Tran 2012; Bui et al. 2013). The procedures can be briefly described as follows:

– The daily rainfall data were collected from three National Hydro-meteorological Stations in Bac Kan province (namely Bac Kan, Cho Ra and Ngan Son), in which the data from 1961 to 2010 was used for calibration, and the data from May to August 2013 was used for validation.

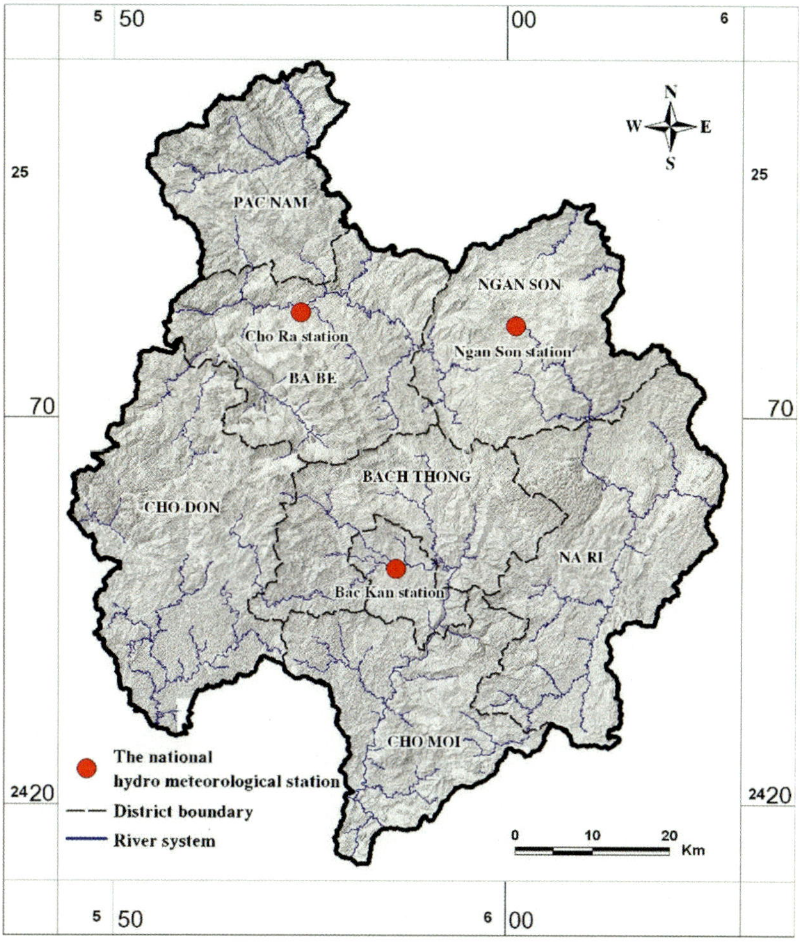

Fig. 1 Sketch map showing Bac Kan province with *red dots* indicating the locations of three national hydro-meteorological stations

- The relationships between the daily rainfall and the occurrences of geohazard events in the study area were analysed by following the procedure as shown in Fig. 2.
- The relationships between the maximum daily rainfall from 1961 to 2010 and their return periods (probability of occurrences) were determined by applying the Gumble method through the analysis of the rainfall data measured at three National Hydro-meteorological Stations in Bac Kan province (Phan Van 2005).
- Daily rainfall values of the dates recorded with geohazard events were analysed to obtain the probability of those rain events as well as the probability of related geohazard events.
- The relationships among the maximum daily rainfall, the daily rainfall on the dates recorded with geohazards, and antecedent rainfall values were calibrated and validated to establish rainfall thresholds by apply the methods developed by Zezere et al. (2005), Marques et al. (2008), Jaiswal et al. (2011), Bui et al. (2013).

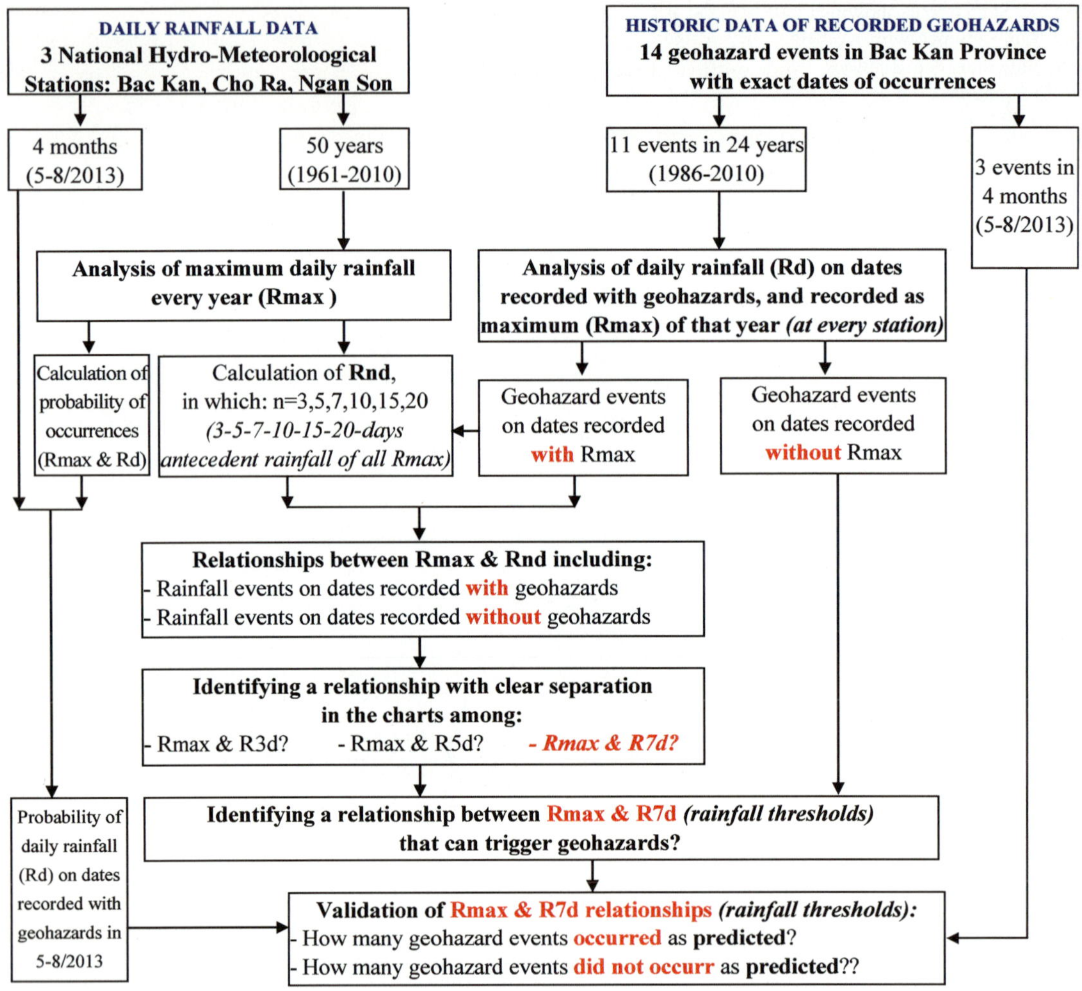

Fig. 2 Procedures of analysing and evaluating the relationships between the rainfall factors (measured at three National Hydro-Meteorological Stations) and the occurrence of geohazard events (recorded by field surveys) in Bac Kan province

3 Results and Discussion

Information on 14 historic geohazard events in the study area was analysed to determine the exact dates of their occurrence for all three stations of Bac Kan province. Then all the geohazards-related values of daily rainfall and 3, 5, 7, 10, 15 and 20-days antecedent were extracted, calibrated and analysed, and their return periods were calculated as shown in Tables 1 and 2; Fig. 3. The field surveys point out that there are many geohazard events that occurred on the days with very low rainfall (e.g. on 17/8/2003 and 31/7/2010), or on the days without rains (or the storms/heavy rains/prolonged rains already finished) (e.g. on 1/7/1986, 27/7/2002, and 10/7/2009). The occurrence of geohazards on those dates can be explained by the influences of the rains on the previous days, which has made the soil and rock mass saturated, or the consequences of flood water from upstreams, etc. Therefore, the antecedent rainfall also plays important to trigger geological disasters.

The daily rainfall (R_t) of 11 geohazard events that occurred from 1986 to 2010 was calibrated in the relation with the 7 days-antecedent rainfall

Table 1 Maximum daily rainfall and antecedent rainfall related to the geohazard events that were recorded in Bac Kan province from 1986 to 2013

Dates recorded with geohazards (dd/mm/year)	Daily rainfall	Return period	Antecedent rainfall					
			3 days	5 days	7 days	10 days	15 days	20 days
Bac Kan station								
01/07/1986			0	0.2	31.1	80.1	131.1	206.7
22/09/1990	193.7	12	183.5	183.5	186.3	206.2	206.2	223.5
05/10/2000	46.2	1.1	1.2	17.9	33.6	91.4	92.8	92.8
27/07/2002			21.6	27	84.6	84.7	113.8	125.2
17/08/2003	0.9	1	67.6	92.5	100.9	190.2	191.4	191.4
17/07/2006	176.6	8.3	7.2	10	67.1	91.8	121.9	201.1
13/04/2008	43.1	1.1	1.1	1.1	1.2	1.3	14.1	14.8
13/08/2008	11.2	1	34.3	137.6	159.6	162.7	175.3	222.1
04/07/2009	53.3	1.1	39.7	42.6	43.8	84.2	120.8	155.7
10/07/2009			1.6	39.4	99.8	132.4	176.9	213.5
31/07/2010	0.5	1	33.4	36	85.1	131.9	194.2	199.7
30/05/2013	134.4	3.5	51	65.8	65.8	82.6	82.6	85.3
29/07/2013	27.8	1	59.4	62.5	101.3	149	185.6	238.8
04/08/2013	15.2	1	101.3	104.6	151.4	191.8	269	313.7
Cho Ra station								
01/07/1986			0.2	0.2	0.4	129.5	183.9	262.3
22/09/1990	21.7	1	101.2	101.2	109	121	121	137
05/10/2000	32.6	1	9.1	31.2	42.6	53.9	68.1	68.1
27/07/2002			21.5	27.1	166.6	166.7	231.9	235.7
17/08/2003	2.8	1	68.2	99.5	104.8	112.9	113.8	115.7
17/07/2006	210.9	112	37	45.1	79.5	119.3	179.5	249
13/04/2008	50.9	1.1	0.6	0.6	0.7	0.9	43.8	45.7
13/08/2008	78.9	1.5	18.5	158.9	166.2	185.4	198,4	210.4
04/07/2009	138.1	8.6	46.9	71.7	79.4	109.5	136.8	157.6
10/07/2009	0.4	1	1.7	47	192.8	232.3	294.6	321.9
31/07/2010	2.6	1	69.2	69.5	118.8	154.4	213.5	220.3
30/05/2013	51.1	1.1	16.3	17.3	20.7	49.5	49.9	57
29/07/2013	25.7	1	65.7	73.1	112.1	145.3	214.1	234.8
04/08/2013	47.2	1	62.9	63	91.8	161.8	209	280
Ngan Son station								
01/07/1986			6.7	13.5	14.9	45	104.7	218.8
22/09/1990	63.2	1.1	249.3	249.3	249.3	256.6	271	271
05/10/2000	110	2.2	41.3	53.3	100.6	193.2	215.9	215.9
27/07/2002			10.9	16.5	79.8	84.7	102.6	119.7
17/08/2003	1.9	1	116.6	125.5	137.5	172.5	172.5	178
17/07/2006	165.6	8.1	0.9	4.9	55.3	75,2	119.4	295.6

(continued)

Table 1 (continued)

Dates recorded with geohazards (dd/mm/year)	Daily rainfall	Return period	Antecedent rainfall					
			3 days	5 days	7 days	10 days	15 days	20 days
13/04/2008	51	1.1	1.2	1.2	1.5	2.6	22.9	24
13/08/2008	33.3	1	19.7	186	209.3	222.5	232.6	346.1
04/07/2009	89.5	1.6						
10/07/2009	0.3	1	4.2	30.1	127.1	153.7	221.7	278.8
31/07/2010	19.2	1	32	85.4	107.8	138.4	232.2	286.2
30/05/2013	50	1.1	38.7	49.2	54.7	86.4	88.5	113.5
29/07/2013	13.8	1	46.4	52.2	70.1	101.2	145.3	177.9
04/08/2013	30	1	118.8	118.9	148.9	179.4	214.8	260.3

(R_{7d}). Results showed that they have correlation through three threshold equations as shown in Fig. 4. To validate those rainfall thresholds, daily rainfall from 01/05/2013 to 31/08/2013 measured at 3 stations were used in combination with geohazard events recorded by the field surveys. The validation as shown in Fig. 5 points out that the actual rainfall observed during this period at 3 stations in Bac Kan province were exceeded many times. Compared with rainfall data during the period from 01/05/2013 to 31/08/2013, we can see that the events occurred on 30/05/2013 and at night of 3/8/2013 and early morning of 04/08/2013 exceeds the thresholds the most.

The correlations between daily rainfall and 3, 5, 7, 10, 15, 20-days antecedent rainfall measured at three Hydro-Meteorological Stations of Bac Kan, Cho Ra and Ngan Son as shown in Table 2 figure out that:

- At Bac Kan Station: there is a clear separation in the distribution of the relationship of 7, 10, 15 and 20-days antecedent rainfall with the daily maximum rainfall on dates recorded *with* geohazard events from that relationship with the ones recorded *without* geohazard events.
- At Cho Ra Station: there is a clear separation in the distribution of the relationship of only 7-days antecedent rainfall with the daily maximum rainfall on dates recorded *with* geohazard events from that relationship with the ones recorded *without* geohazard events.
- At Ngan Son Station: there is no separation in the distribution of relationship of only 7-days antecedent rainfall with the daily maximum rainfall on dates recorded *with* geohazard events from that relationship with the ones recorded *without* geohazard events.

The obtained rainfall thresholds were mainly based on only daily rainfall of three different rain gauge stations. Therefore, the thresholds established for this station/one area cannot be affirmed to be applicable to other station/areas if those stations/areas have different characteristics. To improve the accuracy of these thresholds, and to use them effectively in prediction models, it is necessary to collect more sufficient data on the past events, particularly the exact dates of occurrences.

The observation of meteorological factors in the area of pilot studies should have a long enough period of monitoring and a full complement of accurate information about events occurring geological hazards in the area research. Therefore, the relationship set for rainfall stations in the study area cannot be affirmed applicable to monitoring stations in other areas or not. The observation of meteorological factors in the area

Table 2 Maximum daily rainfall and 3, 5, 7, 10, 15 and 20 days-antecedent rainfall measured at three National Hydro-Meteorological Stations of Bac Kan province from 1986 to 2010, in which: green-dot notations for the values with information of geohazards, and the red triangle notations for the values without information of geohazards

Measured rainfall	Bac Kan station	Cho Ra station	Ngan Son station
(a) Daily rainfall versus 3 days-antecedent rainfall	(a1) 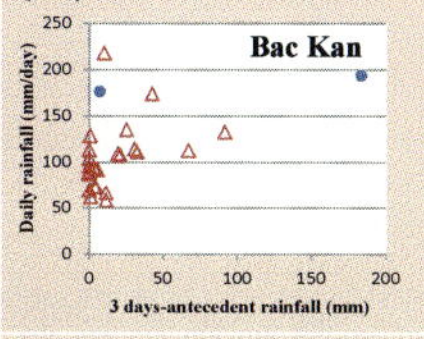	(a2) 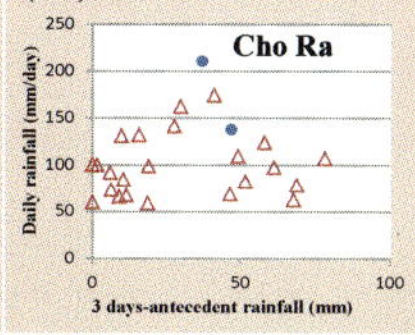	(a3) 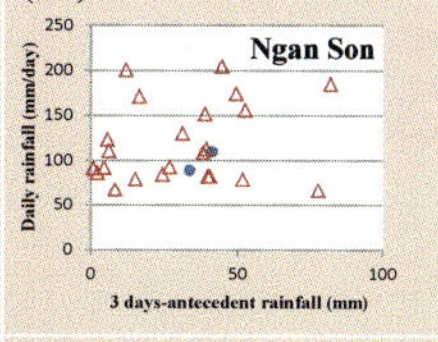
(b) Daily rainfall versus 5 days-antecedent rainfall	(b1) 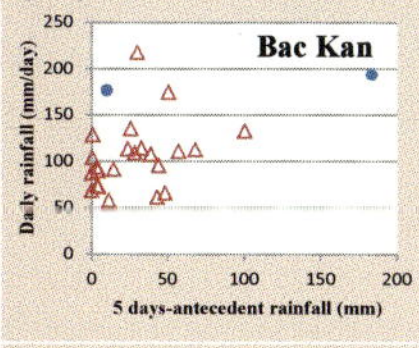	(b2) 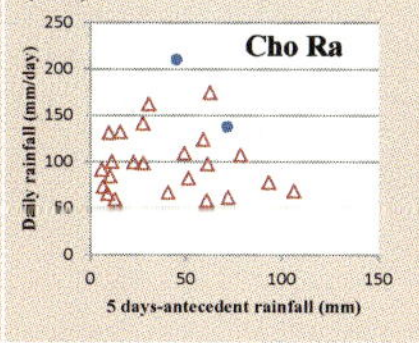	(b3) 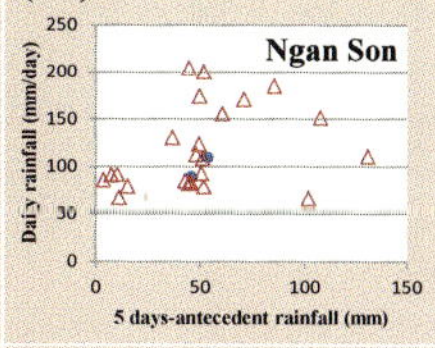
(c) Daily rainfall versus 7 days-antecedent rainfall	(c1) 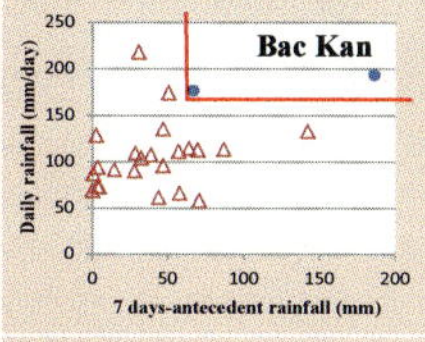	(c2) 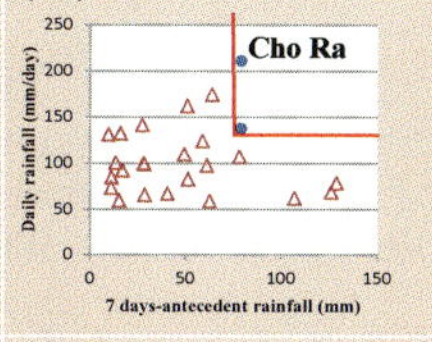	(c3) 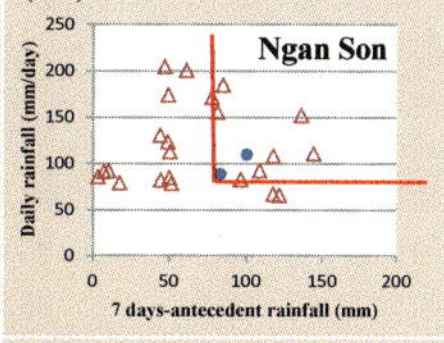
(d) Daily rainfall versus 10 days-antecedent rainfall	(d1) 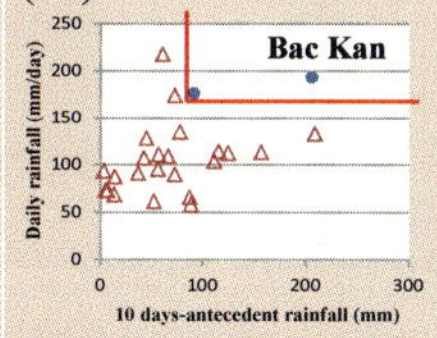	(d2) 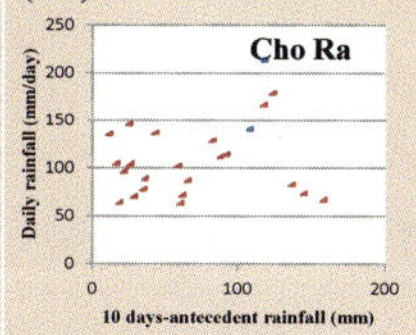	(d3) 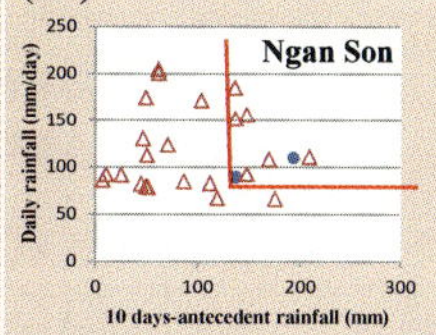
(e) Daily rainfall versus 15 days-antecedent rainfall	(e1) 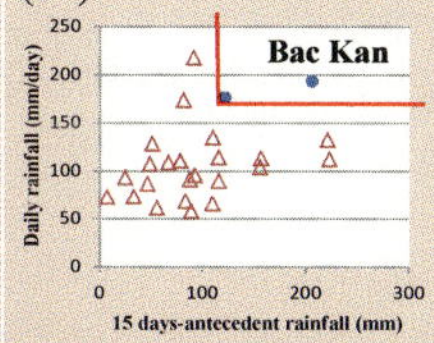	(e2) 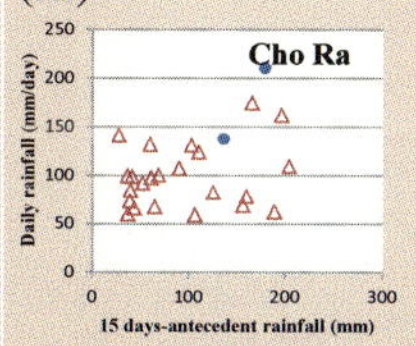	(e3) 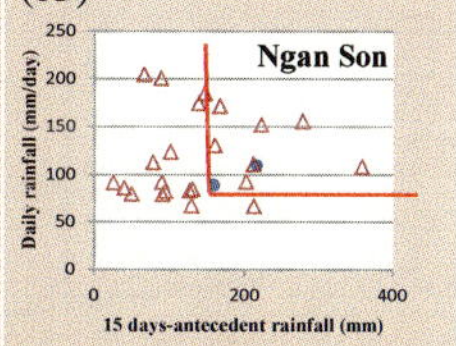
(f) Daily rainfall versus 20 days-antecedent rainfall	(f1) 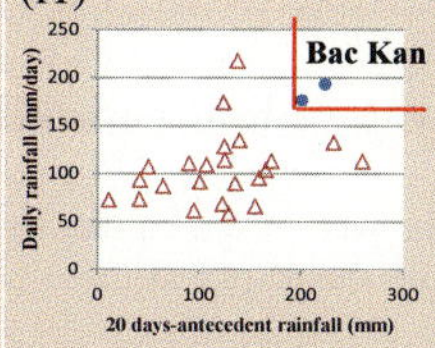	(f2) 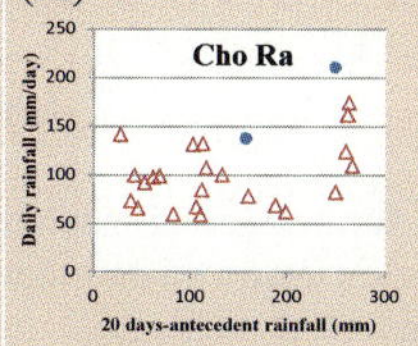	(f3) 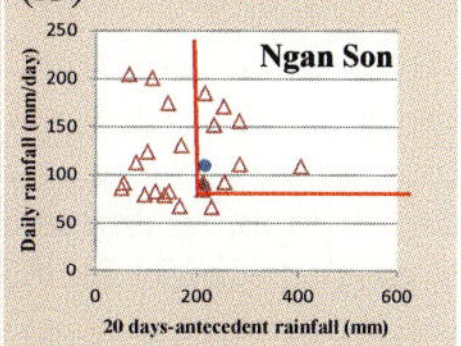

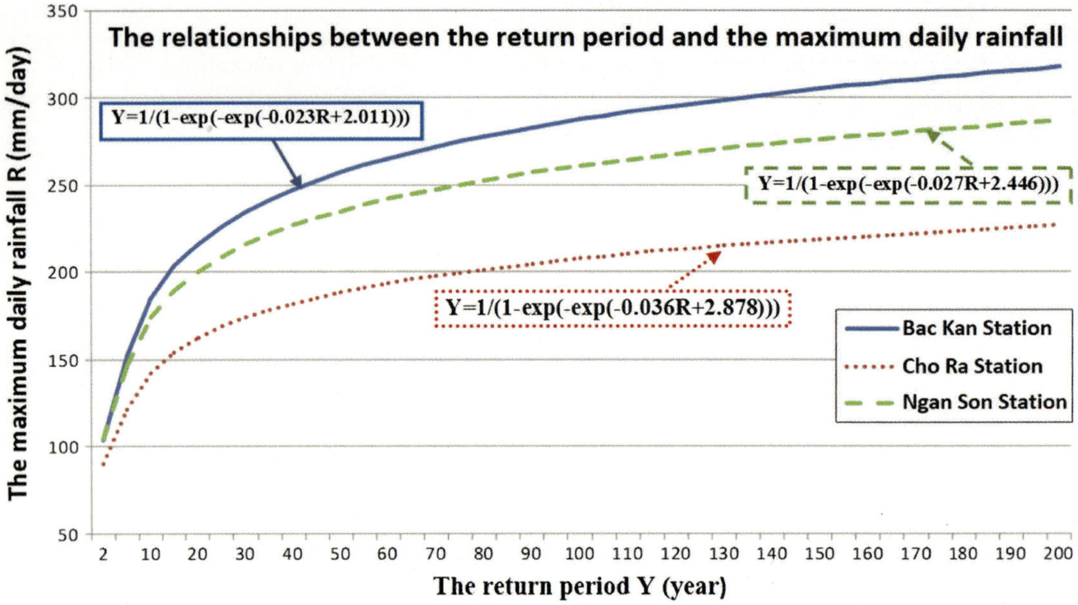

Fig. 3 Graphs showing the relationships between the return period (Y) and the maximum daily rainfall (R) measured at three National Hydro-Meteorological Stations: Bac Kan, Cho Ra and Ngan Son

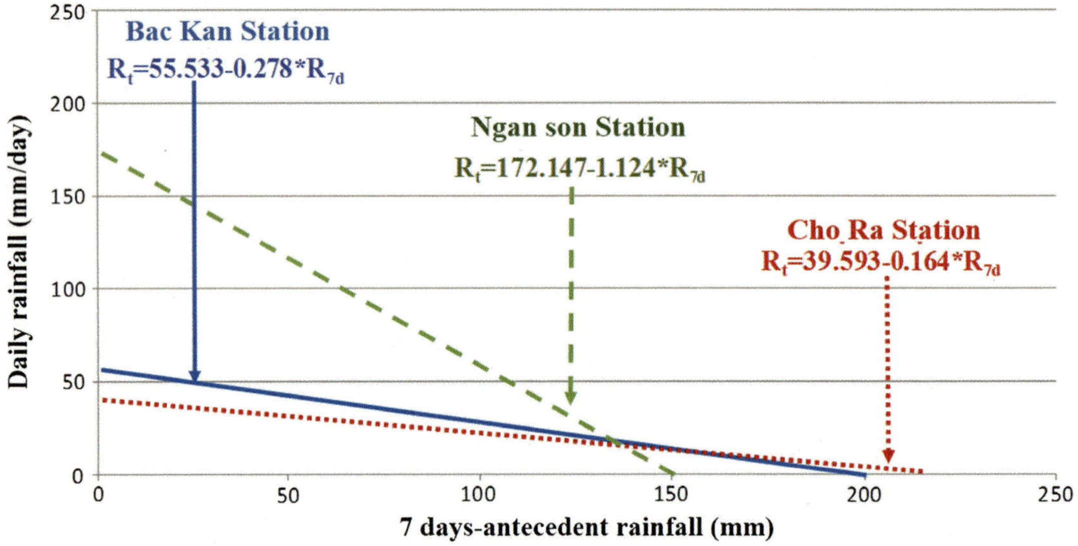

Fig. 4 Rainfall thresholds for triggering geohazards were established by correlating daily rainfall (Rt) with the 7 days-antecedent rainfall from 1986 to 2010 that were measured at three National Hydro-Meteorological Stations

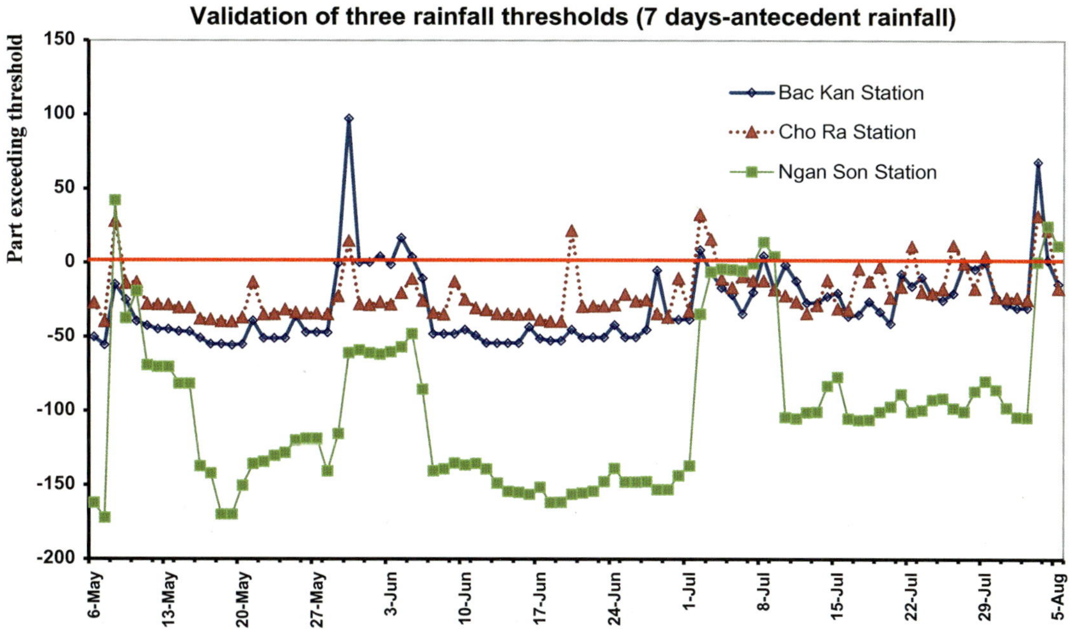

Fig. 5 Validation of three rainfall thresholds for triggering geohazards by using rainfall data from 1st May to 31st August 2013 that were measured at three National Hydro-Meteorological Stations (namely Bac Kan, Cho Ra and Ngan Son Stations)

of pilot studies should have a long enough period of monitoring and a full complement of accurate information about events occurring geological hazards in the area research.

4 Conclusion

Timely forecasting and warning of any type of geohazards play an important role in the prevention, mitigation and reduction of geohazard risk. The determination of geohazard-triggered rainfall thresholds for any specific area will improve the efficiency of forecasting and disaster warnings. This study presents an integration and adjustment of developed methods to establish the thresholds for the testing area in Bac Kan province. The results showed that the daily rainfall is rather correlated with the 7 days-antecedent rainfall to trigger a series of geohazard events in Bac Kan during the period of 1986–2010. This result is in accordance with the actual situations of the geohazard occurrences in Bac Kan province. Thus, the methodology can be improved to obtain more accurate thresholds for this case study area with more supplementary data, or can be applied to other places with suitable adjustments.

Acknowledgements This study is the part of the "Research and application of WebGIS technology, high-resolution RADAR image analysis and GIS models for the establishment of a warning system on geohazards and natural disasters in Vietnam. A case study in Bac Kan". It was carried out under the Contract for Scientific Research and Technological Development with the number of TNMT.03.15/HDKHCN between the Minister of Natural Resources and Environment and Director of the Institute of Geosciences and Mineral Resources.

References

Bui DT, Pradhan B, Lofman O, Revhaug I, Dick OB (2013) Regional prediction of landslide hazard using probability analysis of intense rainfall in the Hoa Binh province, Vietnam. Nat Hazards 66(2):707–730. doi:10.1007/s11069-012-0510-0

Guzzetti F, Peruccacci S, Rossi M, Stark CP (2007) Rainfall thresholds for the initiation of landslides in central and southern Europe. Meteorol Atmos Phys 98 (3):239–267. doi:10.1007/s00703-007-0262-7

Jaiswal P, Jetten VG, van Westen CJ, Ayyasami K (2011) Landslide risk quantification along transportation corridors based on historical information. PhD Dissertation, University of Twenty Faculty of Geo-Information and Earth Observation ITC, Enschede, p 243

Le QH et al (2014) Project report research and application of WebGIS technology, high-resolution RADAR image analysis and GIS models for the establishment of a warning system on geohazards and natural disasters in Vietnam. A case study in Bac Kan (in Vietnamese). Vietnam Institute of Geosciences and Mineral Resources. Hanoi, p 182

Marques R, Zêzere J, Trigo R, Gaspar J, Trigo I (2008) Rainfall patterns and critical values associated with landslides in Povoação County (São Miguel Island, Azores): relationships with the North Atlantic Oscillation. Hydrol Process 22(4):478–494. doi:10.1002/hyp.6879

Nguyen VT et al (1995) Development of methods for flash flood warning. Under the state-funded project research on the mechanisms and prevention measures of flash floods (in Vietnamese). General Department of Meteorology and Hydrology, National Center for Hydro-Meteorological Forecasting. Hanoi, p 122

Phan Van T (2005) Statistical methods in climate studies (in Vietnamese). Publishing House of VNU, p 260

Tran T, La TH (2012) Handbook flash flood—Basic concept and study methods. Publishing House of Sciences and Technology, Hanoi, p 267

Tran Van T et al (2002) Project report assessment on geological hazards in the eight coastal provinces of Central Vietnam (from Quang Binh to Phu Yen): present situation, causes, prediction and recommendation of remedial measures (2 volumes in Vietnamese). Vietnam Institute of Geosciences and Mineral Resources. Hanoi, p 252

Vu MC et al (2007a) Project report research on the mechanisms and prevention measures of flash floods for the Northeast of Vietnam (Cao Bằng, Bắc Kạn, Thái Nguyên) (in Vietnamese). Water Resources University of Hanoi. Ministry of Agriculture and Rural Developement. Hanoi. p 259

Vu TT et al (2007b) Project report investigation on geohazards in some critical areas of Northeastern Vietnam for socio-economic planning and development (in Vietnamese). Vietnam Institute of Geoscience and Mineral Resources. Hanoi, p 270

Zêzere JL, Trigo RM, Trigo IF (2005) Shallow and deep landslides induced by rainfall in the Lisbon region (Portugal): assessment of relationships with the North Atlantic Oscillation. Nat Hazards Earth Syst Sci 5 (3):331–344. doi:10.5194/nhess-5-331-2005

TXT-tool 2.039-1.2 Rainfall Thresholds for the Possible Initiation of Landslides in the Italian Alps

Maria Teresa Brunetti, Silvia Peruccacci, Michela Rosa Palladino, Alessia Viero and Fausto Guzzetti

Abstract

Rainfall-induced landslides are frequent in the Italian Alps, where they cause severe economic damages and loss of life. The prediction of rainfall-induced slope failures is of utmost importance for civil protection purposes and relies upon the definition of physically based or empirical rainfall thresholds. Reliable empirical rainfall thresholds require a large amount of information on the geographical and temporal location of rainfall events that caused past mass movements. We have compiled a catalogue listing 453 rainfall events that have triggered 509 landslides in the Italian Alps in the 13-year period 2000–2012. For the purpose, we searched national and local newspapers, blogs, technical reports, historical databases, and scientific journals. In the catalogue, for each rainfall event that triggered one or more failures, the information includes: (i) landslide geographical position, (ii) date of the landslide occurrence, (iii) landslide type (if available from the source of information), and (iv) rainfall information. Using the available information, we calculated the cumulated amount (E) and the duration (D) of the rainfall that likely caused the documented slope failures. We exploited the catalogue to calculate new ED threshold curves and their associated uncertainties for the Italian Alps adopting a frequentist approach. To define seasonal rainfall thresholds, we also investigated the monthly distribution of the landslides. The new

M.T. Brunetti (✉) · S. Peruccacci · F. Guzzetti
Istituto di Ricerca per la Protezione Idrogeologica, Consiglio Nazionale delle Ricerche, Via Madonna Alta 126, 06128 Perugia, Italy
e-mail: mariateresa.brunetti@irpi.cnr.it

S. Peruccacci
e-mail: silvia.peruccacci@irpi.cnr.it

F. Guzzetti
e-mail: fausto.guzzetti@irpi.cnr.it

M.R. Palladino
Istituto di Ricerca per la Protezione Idrogeologica, Consiglio Nazionale delle Ricerche, Strada delle Cacce 73, 10135 Turin, Italy
e-mail: michela.palladino@irpi.cnr.it

A. Viero
Istituto di Ricerca per la Protezione Idrogeologica,

K. Sassa et al. (eds.), *Landslide Dynamics: ISDR-ICL Landslide Interactive Teaching Tools*, https://doi.org/10.1007/978-3-319-57774-6_26

thresholds are compared with similar curves in the same general area. We expect the results of our study to improve the ability to forecast landslides in the Italian Alps and, more generally, in the wider Alpine region.

Keywords
Landslides • Italian Alps • Rainfall thresholds>

Contents

1 Introduction ... 362
2 Catalogue of Alpine Events ... 362
3 Frequentist Method ... 364
4 Rainfall Thresholds for the Italian Alps ... 364
5 Seasonal Thresholds ... 365
6 Climatic Thresholds ... 367
7 Discussion and Conclusions ... 368
References ... 369

1 Introduction

The Italian Alpine area is located in northern Italy, approximately in an east-west direction. The territory covers 53,000 km^2, and the highest peak (Mont Blanc) rises 4810 m above sea level. The study area (Fig. 1) extends across seven Italian administrative regions Piedmont, Liguria, Aosta Valley, Lombardy, Trentino-Alto Adige, Veneto and Friuli-Venezia Giulia.

In the Alpine area rainfall-induced landslides, including soil slips and debris flows, are frequent (Fig. 2). Landslides can cause damage with human and economic consequences. To foster the currently limited ability to forecast the occurrence of landslides in Italy, the Italian National Department for Civil Protection (DPC) is funding a research project aimed at defining the regional and sub-regional rainfall conditions that can result in landslides in Italy. In particular, the research activity is focused on the use of empirical rainfall thresholds in a nation-wide landslide warning system.

Empirical rainfall thresholds are calculated analysing past rainfall events that are known or presumed to have triggered slope failures. In this work, the calculation of reliable rainfall thresholds is based on a frequentist approach (Brunetti et al. 2010; Peruccacci et al. 2012), which requires a large amount of information on the geographical and temporal location of the landslides, and on the rainfall conditions responsible for the failures. For this purpose, we are compiling a national catalogue of past rainfall events that have triggered landslides in Italy.

2 Catalogue of Alpine Events

The national catalogue of rainfall events that have triggered landslides in Italy includes 453 rainfall events that initiated 509 landslides (yellow dots in Fig. 1) in the Italian Alps in the 13-year period 2000–2012. We searched digital or printed national and local newspapers, blogs, technical reports and historical archives, and scientific papers to collect information on the occurrence of rainfall-induced landslides in the study area. Almost all the information sources report news on landslides that affected the population and/or the infrastructures (e.g., houses, roads) and had impact on the anthropic activities. Mass movements that occurred in isolated areas are not included in the catalogue, which is likely not representative for the population of rainfall-induced landslides. Nevertheless, we maintain that the catalogue is a representative sample of failures that had significant consequences for the population and is therefore suitable for civil protection purposes.

Hourly rainfall measurements from more than 650 rain gauges were obtained from the DPC,

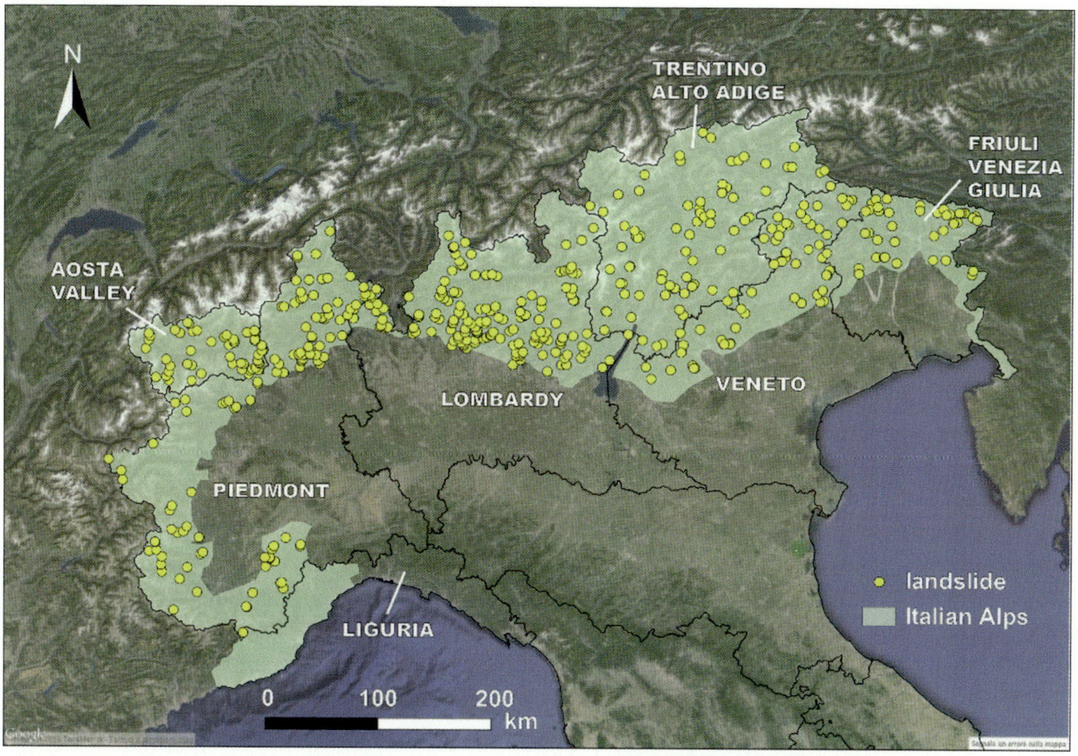

Fig. 1 Study area located in the Italian Alps. *Yellow dots* are the documented landslides

Fig. 2 Examples of landslides in the Italian Alps, (**a**) in the Soana Valley, Piedmont, and (**b**) in the Aosta Valley. Photos by L. Turconi, IRPI Torino

and from Italian Regional Agencies for the Environment Protection.

Information in the catalogue includes: (i) the geographical position of each landslide (i.e., latitude, longitude, municipality, province, region, and position precision); (ii) temporal information (i.e., day, month, year, time or inferred time, date, temporal precision);

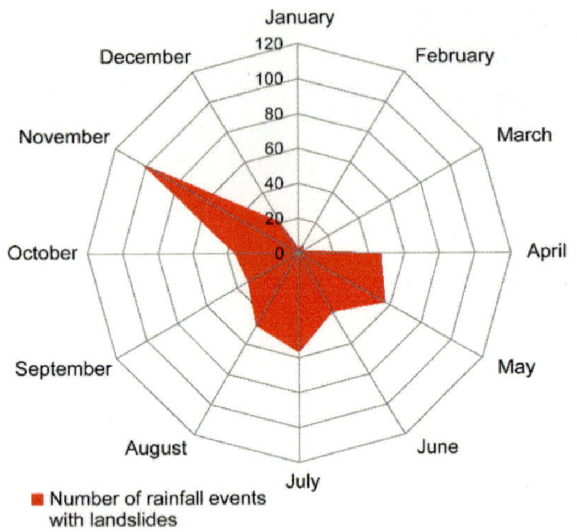

Fig. 3 Monthly distribution of rainfall events with landslides in the Italian Alps

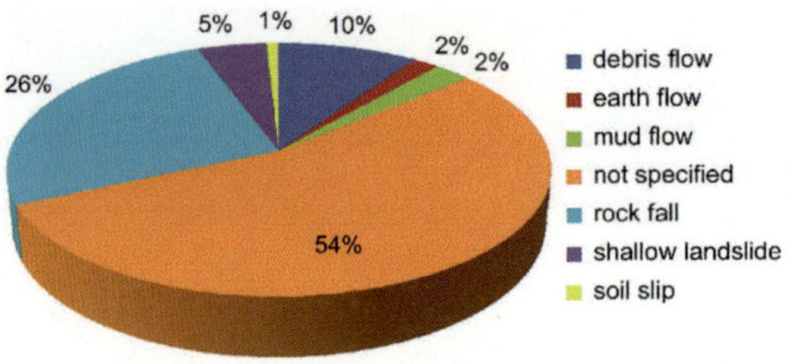

Fig. 4 Distribution of landslide types in the Alpine dataset. For the largest percentage (54%) the mass movement type is not reported by the source of information

(iii) landslide type (where available from the information source); and (iv) rainfall information (i.e., representative rain gauge, rainfall duration D, mean intensity I, cumulated event rainfall E). Hourly rainfall measurements were used to reconstruct the cumulated amount (E in millimetres) and the duration (D in hours) of the rainfall events presumably responsible for the documented slope failures.

Figure 3 shows the monthly distribution of the rainfall events that induced landslides in the Alpine data set.

About 24% of the rainfall events with landslides occurred in November, while less than 3% were registered between January and March, when the precipitation is mainly snow. Landslides due to snowmelt are not considered in this work.

In most cases the sources of information do not describe or specify the type of mass movement. For this reason, 54% of rainfall-induced landslides is classified as "not specified" in the catalogue (Fig. 4). For the remaining events rock falls (26%) and debris flows (10%) are by far the most frequent types of landslide in the Italian Alps.

3 Frequentist Method

Empirical rainfall thresholds and associated uncertainties for the Alpine area were defined using a frequentist probabilistic approach (Brunetti et al. 2010; Peruccacci et al. 2012). The method allows obtaining objective and reproducible thresholds of the form:

$$E = (\alpha \pm \Delta\alpha) \cdot D^{(\gamma \pm \Delta\gamma)}$$

where α is a scaling parameter, γ is the slope of the power law threshold curve, and $\Delta\alpha$ and $\Delta\gamma$ are the resultant uncertainties. Within this approach it is possible to define thresholds corresponding to different exceedance probabilities (e.g., 1, 5, 10%, etc.). This allows to estimate the percentage of rainfall events that are expected to trigger landslides under a given threshold in a sample of rainfall events with landslides.

4 Rainfall Thresholds for the Italian Alps

We analysed the Alpine data set shown in the map of Fig. 1 to define reliable rainfall thresholds for the regional area. The distribution of the cumulated event rainfall E (in mm) as a function of the duration D (in h) for rainfall events that induced slope failures in the alpine area is shown

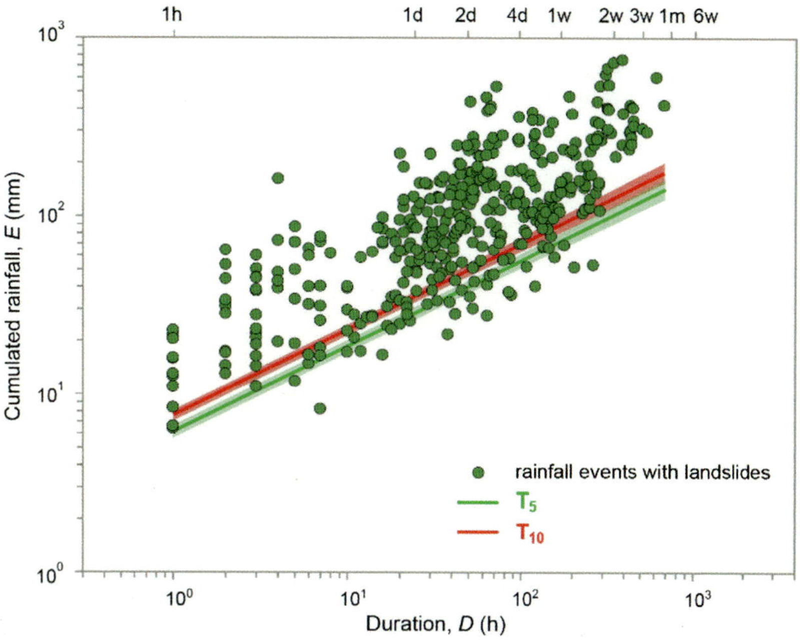

Fig. 5 Distribution of rainfall events with landslides in the Italian Alps (*green dots*) and the new rainfall thresholds T_5 (*green line*) and T_{10} (*red line*) for the Italian Alps. The *green* and *red shaded areas* represent the uncertainties associated to the thresholds parameters of T_5 and T_{10}, respectively

in Fig. 5. In the same Figure are drawn the new *ED* rainfall thresholds at 5% (green line) and 10% (red line) of exceedance probability, T_5 and T_{10}, respectively.

The equations of the new rainfall thresholds are:

$$T_5{:}E = (6.2 \pm 0.5) \cdot D^{(0.48 \pm 0.02)} \quad (1)$$

$$T_{10}{:}E = (7.5 \pm 0.5) \cdot D^{(0.48 \pm 0.02)} \quad (2)$$

In Fig. 6 the T_{10} threshold (red line) is compared with similar thresholds (Marchi et al. 2002; Aleotti 2004) defined in the same general area.

In particular, Aleotti (2004) defined a threshold at the same 10% exceedance probability for the Piedmont region. Marchi et al. (2002) calculated a threshold for debris flows in the Moscardo Torrent basin (Eastern Italian Alps) analysing 15 storms. For duration $D < 30$ h, the T_{10} threshold is lower than the other two.

5 Seasonal Thresholds

We investigated the seasonal distribution of rainfall events that triggered landslides in the Italian Alps. For this purpose, we selected two periods from June to September (J–S) and from October to May (O–M), and assigned each rainfall event with landslides to a period.

The resulting distribution is shown in Fig. 7. The inset in the figure displays the percentage of rainfall events in the two periods as a function of the rainfall duration *D*. It turns out that the warm season (J–S) is characterized by rainfall events with a mean duration shorter than that of rainfall events in the cold season (O–M). This is probably related to the climate, which is characterized by distinct meteorological events in the warm and cold seasons.

In the (warmer) summer season, rainfall is frequently associated with convective rainfall events characterized by short duration and high

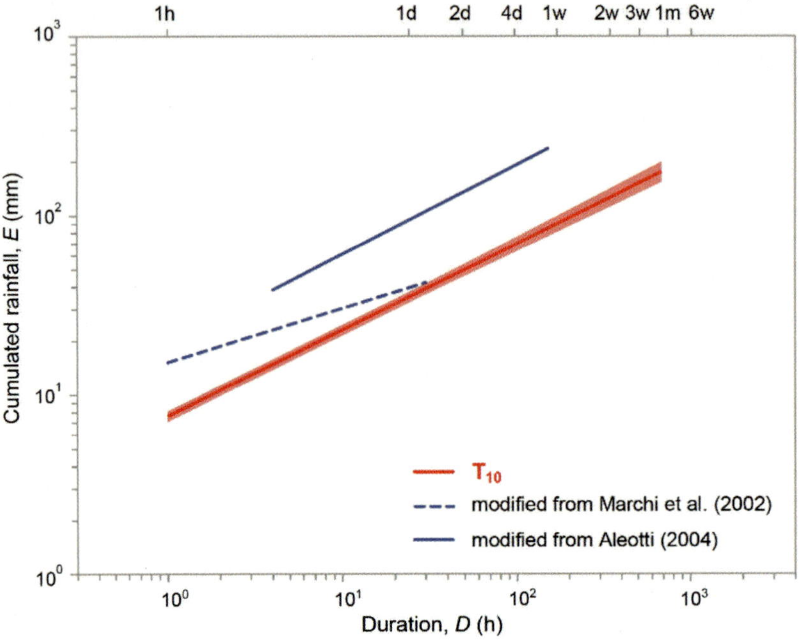

Fig. 6 Comparison between the rainfall threshold for the Italian Alps (*red line*) and thesholds obtained by other authors in the same area. The *red shaded area* represents the uncertainty associated to the thresholds parameters

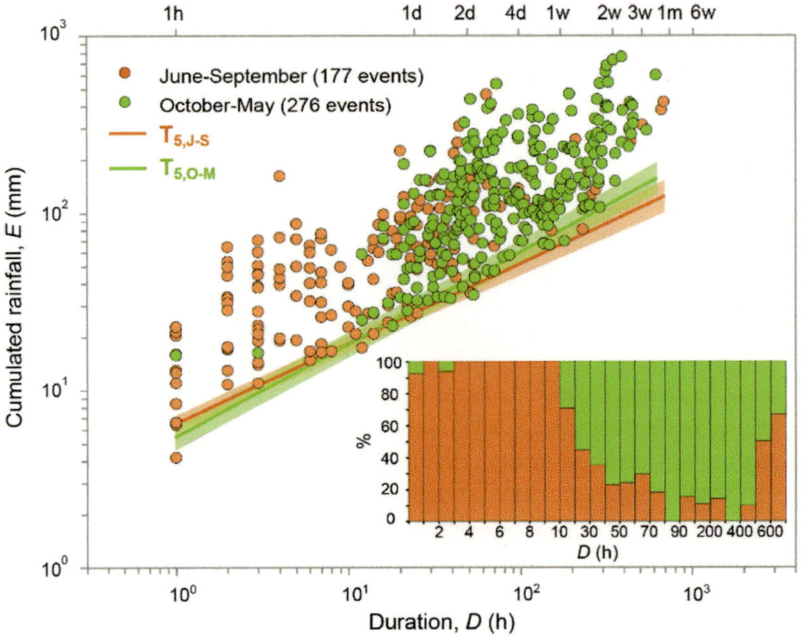

Fig. 7 Distribution of rainfall events with landslides in the two periods June–September (*orange dots*) and October–May (*light green dots*) and the new seasonal rainfall thresholds $T_{5,J\text{–}S}$ (*orange line*) and $T_{5,O\text{–}M}$ (*light green line*). The *orange* and *light green shaded area* represent the uncertainty associated to the thresholds parameters. The inset shows the percentage of rainfall events in the two periods as a function of the event duration *D*

intensity rates. In the two (colder) autumn and spring seasons, abundant rainfall is associated with regional frontal systems, characterized by long duration and low mean intensity. Winter is a relatively dry season, with a proportion of the precipitation falling as snow.

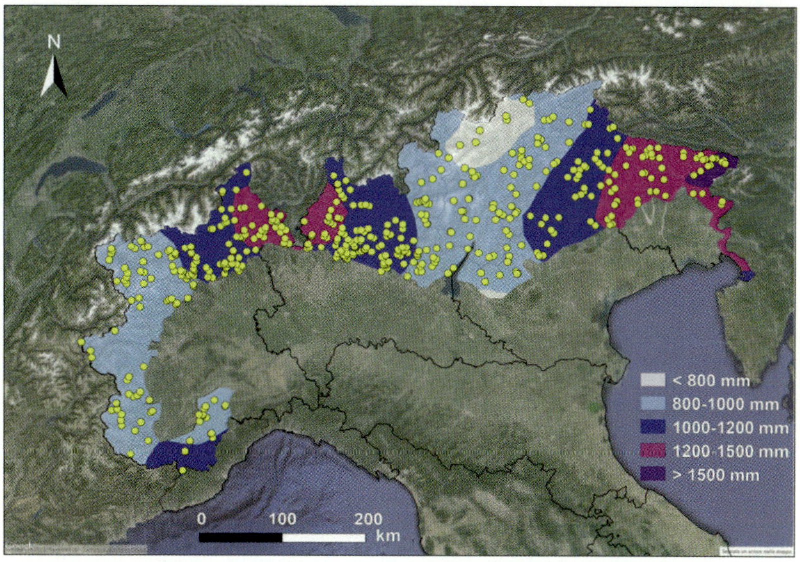

Fig. 8 Subdivision of the alpine landslides (*yellow dots*) among the separated classes identified by the MAP chart of the Italian Alps obtained using the mean annual rainfall data from Centro Nazionale di Cartografia Pedologica (www.soilmaps.it)

In Fig. 7, the rainfall thresholds for the J–S and O–M seasons at 5% of exceedance probability are $T_{5,J\text{–}S}$ (orange line) and $T_{5,O\text{–}M}$ (light green line), and their associated uncertainties (coloured shaded areas).

The corresponding equations are:

$$T_{5,J-S}{:}E = (6.5 \pm 0.6) \cdot D^{(0.45 \pm 0.03)} \quad (3)$$

$$T_{5,O-M}{:}E = (5.5 \pm 0.8) \cdot D^{(0.52 \pm 0.03)} \quad (4)$$

Despite the different slope of the two curves, the thresholds are very similar at low durations. For $D = 120$ h, the cumulated rainfall of $T_{5,J\text{–}S}$ is about 52 mm, and that of $T_{5,O\text{–}M}$ is about 66 mm.

6 Climatic Thresholds

Long-term climatic features, and in particular rainfall regimes, are known to influence rainfall conditions responsible for landslides in various environmental settings (Wieczorek and Guzzetti 2000; Guzzetti et al. 2007). The mean annual precipitation (MAP) is used here to identify climatically homogeneous areas (Govi and Sorzana 1980; Aleotti 2004). The MAP chart (Fig. 8) is obtained using the mean annual rainfall data from Centro Nazionale di Cartografia Pedologica (www.soilmaps.it).

The map is classified by ranking the MAP values into five interval classes: (i) Very Low (VL, MAP < 800 mm), (ii) Low (L, MAP = 800 1000 mm), (iii) Medium (M, MAP = 1000 1200 mm), (iv) Medium-High (MH, MAP = 1200 ÷ 1500 mm), and (v) High (H, MAP > 1500 mm).

The figure puts into evidence a noticeable variability of the MAP within the alpine region, with the minimum values reached locally in the north-central part of the alpine chain (<800 mm per year) and the maxima characterizing the eastern part (>1500 mm per year). Figure 8 shows also the location of rainfall-induced landslides within the study area (yellow dots).

In order to associate a landslide to a MAP class, a circular buffer was associated to each failure according to the mapping accuracy level of each landslide, (Peruccacci et al. 2012). When the intersection between the landslide circular buffer and a given MAP class is at least 75% of

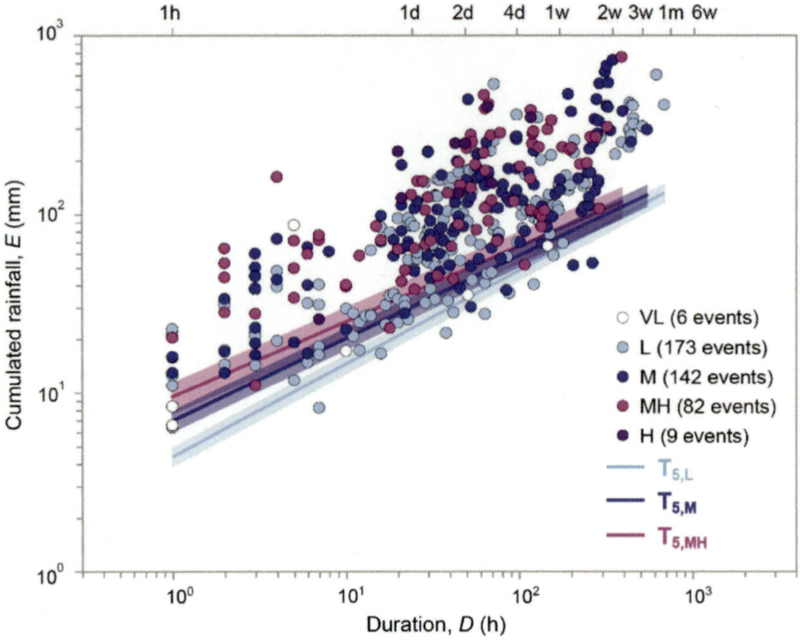

Fig. 9 Distribution of rainfall events with landslides in five MAP classes: very low (VL *white dots*), low (L *light blue dots*), medium (M *blue dots*), medium-high (MH *fucsia dots*), and high (H *purple dots*), and the new climatic rainfall thresholds $T_{5,L}$ (*light blue line*), $T_{5,M}$ (*blue line*) and $T_{5,MH}$ (*fucsia line*). The *coloured shaded areas* represent the uncertainty associated to the thresholds parameters

the buffer area, this MAP class is assigned to the landslide. For lower percentages, the failure is not associated. Figure 9 shows the distribution of (*D*, *E*) pairs corresponding to rainfall events that triggered landslides in the five MAP classes (VL, L, M, MH and H).

We calculated the *ED* thresholds for three MAP classes (L, M, and MH), since the number of empirical data points falling within the VL, and H classes is not sufficient to calculate rainfall thresholds. The 5% *ED* thresholds for the three classes, and their associated uncertainties (coloured shaded areas). are plotted in Fig. 9.

The equations of the new climatic thresholds are:

$$T_{5,L}{:}E = (4.4 \pm 0.5) \cdot D^{(0.52 \pm 0.02)} \qquad (5)$$

$$T_{5,M}{:}E = (7.0 \pm 0.9) \cdot D^{(0.46 \pm 0.03)} \qquad (6)$$

$$T_{5,MH}{:}E = (9.5 \pm 1.8) \cdot D^{(0.42 \pm 0.04)} \qquad (7)$$

7 Discussion and Conclusions

We defined new *ED* rainfall thresholds for the Italian Alps using a catalogue of 453 rainfall events that triggered 509 landslides. The study area is located in the northern Italy (Fig. 1). We defined *ED* thresholds using the entire Alpine data set (Eqs. 1 and 2).

The new Alpine thresholds are generally lower than similar curves found by previous authors in two Alpine sub areas (Fig. 6). The result is that landslides affecting the population and infrastructures in the study area are triggered by less severe rainfall conditions than previously recognized in the same general area. The reason can be related to the source of information used to compile the Alpine data set. The catalogue of rainfall events with landslides used in this work was built primarily searching national and local newspapers. These sources mainly include information on low to moderate rainfall events that have resulted in single landslides, or a few landslides. In most cases, the failures have caused minor damage to the transportation network and to the built-up areas. As a result, the relative proportion of large rainfall events with landslides is probably lower than that found in previous studies.

The seasonal classification of the rainfall events in the Alpine data set (Fig. 7) shows that in the cold O–M season landslides are preferably initiated by rainfall events with duration $D > 10$ h, while in the warm J–S period landslides occur also at shorter rainfall durations. Moreover the two distributions have a similar scattering in the (D, E) plane but slightly different slopes, as confirmed by the rainfall thresholds (Eqs. 3 and 4). However, the two curves are statistically indistinguishable due to the uncertainty, $\Delta\alpha$ and $\Delta\gamma$, in the threshold parameters. The uncertainty around the threshold curves can be reduced by increasing the number of events in the data set (Peruccacci et al. 2012).

The climatic classification of the rainfall events with landslides in the Italian Alps is based on the categorization of MAP values into 5 classes. Figure 9 reveals that as the MAP increases the corresponding threshold increases too. This result is in agreement with previous findings (Giannecchini 2006; Chen et al. 2015) confirming that rainfall thresholds are higher where the average rainfall conditions are more severe. This outcome could be an effect of the physical resilience of a territory, and of the ability to cope with local environmental settings.

We expect that the rainfall thresholds determined for the Italian Alps will help to forecast the occurrence of landslides, and to prevent casualties and economic damage in this mountainous area.

Acknowledgements Research conducted with funding provided by the Italian National Department for Civil Protection (DPC). MRP and AV were supported by DPC grants. The DPC provided access to the national rainfall database used in the study. Additional rainfall information was made available by the Regional Agencies for the Environment Protection.

References

Aleotti P (2004) A warning system for rainfall-induced shallow failures. Eng Geol 73:247–265

Brunetti MT, Peruccacci S, Rossi M, Luciani S, Valigi D, Guzzetti F (2010) Rainfall thresholds for the possible occurrence of landslides in Italy. Nat Hazards Earth Syst Sci 10:447–458

Chen CW, Saito H, Oguchi T (2015) Rainfall intensity–duration conditions for mass movements in Taiwan. Progr Earth Planet Sci 2:14. doi:10.1186/s40645-015-0049-2

Giannecchini R (2006) Relationship between rainfall and shallow landslides in the southern Apuan Alps (Italy). Nat Hazards Earth Syst Sci 6:357–364. doi:10.5194/nhess-6-357-2006

Govi M, Sorzana PF (1980) Landslide susceptibility as a function of critical rainfall amount in Piedmont basins (North–Western Italy). Stud Geomorphol Carpatho Balcanica 14:43–61

Guzzetti F, Peruccacci S, Rossi M, Stark CP (2007) Rainfall thresholds for the initiation of landslides in central and southern Europe. Meteorol Atmos Phys 98:239–267

Marchi L, Arattano M, Deganutti AM (2002) Ten years of debris-flow monitoring in the Moscardo Torrent (Italian Alps). Geomorphology 46:1–17

Peruccacci S, Brunetti MT, Luciani S, Vennari C, Guzzetti F (2012) Lithological and seasonal control of rainfall thresholds for the possible initiation of landslides in central Italy. Geomorphology 139–140:79–90

Wieczorek GW, Guzzetti F (2000) A review of rainfall thresholds for triggering landslides. In: Claps P, Siccardi F (eds) Mediterranean storms, proceedings plinius conference '99, Maratea, GNDCI pub. n. 2012. Editoriale Bios, Cosenza, p 407–414, 14–16 Oct 1999

TXT-tool 2.039-1.3 Topographic and Pedological Rainfall Thresholds for the Prediction of Shallow Landslides in Central Italy

Silvia Peruccacci, Maria Teresa Brunetti and Fausto Guzzetti

Abstract

In Italy, rainfall-induced shallow landslides are frequent phenomena that cause casualties, damages and economic losses every year. At the national and regional scales, empirical rainfall thresholds can predict the occurrence of multiple rainfall-induced shallow landslides. In this work, we updated an historical catalogue listing 553 rainfall events that triggered 723 landslides in the Abruzzo, Marche and Umbria regions, central Italy, between February 2002 and March 2011. For each event, the rainfall duration (*D*) and the cumulated event rainfall (*E*) responsible for the failure are known, together with the exact or approximate location of the landslide. Landslides were mapped as single points and were attributed a level of mapping accuracy *P*, in 3 classes. To analyse the influence of topography and soil characteristics on the occurrence of rainfall-induced shallow landslides, we subdivided the study area in three topographic divisions, and eight soil domains. We analysed the (*D*, *E*) rainfall conditions that resulted in the documented shallow landslides in each topographic division and regional soil domain, and we computed *ED* rainfall thresholds at 5% exceedance probability level for the two subdivisions. We expect that the new topographic and pedological thresholds will contribute to forecast shallow landslides in central Italy, and in areas characterized by similar morphological and soil settings.

Keywords

Shallow landslides · Central Italy · Rainfall thresholds

Contents

1 **Introduction** ... 372
2 **Catalogue of Rainfall Events with Landslides** 372
3 **Definition of Rainfall Thresholds** ... 374
4 **Rainfall Thresholds for Central Italy** ... 375

S. Peruccacci (✉) · M.T. Brunetti · F. Guzzetti
Istituto di Ricerca per la Protezione Idrogeologica, Consiglio Nazionale delle Ricerche, via Madonna Alta 126, 06128 Perugia, Italy
e-mail: silvia.peruccacci@irpi.cnr.it

M.T. Brunetti
e-mail: mariateresa.brunetti@irpi.cnr.it

F. Guzzetti
e-mail: fausto.guzzetti@irpi.cnr.it

K. Sassa et al. (eds.), *Landslide Dynamics: ISDR-ICL Landslide Interactive Teaching Tools*, https://doi.org/10.1007/978-3-319-57774-6_27

4.1 Topographic Thresholds 375
4.2 Pedological Thresholds.................................... 376
5 Discussion and Conclusion 378
References.. 380

1 Introduction

In Italy, shallow landslides triggered by intense and/or prolonged rainfall cause casualties, damages and economic losses every year, and represent a serious hazard to the population. At the national and regional scales, the prediction of rainfall-induced shallow failures can be achieved by defining empirical rainfall thresholds. Rainfall thresholds are relationship linking rainfall events to the occurrence of landslides. In the literature, empirical thresholds are established through the analysis of past rainfall events that have resulted in documented landslides (for an extensive review see e.g. Guzzetti et al. 2007).

In this work, we updated an historical catalogue of 553 rainfall events that have resulted in 723 landslides in the Abruzzo, Marche and Umbria regions, central Italy, between 2002 and 2011. We exploited this catalogue to determine new cumulated event rainfall—rainfall duration (*ED*) thresholds for the possible occurrence of slope failures in central Italy. To define objective and reproducible thresholds, we adopted the frequentist statistical method proposed by Brunetti et al. (2010) and refined by Peruccacci et al. (2012) to determine regional, lithological and seasonal thresholds in the same study area. It is known that the local topographic (morphological), lithological, soil, and climatic settings influence the occurrence and the spatial distribution of shallow landslides in an area. In this work, we investigated the role of topography and soil types in the cumulated amount (*E*) and the duration (*D*) of the rainfall responsible for the known shallow landslides in our study area. First, we subdivided the study area in 3 physiographic provinces, and we defined *ED* rainfall thresholds for the topographic terrain classification. Next, using a small-scale pedological map, we obtain information on the main regional soil domains where landslides were reported, and we calculated *ED* thresholds for three soil regions. We conclude discussing the results obtained.

2 Catalogue of Rainfall Events with Landslides

In a previous work (Peruccacci et al. 2012), we compiled a catalogue of 442 rainfall events that triggered 573 known shallow landslides in the Abruzzo, Marche, and Umbria regions, central Italy, between February 2002 and August 2010. In the Umbria region, the landslide information was obtained searching national, regional, and local newspapers, and reports of landslide events compiled by fire brigades. In the Marche and the Abruzzo regions, newspapers were the only source of information.

A large amount of information on rainfall-induced landslides is required to define reliable rainfall thresholds. In order to collect additional information, we began a collaboration with: (i) the Marche Regional Department for Civil Protection, and (ii) provincial fire brigades of the Marche region. As a result, we found information on 111 rainfall events that resulted in 150 documented slope failures in the Marche region. Merging the data, we obtained a new catalogue listing 553 rainfall events in central Italy that have caused 723 landslides in the period between February 2002 and March 2011 (Fig. 1).

Figure 2a summarizes the total number of rainfall events (purple bars) and of landslides (yellow bars) in the Abruzzo, Marche and Umbria regions. Figure 2b portrays the monthly distributions of rainfall events (in purple), and of landslides (in yellow): rainfall induced failures in central Italy occur mostly in December, but are also frequent in March. Depending on the quality of the available information, we mapped the observed landslides as single points using Google Earth®, assigning to each failure a level

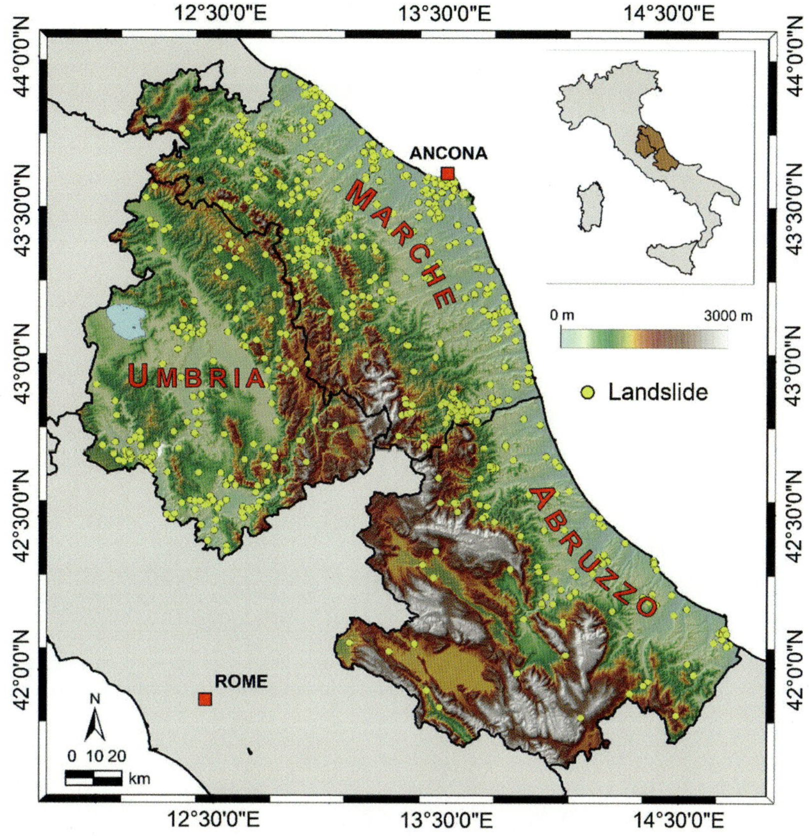

Fig. 1 Map of the study area, including Abruzzo, Marche and Umbria regions (central Italy). *Yellow dots* show location of 723 rainfall induced landslides

of mapping accuracy P, in three classes: high, $P_1 < 1$ km^2; medium, $1 \leq P_{10} < 10$ km^2; and low, $10 \leq P_{100} < 100$ km^2 (Peruccacci et al. 2012).

For each rainfall event the information includes: the landslide type, the timing of the slope failure, and the geographical accuracy for the landslide location. Landslide types were classified as rock fall (31, 4.3%), debris flow (5, 0.7%), earth flow–mud flow (34, 4.7%), and shallow landslide (653, 89.9%, Fig. 3a). Most of the slope failures are classified as generic shallow landslides, due to a lack of a detailed description of the failure type in the source of information. Landslide information in the database comprises the exact time, a part of the day, or the date of occurrence of the landslides. The time of the failure is known for a large portion (231, 41.8%) of the rainfall-induced landslides in the catalogue (Fig. 3b). The majority of the landslides have a mapping accuracy level P_1 (350, 48.4%) followed by the P_{10} level (281, 38.9%) and the P_{100} level (92, 12.7%) (Fig. 3c). Bar charts in Fig. 3d show abundance of information for the Abruzzo, Marche and Umbria regions.

For each rainfall event, the duration *D* (in hours) and the total event rainfall *E* (in mm) were determined using hourly rainfall measurements from a network of 150 rain gauges. Generally, the rain gauge used to reconstruct the rainfall responsible for the failure (i.e. *D* and *E*) was selected considering the geographic distance to the landslide, and its elevation with respect to the elevation of the landslide. In the updated catalogue, rainfall events that caused landslides in central Italy are in the range of duration $1 < D < 1212$ h, and in the range of cumulated rainfall $7.0 < E < 457.2$ mm.

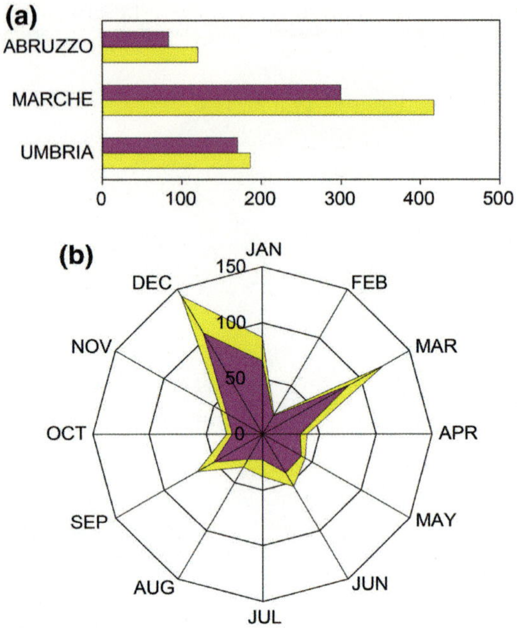

Fig. 2 **a** Number of rainfall events (*purple bars*) and landslides (*yellow bars*) in the Abruzzo, Marche and Umbria regions. **b** Number of rainfall events (*in purple*) and landslides (*in yellow*) subdivided by month

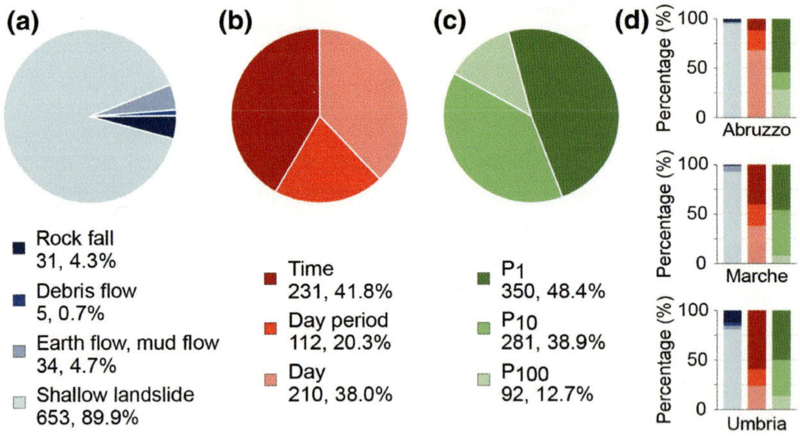

Fig. 3 Statistics obtained from the catalogue of rainfall events with landslides in central Italy. **a** Number and percentage of landslides of different types. **b** Number and percentage of landslides with different level of temporal accuracy. **c** Number and percentage of landslides with different level of mapping accuracy. **d** *Bar charts* show abundance of information for each region

3 Definition of Rainfall Thresholds

Guzzetti et al. (2007, 2008) highlighted the need to define objective and reproducible empirical rainfall thresholds for the possible initiation of landslides. In this work, we used a frequentist statistical method (Brunetti et al. 2010; Peruccacci et al. 2012) to calculate new *ED* thresholds in central Italy. The method avoids subjective criteria in the determination of the thresholds, that are common in many of the published rainfall thresholds for the possible initiation of landslides (Guzzetti et al. 2007, 2008). In particular, the method assumes the threshold curve is a power law:

$$E = (\alpha \pm \Delta\alpha) \times D^{(\gamma \pm \Delta\gamma)} \quad (1)$$

where, E is the cumulated (total) event rainfall (in mm), D is the duration of the rainfall event (in h), "α" is a scaling parameter (the intercept), "γ" is the slope of the power law threshold curve, and "$\Delta\alpha$" and "$\Delta\gamma$" are the uncertainties associated to α and γ, respectively.

Specifically, the ensemble of the rainfall *ED* conditions responsible for the slope failures is fitted with a power law. For each event, the difference between the cumulated event rainfall and the fit is calculated. The probability density of the distribution of the differences is determined through a Kernel Density approach, and the result modelled with a Gaussian function. Using the modelled distribution, thresholds corresponding to different exceedance probability levels are calculated. As an example, assuming the set of the empirical (D, E) data points is sufficiently complete and representative of the conditions that led to slope failures in a study area, the 5% threshold is the curve that should leave 5% of the rainfall events with landslides below the threshold.

A consolidated bootstrap statistical technique is used to determine the uncertainty associated with the parameters that define the power law threshold curve.

4 Rainfall Thresholds for Central Italy

It is widely believed that landslide occurrence is affected by the local topographic (morphological), lithological, soil, and climatic settings. For this work, we exploited the new catalogue of rainfall events with landslides in central Italy to investigate the role of topography and soil types on rainfall thresholds for the possible initiation of landslides. We defined *ED* thresholds for 5% exceedance probability level (i) for the three individual physiographic provinces in the study area, and (ii) for three main soil regions in the study area.

4.1 Topographic Thresholds

To analyse the influence of topography on the rainfall conditions responsible for the slope failures, we used the topographic divisions of Italy proposed by Guzzetti and Reichenbach (1994). They divided the entire Italian territory into eight major physiographic provinces reflecting physical, geological, and structural changes of the Italian landscape.

According to this topographic division, in the study area we identified three physiographic provinces, namely: (i) Apennines Mountain System, AM; (ii) Tyrrhenian Lowland, TL, and (iii) Adriatic Lowland, AL (Fig. 4).

Using the procedure adopted by Peruccacci et al. (2012), we attributed each landslide to a specific physiographic province. In particular, each landslide, mapped as a single point, was represented by a circle, with the area of the circle, A, dependent on the mapping accuracy (P_1, A = 0.5 km^2; P_{10}, A = 5 km^2; P_{100}, A = 50 km^2). The landslide circles were then intersected with the topographic division, and a physiographic province was attributed to each landslide if it covered 75% of more of the circle. The histogram in Fig. 4 shows the number of failures in each physiographic province. AL has the largest number of landslides (351, 48.5%); in AM and TL landslides were 221 (30.6%) and 151 (20.9%), respectively.

Figure 5 shows, in logarithmic coordinates, the distribution of the rainfall conditions (D, E) that caused landslides in AM, Apennines Mountain System (175 light green dots), TL, Tyrrhenian Lowland (136 red dots), and AL, Adriatic Lowland (253 dark green dots). Coloured lines are the 5% *ED* power law thresholds, $T_{5,AM}$, $T_{5,TL}$, $T_{5,AL}$ (Table 1). Figure 5 also shows the proportion of rainfall events for different rainfall durations.

Figure 6 shows $T_{5,AM}$, $T_{5,TL}$, and $T_{5,AL}$ in linear coordinates in the range $1 \leq D \leq 120$ h,

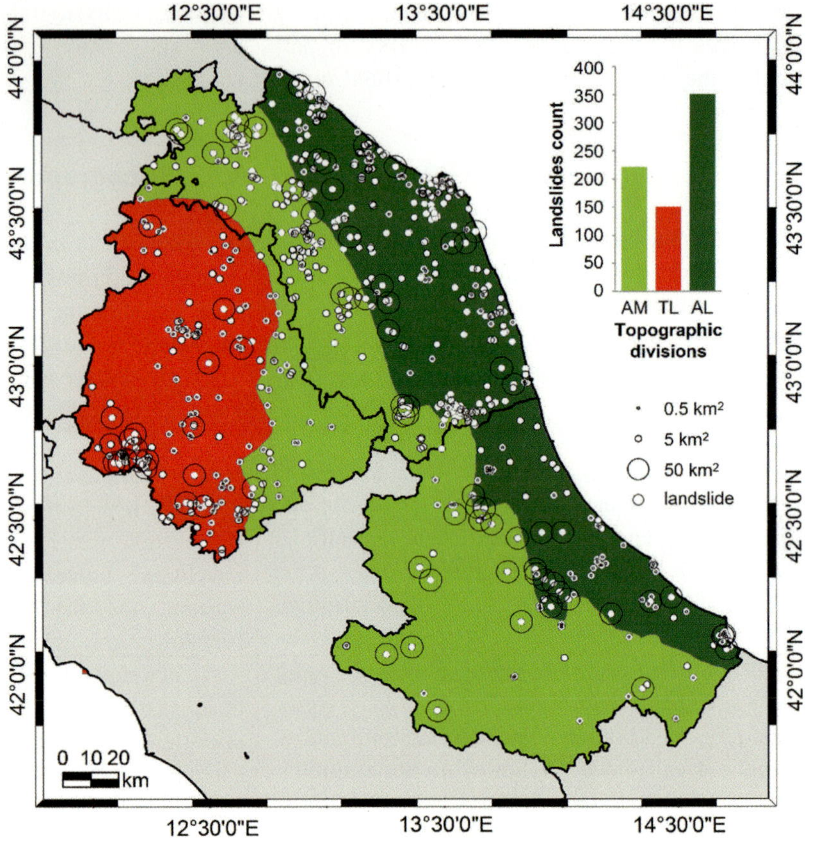

Fig. 4 Map of the study area, including Abruzzo, Marche and Umbria regions (central Italy). *White dots* show location of 723 rainfall induced landslides. *Black circles* show uncertainty in the landslide location, in three classes. *Histogram* shows the number of landslides in each physiographic province

a typical range of rainfall duration used to forecast rainfall-induced slope failure in Italy (Brunetti et al. 2009).The shaded areas around the threshold curves portray the uncertainty associated with the different topographic thresholds. The uncertainty depends on the number and distribution of the data points.

Inspection of Figs. 5 and 6 reveals that $T_{5,AM}$, $T_{5,TL}$, and $T_{5,AL}$ are substantially equivalent and statistically indistinguishable in our study area.

4.2 Pedological Thresholds

To investigate the relationship between soil type and rainfall thresholds, we obtained the pedological information from a synoptic map published by Costantini et al. (2004). They divided the Italian territory into 34 soil regions reflecting morphology, soil water, temperature regime, and more diffused soil types (according to FAO 1990). We partitioned the study area in eight soil regions from S_1, to S_8 (Table 2; Fig. 7), and we attributed each landslide to a specific soil region. The histogram in Fig. 7 portrays the number of failures in each soil region. S_3 has the largest number of landslides (270, 37.3%); in S_1 and S_2 landslides were 156 (21.6%) and 127 (17.6%), respectively.

We used this information to determine *ED* thresholds for three soil regions (S_1, S_2, and S_3). For the remaining soil regions, the number of events was insufficient to determine reliable thresholds.

Figure 8 shows, in log-log coordinates, the ensemble of the rainfall conditions that have resulted in slope failures in S_1 (128 dark blue dots), in S_2 (94 magenta dots), and in S_3 (198 grey dots) together with the 5% *ED* power law thresholds $T_{5,S1}$, $T_{5,S2}$, and $T_{5,S3}$ (Table 1). Figure 8 also shows the proportion of rainfall events for different rainfall durations.

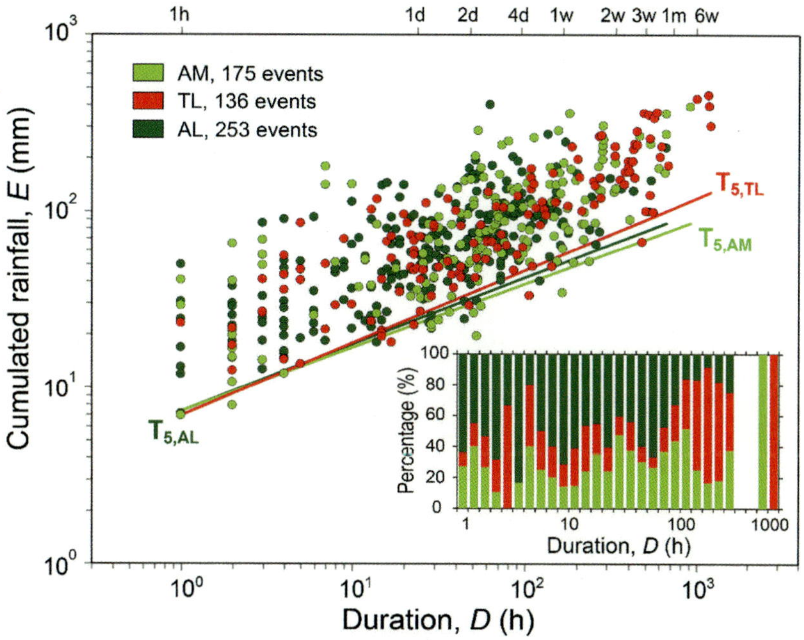

Fig. 5 Rainfall duration D (x-axis) and cumulated event rainfall E (y-axis) conditions that have produced landslides in Apennines mountain system (AM, *light green dots*), Tyrrhenian lowland (TL, *red dots*), and Adriatic lowland (AL, *dark green dots*), in the study area. *Coloured lines* are 5% power law thresholds ($T_{5,AM}$, $T_{5,TL}$, $T_{5,AL}$). Data shown in logarithmic coordinates. *Bar chart* shows proportion of rainfall events for different rainfall durations

Table 1 Rainfall *ED* thresholds for the possible initiation of landslides in central Italy

#	Label	Area	Events	Threshold	Range	Uncertainty
1	$T_{5,AM}$	Apennines mountain system	175	$E = 7.4 \times D^{0.36}$	$1 < D < 918$	$\Delta\alpha = 1.1$, $\Delta\gamma = 0.03$
2	$T_{5,TL}$	Tyrrhenian lowland	136	$E = 7.0 \times D^{0.41}$	$1 < D < 1212$	$\Delta\alpha = 0.8$, $\Delta\gamma = 0.02$
3	$T_{5,AL}$	Adriatic lowland	253	$E = 7.8 \times D^{0.37}$	$1 < D < 663$	$\Delta\alpha = 0.7$, $\Delta\gamma = 0.02$
4	$T_{5,S1}$	Cambisol–leptosol region with luvisols of the central Apennine	128	$E = 7.2 \times D^{0.36}$	$1 < D < 1176$	$\Delta\alpha = 1.2$, $\Delta\gamma = 0.04$
5	$T_{5,S2}$	Cambisol–regosol region, with luvisols of eastern italy	94	$E = 7.1 \times D^{0.42}$	$1 < D < 315$	$\Delta\alpha = 1.2$, $\Delta\gamma = 0.04$
6	$T_{5,S3}$	Cambisol–regosol region with vertisols of central and southern Italy	198	$E = 7.9 \times D^{0.37}$	$1 < D < 662$	$\Delta\alpha = 0.8$, $\Delta\gamma = 0.03$
7	$T_{5,PO}$	Post-orogenic sediments	211	$E = 7.4 \times D^{0.37}$	$1 < D < 1188$	$\Delta\alpha = 0.6$, $\Delta\gamma = 0.02$
8	$T_{5,FD}$	Flysch deposits	82	$E = 10.3 \times D^{0.37}$	$1 < D < 1212$	$\Delta\alpha = 1.5$, $\Delta\gamma = 0.03$
9	$T_{5,CC}$	Carbonate complex	88	$E = 9.0 \times D^{0.3}$	$1 < D < 663$	$\Delta\alpha = 1.5$, $\Delta\gamma = 0.04$

Label: label of the thresholds. Area: area for which the threshold was defined. *Equation D*, rainfall duration, in hour; *E*, cumulated event rainfall, in mm. Range: range of validity for the threshold. Uncertainty: uncertainty associated with the intercept α and the slope γ of the threshold curve. Source: # 1–6, this work; # 7–9, Peruccacci et al. (2012)

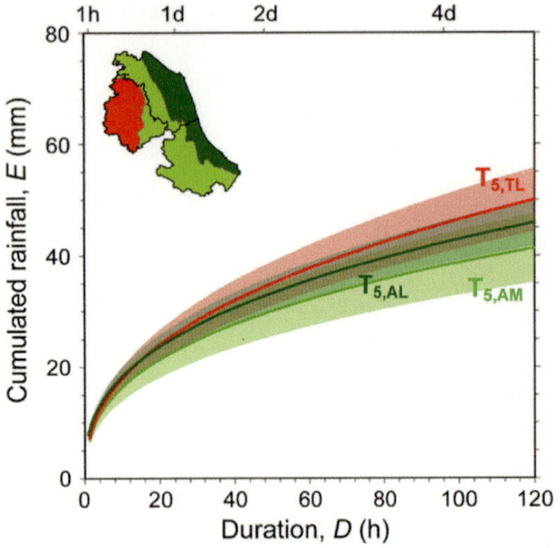

Fig. 6 Cumulated event rainfall–rainfall duration (*ED*) thresholds for *AM* Apennines mountain system; *TL* Tyrrhenian lowland; *AL* Adriatic lowland. *Solid coloured lines* are 5% power law thresholds ($T_{5,AM}$, $T_{5,TL}$, $T_{5,AL}$) in linear coordinates. *Shaded areas* show uncertainty associated with the threshold curves

Figure 9 portrays $T_{5,S1}$, $T_{5,S2}$, and $T_{5,S3}$ in linear coordinates in the range $1 \leq D \leq 120$ h, with the shaded area showing the uncertainty associated with the thresholds.

Inspection of Figs. 8 and 9 reveals that $T_{5,S1}$, $T_{5,S2}$, and $T_{5,S3}$ are similar. Further inspection of Figs. 8 and 9 indicates that $T_{5,S2}$ is slightly steeper and higher than $T_{5,S1}$ (i.e., their uncertainties do not overlap completely). This could mean that, increasing rainfall duration, a smaller amount of rainfall is necessary to trigger landslides in S_1, rather than in S_2. The trend for each soil region has to be confirmed by collecting a larger number of events.

5 Discussion and Conclusion

We updated an existing catalogue of rainfall events that resulted in landslides in central Italy. The final catalogue listed information on 553 rainfall events that produced 723 slope failures in the period between February 2002 and March 2011.

Peruccacci et al. (2012) investigated the relationship between lithology and rainfall thresholds in central Italy. They defined *ED* thresholds and their associated uncertainty for the 3 main lithological domains: post-orogenic sediments PO, flysch deposits FD, and carbonate rocks CC, $T_{5,PO}$, $T_{5,FD}$ and $T_{5,CC}$, respectively (Table 1). In the study area, they found that: (i) threshold for flysch terrain was definitively higher than those for post-orogenic and carbonate sediments, and (ii) thresholds for post-orogenic sediments and for carbonate rocks were statistically indistinguishable.

To investigate the influence of other environmental factors on the rainfall conditions responsible for shallow failures in the same study area, we segmented the catalogue of rainfall event with landslides on topography and soil regions to determined objective cumulated event rainfall—rainfall duration (*ED*) thresholds in central Italy. The analysis of the segmented catalogue reveals

Table 2 Soil regions identified in the study area using the synoptic map published by Costantini et al. (2004)

Code	Soil region
S_1	16.4: Cambisol–leptosol region with luvisols of the central Apennine
S_2	61.1: Cambisol–regosol region, with luvisols of eastern Italy
S_3	61.3: Cambisol–regosol region with vertisols of central and southern Italy
S_4	56.1: Cambisol–andosol region with regosols of central Italy
S_5	59.7: Cambisol–leptosol region with luvisols of the southern Apennine
S_6	64.4: Cambisol–fluvisol region with luvisols and vertisols of Arno and Tevere rivers and coastal plains in central Italy
S_7	78.1: Regosol–cambisol region with calcisols of the northern Apennine
S_8	78.2: Regosol–cambisol region with calcisols of the middle Apennine

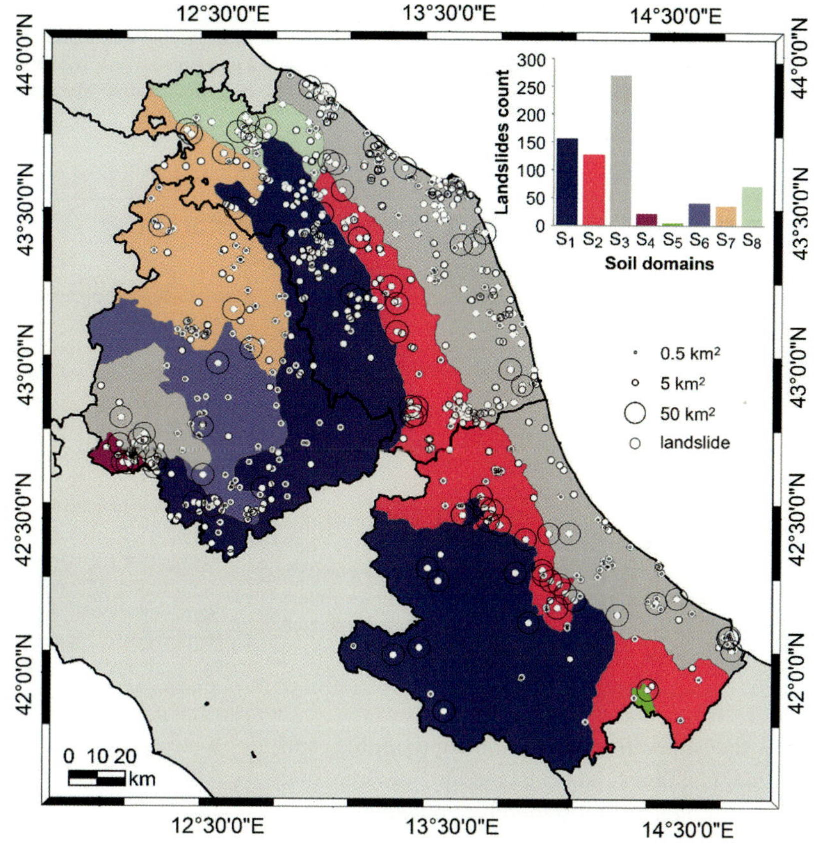

Fig. 7 Simplified pedological division of the study area (from S_1, to S_8). White dots show location of rainfall-induced landslides. *Black circles* show uncertainty in the landslide location, in three classes. *Histogram* shows number of landslides in each soil region

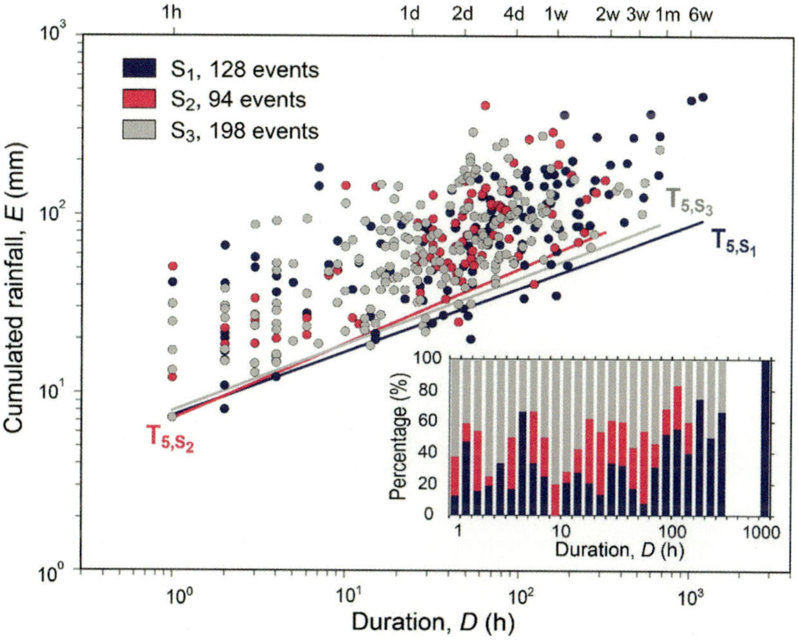

Fig. 8 Rainfall duration D (x-axis) and cumulated event rainfall E (y-axis) conditions that have triggered landslides in S_1 (*dark blue dots*), S_2 (*magenta dots*), and S_3 (*grey dots*), in the study area. *Coloured lines* are 5% power law thresholds ($T_{5,S1}$, $T_{5,S2}$, $T_{5,S3}$). Data shown in logarithmic coordinates. *Bar chart* shows proportion of rainfall events for different rainfall durations

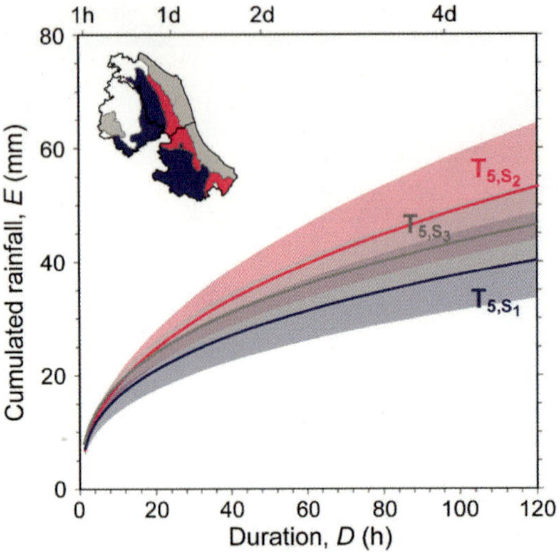

Fig. 9 Cumulated event rainfall–rainfall duration (*ED*) thresholds for S_1, S_2, and S_3. *Solid coloured lines* are 5% power law thresholds ($T_{5,S1}$, $T_{5,S2}$, $T_{5,S3}$) in linear coordinates. *Shaded areas* show uncertainty associated with the threshold curves

that the differences between topographic and pedological thresholds exist but are not large, and the thresholds are substantially equivalent.

Although the size of the data set is significant (553 rainfall events with landslides), we acknowledge that it may be insufficient to identify (subtle) differences in the rainfall conditions required to initiate slope failures due to the different topographic and pedological settings.

Acknowledgements Research conducted with funding provided by the Italian National Department of Civil Protection (DPC). DPC provided access to the national rainfall database used in the study. Additional rainfall information was made available by the Marche Regional Department for Civil Protection. We thank provincial fire brigades of the Marche region for providing access to their database on rainfall-induced landslides.

References

Brunetti MT, Peruccacci S, Rossi M, Guzzetti F, Reichenbach P, Ardizzone F, Cardinali M, Mondini A, Salvati P, Tonelli G, Valigi D, Luciani S (2009) A prototype system to forecast rainfall induced landslides in Italy. In: Proceedings of the 1st Italian workshop on landslides, rainfall-induced landslides: mechanisms, monitoring techniques and now-casting models for early warning systems, vol 1. Doppiavoce Edizioni, Naples, pp 157–161

Brunetti MT, Peruccacci S, Rossi M, Luciani S, Valigi D, Guzzetti F (2010) Rainfall thresholds for the possible occurrence of landslides in Italy. Nat Hazards Earth Syst Sci 10:447–458

Costantini EAC, Urbano F, L'Abate G (2004) Soil regions of Italy. http://www.soilmaps.it/download/csi-BrochureSR_a4.pdf. Last accessed 7 Aug 2013

Guzzetti F, Reichenbach R (1994) Towards a definition of topographic divisions for Italy. Geomorphology 11:57–74

Guzzetti F, Peruccacci S, Rossi M, Stark CP (2007) Rainfall thresholds for the initiation of landslides in central and southern Europe. Meteorol Atmos Phys 98:239–267

Guzzetti F, Peruccacci S, Rossi M, Stark CP (2008) The rainfall intensity-duration control of shallow landslides and debris flows: an update. Landslides 5(1):3–17

Peruccacci S, Brunetti MT, Luciani S, Vennari C, Guzzetti F (2012) Lithological and seasonal control on rainfall thresholds for the possible initiation of landslides in central Italy. Geomorphology 139–140:79–90. doi:10.1016/j.geomorph.2011.10.005

TXT-tool 2.039-4.1 FLaIR Model (Forecasting of Landslides Induced by Rainfalls)

Pasquale Versace, Giovanna Capparelli and Davide Luciano De Luca

Abstract

Mathematical models for landslide forecasting constitute an important component for Early Warning Systems. This teaching tool focuses on the empirical model named FLaIR (Forecasting of Landslides Induced by Rainfalls), developed at Laboratory of Environmental Carthography and Hydraulic and Geological Modelling (CAMILab) of University of Calabria (Italy). FLaIR is a general framework for many empirical models proposed in technical literature: in particular, it reproduces as particular cases all the ID (Intensity-Duration) schemes (Capparelli and Versace 2011). From the website www.camilab.unical.it it is possible to download the software FLaIR.exe, together with a user guide.

Keywords

Landslide forecasting · Early warning systems · Empirical models

Contents

1 **Introduction** ... 381
2 **Mathematical Background of FLaIR Model** .. 382
3 **Parameter Estimation** ... 384
3.1 Ranking Method ... 384
4 **Real-Time Forecasting** ... 386
4.1 Example of Application ... 387
References ... 388

1 Introduction

Prediction of landslides induced by rainfall constitutes an important topic for the development of Early Warning Systems (EWS), and the most adopted models are based on empirical approaches. Although they are not so adequate to model hydraulic and geotechnical aspects of landslide triggering, empirical models are preferred for EWS because of their simplicity: in fact, they can easily be calibrated on the basis of only information about historical movements, and can be

P. Versace (✉) · G. Capparelli · D.L. De Luca
Department of Informatics, Modelling, Electronics and System Engineering, University of Calabria, Ponte Pietro Bucci, 41B Building 5th Floor, 87036 Arcavacata, CS, Italy
e-mail: linoversace@libero.it; pasquale.versace@unical.it

G. Capparelli
e-mail: giovanna.capparelli@unical.it

D.L. De Luca
e-mail: davide.deluca@unical.it

K. Sassa et al. (eds.), *Landslide Dynamics: ISDR-ICL Landslide Interactive Teaching Tools*, https://doi.org/10.1007/978-3-319-57774-6_28

straightforwardly used in real time as only knowledge about rainfall time series (predicted or observed in real time) is required to evaluate the exceedance of critical conditions.

Several empirical models were proposed in technical literature, and the most adopted belong to the class of Intensity-Duration (ID) schemes (Caine 1980; Cannon and Gartner 2005; Corominas et al. 2005; Crosta and Frattini 2001; Godt et al. 2006; Guzzetti et al. 2007, 2008; Nadim et al. 2009).

This teaching tool focuses on the model named FLaIR (Forecasting of Landslides Induced by Rainfalls, Sirangelo and Versace 1996; Capparelli et al. 2009; Capparelli and Versace 2011), which was developed at Laboratory of Environmental Carthography and Hydraulic and Geological Modelling (CAMILab) of University of Calabria (Italy). FLaIR has many advantages compared with other empirical schemes; in particular (i) it considers the real pattern of rainfall input and not average values along assigned durations, and thus it is possible to discriminate the influence of rainfall heights, on the basis of their lag-time from the current instant; (ii) it provides an unique threshold value and not a reference function like an ID scheme, and consequently it allows for a simpler check about exceedance or not of critical conditions for landslide forecasting, (iii) it is more flexible as it is suitable for both shallow landslides, induced by recent rainfall events, and deep movements, triggered by precipitation aggregated on longer durations. Moreover, FLaIR constitutes a general framework for many empirical models, as it reproduces as particular cases all the ID schemes proposed in technical literature (Capparelli and Versace 2011).

A short description of FLaIR mathematical background is reported in Sect. 2; parameter estimation is discussed in Sect. 3, while the use for real-time forecasting is shown in Sect. 4.

From the website www.camilab.unical.it it is possible to download the software FLaIR.exe, together with a user guide.

2 Mathematical Background of FLaIR Model

In empirical models it is possible to identify a **mobility function** $Y(t)$, which is a generic function of the rainfall that can be correlated with landslide occurrence. In details, if $P[E_t]$ is the occurrence probability of a landslide at time t, and assuming that it depends only on $Y(t)$, then $P[E_t]$ can be expressed as:

$$P[E_t] = \begin{cases} 0 & if \quad Y(t) < Y_1 \\ g[Y(t)] & if \quad Y_1 \le Y(t) \le Y_2 \\ 1 & if \quad Y(t) > Y_2 \end{cases} \qquad (1)$$

where $g[\cdot]$ is a non-decreasing generic function that can take values between [0;1] in the interval $[Y_1;Y_2]$;Y_1 is the value of $Y(t)$ for which mobilization is impossible; and Y_2 is the value of $Y(t)$ for which mobilization is certain (Fig. 1).

A simple version of Eq. (1) is obtained by assuming $Y_1 = Y_2 = Y_{cr}$, where Y_{cr} represents a threshold value of $Y(t)$ which separates the condition "impossible mobilization" from "certain mobilization"; that is (see also Fig. 2):

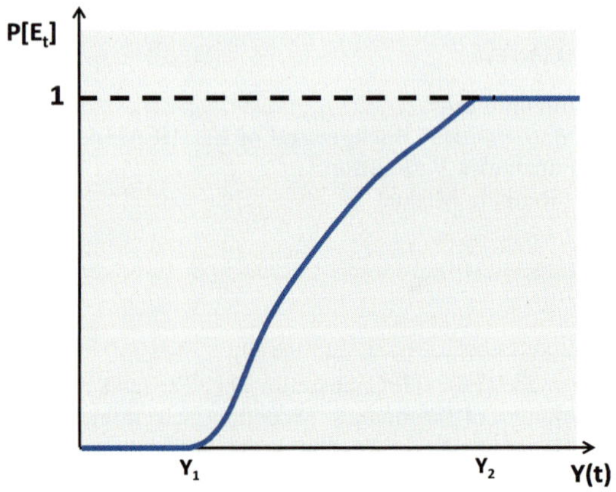

Fig. 1 Relationship between the occurrence probability $P[E_t]$ of a landslide at time t and the mobility function $Y(t)$, on the basis of Eq. (1)

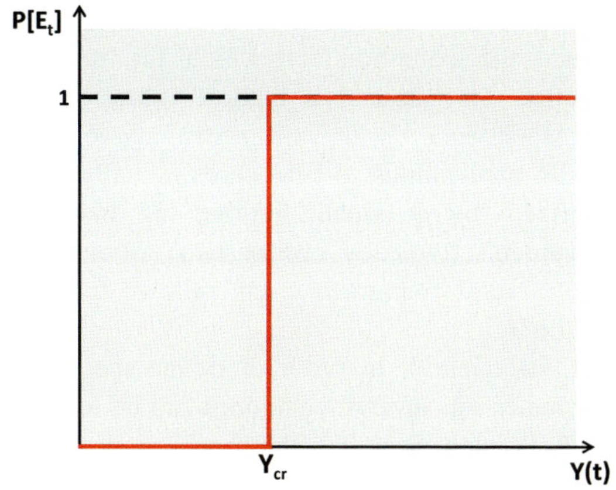

Fig. 2 Relationship between the occurrence probability $P[E_t]$ of a landslide at time t and the mobility function $Y(t)$, on the basis of Eq. (2)

$$\begin{cases} P[E_t] = 0 & \textit{if } Y(t) < Y_{cr} \\ P[E_t] = 1 & \textit{if } Y(t) \geq Y_{cr} \end{cases} \tag{2}$$

The choice of criteria adopted by different authors for defining threshold values is different: e.g., Cannon and Ellen (1985) assume that the threshold corresponds to "abundant landslides", while for Wieczorek (1987), the threshold corresponds to "one or more than one landslides." Nevertheless, the threshold approach remains the most widely used in rainfall–landslide studies, because of the difficulty in function identification and parameter calibration in Eq. (1) owing to the lack of experimental data.

In FLaIR model (Forecasting of Landslides Induced by Rainfall, Capparelli and Versace 2011), the mobility function $Y(t)$ is computed as a convolution between the rainfall infiltration rate $I(\cdot)$ and a filter function $\psi(\cdot)$:

$$Y(t) = \int_0^t \psi(t - \tau) I(\tau) d\tau \tag{3}$$

and a landslide trigger is predicted by the model when $Y(t)$ exceeds a critical threshold Y_{cr}, which is a parameter to be estimated (see Sect. 3).

The infiltration rate $I(\tau)$ is assumed proportional to the rainfall intensity $P(\tau)$, according to the following simple relationship:

$$I(\tau) = \begin{cases} P(\tau) & \textit{when} \quad P(\tau) \leq P_0 \\ P_0 & \textit{when} \quad P(\tau) > P_0 \end{cases} \tag{4}$$

where P_0 depends on soil characteristic, and it is a parameter to be estimated. In the simplest version, P_0 is set equal to $+\infty$, and then:

$$Y(t) = \int_0^t \psi(t - \tau) P(\tau) d\tau \tag{5}$$

The function $\psi(\cdot)$ is typical for each case study and it plays a central role in mobility function evaluation. It can assume different expressions (Iiritano et al. 1998), like rectangular:

$$\psi(t) = \begin{cases} 1/t_0 & \text{if } 0 < t \leq t_0 \\ 0 & \text{elsewhere} \end{cases} \tag{6}$$

exponential:

$$\psi(t) = \frac{1}{k} e^{-\frac{t}{k}} \quad t \geq 0,\ k > 0 \tag{7}$$

gamma:

$$\psi(t) = \frac{1}{\beta^\alpha \Gamma(\alpha)} t^{\alpha - 1} e^{-\frac{t}{\beta}} \quad t \geq 0,\ \alpha > 0,\ \beta > 0 \tag{8}$$

power:

$$\psi(t) = m t^{-q} \quad 0 < t < T,\ m > 0,\ 0 < q < 1 \tag{9}$$

mixture of two exponential functions:

$$\psi(t) = \omega \beta_1 \exp(-\beta_1 t) + (1 - \omega) \exp(-\beta_2 t)$$
$$t \geq 0,\ \beta_1 \geq \beta_2 > 0,\ 0 \leq \omega \leq 1 \tag{10}$$

In particular, Eq. (10) is very flexible: the first addendum reproduces the effect of the most recent rainfall (short-term component) while the second addendum is referred to the influence of antecedent precipitation (long-term component). The terms ω and $(1 - \omega)$ are the weights of the two components.

Figure 3 shows the differences in mobility function when different filter functions are adopted to transform rainfall time series.

3 Parameter Estimation

FLaIR calibration is carried out by: (i) parameter estimation referred to the chosen filter function $\psi(\cdot)$; (ii) assessment of the critical threshold Y_{cr}. With this goal, dates of historical landslide occurrences and rainfall database are used.

The estimated parameters for $\psi(\cdot)$ must ensure the condition for which the mobility function $Y(t)$ attains its highest values just in correspondence with historical movements. FLaIR applications to real cases, usually characterized by a small number of historical movements (typically one or two), often show that several parameter sets allow for this condition.

Consequently, in order to define the "best" parameter set, several techniques can be used. In the following Sect. 3.1 the *ranking method* is only discussed.

3.1 Ranking Method

Concerning a specific filter $\psi(\cdot)$, application of ranking method consists of the identification of the admissibility region, defined as the ensemble

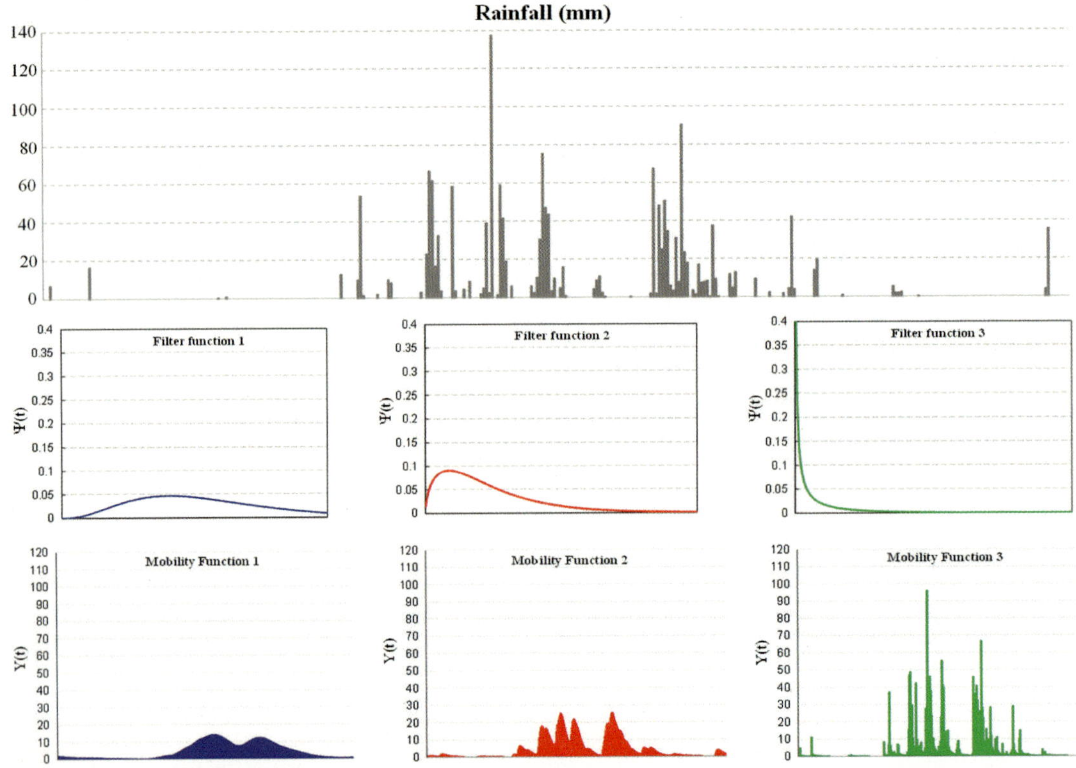

Fig. 3 Example of differences in mobility function when different filter functions are adopted to transform rainfall time series. Adapted from Capparelli and Versace (2011)

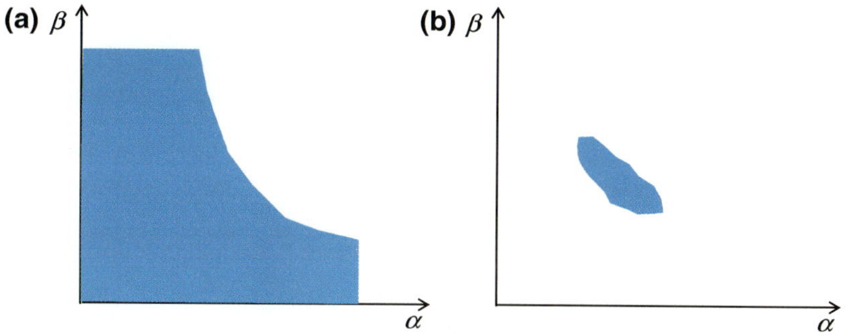

Fig. 4 **a** Example of large admissibility region; **b** example of small admissibility region

of all the parameter sets $\underline{\theta} = (\theta_1, \theta_2, \ldots, \theta_n)$ for which the mobility function $Y(t;\underline{\theta})$ assumes its k highest values in correspondence with the k known historical movements of the landslide.

Depending on the size of admissibility region, the *informative content* of landslide events can be defined. Figure 4 shows examples of two different admissibility region, related to gamma filter (Eq. 8):

- in the former (Fig. 4a), a lot of parameter sets allow for the highest values of $Y(t;\underline{\theta})$ when historical landslides occurred. This situation is typical when historical movements are triggered by only heavy rain events. In this case the informative content is very low, and then using the rainfall histogram (and not FLaIR model) produces the same results in terms of forecasting of landslide triggering;
- in the latter (Fig. 4b), $Y(t;\underline{\theta})$ presents its highest values in correspondence with historical landslides only for few parameter sets. In this case the informative content is high.

For each admissible parameter set $\underline{\theta}$ it is possible to define a lower limit function (indicated as f_L) and an upper limit function (indicated as f_U) for $Y(t;\underline{\theta})$; they represent, respectively, the highest value that did not produce any movement and the lowest value for which movement occurred. Figure 5 refers to a simple case with only one historical movement, but it can be straightforwardly generalized to multiple movements. It represents the mobility function, evaluated on the basis of a rain data sample with a time resolution equal to Δt. A tolerance interval of length $(u_1 + u_2)\Delta t$ is considered, in order to take into account the uncertainty in the information about the date of landslide movement. Finally, to avoid the possibility of estimating the lower limit function in the same rainfall event that caused the landslide, a disjunction interval, of length equal to $(v_1 + v_2)\Delta t$, is introduced.

Starting from the ensemble of all the admissible parameter sets $\underline{\theta}$, the best set is chosen by maximizing the difference:

$$r = f_U - f_L \tag{11}$$

that means to allow for the largest possible gap between the mobility function $Y(\cdot)$ values related to no movement and those related to a movement.

Finally, estimation of Y_{cr} has to be carried out, and this requires particular care, especially in the frequent case of only one historical movement:

- if $Y_{cr} = f_L$ then the model could forecast very frequently a landslide occurrence;
- if $Y_{cr} = f_U$ then the model could not be able to identify situations of real hazard, because only events larger than the one that induced the historical occurrence would be believed able to trigger a second movement.

On the basis of the above aspects, it could be preferable to use a critical threshold equal to the lower limit function.

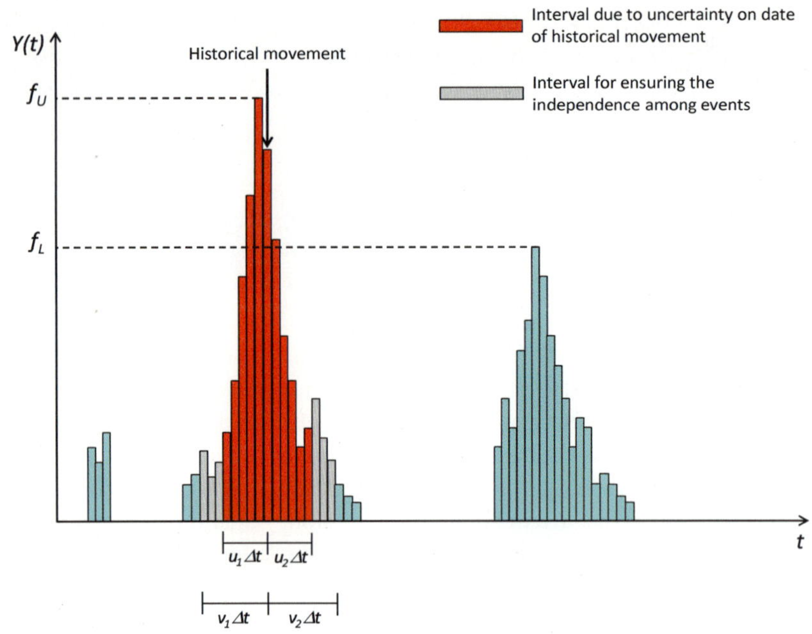

Fig. 5 Evaluation of upper limit function f_U and lower limit function f_L

4 Real-Time Forecasting

Use of FLaIR model for real-time forecasting consists in evaluating, with suitable early time, the probability that at time *t*, the mobility function $Y(t)$ exceeds the critical value Y_{cr}, estimated on the basis of historical information (Sect. 3), or its percentages which are indicated as Levels of Criticality (*LC*) and are defined in the following way:

$$LC_1 = \xi_1 Y_{cr} \quad (12)$$

$$LC_2 = \xi_2 Y_{cr} \quad (13)$$

$$LC_3 = \xi_3 Y_{cr} \quad (14)$$

with $\xi_1 < \xi_2 < \xi_3 < 1$. LC_1, LC_2 and LC_3 are also named ordinary, moderate and high criticality, respectively.

In details, setting τ as the current time, the mobility function evaluated at τ and related to the future time *t* (with clearly $\tau < t$) is indicated as $Y_\tau(t)$, which can be written as sum of two terms:

$$Y_\tau(t) = \int_0^\tau \psi(t-u)P(u)du + \int_\tau^t \psi(t-u)P(u)du \quad (15)$$

The first term is evaluated on the basis of observed rainfall depth until time τ and it can be considered as the ***deterministic*** component of $Y_\tau(t)$.

The second term is valuable by considering rainfall nowcasting for $P(\cdot)$ in the interval [τ; *t*], which is usually carried out with a Probabilistic Quantitative Precipitation Forecast (**PQPF**), that means a large number *N* of future rainfall scenarios (Fig. 6a). PQPF can be obtained by using stochastic or meteorological models, and the second addendum of Eq. (15) is indicated as the ***probabilistic*** component. From *N* predicted rainfall scenarios it is possible to obtain *N* realizations of $Y_\tau(t)$ (Fig. 6b).

For each predicted scenario of $Y_\tau(t)$, the dimensionless index $\xi(t) = Y_\tau(t)/Y_{cr}$ can be defined, and the maximum value ξ_{max} in the interval [τ; *t*] is evaluated: consequently, the

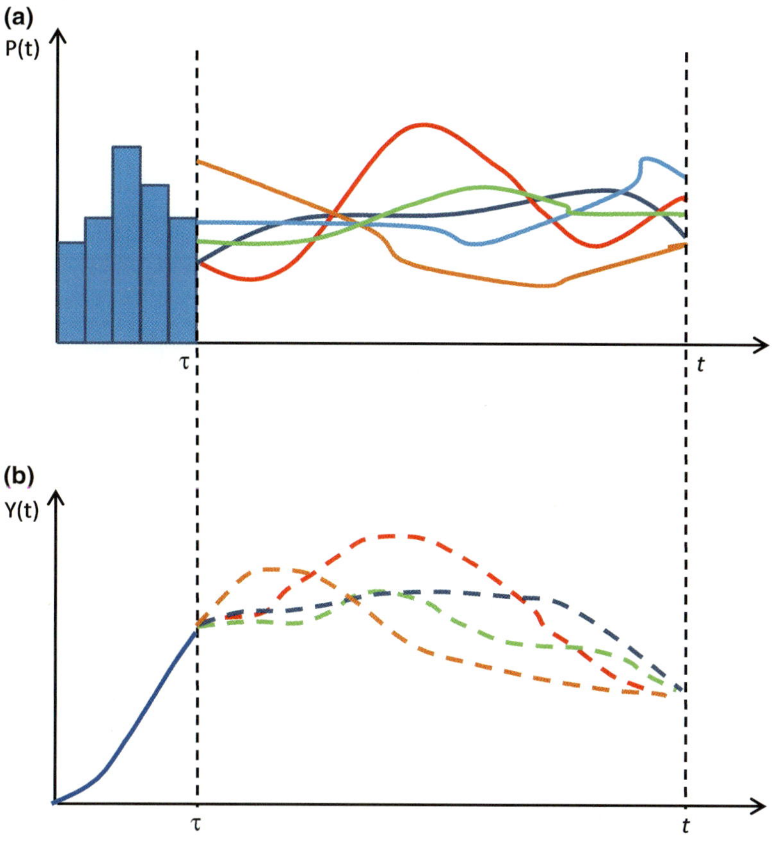

Fig. 6 Real-time forecasting in the interval [τ; t]: **a** rainfall nowcasting; **b** FLaIR scenarios

exceedance probabilities can be computed in the following way:

$$P[\xi(t) > \xi_1] = \frac{N_{[\xi_{\max} > \xi_1]}}{N} \quad (16)$$

$$P[\xi(t) > \xi_2] = \frac{N_{[\xi_{\max} > \xi_2]}}{N} \quad (17)$$

$$P[\xi(t) > \xi_3] = \frac{N_{[\xi_{\max} > \xi_3]}}{N} \quad (18)$$

where $N_{[\xi_{\max} > \xi_1]}, N_{[\xi_{\max} > \xi_2]}$ and $N_{[\xi_{\max} > \xi_3]}$ indicate the number of predicted scenarios for which $\xi_{\max}$ is greater than ξ_1, ξ_2 and ξ_3, respectively.

4.1 Example of Application

FLaIR model was applied in several areas of Italy, in the context of agreements with Italian National Department of Civil Protection and with Regional Administrations.

As an example, the case study of Montenero di Bisaccia, located in Molise region (southern Italy), is shown, where a landslide occurred on 1st March 2006.

The stochastic model named PRAISE (Prediction of Rainfall Amount Inside Storm Events, Sirangelo et al. 2007) was used for rainfall prediction in real time. PRAISE model was calibrated by considering the hourly time series of Palata raingauge (close to Campobasso city).

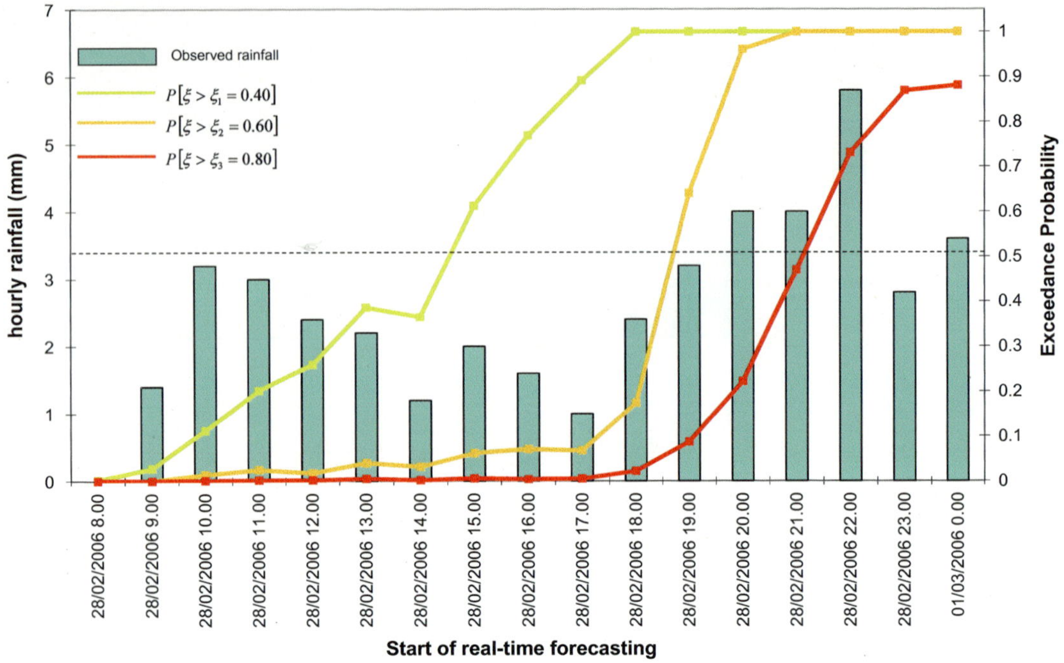

Fig. 7 Case study of Montenero di Bisaccia (southern Italy). Temporal behavior of exceedance probability referred to the thresholds ξ_1, ξ_2 and ξ_3 for a lapse time equal to 6 h successive to the beginning of the forecast

Gamma function (Eq. 8) was used as filter $\psi(\cdot)$, and parameter estimation provided $\alpha[-] = 0.9, \beta[\text{days}] = 20$ and $Y_{cr}[\text{mm/day}] = 5$.

The following threshold values for the dimensionless index were considered: $\xi_1 = 0.4, \xi_2 = 0.6$ and $\xi_3 = 0.8$.

Figure 7 represents on horizontal axis the time in which rainfall occurred and mobility function forecasting is performed; hourly rainfall heights measured by raingauge are represented on the left vertical axis, while in the right vertical axis there is the probability evaluation that the mobility function could exceed threshold values in the successive six hours. From the figure it is clear that approaching to the landslide date, $P[\xi(t) > \xi_3]$ also increases; consequently, using FLaIR coupled with a rainfall predictor allows to provide, with sufficient advance, the exceedance of the several thresholds.

References

Caine N (1980) The rainfall intensity-duration control of shallow landslides and debris flows. Geogr Ann A 62:23–27

Cannon SH, Ellen SD (1985) Rainfall conditions for abundant debris avalanches, San Francisco Bay region, California. Cal Geol 38(12):267–272

Cannon SH, Gartner JE (2005) Wildfire-related debris flow from a hazards perspective. In: Jakob M, Hungr O (eds) Debris flow hazards and related phenomena. Springer, Berlin, pp 363–385

Capparelli G, Versace P (2011) FLaIR and SUSHI: two mathematical models for early warning systems for rainfall induced landslides. Landslides 8:67–79. doi:10.1007/s10346-010-0228-6

Capparelli G, Biondi D, De Luca DL, Versace P (2009) Hydrological and complete models for forecasting landslides triggered by rainfalls. In: Proceedings of the first Italian workshop on landslides, 8–10 June 2009, Napoli (Italy), pp 162–173

Corominas J, Ayala FJ, Cendrero A, Chacón J, Díaz de Terán JR, Gonzáles A, Moja J, Vilaplana JM (2005)

Impacts on natural hazard of climatic origin. In: ECCE final report: a preliminary assessment of the impacts in Spain due to the effects of climate change. Ministerio de Medio Ambiente

Crosta GB, Frattini P (2001) Rainfall thresholds for triggering soil slips and debris flow. In: Mugnai A, Guzzetti F, Roth G (eds) Proceedings 2nd EGS Plinius conference on Mediterranean storms, Siena, Italy, pp 463–487

Godt JW, Baum RL, Chleborad AF (2006) Rainfall characteristics for shallow landsliding in Seattle, Washington, USA. Earth Surf Proc Land 31:97–110

Guzzetti F, Peruccacci S, Rossi M, Stark CP (2007) Rainfall thresholds for the initiation of landslides in central and southern Europe. Meteorol Atmos Phys 98:239–267

Guzzetti F, Peruccacci S, Rossi M, Stark CP (2008) The rainfall intensity-duration control of shallow landslides and debris flows: an update. Landslides 5:3–17

Iiritano G, Versace P, Sirangelo B (1998) Real-time estimation of hazard for landslides triggered by rainfall. Environ Geol 35(2–3):175–183

Nadim F, Cepeda J, Sandersen F, Jaedicke C, Heyerdahl H (2009) Prediction of rainfall-induced landslides through empirical and numerical models. In: Proceeedings of IWL—The first Italian workshop on landslides, vol 1. Naples, 8/10 June 2009, pp 206—215

Sirangelo B, Versace P (1996) A real time forecasting model for landslides triggered by rainfall. Meccanica 31(1):73–85

Sirangelo B, Versace P, De Luca DL (2007) Rainfall nowcasting by at site stochastic model P.R.A.I.S.E. Hydrol Earth Syst Sci 11:1341–1351

Wieczorek GF (1987) Effect of rainfall intensity and duration on debris flows in central Santa Cruz Mountains, California. Geol Soc of Am Rev in Eng Geol 7:93–104

TXT-tool 2.386-2.1 A System to Forecast Rainfall-Induced Landslides in Slovenia

Mateja Jemec Auflič, Jasna Šinigoj, Tina Peternel, Martin Podboj, Matija Krivic and Marko Komac

Abstract

Nowadays, an increasingly important area of work in landslides prediction is the development of tools to generate warnings to potentially affected communities. For the last 20 years many countries and cities have been affected by heavy rainfall leading to landslides and Slovenia is no exception. Therefore, public awareness of extreme events has notably increased across the world in different sectors. In order to provide a solution to the problem a national landslide warning system, Masprem, has been developed based on three main raster datasets: landslide-triggering rainfall threshold values, precipitation forecasts, and landslide susceptibility maps. At the moment, a landslide warning system predicts landslide probability for the area of Slovenia in the scale 1:250,000 twice a day. A Masprem system also calculates landslide predictions on a local level, including maps of exposure of inhabitants, buildings and different types of infrastructures to potential landslide occurrence at a scale of 1:25,000 for five selected municipalities.

M. Jemec Auflič (✉) · J. Šinigoj · T. Peternel · M. Podboj · M. Krivic
Geological Survey of Slovenia, Dimičeva Ulica 14, 1000 Ljubljana, Slovenia
e-mail: mateja.jemec@geo-zs.si

J. Šinigoj
e-mail: jasna.sinigoj@geo-zs.si

T. Peternel
e-mail: tina.peternel@geo-zs.si

M. Podboj
e-mail: martin.podboj@geo-zs.si

M. Krivic
e-mail: matija.krivic@geo-zs.si

M. Komac
Marko Komac s.p., Mesarska cesta 12, Ljubljana, Slovenia
e-mail: m.komac@telemach.net

K. Sassa et al. (eds.), *Landslide Dynamics: ISDR-ICL Landslide Interactive Teaching Tools*, https://doi.org/10.1007/978-3-319-57774-6_29

In the present Teaching Tool a design of the system for modelling a probability of landslide through time in Slovenia is highlighted.

Keywords
Landslides • Rainfall • Warning system • System architecture • Slovenia

Contents

1 **Introduction** ... 392
2 **Methodology** ... 393
2.1 Dynamic Forecast Modelling Module . 393
2.2 Real-Time Rainfall Data ... 395
2.3 Rainfall-Triggering Values ... 396
2.4 Landslide Susceptibility Model ... 396
2.5 Modelling Exposure Maps on a Local Level ... 397
3 **System Architecture** ... 399
3.1 Application System of Masprem ... 400
3.2 Masprem Web Application ... 400
4 **Conclusion** ... 401
References ... 404

1 Introduction

The development of information systems has led to progress in the design of landslide forecasting systems that, through the input rainfall data, reflect possible scenarios or situations that may happen in nature. The developed models serve as a useful tool for the prediction of landslide areas where rainfall exceeds a defined threshold and as a warning for the population of an increased probability of landslides occurring.

One of the first landslide warning systems was designed for the San Francisco Bay Region (Keefer et al. 1987). It forecasts debris flow and landslide initiation based on a defined rainfall threshold. In a few places of the world (Japan, China, New Zealand, Central America), rainfall thresholds are a part of operational landslide warning systems that compare real-time rainfall measurements with established thresholds. When pre-established values are exceeded, alarm messages are issued. Similar systems warning of susceptibility to the occurrence of landslides are also active in areas of Hong Kong (Chan and Pun 2004) and in Rio de Janeiro, Brazil (Ortigao et al. 2002). In areas with numerous landslide occurrences, landslide early warning systems in a real-time mode have also been developed in several countries in Europe systems: ILEWS (Thiebes 2012), AlpEWAS (Thuro et al. 2009), DORIS (Manunta et al. 2012).

The territory of Slovenia is relatively highly exposed to exogenous slope processes due to its geological, geomorphological and climate diversity. Such slope processes include slope mass movements that can be further categorised into landslides, soil slips, debris flows and rockfalls. Recently, triggered by increasingly frequent extreme weather events and the country's precipitation regime, slope mass movements in Slovenia have become common and have caused considerable damage. According to Komac and Ribičič (2006) a quarter of Slovenian territory is exposed to landslides. In Slovenia, intense short and less intense long duration rainfall is a primary cause of shallow landslides that, in some estimates, sum to the number of 10,000 or approximately one landslide per two square kilometers. In the past years rainfall has triggered numerous shallow landslides and caused extensive damage to buildings, infrastructure, roads, and unfortunately also caused deaths. These events could be identified and to some extent also minimized with better knowledge of the relation between landslides and rainfall. To tackle the problem from a prevention perspective, a system to forecast probability of landslide occurrences (acronym is Masprem, A National Warning System for Rainfall-Induced

Landslides in Slovenia) has been developed in 2013 for the whole country, which spreads over an area of 20,000 km^2, and financed by the Slovenian disaster relief office and Ministry of Defence (Jemec Auflič et al. 2016). The designed operational system is based on precipitation forecasts, rainfall threshold values and landslide susceptibility maps and informs inhabitants of an increased probability of landslide occurrences as a consequence of heavy precipitation that exceeds the rainfall thresholds. A near real-time system to forecast probability of landslide occurrences is currently operational in the validation phase (Šinigoj et al. 2015).

The purpose of this teaching tool is to present the researches and students a landslide forecast system and a model of probability of landslides through time using a real-time forecast at a regional level.

2 Methodology

A national system to forecast probability of landslide occurrences in Slovenia is built from three main raster datasets: (1) a rainfall forecast model for the 24-h period ALADIN-SI (Pristov et al. 2012); (2) landslide-triggering rainfall threshold values, and (3) landslide susceptibility maps. At the moment, a landslide warning system forecast probability of landslide is calculated for the area of Slovenia at a 1:250,000 scale and for five selected municipalities at a 1:25,000 scale, where the exposure of inhabitants, buildings and infrastructure is calculated. The system is operational from September 2013 and runs in 12-h cycling mode for the 24 h period ahead, yet due to the testing phase of model prediction, the results need to be treated with care and within their reliability. The model results of probability of landslides are classified into six classes, with values ranging from one to six, where class one represents areas with negligible landslide probability and class six areas with very high landslide probability. Landslide forecast models are automatically transferred to Administration for civil protection and disaster relief to inform them of the increased probability of landslide occurrences as a consequence of heavy precipitation that could exceed the rainfall threshold.

Input datasets (landslide-triggering rainfall threshold values, precipitation forecasts, and landslide susceptibility maps) are integrated into the system through separate modules (Fig. 1). Calculation of forecast models is performed through the dynamic forecast modelling module, as shown in Fig. 1 (Šinigoj et al. 2015).

2.1 Dynamic Forecast Modelling Module

The dynamic forecast modelling module is one of the key elements of the system because it contains an algorithm for calculating predictions.

The use of fuzzy logic method allows for a gradual transition between the variables. In the case of determination of rainfall thresholds that are continuous variables, the use of fuzzy logic is very reasonable. Because of different rainfall patterns and intensities that vary both spatially and temporally, the precise boundaries of the rainfall threshold over which a landslide always occurs is very difficult to define. Therefore, the model considers continuous rainfall threshold values for each engineering-geological unit. Engineering geological units were created based on lithology units of Slovenia according to EG characteristics on three levels. The first one is the basic separation into soils, soft rocks and rocks. The second level is a more detailed separation on the basis of their origin and the third one on the basis of the composition, rock strength and particle size ranges (Ribičič et al. 2003).

The diagram of three separate modules of the Masprem system required for dynamic modelling of rainfall induced landslides is shown in Fig. 1. The three input models were combined using the following IF-THEN algorithm (modified after Glade et al. 2000; Li et al. 2010):

IF ([forecasted precipitation value (RT(x, y))] > [rainfall triggering value (R_{FALL} (x,y))]) AND [landslide susceptibility value] = 1–5

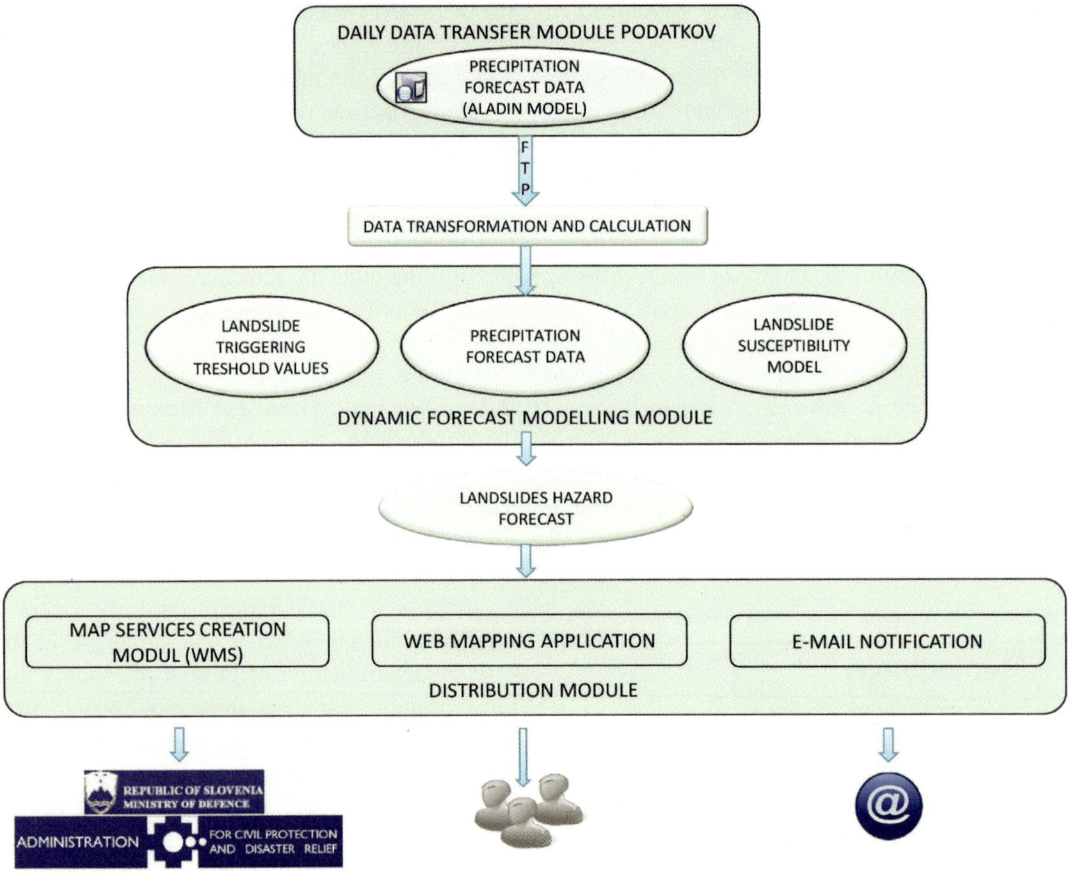

Fig. 1 Three separate modules of the Masprem system

THEN [forecasted rainfall induced landslide value] = 1–5.

The final model values are classified into six classes—none/negligible (0), very low (1), low (2), moderate (3), high (4), and very high (5).

By using the threshold approach, two major thresholds can be defined: minimum threshold and maximum threshold, which identify the lower and upper boundaries of the threshold's probability range (White et al. 1996). The minimum threshold (R_{TMIN}) defines the lowest level, below which a landslide does not occur. The maximum threshold (R_{TMAX}) is defined as the level above which a landslide always occurs. Below a certain value (R_{TMIN}) the probability of the event triggering is almost none (0), while above a certain value (R_{TMAX}) the probability of the event triggering is almost certain (1). Between the two values the probability of triggering rises from 0 to 1, depending on the function that defines the transition. The difference between the R_{TMIN} and R_{TMAX} is set to 30 mm to account for the classification error. R_{SUM} is the total amount of forecasted precipitation and rainfall threshold together. It follows that landslide triggering rainfall threshold (R_{FALL}) for each location (cell) x, y in the time interval [0, t] is:

$$R_{FALL}(x,y) = \begin{cases} 0 & if\ Rsum(x,y) < R_{TMAX}(x,y) \\ s, s \in (0,1) & if\ R_{TMIN}(x,y) \leq Rsum(x,y) \leq R_{TMAX}(x,y) \\ 1 & if\ Rsum(x,y) > R_{TMAX}(x,y) \end{cases}$$

Methods of deriving the three input data sets, being an important part of the system, will be briefly described below in a separate section.

2.2 Real-Time Rainfall Data

Different types of numerical weather forecast models exist: (1) a short-range forecast where a weather forecast is made for a time peri of up to 48 h; (2) medium-range forecasts are for a period extending from about three days to seven days in advance, and (3) long-range forecasts are for a period greater than seven days in advance, but there are no absolute limits to the period (Jemec Auflič and Komac 2013; Rosi et al. 2016). Due to the chaotic nature of the atmosphere short-range forecasts are generally more accurate than the other types of forecasts (Mercogliano et al. 2010). ALADIN (Aire Limitee Adaptation dynamique Development INternational) is a regional mesoscale model for numerical forecast of weather that simulates events in the atmosphere over much of continental Europe (Bubnová et al. 1995). The ALADIN system has been used in Slovenia as an operational weather forecast system since 1997 (Pristov et al. 2012). A regional ALADIN model for Slovenia predicts the status of the atmosphere over the area of Slovenia up to 72 h ahead. A model simulates the precipitation (kg/m^2), snowfall, water in snow pack, and air temperature data. ALADIN/SI is a grid point model (439 × 2421 × 43), where the horizontal distance between the grid points is 4.4 km and it runs in a 6-h cycling mode for the next 54 h by the Environmental Agency of Republic of Slovenia

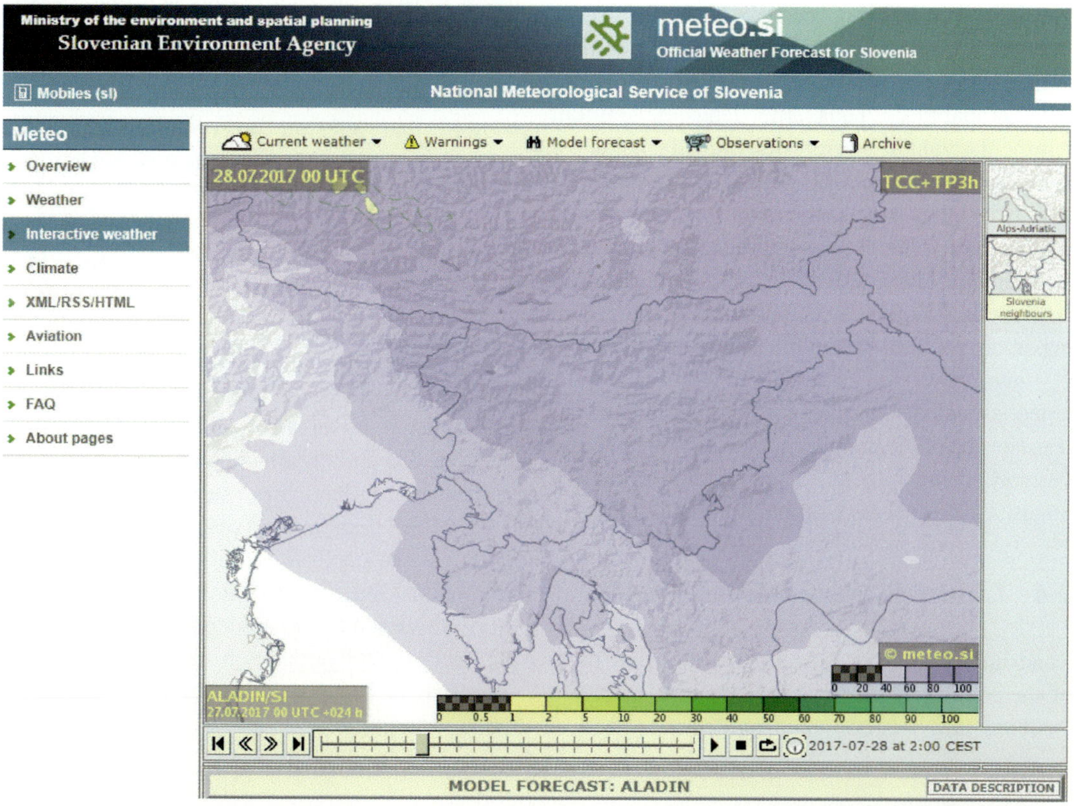

Fig. 2 Numerical meteorological model ALADIN/SI, forecast for precipitation (ARSO, http://meteo.arso.gov.si/)

(ARSO 2015). In Fig. 2, a numerical meteorological model ALADIN/SI is shown, as an example for precipitation forecast as a real-time rainfall data for modelling probability of landslides through time.

2.3 Rainfall-Triggering Values

To analyse rainfall patterns that govern slope failure, it is essential to understand the relationship between rainfall and landslide occurrence.

Analyses of landslide occurrences in Slovenia have shown that in areas where intensive rainstorms occur (maximal daily rainfall for the 100-year period), and where the geological settings are favorable (landslide-prone), an abundance of landslides can be expected. This clearly indicates the spatial and temporal dependence of landslide occurrence upon intensive rainfall. For defining rainfall thresholds, the frequency of spatial occurrence of landslide per spatial unit was correlated with lithological unit, and 24-h maximum rainfall data with a return period of 100 years. The result of frequency of landslide occurrence and rainfall data provides a good basis for determining the critical rainfall threshold over which landslides occur with high probability. Thus, the landslide rainfall threshold values were determined using a non-parametric statistical method chi-square (χ^2) for each lithological unit. Then we separately cross-analysed the occurrence of landslides within each unique class derived from the spatial cross analysis of lithological units and classes of 24-h maximum rainfall. Maximum daily rainfalls above 100 mm proved to be critical for landslide occurrence, especially in more loose soils and in less resistant rocks (Quaternary, Tertiary, Triassic, and Permo-Carbonian rocks). The critical 24-h rainfall intensities (thresholds) can be found in Table 1.

2.4 Landslide Susceptibility Model

Landslides in Slovenia occur almost in all parts of the country. Based on the extensive landslide database that was compiled and standardized at a national level, and based on analyses of landslide spatial occurrence, a landslide susceptibility map of Slovenia at a scale of 1:250,000 was produced (Komac and Ribičič 2006; Komac 2012) (Fig. 3). Altogether more than 6,600 landslides were included in the national database, of which roughly half are on known locations. Of 3,241 landslides with known location, a random but representative 67% were selected (landslide learning set) and used for the univariate statistical analyses (χ^2) to analyse the landslide occurrence in relation to the spatio-temporal precondition factors (lithology, slope inclination, slope curvature, slope aspect, distance to geological boundaries, distance to structural elements, distance to surface waters, flow length, and land-cover type). The analyses were conducted using GIS in raster format with the 25 × 25 m pixel size. The results of the analyses were later used to develop a weighted linear susceptibility model from which more than 156,000 automatically calculated models with random weight combinations were derived. The landslide testing subset (33% of landslides) and representative areas with no landslides were used for the validation of all models developed. The results showed that relevant precondition factors for landslide occurrence are (with their weight in a linear model): lithology (0.33), slope inclination (0.23), land-cover type (0.27), slope curvature (0.08), distance to structural elements (0.05), and slope aspect (0.05). Roughly 8% of Slovenia is extremely susceptible to potential hazards posed by landsliding. 11% of the population lives in the areas of high landslide susceptibility that spread over 16% of Slovenia. 5.7% of Slovenia's inhabitants live in the areas of moderate landslide susceptibility (10%). 6.7% of the population lives in the areas of low landslide susceptibility (20%), 3.7% of the population lives in the areas of insignificant (very low) landslide susceptibility (18%), and the rest of the population (65%) lives in the areas where the possibility of landslide occurrence is negligible (28%).

Table 1 Landslide rainfall threshold values for lithostratigraphic units

Lithostratigraphic units	Critical 24-h rainfall intensities (mm)
Predominantly clay soils (soils)	–
Marsh and lake sediments (clay, silt, peat) (soils)	–
Alluvium, fluvial loose sediments in terraces (soils)	–
Clayey—diluvial, proluvial (soils)	<120
Gravely with a clayey component (soils)	150–180
Gravely (predominantly thick fraction), moraines (soils)	210–240
Clayey (soils)	120–150
Alternation of fine and coarse grain soils (soils)	<120
Pebbly (soils)	<120
Mine tailings—gangues (soils)	120–150
Clayey, marly rocks (soft rocks)	120–150
Clayey, marly and limestone (soft rocks)	–
Alternation of different materials (marl, sand, sandstone, conglomerate pebble, clay) (soft rock)	120–150
Conglomerate (soft rock)	120–150
(slaty) Claystones with inclusion of other rocks (rocks)	120–150
Marl and sandstone (flysch) with inclusions of other rocks (rocks)	210–240
Sandstones and conglomerates with inclusionsof other rocks (rocks)	150–180
Stratified and cliff limestones (rocks)	–
Flat limestones (rocks)	–
Limestones and dolomites (rocks)	–
Dolomites (rocks)	–
Limestones with marls (rocks)	<120
Limestones with inclusions of other rocks (rocks)	210–240
Limestone conglomerates and breccia (rocks)	150–180
Phyllites, schists and slate (rocks)	180–210
Amphibolite and gneiss (rocks)	120–150
Diabase and other magmatic rocks with tuff (rocks)	120–150
Amphibolites, serpentinites, diaphthorites (rocks)	120–150
Tonalite, dacite, granodiorite (rocks)	–

Critical 24-h rainfall intensities are presented only for litostratigraphic units for which the number of observed landslides were statistically higher than the number of expected landslides

2.5 Modelling Exposure Maps on a Local Level

A Masprem system also calculates landslide prediction on a local level, including maps of exposure of inhabitants, buildings and different types of infrastructure to potential landslide occurrence at a scale of 1:25,000 for five selected municipalities. The probability of landslide occurrence on a local scale is calculated very similarly to the calculation probability of landslide occurrences in national scale by difference in the scale of landslide susceptibility map (1:25,000).

For the five selected Slovenian municipalities, exposure maps of inhabitants, buildings and different types of infrastructure to potential

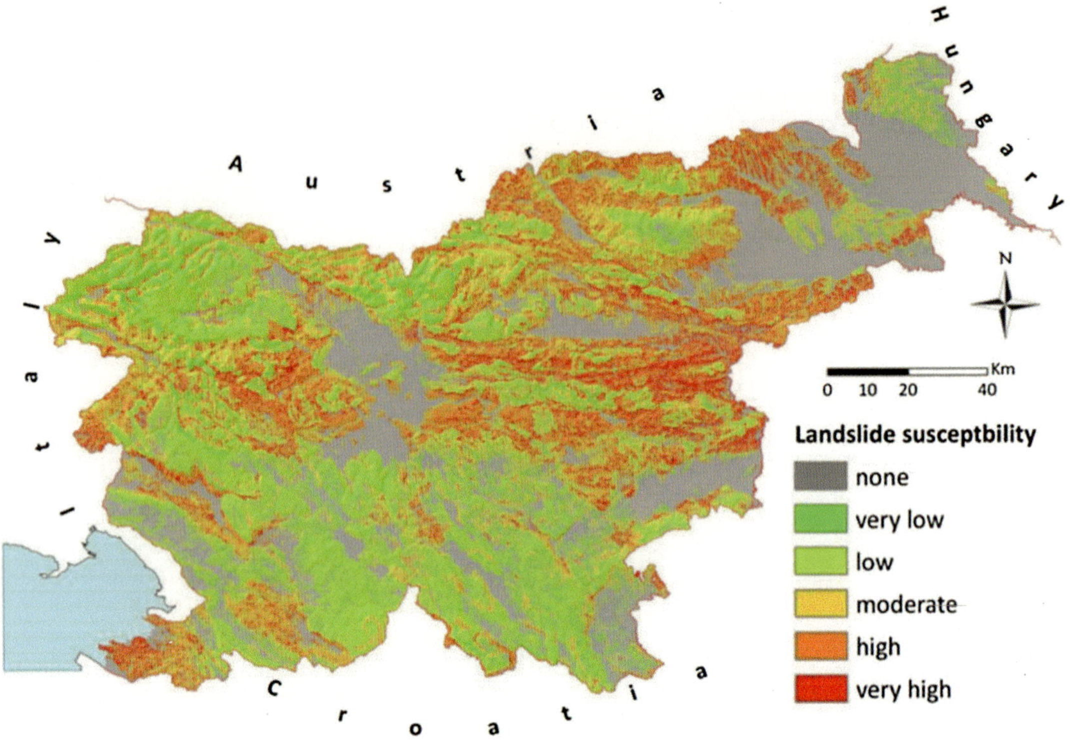

Fig. 3 Landslide susceptibility map of Slovenia (Komac 2012)

landslides as a consequence of heavy precipitation that would exceed the landslide triggering values were produced (Peternel et al. 2014). Exposure maps were elaborated based on a synthesis of analysis of event-based landslide inventory and field investigations. The exposures of inhabitants, buildings and different types of infrastructure to potential landslides were calculated with a cross-tabulation approach, using the susceptibility map at scale of 1:25,000 and a map of distribution of inhabitants, buildings and different types of infrastructure. The data on socio-economic elements (inhabitants, buildings and different types of infrastructure) were obtained from The Surveying and Mapping Authority of Republic of Slovenia (Ministry of Infrastructure and Spatial Planning). All the analyses were conducted in GIS, using ArcGIS software tools with a raster format 5 × 5 m. The levels of exposure were ranked into six classes, where level 1 represented areas with negligible exposure to potential landslides and level 6 represented areas with the highest level of exposure to potential landslides. The results of analyses are exposure maps of inhabitants, buildings and different types of infrastructure to potential landslides in the five Slovenian municipalities; they represent simplified risk assessment without inclusion of the economic loss. Exposure maps to potential landslides present basics for landslide risk assessment and later for the management of the landslide risk. Exposure maps should be used for guidance in the regional and local spatial planning process and could present preliminary warnings for the local authorities and the civil population. Exposure maps are not intended for direct use in the spatial planning process, as they only represent the current status of potential damage, and lack the information on the spatial distribution of the hazardous zones. Figure 4 shows factors influencing the development of exposure maps.

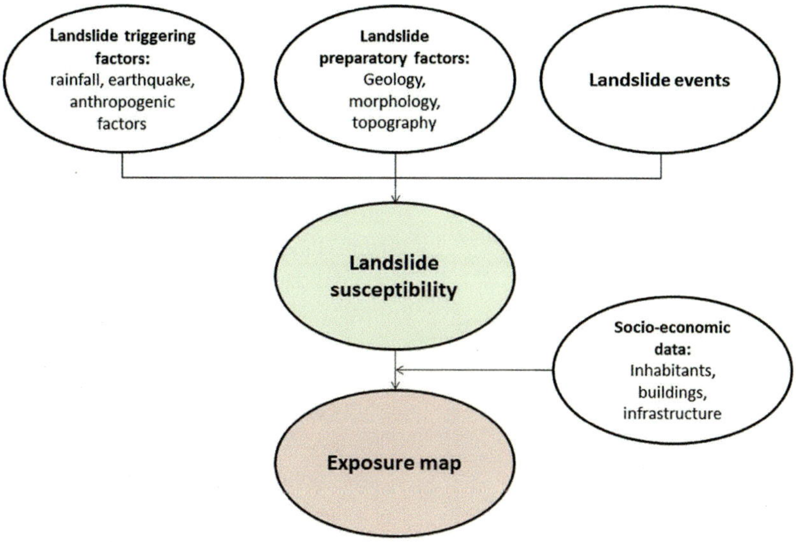

Fig. 4 Factors influencing the development of exposure maps

3 System Architecture

The Masprem system is used for calculating and displaying landslide prediction maps. It is a fully automated system developed and implemented by the Geological Survey of Slovenia. It is based on open source software (Šinigoj et al. 2015). Figure 5 demonstrates the concept of the architecture of the Masprem system.

The Masprem system includes: spatial data stored in a PostgreSQL database and spatial

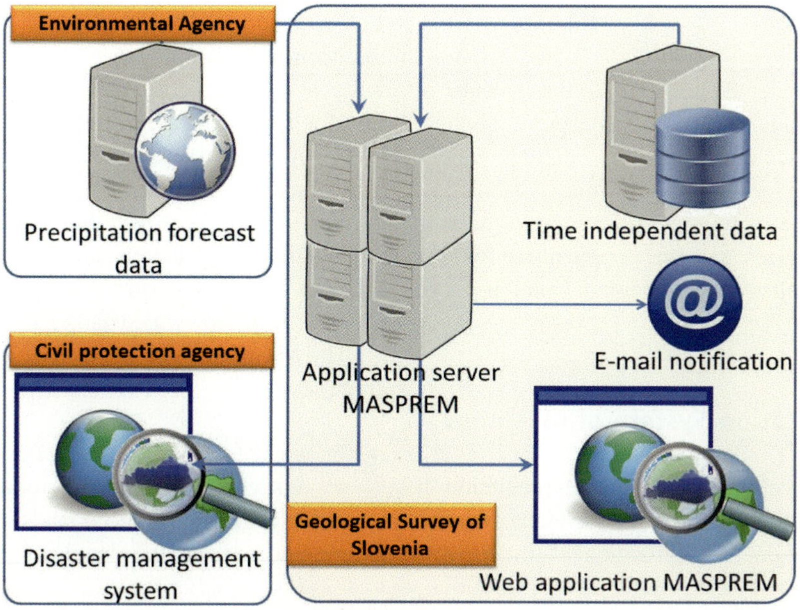

Fig. 5 The concept of the architecture of the Masprem system

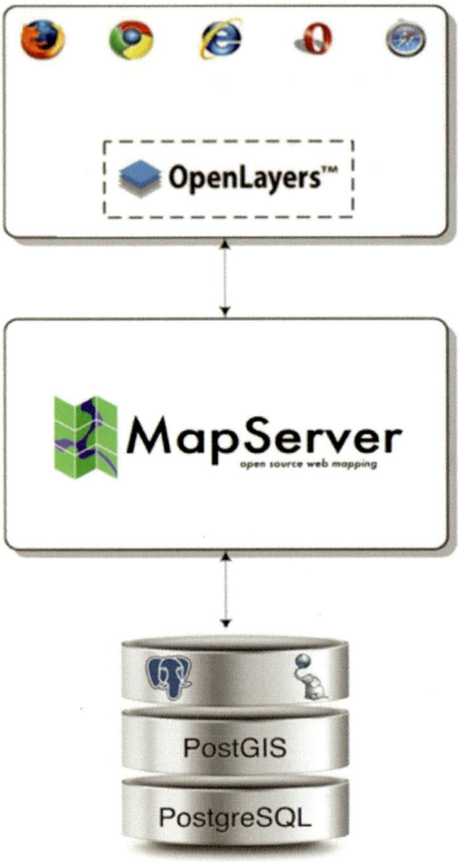

Fig. 6 Architecture of the Masprem system

database extender PostGIS (input data, results of calculated models—rasters), Masprem application for calculating various models (Java, GDAL), MapServer for publishing spatial data and interactive mapping applications to the web as WMS services and web application for displaying results (Implemented on top of OpenLayers) (Fig. 6).

3.1 Application System of Masprem

Dynamic modelling for landslide prediction is performed based on three independent datasets. The first input data are dynamic real-time rainfall derived from forecast of Aladin-SI twice a day. When Aladin-SI models are transferred successfully to the Geological Survey of Slovenia (GeoZS) server the conversion process to raster data starts and stores data in a PostgreSQL database. The same procedure is repeated with the remaining two rasters data or static input data sets landslide triggering threshold values and landslide susceptibility maps. When all input raster data are stored a forecast of rainfall-induced landslides is calculated and stored in database. Based on the final result, the WMS service that is responsible for the distribution of data through the service for download and review of results in a web application is created (Fig. 7). When the probability of landslide occurrences is increased, the system automatically sends an email to people responsible for disaster management at the Civil Protection Agency of Slovenia and to landslide experts (Masprem team) at the Geological Survey of Slovenia.

3.2 Masprem Web Application

After the calculation of the most plausible model for the probability of landslides, results have to be relayed to the end user, whether the decision-makers or general public. The most suitable tool for communication is the internet and for this purpose an interface was developed using open-source software. Programming was performed in Java and the results were transformed into the Web Map Service standard, used for displaying GIS-related raster images/data, and put on the internet via a web application (source: http://www.geo-zs.si/).

The Masprem Web application is presented in three-level:

- National level at the scale 1:250,000 (Fig. 8 a); currently, results of probabilistic modelling of rainfall induced landslides are shown generalized at the level of municipalities

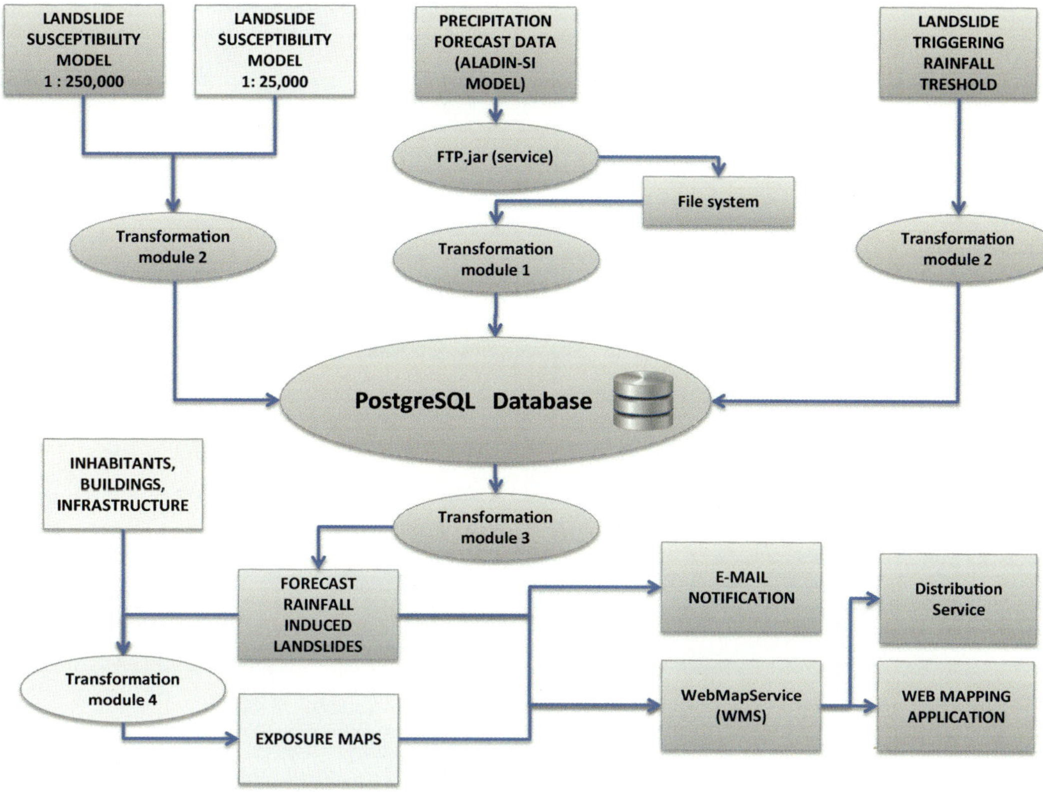

Fig. 7 Diagram of dynamic modelling of rainfall-induced landslides (Šinigoj et al. 2015)

(Fig. 8b). Due to the different degree of susceptibility of areas within municipalities, a new, more efficient display prediction is required in the future.

- The level of municipalities at the scale 1:25,000 (Fig. 9a).
- Model predictions of population exposure, buildings and infrastructure at the scale 1:25,000 (Fig. 9b).

4 Conclusion

Society strives towards self-protection and self-preservation, or at least mitigation of hazard consequences if they cannot be prevented. To manage the hazards associated with shallow landslides, accurate predictive warning systems for rainfall-induced landslides are needed. In the present Teaching Tool we introduced the design of the system for modelling the probability of landslides through time in Slovenia (Masprem). A system to forecast rainfall-induced landslides is based on real-time rainfall data, landslide-triggering rainfall threshold values and landslide susceptibility maps. The importance of their inclusion and development of system architecture is highlighted.

The development of a real-time landslide forecast system certainly brings benefits for the owners and planners of various infrastructure (roads, railways, gas/oil pipelines, buildings, electrical power systems), civil agencies, local

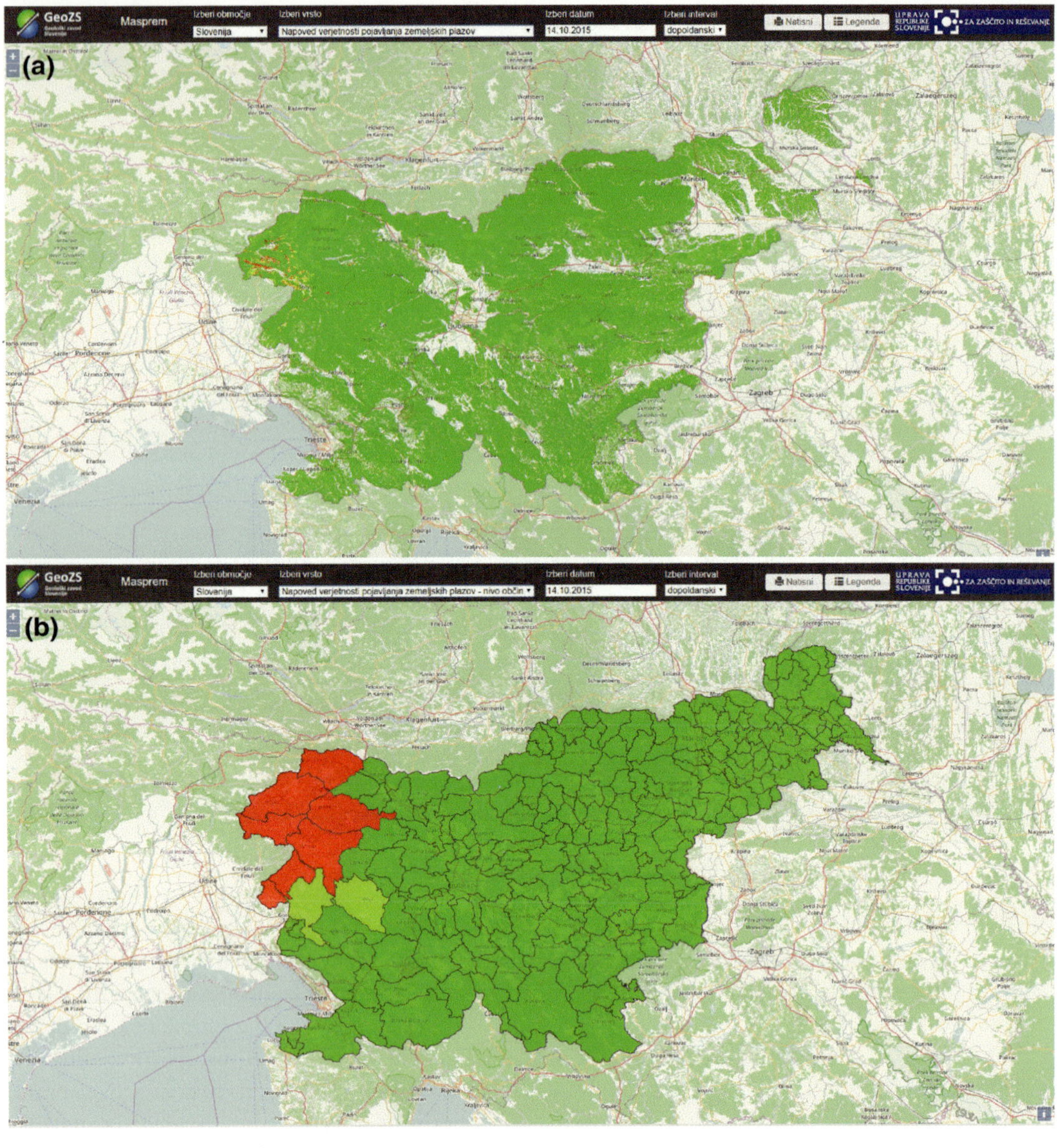

Fig. 8 **a** Web application of Masprem system at the national level in the scale 1:250,000. **b** Generalized results of probabilistic modelling of rainfall induced landslides at the national level in the scale 1:250,000

authorities, relevant government agencies and the citizens in the most exposed areas. For the next step, a validation has to be performed comparing the results of model runs with corresponding landslide events. When the validation phase is finished and the certainty of the system is high enough, the system will be able to inform infrastructure owners, civil agencies and operators of an increased probability of landslides as a consequence of heavy precipitation exceeding landslide-triggering rainfall values and enable them to mitigate risks. Furthermore this will improve the quality of life of citizens living in hazard areas, as well as their property and lives. First responders will have an early warning system that will give them enough time to save property and lives and also predict negative effects of landslides on

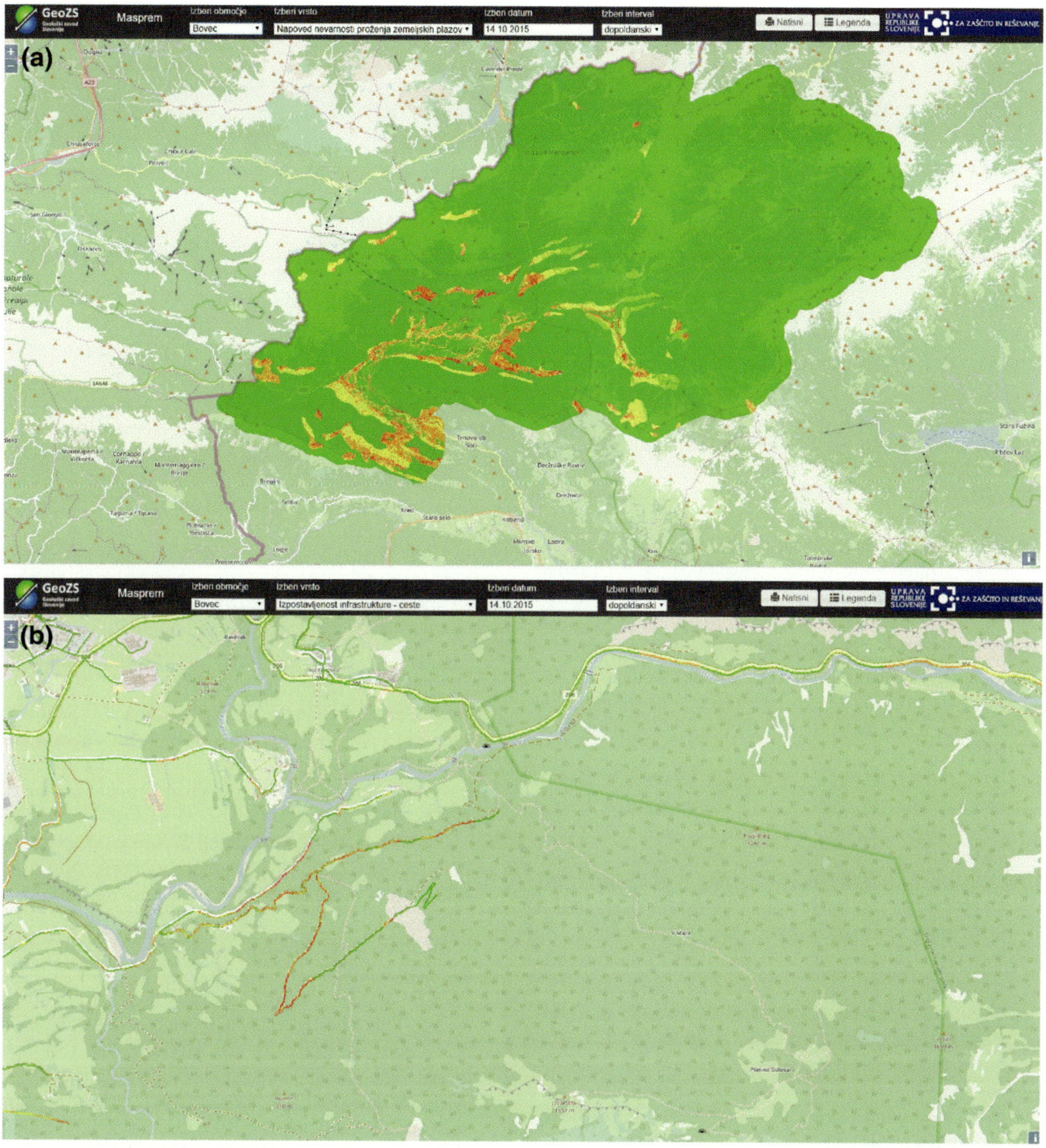

Fig. 9 **a** Forecast map of rainfall-induced landslides for the level of municipality at the scale 1:25,000. **b** Model prediction of exposure of infrastructures at the scale 1:25,000 when forecasted precipitation exceeds the rainfall threshold

infrastructure. Planners will have a susceptibility map developed to assist them when planning new infrastructure. Insurers and risk managers can use this system to assess the risk in a portfolio of exposures. This might help to guide an insurer's underwriting strategy or to decide how much reinsurance to purchase.

Acknowledgments The authors would like to thank the Administration for Civil Protection and Disaster Relief and Ministry for Defence for financing the project Masprem, the Slovenian Environment Agency (ARSO) for providing ALADIN-SI data and DG Information Society at the European Commission for financing the project InGeoClouds. Special thanks are also directed to Infrastructure programme of Geological Survey of Slovenia for support and implementing the project

Masprem and assistance in managing spatial data. The authors would like to thank colleagues with whom they conducted the research and worked on the project.

References

ARSO Ministry for Environment and Spatial Planning, Environmental Agency of the Republic of Slovenia (2015) National meteorological service of Slovenia. URL: http://meteo.arso.gov.si/met/en/app/webmet/. Last accessed 20 Dec 2015

Bubnová RG, Hello P, Béenard P, Geleyn JF (1995) Integration of the fully elastic equations cast in the hydrostatic pressure terrain following coordinate in the framework of the ALADIN NWP system. Mon Wea Rev 123:515–535

Chan RKS, Pun WK (2004) Landslip warning system in Hong Kong. Geotech News 22(4):33–34

Glade T, Crozier MJ, Smith P (2000) Applying probability determination to refine landslide-triggering rainfall thresholds using an empirical "Antecedent Daily Rainfall Model". Pure appl Geophys 157(6/8):1059–1079. doi:10.1007/s000240050017

Jemec Auflič M, Komac M (2013) Rainfall patterns for shallow landsliding in perialpine Slovenia. Nat Hazards 67(3):1011–1023. doi:10.1007/s11069-011-9882-9

Jemec Auflič M, Šinigoj J, Krivic M, Podboj M, Peternel T, Komac M (2016) Landslide prediction system for rainfall induced landslides in Slovenia (Masprem) = Sistem opozarjanja na nevarnost proženja zemeljskih plazov v Sloveniji (Masprem). Geologija 59 (2):259–271. doi:10.5474/geologija.2016.016. ISSN 0016-7789. [Tiskana izd.]

Keefer DK, Wilson RC, Mark RK, Brabb EE, Brown WM, Ellen SD, Harp EL, Wieczorek GF, Alger CS, Zatkin RS (1987) Real-time landslide warning during heavy rainfall. Science 238:921–925. doi:10.1126/science.238.4829.921

Komac M (2012) Regional landslide susceptibility model using the Monte Carlo approach—the case of Slovenia. Geol Q 56(1):41–54

Komac M, Ribičič M (2006) Landslide susceptibility map of Slovenia at scale 1:250,000. Geologija 49(2):295–309. doi:10.5474/geologija.2006.022

Li CJ, Ma TH, Zhu XS (2010) aiNET-and GIS-based regional prediction system for the spatial and temporal probability of rainfall triggered landslides. Nat Hazards 52:57–78

Manunta M, Calo F, Paglia L, Bonano M, Lanari R (2012) DORIS FP7-EU project: exploitation of 20 years DInSAR Data archive for landslide monitoring. In: Proceedings of fringe 2011, held 19–23 Sept, 2011 in Frascati, Italy. ESA SP-697. p 61

Mercogliano P, Schiano P, Picarelli L, Olivares L, Catani F, Tofani V, Segoni S, Rossi G (2010) Short term weather forecasting for shallow landslide prediction. In: Malet JP, Glade T, Casagli N (eds) International Conference on mountain risks. Bringing Science to Society, Firenze, pp 525–530

Ortigao JAR, Justi MG, D'Orsi R, Brito H (2002) Rio-watch 2001: the Rio de Janeiro landslide alarm system. In: Ho KKS, Li KS (eds) Proceedings, 14th Southeast Asian geotechnics conference—geotechnical engineering: meeting society's needs, pp 237–241

Peternel T, Šinigoj J, Komac M, Jemec Auflič M, Krivic M (2014) Exposure of inhabitants, buildings and infrastructure to landslides: a case of five Slovenian municipalities. Geologija 57(2):193–202. doi:10.5474/geologija.2014.017

Pristov N, Cedilnik J, Jerman J, Strajnar B (2012) Priprava numerične meteorološke napovedi ALADIN-SI. Vetrnica, pp 17–23

Ribičič M, Šinigoj J, Komac M (2003) New general engineering geological map of Slovenia. Geologija 46/2:397–404 (Ljubljana)

Rosi A, Peternel T, Jemec Auflič M, Komac M, Segoni S, Casagli N (2016) Rainfall thresholds for rainfall-induced landslides in Slovenia. Landslides: J Int consortium landslides 13(6):1571–1577. doi:10.1007/s10346-016-0733-3

Šinigoj J, Jemec Auflič M, Kumelj Š, Krivic M, Požar M, Podboj M, Tukić M, Peternel T, Prkić N (2015) Nadgradnja sistema za obveščanje in opozarjanje v primeru proženja zemeljskih plazov—Masprem2: poročilo ob prvem mejniku. Geološki zavod Slovenije, Ljubljana, p 42

Thiebes B (2012) Landslide analysis and early warning systems: local and regional case study in the Schwabian Alb, Germany. Springer, Heidelberg, p 257 (Springer theses, recognizing outstanding Ph.D. research)

Thuro K, Wunderlich Th, Heunecke O, Singer J, Wasmeier P, Schuhback S, Festl J, Glabsch J (2009) alpEWAS—the Aggenalm landslide—a testing site for an integrative 3D early warning system. In: Geotechnologien science report, Statusseminar 12–13 Oct 2009. Technical University Munich, Munich, pp 33–48

White ID, Mottershead DN, Harrison J (1996) Environmental systems, 2nd edn. Chapman & Hall, London, 616 p. http://www.geo-zs.si

TXT-tool 2.886-1.1 Early Warning Criteria for Debris Flows and Their Application in Taiwan

Chyan-Deng Jan, Ji-Shang Wang, Yi-Chao Zeng and Feng-Hao Kuo

Abstract

As a result of the heavy rains that accompany typhoons, steep topography, young and weak geological formations, strong earthquakes, loose soils, and land development in mountainous terrain, many areas in Taiwan are susceptible to landslides and debris flows. For evacuation in case of emergency, it will be necessary to enhance public awareness of many debris-flow hazards and educate people on how to react to these hazards. A rainfall-based debris-flow warning model and system is needed for preventing or mitigating debris-flow hazards. A rainfall-based warning system for debris flows has been developed and applied by the Soil and Water Conservation Bureau (SWCB) in Taiwan. The rainfall-based warning model used to predict the occurrence potential of a debris flow and its applications in Taiwan are presented herein.

Keywords

Debris flow · Landslide · Rainfall-based warning model · Taiwan

Contents

1 **Introduction** 406
2 **Definition of a Rainfall Event and Related Rainfall Parameters** 406
3 **A Rainfall-Based Warning Model** 407
4 **Applications** 409
5 **Case Studies of Debris Flow Warnings** 410
5.1 Debris Flows Triggered by Typhoon Mindulle 410
5.2 Debris Flows Triggered by Typhoon Haitang. 411
5.3 Debris Flows Triggered by Typhoon Morakot 413
6 **Conclusion** 414
References 414

C.-D. Jan (✉)
Department of Hydraulic and Ocean Engineering, National Cheng Kung University, No. 1 University Road, Tainan 70101, Taiwan
e-mail: cdjan@mail.ncku.edu.tw

J.-S. Wang · Y.-C. Zeng
Soil and Water Conservation and Ecological Engineering Research Center, National Cheng Kung University, No. 1 University Road, Tainan 70101, Taiwan
e-mail: rheo.js@gmail.com

Y.-C. Zeng
e-mail: zengyichao@gmail.com

F.-H. Kuo
Department of Engineering, Cheng Yao Construction Company, 12F-1, No. 93, Jiangong 1st Rd., East Dist., Hsinchu 300, Taiwan
e-mail: hugebig.tw@yahoo.com.tw

K. Sassa et al. (eds.), *Landslide Dynamics: ISDR-ICL Landslide Interactive Teaching Tools*, https://doi.org/10.1007/978-3-319-57774-6_30

1 Introduction

Taiwan island was formed by the collision of an island arc with the Asian continental margin, and still has active mountain formation. As a result, two-thirds its area of 36,000 km^2 is covered by high mountains and hills. With a warm and humid climate throughout the year, Taiwan receives an average annual rainfall of 2500 mm, and rainfall may exceed 5000 mm in some mountainous regions.

Abundant loose soils, steep slopes, and large amounts of water are three basic conditions for the occurrence of a debris flow. Large amounts of loose soils provide the basic materials for the body of the debris flow, the steep slope provides the energy for debris flow movement, and water, especially rainfall, plays the role of a triggering factor to initiate debris flow movement. In Taiwan, water mainly comes from some severe rainfalls, especially in rain brought by typhoons. For a specified watershed of a debris-flow gully, the changes in topographic and geological conditions, as well as loose soil conditions in a period of normal weather are small compared with the changes caused by rainfall. Therefore, if one can analyze the rainfall characteristics and their relation with debris flow occurrence, one can estimate the potential of debris flow occurrence in any given rainfall event, based on the rainfall conditions (Chen et al. 2012, 2013; Jan et al. 2003; Jan and Chen 2005). This paper introduces the early warning rainfall criteria for debris flows and their application in Taiwan.

2 Definition of a Rainfall Event and Related Rainfall Parameters

Many researchers use different methods to define a rainfall event for different purposes, and this may result in different values for rainfall parameters, even for the same rainfall event. Herein, a rainfall event used to study its effect on debris flow initiation is defined so that the starting-time for the rainfall event is the time at which the hourly rainfall just begins to exceed 4 mm, and the ending-time for the rainfall event is the time when the hourly rainfall has dropped below 4 mm and lasts over 6 h, as schematically indicated in Fig. 1.

The common rainfall parameters used to set up a rainfall-based debris-flow warning model are the rainfall intensity I, the rainfall duration T and the accumulated rainfall R, and also the antecedent rainfall P of the given rainfall event. Previous researchers have used the rainfall intensity I and the rainfall duration T to derive the critical volume of water for triggering a landslide or debris flow. However, an important contribution to the occurrence of a debris flow is not only the rainfall event but also its antecedent rainfall. Except for the given rainfall event itself, the closer the antecedent rainfall is to the main rainfall, the greater its contribution to triggering a debris flow. An effective accumulated rainfall contributing to the occurrence of a debris flow should thus include both the given rainfall event and its antecedent rainfall. In the present model, the effective accumulated rainfall $R(t)$ at time t in

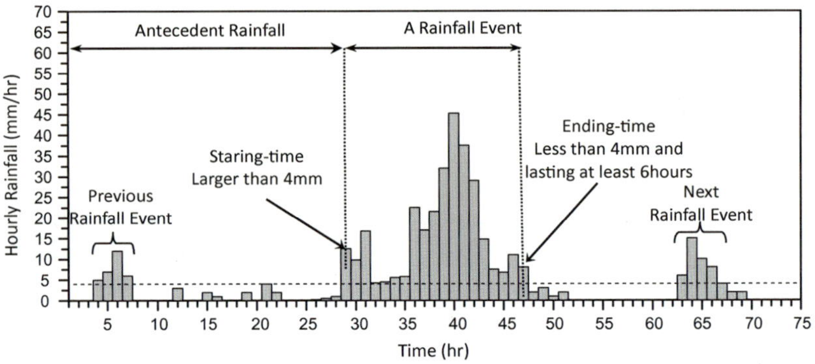

Fig. 1 Definition of a rainfall event used for debris-flow-event analysis

a rainfall event is defined as the sum of the accumulated rainfall $R(t)$ at time t of the rainfall event, and the seven-day effective antecedent rainfall $R_t(t)$ is defined as,

$$R_t(t) = R(t) + \sum_{i=1}^{7} \alpha^i R_i \qquad (1)$$

in which $R(t)$ is the amount of accumulated rainfall at time t in the given rainfall event; R_i is the amount of the antecedent i day's rainfall; and α is a weighting factor. We previously used $\alpha = 0.8$, but 0.7 is used currently. An instant rainfall triggering index (*RTI*) at time t for estimating the potential for debris flow occurrence is defined as the product of hourly rainfall intensity $I(t)$ and the effective accumulated rainfall $R_t(t)$ at time t, i.e.,

$$RTI(t) = R_t(t) \times I(t) \qquad (2)$$

The volume of water needed for debris flow movement depends on physiographic conditions in the watershed. With conditions of higher porosity or water content of a soil on a steeper slope, less rainfall is usually needed to trigger a debris flow. However, physiographic factors and rainfall conditions for a specified watershed of a debris flow gully usually include a lot of uncertainties, or may even be unknown. Thus the critical rainfall for triggering debris flows in different rainfall events for the same area usually has some uncertainty that should be considered in debris-flow warning practice.

3 A Rainfall-Based Warning Model

The rainfall parameters for all rainfall events that have ever triggered debris flows are very important data for establishing a debris flow warning model. Rainfall conditions before the debris flow event represent the amount of water that reaches the critical trigger for debris flow occurrence; while the rainfall after the debris flow occurs mainly affects the magnitude of the debris flow. However, only some debris-flow events have been observed to be initiated, and most of them were about at the time of the peak of hourly rainfall in a rainfall event.

In most instances, the exact time for debris flow initiation in a rainfall event is unclear due to the fact that debris flows usually occur in mountainous areas, so there is a lack of accurate recordings. If the time of the debris flow occurrence is unknown, the time of the peak hourly rainfall in a rainfall event is assumed to be the time of debris-flow initiation in the parameter analysis of a rainfall event, as shown in Fig. 1. The accumulated rainfall is defined as the total rainfall between the starting-time of a rainfall event to the time of debris flow initiation. The hourly rainfall intensity I is defined as the hourly rainfall intensity at the time of debris flow occurrence.

The representative value of the rainfall triggering index (*RTI*) for each historical rainfall event, no matter whether it triggered a debris flow or not, is calculated and used to determine the critical values of *RTI* for debris flow warnings in a specified debris-flow-prone area. The value of *RTI* for a historical rainfall event is defined as the product of the hourly rainfall intensity I (mm/h) and the effective accumulated rainfall R_t (mm) at the time of debris-flow occurrence, if available; otherwise, *RTI* is the product of the peak rainfall intensity and its corresponding effective accumulated rainfall R_t at the time of the peak rainfall intensity in the rainfall event. A scatter-plot of the rainfall triggering index *RTI* (mm^2/h) of historical rainfall events versus rainfall event consecutive number N is shown in Fig. 2.

The symbols of RTI_j shown in the figure refer to the value at which j percent (for example, 10% or 90%) of *RTI*-values for historical rainfall events in a given debris-flow watershed during a given period is smaller than RTI_j. If there are a lot of debris flow events (i.e., more then 10) in the debris-flow watershed, the empirical lower critical line of *RTI* could be defined as the lowest *RTI*-values of rainfall events that had triggered debris flows. Otherwise we use $RTI = RTI_{10}$ as the lower critical line. The upper critical warning line of *RTI* is defined as that 90% of *RTI*-values for the historical rainfall events (no matter

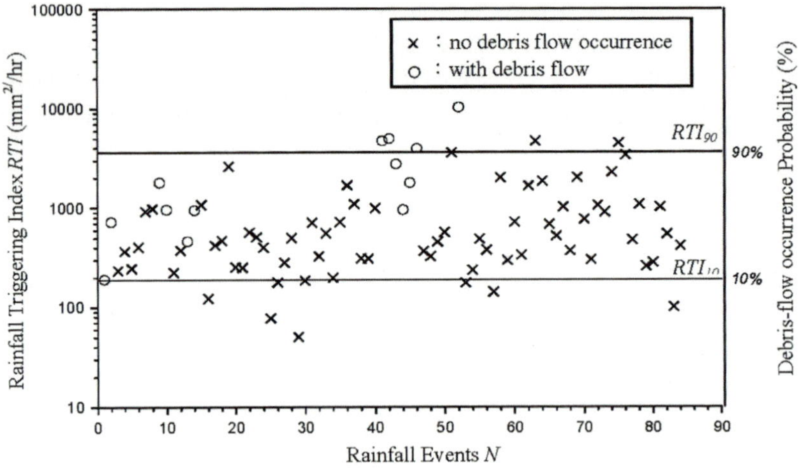

Fig. 2 Distribution of *RTI*-values for corresponding historical rainfall events that occurred in a debris-flow watershed

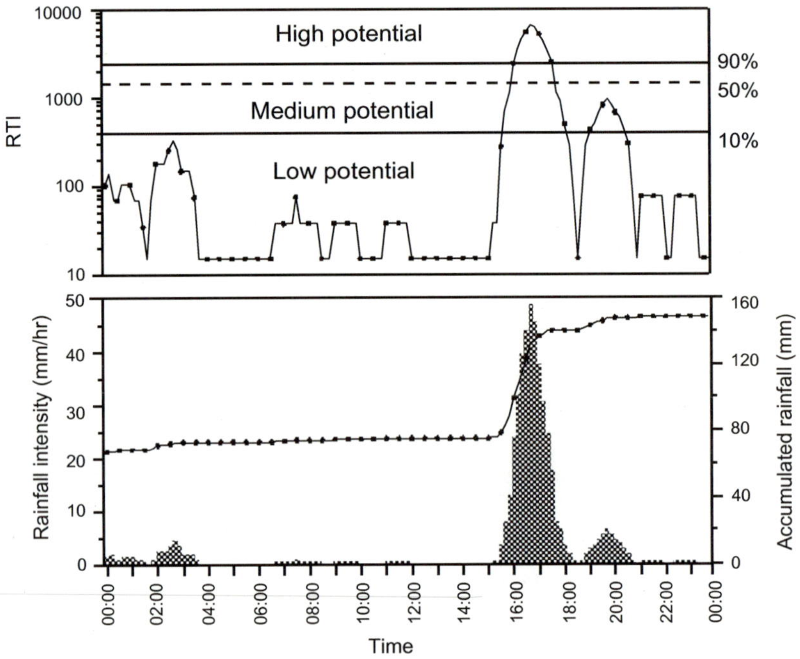

Fig. 3 Schematic diagrams of temporal variations of rainfall intensity, effective accumulated rainfall, *RTI*-values and corresponding debris flow potentials during a rainfall-time series

whether they triggered debris flows or not) is smaller than it (i.e., $RTI = RTI_{90}$). The lower and upper critical warning lines divide the *RTI*-plot into three parts. Once the two critical *RTI*-values are determined, one uses *RTI* values as a vertical coordinate, while the time considered as a horizontal coordinate, as shown in Fig. 3. The location of the value of instant $RTI(t)$ at time t on the *RTI* figure will indicate the debris-flow occurrence potential.

The *RTI*-values in the lower part are all smaller than the lower critical value (RTI_{10}). When the value of *RTI* is smaller than RTI_{10}, the probability of debris flow occurrence is small. The *RTI*-values in the higher part are all larger than the upper critical value, and so the probability of debris-flow occurrence is high. The *RTI*-values in the middle part are between RTI_{10} and RTI_{90} and thus the probability of debris-flow occurrence is medium (may be between 10 and

90%). The relation of the probability of debris-flow occurrence P and rainfall triggering index RTI between the lower and the upper critical line is assumed to be a linear relation, given as

$$P(RTI) = 0.1 + 0.8\left(\frac{RTI - RTI_{10}}{RTI_{90} - RTI_{10}}\right) \quad (3)$$

Figure 3 shows schematic diagrams of temporal variations of rainfall intensity, effective accumulated rainfall, RTI-values and corresponding debris flow potentials during a rainfall-time series. Based on the empirical lower and upper critical lines of RTI, the temporal variations of debris-flow potential during a rainfall event can be estimated, so as to provide debris flow warning for people in mountainous areas.

4 Applications

The Soil and Water Conservation Bureau (SWCB) of the Council of Agriculture (COA) is responsible for all debris flow hazards prevention and mitigation using engineering or non-engineering works in Taiwan. The web-based decision support system for debris-flow disaster management has been developed. The system integrates various kinds of information and provides the most important decision-making support for various levels of governments in Taiwan (Liu et al. 2013; Yin et al. 2015).

Structural countermeasures used for debris flow mitigation include the construction of debris barriers, debris breakers, debris basins, deflection berms, slit dams, check dams, and/or Sabo dams in debris-flow gullies and on alluvial fans. In addition to such debris-flow control structures, other countermeasures include the installation of rainfall-based warning and monitoring systems and instrument-operated warning systems. The debris-flow warning model proposed in this paper is adopted by the Soil and Water Conservation Bureau and the critical RTI-value for debris flow warning for evacuation is set at $RTI = RTI_{70}$ in practice. The rainfall triggering index RTI is based on both the accumulated rainfall and the rainfall intensity. For engineers and officers in debris-flow management, it is not difficult to understand the meanings of RTI. However, the general public can easily understand the meaning of the accumulated rainfall but not the rainfall intensity. Therefore, we simplify it by using the accumulated rainfall as the single rainfall parameter presented in debris-flow warnings for the public. For this purpose, we use RTI_{70} as a critical value and then divide RTI_{70}, using I = 10 mm/h as the value to obtain the corresponding accumulated rainfall R_t (that was rounded by using 50 mm as an interval), as shown in Fig. 4 and Table 1.

The use of I = 10 mm/h as a divisor is generally based on the trial-and-error method, with the consideration that most debris flows occur at a rainfall intensity of greater than 10 mm/h in Taiwan. The accumulated rainfall-based critical value for practical debris-flow warning is basically classified into 9 levels between 200 and 600 mm, with 50 mm as an interval. The nine categories of critical accumulated rainfall for different debris-flow streams thus are R_c = 200, 250, 300, 350, 400, 450, 500, 550 and 600 mm, as shown in Table 1. These R_c-values have been

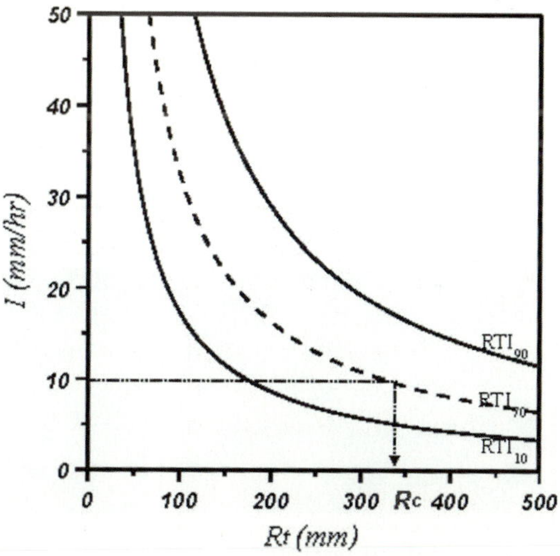

Fig. 4 The method used to assess the critical accumulated rainfall for debris flow warning, based on the curve of RTI_{70} under the rain intensity of 10 mm/h

Table 1 Nine categories of critical accumulated rainfall for debris-flow warnings used in Taiwan

Category	A	B	C	D	E	F	G	H	I
R_c (mm)	200	250	300	350	400	450	500	550	600

currently adopted by the Soil and Water Conservation Bureau for debris flow warning in Taiwan.

If the predicted rainfall has an effective accumulated rainfall R_t larger than R_c, the debris flow warning level is yellow and in this situation, people in the potential areas of debris flow flooding should be prepare to leave or to be evacuated. If the actual rainfall has an R_t value larger than R_c, the warning level is red and people in the potential debris-flow flooding areas should be forcibly evacuated. In areas that have lower values of R_c, the physiographic conditions are poor and debris flows would easily occur in these areas; areas with larger values of R_c have physiographic conditions that are better and more rainfall is need to trigger debris flows, as shown in Fig. 5.

The debris-flow warning criteria for different areas also vary with their corresponding physiographic conditions. The lower debris flow warning criteria usually apply to areas that are more landslide-prone or in more fragile geological condition in the watershed. Figure 5 shows three examples of the physiographic conditions in the watersheds of streams prone to debris-flows, having R_c of 200, 400 and 600, respectively. These three debris-flow streams are named as Kaohsiung DF007, Tainan DF033 and Pintung DF055, respectively.

5 Case Studies of Debris Flow Warnings

The warning model described in the previous section has been adopted by the Soil and Water Conservation Bureau to estimate debris-flow occurrence potential, to provide debris flow warnings for people in mountainous areas in Taiwan. Several examples of the application of this debris-flow warning model in Taiwan are discussed below.

5.1 Debris Flows Triggered by Typhoon Mindulle

Typhoon Mindulle, which landed on Hulien in Eastern Taiwan on July 1, 2004 and then left on July 2, brought heavy rain in the eastern and southern parts of Taiwan while moving through the country from July 1–2. However, additional torrential rain fell after Typhoon Mindulle left, especially on the central and

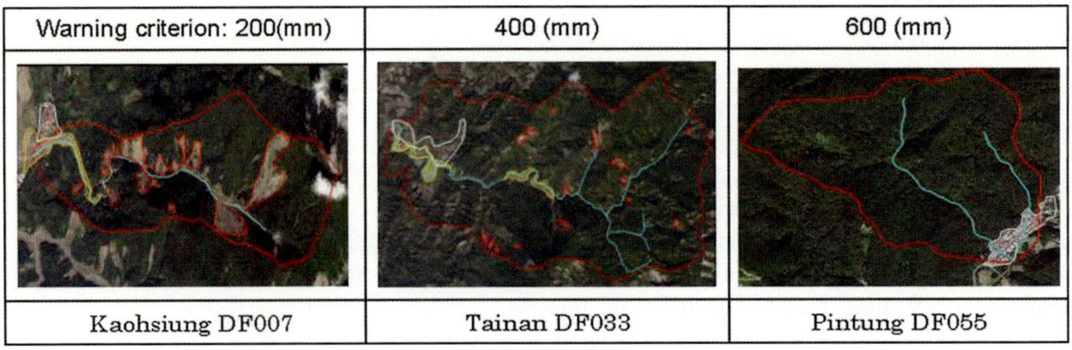

Fig. 5 Physiographic conditions corresponding to three different debris-flow warning criteria

southern areas of Taiwan. The three-day rainfalls between July 2 and 4 exceeded 2000 mm. The heavy rainfalls caused numerous floods, landslides and debris flows, and killed 41 people, mainly by flooding, rather than by debris flows or landslides.

Four examples of debris-flow warnings in central Taiwan are discussed here. They are at Heping and Dongshi in Taichung County, and at Renai and Guosing in Nantou County, respectively. Based on the debris-flow warning model, as well as the predicted and immediate rainfall data from the rainfall stations, debris flow warnings had been issued at various times on July 2 for the above-mentioned four areas.

The time of debris-flow occurrence and the debris-flow warning time for each of the four areas are listed in Table 2. From this table one can see that the warning times of debris flow occurrence were about 15–28 h before the time the debris flows occurred. The warning time provided local government and people enough time to do some easy mitigation works and to carry out evacuations from areas threatened by debris flows.

The temporal variations of rainfall intensity, accumulated rainfall, rainfall triggering index and its corresponding probability of debris flow occurrence at Heping, on July 3, 2004 are shown in Fig. 6. The first warning of potential debris-flow occurrence at Heping was issued by the Soil and Water Conservation Bureau at 10:40 a.m. on July 2. After 02:10 a.m. on July 3, heavy rain was continuously falling, so the probability of debris-flow occurrence was consecutively higher than 90% for many hours. Finally the debris flow occurred at 08:30 on July 3; at this time the rainfall intensity and effective accumulated rainfall were 81.5 mm/h and 472 mm, respectively. The peak hourly rainfall of 125.5 mm occurred at 9:00 a.m. and its corresponding effective accumulated rainfall was 527 mm. After the debris flow initiated, heavy rains still continuously fell for hours, and this enlarged flooding areas.

The data for debris flows in the Typhoon Mindulle event show that the hourly rainfall intensity at the time of debris-flow occurrence was 34–125 mm/h and its corresponding accumulated rainfall was 380–472 mm. The difference between the time of debris flow occurrence and the time of peak rainfall intensity was within 1 h. In addition, at the time the debris flow occurred, the probabilities of debris-flow occurrence estimated by the proposed model were all higher than 90% for these four examples.

5.2 Debris Flows Triggered by Typhoon Haitang

Typhoon Haitang formed to the northwest of Guam on July 12, 2005, and passed through Taiwan at 22:00 on July 18. It brought abundant rain and caused severe disasters in Taiwan,

Table 2 Comparison of the time of debris-flow warning and the time of debris-flow occurrence in four debris-flow prone areas during heavy rainfalls brought by Typhoon Mindulle in 2004

Location	Time of debris flow occurrence	Rainfall intensity and effective accumulated rainfall at the time of debris-flow occurrence		Debris-flow warning time by the SWCE based on the proposed model	Length of warning time ahead the time of debris-flow occurrence
		I (mm/h)	R_t (mm)		
Heping	07/03 08:30	81.5	472	07/02 10:40	26 h 50 min
Dongshi	07/03 08:00	34.0	421	07/02 16:10	15 h 50 min
Renai	07/03 09:10	55.5	380	07/02 09:50	28 h 20 min
Guosing	07/03 08:00	125.0	419	07/02 10:50	21 h 10 min

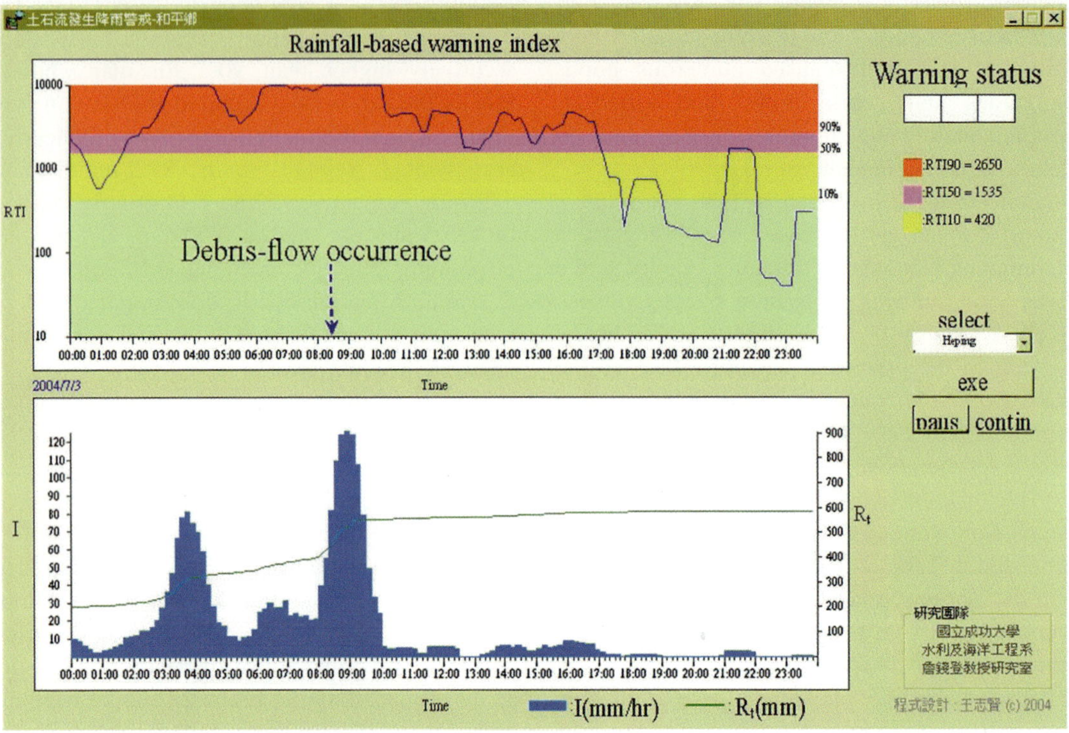

Fig. 6 Temporal variations of the rainfall intensity, the effective accumulated rainfall, the rainfall triggering index and its corresponding debris-flow occurrence probability at Heping, Taichung county on July 3, 2004

especially in mountainous areas in north-eastern and southern Taiwan, where the 24-h rainfall exceeded 300 mm. The heavy rainfalls caused 11 debris flows in Taiwan; four of them occurred in southern Taiwan. However, no people were killed by debris flows during Typhoon Haitang, due to the contribution of accurate debris flow warnings. The rainfall-based debris flow warning criteria adopted by the Soil and Water Conservation Bureau were applied to debris flow warning practice at Kaohsiung A031 in Liouquei during Typhoon Haitang. The temporal variations in rainfall intensity, effective accumulated rainfall and rainfall triggering index at Liouquei during July 18–20 are shown in Fig. 7. From the figure, it can be seen that the accumulated rainfall exceeded the critical value at 19:00 p.m., and Bureau informed the Administration Offices of Liouquei of the need to evacuate residents at 20:00 p.m. on July 18.

About 40 village residents were quickly evacuated to a safer place at night. The debris flow (named Kaohsiung A031) at Liouquei occurred at 14:00 on July 19, at which time the hourly rainfall and accumulated rainfall were 20.5 mm/h and 867 mm, respectively. This debris flow event buried five houses and severely damaged a main road. The amount of soil and rock brought down by the debris flow was about 500 m^3 and the deposition depth was about 2–3 m. Fortunately, there were no injuries or deaths in the debris flow event. The evacuation time was about 18 h ahead of the time the debris flow occurred, and this allowed the officers in charge of debris flow management enough time to reconfirm rainfall conditions, and to inform the local government and villages, and the local residents had enough time to leave or be evacuated before the debris flow occurred.

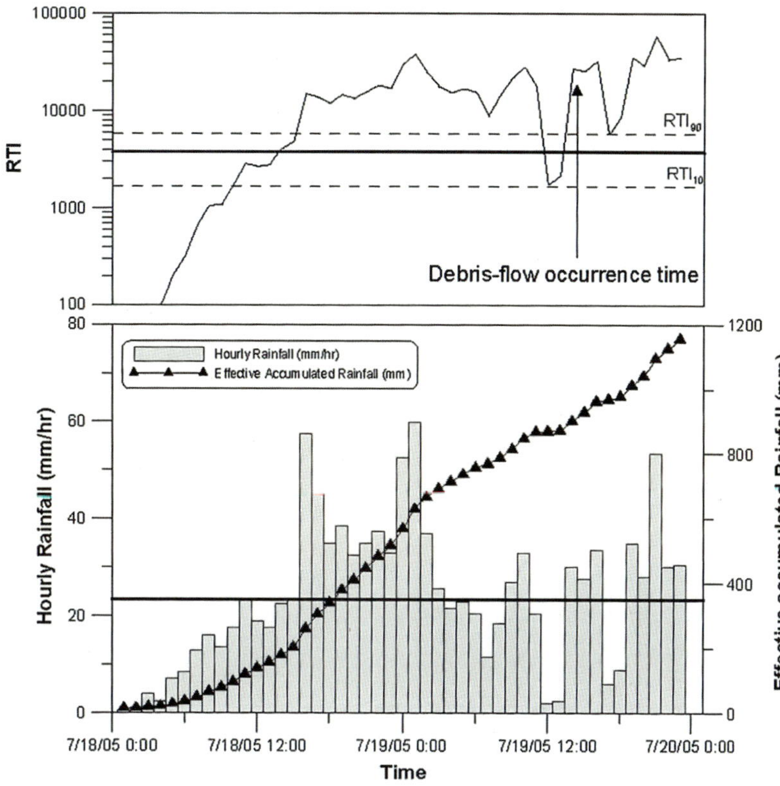

Fig. 7 Temporal variations of the rainfall intensity, the effective accumulated rainfall and the rainfall triggering index during Typhoon Haitang at Liouquei, Kaohsiung in Taiwan

5.3 Debris Flows Triggered by Typhoon Morakot

Typhoon Morakot struck Taiwan during August 7–10, 2009 and brought a copious amount of rainfall, with a peak up to 3000 mm, surpassing the previous record of 1987 mm caused by Typhoon Herb in 1996. The extreme amount of rain triggered enormous mudslides and debris flows, and caused severe flooding throughout southern Taiwan. One deep-seated landslide that occurred in Xiao-lin buried about 400 people in the village. The Xin-kai tribe was destroyed by a huge debris flow with a sediment volume of over one million cubic meters. The debris flow buried 38 buildings, and killed four people, with 24 people missing. Nevertheless, the debris-flow warning still made a good contribution, helping people from about 20 houses to leave, thus avoiding disaster. Table 3 shows the warning time and the time of debris-flow/landslide occurrence in Southern Taiwan.

Table 3 The time of sediment-related hazards and the corresponding time of debris-flow warnings in Southern Taiwan during Typhoon Morakot

Site	Hazard type	Occurrence time	Warning time
Xi-an	Debris flow	20:00, August 8	23:00, August 7
Dong-an	Debris flow	20:00, August 8	23:00, August 7
Xin-kai	Debris flow	21:00, August 8	23:00, August 7
Xin-long	Landslide	04:00, August 9	23:00, August 7
San-lin	Debris flow	05:00, August 9	08:00, August 8
Xiao-lin	Landslide	06:00, August 9	23:00, August 7

6 Conclusion

Aggressive activities on debris-flow disaster reduction have been conducted since 1996 after severe debris flows caused by Typhoon Herb. The web-based decision support system for debris-flow disaster management and the rainfall-based debris flow warning system have been developed and proved to be effective for evacuation operation and disaster reduction. Education and evacuation practice for the local community residents are very important in managing debris flow disaster reduction. Evacuation operations sometimes may cause complaints and criticism from the public due to false alarms. To improve the performance of warnings, the rainfall-based warning criteria and the system should be regularly revised based on the feedback from new debris flow events and the local public opinion.

Acknowledgements Special thanks are directed to the Ministry of Science and Technology as well as to the Soil and Water Conservation Bureau of the Council of Agriculture in Taiwan for their long-term supports to the authors in the studies of debris-flow warning and debris-flow disaster reduction.

References

Chen JC, Huang WS, Jan CD, Yang YH (2012) Recent changes in the number of rainfall events related to debris-flow occurrence in the Chenyulan stream watershed, Taiwan. Nat Hazards Earth Syst Sci 12:1539–1549

Chen JC, Jan CD, Huang WS (2013) Characteristics of rainfall triggering of debris flows in the Chenyulan watershed, Taiwan. Nat Hazards Earth Syst Sci 13:1015–1023

Jan CD, Chen CL (2005) Debris flow caused by Typhoon Herb in Taiwan (chapter 21). In: Jakob M, Hungr O (eds) Debris-flow hazards and related phenomena. Springer, Berlin, pp 539–563

Jan CD, Lee MH, Huang TH (2003) Effect of rainfall on debris flows in Central Taiwan. In: International conference on slope engineering, Hong Kong. pp 751–756

Liu KF, Jan CD, Lin PS, Li HC (2013) Advance of geo-disaster mitigation technologies in Taiwan. In: Wang F et al (eds) Progress of geo-disaster mitigation technology in Asia. Springer, Berlin, pp 77–103

Yin HY, Lee CY, Jan CD (2015) A Web-based decision support system for debris flow disaster management in Taiwan. In: Lolino G et al (eds) Engineering geology for society and territory, vol 3. Springer, Berlin

TXT-tool 2.081-5.1 High-Resolution Rainfall Prediction for Early Warning of Landslides

Ryo Onishi, Keigo Matsuda and Keiko Takahashi

Abstract

Most of tragic landslides are caused by heavy rainfall. Reliable rainfall prediction is thus invaluable for early warning. This paper describes a recent development of our next generation numerical weather prediction model –the Multi-Scale Simulator for the Geoenvironment (MSSG)—after introducing general information about numerical weather prediction models. The MSSG can be categorized as a global cloud-resolving model that can be used for global simulation with high-resolutions without the aid of cumulus parameterization. MSSG is thus expected to have a good skill for local rainfall prediction as well as for long-term prediction. This paper also shows a couple of examples that show a good prediction skill of MSSG and a powerful physics scheme implemented in MSSG for the research on local rainfalls.

Keywords

Multi-scale weather simulation • Earth simulator • Orographic precipitation • Turbulence enhancement of cloud development

Contents

1 **Introduction** ... 416
2 **Numerical Weather Prediction Models** ... 416
3 **Multi-scale Weather Prediction Model—Multi-scale Simulator for the Geoenvironment (MSSG)** ... 417
3.1 Overview ... 417
3.2 Dynamical Core of the Atmosphere Component ... 417
3.3 Cloud Microphysics ... 417
4 **Rainfall Predictions** ... 419
4.1 Regional Numerical Weather Prediction ... 419
4.2 Turbulence Influence on Orographic Precipitation ... 419
5 **Conclusions** ... 422
References ... 422

R. Onishi (✉) · K. Matsuda · K. Takahashi
Center for Earth Information Science and Technology, Japan Agency for Marine-Earth Science and Technology, 3173-25 Showamachi Kanazawa-ku, Yokohama 236-0001, Kanagawa, Japan
e-mail: onishi.ryo@jamstec.go.jp

K. Takahashi
e-mail: takahasi@jamstec.go.jp

K. Sassa et al. (eds.), *Landslide Dynamics: ISDR-ICL Landslide Interactive Teaching Tools*, https://doi.org/10.1007/978-3-319-57774-6_31

1 Introduction

Most of tragic landslides are caused by heavy rainfall. For example, the most tragic landslide disaster for recent Japan happened in Hiroshima in August, 2014. A series of landslides following local heavy rain reportedly killed seventy-four people. An automated observatory recorded over 200 mm rainfall accumulation within 3 h. This intense rainfall within a short period made it difficult to issue an early warning and, unfortunately, the evacuation advisory by Hiroshima prefecture was indeed too late.

Reliable rainfall prediction is of course invaluable for early warning. There are several numerical weather prediction models available in the world. Examples are WRF (www.wrf-model.org), MRI-NHM (www.mri-jma.go.jp/Project/mrinpd/INDEXE.htm), and COSMO (www.cosmo-model.org). These are called regional models as they target regional- (or meso-) scale wearther with O(1 km) horizontal resolution. Such regional models are usually used for short-term forecasts, e.g., shorter than 2 days. However, it is often valuable for early warning if the chance of heavy rain is predicted more than 2 days before. The regional models need, for forecast, boundary conditions provided by global models. Global models are generally run with O (10 km) horizontal resolutions and used for guidance in long-term forecasts (usually more than 2 days). Due to the coarse horizontal grid of O(10 km), the global models need the aid of cumulus parameterization and are usually not good at local heavy rain predictions.

Recent development of supercomputer systems allows us to run global simulations with O (1 km) horizontal resolutions without the aid of cumulus parameterizations. The Multi-Scale Simulator for the Geoenvironment (MSSG; Takahashi et al. 2006, 2013 and references therein) can be used for such simulations and is called a global cloud-resolving model. The global cloud-resolving model is expected to have good skills in local rainfall predictions as well as in long-term predictions.

In the following subsection, the numerical weather prediction models are briefly reviewed. In Sect. 3, we explain the global cloud-resolving model MSSG. Some examples from MSSG simulations are shown in Sect. 4. This paper is concluded in Sect. 5 followed by an Acknowledgement.

2 Numerical Weather Prediction Models

Two types of models are in use for numerical weather prediction (NWP): global and regional models. Global models are generally run with O(10 km) horizontal resolutions and used for guidance in long-term forecasts (usually more than 2 days). Regional models are used for shorter-term forecasts, and are run with O(1 km) horizontal resolutions. For the global models, some meteorological processes are too small to be explicitly included. Then parameterization is a way for representing these processes by relating them to grid-scale variables. For example, a typical cumulus cloud has a scale of O(1 km), which is much smaller than the horizontal resolutions of global models. The convection process by cumulus clouds should be parameterized and the parameterization is called cumulus parameterization, on which most of global models rely.

Recent development of supercomputer systems allows us to run global simulations with O (1 km) horizontal resolutions without the aid of the cumulus parameterizations. The Multi-Scale Simulator for the Geoenvironment (MSSG; Takahashi et al. 2006, 2013 and references therein) can be used for such simulations and is called a global cloud-resolving model. The global cloud-resolving model is expected to have good skills in local rainfall predictions as well as in long-term predictions.

3 Multi-scale Weather Prediction Model—Multi-scale Simulator for the Geoenvironment (MSSG)

3.1 Overview

The MSSG is an atmosphere-ocean coupled non-hydrostatic model aimed at seamless simulations from global to local scales (Takahashi et al. 2006, 2013 and references therein) (Fig. 1). MSSG consists of atmospheric (MSSG-A) and ocean (MSSG-O) components. MSSG adopts a conventional latitude-longitude grid system for regional simulations and the Yin-Yang grid system (Kageyama and Sato 2004; Baba et al. 2010), which consists of two overlapping latitude-longitude grids thus avoiding the polar singularity problem, for global simulations. MSSG has been used for a wide range of applications. A cloud-system-resolving global ocean-atmosphere coupled MSSG successfully simulated an observed MJO propagation (Sasaki et al. 2016). A global atmosphere-ocean coupled experiment at an 11 km horizontal resolution with a nested region at a 2.7 km horizontal resolution simulated sea surface cooling induced by a TC along the track (Takahashi et al. 2013). High-resolution regional atmospheric simulations have been conducted to investigate the influences of the choice of cloud microphysics scheme and that of in-cloud turbulence on the development of clouds (Onishi et al. 2015; Seifert and Onishi 2016). MSSG-O has been used to investigate the dispersion of radionuclides released from the Fukushima Dai-ichi nuclear power plant with 2 km horizontal resolution (Choi et al. 2013) and the effect of wind on long-term summer water temperature trends in Tokyo Bay, Japan with 200 m horizontal resolution (Lu et al. 2015). MSSG-A has been applied to building-resolving urban atmosphere simulations with 5 m spatial resolutions to clarify heat environments at streets (Takahashi et al. 2013).

3.2 Dynamical Core of the Atmosphere Component

This paper describes the atmospheric component of MSSG (MSSG-A). The dynamical core of MSSG-A is based on the nonhydrostatic equations and predicts the three wind components, air density, water vaper mixing ratio and pressure.

The third-order Runge-Kutta scheme is used for time integrations. The fast terms relating to acoustic and gravity waves are calculated separately with a shorter time step, respectively (Wicker and Skamarock 2002). A fifth-order upwind scheme (Wicker and Skamarock 2002) was chosen for momentum advection and the second-order weighted average flux (WAF) scheme with the Superbee flux limiter (Toro 1989) for scalar advections. For turbulent diffusion, the Mellor-Yamada-Nakanishi-Niino level 2.5 scheme (Nakanishi and Niino 2009) was used. The MSSG-Bulk model (Onishi and Takahashi 2012), which is a six-category bulk cloud microphysics model (see the following subsection for detail), was used for explicit cloud physics and Model Simulation radiation TRaNsfer code (MstrnX; Sekiguchi and Nakajima 2008) was used for calculating longwave and shortwave radiation transfer.

3.3 Cloud Microphysics

The bulk cloud microphysics model in MSSG (MSSG-Bulk model; Onishi and Takahashi 2012) prognoses mixing ratios of water vapor, cloud water, rain, cloud ice, snow and graupel, and in addition the number density of cloud ice particles. Thus, the bulk model is a one-moment model for warm rain processes and a partial two-moment model for cold rain processes.

Figure 2 shows the cloud microphysical processes considered in MSSG-Bulk. The prognostic variables are the mixing ratios of water vapor, cloud, rain, cloud ice, snow, and graupel, and the number concentration of cloud ice; Q_v, Q_c, Q_r, Q_i, Q_s, Q_g and N_i, respectively.

MSSG also has the warm-bin/cold-bulk hybrid cloud microphysical model named MSSG-Bin (Onishi and Tahakashi 2012). In that hybrid model, a spectral bin scheme is used for liquid droplets, while a bulk scheme is used for solid (ice) particles. That is, the expensive but more reliable spectral bin scheme treats the

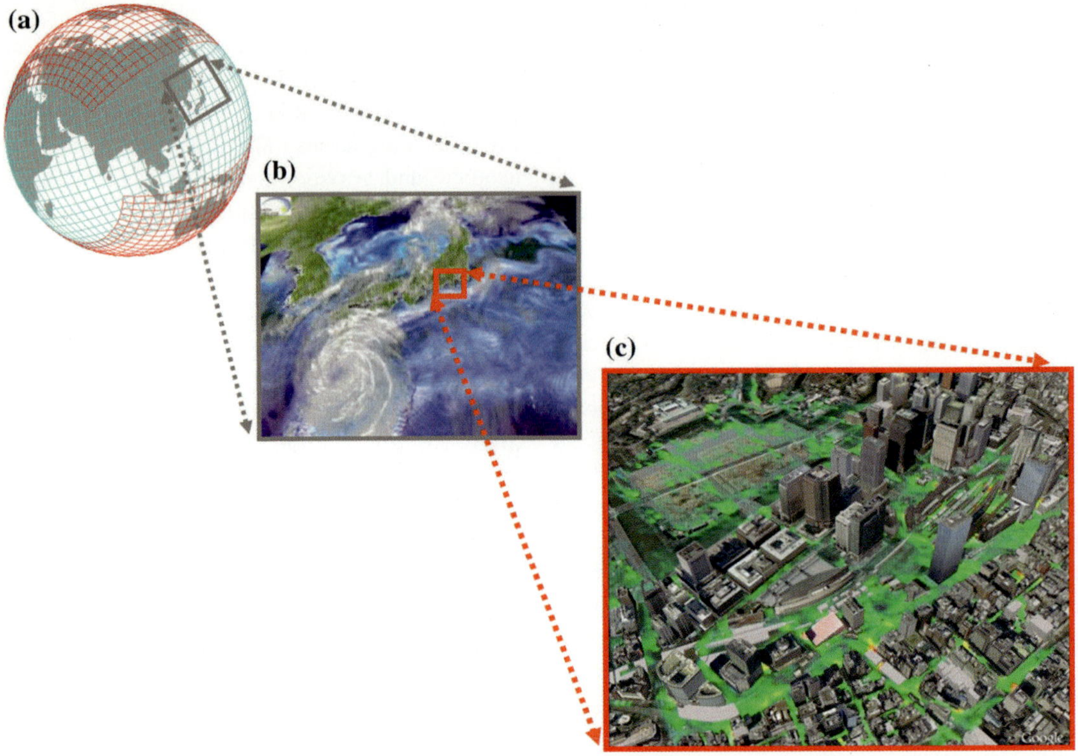

Fig. 1 MSSG is designed to be applicable to **a** global scale, **b** meso scales and up to **c** urban scales. The Yin-Yang grid system, which consists of two overlapping latitude-longitude grids indicated in *blue* and *red*, is adopted for global simulations

Fig. 2 Cloud microphysical processes in MSSG-Bulk

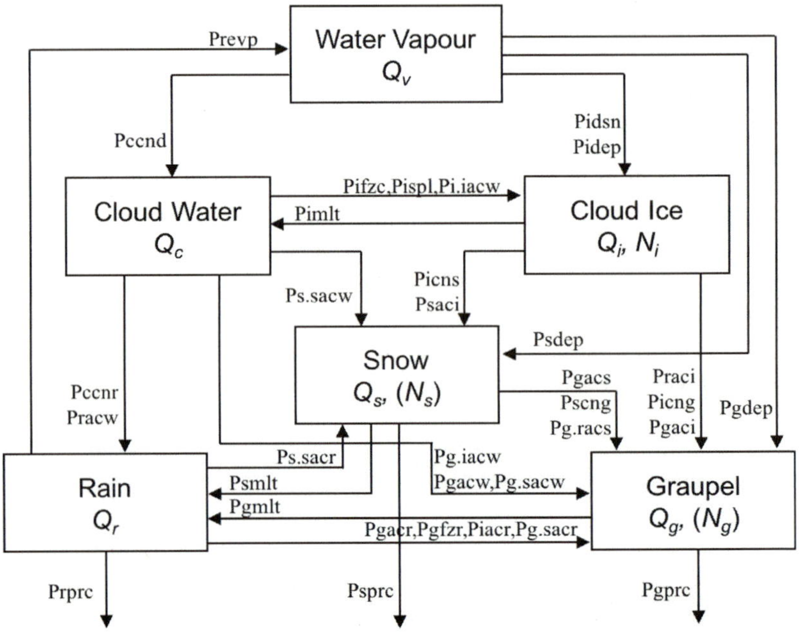

relatively well-understood physics of the liquid phase, and the computationally efficient but less robust bulk scheme is used to treat the poorly understood physics of the ice phase.

4 Rainfall Predictions

4.1 Regional Numerical Weather Prediction

On 4th September 2005, local heavy rainfall associated with mesoscale convective systems caused a serious urban flooding in Suginami ward, in the heart of Tokyo metropolitan city.

MSSG was run on 1 km horizontal resolution and 32 vertical layers for the computational domain shown in Fig. 3. The initial data were obtained by interpolating the Grid Point Value (GPV) data at 06:00UTC on 4th September 2005 provided by Japan Meteorological Business Support Center. Figure 4 shows the surface precipitation at 00:00 on 5th September 2005, when a strong rain band lay on the east of Tokyo and Kanagawa. The local maximum precipitation was underestimated, but the south-north rain band was reproduced well by MSSG.

4.2 Turbulence Influence on Orographic Precipitation

The turbulence-enhanced collision rate of cloud droplets may be able to explain the rapid growth of cloud droplets, which often result in fast rain initiation in the early stages of cloud development (Falkovich and Pumir 2007; Grabowski and Wang 2013).

The change rate of particle number density, $n_f(r, x_i, t)$, by the stochastic collision-coalescence process is represented by

$$\left(\frac{dn_f(r)}{dt}\right)_{col} = \frac{1}{2}\int_0^r K_c(r', r'')n_f(r')n_f(r'')dr' - \int_0^r K_c(r, r')n_f(r)n_f(r')dr', \quad (1)$$

where $r'' = (r^3 - r'^3)^{1/3}$ and $K_c(r_1, r_2)$ is the collision kernel describing the rate at which a particle of radius r_1 is collected by a particle of radius r_2. The conventional collision kernel model is the hydrodynamic kernel model, which describes the collision due to the settling velocity difference between two particles with different sizes

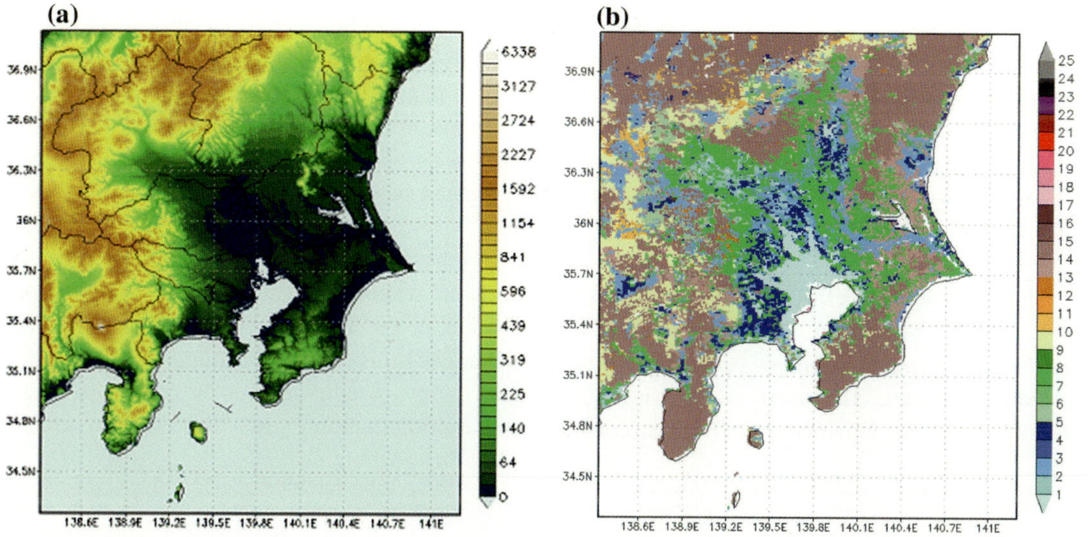

Fig. 3 Computed area. **a** *Left figure* shows the orographic elevation and **b** *right* shows the land use index

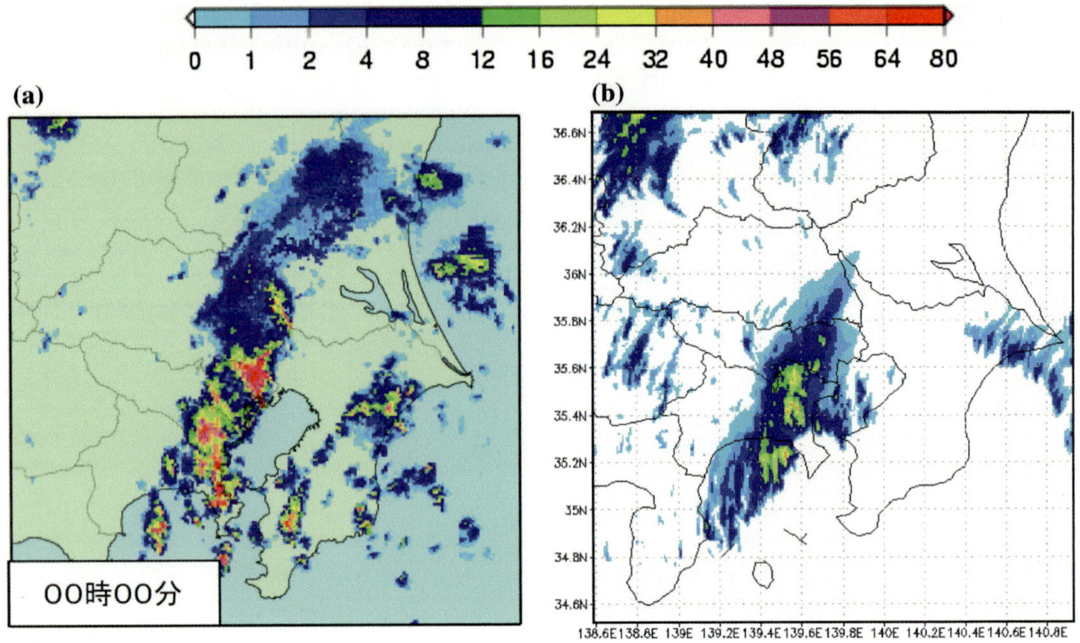

Fig. 4 Surface precipitation at 00:00 JST on 4 September 2005 from **a** the radar observation (Tokyo district meteorological observatory) and **b** the MSSG simulation

$$\left\langle K_{c,hydr,12} \right\rangle = R_{12}^2 \left| V_{p,1} - V_{p,2} \right|, \tag{2}$$

where $\langle\ \rangle$ denotes an ensemble average, $R_{12}(= r_1 + r_2)$ is the collision radius and $V_{p,i}$ is the settling velocity of particles with radius r_i. This hydrodynamic kernel cannot describe the collisions due to turbulence since it contains no flow parameters.

The turbulent collision kernel that involves the turbulence effects can be written in the following form Wang et al. (1998).

$$\left\langle K_{c,turb,12} \right\rangle = 2\pi R_{12}^2 \langle |w_r(x = R_{12})| \rangle g_{12}(x = R_{12}). \tag{3}$$

Here $w_r(x = R_{12})$ is the radial relative velocity at contact separation and describes the turbulence-enhanced relative velocities of two colliding particles. The term of $g_{12}(x = R_{12})$ is the radial distribution function (RDF) at contact separation, which is so-called the "accumulation effect" and measures the effect of particle preferential distributions. Currently, there are a couple of models available for the turbulent kernel based on Eq. (3): Examples are Ayala-Wang model (Ayala et al. 2008; Wang et al. 2008) and our developed model (Onishi model; Onishi et al. 2015). This study adopted the latter model for the investigation of the turbulent collisions on cloud development.

Figure 5 shows the computational domain for mesoscale orographic flow over Mt. Hiei, which is located between Kyoto and Shiga prefectures. The MSSG simulation with MSSG-Bin cloud microphysics model was performed for the domain whose domain was 40 km × 20 km 15 km in the x-(streamwise), y-(spanwise) and z-(vertical) directions. The number of numerical grid points used was 400 × 200 × 48. Horizontal computational grids were regular and the

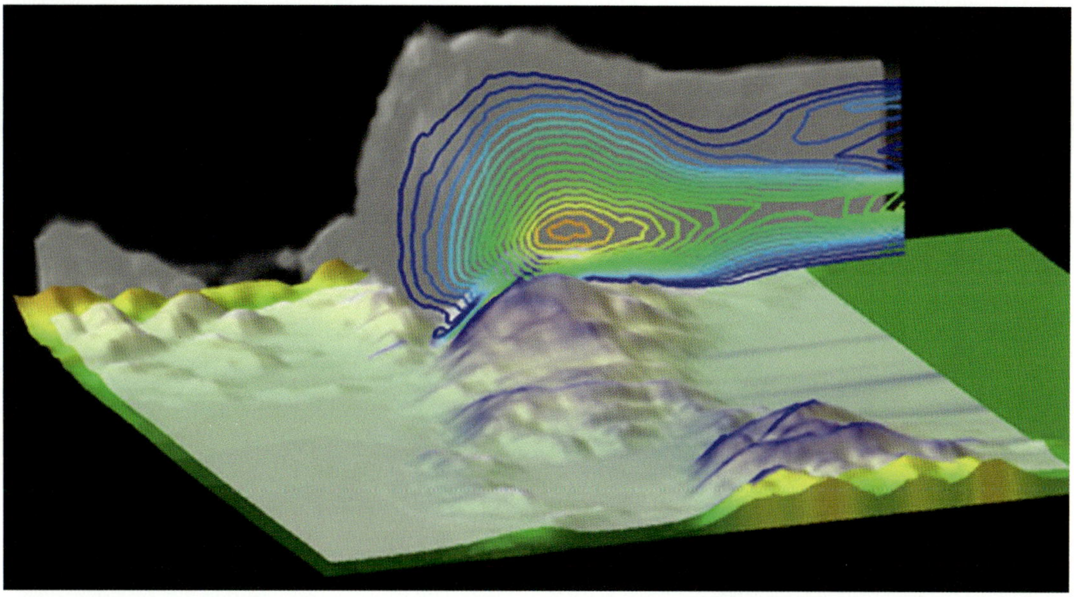

Fig. 5 Bird-eye's view of the computational domain. A snap shot of three-dimensional cloud distribution (*in white*) and the surface precipitation (*in blue*) are also drawn

Table 1 Collision kernel models used in numerical simulations

RUN	Collision kernel model	Turbulence effect
RUN-NoT	Hydrodynamic model	Not considered
RUN-T	our developed model	Considered

height-based, terrain-following-coordinate system has been chosen for vertical direction. The minimum vertical spacing was 10 m in the vicinity of the surface and the maximum one was 500 m in the top of the domain. The moist air, whose relative humidity was 95%, flew into the domain with the streamwise velocity U_0 = 15 m/s toward east. In MSSG-Bin model, 33 bins (classes) were used for so-called mass-doubling size resolution. The droplet collision growth was calculated based on Eq. (1) using the collision kernel models listed in Table 1.

Figure 6 shows the surface precipitation obtained from RUN-NoT and RUN-T. It shows large part of surface precipitation in windward slopes. This means the present precipitation process is governed by the local orography. Intense precipitation, over 80 mm/h, is observed in RUN-T, indicating that the precipitation is significantly enhanced by turbulent collisions of cloud droplets.

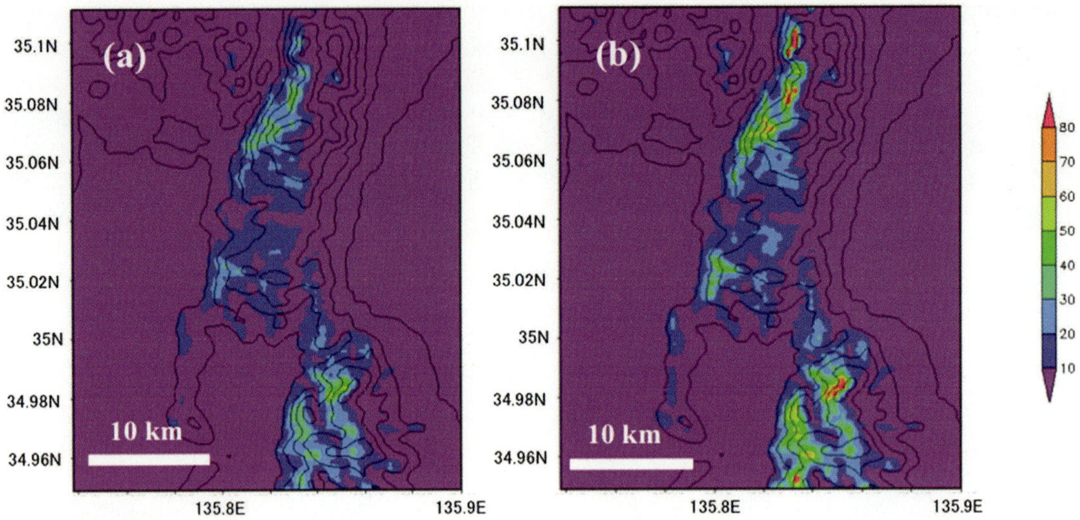

Fig. 6 Surface precipitation for **a** RUN-NoT and **b** RUN-T. Large part of surface precipitation was seen in windward slope. It was clearly shown that the precipitation is enhanced by turbulent collisions of cloud droplets

5 Conclusions

We have described a recent development of our next generation numerical weather prediction model—the Multi-Scale Simulator for the Geoenvironment (MSSG). MSSG is categorized as a global cloud-resolving model that can be used for global simulation with high resolution without the aid of cumulus parameterization. The examples have shown that MSSG has good prediction skill and is a promising tool for investigation of detail orographic local precipitation. Thus MSSG can be a promising numerical weather prediction model for early warning of landslides.

Acknowledgements The numerical simulations were performed on the Earth Simulator of the Japan Agency for Marine-Earth Science and Technology (JAMSTEC).

References

Ayala O, Rosa B and Wang LP (2008) Effects of turbulence on the geometric collision rate of sedimenting droplets. Part 2. Theory and parameterization. New J Phys 10:075016

Baba Y, Takahashi K, Sugimura T, Goto K (2010) Dynamical core of an atmospheric general circulation model on a Yin–Yang grid. Mon Weather Rev 138:3988–4005

Choi Y, Kida S, Takahashi K (2013) The impact of oceanic circulation and phase transfer on the dispersion of radionuclides released from the Fukushima Dai-Ichi Nuclear Power Plant. Biogeosciences 10:4911–4925

Falkovich G, Pumir A (2007) Sling Effect in collisions of water droplets in turbulent clouds. J Atmos Sci 64:4497–4505

Grabowski WW, Wang L-P (2013) Growth of cloud droplets in a turbulent environment. Annu Rev Fluid Mech 45:293–324

Kageyama A, Sato T (2004) The Yin-Yang grid: an overset grid in spherical geometry. Geochem Geophys Geosyst 5:Q09005

Lu L-F, Onishi R, Takahashi K (2015) The effect of wind on long-term summer water temperature trends in Tokyo Bay, Japan. Ocean Dyn 65:919–930

Nakanishi M, Niino H (2009) Development of an improved turbulence closure model for the atmospheric boundary layer. J Meteor Soc Jpn 87:895–912

Onishi R, Takahashi K (2012) A warm-bin–cold-bulk hybrid cloud microphysical model. J Atmos Sci 69:1474–1497

Onishi R, Matsuda K, Takahashi K (2015) Lagrangian tracking simulation of droplet growth in turbulence–turbulence enhancement of autoconversion rate. J Atmos Sci 72:2591–2607

Sasaki W, Onishi R, Fuchigami H, Goto K, Nishikawa S, Ishikawa Y, Takahashi K (2016) MJO simulation in a cloud-system-resolving global ocean-atmosphere coupled model. Geophys Res Lett 43:9352–9360

Seifert A, Onishi R (2016) Turbulence effects on warm-rain formation in precipitating shallow convection revisited. Atmos chem Phys 16:12127–12141

Sekiguchi M, Nakajima T (2008) A k-distribution-based radiation code and its computational optimization for an atmospheric general circulation model. J Quant Spectrosc Radiat Transf 109:2779–2793

Takahashi K, Peng X, Onishi R, Ohdaira M, Goto K, Fuchigami H and Sugimura T (2006) Multi-scale weather/climate simulations with multi-scale simulator for the geoenvironment (MSSG) on the earth simulator. Annual report of the earth simulator center, Apr 2005–Mar 2006, pp 31–39

Takahashi K, Onishi R, Baba Y, Kida S, Matsuda K, Goto K, Fuchigami H (2013) Challenge toward the prediction of typhoon behaviour and down pour. J Phys Conf Ser 454(012):072

Toro EF (1989) A weighted average flux method for hyperbolic conservation laws. Proc R Soc Lond A423:401–418

Wang L-P, Wexler AS, Zhou Y (1998) Statistical mechanical descriptions of turbulent coagulation. Phys Fluids 10:2647–2651

Wang L-P, Ayala O, Rosa B, Grabowski WW (2008) Turbulent collision efficiency of heavy particles relevant to cloud droplets. New J Phys 10(075):013

Wicker LJ, Skamarock WC (2002) Time-split-ting methods for elastic models using forward time schemes. Mon Weather Rev 130:2088–2097

TXT-tool 2.386-1.1 Intensity-Duration-Frequency Curves for Rainfall-Induced Shallow Landslides and Debris Flows Using Copula Functions

Nejc Bezak, Mitja Brilly, Mojca Šraj and Matjaž Mikoš

Abstract

Geological hazards such as debris flows or deep-seated landslides can cause major economic damage and endanger human lives. In order to warn people living in a hazardous area during extreme meteorological events that may trigger debris flows or landslides, in many parts of the world different early warning systems (EWS) have been developed and installed. Empirical rainfall thresholds can be applied as a part of advanced early warning systems. This study combines two concepts: (i) empirical rainfall thresholds and (ii) intensity-duration-frequency (IDF) curves, that both include information about rainfall duration and rainfall intensity. The IDF curves were constructed using a copula function. Post-event analysis of several extreme events in Slovenia, Europe, in the last 25 years was carried out. The triggering conditions responsible for the initiation of the extreme events were investigated, and several empirical rainfall thresholds were tested. It was found that different meteorological conditions were responsible for triggering deep-seated and shallow landslides. A comparison between empirical rainfall thresholds and IDF curves indicated that some thresholds correspond to a return period significantly smaller than two years. If the IDF curves are incorporated into the early warning systems, this will mean that warnings can be issued with a corresponding probability.

Keywords

Debris flows · Early warning systems · Empirical rainfall thresholds
Extreme events · Intensity-duration-frequency (IDF) curves
Landslides

N. Bezak (✉) · M. Brilly · M. Šraj · M. Mikoš
Faculty of Civil and Geodetic Engineering, University of Ljubljana, Jamova Cesta 2, 1000 Ljubljana, Slovenia
e-mail: nejc.bezak@fgg.uni-lj.si

M. Brilly
e-mail: mitja.brilly@fgg.uni-lj.si

M. Šraj
e-mail: mojca.sraj@fgg.uni-lj.si

M. Mikoš
e-mail: matjaz.mikos@fgg.uni-lj.si

K. Sassa et al. (eds.), *Landslide Dynamics: ISDR-ICL Landslide Interactive Teaching Tools*, https://doi.org/10.1007/978-3-319-57774-6_32

Contents

1 **Introduction** 426
2 **Data and Methods** 426
3 **Results and Discussion** 429
3.1 Evaluation of the Selected Empirical Rainfall Thresholds 429
3.2 IDF Curves Using Copula Functions 429
4 **Conclusions** 429
References 431

1 Introduction

Different empirical rainfall thresholds can be used in an early warning system (EWS) for shallow landslides and debris flows (e.g., Segoni et al. 2015). Many different empirical rainfall thresholds can be found in the literature, and the thresholds were developed using field data from various parts of the world. Guzzetti et al. (2007) provided an overview of several empirical rainfall thresholds for debris flows, soil slips, and shallow landslides. On the other hand, the intensity-duration-frequency (IDF) curves also connect rainfall intensity and rainfall duration (e.g., Koutsoyiannis et al. 1998). Thus, a comparison of both types of curves can yield an interesting relationship (Bezak et al. 2016). IDF curves are constructed using a univariate approach in which rainfall durations are pre-defined, and rainfall intensities are used in the probabilistic approach. To simultaneously study both rainfall intensity and actual rainfall duration, copula functions can be applied to construct the IDF curves (e.g., Ariff et al. 2012; Bezak et al. 2016; Singh and Zhang 2007).

The main aim of this study is to construct IDF curves using copula functions for several rainfall stations in Slovenia, and compare IDF curves and several empirical rainfall thresholds for selected extreme meteorological events that have occurred in Slovenia in the last 25 years. The post-event analysis of the selected extreme meteorological events was also carried out.

2 Data and Methods

Six extreme meteorological events (case studies) in the last 25 years in Slovenia were selected to perform a post-event analysis of shallow landslides, deep-seated landslides and debris flows triggered by the meteorological events. Table 1 lists the basic characteristics of the selected case studies, and Fig. 1 shows the location of the events. The selected extreme events caused 10 casualties and approximately €100 million in economic damage. Bezak et al. (2016) has provided more detailed information on the journal papers that have analysed some of the investigated extreme meteorological events (e.g., Logar et al. 2005; Mikoš et al. 2005; Rusjan et al. 2009; Petkovšek et al. 2011).

For each of the selected events, at least one rainfall station was selected based on the data availability and the geographical location of the nearest rain gauge (Fig. 1). Sixteen rain gauges in Slovenia were selected. Rainfall records ranged from 11 to 66 years, with a mean of 34 years. The 5-minute rainfall measurements were cumulated to 1-hour periods, which were

Table 1 Main characteristics of the selected extreme meteorological events

Event name	Event location	Year	Type of event
Solčava	Macesnik	1989 or 1990	Deep-seated landslide
Log pod Mangartom	Stože	2000	Deep-seated landslide and debris flow
Lokavec	Slano Blato	2000	Deep-seated landslide
SE Slovenia	SE area	2005	Shallow landslides
Železniki	Železniki	2007	Shallow landslides
Lokavec	Stogovce, Znosence	2010	Deep-seated landslide

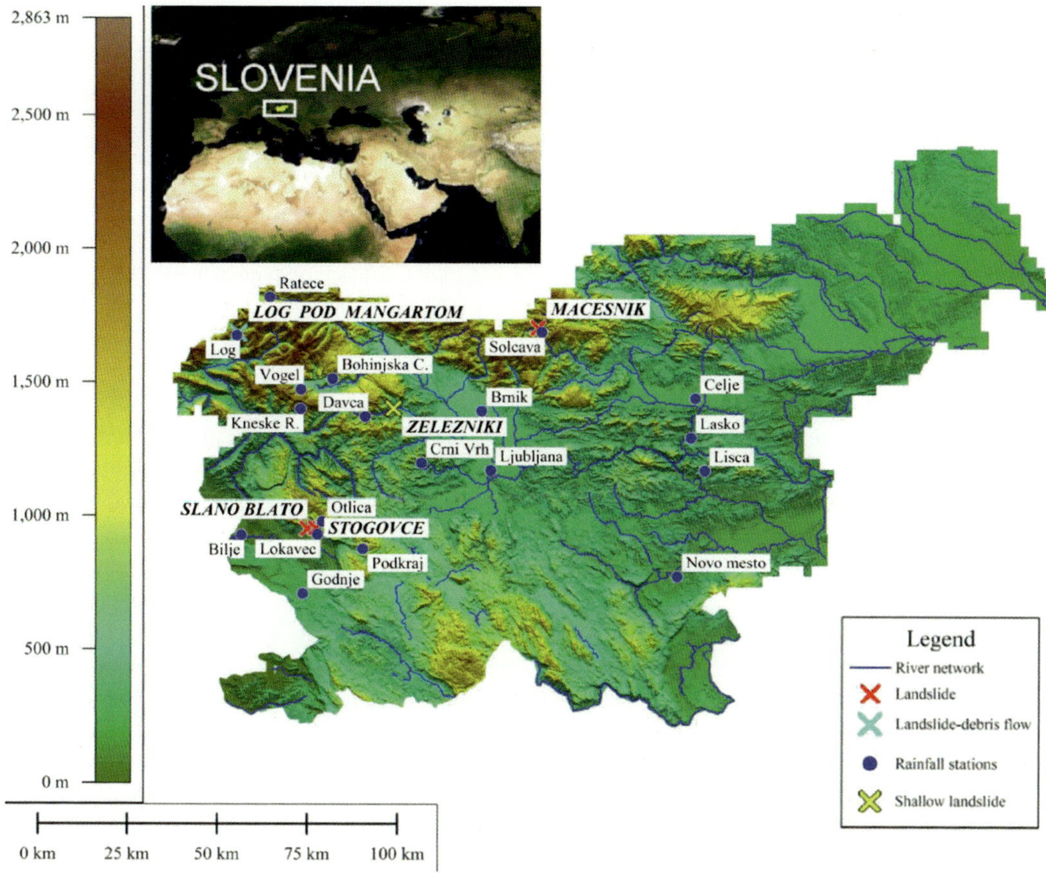

Fig. 1 Location of the selected case studies on a topographic map of Slovenia

used in the study. The completeness of the rainfall records was 93.7%.

In the first step of the study (data pre-processing phase), the rainfall data was divided into individual rainfall events. For the purpose, the inter-event time had to be defined. Inter-event is the duration of the period without rainfall between two consecutive rainfall events. The inter-event time selection can have a notable influence on the results of the rainfall analysis, and should be considered carefully (e.g., Bezak et al. 2016; Segoni et al. 2015).

Three different types of empirical rainfall thresholds were selected for the post-event analysis for the selected extreme meteorological events. The selected empirical rainfall thresholds, and the authors who proposed the threshold are listed in Table 2.

In the next step of the study the IDF curves were constructed using copula functions, which can be used to study jointly two or more related variables. One of the properties of the copula function is that the marginal distributions are modelled independently of the dependence structure among the selected variables (e.g., Salvadori et al. 2007). More information about copula functions can be found in copula text-books, such as Salvadori et al. (2007). In the following, we describe the methodology for the construction of the IDF curves using copula functions (Bezak et al. 2016):

- Subsample definition (e.g., annual maximum method, peak-over threshold method or some other method);
- Selection of suitable marginal distribution functions to describe the rainfall intensity and

Table 2 Overview of the selected empirical rainfall thresholds

Author	Equation	Parameters
Caine (1980)	$I = 14.82D^{-0.39}$	*I*–mean rainfall intensity; *D*–rainfall duration
Clarizia et al. (1996)	$I = 10D^{-0.77}$	Same as above
Guzzetti et al. (2008)	$I = 2.2D^{\beta-0.44}$	Same as above
Aleotti (2004) (I)	$NCR = 11.5e^{-0.08NAR}$	*NCR*–normalized critical rainfall; *NAR*–10 day normalized antecedent rainfall
Aleotti (2004) (II)	$I_{MAP} = 0.76D^{-0.33}$	I_{MAP}–normalized mean rainfall intensity; *D*–rainfall duration

rainfall duration and marginal distributions parameters estimation;

- Copula function selection and estimation of the copula parameter(s);
- Determination of the relationship between the design value and the return period.

In our study, the annual maximum (AM) series method was used for subsample definition. Two different types of subsamples were used:

- Subsample 1: Inter-event was 6 h, and AM events were defined based on the maxima of the rainfall intensities;
- Subsample 2: Inter-event was 18 h, and AM events were defined based on the maxima of the rainfall amount.

In the next step the marginal distribution functions were selected. The Gumbel distribution was used to describe rainfall intensity and the Gamma distribution was selected to describe the rainfall duration. More information about the marginal distribution functions can be found in Hosking and Wallis (2005). The Kolmogorov-Smirnov statistical test was also used to check the suitability of the selected marginal distribution functions. To link the univariate marginal distribution functions with the multivariate probability distribution, a copula function was used. The Frank copula function from the Archimedean family was selected to construct the IDF relationship. This one-parameter copula was used previously in similar studies (e.g., Ariff et al. 2012; Singh and Zhang 2007). More information about the Frank copula can be found in copula textbooks (e.g., Salvadori et al. 2007). To test the adequacy of the Frank copula, the Cramer-von Mises statistical test was used. This test is included in the R package *copula* (Kojadinovic and Yan 2010) that was used for copula analysis. To construct the IDF relationship, the conditional copula has to be defined (e.g., Singh and Zhang 2007):

$$C(u|V = v) = \frac{[\exp(-\theta u) - 1][\exp(-\theta v)]}{[\exp(-\theta u) - 1][\exp(-\theta v) - 1] + [\exp(-\theta) - 1]} = 1 - 1/T, \quad (1)$$

where θ is the Frank copula parameter, T is the return period [years], U and V are the marginal distribution functions that are used to model rainfall intensities and rainfall durations, respectively, and C is the selected copula function. The IDF curves can be constructed by solving Eq. 1 for a given rainfall duration and return period T values.

3 Results and Discussion

3.1 Evaluation of the Selected Empirical Rainfall Thresholds

In the first step of the study, the evaluation of the selected empirical rainfall thresholds was carried out. Figure 2 shows post-event analysis for the selected case studies. For some extreme events more than one rain gauge was used, and for all case studies two sub-samples (shown in Sect. 2) were tested (Fig. 1). One can see that for some extreme events (e.g., Log pod Mangartom) a significant difference exists in the position of the event in the rainfall intensity-duration space due to the different sub-sample methods used (Fig. 2).

An important issue related to the use of empirical rainfall thresholds as part of an EWS is the appropriate density of rain gauges. From Fig. 2 it can be seen that rain gauges too far from the location of the event can be non-representative. This is particularly important where the meteorological conditions are variable. A combination of different empirical rainfall thresholds (e.g., Aleotti 2004; Caine 1980; Clarizia et al. 1996; Guzzetti et al. 2008) would be an interesting option for the use of empirical rainfall thresholds as part of an EWS. This would enable one to consider different variables (e.g., *I, D, NAR, NCR* and *IMAP*) in the EWS, and not just the rainfall intensity and rainfall duration.

3.2 IDF Curves Using Copula Functions

In the next step of the study, the IDF curves were constructed using copula functions. The procedure described in Sect. 2 was used to define the IDF curves for the stations shown in Fig. 1 and for both types of sub-samples (sub-sample 1 and 2) (Bezak et al. 2016). Figure 3 shows the IDF curves defined using the Frank copula from the Archimedean family for four rainfall stations in Slovenia (using sub-sample 1).

A comparison between the IDF curves and the empirical rainfall thresholds was carried out by Bezak et al. (2016). A relevant conclusion of this study was that some empirical rainfall thresholds (e.g., Clarizia et al. 1996; Guzzetti et al. 2008) correspond to rainfall events with return periods significantly smaller than two years (Bezak et al. 2016). This means that debris flows and shallow landslides should occur several times a year, but this number of occurrences is not the case for Slovenia (Bezak et al. 2016). Moreover, the Caine (1980) threshold intersects with the rainfall intensity-rainfall duration points with the 2-year return period (Bezak et al. 2016). More results about the IDF curves that were constructed using copulas can be found in Bezak et al. (2015, 2016).

4 Conclusions

This paper presents post-event analysis of selected extreme meteorological events that occurred in Slovenia, Europe, in the last 25 years. For more than 10 rain gauges, the intensity-duration-frequency (IDF) curves were defined using copula functions, and a comparison between IDF curves and selected empirical rainfall thresholds was performed. The main conclusions can be summarized as follows (Bezak et al. 2016):

- rainfall triggering conditions responsible for the initiation of shallow landslides and deep-seated landslides were different;
- the inter-event selection can have significant impact on the rainfall analysis results;
- not all the tested empirical rainfall thresholds are suitable for the Slovenian meteorological conditions, where convective events are frequent;
- different empirical rainfall thresholds should apply for different parts of Slovenia;
- IDF curves can be constructed using copula functions.

The comparison between empirical rainfall thresholds and IDF curves showed that the IDF curves, constructed using copulas or using univariate approaches, can potentially be included in the Early Warning Systems (EWS). In Slovenia,

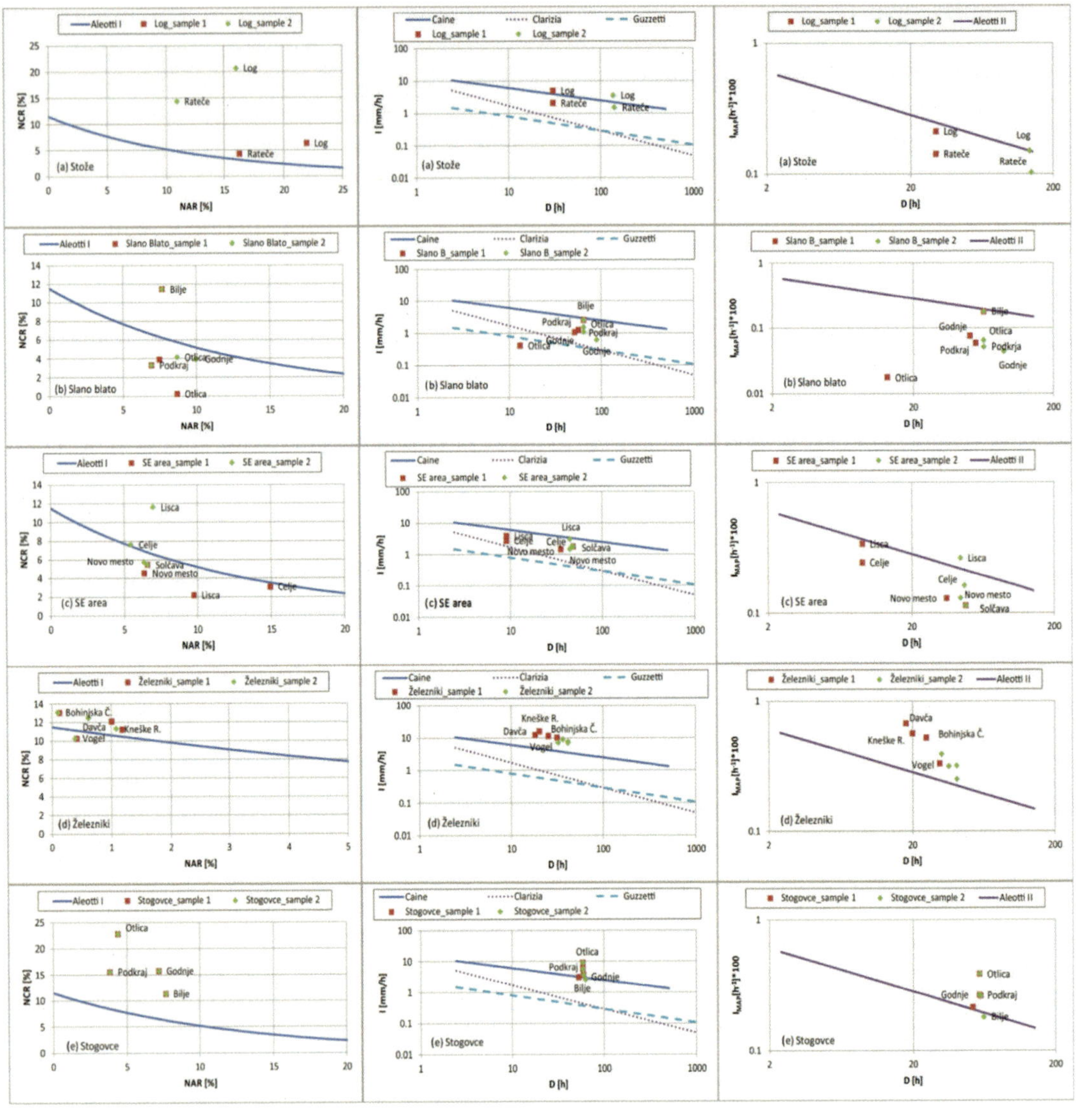

Fig. 2 Evaluation of different empirical rainfall thresholds for several case studies

an Early Warning System for landslides has been recently developed (for a detailed description see TXT-Tool 2.386-2.2 2016) that takes into account 100-year 24-hour rainfall estimates determined from measured values in rain gauges. Such rainfall threshold values are deterministic; the IDF values determined using Copula functions are not.

The IDF curves shown in Fig. 3 represent probability bands (levels), and if this information is included in the Early Warning System, a warning (or critical state) with corresponding probability can be issued. Colours (green, yellow, orange and red) can be used to represent different probability levels to present the probabilities to the general public. A similar approach is used in early warning systems for floods where critical states are issued using different colours. However, the methodology discussed in this tool has to be tested using data (shallow landslides, debris flows and deep-seated landslides) from different parts of the world, and a comprehensive

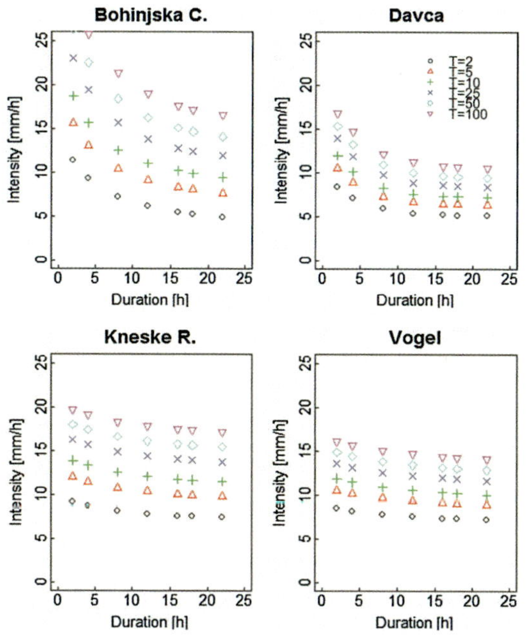

Fig. 3 Intensity-duration-frequency (IDF) relationships constructed using a copula function approach for the four rainfall stations using subsample 1 methodology (Bezak et al. 2015)

post-event analysis has to be carried out. We expect that this methodology will be further tested and proposed in different areas.

Acknowledgements Authors wish to thank the Slovenian Environment Agency (ARSO) for rainfall data provision. The results of the study are part of the Faculty of Civil and Geodetic Engineering (UL FGG) work on the European research project SedAlp (www.sedalp.eu), jointly financed by the European Union through the Alpine Space program, and by the Slovenian Research Agency (ARRS) through the Research Programme P2-0180 Water science and technology, and Geotechnics.

References

Aleotti P (2004) A warning system for rainfall-induced shallow failures. Eng Geol 73:247–265

Ariff NM, Jemain AA, Ibrahim K, Zin WZW (2012) IDF relationships using bivariate copula for storm events in Peninsular Malaysia. J Hydrol 470:158–171

Bezak N, Šraj M, Brilly M, Mikoš M (2015) Return Period Determination for Several Extreme Rainfall-Induced Events Using the IDF Relationship Obtained via Copulas. In: 12th International Conference on Applications of Statistics and Probability in Civil Engineering, ICASP12 (324), pp 1–8

Bezak N, Šraj M, Mikoš M (2016) Copula-based IDF curves and empirical rainfall thresholds for flash floods and rainfall-induced landslides. J Hydrol. doi:10.1016/j.jhydrol.2016.02.058

Caine N (1980) The rainfall intensity-duration control of shallow landslides and debris flows. Geografiska Annaler Ser A Phys Geogr 62:23–27

Clarizia M, Gullà G, Sorbino G (1996) Sui meccanismi di innesco dei soil slip. In: international conference prevention of hydrogeological hazards: the role of scientific research vol 1, pp 585–597

Guzzetti F, Peruccacci S, Rossi M, Stark CP (2007) Rainfall thresholds for the initiation of landslides in central and southern Europe. Meteorol Atmos Phys 98:239–267

Guzzetti F, Peruccacci S, Rossi M, Stark CP (2008) The rainfall intensity-duration control of shallow landslides and debris flows: an update. Landslides 5:3–17

Hosking JRM, Wallis JR (2005) Regional frequency analysis: an approach based on L-moments. Cambridge University Press, Cambridge, 244 p

Kojadinovic I, Yan J (2010) Modeling multivariate distributions with continuous margins using the copula R package. J Stat Softw 34(9):1–20

Koutsoyiannis D, Kozonis D, Manetas A (1998) A mathematical framework for studying rainfall intensity-duration-frequency relationships. J Hydrol 206:118–135

Logar J, Bizjak KF, Kočevar M, Mikoš M, Ribičič M, Majes B (2005) History and present state of the Slano Blato landslide. Nat Hazards Earth Sys Sci 5:447–457

Mikoš M, Fazarinc R, Pulko B, Petkovšek A, Majes B (2005) Stepwise mitigation of the Macesnik landslide, N Slovenia. Nat Hazards Earth Sys Sci 5:947–958

Petkovšek A, Fazarinc R, Kočevar M, Maček M, Majes B, Mikoš M (2011) The Stogovce landslide in SW Slovenia triggered during the September 2010 extreme rainfall event. Landslides 8:499–506

Rusjan S, Kobold M, Mikoš M (2009) Characteristics of the extreme rainfall event and consequent flash floods in W Slovenia in September 2007. Nat Hazards Earth Sys Sci 9:947–956

Salvadori G, De Michele C, Kottegoda NT, Rosso R (2007) Extremes in nature: an approach using copulas. Springer, Berlin, 292 p

Segoni S, Battistini A, Rossi G, Lagomarsino D, Catani F, Moretti S, Casagli N (2015) Technical Note: An operational landslide early warning system at regional scale based on space–time-variable rainfall thresholds. Nat Hazards Earth Sys Sc 15:853–861

Singh VP, Zhang L (2007) IDF curves using the Frank Archimedean copula. J Hydrol Eng 12:651–662

TXT-tool 2.386-2.2 2016. A system to forecast rainfall-induced landslides in Slovenia (this book)

TXT-tool 2.039-1.5 An Algorithm for the Objective Reconstruction of Rainfall Events Responsible for Landslides

Massimo Melillo, Maria Teresa Brunetti, Silvia Peruccacci, Stefano Luigi Gariano and Fausto Guzzetti

Abstract

The primary trigger of damaging landslides in Italy is intense or prolonged rainfall. Definition of the rainfall conditions responsible for landslides is a crucial issue and may contribute to reducing landslide risk. Criteria for identifying the rainfall conditions that could initiate slope failures are still lacking or uncertain. Expert investigators usually reconstruct rainfall events manually. In this paper, we propose an algorithm for the objective and reproducible definition of rainfall conditions responsible for landslides, from a series of hourly rainfall data. The algorithm, which is implemented in R (http://www.r-project.org), performs a series of actions: (i) removes isolated events with negligible amount of rainfall and random noise generated by the rain gauge; (ii) aggregates rainfall measurements in order to obtain a sequence of distinct rainfall events; (iii) identifies single or multiple rainfall conditions responsible for the slope failures. The result is the objective reconstruction of the duration, *D*, and the cumulated rainfall, *E*, for rainfall events, and for rainfall conditions that have resulted in landslides. We tested the algorithm using rainfall and landslide information for the period between January 2002 and December 2012 in Sicily, Southern Italy. The algorithm reconstructed 13,537 rainfall events and 343 rainfall conditions as possible triggers using the information on 163 documented landslides. The comparison between automatic and manually method highlights that most (87.7%) of the rainfall conditions obtained manually were reconstructed accurately. Use of the algorithm

M. Melillo (✉) · M.T. Brunetti · S. Peruccacci · S.L. Gariano · F. Guzzetti
CNR Research Institute for Geo-Hydrological Protection (IRPI), via Madonna Alta, 126-06128 Perugia, Italy
e-mail: massimo.melillo@irpi.cnr.it

M.T. Brunetti
e-mail: mariateresa.brunetti@irpi.cnr.it

S. Peruccacci
e-mail: silvia.peruccacci@irpi.cnr.it

S.L. Gariano
e-mail: gariano@irpi.cnr.it

F. Guzzetti
e-mail: fausto.guzzetti@irpi.cnr.it

S.L. Gariano
Department of Physics and Geology, University of Perugia via Pascoli, 06123 Perugia, Italy

K. Sassa et al. (eds.), *Landslide Dynamics: ISDR-ICL Landslide Interactive Teaching Tools*, https://doi.org/10.1007/978-3-319-57774-6_33

should contribute to reducing the current subjectivity inherent in the manual treatment of the rainfall and landslide data.

Keywords
Algorithm • Landslide • Rainfall • Rainfall event • Threshold

Contents

1 **Introduction** ... 434
2 **The Algorithm** ... 434
2.1 Reconstruction of the Rainfall Events ... 436
2.2 Selection of the Rainfall Events Responsible for Landslides ... 439
3 **Software** ... 441
4 **Test Case** ... 441
5 **Discussion** ... 443
6 **Conclusion** ... 445
References ... 445

1 Introduction

Landslides are widespread phenomena that cause casualties and economic damage worldwide (Brabb and Harrod 1989; Keefer and Larsen 2007; Salvati et al. 2010; Petly 2012). In Italy, rainfall is a primary trigger of landslides. The prediction of rainfall-induced landslides relies largely on the definition of empirical rainfall thresholds. Generally, rainfall thresholds are determined using empirical distributions of rainfall conditions that have resulted in landslides, including rainfall intensity, I, versus rainfall duration, D, (e.g., Caine 1980; Aleotti 2004; Guzzetti et al. 2007, 2008; Cannon et al. 2008; Martelloni et al. 2012; Staley et al. 2012; Rosi et al. 2012, 2015; Lee et al. 2014) and cumulated event rainfall, E, versus rainfall duration, D, (e.g., Innes 1983; Cannon and Ellen 1985; Wieczorek 1987; Crosta and Frattini 2001; Kanji et al. 2003; Vennari et al. 2014; Gariano et al. 2015).

Rainfall thresholds are affected by uncertainties that limit their use in modern landslide warning systems (Aleotti 2004; Godt et al. 2006; Guzzetti et al. 2008; Bach-Kirschbaum et al. 2012; Komac et al. 2014; Segoni et al. 2015). A specific source of uncertainty lays in the characterization of the rainfall event responsible for the landslides (Guzzetti et al. 2007). An objective definition of the rainfall conditions responsible for the failure does not exist (Guzzetti et al. 2007) or is poorly formalized and ambiguous (Aleotti 2004). Criteria for establishing the duration of an event, or for deciding the conditions to separate (or to combine) successive events (Reichenbach et al. 1998; Guo 2002; Guzzetti et al. 2007; Saito et al. 2010; Shamsudin et al. 2010; Segoni et al. 2014a, b) are also lacking. Generally, investigators do not specify how individual rainfall events are determined. Subsequently, the identification and measurement of the rainfall conditions responsible for landslides remains subjective.

In this work, we propose an algorithm for the objective definition of rainfall events, and for the quantitative measurement of the rainfall conditions that characterize a rainfall event. The algorithm systematizes the actions performed and the decisions taken, by an expert investigator that defines heuristically rainfall events from a typical rainfall record and information on landslide occurrence (Guzzetti et al. 2007, 2008; Brunetti et al. 2010; Berti et al. 2012). Use of the algorithm will contribute to reducing uncertainty in the definition of landslide-triggering rainfall events, to compiling large catalogues of rainfall events with landslides and to determining reliable rainfall thresholds for possible landslide occurrence.

2 The Algorithm

The framework shown in Fig. 1 represent the structure of the proposed algorithm described in this section. Using a standard record of rainfall

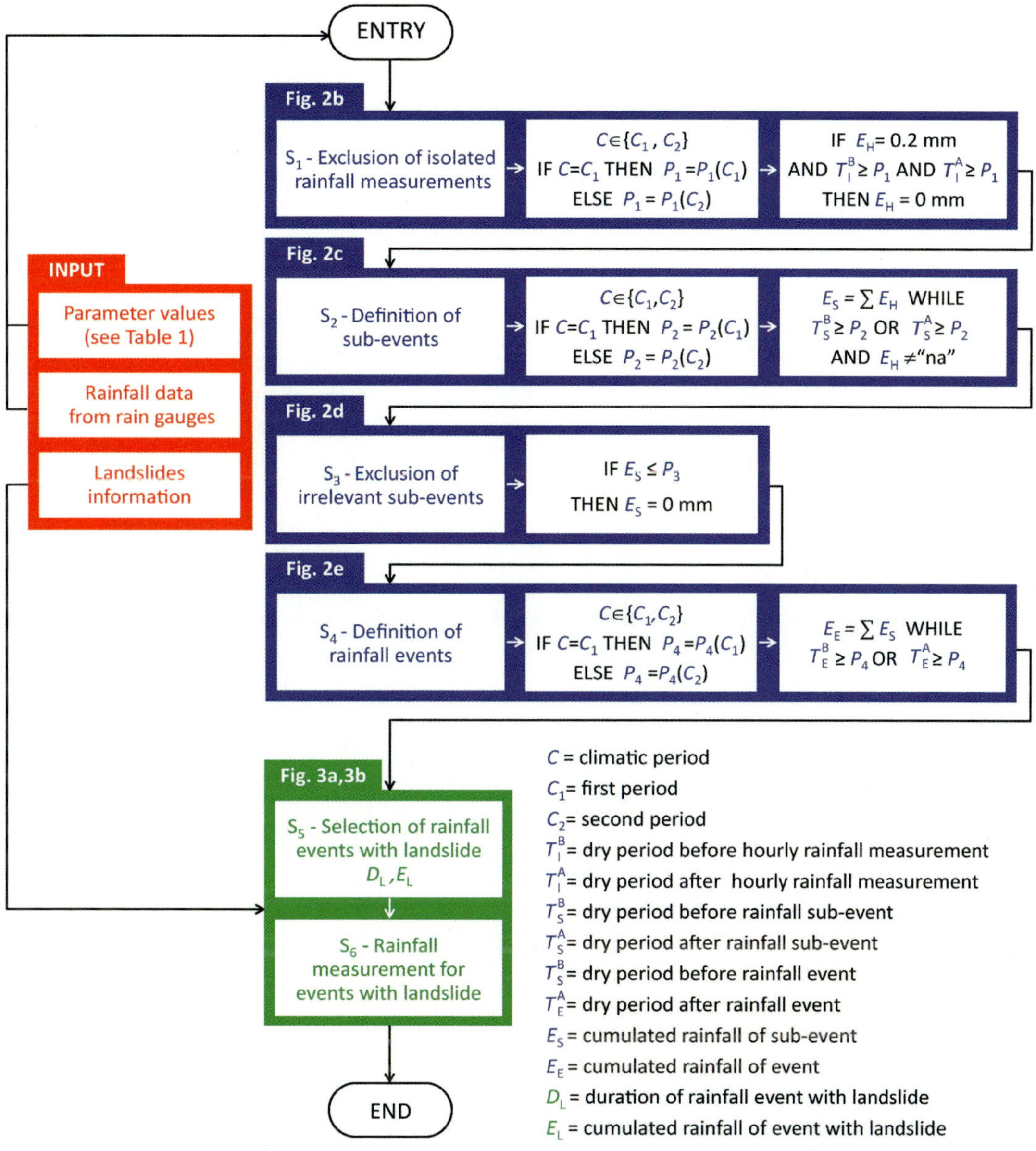

Fig. 1 Structure of the algorithm proposed for the objective reconstruction of rainfall events and of rainfall conditions responsible for landslides

measurements and a limited number of parameters, the procedure (1) detects rainfall events from a standard record of rainfall measurements, (2) determines the rainfall duration (D_E) and the cumulated (total) rainfall (E_E) for the detected events, (3) associates landslide information (or the lack of landslide information) to the detected rainfall events, and (4) measures the rainfall duration (D_L) and the cumulated rainfall (E_L) responsible for the landslide.

The structure of the algorithm includes two main logical blocks (Fig. 1). The first performs the automatic reconstruction of the rainfall events. The second selects the rainfall events responsible for the landslides. A rainfall event is a period of continuous rainfall or a chronological ensemble of periods of continuous rainfall in a rainfall record, separated from preceding and successive rainfall events by periods with no rainfall (i.e., dry periods, *T*). The length of the dry periods may vary, depending, e.g., on seasonal, meteorological or climatic conditions. To illustrate the algorithm, we use a record of hourly rainfall measurements obtained for a single rain gauge in Sicily, southern Italy. We then test the algorithm in the same geographical region using rainfall measurements obtained from a network of 105 rain gauges.

2.1 Reconstruction of the Rainfall Events

For the reconstruction of the individual rainfall events from a record of rainfall measurements, the algorithm perform a series of five steps including a pre-processing step (Fig. 1).

Step 0 *Pre-processing of rainfall data*

From a single rain gauge, the algorithm works on a continuous (hourly) record of rainfall measurements obtained in a period. These measurements are generally discontinuous (incomplete), with individual or multiple rainfall measurements missing in the record due to technical and operational problems. The gaps in the record can cover periods in a variable range from a minimum of 1 h to several days or weeks and are typically marked by specific "tags" in the record. In some cases, tags are missing in the rainfall record, and it is difficult—or impossible—to single out measurement gaps in the record. The algorithm checks the continuity of the record and detects the gaps. More specifically, the algorithm searches the rainfall record for tagged and untagged missing measurements and replaces them with the "na" tag (measurement not available). In addition, a rainfall record may contain hourly rainfall measurements, E_H, that are lower than the instrumental sensitivity of a rain gauge (e.g., $G_S = 0.2$ mm), $E_H < G_S$. In this case these measurements are considered noise in the rainfall record, and the algorithm sets the measurements to $E_H = 0.0$ mm. After these preliminary operations, the corrected rainfall record is ready to be processed to reconstruct the rainfall events. Figure 2a shows an example of a corrected rainfall record.

Step 1 *Detection and exclusion of isolated rainfall measurements*

The algorithm starts by searching for isolated hourly rainfall measurements in the corrected rainfall record (S_1 in Fig. 1). An isolated rainfall measurement is defined as an hourly measurement separated from the immediately preceding and the immediately following rainfall measurements by dry periods (T_I^B before, and T_I^A after) that exceed a given length P_1. The length of the dry period (P_1) depends on the seasonal or the local climatic conditions, i.e., $P_1 = P_1(C)$.

In Sicily, as in the whole Mediterranean area, two seasonal periods can be identified for landslide initiation: (i) a "warm" spring-summer period C_W, and (ii) a "cold" autumn-winter period C_C. For the C_W warm period the dry interval separating isolated rainfall measurements is $P_1 = 3$ h, and for the C_C cold period the dry interval is $P_1 = 6$ h (Table 1).

Study of the mean annual evapotranspiration (ETR) in Italy (Melillo 2009) using the Thornthwaite-Mather method (Thornthwaite and Mather 1957) revealed that the evapotranspiration in the warm period is about twice the evapotranspiration in the cold period, ETR $(C_W) \cong 2 \cdot \mathrm{ETR}(C_C)$. Here, we assume that the evapotranspiration is inversely proportional to the time necessary to dry the soil, and we set a factor of two between all relevant parameters in the C_W and C_C periods. Once the isolated rainfall measurements are identified, their individual relevance for the reconstruction of the rainfall conditions responsible for possible landslide occurrence is evaluated. We consider relevant the isolated hourly rainfall measurements that exceed a minimum value E_R (e.g., $E_R = 0.2$ mm), $E_H > E_R$, and irrelevant the measurements with $E_H = E_R$. The later measurements, shown by red

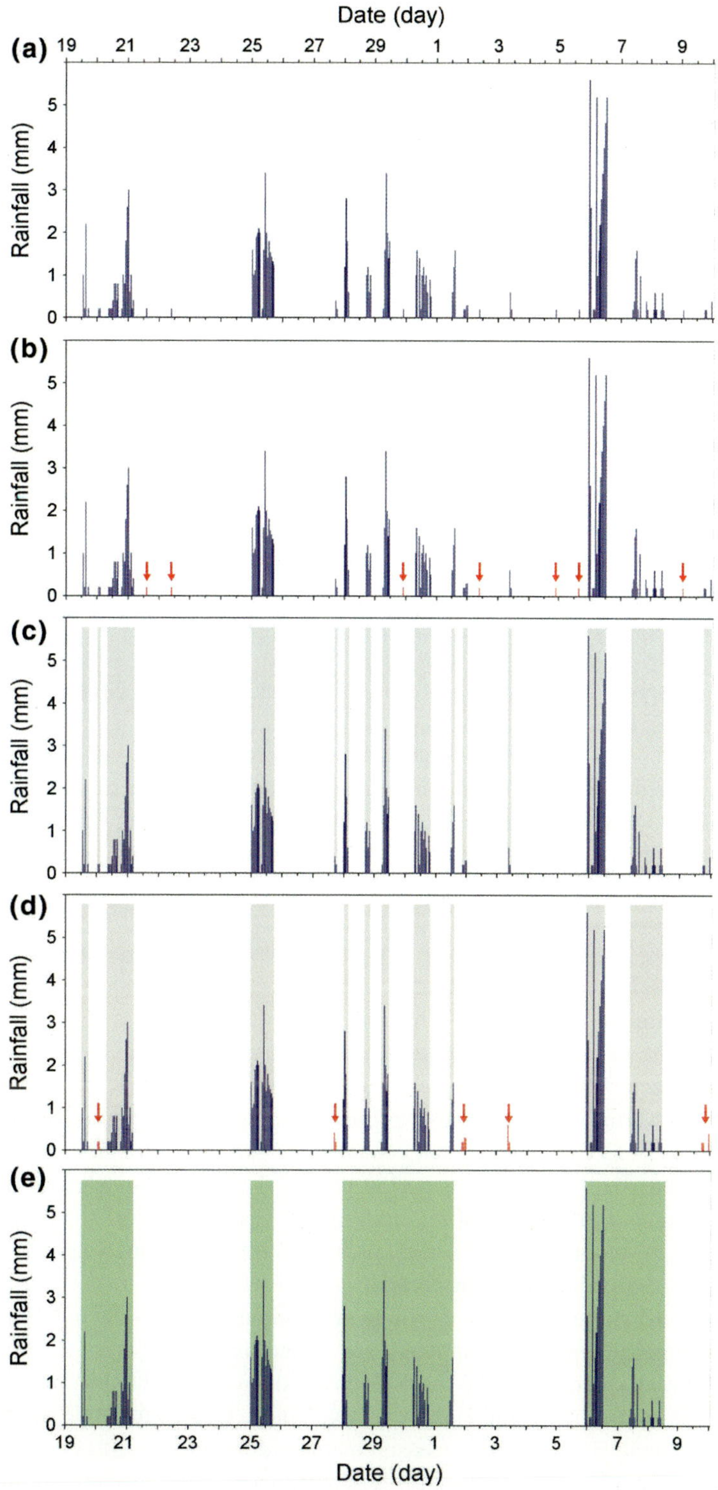

Fig. 2 Example of the application of the algorithm for the reconstruction of rainfall events. **a** *Blue bars* show hourly rainfall measurements obtained by the Aragona-Torre Salto rain gauge, Sicily, southern Italy, from 19 September to 10 October 2006. **b** Selection of the isolated hourly rainfall measurements (*red bars*), shown by *red arrows*. **c** Identification of the rainfall sub-events, highlighted by *grey-shaded* areas. **d** Selection of irrelevant sub-events (*red bars*), shown by *red arrows*. **e** Identification of rainfall events, highlighted by *green-shaded* areas

Table 1 Parameters used by the algorithm

Step	Parameter name	Parameter value		Unit
		$P(C_W)$	$P(C_C)$	
S_0	G_S	0.2	0.2	mm
S_1	E_R	0.2	0.2	mm
S_1	P_1	3	6	h
S_2	P_2	6	12	h
S_3	P_3	1	1	mm
S_4	P_4	48	96	h

The first column lists the step in the logical framework of the algorithm where the parameter is used (Fig. 1). Two climatic periods are considered: C_W a "warm" spring-summer period; and C_C a "cold" autumn-winter period

bars in Fig. 2b, contribute a negligible (irrelevant) amount of rain to the rainfall event (e.g., due to the presence of fog and/or humidity in the air). For the purpose of the analysis, the algorithm sets the isolated, irrelevant measurements to $E_H = 0.0$ mm.

Step 2 *Identification of rainfall sub-events*

After first step, the algorithm proceeds by searching for individual rainfall sub-events (S_2 in Fig. 1), where a rainfall sub-event is a period of continuous rainfall separated from the immediately preceding and the immediately following sub-events by dry periods with no rain. As before, the length P_2 of the dry period may vary, depending on the seasonal and the climatic conditions, $P_2 = P_2(C)$. The separation depends on the meteorological conditions and the rainfall characteristics in the two climatic periods. In the C_W warm period, rainfall is primarily brought to the study area by local convective storms, whereas in the C_C cold period rainfall is most commonly the result of regional frontal systems. When reconstituting a rainfall sub-event, the algorithm checks for the continuity of the rainfall record in the sub-event. If single or multiple "na" measurements (interruptions) are found in the rainfall record in the period covered by the sub-event, the sub-event is excluded from the analysis. If no "na" measurements are found, the sub-event is defined (grey shaded areas in Fig. 2c), and rainfall metrics are computed for the sub-event, including: (i) the sub-event duration D_S, computed by summing the number of hours in the sub-event, and (ii) the sub-event total rainfall E_S, computed by summing the hourly rainfall measurements in the sub-event, $E_S = \sum E_H$.

Step 3 *Exclusion of irrelevant rainfall sub-events*

Next, the algorithm searches for sub-events that can be considered irrelevant for the reconstruction of rainfall events responsible for landslide occurrence (S_3 in Fig. 1). For the purpose, a sub-event is considered irrelevant if the cumulated (total) rainfall for the sub-event is lower than a given threshold value $E_S \leq P_3$, regardless of the duration of the sub-event. In a Mediterranean climate, $P_3 = 1$ mm (Table 1) is a reasonable threshold to exclude sub-events whose contribution can be considered irrelevant for the possible initiation of rainfall-induced landslides. Irrelevant sub-events (red bars in Fig. 2d) are excluded from the subsequent analysis.

Step 4 *Identification of rainfall events*

In this step, the algorithm aggregates single or multiple sub-events to obtain single rainfall events (S_4 in Fig. 1). The single rainfall event is defined as a period of continuous rainfall, or an ensemble of periods of continuous rainfall, separated from the preceding and the successive events by dry periods. Again, the minimum length P_4 of the inter-event dry periods may vary, depending on meteorological and seasonal conditions i.e., $P_4 = P_4(C)$. As an example, to identify the rainfall events (Fig. 2c), we used a minimum dry period $P_4 = 48$ h for the C_W warm period, and a minimum dry period of $P_4 = 96$ h for the C_C cold period (Table 1). After the reconstruction of the rainfall events (green shaded areas in Fig. 2e), the algorithm calculates rainfall metrics for each of the detected rainfall events, including: (i) the event duration D_E, computed summing the number of hours in the rainfall event (including hours for which $E_H = 0$), and (ii) the event total cumulated rainfall E_E, computed by summing the sub-event rainfall $E_E = \sum E_S$.

2.2 Selection of the Rainfall Events Responsible for Landslides

In two additional steps, described in the following, the algorithm combines independent information on the temporal occurrence of landslide(s) with the information on the rainfall events obtained before (Fig. 1).

Step 5 *Selection of rainfall events with landslides*

In the first additional step, the algorithm selects the rainfall events for which information on landslide occurrence is available. We assign to each landslide a record of rainfall measurements obtained from a single rain gauge. Criteria to select the rain gauge include proximity, the elevation difference between the rain gauge and the landslide, and the local morphological setting (Brunetti et al. 2010; Peruccacci et al. 2012). The algorithm compares the dates (start date and end date) of the rainfall events identified by the first logical block, with a record listing the date (day and time) of occurrence of the landslides. Each landslide in the temporal record (Fig. 3a) is associated to a single rainfall event (Fig. 3b).

Step 6 *Rainfall measurements for events with landslides*

In this step the algorithm calculates the rainfall metrics responsible for the failure, namely: (i) the rainfall duration D_L, and (ii) the cumulated rainfall E_L. Note that the rainfall duration D_L and the cumulated rainfall E_L responsible for landslide occurrence are not necessarily the same as the rainfall duration D_E and the cumulated event rainfall E_E defined at the end of the first logical block for the entire rainfall event (Fig. 3b).

Most commonly, landslides occur before (and sometimes well before) the end of a rainfall period, and the rainfall after the landslide occurrence cannot be considered relevant for the initiation of the slope failure. In this case, $E_L < E_E$ and $D_L < D_E$. Occasionally, landslides fail after the end of the rainfall event (Guzzetti et al. 2004). In this case, the cumulated rainfall responsible for the landslide corresponds to the cumulated event rainfall, $E_L = E_E$, and the rainfall duration is $D_L = D_E$. The algorithm considers the different conditions, and calculates the correct values for D_L and E_L.

For complex rainfall events (Fig. 3b)—which are the majority in a typical rainfall record—it is often difficult (or impossible) to decide a single duration, and the corresponding cumulated amount of rain responsible for the landslides. In this case, the algorithm reconstructs multiple aggregations of rainfall sub-events that are likely to trigger landslides. In Fig. 3c, d and e we show that the complex rainfall event portrayed in Fig. 3b is characterized by three sub-events with significantly different rainfall durations (D_L = 28, 91 and 178 h) and cumulated rainfall amounts (E_L = 104.2, 218.4 and 263.2 mm). Without external information, the three sub-events identified by the algorithm are equally probable as possible landslide triggers.

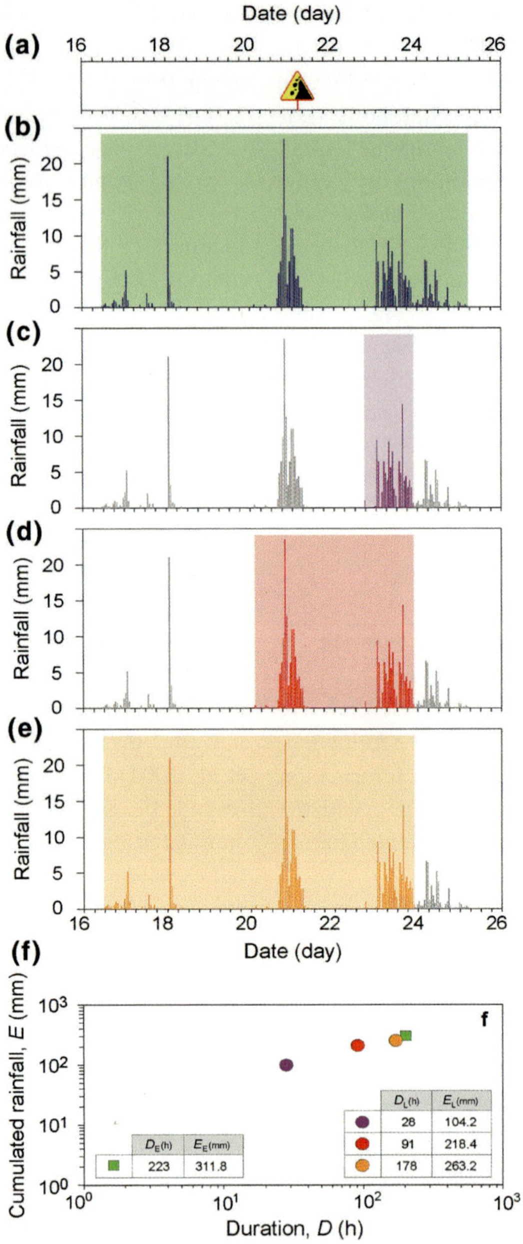

Fig. 3 Application of the algorithm for the reconstruction of rainfall events that have resulted in landslides. **a** Time period from 16 to 26 December 2006. *Traffic sign* shows time of occurrence of a landslide near to the Riposto-Praiola rain gauge, Sicily, southern Italy. **b** *Blue bars* show hourly rainfall measurements obtained by the Riposto-Praiola rain gauge between 16 and 26 December 2006. The *green-shaded* area highlights the rainfall event identified by the first logical block of the algorithm. *Purple*, *red*, and *orange bars* in (**c**)–(**e**) show rainfall measurements that represent the first, second, and third sub-events, respectively. **f** Rainfall conditions for the identified events. *Green square* shows the rainfall duration, D_E, and the cumulated rainfall, E_E, for the rainfall event considered in (**b**). *Purple*, *red*, and *orange dots* show the D_L and E_L conditions that have resulted in landslides for the three periods considered in (**c**)–(**e**) respectively

3 Software

We developed a code for the proposed algorithm using the R open-source software for advanced statistical computing and graphics, release 2.15.2 (http://www.r-project.org). The software uses two text files as input: one listing the rainfall record, and a second file with the necessary landslide information. The software is available from: http://geomorphology.irpi.cnr.it/tools/rainfall-events-and-landslides-thresholds/definition-of-rainfall-events-and-rainfall-events-with-landslides/algorithm/, together with the rainfall and landslide data used to prepare Figs. 2 and 3

4 Test Case

We tested the proposed algorithm using rainfall and landslide data available in Sicily, Southern Italy (Fig. 4). The rainfall data consisted of hourly rainfall measurements collected in the 11-year period from 1 January 2002 to 31 December 2012 by a network of 105 rain gauges operated by the Sistema Informativo Agrometeorologico Siciliano (SIAS). Figure 4 shows the geographical distribution of the 105 rain gauges (white and red squares). The landslide information consisted of the geographical location and the occurrence time of 163 rainfall induced landslides in the period from July 2002 to November 2011 (yellow dots in Fig. 4). The information on landslide occurrence was collected from digital archives of national and local newspapers and blogs, and from technical reports provided by local Fire Brigades in Sicily.

We selected a subsample of 59 rain gauges in the vicinity of the single landslides (red squares in Fig. 4) to reconstruct the rainfall sub-events responsible for the failures.

Prior to the development of the algorithm, we had determined the rainfall duration D_L^* and the corresponding cumulated event rainfall E_L^* responsible for the ensemble of 163 rainfall-induced landslides in Sicily. For the purpose, we used the heuristic approach proposed by Brunetti et al. (2010) and updated by Peruccacci et al. (2012).

To separate two rainfall events with the heuristic approach we used a dry (no rain) period of two days (48 h) between April and October, and a dry period of four days (96 h) from November to March. Due to the Mediterranean climate (Köppen 1931; Trewartha 1968) in Sicily the warm period is longer (7 months from April to October) than the cold period (5 months from November to March). We determined the rainfall conditions for a

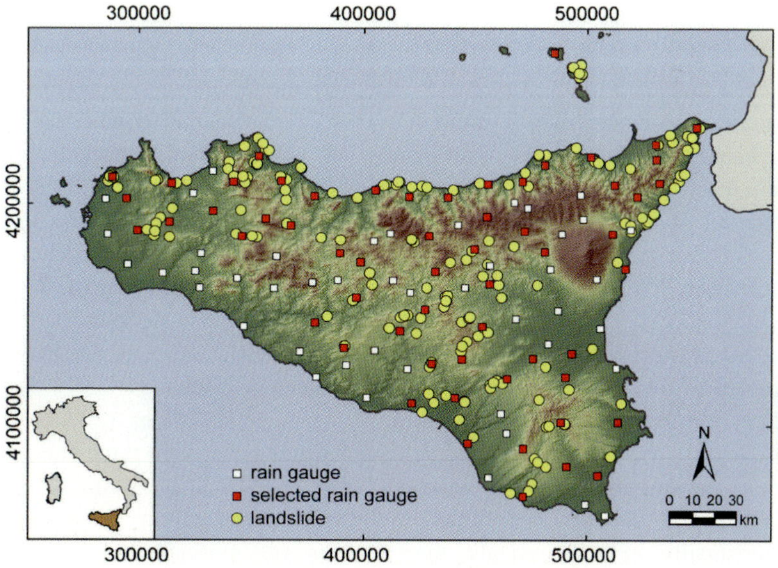

Fig. 4 Map showing the geographical location of 105 rain gauges in Sicily (*squares*). *Yellow dots* show the location of 163 rainfall-induced landslides in the period from July 2002 to November 2011. *Red squares* show the locations of the 59 rain gauges used in this study

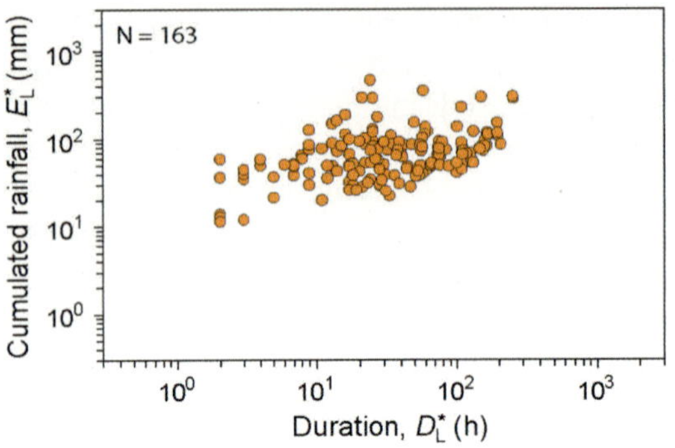

Fig. 5 Rainfall duration, D_L^*, and cumulated event rainfall, E_L^*, conditions (N = 163) responsible for the 163 landslides. Expert investigators determined the rainfall conditions using a heuristic approach

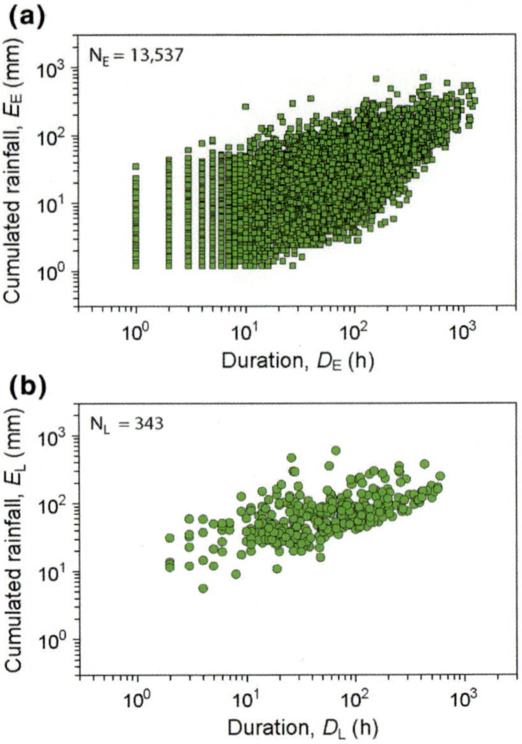

Fig. 6 **a** *Green squares* show the rainfall duration, D_E, and cumulated event rainfall, E_E, events (N_E = 13,537) reconstructed by the algorithm in the 11-year period from 2002 to 2012 using rainfall measurements obtained by 59 rain gauges (*red squares* in Fig. 4) and the parameters listed in Table 1. **b** *Green circles* show rainfall duration, D_L, and cumulated event rainfall, E_L, conditions for the subset of N_L = 343 rainfall events responsible for the 163 landslides shown by the *yellow dots* in Fig. 4

number of ambiguous cases based on experience. For each triggered landslide, a single set of rainfall (D_L, E_L) conditions considered responsible for the landslide was determined. This is a typical result obtained when deciding heuristically the rainfall conditions for possible landslide occurrence (Guzzetti et al. 2007, 2008; Brunetti et al. 2010; Peruccacci et al. 2012). Figure 5 portrays the N = 163 rainfall (D_L^*, E_L^*) conditions responsible for slope failures in Sicily reconstructed manually through expert judgment.

We then applied the algorithm to the same rainfall data and landslide information. To be consistent with the expert-based, heuristic method, we set the warm period C_W from April to October, and the cold period C_C from November to March. Table 1 lists the values for the parameters P_1, P_2, P_3, and P_4 used by the algorithm to reconstruct the rainfall events.

Figure 6a shows the distribution of the cumulated rainfall E_E as a function of the event duration D_E for N_E = 13,537 rainfall events in the 11-year period from 1 January 2002 to 31 December 2012 reconstructed by the algorithm using the subset of 59 rain gauges in the vicinity of the single landslides, and Fig. 6b shows the sub-set of N_L = 343 rainfall events with landslides reconstructed by the algorithm.

Because the heuristic method defined a single set of rainfall conditions for each landslide, the number of rainfall conditions reconstructed heuristically (Fig. 5) equals the number of landslides (N = 163). For each landslide, the algorithm reconstructed a variable number of rainfall conditions (from one to six) as possible landslide triggers. As a result, the number of rainfall conditions responsible for landslides identified by the algorithm is larger, N_L = 343 (Fig. 6b). Inspection of the reconstructed events revealed that for 71 landslides (43.6%) the algorithm reconstructed a single rainfall event, for 40 landslides (24.5%) two rainfall events, and for four landslides (2.5%) six events. For the remaining 48 landslides (29.5%) the algorithm reconstructed between three and five events. We stress that the multiple rainfall conditions identified by the algorithm for a single landslide are all equally probable as possible triggers of the landslide.

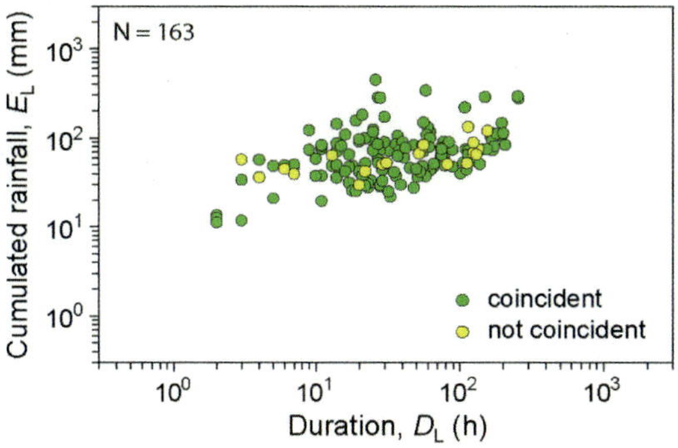

Fig. 7 Comparison of rainfall (D_L, E_L) conditions reconstructed by the algorithm or obtained manually by an expert investigator. *Green dots* (N_L = 143, 87.7%) show events for which the durations decided by the algorithm and by the expert investigator differed by <10% and were considered corresponding (equal) events. *Yellow dots* (N_L = 20, 12.3%) are "not coincident" events for which the durations decided by the algorithm and by the investigator differed by 10% or more

To study the differences between the events selected by the algorithm and those identified by the heuristic method, we selected the 163 rainfall events (one event for each landslide) for which the rainfall duration identified by the algorithm D_L was most similar to the corresponding duration defined heuristically by the expert investigator, D_L^*. We decided that two paired events have the same duration when their values for D_L and D_L^* differ by less than 10%. This is a reasonable assumption considering the uncertainties associated to the definition of the time of occurrence of a landslide. We further verified that for all the corresponding events, the values for the cumulated rainfall measured by the algorithm E_L and by the expert investigator E_L^* differ much less than 10%. We found that for 143 events (87.7%) (green dots in Fig. 7) the algorithm and the expert investigator provided coincident results. We consider this a measure of the ability of the algorithm to reproduce consistently the results obtained by the expert investigator. We investigated the 20 non-coincident events (12.3%) (yellow dots in Fig. 7), and found that the differences were due to subjective interpretations made by the investigator that resulted in the definition of systematically shorter rainfall events, i.e., $D_L^* < D_L$.

5 Discussion

The algorithm here presented applies the decisions taken by an expert investigator that reconstructs manually (i.e., heuristically) the rainfall events, and measures the rainfall conditions that have resulted in landslides.

The algorithm has several advantages over the traditional, manual methods. The first advantage is the fact that the algorithm performs an objective and reproducible reconstruction of the rainfall events. Experience gained in a national project for the collection of information on the rainfall (D_L, E_L) conditions that have resulted in landslides in Italy (Gariano et al. 2012) indicates that it is difficult for an investigator who has to analyse hundreds of landslides and multiple rain gauges to be consistent in the identification of the rainfall events, and in measuring the rainfall conditions responsible for the landslides. The problem is exacerbated when the number of investigators increases, making it difficult to prepare accurate catalogues of rainfall events with landslides covering large geographical areas (e.g., a nation). Use of the algorithm significantly reduces the uncertainty (operational variability) introduced by the presence of multiple investigators.

Another obvious advantage is the fact that use of the code reduces significantly the time necessary to determine the rainfall events, to associate the landslide information to a rainfall event, and to determine the rainfall duration (D_L) and the cumulated rainfall (E_L) responsible for landslide occurrence. We estimate that the time required by the expert investigators to determine the rainfall conditions responsible for the 163 landslides in Sicily between July 2002 and November 2011 (yellow dots in Fig. 4), and to search the results for possible errors, was about one month. Using the algorithm, a single investigator completed the equivalent operations in three hours.

Furthermore, the automation becomes a particular advantage where multiple rainfall records have to be tested for the same landslide, i.e., where multiple rain gauges exist near a landslide. The algorithm can be easily improved to select automatically or semi-automatically the most representative rain gauge for a specific landslide from a pool of rain gauges, and to evaluate the influence of the selection of different rain gauges in the definition of the rainfall conditions responsible for the landslide. This leads to the possibility of quantifying the uncertainty related to the selection of the rain gauges in the definition of rainfall thresholds for possible landslide occurrence, a problem currently unresolved (Guzzetti et al. 2007, 2008).

To explain how the algorithm operates, we have used rainfall measurements cumulated over a period of one hour. However, the algorithm is independent of the temporal resolution of the rainfall record, and it is applicable to sub-hourly rainfall data (with measurements every, e.g., 5, 10, 20, or 30 min), to rainfall cumulated over more than one hour (e.g., every 2, 3, 6, 12 h), and even to daily rainfall measurements. To operate, the algorithm uses six parameters, listed in Table 1. These parameters can be changed and adjusted to different physical (e.g., climatic, meteorological) or operational (e.g., type of rain gauges) conditions. The algorithm is independent from the local or regional climatic conditions, and from the operational settings of the rain gauge network.

The algorithm checks the rainfall record for missing measurements. This allows singling out incomplete rainfall events, which can be eliminated from the subsequent analyses reducing the uncertainty associated to the definition of the rainfall conditions responsible for the landslides. Manual check of a rainfall record is a time consuming, and error-prone operation. Due to the inherent lack of consistency of an investigator, the manual operation might not detect all the gaps in a rainfall record. This is a further advantage of the software code.

As explained before, when applied to a typical record of rainfall measurements, the algorithm identifies multiple rainfall (D_L, E_L) conditions that can be responsible for landslides (Fig. 3f). Conversely, due to practical and operational constrains, when an investigator searches a rainfall record manually to define the rainfall conditions responsible for landslides, the investigator identifies only a single set of rainfall conditions. Nevertheless, in the absence of external information, all the events are equally probable as possible landslide triggers. This implies that all the events should be considered equally for the statistical analysis of the events, or for reconstruction of rainfall thresholds. This is an advantage over the existing manual procedures.

To reconstruct the rainfall events, the algorithm uses information on the separation between successive events decided by the investigator. This is the same information used—explicitly or implicitly—by an investigator that searches a rainfall record and separates two successive events manually. The problem of the manual method is twofold: (i) the criteria for the separation of the events are often not clear or explicit, or are not applied consistently by the investigator, and (ii) for operational and practical problems, only a single set of criteria is used to separate successive events manually. This can condition the subsequent analyses. The software allows changing the separation criteria, and rapidly reconstructing the rainfall events and the rainfall conditions responsible for landslides using different parameter values (Table 1). This is an advantage over manual methods that opens to the possibility of evaluating the uncertainty introduced by selecting different criteria to separate (or to combine) successive rainfall events.

An additional advantage of the proposed algorithm is the fact that the rainfall events are defined independently from the landslide information, using only the rainfall record and a set of event separation criteria. Indeed, the landslide information is associated with a rainfall event only after the event has been identified. This is a significant advantage over traditional methods that—with a few exceptions (e.g., Onodera et al. 1974; Lumb 1975; Jibson 1975; Corominas and Moya 1999; Biafiore et al. 2002; Marchi et al. 2002; Zezere and Rodriquez 2002; Pedrozzi 2004; Giannecchini 2005; Berti et al. 2012; Segoni et al. 2014a, b)—consider the rainfall events that have resulted in landslide and ignore all the other events, which are the majority in a rainfall record. The ability to reconstruct independently the rainfall events and the rainfall events that have resulted in landslides is important because it allows using conditional probability and Bayesians inference to establish rainfall thresholds for possible landslide occurrence (Berti et al. 2012). This is a significant improvement for the application of rainfall thresholds in modern landslide warning systems (Aleotti 2004; Godt et al. 2006; Guzzetti et al. 2008; Bach-Kirschbaum et al. 2012; Rossi et al. 2012; Segoni et al. 2015). Further, the ability to reconstruct rainfall events independently from the landslide information opens to the possibility of using the algorithm to investigate processes and hazards different from landslides, including. e.g., high-intensity rainstorms, flash flooding, and droughts.

6 Conclusion

We developed and tested an algorithm for the objective and reproducible reconstruction of rainfall events, and of rainfall events that have resulted in landslides. The algorithm exploits a continuous record of rainfall measurements, and information on the time of occurrence of landslides, to determine: (i) the duration and the cumulated rainfall of rainfall events, and (ii) the duration and the cumulated rainfall of single or multiple rainfall conditions responsible for the landslide initiation. Use of the algorithm accelerates considerably the slow and tedious process of the definition of the rainfall conditions responsible for landslides, and reduces the subjectivity inherent in the manual treatment of the rainfall and landslide data. This decreases the uncertainty associated to the definition of the rainfall events.

We expect that the proposed algorithm, and the software that implements the algorithm that is made publicly available, will be used for the objective and reproducible definition of large sets of rainfall conditions that have resulted in landslides in different geographical areas, and will contribute to reduce at least part of the uncertainty associated with the definition of rainfall thresholds for possible landslide occurrence. Rainfall thresholds characterized by a reduced uncertainty, or for which the uncertainty is known, will contribute to more reliable landslide warning systems.

Acknowledgements The work was supported by a grant of the Italian national Department for Civil Protection (DPC). MM and SLG were supported by DPC grants. We thank Provincial Fire Brigades in Sicily for providing the landslide information.

References

Aleotti P (2004) A warning system for rainfall-induced shallow failures. Eng Geol 73:247–265

Bach Kirschbaum D, Adler RF, Hong Y, Kumar S, Peters-Lidard C, Lerner-Lam A (2012) Advances in landslide nowcasting: evaluation of a global and regional modeling approach. Environ Earth Sci 66:1683–1696. doi:10.1007/s12665-011-0990-3

Berti M, Martina MLV, Franceschini S, Pignone S, Simoni A, Pizziolo M (2012) Probabilistic rainfall thresholds for landslide occurrence using a Bayesian approach. J. Geohys Res 117:F04006. doi:10.1029/2012JF002367

Biafiore M, Braca G, De Blasio A, Martone M, Onorati G, Tranfaglia G (2002) Il monitoraggio ambientale dei territori campani a rischio di frane e di alluvioni: lo sviluppo della rete idropluviometrica del Servizio Idrografico e Mareografico Nazionale. Unpublished report (in Italian)

Brabb EE, Harrod BL (1989) Landslides: extent and economic significance. AA Balkema Publisher, Rotterdam

Brunetti MT, Peruccacci S, Rossi M, Luciani S, Valigi D, Guzzetti F (2010) Rainfall thresholds for the possible occurrence of landslides in Italy. Nat Hazards Earth Syst Sci 10:447–458

Caine N (1980) The rainfall intensity-duration control of shallow landslides and debris flow. Geogr Ann A 62:23–27

Cannon SH, Ellen SD (1985) Rainfall conditions for abundant debris avalanches in the San Francisco Bay region, California. Calif Geol 38(12):267–272

Cannon SH, Gartner JE, Wilson RC, Bowers JC, Laber JL (2008) Storm rainfall conditions for floods and debris flows from recently burned areas in southwestern Colorado and southern California. Geomorphology 96 (3–4):250–269

Corominas J, Moya J (1999) Reconstructing recent landslide activity in relation to rainfall in the Llobregat River basin, Eastern Pyrenees, Spain. Geomorphol 30:79–93

Crosta GB, Frattini P (2001) Rainfall thresholds for triggering soil slips and debris flow. In: Proceedings of the 2nd EGS Plinius Conference on Mediterranean

Storms, Siena, 2000, CNR GNDCI publication 2547, pp 463–487

Gariano SL, Iovine GR, Brunetti MT, Peruccacci S, Luciani S, Bartolini D, Palladino M, Vessia G, Viero A, Vennari C, Antronico L, Deganutti AM, Luino F, Parise M, Terranova OG, Guzzetti F (2012) Populating a catalogue of rainfall events that triggered shallow landslides in Italy. 86° Congresso della Società Geologica Italiana, Arcavacata di Rende, 18-20 Settembre 2012. Rendiconti Online della Società Geologica Italiana 21:396–398

Gariano SL, Brunetti MT, Iovine G, Melillo M, Peruccacci S, Terranova O, Vennari C, Guzzetti F (2015) Calibration and validation of rainfall thresholds for shallow landslide forecasting in Sicily, southern Italy. Geomorphol 228:653–665

Giannecchini R (2005) Rainfall triggering soil slips in the southern Apuane Alps (Tuscany, Italy). Adv Geosci 2:21–24

Godt JW, Baum RL, Chleborad AF (2006) Rainfall characteristics for shallow landsliding in Seattle, Washington, USA. Earth Surf Proc Land 31:97–110

Guo J (2002) Overflow risk analysis for stormwater quality control basins. J Hydrol Eng 7(6):428–434

Guzzetti F, Cardinali M, Reichenbach P, Cipolla F, Sebastiani C, Galli M, Salvati P (2004) Landslides triggered by the 23 November 2000 rainfall event in the Imperia Province, Western Liguria, Italy. Eng Geol 73:229–245

Guzzetti F, Peruccacci S, Rossi M, Stark CP (2007) Rainfall thresholds for the initiation of landslides in central and southern Europe. Meteorol Atmos Phys 98:239–267

Guzzetti F, Peruccacci S, Rossi M, Stark CP (2008) The rainfall intensity-duration control of shallow landslides and debris flows: an update. Landslides 5(1): 3–17

Innes JL (1983) Debris flows. Prog in Phys Geogr 7:469–501

Jibson RW (1975) Debris flow in southern Porto Rico. Geol Soc of America special paper 236:29–55

Kanji MA, Massad F, Cruz PT (2003) Debris flows in areas of residual soils: occurrence and characteristics. In: International workshop on occurrence and mechanisms of flows in natural slopes and earthfills. Iw-Flows, Sorrento, Associazione Geotecnica Italiana 2, pp 1–11

Keefer DK, Larsen MC (2007) Assessing Landslide Hazards. Science 316:1136–1138

Komac M, Šinigoj J, Jemec Auflič M (2014) A national warning system for rainfall-induced landslides in Slovenia. In: Sassa K, Canuti P, Yin Y (eds) Landslide science for a safer geoenvironment: methods of landslide studies. Springer, Berlin, pp 577–582

Köppen WP (1931) Grundriss der Klimakunde. Walter de Grayter, Berlin

Lee ML, Ng KY, Huang YF, Li WC (2014) Rainfall-induced landslides in Hulu Kelang area. Malaysia Nat Hazards 70(1):353–375

Lumb P (1975) Slope failure in Hong Kong. Q J Eng Geol 8:31–65

Marchi L, Arattano M, Deganutti AM (2002) Ten years of debris-flow monitoring in the Moscardo Torrent (Italian Alps). Geomorphol 46:1–17

Martelloni G, Segoni S, Fanti R, Catani F (2012) Rainfall thresholds for the forecasting of landslide occurrence at regional scale. Landslides 9:485–495

Melillo M (2009) Modellizzazione del Lago Trasimeno. PhD thesis (in Italian). http://www.scribd.com/doc/187518230/Modellizzazione-Lago-Trasimeno. Accessed November 2013

Onodera T, Yoshinaka R, Kazama H (1974) Slope failures caused by heavy rainfall in Japan. In: Proceedings 2nd International Congress of the International Assesment Engineering Geology, San Paulo, vol 11, pp 1–10

Pedrozzi G (2004) Triggering of landslides in Canton Ticino (Switzerland) and prediction by the rainfall intensity and duration method. Bull Eng Geol Environ 63(4):281–291

Peruccacci S, Brunetti MT, Luciani S, Vennari C, Guzzetti F (2012) Lithological and seasonal control on rainfall thresholds for the possible initiation of landslides in central Italy. Geomorphol 139–140:79–90. doi:10.1016/j.geomorph.2011.10.005

Petly (2012) Global patterns of loss of life from landslides. Geology 40:927–930

Reichenbach P, Cardinali M, De Vita P, Guzzetti F (1998) Regional hydrological thresholds for landslides and floods in the Tiber River Basin (central Italy). Environ Geol 35(2–3):146–159

Zezere JL, Rodrigues, ML (2002) Rainfall thresholds for landsliding in Lisbon Area (Portugal) In: Rybar J, Stemberk J,Wagner P (eds) Landslides. A.A. Balkema, Lisse, pp 333–338

Rosi A, Segoni S, Catani F, Casagli N (2012) Statistical and environmental analyses for the definition of a regional rainfall threshold system for landslide triggering in Tuscany (Italy). J Geogr Sci 22(4):617–629

Rosi A, Lagomarsino D, Rossi G, Segoni S, Battistini A, Casagli N (2015) Updating EWS rainfall thresholds for the triggering of landslides. Nat Hazard 78:297–308

Rossi M, Peruccacci S, Brunetti MT, Marchesini I, Luciani S, Ardizzone F, Balducci V, Bianchi C, Cardinali M, Fiorucci F, Mondini AC, Reichenbach P, Salvati P, Santangelo M, Bartolini D, Gariano SL, Palladino M, Vessia G, Viero A, Antronico L, Borselli L, Deganutti AM, Iovine G, Luino F, Parise M, Polemio M, Guzzetti F (2012) SANF: a national warning system for rainfall-induced landslides in Italy. In: Eberhardt et al (eds) Landslides and engineered slopes: protecting society through improved understanding. Taylor & Francis Group, London, pp 1895–1899

Saito H, Nakayama D, Matsuyama H (2010) Relationship between the initiation of a shallow landslide and rainfall intensity-duration thresholds in Japan.

Geomorphol 118:167–175. doi:10.1016/j.geomorph.2009.12.016

Salvati P, Bianchi C, Rossi M, Guzzetti F (2010) Societal landslide and flood risk in Italy. Nat Hazards Earth Syst Sci 10:465–483

Segoni S, Rossi G, Rosi A, Catani F (2014a) Landslides triggered by rainfall: a semiautomated procedure to define consistent intensity-duration thresholds. Comput Geosci 63:123–131

Segoni S, Rosi A, Rossi G, Catani F, Casagli N (2014b) Analysing the relationship between rainfalls and landslides to define a mosaic of triggering thresholds for regional scale warning systems. Nat Hazards Earth Syst Sci 14:2637–2648

Segoni S, Battistini A, Rossi G, Rosi A, Lagomarsino D, Catani F, Moretti S, Casagli N (2015) Technical note: an operational landslide early warning system at regional scale based on space–time variable rainfall thresholds. Nat Hazards Earth Syst Sci 15:853–861

Shamsudin S, Dan'azumi S, Aris A (2010) Effect of storm separation time on rainfall characteristics—A case study of Johor, Malaysia. Eur J Sci Res 45(2):162–167

Staley DM, Kean JW, Cannon SH, Schmidt KM, Laber JL (2012) Objective definition of rainfall intensity–duration thresholds for the initiation of postfire debris flows in southern California. Landslides 10:547–562. doi:10.1007/s10346-012-0341-9

Vennari C, Gariano SL, Antronico L, Brunetti MT, Iovine G, Peruccacci S, Terranova O, Guzzetti F (2014) Rainfall thresholds for shallow landslide occurrence in Calabria, southern Italy. Nat Hazards Earth Syst Sci 14:317–330

Wieczorek GF (1987) Effect of rainfall intensity and duration on debris flows in central Santa Cruz Mountains. In: Costa JE, Wieczorek GF (eds) Debris flow/avalanches: process, recognition, and mitigation. Geological Society of America, Reviews in Eng Geology, vol 7, pp 93–104

Thornthwaite CW, Mather JR (1957) Instructions and tables for computing potential evapotranspiration and the water balance. Publ Climatol Lab Climatol Drexel Inst Technol 10(3):185–311

Trewartha GT (1968) An introduction to climate, 4th edn. McGraw-Hill, New York

TXT-tool 2.385-1.1 A Comprehensive Landslide Monitoring System: The Kostanjek Landslide, Croatia

Snježana Mihalić Arbanas, Martin Krkač, Sanja Bernat Gazibara, Marko Komac, Marin Sečanj and Željko Arbanas

Abstract

In this paper the general design of the integrated monitoring system of the Kostanjek landslide (Zagreb, Croatia) will be briefly presented. A comprehensive automated real-time monitoring system has been installed as a part of the research activities in the Croatian–Japanese SATREPS FY2008 scientific project on landslides in 2011–2013. The monitoring system consist of sensors for geodetic, hydrological and geotechnical monitoring to monitor landslide movement and landslide causal factors. Equipment for landslide monitoring at the surface and in the underground include 15 precise GNSS stations, long- and short-span extensometers, pore pressure gauges in boreholes, water level gauges in wells, rain gauge, weather station and accelerometers. All monitoring equipment is connected in the system entitled the Kostanjek Landslide Observatory. Monitoring of this deep and large landslide during multiple extreme hydro-meteorological events from January 2013 to January 2015 enabled collection of data for the analysis of landslide movement both on and below the surface as well as analysis of the relationship between landslide reactivations and their triggers, i.e., changes in groundwater level and precipitation. Monitoring results have proven established monitoring system as a sound tool for long-term monitoring as well as for landslide characterization. Preliminary application of landslide characterization (i.e., determination of landslide

S. Mihalić Arbanas (✉) · M. Krkač · S. Bernat Gazibara · M. Sečanj
Department of Geology and Geological Engineering, University of Zagreb, Pierottijeva 6, 10000 Zagreb, Croatia
e-mail: smihalic@rgn.hr

M. Krkač
e-mail: martin.krkac@rgn.hr

S. Bernat Gazibara
e-mail: sanja.bernat@rgn.hr

M. Sečanj
e-mail: marin.secanj@rgn.hr

M. Komac
University of Nova Gorica, Nova Gorica, Slovenia
e-mail: m.komac@telemach.net

Ž. Arbanas
Department of Hydrotechnics and Geotechnics, University of Rijeka, Radmile Matejčić 3, 51000 Rijeka, Croatia
e-mail: zeljko.arbanas@gradri.uniri.hr

K. Sassa et al. (eds.), *Landslide Dynamics: ISDR-ICL Landslide Interactive Teaching Tools*, https://doi.org/10.1007/978-3-319-57774-6_34

geometry) was presented in the form of numerical modeling to access changes on the landslide surface and run-out length for hazard and risk mapping purposes.

Keywords
Landslide monitoring · Kostanjek landslide · GNSS · Extensometer Accelerometer · Rain gauge · SATREPS project · City of Zagreb

Contents

1 Introduction ... 450
2 Framework for Monitoring Establishment ... 451
2.1 The Japanese-Croatian SATREPS FY2008 Project ... 452
2.2 Historical Data from Landslide Investigation and Existing Landslide Model ... 453
3 Monitoring System Development and Sensor Network ... 455
3.1 Sensors for Displacement Measurement ... 455
3.2 Sensors for Hydrological Measurement ... 457
3.3 Sensors for Geophysical Measurement ... 457
4 Monitoring Results and Its Preliminary Application ... 458
4.1 Analysis of Monitoring Results ... 458
4.2 Preliminary Application of Monitoring Results 459
5 Discussion and Conclusions ... 460
References ... 463

1 Introduction

Kostanjek landslide is an example of a reactivated deep-seated large translational landslide formed in soft rock-hard soil, i.e., marls. The landslide is one of numerous landslides within the borders of the City of Zagreb (Croatia) but is specific according to its size, as interpreted by Ortolan (1996) and Stanić and Nonveiller (1996): the landslide volume is 32.6×10^6 m^3 and the depth of displaced mass is 90 m. Landslide velocities have been changing over the last 50 years, from landslide activation in 1963 until today, in a range from extremely slow to very slow. The initial landslide was developed as a consequence of loss of global stability of gentle to steep slopes above an open pit mine of marl and a cement factory 'Sloboda' (Fig. 1). Slope movements were caused by mining activities, i.e., undercutting of the slope toe and uncontrolled massive blasting. Following the initial slow movements that caused settlement and fractures of industrial cement factory buildings in 1963, and damaging numerous private houses within an area of approximately 1 km^2 in a very short period, attention shifted to the unstable slopes above the cement factory known as Kostanjek landslide.

Although numerous surface exploration and visual studies were undertaken between the 1966 and 2010 (Ferić et al. 2010) the rudimentary nature of the monitoring undertaken did not provide conclusive evidence regarding the rate and extent of the movement of the Kostanjek landslide. Recently, more detailed assessments were carried out in the framework of the Japanese-Croatian scientific bilateral project 'Risk Identification and Land-Use Planning for Disaster Mitigation of Landslides and Floods in Croatia' (Mihalić and Arbanas 2012) to assess the status of the historical monitoring points at the area of the landslide, to assess the hazard of further slope movement, and review the options for updated instrumentation to monitor movements of the landslide.

The purpose of this teaching material is to provide the researchers and students with a synthesized review of the results of bilateral scientific project joint research activities of the deep-seated landslide in an urban area as a stimulus for further research. The following

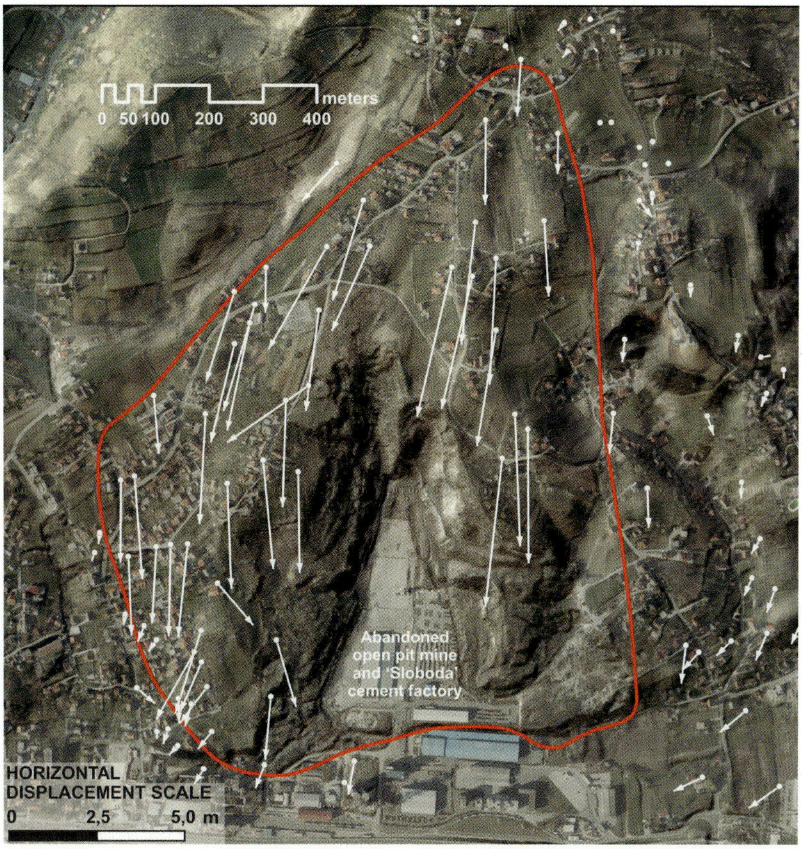

Fig. 1 Horizontal displacements (*white arrows*) at the Kostanjek landslide area in the period 1963–1988 based on interpretation of stereo pairs of aerial photographs from 1963, 1979, 1981, 1985 and 1988 (Ortolan and Pleško 1992). Red line depicts the outline of the Kostanjek landslide according to interpretation by Ortolan (1996). In the background are maps from March 2012: hillshade map (with an azimuth of 315° and a sun angle of 45°) generated from 2 × 2 m bare earth DEM overlain by an orthophoto image. Devastated slopes of the abandoned open pit mine are clearly expressed by rough relief forms in the middle of the Kostanjek landslide. The former cement factory 'Sloboda' was placed in the plain area in the bottom middle part of the landslide

sections briefly describe the study area with summarized overviews of historical data related to the landslide investigation, monitoring and modelling. This section is followed by a short overview of monitoring results and presentation of numerical modeling results. After discussing the potential advantages and the current limitations of the monitoring system, the conclusions present a perspective on the future analysis and application of continuous observations of the Kostanjek landslide to the prognosis of landslide hazard and risk.

2 Framework for Monitoring Establishment

By definition, landslides are characterized by movement. Knowledge of the movement magnitude and velocity, that is, movement distribution along the slope, are the most important data for all landslide analysis. Monitoring is required to observe the changing conditions that may lead to total failure of the slope where slope movement is occurring, where safety factors against sliding are low, or where high risk is present from a possible slope failure (Mihalić Arbanas and Arbanas 2015). Landslide movement monitoring expressed via ground surface displacements and deformation of structures (including the landslide body) related to landslides can be accomplished using different types of monitoring systems and techniques. Baroň et al. (2012) summarized the state of the art of the monitoring parameters and presented some examples of prominent European landslides.

Integrated monitoring systems are usually established at landslides for long-term landslide monitoring, which presents high risks for particular societies. Examples of landslides that are equipped with integrated monitoring systems include Ancona (Cotecchia 2006), Corvara (Corsini et al. 2005), Tessina (Petley et al. 2005) and Valoria (Castagnetti et al. 2013) in Italy; Gradenbach in Austria (Brückl et al. 2006); Åknes in Norway (Blikra 2012); Turtle Mountain in Canada (Froese et al. 2012); and Grohovo in Croatia (Arbanas et al. 2012, 2016). The planning and design of an integrated monitoring system of the deep-seated Kostanjek landslide began in 2010 in the framework of the bilateral scientific Japanese-Croatian SATREPS FY2008 project 'Risk Identification and Land-Use Planning for Disaster Mitigation of Landslides and Floods in Croatia' (Mihalić and Arbanas 2012). The selection of measurement instrumentation and methods was performed based on analysis of historical data related to investigation, monitoring and modelling of the Kostanjek landslide.

2.1 The Japanese-Croatian SATREPS FY2008 Project

In the framework of the Japanese-Croatian joint research from the period 2009–2011, the Kostanjek landslide monitoring project activities were initiated; this included planning and design of an monitoring system in a period 2009–2010, installation of the instrumentation in a two-year period 2011–2013 and analysis of monitoring results (Krkač et al. 2013, 2014). The objective of the monitoring system was to continuously observe the parameters of landslide movement in real-time and monitor the landslide's causal factors, i.e., processes that trigger landslide reactivations, such as precipitation, groundwater changes and earthquakes. Continuous data series, collected by sensor networks, aimed for the analysis of landslide movement and its causes and the analysis of the interrelations between the dynamics of landslide and causal processes.

The Kostanjek landslide monitoring project activities were administered and developed by Project's Working Group 1, with working group members from the Faculty of Mining, Geology and Petroleum Engineering (University of Zagreb), Faculty of Civil Engineering (University of Rijeka), Research Institute for Natural Hazards and Disaster Recovery (Niigata University), International Consortium on Landslides and Disaster Prevention Research Institute (Kyoto University). It involved an international team of university research groups working together with consultants and contractors to implement a near–real-time monitoring network and conduct studies to better understand the geological settings of the area and mechanisms of failure. Prior to the completion of the Kostanjek landslide monitoring system in January, 2011, the Government of the City of Zagreb determined that the continued long-term monitoring of the Kostanjek landslide was important for public safety of the residents. The local authorities from the City of Zagreb must find cost-effective solutions for this human and environmental threat. Potential hazards in the form of large movements of reactivated landslide can become disastrous because there are approximately 290 endangered buildings on the landslide surface.

The first priority of the monitoring system is to provide an early warning to residents of the potential for dangerous sliding originating from slope movements in the area of the Kostanjek landslide. The secondary priority is to provide an opportunity for the research community to test and develop instrumentation and monitoring technologies and to better understand the mechanics of slowly moving masses of soft rock-hard soil. Hence, the working name for the whole monitoring system is Kostanjek Landslide Observatory (KLO). It is the intention of the Faculty of Mining, Geology and Petroleum Engineering (University of Zagreb) to make all data from the KLO available to the research community and to work with researchers and companies to test and develop new monitoring technologies applicable to similar phenomena. This ongoing research will aid in understanding the movements of the entire Kostanjek landslide mass, including the lower artificial slopes in the abandoned open pit as well as urbanized slopes which are populated (Fig. 1), in order to provide a better model for prediction of future movements.

2.2 Historical Data from Landslide Investigation and Existing Landslide Model

Since its activation in 1963, there have been various studies and interpretations of the materials, mechanism, causal factors and remedial measures of the Kostanjek landslide. Initial studies from the period 1966–1970 focused on the monitoring of displacements of industrial buildings in the toe part of the landslide and interpretation of subsidence and uplifting of the ground. Swelling of the unloaded marl layers caused by the excavation was identified as a possible cause of the damage (Pehnec 1967). Several years later Nonveiller (1976) highlighted the potential for sliding on the basis of analysis of incurred movements. Excavation in the quarry was stopped in 1988 after mining activities and excavation were identified as the main triggering factors of the landslide. A total volume of 5.3×10^6 m^3 of rock was excavated.

More recently, more detailed investigation, studies and interpretations of the geological settings and landslide model were undertaken in the period 1984–1994, and results were published in Ortolan et al. (1987), Ortolan and Pleško (1992), Stanić and Nonveiller (1995, 1996) and Ortolan (1996). It has been widely recognized that the 1963 displacements originated in the foot part of the large landslide with the following dimensions: total landslide area of approximately 1 km^2; volume of moving landslide mass of 32×10^6 m^3; depth of displaced mass 90 m. Interpretation of the landslide model was done on the basis of results of investigations of the Kostanjek landslide from the period 1988–1989, including surface exploration and engineering geological mapping; subsurface investigation; geodetic and geotechnical monitoring; and laboratory analysis. The main disadvantage of this landslide model is the high level of uncertainty of interpretation caused by (1) a lack of clearly expressed landslide features, especially main scarp, flanks and accumulation at the toe; and (2) a lack of subsurface investigations and monitoring (e.g., only 6 boreholes were drilled inside or near the landslide area). For the interpretation of landslide movement, the most important data from this period of investigation are traces of sliding surfaces and vectors of displacements. During engineering geological mapping in 1987, dug wells cut by sliding were systematically recorded, providing evidence of sliding surfaces along the western, north-western and northern boundary of the landslide. According to the photo interpretation of aerial stereo pairs from 1963, 1979, 1981, 1985 and 1988, horizontal displacements of the ground surface in the period 1963–1988 were detected in a range 3–6 meters (average 12–24 cm per year), as is depicted in Fig. 1.

A geotechnical report prepared by Croatian Civil Engineering Institute (IGH) in 2008 provides a comprehensive review of all historical investigations at the area of the Kostanjek landslide, with the presentation of the historical landslide model, its geometry, mechanism and contributing factors. With respect to an interpreted historical movement of the Kostanjek landslide, as the specific mechanism of recent movement is not known, there is need to better define the subsurface conditions and contours of landslide bodies contributing to the movement.

Recent field studies have been carried out in the framework of the Croatian-Japanese SATREPS FY2008 scientific joint research project (in the period 2009–2012) by Furuya et al.

(2011) to characterize the landslide contour patterns at the surface and by Vrsaljko et al. (2012) to interpret geological units in the landslide body. Field mapping has also been supplemented by reviews of morphology derived from the available 1-m digital elevation model (DEM) and by monitoring of recent movements from the period 2010–2012 at the 35 stable geodetic points shown in Fig. 2 (Županović et al. 2012). Subsequent mapping of the Kostanjek landslide area has highlighted the similar geological conditions.

Although undercutting of slopes by excavation in an open pit mine was considered to be the main driver for historical sliding in the central part of the area, other factors are believed to have contributed to the actual movements, including:

- weakening of the marl due to blast-induced seismic activity in 1963 in the wider area,
- creeping of the superficial deposits due to daily and seasonal changes of soil moisture,
- water accumulation in cracks at the locations of damaged sewage utilities,
- decrease of slope stability due to mining activities (excavation and fills) in the area of the abandoned open pit mine,
- expansion of marl due to unloading.

Future work is expected to utilize new data from: chemical, isotope and mineralogical analysis of rock samples (Martinčević et al. 2013) and groundwater samples (Watanabe et al. 2011; Yamamoto et al. 2013), hydrological investigations (Krkač et al. 2011; Krkač and Rubinić 2013), laboratory testing of mechanical properties (Oštrić et al. 2012) and additional surface geophysical investigations to further improve understanding of the subsurface structure of the landslide area, including groundwater, soil and rock properties. The most important data were collected by continuous monitoring of landslide movement and landslide causal factors,

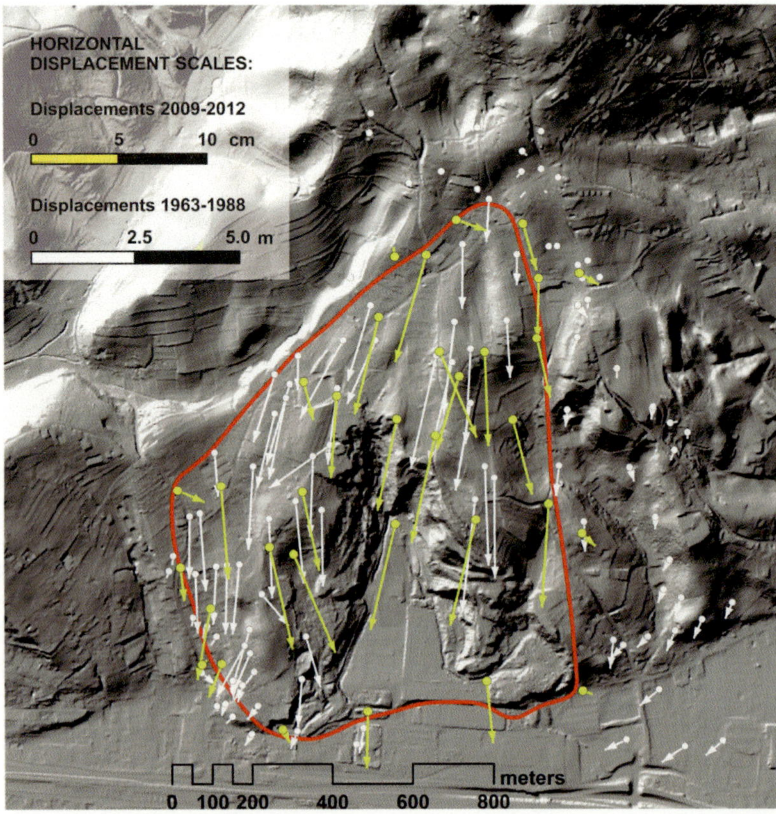

Fig. 2 Recent horizontal displacements (*yellow arrows*) at the Kostanjek landslide area in the period 2010–2012 (Županović et al. 2012) compared with historical horizontal displacement (*white arrows*) for the period 1963–1988 (Ortolan and Pleško 1992). *Red line* depicts Kostanjek landslide contour according to Ortolan (1996). The background is a hillshade map (with an azimuth of 315° and a sun angle of 45°) generated from a 2 × 2 m LiDAR bare earth DEM (scanned in March 2012)

precipitation and earthquakes. Integration of real-time GNSS monitoring data with other sensor data in the frame of GIS software is shortly presented in the paper by Baučić et al. (2013).

3 Monitoring System Development and Sensor Network

Based on the joint research in the framework of Croatian-Japanese project, the monitoring system on the Kostanjek landslide was designed to include a number of different types of instruments communicating in near–real time to a data acquisition-processing center located at the Faculty of Mining, Geology and Petroleum Engineering, University of Zagreb (UNIZG-RGNF). The system, whose conceptual design was outlined by Nagai et al. (2011), is meant to improve or influence public safety, public education, scientific research, and university education.

The system is designed to measure changes in conditions that affect the potential for a reactivation of sliding from slope cuts of abandoned open pit mines, and to provide early warning of extreme conditions to authorities responsible for emergency preparedness. The public education role involves raising the level of awareness of the general public regarding natural hazards and their potential impacts. The scientific research role of the system is to provide long-term monitoring data that can be used to gain a better understanding of the mechanisms associated with landslide in hard soil-soft rock (Pannonian and Sarmatian marl), and to advance the development of technology in landslide monitoring. Finally, a monitoring system that is housed at the UNIZG-RGNF has the potential to increase the educational potential of the University of Zagreb, and thereby benefit the national educational capacities.

The design process involved defining the preliminary data requirements and reviewing options for instrument types and locations, measurement frequency, and equipment required for data acquisition and management. The monitoring framework should provide complementary types of instruments with varying sensitivities to movement and climatic influences, and also have enough redundancy built into the system to be able to distinguish actual movements. In considering the types of sensors most suitable for providing early warning for impending slope movements, the sensors were grouped as follows:

(1) Sensors for displacement measurement (GNSS receivers, short- and long-span extensometers, inclinometer),
(2) Sensors for hydrological measurements (rain gauge, meteorological station, outflow weirs, water level sensors, piezometers),
(3) Sensors for geophysical measurements (accelerometers).

The sensor network installed at the Kostanjek landslide area encompasses approximately 40 sensors for the monitoring of landslide movement and landslide causal factors. Figure 3 provides the layout of the sensor network.

3.1 Sensors for Displacement Measurement

The sensors for displacement measurements are those that provide a reliable data stream on a year-round basis and measure easily interpretable parameters of the superficial and subsurface movements. The displacement measurement sensor network includes 15 Global Navigation Satellite System (GNSS) receivers, 9 extensometers and 1 inclinometer.

A Trimble GNSS monitoring system consists of fifteen double-frequency NetR9 TI-2 GNSS reference stations with Zephyr Geodetic 2 GNSS antennas installed on each of GNSS reference points. GNSS receivers are fixed to 4 m high poles with 1 m deep reinforced foundations. They collect GNSS raw data and deliver this data in real-time, over communication lines (using routers), to Trimble 4D Control software installed on an application/data server in a data center at UNIZG-RGNF. GNSS receivers provide data

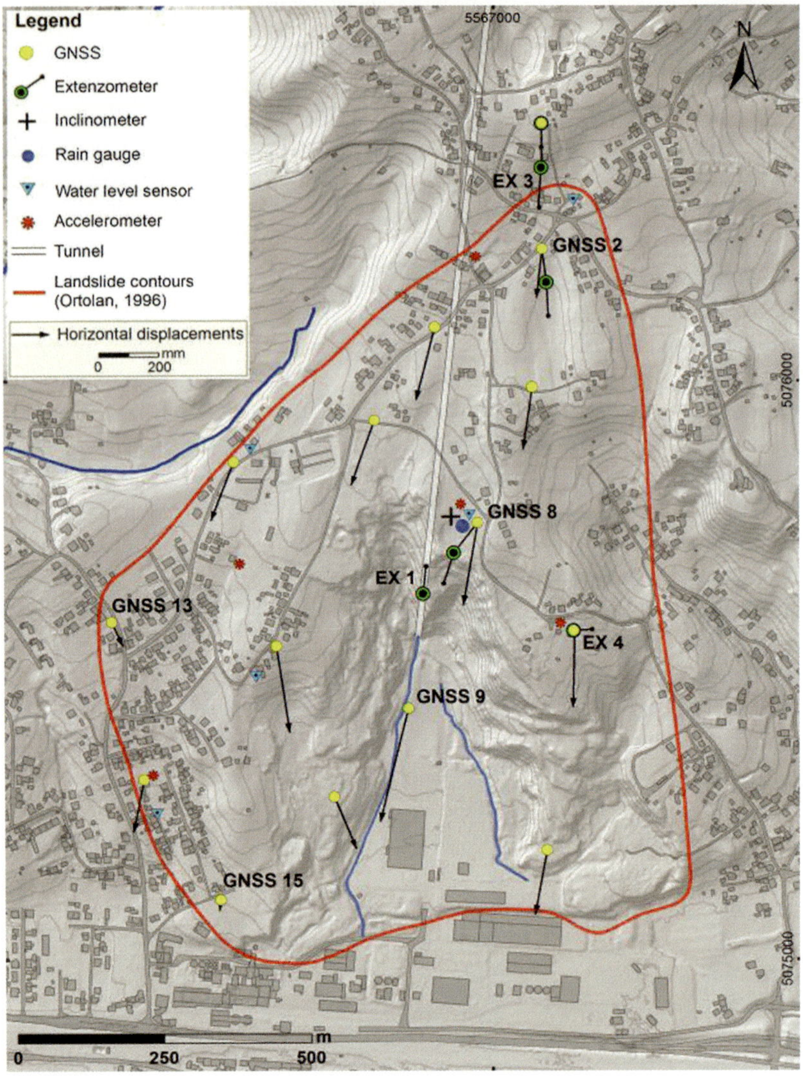

Fig. 3 Locations of the monitoring sensors at the Kostanjek landslide and the cumulative horizontal displacements that were recorded by the GNSS sensor network from January 2013 to January 2015 (according to Krkač 2015)

on absolute positions of surficial points with precisions for measuring movement in the range of cm to mm. All monitoring stations are supplied with electricity from public network.

The locations of GNSS receivers (Fig. 3) can be grouped as follows: above the landslide crown, i.e., outside the landslide area to check the assumption that this area is stable (1 GNSS receiver); near the top of the abandoned slope cuts (4 GNSS receivers); inside a slope cut/fill in the eastern part of the landslide (1 GNSS receiver); inside a slope cut in the western part of the landslide (1 GNSS receiver); inside a flat area in the central part of the abandoned open pit mine with the evidence of uplifting (1 GNSS receiver); along the western and north-western landslide border (5 GNSS receivers); and in northern part of landslide (2 GNSS).

To be able to calculate very precise coordinates of 15 GNSS reference stations at the landslide area, the system needs at least one GNSS reference stations outside of the landslide zone. For that purpose, the system is using data from a permanent GNSS reference station in Gornji Stupnik, which is 7 km away from Kostanjek, and for T4DC monitoring software it will be the 16th GNSS reference station. T4DC monitoring software installed on an application/data server collects GNSS raw data and synchronizes, calculates and adjusts the collected data, then operates an alarm if needed and analyzes measurement results.

The long- and short-span wire extensometers, type NetLG-501E Osasi, provide data on absolute deformation with submillimetre-level precision. Five long-span extensometers are placed from the top of the most stable point above the landslide, perpendicular to the main scarp, and in the direction of sliding. One short-span extensometer is placed perpendicular to the fracture in the eastern part of the landslide, where a scarp with steep displacement is clearly visible. Two long-span extensometers are installed to cross the crown of an artificial steep slope where the highest magnitude of displacement is expected. One short-span extensometer is installed in the underground, in a tunnel which crosses the sliding surface. Extensometers with data loggers are fixed on 4 m high poles with reinforced foundations and some are on the same poles as the GNSS routers. Data collection from extensometers is manual, but data transfer using routers is currently underway.

One inclinometer casing is installed in a 100 m deep vertical borehole in the middle of the landslide for measurements of the inclination of the pipe by a high-precision probe in even distances, for example, every 50 cm. The depth of the present-day major active shear surface is at 62.5 m (on the basis of two measurements, in May 2012 and in February 2013). The deep sliding surface is considered very important for the appraisal of future scenarios of the evolution of mass movement.

3.2 Sensors for Hydrological Measurement

Sensors for hydrological measurements are necessary to provide background data that are useful in the interpretation of displacement measurement results. The planned hydrological measurement sensors include a rain gauge and meteorological station (rain, wind, temperature, barometric pressure). The rain gauge has been installed in the middle part of the landslide from 2011. The meteorological station was installed in 2014 because of the limitations of the rain gauge in terms of precipitation measurements during snowy periods (Krkač et al. 2014). The data recorded from these sensors is being used to correlate observed displacements with meteorological changes. Two outflow weirs (at the mouth of the abandoned tunnel and on a spring outside the landslide area) are equipped by two water level gauges to assess delays in superficial and groundwater discharge and pore pressure after precipitation. Two water level gauges were installed in the central part of the landslide, while three water level gauges were installed on the western, northwestern and northern landslide boundaries (two in dug wells and one in a borehole). Three piezometers with pore pressure gauges were installed in the central part of the landslide body, one is in the zone of the sliding surface and two are in the landslide mass, to measure pore pressure and groundwater levels at two depths inside the landslide mass and one below the landslide.

All hydrological sensors have a power supply from batteries because of their low power consumption. Data collection from water level gauges is manual, but data transfer using routers is enabled for a rain gauge and meteorological station as well as for piezometers with pore pressure gauges.

3.3 Sensors for Geophysical Measurement

The geophysical measurement sensor network encompasses seven accelerometers installed inside the landslide area for the purpose of (i) monitoring local micro-earthquake activity in the landslide area; (ii) monitoring regional earthquake activity, including strong motion; and (iii) monitoring of any ground tremors associated with the landslide, including possible ground inclination. This is a low cost and hi-fidelity broad-band monitoring system consisting of three-component MEMS accelerometer and three-channel autonomous broadband digital recorders with GPS to keep accurate synchronization between each other. A Seismic Source DAQ3-3 3CH high-fidelity digital logger with accurate GPS clocking enables continuous recording, with data harvesting by the attached USB memory every three weeks. Power is supplied using rechargeable batteries.

Three accelerometers (JGI-SVAC-3C MEMS, Colibrys 1600) are installed in three boreholes at the central monitoring station in the middle of landslide: one is in a deep borehole below the landslide (90 m depth); one is in a borehole at depth inside the landslide body (20 m); and one is installed in a shallow borehole near the surface. Four accelerometers (Colibrys SF3000L 3-C MEMS) are installed near the surface, in shallow boreholes at depths of approximately 1.5 m. They are spatially arranged to cover all parts of the landslide area that are supposed to be separate landslide bodies: the upper part of the landslide, the left (eastern) landslide flank and the adjacent valley with shallow creeping phenomena.

4 Monitoring Results and Its Preliminary Application

Monitoring can be generally defined as the systematic repetition of observations of a particular object or area (Stumpf et al. 2011). Landslide monitoring in particular comprises a number of different tasks that will influence the choice of the optimal technique: detection, rapid mapping, fast characterization and long-term monitoring. Kostanjek Landslide Observatory is a long-term monitoring system aimed at processing data for retrieving deformation patterns and time series. The following sections contain monitoring results of different sensors installed at the Kostanjek landslide and graphical presentation of representative time series. Detailed information about their applicability with respect to interpretation of landslide movement phases and corresponding triggering conditions can be found in Krkač (2015). Results of numerical modeling of the Kostanjek landslide using LS-RAPID software (Integrated Landslide Simulation Model) (Sassa and He 2013) are presented to illustrate importance of landslide characterization (i.e., determination of landslide geometry) to access changes on the landslide surface and run-out length which are input data for evaluation of hazard and risk levels for zonation purposes.

4.1 Analysis of Monitoring Results

Landslide movement sensors recorded multiple reactivations of the Kostanjek landslide from external triggers in 2013 and 2014. Krkač (2015) identified five periods of faster movement (landslide reactivations) based on analyses of observations from the GNSS network. The cumulative horizontal displacements that were recorded by the GNSS sensor network over these two years are shown in Fig. 3. The maximal displacement of 426 mm was measured in the lower central part of the landslide (GNSS 9), while the minimal displacement of 38 mm (GNSS 15) was measured along the southwestern landslide boundary. The amount of displacement in the central part of the landslide and in the foot part is approximately two to four times higher than the displacement near the landslide's boundaries (Fig. 3).

Figure 4 shows the evolution of the 2D surface displacement that was registered by four GNSS sensors. During the monitoring period, the first and most intensive period of faster movement occurred from 22nd January to 24th May 2013. During this period, the horizontal displacements ranged from 36.9 mm at the northwestern landslide boundary (GNSS 13) to 168 mm in the lower central part of the landslide (GNSS 9). The maximum recorded velocities were from 1.8 mm/day (GNSS 2, located at the northern boundary) to 4.8 mm/day (GNSS 9). The maximum rates of displacement, with velocities of 2–4.8 mm/day, occurred during the first week of April. The first period of faster movement was caused by a total precipitation amount of 455.2 mm from December 2012 to March 2013 during extreme hydro-meteorological conditions (Bernat et al. 2014, 2015).

During the monitoring period, three of the seven installed extensometers showed significant displacement, namely, the long-span extensometer (EX 3) that crosses the main scarp, the short-span extensometer (EX 4) that crosses the fracture in the eastern part of the landslide and the short-span extensometer (EX 1) that crosses the sliding surface in the abandoned tunnel (Figs. 3 and 5). The maximum recorded relative displacement was 142 mm, which was measured by the extensometer that crosses the main scarp (EX 3). During the first period of faster movement, all three extensometers displayed extension, but the amount of measured relative displacement varied from 40 mm along the fracture on the eastern side of the landslide body to 72.5 mm on the main scarp and 97 mm in the central part, where the sliding surface intersects the tunnel (Fig. 5).

The inclinometric profile, which was obtained based on measurements from March 2012 to February 2013, indicates that the failure occurred in a thin basal shear zone at a depth of 62.5 m (Fig. 6). Deformation above 62.5 m can be considered negligible in terms of landslide mechanisms. Deformation at depths of 15–30 m is probably a consequence of borehole casing deformation from improper inclinometer tube installation and voids between the inclinometer casing and the in situ ground (Krkač et al. 2014).

4.2 Preliminary Application of Monitoring Results

Gradiški et al. (2013) presented simulation of the behavior of the Kostanjek landslide (Fig. 7) with the LS-RAPID software (Sassa and He 2013). An existing landslide model from Ortolan (1996) was modified based on the monitoring results by creating an ellipsoidal sliding surface with a maximum depth of 65 m in the central part of the landslide body (according to the maximum displacements). The parameters that were used for these analyses were determined from drained test samples in a ring shear apparatus, which was also performed in the framework of the Japanese-Croatian project. Simulation was performed for the assumed pore pressure on the sliding surface. According to the results of the simulation, the most unstable part of the landslide is the central part of the landslide body, the slopes of the abandoned open marl pit. In the simulation, the movements started in the central part of the landslide body, and the failure area expanded around the initial failure zone. At the end of the simulation, the area of the entire landslide mass corresponded to the landslide contour from the historical landslide model according to Ortolan (1996). This is also in accordance with the new surface deformations (cracks, bulging, and subsidence) that developed during very recent landslide movement in 2013–2014, as shown by the arrows in Fig. 3.

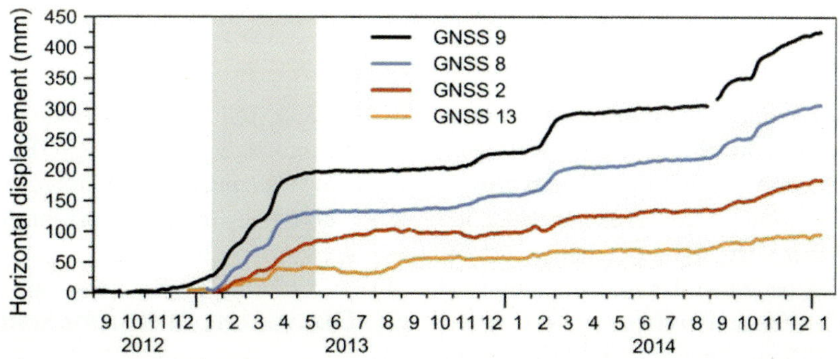

Fig. 4 Cumulative horizontal displacements versus time for GNSS 2 and 13 (near the landslide boundary), GNSS 8 and GNSS 9 (in the central part of the landslide). The gray area indicates the first period of faster movement (according to Krkač 2015)

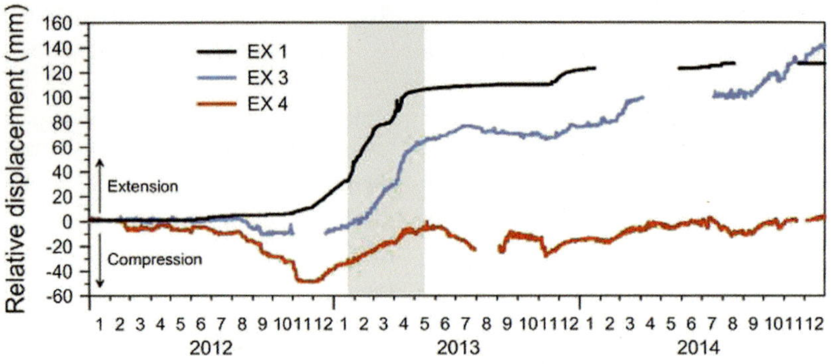

Fig. 5 Relative displacements versus time for EX 1 (extensometer below ground installed across the sliding surface in the abandoned tunnel), EX 3 (extensometer at the surface installed across the main scarp) and EX 4 (extensometer at the surface installed across a fracture in the eastern part of the Kostanjek landslide). The *gray area* indicates the first period of faster movement (Mihalić Arbanas et al. 2016)

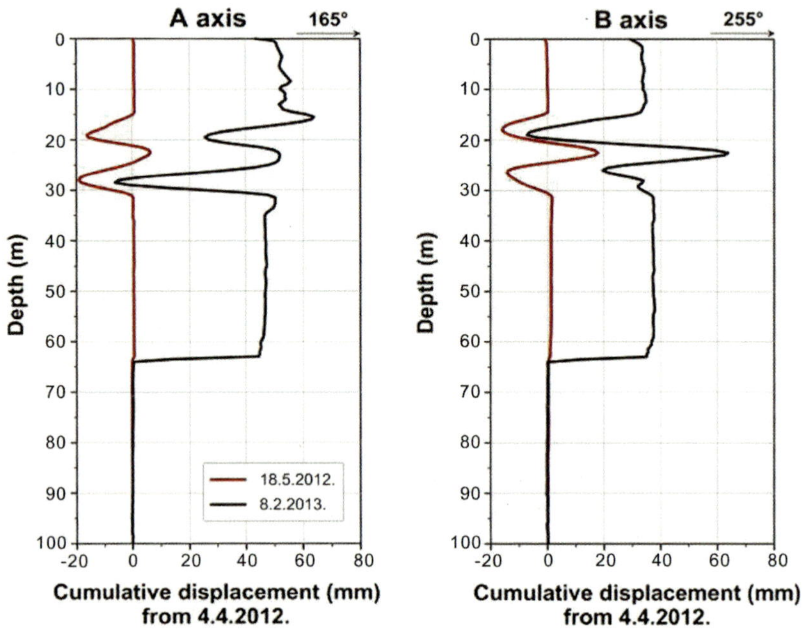

Fig. 6 Cumulative inclinometer displacements for the A and B axes (Krkač et al. 2014)

More reliable simulations of the Kostanjek landslide behavior require defining the following factors more precisely: the sliding surface, with more correct positions along particular landslide cross sections; the groundwater table surface, which needs to be derived based on measured groundwater levels; and soil parameters, determined from additional undrained ring shear tests. Additional subsurface investigations, necessary to interpret more reliably the sliding surface depths, include drilling of boreholes with depths of 10–95 m, followed by soil/rock determination and inclinometer monitoring. The groundwater table surface must be monitored and measured continuously by at least seven piezometers.

5 Discussion and Conclusions

This paper describes integrated landslide monitoring system that was established in the framework of the scientific Japanese-Croatian

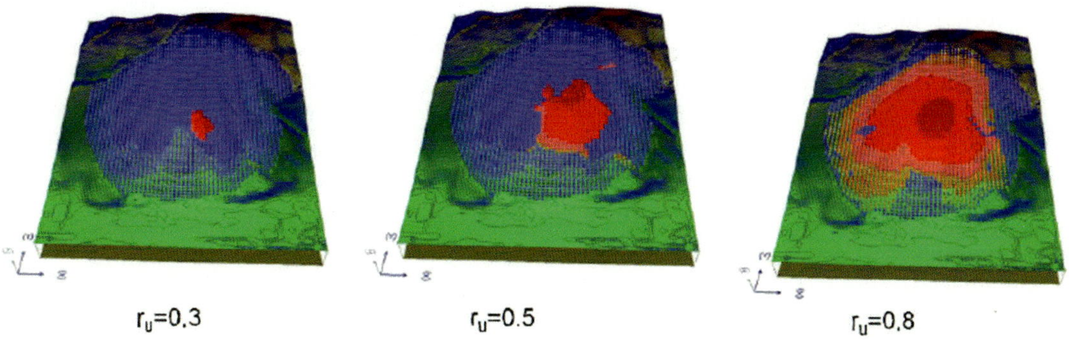

Fig. 7 Simulations of the Kostanjek landslide behavior for r_u = 0.3, r_u = 0.5, and r_u = 0.8 (according to Gradiški et al. 2013)

SATREPS FY2008 project (2009–2014) at the area of the deep-seated Kostanjek landslide, located in the western part of the hilly area, in an urban part of the City of Zagreb in Croatia. The methodology for landslide monitoring is based on new advanced technologies, which enable continuous, semi-automated monitoring in near–real-time necessary for development of early warning system. The general objective of the research was to find out method to efficiently gather more reliable data for risk mitigation to satisfy the requirements of land use planning, construction and civil protection at the area of large and deep-seated Kostanjek landslide. This discussion presents the main advantages and disadvantages of the established monitoring system.

Present-day movements of the Kostanjek landslide in the period from January 2013 to January 2015, ranging from 38 mm to 426 mm at the surface and at depth, cause, on a yearly basis, damage to private houses and infrastructure. Moreover, a very attractive city area over the last 50 years in the area of the Kostanjek landslide has been situated near the toe of the landslide, which has prevented development of this part of the city. The Kostanjek landslide presents a risk for approximately 290 buildings (mostly residential houses) in an area of one square kilometre in the urban part of the City of Zagreb. The automated continuous monitoring sensor network provided good quality data for the establishment of relationships between landslide movement parameters and the triggering causal factor parameters to predict faster movements, accelerated movement and/or large displacements. The monitoring sensors at the Kostanjek landslide recorded multiple landslide reactivations from external triggers, such as during the winter of 2012/2013 and the spring of 2013.

The advantages of the monitoring system at the Kostanjek Landslide Observatory are as follows: (i) using multiple sensors for the purpose of movement monitoring at the same position in the central part of the landslide (consisting of a GNSS station, extensometer, inclinometer and vertical extensometer), guarantees redundant measurements and can prevent data loss if one instrument fails; (ii) using the same position for different types of movement monitoring sensors in the central part of a landslide (consisting of the aforementioned sensors) also enables the spatial correlation of measurement data on the landslide surface and within the landslide profile; and (iii) using different geodetic and geotechnical sensors in combination with hydrological monitoring equipment (measuring precipitation and pore pressures in the landslide profile) allows a reconstruction of the relationships between rainfall, groundwater level, and consequent landslide behavior as a basis for establishing an early warning system. However, the main disadvantages of the Kostanjek monitoring system are related to missing sensors in all parts of this large and deep-seated landslide. Additional inclinometers in different parts of the landslide are necessary to provide reliable data regarding the geometry of the entire sliding surface. Additional piezometers are also necessary in different parts of the landslide to provide measurements of pore water pressure to better understand the landslide behavior and correlation with causal factors.

The greater displacements in the central part of the landslide compared to the displacement along the landslide boundaries implies the necessity of identifying zones within the landslide body with different hazard and risk levels. This can be achieved by combining the described series of monitoring data from the Kostanjek landslide with other advanced remote sensing technologies. Tofani et al. (2013) and Komac et al. (2015) integrated Persistent Scatterer Interferometry (PSI) data with in situ monitoring instrumentation and showed that the joint use of satellite and ground-based data with the geological interpretation of a landslide allows a better understanding of landslide geometry and kinematics. The integration of ground-based monitoring data and PSI data are potentially providing sound results for the characterization of the slow-moving Kostanjek landslide. Defining the state of activity of the large Kostanjek landslide, would be useful when dealing with landslide mapping, monitoring and hazard analysis. In particular, satellite SAR (Synthetic Aperture Radar) interferometry (Tofani et al. 2013) has proven to be a sound tool to assess changes on the Earth's surface for landslide mapping and monitoring purposes. PSI represents a powerful tool to measure landslide displacement because it offers a synoptic view that can be repeated at different time intervals and at various scales.

As the specific and total risk arising from this situation is potentially significant, both in relation to the extent and magnitude of the landslide and the socioeconomic vulnerability and value of the elements at risk, monitoring has been carried out in close collaboration with end-users from city government, within the scientific Croatian-Japanese research project. In principle, a landslide hazard can be fully assessed only by defining where, how and when slope instability will take place, thus defining the spatial, kinematical and temporal components of the hazard respectively. The monitoring data presented in this paper are principally aimed at assessing the magnitude of the Kostanjek landslide that, expressed in terms of volume, velocity or energy, makes up the kinematical component of the hazard. Nonetheless, the same monitoring data can also support a full assessment of the spatial and temporal component of the hazard. The spatial component of the hazard, for instance, depends upon how prone the slope is to fail in relation to various causal factors and is related to the expected extent of the phenomenon, including possible retrogression, enlargement or advancement of the landslide. This, in the specific case of the Kostanjek landslide, will have to be estimated through the use of slope stability models built up and validated using geomorphological evidence and monitoring data. On the other hand, the temporal component of the hazard is related to the frequency in time with which a certain damaging landslide scenario with given spatial and kinematics characteristics is expected to occur. It can in principle be assessed using data from the inclinometers, piezometers, the meteorological station and accelerometers installed on the Kostanjek landslide. However, this is only possible if a close relationship between movements and triggering factors, such as critical rainfall or groundwater levels or earthquake shaking, can be highlighted. Establishment of an early warning system and defining of alarm thresholds will be based on existing knowledge of the Kostanjek landslide behavior so far, based on collected consequent comprehensive monitoring data.

Acknowledgements Results presented herein have been obtained with the financial support from JST/JICA's SATREP Program (Science and Technology Research Partnership for Sustainable Development). Ministry of Science, Education and Sports of the Republic of Croatia have been financing preparatory and installation works. City Office for Physical Planning, Construction of the City, Utility Services and Transport, City of Zagreb enabled and financed supply of the system with electricity from public network. These supports are gratefully acknowledged.

We would like to express our very great appreciation to the company Geomatika- Smolčak Ltd. for sponsoring the Kostanjek Landslide Observatroy development with consultations during the planning and development of the system. Our special thanks are extended to our colleagues Mr. Kristijan Špehar for his support in the site and cabinet work. Authors want to thank the Emergency Management Office of the City of Zagreb for continuous collaboration. We are also thankful to the City Office for the Strategic Planning and Development of the City, City of Zagreb for enabling us use digital data, orthophoto maps and LiDAR DEM. We wish to thank numerous citizens and the City of Zagreb for their contribution to this project, by enabling use of private and City's land for measurement stations establishment.

References

Arbanas Ž, Sassa K, Marui H, Mihalić S (2012) Comprehensive monitoring system on the Grohovo Landslide, Croatia. In: Proceedings of the 11th international & 2nd North American symposium on landslides. Banff, Canada, pp 1441–1447

Arbanas Ž, Mihalić Arbanas S, Vivoda Prodan M, Peranić J, Jagodnik V, Dugonjić Jovančević S (2016) TXT-tool 2.385-1.1 a comprehensive landslide monitoring system: the Grohovo landslide, Croatia. In: Sassa K et al (eds) Landslide dynamics: ISDR-ICL landslide interactive teaching tools. Springer, Heidelberg (in press)

Baroň I, Supper R, Ottowitz D (eds) (2012) The Safeland project, deliverable 4.6. In: Report on evaluation of mass movement indicators. Unpublished report. Geological Survey of Austria (GSA), p 328

Baučić M, Mihalić Arbanas S, Krkač M (2013) Geographic information system of landslide Kostanjek: integration of real-time GNSS monitoring data with other sensor data. In: Proceedings of the 1st regional symposium on landslides in the Adriatic-Balkan region. Zagreb, Croatia, pp 12–13

Bernat S, Mihalić Arbanas S, Krkač M (2014) Inventory of precipitation triggered landslides in the winter of 2013 in Zagreb (Croatia, Europe). In: Proceedings of the 3rd world landslide forum, landslide science for a safer geoenvironment, vol 2. Beijing, China, pp 829–836

Bernat S, Mihalić Arbanas S, Krkač M (2015) Catalog of precipitation events that triggered landslides in northwestern Croatia. In: Proceedings of the 2nd regional symposium on landslides in the Adriatic-Balkan region. Beograd, Serbia (in press)

Blikra LS (2012) The Åknes rockslide, Norway. In: Clague JJ, Stead D (eds) Landslides—types, mechanisms and modeling. Cambridge University Press, Cambridge, pp 323–334

Brückl E, Brunner FK, Kraus K (2006) Kinematics of a deep-seated landslide derived from photogrammetric, GPS and geophysical data. Eng Geol 88:149–159

Castagnetti C, Bertacchini E, Corsini A, Capra A (2013) Multi-sensors integrated system for landslide monitoring: critical issues in system setup and data management. Eur J Rem Sens 46:104–124

Corsini A, Pasuto A, Soldati M, Zannoni A (2005) Field monitoring of the Corvara landslide (Dolomites, Italy) and its relevance for hazard assessment. Geomorphology 66:149–165

Cotecchia V (2006) The second hans cloos lecture. Experience drawn from the great Ancona landslide of 1982. Bull Eng Geol Env 65:1–41

Ferić P, Mihalić S, Krkač M, Arbanas Ž, Podolszki L (2010) Kostanjek landslide: current state and planned project activities. In: Proceedings 1st workshop of the project risk identification and land-use planning for disaster mitigation of landslides and floods in Croatia. Dubrovnik, Croatia, p 6

Froese CR, Charrière M, Humair F, Jaboyedoff M, Pedrazzini A (2012) Characterization and management of rockslide hazard at Turtle Mountain, Alberta, Canada. In: Clague JJ, Stead D (eds) Landslides—types, mechanisms and modeling. Cambridge University Press, Cambridge, pp 310–322

Furuya G, Miyagi T, Hamasaki E, Krkač M (2011) Geomorphologic mapping and 3D modeling of the Kostanjek landslide, Zagreb. Proceedings of the 2nd workshop of the project risk identification and land-use planning for disaster mitigation of landslides and floods in Croatia. Dubrovnik, Croatia, pp 21–23

Gradiški K, Krkač M, Mihalić Arbanas S, Bernat S (2013) Slope stability analyses of the Kostanjek Landslide for extreme rainfalls in the winter of 2013. In: Proceedings of the 4th workshop of the project risk identification and land-use planning for disaster mitigation of landslides and floods in Croatia. Split, Croatia, pp 15–16

Komac M, Holley R, Mahapatra P, Van der Marel H, Bavec M (2015) Coupling of GPS/GNSS and radar interferometric data for a 3D surface displacement monitoring of landslides. Landslides 12(2):241–257

Krkač M (2015) A phenomenological model of the Kostanjek landslide movement based on the landslide monitoring parameters. PhD thesis. Faculty of mining, geology and petroleum engineering. University of Zagreb, Zagreb (in Croatian)

Krkač M, Rubinić J (2013) Analysis of water fluctuation dynamics in the wider area of the Kostanjek landslide. In: Proceedings of the 1st regional symposium on landslides in the Adriatic-Balkan region. Zagreb, Croatia, p 10

Krkač M, Rubinić J, Mihalić S (2011) Kostanjek landslide—analyses of groundwater discharge as a basis for the new hydrological monitoring. Proceedings of the 2nd workshop of the project risk identification and land-use planning for disaster mitigation of landslides and floods in Croatia. Dubrovnik, Croatia, pp 17–20

Krkač M, Mihalić Arbanas S, Nagai O, Arbanas Ž (2013) The Kostanjek landslide—monitoring system development and sensor network. In: Proceedings of the 1st regional symposium on landslides in the Adriatic-Balkan region. Zagreb, Croatia, p 11

Krkač M, Mihalić Arbanas S, Arbanas Ž, Bernat S, Špehar K (2014) The Kostanjek landslide in the City of Zagreb: Forecasting and protective monitoring. In: Proceedings of the XII IAEG congress, engineering geology for society and territory, vol 5. Torino, Italy, pp 715–719

Martinčević J, Mihalić Arbanas S, Krkač M, Bernat S, Miklin Ž, Podolszki L (2013) Mineralogical composition of the Kostanjek landslide sediments and its possible influence on the sliding and swelling processes. In: Proceedings of the 1st regional symposium on landslides in the Adriatic-Balkan region. Zagreb, Croatia, pp 7–8

Mihalić S, Arbanas Ž (2012) The Croatian–Japanese joint research project on landslides: activities and public benefits. In: Sassa K et al (eds) Landslides: global risk preparedness. Springer, pp 335–351. doi 10.1007/978-3-642-22087-6_24

Mihalić Arbanas S, Arbanas Ž (2015) Landslides—a guide to researching landslide phenomena and processes. In: Gaurina-Međimurac N (ed) Handbook of research on advancements in environmental engineering. IGI Global-Hershey, pp 474–510

Mihalić Arbanas S, Krkač M, Bernat S (2016) Application of advanced technologies in landslide research in the area of the City of Zagreb (Croatia, Europe). Geol Croat 69(2):231–243

Nagai O, Krkač M, Mihalić S (2011) Introduction of one of methods to predict failure time of a slope widely used in Japan and application to the Kostanjek landslide. In: Proceedings of the 2nd workshop of the project risk identification and land-use planning for disaster mitigation of landslides and floods in Croatia. Dubrovnik, Croatia, pp 46–50

Nonveiller E (1976) Analysis of ground displacement causes at the area of the "Sloboda" cement factory in Podsused. Croatian Civil Engineering Institute, Zagreb (in Croatian). Geotech Rep

Ortolan Ž (1996) Development of 3D engineering geological model of deep landslide with multiple sliding surfaces (example of the Kostanjek landslide). PhD thesis. Faculty of Mining, Geology and Petroleum Engineering, University of Zagreb, Zagreb (in Croatian)

Ortolan Ž, Pleško J (1992) Repeated photogrammetric measurements at shaping geotechnical models of multi-layer landslides. Rudarsko—Geološko—Naftni Zbornik 4:51–58

Ortolan Ž, Stanić B, Nonveiller E, Pleško J (1987) Consequences of mining in the marl quarry Kostanjek—Podsused. In: Proceedings of the 9th Yugoslavia symposium on hydrology and engineering of geology, vol 2, pp 117–128 (in Croatian)

Ortolan Ž, Mihalinec Z, Stanić B, Pleško J (1995) Application of photogrammetric measurements at shaping geotechnical repeated models of multi-layer landslides. Proceedings of the 6th international symposium on landslides, Christchurch, New Zealand, pp 1685–1691

Oštrić M, Ljutić K, Krkač M, Setiawan H, He B, Sassa K (2012) Undrained ring shear tests performed on samples from Kostanjek and Grohovo landslide. In: Proceedings of the ICL symposium, Kyoto, Japan, pp 47–52

Pehnec V (1967) Damage of the factory buildings in Podsused due to swelling of marl. Građevinar 6:197–201 (in Croatian)

Petley DN, Mantovani F, Bulmer MN, Zannoni A (2005) The use of surface monitoring data for the interpretation of landslide movement patterns. Geomorphology 66(1–4):133–147

Sassa K, He B (2013) TXT-tool 3.081–1.2 landslide dynamics. In: Sassa K, He B, McSaveney M, Osamu N (eds) ICL landslide teaching tools. International Consortium on Landslides-Kyoto, pp 215–237

Stanić B, Nonveiller E (1995) Large-scale landslide in Kostanjek area. Građevinar 47(4):201–209 (in Croatian)

Stanić B, Nonveiller E (1996) The Kostanjek landslide in Zagreb. Eng Geol 42:269–283

Stumpf A, Kerle N, Malet J-P (eds) (2011): The Safeland project, deliverable 4.4. In: Guidelines for the selection of appropriate remote sensing technologies for monitoring different types of landslides. Unpublished report. Faculty for Geo-information Science and Earth Observation—ITC, University of Twente, United Nations University, p 91

Tofani V, Raspini F, Catani F, Casagli NM (2013) Persistent scatterer interferometry (PSI) technique for landslide characterization and monitoring. Remote Sens 5:1045–1065

Vrsaljko D, Mihalić S, Bošnjak M, Krkač M (2012) Lithostratigraphical investigations at the wider area of the Kostanjek landslide: review of existing data and planned activities. In: Proceedings of the 2nd project workshop of the Croatia - Japan project on risk identification and land-use planning for disaster mitigation of landslides and floods in croatia. Rijeka, Croatia, pp 9–13

Watanabe N, Krkač M, Furuya G, Wang C, Mihalić S (2011) Hydrochemical characteristics of groundwater from the Kostanjek Landslide in Croatia. In: Proceedings of the 2nd workshop of the project risk identification and land-use planning for disaster mitigation of landslides and floods in Croatia. Dubrovnik, Croatia, pp 14–16

Yamamoto S, Watanabe N, Krkač M, Furuya G, Wang C, Mihalić Arbanas S (2013) Geochemical constraints on the origins of groundwater from the Kostanjek landslide in the western part of Zagreb, Croatia. In: Proceedings of the 1st regional symposium on landslides in the Adriatic-Balkan region. Zagreb, Croatia, pp 9–10

Županović LJ, Opatić K, Bernat S (2012) Determination displacements of landslides Kostanjek with relative static method. Ekscentar 15:46–53 (in Croatian)

TXT-tool 2.385-1.2 Landslide Comprehensive Monitoring System: The Grohovo Landslide Case Study, Croatia

Željko Arbanas, Snježana Mihalić Arbanas, Martina Vivoda Prodan, Josip Peranić, Sanja Dugonjić Jovančević and Vedran Jagodnik

Abstract

The Grohovo Landslide is located on the north-eastern slope of the Rječina Valley, Croatia and it was reactivated in 1996 at the location of the landslide from 19th century. In 2009 the Croatian-Japanese joint research project "Risk identification and Land-Use Planning for Disaster Mitigation of Landslides and Floods in Croatia" was initiated and the Grohovo Landslide was chosen as a pilot area for monitoring system development. A comprehensive monitoring system was designed and installed. Integrated monitoring system consists of survey using GPS and robotic total station so as geotechnical monitoring using long span and sort span wire extensometers, inclinometers, pore pressure gauges and rain gauges. All measurements will be integrated in GIS for landslide risk management and early warning system. The monitoring results should provide a basis for develop and validate confidential numerical models and adequate hazard management.

Keywords

Landslide • Monitoring system • Early warning system

Contents

1 **Introduction** 466
2 **Landslide, History and Present State** 466
2.1 Geological Settings of the Rječina River Valley 466
2.2 History of the Grohovo Landslide 467
2.3 Description of the Grohovo Landslide 468

Ž. Arbanas (✉) · M. Vivoda Prodan · J. Peranić · S. Dugonjić Jovančević · V. Jagodnik
Faculty of Civil Engineering, University of Rijeka, Radmile Matejčić 3, 51000 Rijeka, Croatia
e-mail: zeljko.arbanas@gradri.uniri.hr

M. Vivoda Prodan
e-mail: martina.vivoda@gradri.uniri.hr

J. Peranić
e-mail: josip.peranic@gradri.uniri.hr

S. Dugonjić Jovančević
e-mail: sanja.dugonjic@gradri.uniri.hr

V. Jagodnik
e-mail: vedran.jagodnik@gradri.uniri.hr

S. Mihalić Arbanas
Faculty of Mining, Geology and Petroleum EngineeringUniversity of Zagreb, Pierottieva 6, 10000 Zagreb, Croatia
e-mail: snjezana.mihalic@rgn.hr

K. Sassa et al. (eds.), *Landslide Dynamics: ISDR-ICL Landslide Interactive Teaching Tools*, https://doi.org/10.1007/978-3-319-57774-6_35

3 Monitoring System on the Grohovo Landslide 470
3.1 Landslide Monitoring Techniques 470
3.2 The Grohovo Landslide Comprehensive Monitoring System 471
3.3 Results of the Grohovo Landslide Monitoring 474
3.4 The Grohovo Landslide Early Warning System Establishment 475
4 Conclusion 476
References 477

1 Introduction

The Grohovo Landslide, the largest active landslide along the Croatian part of the Adriatic coast, is located on the north-eastern slope of the Rječina Valley. It was noted that during 19th and 20th centuries considerable instabilities on the Rječina Valley slopes occurred. The last complex retrogressive landslide was reactivated in December 1996, after long dormant period and about 3.0×10^6 m^3 moved down the slope and buried the Rječina riverbed. After initial movements, the landslide retrogressively developed up the slope. Slip surfaces are considered to be on the contact of superficial deposits and flysch bedrock. After removing displaced mass from the landslide foot, further significant movements stopped, and monitoring indicated that only small displacements developed in the upper part of the slope. In 2009 the Croatian-Japanese research joint project "Risk identification and Land-Use Planning for Disaster Mitigation of Landslides and Floods in Croatia" was initiated and the Grohovo Landslide was chosen as a pilot area for monitoring system development. A comprehensive monitoring system was designed consisting of geodetical and geotechnical monitoring. Installation of monitoring equipment started in March 2011 and first stage was completed to the end of 2011. Integrated monitoring system will consist of surveying system using GPS and robotic total station, so as geotechnical monitoring using pore pressure gauges, inclinometers, extensometers and seismographs. All measurements will be integrated in GIS for landslide risk management and early warning system. Detailed description of the Grohovo Landslide occurrence and possibilities and consequences of landslide development will be analyzed. The integrated monitoring system and used monitoring equipment are described (Arbanas et al. 2012a, 2014a, b).

2 Landslide, History and Present State

2.1 Geological Settings of the Rječina River Valley

The Grohovo Landslide is located on the north-eastern slope of the Rječina Valley outside the City of Rijeka, the largest Croatian port on the northeastern Adriatic coast. This area is part of a dominant morphostructural unit that strikes in the northwest-southeast direction along the Rječina Valley, Sušačka Draga Valley, Bakar Bay and the Vinodol Valley (Velić and Vlahović 2009).

The Rječina River is 18.7 km long, and the mouth is located in the center of the City of Rijeka. The Rječina River is a typical karstic river originating from a strong karstic spring at the foot of the Gorski Kotar Mountains. The annual average profusion of the Rječina spring is 7.76 m^3/s with maximum flow rates ranging from 0 to over 100 m^3/s (Karleuša et al. 2003). Some of the water from the Rječina Spring is used for water supply of Rijeka City, while some of the water from the Valići Lake is used for electric power production in the Rijeka Hydropower Plant.

The kinematics of the structural elements in the investigated part of the Rječina Valley are based on the relationship between relatively rigid carbonate rocks and ductile siliciclastic rocks during simultaneous deformations. The Cretaceous and Paleogene limestones are situated at the top of the slopes, while the Paleogene siliciclastic rocks and flysch are situated on the lower slopes and the bottom of the valley (Fig. 1). The flysch complex is a block that has been squeezed between the limestone rock blocks on the northeastern and southwestern sides. The effects of the deformation are most distinctive at the contacts between the two rocky complexes, where the relatively rigid limestone is pushed into the more ductile siliciclastic rocks. In this way, the formerly straight tectonic contact deformed into its present toothed appearance (Blašković 1999; Benac et al. 2005).

The siliciclastic or flysch bedrock is lithologically heterogeneous, with common vertical and lateral alternations of different lithological sequences. Microscopic petrological analysis of the bedrock has shown the presence of silty marl, laminated silt to silty shale and fine-grained sandstones (Benac et al. 2005, 2011). Unlike the limestone, the flysch is prone to weathering, which causes the formation of a clayey weathering zone on the flysch bedrock. Over time, coarse-grained fragments originating from rockfalls were mixed with clay from the weathered flysch zone, forming slope superficial deposits a few meters thick.

2.2 History of the Grohovo Landslide

Numerous historical descriptions, figures and maps describing landslides were found in the Croatian State Archive (Anon 2011) in Rijeka which provides evidence of the occurrence of landslides in the Rječina Valley near the Grohovo Village. Sliding was first documented in 1767, when numerous landslides and rockfalls in the Rječina River Valley were caused by the 1750 earthquake, which had an epicenter in the City of Rijeka. Large landslides triggered by rainfall and floods were noticed on both river banks near the Grohovo Village at the end of 19th century. A large landslide occurred on the southwestern slope in 1870, and after reactivation of the slide in 1885 part of the Grohovo Village was buried by a rock avalanche. A large landslide occurred in 1893 on the northeastern slope of the Rječina River Valley at the location of the recent landslide, and the Rječina River channel was shifted to the south by approximately 50 m. Because of these landslides and the resulting damage, the Ministry of Agriculture of the Hungarian Kingdom designed the Rječina Recovery Project in 1894, and restoration began in 1889. The designer of the Rječina Recovery Project, civil engineer Bela Pech, mapped all the landslides on the 1894 topographic map, Fig. 2. Restoration of the Rječina River banks was finished in 1908, and during construction sliding had to be stabilized using drain trenches and dikes.

Fig. 1 Aerial view of the Rječina River Valley

Numerous landslides occurred during the first half of the 20th century but did not cause significant damage to structures on the river banks. New problems with landslides in the Rječina valley occurred during construction of the Valići Dam in 1960, when landslides appeared on the northeast slope near the dam. This landslide was stabilized during the dam's construction.

Construction of the Valići Dam as part of the Rijeka Hydropower Plant did not significantly influence the water regime of the Rječina River downstream of the dam, especially during extended periods of heavy rain. Because of the relatively small volume of the Valići Lake, the dam cannot accumulate all the inflow during heavy rain and a relatively large quantity of water must be released into the downstream part of the Rječina River. In these situations, erosion along

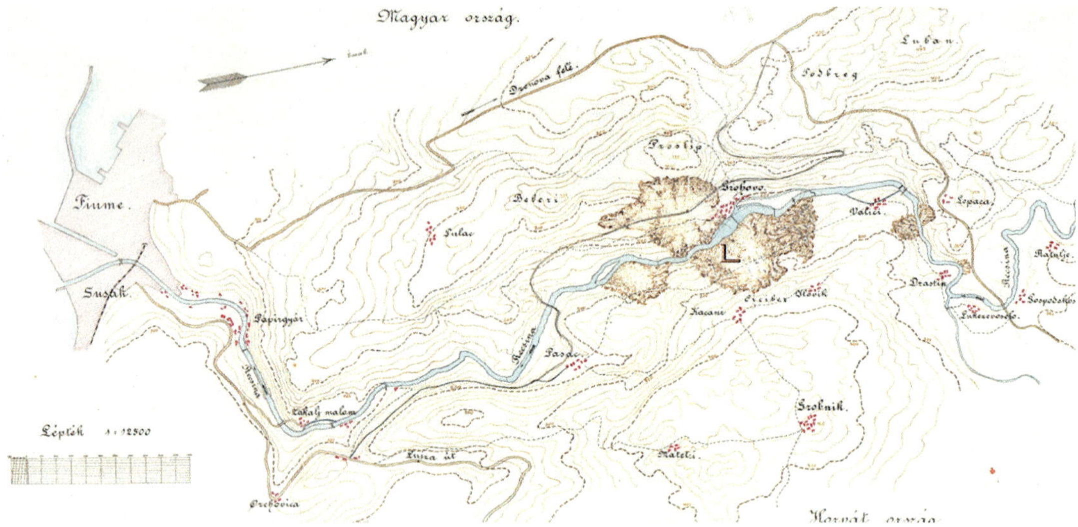

Fig. 2 Topographic map of the Rječina River Valley showing landslides occurred in 1894 (Anon 2011)

the Rječina River presents a hazard to the possible landslide (Mihalić and Arbanas 2013).

The last large Grohovo Landslide occurred in December 1996 (Fig. 3) at the location of the landslide from 19th century on the northeast slope. Comparison with old topographic map from 1894 shows that the landslide mass had a significantly smaller volume than the older sliding. The slide mass completely buried the Rječina River channel, and a landslide foot formed a dam and the lake behind. Immediately after sliding, this landslide mass was removed from the river channel, eliminating the risk of dam collapse and the resulting water wave that could have caused serious damage to the City of Rijeka. Comprehensive landslide remediation projects were never conducted, but further movements of the landslide were reduced (Benac et al. 2010).

2.3 Description of the Grohovo Landslide

The results of investigation works on the Grohovo Landslide after last sliding occurrence in December 1996 have shown the formation of a complex land-slide with thirteen sliding bodies including movements of the carbonate mega-blocks at the top of the slope (Benac et al. 2005). Boundaries between these bodies are mostly clearly expressed. Sliding bodies represent different types of mass movements. The whole reactivated sliding mass has a volume of approximately 3.0×10^6 m^3. On the basis of geological mapping and geophysical surveys results, thickness of the displaced sliding mass could be estimated and the position of sliding surfaces could be determined. Morphometric characteristics of the total complex Grohovo Landslide, described according to WP/WLI Suggested Nomenclature for Landslides (IAEG 1990), are presented in Table 1.

Slip surface was formed on the contact of slope deposits and the flysch bedrock. Initial landslide body was the most moved one, what is visible from the range of displaced material in the Rječina riverbed, and by the tilted trees. Due to the magnitude of displacement (up to 20 m) primary geological settings was completely disturbed.

According to accepted classifications, the investigated landslide is a complex composite landslide. It is a complex and retrogressive landslide (Skempton and Hutchinson 1969). Movements of mixed rock and soil material in

Fig. 3 A view on the Grohovo landslide

Table 1 Dimension of the Grohovo landslide occurred 1996 according WP/WLI suggested nomenclature for landslides (IAEG 1990)

Characteristic	Value
Total length (L)	425 m
Length of the displaced mass (L_d)	420 m
Length of the rupture surface (L_r)	405 m
Width of the displaced mass (W_d)	200 m
Width of the rupture surface (W_r)	200
Depth of the displaced mass (D_d)	6–9 m
Depth of the rupture surface (D_r)	6–9 (20) m
Total height (ΔH)	165 m

initial landslide body have characteristics of debris avalanches, according to velocity of movements (Varnes 1978; Cruden and Varnes 1996). Block sliding of a rock mass is a special phenomenon. Due to the fact that the position of the slip surface was predisposed by geological settings, the landslide can also be considered as a consequent translational, as well as the blocky slide type IIIb (Antoine and Giraud 1995). According to landslide activity it could be considered a reactivated landslide on unstable slope, type Ib (Crozier 1984).

Based on existing monitoring results (Benac et al. 2005, 2010), stability analyses of wider landslide area (Arbanas et al. 2010) so as detailed 3D analyses of the reactivated landslide (Wang et al. 2012) it was identified real hazard of further sliding in the Rječina River valley. Landslide occurrence is very probable during unfavorable hydrogeological conditions. These landslides also have an indirect threat. The main risk of sliding is the possible covering of the Rječina River channel to form a lake. During heavy rainfall, the lake level rises and the accumulated water can overflow the landslide mass. After dam collapse, the water wave can cause fatalities and serious damage to structures in the City of Rijeka situated on the mouth of the Rječina River. Widening of the existing Grohovo Landslide can also destabilize carbonate mega-blocks on the limestone scarp at the top of the landslide, and the villages at the top of the slope can be endangered. Another threat is forming of landslides on a slope over the Valići Lake and filling the lake with the landslide mass. In addition, other sliding phenomena can cause damage to buildings, roads and other facilities in different parts of the valley (Arbanas et al. 2010).

3 Monitoring System on the Grohovo Landslide

3.1 Landslide Monitoring Techniques

By definition, landslides are characterized by movement. Knowledge of the movement magnitude and velocity distribution along the slope, are the most important data for all landslide analysis (Mihalić Arbanas and Arbanas 2014). Monitoring is required to observe the changing conditions that may lead to total failure of the slope where slope movement is occurring, where safety factors against sliding are low, or where high risk is present from a possible slope failure. Landslide movement monitoring expressed via ground surface displacements and deformation of structures (including the landslide body) related to landslides can be accomplished using different types of monitoring systems and techniques that are classified according to Savvaidis (2003) as follows: satellite and remote sensing techniques, photogrammetric techniques, geodetic or observational techniques, and geotechnical or instrumentation or physical techniques. The selection of measurement instrumentation and methods or the planning and design of a desirable monitoring system depends on the movement types and deformation as well as on the role and purpose of the monitoring campaign (Savvaidis 2003).

All satellite and remote sensing techniques (photography and imagery ranging from ground-based mobile units to airborne or satellite platforms using LiDAR, optical, and radar sensors) used for landslide mapping also can be used for landslide monitoring if multi-temporal images are available.

Photogrammetry is a three-dimensional measurement process that enables the determination of 3D motion vectors in long-term landslide monitoring (Brückl et al. 2006). Images can be acquired by satellites, aircraft, helicopters, or remote unmanned aerial vehicles (UAV) as well as from a ground-based observation points using such equipment as film cameras, digital cameras, scanners, among others. The main advantages of the photogrammetry technique are the reduced time of fieldwork, simultaneous three-dimensional coordinates, and in principle, unlimited number of points (targets) that can be monitored.

Conventional ground-based geodetic techniques have been used for surface displacement monitoring of landslides. Savvaidis (2003) differentiates between two basic methods for the design of a geodetic survey: (i) a horizontal or vertical control benchmark network established in the landslide area under monitoring with control points located in the landslide body and (ii) the total station instruments used to measure angles and distances to target prisms located at the moving landslide mass.

The Global Positioning System (GPS) is often used as a surveying tool in landslide monitoring by positioning the 3D coordinate time series of displacements at discrete points on the landslide surface (Gili et al. 2000). The GPS positioning is based on measuring the transit time of radio signals emitted by stationary orbiting satellites and computing a receiver position, the unit must be in view of at least four satellites. The GPS is

U.S. (military) system; other satellite-related positioning systems are the Russian GLONASS and the European Satellite System GALILEO complemented by the civil Global Navigation Satellite System (GNSS). Current GPS positioning techniques for landslide monitoring typically include the use of either episodic techniques or continuous monitoring and static, rapid-static, or real-time kinematic GPS surveying techniques (Savvaidis 2003). Static or Relative GPS (RGPS) measurements are based on positioning of relative coordinates of the rover stations relative to the reference stations and precise determination of baseline distances between rover stations located on landslide surfaces and reference stations located outside the landslide body (e.g., Gili et al. 2000; Mora et al. 2003; Brückl et al. 2006). Real Time Kinematic GPS (RTK GPS) uses one or even a network of reference stations (Wang 2011; Wang et al. 2014) set up as known points with fixed coordinates. The baseline distances between rover stations located on a landslide surface and fixed reference stations can be calculated quite accurately, but the absolute position of the rovers also can be found relative to the coordinates of the fixed reference stations.

Geotechnical sensors are used extensively in landslide monitoring to ensure efficient data for landslide behaviour prediction and landslide stability analysis. The main geotechnical sensors and instruments used for landslide monitoring include inclinometers, extensometers, crack meters, piezometers, deformeters, tiltmeters, clinometers, load and pressure cells, and geophones. Most of those geotechnical sensors store the measured data using internal loggers awaiting download or the measurements can be automatically logged to a connected computer unit (Savvaidis 2003). An additional group of sensors necessary to complete the group of geotechnical monitoring sensors and enable analyses of measured data is the group of meteorological sensors that are usually integrated in a compact but efficient weather station.

Usually, different types of monitoring techniques and instrumentation are used in different combinations and connected in a unique comprehensive landslide monitoring system. Because of the variability in landslide types and processes, targets of landslide investigation, field conditions, and ongoing technological development of monitoring sensors, no standardization can be adopted as a universal solution for landslide monitoring system setups. The use of multiple sensors for the same purpose (equipment fusion) and at the same position is recommended to guarantee redundancy of measurements and can prevent loss of data if one instrument fails (Arbanas et al. 2014a, b; Mihalić Arbanas and Arbanas 2014). Using the same position for different types of monitoring sensors also enables spatial correlation of measurement data on the landslide surface and inside the landslide profile. Use of different geodetic and geotechnical sensors in combination with hydrological monitoring equipment, that is, to measure data on precipitation and pore pressures in the landslide profile, allows reconstruction of the relationships between rainfall, ground water level, and consequent landslide behavior as a base for establishing an early warning system. Comprehensive monitoring systems have been established on numerous landslides throughout the world and certain of them are described in detail in numerous papers in journals and conference proceedings.

3.2 The Grohovo Landslide Comprehensive Monitoring System

In the frame of Croatian-Japanese joint research SATREPS project "Risk identification and Land-Use Planning for Disaster Mitigation of Landslides and Floods in Croatia the Grohovo Landslide was chosen as a pilot area for comprehensive integrated real-time monitoring system development and an advanced comprehensive monitoring system was designed and installed and shown in Fig. 4. Installation of the monitoring equipment started in May 2011 and completed in October 2013, but further improvements of the system are still on-going. The monitoring system was designed to consist

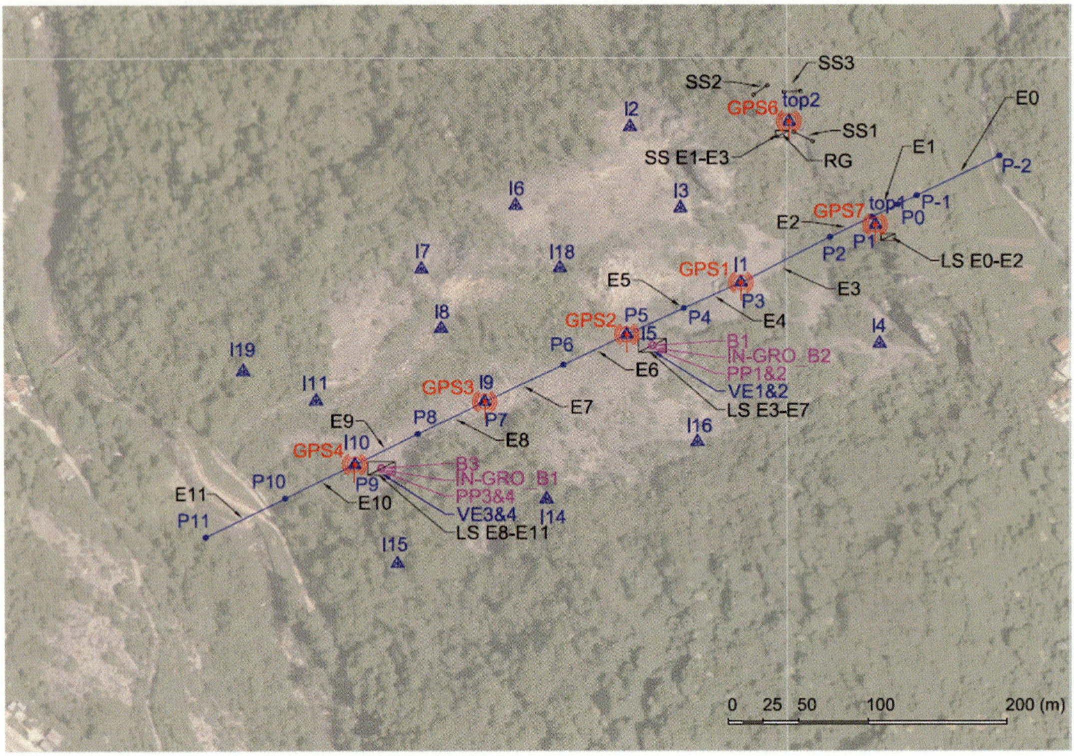

Fig. 4 Installed sensors at the Grohovo landslide: *GPS* GPS rover; *I* prism; *E* long span extensometer wire, *P* extensometer pole; *SS* short span extensometer; *B* position of borehole; *IN* inclinometer casing in borehole; *VE* vertical extensometer in borehole; *PP* pore pressure gauge in borehole; *LS* long span extensometer data logger; *RG* rain gauge

of geodetic and geotechnical monitoring (Arbanas et al. 2012b).

Geodetic monitoring consists of a robotic total station (Leica TM30) measuring 25 geodetic benchmarks (prisms) and a Global Positioning System (GPS) master unit (Leica GMX901) with 9 GPS receivers (rovers). The robotic total station and the GPS master unit are located in a relatively stable area, on the top of the slope opposite to the landslide, together with a meteorological sensor, and web cam (Fig. 5). The robotic station measures 25 benchmarks (prisms) located on the landslide body, on the top of the main scarp and, as reference points, around the landslide, every 30 min. The GPS network is composed of the GPS master unit, which is a reference station for four single frequency GPS rovers located in the landslide body, three single frequency GPS rovers located on top of the limestone scarp above the

Fig. 5 The master unit consisted of the robotic total station, single frequency GPS master unit, meteorological sensor, and web cam

landslide, a single frequency GPS rover located on a dam near the landslide, and 1 reference single frequency GPS rover located on the roof of the Faculty building in University Campus, chosen as a stable reference rover.

Data are transmitted by a Wi-Fi system from each rover to the GPS master unit connected to the field master unit PC, in which the GNSS Spider software creates measurement data files. The master unit PC can be reached remotely by an UMTS module from the main working station at the Faculty of Civil Engineering. The robotic total station, master GPS rover and field PC are powered from a windmill and solar panel mini power plant, while other GPS rovers in the landslide site use their own solar panel installations (Arbanas et al. 2014a, b).

The geotechnical monitoring system includes two vertical inclinometers, and four wire extensometers; 13 long- and three short-span extensometers, four pore-pressure gauges, a weather station and a rain gauge. Pore-pressure gauges, inclinometers and vertical extensometers are installed at two locations inside the central part of the landslide body at the same locations as the GPS rovers and prisms. Long-span extensometers (NetLG 501E Osasi, 13 pcs) are installed in a continuous line from the Rječina riverbed to the limestone mega-blocks at the top of the slope, while short-span extensometers (NetLG 501E Osasi, 3 pcs) are installed over the open cracks at the top of the landslide (Arbanas et al. 2014a). To obtain the measured data from the geotechnical monitoring equipment, it is necessary to download data directly from the installed sensors. After installation in summer 2011, geodetic monitoring and data collection were started in September 2011.

All installed equipment (except the vertical inclinometers) would be connected in one monitoring system with continuous measured data transmitting to the central computer unit at the Faculty of Civil Engineering in Campus of University of Rijeka (Fig. 6). This is a necessary requirement for establishing a comprehensive monitoring system and early warning system and landslide risk management.

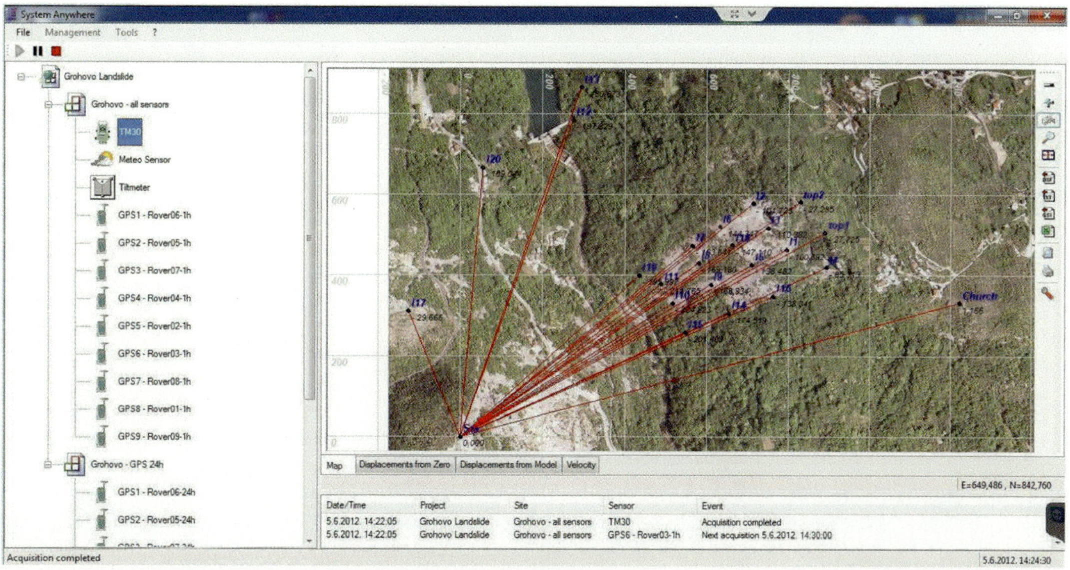

Fig. 6 Screen of System Anywhere software with orthophoto map of the Grohovo landslide and TPS data collection

3.3 Results of the Grohovo Landslide Monitoring

From obtained measurement results, it was very clearly visible that the collected data are liable to numerous significant influences such as daily, monthly and yearly temperature and humidity variation and local disturbing effects caused by deformations of poles on whose robotic total station and GPS master unit receiver are installed. In order to eliminate local disturbing effects, a bi-axis clinometer type Nivel 220 Leica and meteorological sensor were installed (Arbanas et al. 2012a, 2014a, b).

Results obtained from bi-axis clinometer about tilting of 4 m high concrete pillar could enable direct correction of measured results, but to appropriate reduction of weather condition influences, it should be necessary to have one year data collection and analysis (Arbanas et al. 2012a). It was also identified that the raw GPS measurement data are too rough and cannot indicate on real landslide behaviour. Post-processing of one hour and 12 h enables more precise GPS rovers' movement control with accuracy less than 2 mm. Movements of GPS rovers in the landslide area presented on Fig. 7.

After reduction of weather condition influences on concrete pillar where the robotic total station is installed, TPS measurement shows very precise and accurate values of prisms' movements with accuracy less than 1 mm. Movements of prisms in the landslide area presented on Fig. 8 for monitoring period from 29 September 2011 to 9 June 2013 with maximum displacements in upper part of the landslide.

Both measurement's techniques (GS, TPS) show on higher landslide activity in the upper part of the slope while the compression of material and heaving of the GPS rovers and

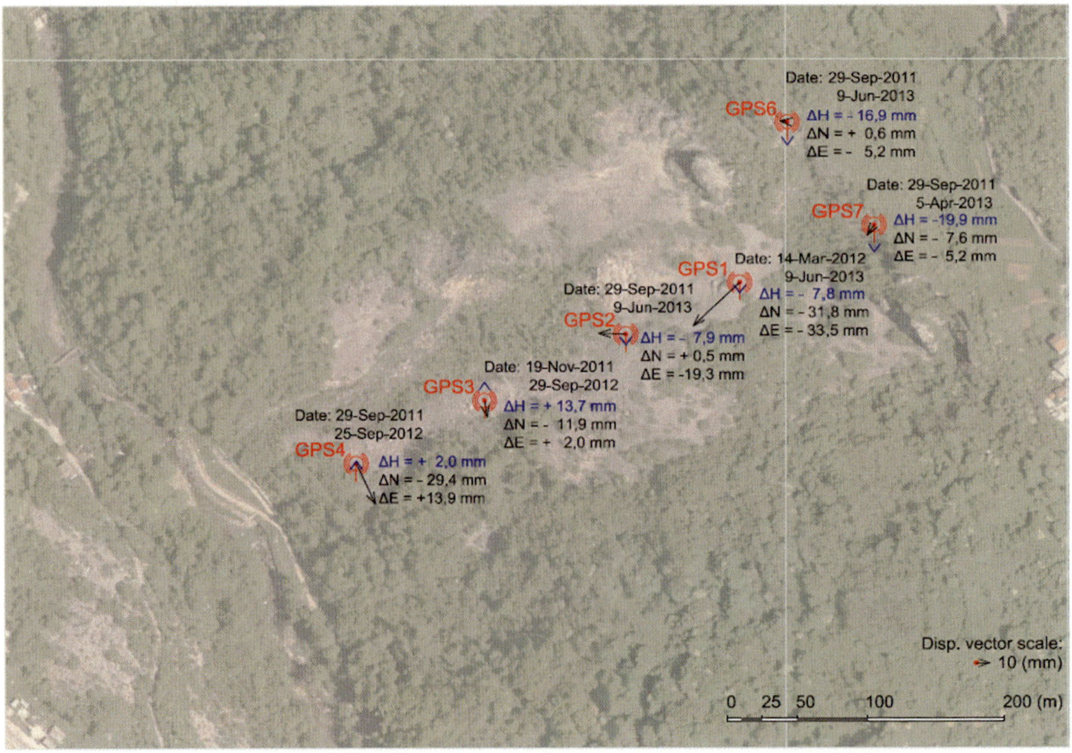

Fig. 7 GPS rovers' displacement vectors for approximately two years monitoring period. Two vectors attached to each GPS rover indicate displacements in 2D and heaving (+) or lowering (−) of GPS rover. Monitoring period is noted near each GPS rover

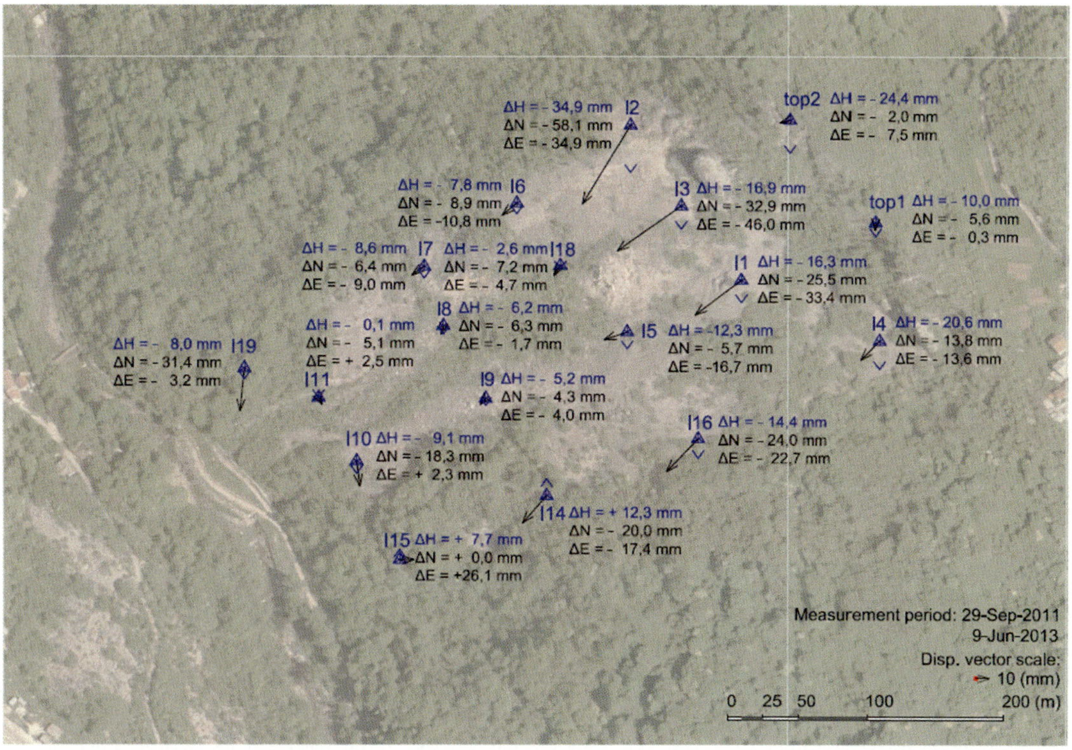

Fig. 8 Prisms' displacement vectors for monitoring period from 29 September 2011 to 9 June 2013. Two vectors attached to each prism indicate displacements in 2D and heaving (+) or lowering (−) of prism

prisms are outlined in the lower part of the landslide body and in the landslide foot. These observations are confirmed by in-line installed long-span wire extensometers' measurements which show extension of in upper part of extensometers' line (E1–E4) and compression in lower part of extensometers' line (E7–E11).

A relationship between rainfall, ground water level, pore pressures and landslide movement is still very hard to establish because of numerous interruptions of in monitoring system working caused by low energy production of mini solar power plant that supply the main unit (robotic total station, main GPS unit, and site PC unit) during the winter period from end of October to early March when the landslide activity is the most expressed. The second reason is very slow landslide activity that requires longer uninterrupted monitoring period.

From previous facts it is possible to conclude that the weakest link in the Grohovo Landslide monitoring system is power supply based on solar devices those cannot produce enough energy during a winter period. The consequences are missing of TPS measurements and loss of GPS measured data because single frequency GPS units have no possibility to store measured data. It should be necessary to improve power supply and optimize energy consumption of main equipment (GPS, TPS, PC).

3.4 The Grohovo Landslide Early Warning System Establishment

The use of different geodetic and geotechnical sensor fusion in combination with hydrological monitoring equipment, which records data about precipitation and pore pressures in the landslide profile, enables reconstruction of relationships between rainfall, groundwater level and consequent landslide behavior as a base for establishing an early warning system. The analyses necessary for an early warning system should link sensor measurements data and possible failure mechanisms with consequences that would follow the sliding occurrence (landslide risk) (Arbanas et al. 2014a, b). In designing the Grohovo Landslide early warning system it was necessary to: (i) Identify real hazards of further sliding and possible direct and indirect threats; (ii) Identify possible movements and landslide widening with high hazard; (iii) Select appropriate equipment relating to position in the field and measurement accuracy as a competent equipment to initiate an alarm and (iv) Define critical limit values (criteria values) that indicating the onset of new sliding and trigger an alarm.

Analyses of the past and prediction of future landslide behavior are the most important steps to consider in development of an early warning system, while equipment selection should depend on measurement of appropriate values with required accuracy. In case of the Grohovo Landslide, the accuracy of the wire extensometer measurements (<0.1 mm) and real-time recording have an advantage over GPS and TPS measurements. Both measurements are still unreliable because of unstable power supply. The accuracy of TPS measurements is <1.0 mm but from the measurement results, it is very clear that the collected data are liable to numerous influences such as daily, monthly and yearly temperature and humidity variations and local disturbing effects caused by deformation of poles on which the robotic total station is installed (Arbanas et al. 2012a). The accuracy of GPS measurement is <2.0 mm but this accuracy is obtained only after 6 h of data post-processing. This fact eliminated GPS data as appropriate equipment to trigger the alarm in an early warning system.

The Grohovo Landslide early warning would be established after completion of the monitoring system and enabling of transmission of real-time extensometers data to the central computer unit with defined critical limit values that indicate the onset of new sliding and trigger the alarm (Arbanas et al. 2014a, b).

4 Conclusion

The Grohovo Landslide, reactivated in December 1996, was chosen as a pilot area for monitoring system development as a part of the Croatian-Japanese joint research SATREPS' project 'Risk Identification and Land-Use Planning for Disaster Mitigation of Landslides and Floods in Croatia' performed from 2009 to 2014. A comprehensive integrated monitoring system consists of a GPS survey network composed of a GPS master unit which is a reference station for 9 single frequency GPS rovers and a robotic total station which measures 25 benchmarks (prisms) every 30 min. The monitoring data collected may be influenced by local disturbing effects caused by deformation of the poles holding the robotic total station and GPS master unit. The geotechnical monitoring system includes 2 vertical inclinometers and 4 wire extensometers; 13 long- and 3 short-span extensometers, 4 pore pressure gauges, a weather station and rain gauge, installed on the same position as certain parts of the geodetic monitoring equipment. Selection of the same position for different types of monitoring equipment enables spatial correlation of measurement data on the landslide. In combination with hydrological measurements it will be possible to construct relationships between rainfall, groundwater level and appropriate landslide behavior, to establish an early warning system. Definition of alarm thresholds will be based on existing knowledge of the Grohovo landslide behavior and on the collected consequent comprehensive monitoring data as well.

Acknowledgements Equipment used in the study was obtained with financial support from the SATREPS (Science and Technology Research Partnership for Sustainable Development) program, financed by the Japan Science and Technology Agency and Japan International Cooperation Agency through the Project Risk Identification and Land-Use Planning for Disaster Mitigation of Landslides and Floods in Croatia. This support is gratefully acknowledged.

This work is an expanded version of the papers published and presented on the 11th International & 2nd North American Symposium on Landslides, held in Banff, Canada in 2012, World Landslide Forum 3, Landslide Science for a Safer Geoenvironment, held in Beijing,

China, and SATREPS Workshop on Landslide Risk Assessment Technology, held in Hanoi, Vietnam in 2014.

References

ANON (2011) City of Rijeka. Technical department (1840–1918). State archive in Rijeka (unpublished documents, in Hungarian)

Antoine P, Giraud A (1995) Typologie des Mouvements de Versants dans un Contexte. Bull IAEG 51:57–62

Arbanas Ž, Benac Č, Dugonjić S (2010) Dynamic and prediction of future behavior of the Grohovo landslide. In: Arbanas Ž, Mihalić S, Ožanić N, Marui H (eds) Proceedings of the 1st workshop of the project risk identification and land-use planning for disaster mitigation of landslides and floods in Croatia, Dubrovnik, 22–24 Nov 2010. JICA, Zagreb, p 5

Arbanas Ž, Sassa K, Marui H, Mihalić S (2012a) Comprehensive monitoring system on the Grohovo landslide, Croatia. In: Eberhardt E, Froese C, Turner K, Leroueil S (eds) Proceedings of the 11th international & 2nd north American symposium on landslides. CRC Press, Vancouver, pp 1441–1447

Arbanas Ž, Jagodnik V, Ljutić K, Dugonjić S, Vivoda M (2012b) Establishment of the Grohovo landslide monitoring system. In: Proceedings of the 2nd workshop of the project risk identification and land-use planning for disaster mitigation of landslides and floods in Croatia, 15–17 Dec 2011. Rijeka, Croatia, pp 29–32

Arbanas Ž, Sassa K, Nagai O, Jagodnik V, Vivoda M, Dugonjić Jovančević S, Peranić J, Ljutić K (2014a) A landslide monitoring and early warning system using integration of GPS, TPS and conventional geotechnical monitoring methods. In: Proceeding of world landslide forum 3, landslide science for a safer geoenvironment. Methods of landslide studies, 13–16 June 2014, vol. 2. Beijing, China, pp 631–636

Arbanas Ž, Mihalić Arbanas S, Vivoda M, Peranić J, Dugonjić Jovančević S, Jagodnik V (2014b) Identification, monitoring and simulation of landslides in the Rječina River Valley, Croatia. In: Proceedings of the SATREPS workshop on landslide risk assessment technology, July 2014, Hanoi, International Consortium on Landslides, pp 200–213

Benac Č, Arbanas Ž, Jurak V, Oštrić M, Ožanić N (2005) Complex landslide in the Rječina River Valley (Croatia): origin and sliding mechanism. Bull Eng Geol Env 64(4):361–371

Benac Č, Dugonjić S, Oštrić M, Arbanas Ž, Đomlija P (2010) Complex landslide in the Rječina River Valley: monitoring results. In: Horvat M (ed) Proceedings of the 4th Croatian geological congress, Šibenik, 14–15 Oct 2010. Croatian Geological Survey, Zagreb, pp 157–158

Benac Č, Dugonjić S, Vivoda M, Oštrić M, Arbanas Ž (2011) Complex landslide in the Rječina Valley: results of monitoring 1998–2010. GeolCroat 64 (3):239–249

Blašković I (1999) Tectonics of part of the Vinodol Valley within the model of the continental crust subduction. Geol Croat 52(2):153–189

Brückl E, Brunner FK, Kraus K (2006) Kinematics of a deep-seated landslide derived from photogrammetric, GPS and geophysical data. Eng Geol 88:149–159

Crozier MJ (1984) Field assessment of slope instability. In: Brunsen D, Prior DB (eds) Slope instability. John Wiley & Sons, New York, pp 103–142

Cruden DM, Varnes DJ (1996) Landslide type and processes. In: Turner AK, Schuster RL (eds) Landslides: investigation and mitigation. Special report 247. National Academy Press, Washington, DC, pp 36–75

Gili JA, Corominas J, Rius J (2000) Using global positioning system techniques in landslide monitoring. Eng Geol 55:167–192

IAEG (1990) Suggested nomenclature for landslides. Bull IAEG 41:13–16

Karleuša B, Oštrić M, Rubinić J (2003) Water management elements in regional planning in Karst, Rječina catchment area—case study. In: Goluža M (ed) Proceedings of the international conference on water in karst area of watersheds Cetina, Neretva and Trebišnjica, Mostar, 25–27 Sept 2003. Mostar, University of Mostar, pp 85–94 (in Croatian)

Mihalić S, Arbanas Ž (2013) The Croatian–Japanese joint research project on landslides: activities and public benefits. In: Sassa K, Rouhban B, Briceño S, McSaveney M, He B (eds) Landslides: global risk preparedness. Springer, Heidelberg, pp 333–349

Mihalić Arbanas S, Arbanas Ž (2014) Landslide mapping and monitoring: review of conventional and advanced techniques. In: Proceedings of the 4th symposium of macedonian association for Geotechnics, 25–28 June 2014. Struga, Macedonia, pp 57–72

Mora P, Baldi P, Casula G, Fabris M, Ghirotti M, Mazzini E, Pesci A (2003) Global positioning systems and digital photogrammetry for the monitoring of mass movements: application to the Ca' di Malta landslide (northern Apennines, Italy). Eng Geol 68:103–121

Savvaidis PD (2003) Existing landslide monitoring systems and techniques. In: Proceedings of the conference from stars to earth and culture. In honor of the memory of Professor Alexandros Tsioumis. Thessaloniki, Greece, pp 242–258

Skempton AW, Hutchinson JN (1969) Stability of natural slopes and embankment foundations, state of the art report. In: Proceedings of the 7th international conference on soil mechanics and foundation engineering, Mexico City, pp 291–340

Varnes DJ (1978) Slope movements, types and processes. In: Turner AK, Schuster RL (eds) Landslides: investigation and mitigation. Special report 247. National Academy Press, Washington, DC, pp 11–33

Velić I, Vlahović I (2009) Geological map of the Republic of Croatia 1:300.000. Croatian Geological Survey, Zagreb (in Croatian)

Wang G (2011) GPS landslide monitoring: single base vs. network solutions—a case study based on the Puerto Rico and Virgin Islands permanent GPS network. J Geodetic Sci 1(3):191–203

Wang C, Arbanas Ž, Mihalić S, Marui H (2012) Three dimensional stability analysis of the Grohovo landslide in Croatia. In: Proceedings of the 2nd world landslide forum, Rome, 3–9 Oct 2011. Springer, Heidelberg (in press)

Wang G, Kearns TJ, Yu J, Saenz G (2014) A stable reference frame for landslide monitoring using GPS in the Puerto Rico and Virgin Islands region. Landslides 11(1):119–129

TXT-tool 2.062-1.2 A Monitoring and Early Warning System for Debris Flows in Rivers on Volcanoes

Teuku Faisal Fathani and Djoko Legono

Abstract

Volcanic eruptions may produce large amounts of pyroclastic material. Such material will then be transported down a volcano's rivers due to slope instability or rainfall, and then settle in alluvial fan areas. Further migration of this deposited sediment along the river may occur through both natural mechanisms (hydraulic phenomena) as well as anthropogenic processes (human interference). This paper presents the application of monitoring and early warning systems to mitigate the impact of debris flows, using adaptive, low-cost, and collaborative-based technology. A long-running and sustainable system for monitoring and early warning of debris flows in the rivers of Mt. Merapi volcano in Indonesia will be used as case study for the implementation of this model. The 2010 Mt. Merapi eruption produced approximately 140 million m^3 of pyroclastic deposits, of which more than 10 million m^3 deposits have a potential to move downstream through Boyong/Code River towards Yogyakarta City and cause damage to the settlement areas. A real-time monitoring and warning system has been developed by considering community wishes in determining the types and placement of monitoring equipment, and maintaining it sustainably. The equipment consists of an automatic rainfall recorder, automatic water level recorder, debris sensor, and interval camera. The information flow for the proposed early warning system has been set up. The central station receives both the results of the real-time monitoring, and information through radio communication from key

T.F. Fathani (✉) · D. Legono
Center for Disaster Mitigation and Technological Innovation (GAMA-InaTEK), Department of Civil and Environmental Engineering, Universitas GAdjah Mada, Jl. Grafika No. 2, Yogyakarta 55281, Indonesia
e-mail: tfathani@ugm.ac.id

D. Legono
e-mail: djokolegono@ugm.ac.id

K. Sassa et al. (eds.), *Landslide Dynamics: ISDR-ICL Landslide Interactive Teaching Tools*, https://doi.org/10.1007/978-3-319-57774-6_36

persons. Afterward, a warning alert is sent to key persons and the debris flow monitoring radio. This newly built system is expected to be integrated with the monitoring system of rivers, not only at Mt. Merapi but also on other volcanoes in Indonesia.

Keywords
Debris flow • Volcanic river • Real-time monitoring
Early warning system • Collaborative-based technology

Contents

1 **Introduction** 480
2 **Debris Flows in Rivers on Volcanoes** 481
3 **Characteristics of Rainfall-Induced Debris Flows** 482
4 **Early Warning System for Debris Flows** 485
5 **Conclusions** 489
References 490

1 Introduction

Mt. Merapi is one of the most active volcanoes in Indonesia. It has a long history of frequent eruptions, with avalanche-type pyroclastic flows caused by collapsing masses of lava. Eruption-induced pyroclastic flows, tephra, lahar flows, and debris flooding from Mt. Merapi have often occurred with a cycle of approximately 9–16 years for larger scale flows and 1–3 years for smaller scale ones. Rivers originating from Mt. Merapi are significant channels for conveying the sediment produced by volcanic activities. Depending upon the volume of sediment accumulating in the upstream areas of the rivers, as well as the amount of rainfall, the rivers may experience debris flows with significant destructive power. The accumulation of sediment in the upper part of the rivers depends on the volcanic activity of Mt. Merapi, and unfortunately, eruption direction is unpredictable. The movement of sediment from the upstream to downstream rivers is dominated by two mechanisms, i.e., natural flow by bed load transport and debris flows, as well as through human-related activities such as sand mining (Wardoyo et al. 2010).

It was reported that the 2006 Mt. Merapi eruption (from April through October 2006) had produced approximately 8 million m^3 of sediment, which has a high potential to move downstream towards the southern part of the mountain (Legono et al., 2008). Subsequently, the 2010 Mt. Merapi eruption (from 25 October through 2 December 2010) produced approximately 140 million m^3 of pyroclastic deposits toward the southern and western part of the mountain (Legono et al. 2011a). Since then, volcanic debris flows have caused severe damage to the property of residents living on the foot slopes.

Several main processes are common at Merapi due to its characteristic style of eruption. First is the raising of a volcanic dome, with pyroclastic debris falls and the generation of hot pyroclastic clouds. As the dome rises due to the force of rising magma, material also will fall down as pyroclastic avalanches and move downslope at high speeds through to the lower points on mountain side. This material consists of dust to block size debris at a very high temperatures. This material will be deposited and create new morphological profiles either on the soil surfaces or in the river beds. When a sufficient rainfall triggers movement of this deposited material, it may then become a secondary natural disaster. The behavior of Mt. Merapi's eruption has changed considerably, particularly in the nature

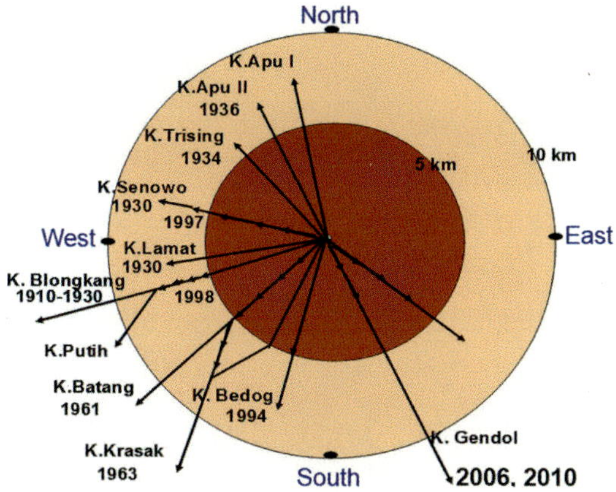

Fig. 1 Dominant directions of Mt. Merapi eruptions

of the location and direction of pyroclastic debris production. The dominant direction of eruption was to the southwest, then to the west in 1998, and the last two eruptions were to the southeast (Fig. 1).

Due to the 2010 Mt. Merapi eruption, a large amount of sediment has been deposited in the upstream reaches of several rivers on the volcano, hence a secondary disaster, rainfall-induced debris flows, may occur over a long period. The presence of river bends also increases the possibility of inundation of some river flood plains in the city area or may create a new stream, causing significant damage to various infrastructure and settlements. Figure 2 shows the collapsed bridge at the west side of Mt. Merapi, while the damage to houses there is shown in Fig. 3. This damage was due to the destructive power of rainfall-induced debris flows.

Because of the large volumes of deposits accumulating in the upstream parts of rivers on volcanoes, there is a need to monitor hydraulic and hydrology parameters, in order to predict the occurrence of debris flows. The monitoring results can be used to both anticipate debris flow disasters and develop the criteria for rainfall-induced debris flows for early warning systems. Therefore, both types of monitoring, i.e., historical and real-time, are very much needed.

2 Debris Flows in Rivers on Volcanoes

The distribution of material from pyroclastic flows towards the southern part of Mt. Merapi threatens the Boyong/Code River area, with the 1994 Mt. Merapi eruption severely affecting the upstream part of the river. Since then, a number

Fig. 2 Collapsed bridge due to the destructive power of a rainfall-induced debris flow

Fig. 3 Damage to houses near a river on the volcano

of sediment control structures have been built to ensure that Yogyakarta City would be protected from debris flows. There are 41 sabo dams, provided with river training structures such as groundsills and retaining walls, along the Boyong River. The 2010 Mt. Merapi eruption potentially can contribute more than 10 million m^3 of pyroclastic deposits to the upstream reaches (Legono et al. 2011a). Triggered by rainfall, this potential source can cause debris flows that may damage the settlement areas surrounding the Boyong/Code River in Yogyakarta City (Fig. 4).

Although local communities in this area have experience with various flood disasters, their preparedness for debris flows is far from sufficient because such debris flows are very rare. Preparedness was inadequate on the afternoon of 29 November 2010, when the first flood after the 2010 Mt. Merapi eruption occurred. The flood carried materials ranging from fine sediment to coarse sand, but no boulders were found. The sediment was assumed to have originated from dust particles falling in the catchment of Boyong/Code River and its accumulation in the upstream areas of the river (Fig. 5). In fact, local communities have initiatives to anticipate debris flow disasters by identifying the occurrence of debris in the upstream parts of the river.

Apparently, during the days after 29 November 2010, rainfall intensity increased, triggering a more intensive debris flow that caused significant damage to infrastructure along some rivers (Pabelan River and Putih River). Fortunately, no casualties were reported. However, there was significant damage to important infrastructure such as roads, bridges, paddy fields, and settlements. A hyper-concentrated flow in Boyong/Code River occurred several times, forcing the local people to evacuate the area (Fig. 6).

3 Characteristics of Rainfall-Induced Debris Flows

Rainfall plays an important role in contributing to massive soil mass movements, both in the form of debris flows and landslides. For many debris flows or landslides, the contribution of rain can be in the form of heavy rain (commonly expressed in rain intensity, in mm/hour), or as relatively long rain (expressed in time units of hours). Depending on the characteristics of the

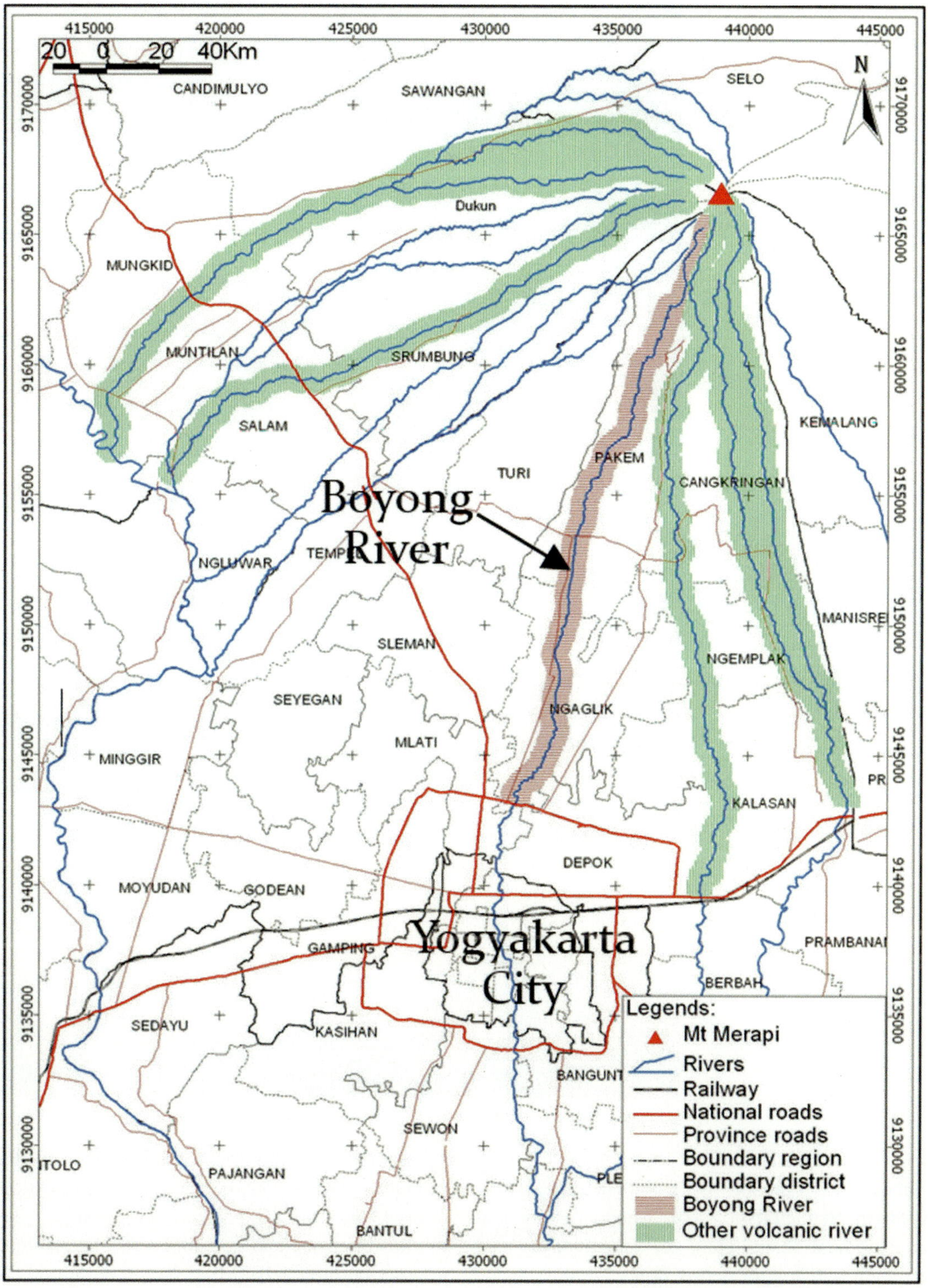

Fig. 4 Volcano rivers toward the southern and western part of Mt. Merapi that have a potential for debris flows. Boyong River is directed toward Yogyakarta City, where its name changes to Code River

Fig. 5 Condition of the Boyong River upstream of Sabo dam BOD4 on 14 January 2011

Fig. 6 Flow conditions of Code River downstream in the Jogoyudan area in Yogyakarta City on 30 November 2010

soil slopes, usually defined in terms of soil mass stability (a function of mass density, shear strength characteristics, soil water content, soil embankment shape, etc.), the rain intensity and duration will trigger debris flows.

Three different approaches can be used to develop criteria for debris flow occurrence as described by MLIT (2004), i.e., based on information on the relationships between the rainfall intensity and rainfall accumulation, rainfall intensity and antecedent rainfall, and rainfall intensity and soil moisture. As an illustration, before 2010 Mt. Merapi eruption, the criterion used as a sign of possible debris flow occurrence in the upstream areas of the rivers on Mt. Merapi was rainfall with an intensity of

25 mm/h (Takahashi 1991) or 50 mm/h with the duration of more than 2 h (Legono et al. 2008). However, at the present, after the 2010 Mt. Merapi eruption, the criteria changed due to the position of the sediment, as instability increased significantly due to a large amount of very loose granular materials forming very steep slopes upstream. Such a condition may cause debris flows to occur even in rainfalls with intensities of less than 50 mm/h and durations of less than 2 h. For two lahar flow occurrences at Kali Putih on 9 January and 23 January 2011, the rainfall intensities upstream were 9 mm/h for 2 h duration and 40.9 mm/h for 2 h duration.

In order to obtain historical and real-time information, the rainfall monitoring should show rainfall intensity values in mm/hour. For debris flows, as the wave propagation of debris flow is very fast (e.g., c. 3–5 m/s), the monitoring of water elevation in stream should have very high accuracy, e.g., at 1.00 cm increments, with a reliably short frequency, e.g., every 10 s, particularly if such equipment is a dedicated part of the development of an early warning system.

As data from the flow monitoring system will be converted into discharge information (in volume units per unit time), the rating curve, which represents the correlation between the water depth and the discharge, should be revised and updated periodically. The necessary calibration is a part of "rule of thumb" of each system development; therefore, it is an obligatory requirement.

4 Early Warning System for Debris Flows

At the moment, the development process for reliable sustainability includes community participation in the form of highly local self-assessment, planning and action (Wisner 2006). In this development of a debris flow early warning system, a collaborative-based process is introduced (Legono et al. 2011b; Fathani and Legono 2013). Although it is similar to the community involvement process, the term "collaborative" is used to emphasize that the community, together with local government, should be involved in solving a problem, i.e., debris flow early warning.

The collaboratively-based debris flow early warning system of Boyong/Code River was developed as follows.

(a) Public consultation meetings on Boyong/Code River debris flows (the people living along the Code River requested information on debris flow occurrence).
(b) Determination of key persons whose task is to receive the earliest news on debris flow phenomenon by SMS blasting (text message).
(c) Construction and installation of integrated monitoring equipment (water elevation sensors, pendulums, camera intervals) and training in its use by key persons.

Figure 7 shows the proposed locations of the instruments for debris flow monitoring and early warning, i.e., one real-time Automatic Rainfall Recorder (ARR) at Turgo, and 4 sets of cross-sectional monitoring equipment (comprising one real-time Automatic Water Level Recorder-AWLR, one real-time debris sensor, one interval camera) at Donoharjo, Rejodani, Krikilan, and Gemawang. A central station was set up in the Hydraulics Laboratory, Department of Civil and Environmental Engineering of Universitas Gadjah Mada, which further addresses maintaining the sustainability of the system.

Figure 8 shows the information flow for the proposed early warning system. The rainfall station in Turgo and cross-sectional monitoring system in four locations along Boyong River will transmit the results of real-time monitoring to the central station. In addition, the central station also receives information via radio communications from the key persons, who monitor debris flows visually. After analyzing the received data, the central station then sends a warning alert to key persons located along Boyong/Code River and to the debris flow monitoring radio using a frequency of 149.940 MHz. This warning can be sent automatically by the system or trigged manually by the operator. The system built for

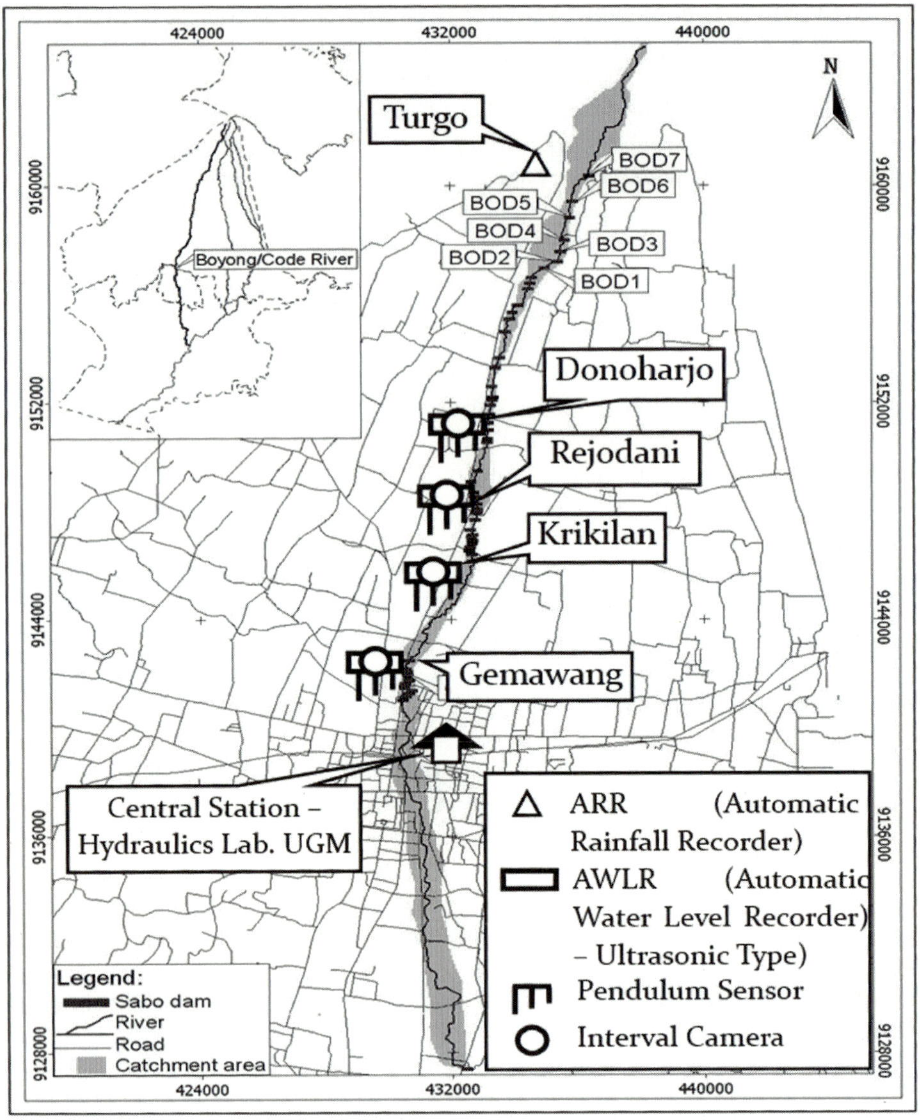

Fig. 7 The proposed location of instruments for the debris flow monitoring and warning system at Boyong/Code River

the Boyong/Code River is expected to be integrated with the monitoring systems of other rivers on Mt. Merapi volcano.

Figure 9 shows the current warning system used by the community surrounding the Code River. The community living near the river has tried to carry out debris flow monitoring in the simplest way. Based on their previous experience, "caution" and "evacuation" elevations are marked on the river slope protection; hence, they can decide the right time to evacuate. The presence of an integrated debris flow early warning system indeed helps the community to increase their awareness and preparedness in facing possible disaster.

As shown in Figs. 7 and 8, the ideal version of an early warning system for debris flows in Boyong/Code Rivers may comprise a water level monitoring system at four locations, i.e., Donoharjo, Rejodani, Krikilan, and Gemawang, with a pendulum-type sensor to confirm the height of the flow, i.e., flow depths equal to

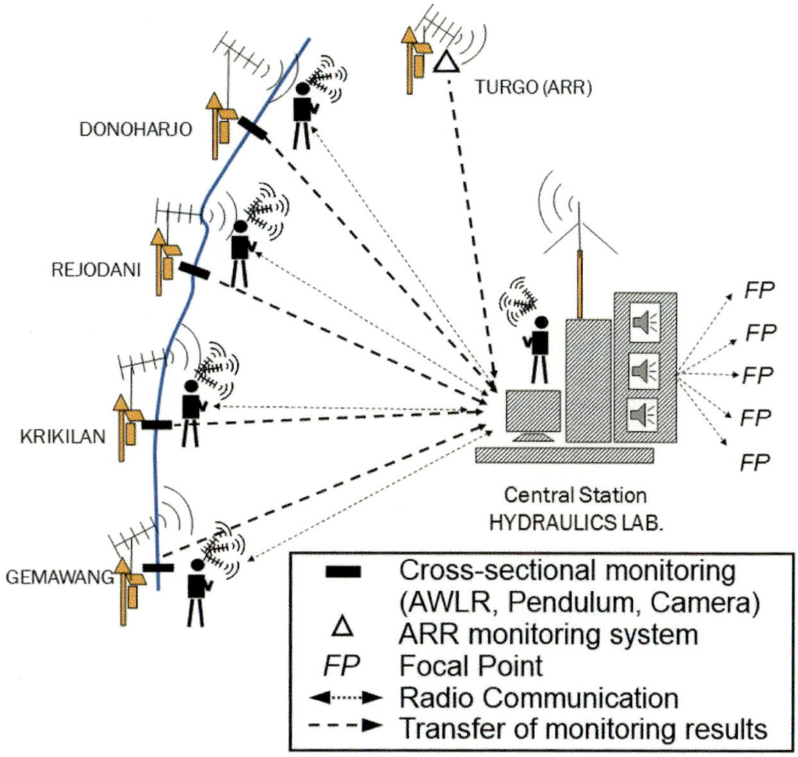

Fig. 8 Proposed information flow for monitoring and early warning of debris flows at Boyong/Code River. Focal Points are key persons

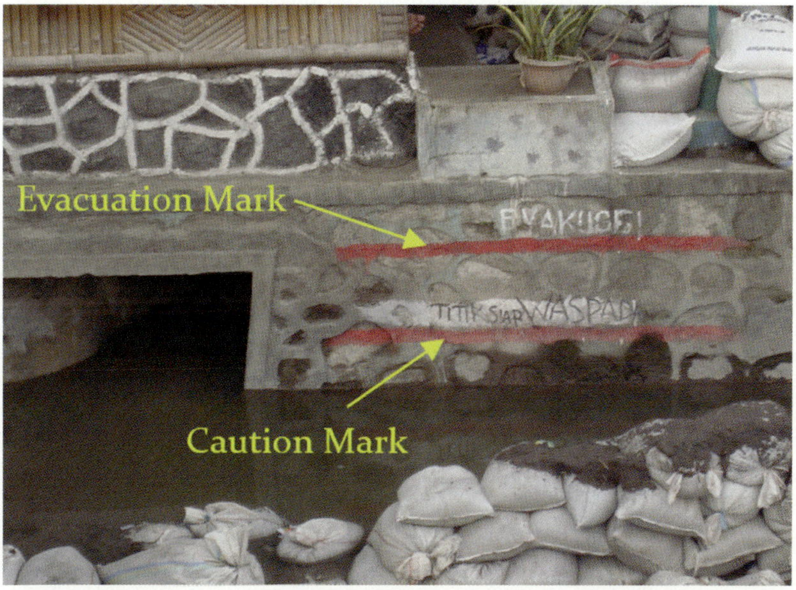

Fig. 9 Bank markings used for existing community warning alerts of debris flows

1.50, 2.25, and 3.00 m. In addition, one additional interval camera was installed to confirm the flow occurrence visually. The sequence of the alert should be such that the debris flow propagation time is long enough to issue the warning to the community, so that they have sufficient time to take necessary actions for evacuation.

Establishment of a collaboratively-based debris flow early warning system consisted of the following (Fathani and Legono 2013):

(a) Installation of an interval camera on the river side, accompanied by the local people (the installation was witnessed by the local people who are also members of an NGO, Jalin Merapi. At the same time, awareness and aspirations of the local people were studied).
(b) After the installation, data on the river flow was collected for several days, and a collaborative review of the data was then conducted (the local people saw the benefits and asked for a higher specification, not the historical/logger type, but the real-time one).
(c) The development and installation of equipment were carried out, taking into account community wishes, in terms of types and locations (the local people showed the best location for the installation). Also, due to the necessity of early debris flow monitoring, the local people were given facilities such as electricity accessibility and key persons for training in the equipment functions and their maintenance).

Figures 10 and 11 show some readings obtained from the installed Automatic Rainfall Recorder (ARR) in Turgo and the Automatic Water Level Recorder (AWLR) in Rejodani, respectively. The Automatic Rainfall Recorder in Turgo recorded rainfall intensity every 6 min in the logger and then sent the data to the central station by radio telemetry. An example of a rainfall intensity of over 50 mm/h that occurred between 14.00–16.00 PM on 13 May 2011 can be viewed at the website (Fig. 10). With this information, key persons should increase their awareness of the possibility of debris flow occurrence in the downstream.

In Fig. 11, the rising of the flow monitored by the ultrasonic equipment every 15 s is shown in real-time on the website. On 9 January 2011, the water elevation relative to a dam crest rose from 39 cm to 130 cm in only 20 s. Further study should be carried out to determine the time needed for the debris to reach Yogyakarta City and the elevation of the flow as it moves downstream.

Figure 12 shows the cross-sectional monitoring system at Rejodani. Ultrasonic monitoring equipment, an interval camera, and three different levels of pendulum are placed in one river cross-section. This system transfers the monitoring results regularly to the central station. The whole system was built by prioritizing the most adaptive and least-cost technology. Through this process, a debris flow monitoring system for

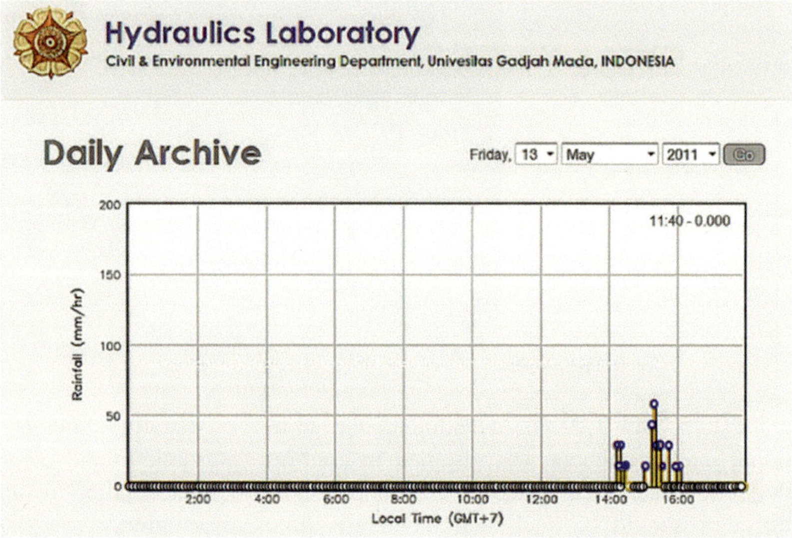

Fig. 10 Example of an online reading obtained from the Turgo Automatic Rainfall Recorder (ARR) monitoring system

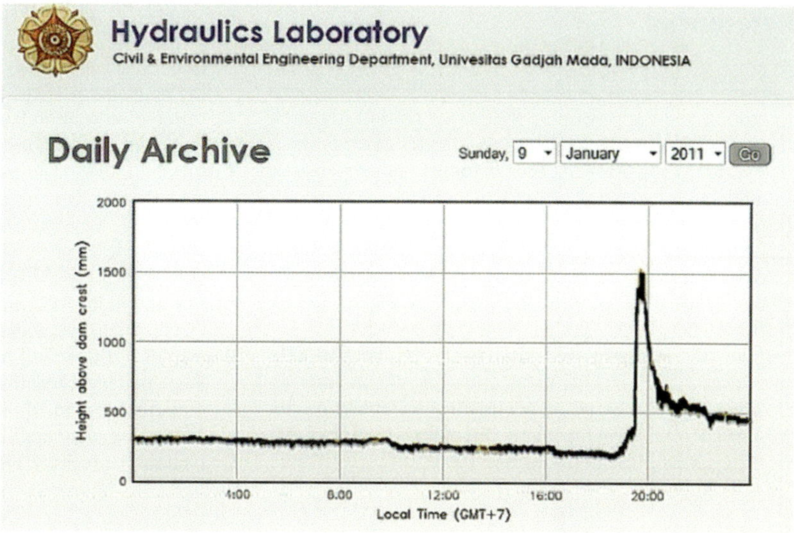

Fig. 11 Example of an online reading obtained from the Rejodani Automatic Water Level Recorder (AWLR) monitoring system

Fig. 12 Cross-sectional monitoring system at Rejodani

Boyong/Code River for community at Yogyakarta City has been implemented. This monitoring system is easy to replicate for rivers threatened by debris flows in other areas surrounding Mt. Merapi.

5 Conclusions

The monitoring of hydrology and hydraulic parameters (including debris flow and rainfall characteristics) should be carried out with consideration of the physical laws of the parameters. Information from the monitoring results should be quantitative and complimentary to the existing system. Both historical logger-type and real-time monitoring systems should be studied further to determine rainfall and debris flow characteristics for warning criteria. Collaborative-based debris flow early warning systems are more sustainable, as communities understand the value of the information they receive. However, support from the local government is still required. More integrated and effective application in the future is still needed, including completion of the monitoring system along Boyong/Code River and also the integration of this system with other debris flow monitoring in the Mt. Merapi area. Further establishment of Standard Operating Procedures utilizing the established early warning system in collaboration with the local community is important. In this way, the sustainability of the effort will be maintained.

References

Fathani TF, Legono D (2013) The application of monitoring and early warning system of rainfall-triggered debris flow at Merapi Volcano, Central Java, Indonesia. In: Wang AF et al (eds) Progress of geo-disaster mitigation technology. Springer, pp 263–275

Legono D, Rahardjo AP, Fathani TF, Prabowo IE (2008) Development of community-based early warning system against debris flow at Mt. Merapi Indonesia. In: Proceeding of the first world landslide forum, 18–21 Nov 2008, Tokyo, Japan, pp 404–406

Legono D, Prabowo IE, Karnawati D, Fathani TF, Rahardjo AP (2011a) Lahar flow disaster monitoring and mitigation based on community collaboration. In: Proceeding of symposium Merapi, 21 Feb 2011, Yogyakarta, Indonesia, pp 11–16

Legono D, Prabowo IE, Fathani TF, Karnawati D, Rahardjo AP (2011b) Development of collaborative-based lahar flow early warning system for Code River at Yogyakarta City. In: Proceeding of international seminar on water related risk management, 15–17 July 2011, Jakarta, Indonesia

Takahashi T (1991) Lahar flow—monograph. IAHR, A. A. Balkema, Rotterdam, The Netherlands

Wardoyo W, Legono D, Fathani TF, Jayadi R (2010) Pyroclastic deposit characteristics of volcanic rivers and its affect to the transport mechanism. In: Proceeding of IAHR-APD international seminar, 21–24 Feb 2010, Auckland, New Zealand

Wisner B (2006) Self-assessment of coping capacity: Participatory, proactive and qualitative engagement of communities in their own risk management. In: Measuring vulnerability to natural hazard. United Nations University Press, Tokyo, pp 316–328

TXT-tool 2.380-1.1 Monitoring and Early Warning System of the Building Constructions of the Livadia Palace, Ukraine

Oleksander Trofymchuk, Iurii Kaliukh and Oleg Klymenkov

Abstract

The present paper briefly presents the general design of integrated monitoring system of the Livadia Palace building constructions placed on active Central Livadia Landslide system. A final part of comprehensive integrated real-time monitoring system has been installing as a part of the research activities by the Institute of Telecommunications and Global Information Space of National Academy of Sciences of Ukraine, a part of the research activities within the framework of the "Landslide protection structures and their development in the Autonomous Republic of the Crimea, Ukraine" IPL Project № 153 (headed by O. Trofymchuk) and a part of the research activities within the "Landslides and Cultural & Natural Heritage" (LACUNHEN) thematic Network of the ICL (headed by C. Margottini). The monitoring systems will consist of sensors for geodetic, hydrological and geotechnical monitoring of the Central Livadia Landslide system and accelerometers-inclinometers aimed at monitoring of the Livadia Palace building constructions. All monitoring equipment was combined in one system with continuous monitoring and data transmitting to the central data unit. Installation of the systems first part was finished in 2001 and final part was finished in 2010.

Keywords

GNSS · Extensometer · Accelerometer · Rain gauge
Landslide and construction monitoring · Central Livadia landslide system
Livadia Palace · Cultural & natural heritage protection

O. Trofymchuk · I. Kaliukh (✉) · O. Klymenkov
Institute of Telecommunications and Global Information Space of the National Academy of Sciences of Ukraine, Kiev, Ukraine
e-mail: kalyukh2002@gmail.com

K. Sassa et al. (eds.), *Landslide Dynamics: ISDR-ICL Landslide Interactive Teaching Tools*, https://doi.org/10.1007/978-3-319-57774-6_37

Contents

1 **Introduction** ... 492
2 **Historical Background. The Livadia Palace** ... 493
3 **Central Livadia Landslide System and Its Activation** ... 495
4 **The Livadia Palace Engineering Survey** ... 499
5 **The Monitoring and Early Warning System of the Livadia Palace Building Constructions** 500
6 **Monitoring Studies Outcomes Analysis** ... 504
7 **Conclusions** ... 506
References ... 507

1 Introduction

The landslide frequency (Lacasse 2013) of about 20 major events per year in Europe is the highest compared to floods, earthquakes and cyclones. Ukraine (Institute of Telecommunications and Global Information Space of National Academy of Sciences of Ukraine) has been a member of the "Landslides and Cultural & Natural Heritage" (LACUNHEN) thematic Network of the ICL since 2012 [head of the LACUNHEN is the Margottini and Vilimek (2014)]. The purpose of the LACUNHEN—International Consortium on Landslides is to create a platform for scientists and practitioners who are ready to contribute to safeguarding relevant endangered Natural and Cultural Heritage sites. LACUNHEN will share and disseminate their respective experience (Margottini and Vilimek 2014), demonstrating how these special "objects" require approaches, techniques and solutions that go far beyond traditional civil engineering perspectives. Within this view, landslides and more generally slope instabilities are an important factor endangering cultural heritage sites and its degradation and requires additional protection measures, creation of the monitoring and early warning systems, etc.

Monitoring and early warning systems are widely used for landslide hazards early warning (Sassa 2005; Lollino 2014; Highland and Bobrowsky 2008; Baum et al. 2005; Mihalic et al. 2013; Karnawati et al. 2013; Reid and LaHusen 1998; etc.). Building constructions early warning and monitoring systems are commonly used by the best practices (Huston 2011; Nagayama and Spencer 2007; etc.). Initially the on-line monitoring system in Ukraine was developed under the direction of Kaliukh (1996) and introduced in 1994 on the south and north vent shafts of the damaged ChNPP Unit 4 with the aim of investigation of building structures consistency degradation issues and development of recommendations as for their further reinforcement.

More than 90% of the territory of Ukraine has complex ground conditions and about 120,000 km^2 of the Ukrainian territory are located in the zone with seismicity of natural origin with a magnitude varying from 6 to 9. Therefore, unpredictable changes of natural geological and man-made factors governing ground conditions may lead to dangerous deformation processes in the Ukraine heritage sites. Periodic exogenic geological processes activation and complex engineering-geological situation within the South Coast of the Crimea (SCC) (Fig. 1) results in landscape transformation and engineering and architecture structures deformations.

Growth of landslides amount is mainly caused by man-made and combined natural and man-made reasons, and less frequently by natural ones purely. Man-made reasons have caused about 600 human-induced landslides on the territory of SCC for the recent 100 years that were registered within SCC Landslides Cadaster (the total amount of landslides in the Crimea by the end of 2009 was 1576).

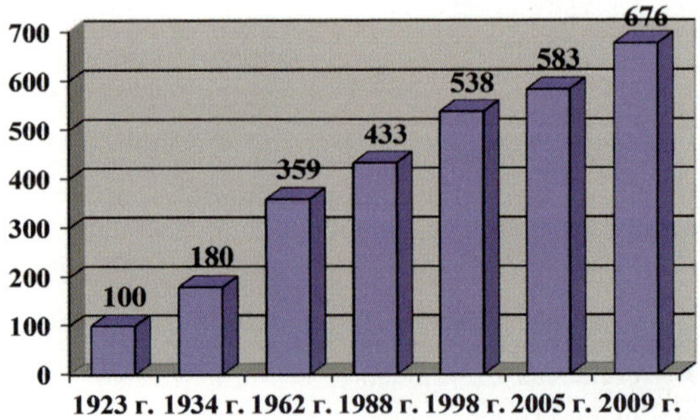

Fig. 1 Dynamics of SCC landslides growth (from Kostel town to Aya Cape)

Thus, about 38% of SCC landslides are man-caused landslides. The most considerable damage is caused by the landslides creating a threat for unique architectural, historic and cultural monuments safety. Over the history of SCC engineering development these landslides have caused more damages than earthquakes. At the present moment there is a threatening situation on the territory of the Livadia Palace and Park Complex.

2 Historical Background. The Livadia Palace

Prior to 1779 a small Greek village was located on the territory of Livadia Palace and Park Complex. The intensive development of Livadia territory refers to the 50-th of the XIX century. Manor houses, church, residential and farm buildings were built. The park with valuable species of subtropical plants: evergreen myrtles, laurels, cedars, pine trees, magnolias and Crimean pine was laid out on the area of 437,000 m^2. In 1860 the royal Romanov's family acquired Livadia and started the reconstruction of one of the houses into the Imperial Palace. Since 1866 Livadia became the royal residence (Fig. 2).

From 1910 to 1911 a new Big Livadia Palace was constructed in place of the old palace built by the famous architect N. Krasnov. It was a two—three storey building, constructed in accordance with the strict architectural forms of Italian Renaissance style with the inclusion of composite elements of other styles (Fig. 3). With this aim a powerful vertical relief planning was performed along with the roads expansion and construction of electric power plant, ice production plant, garage and green theatre. During the Palace construction a collection gallery, drainage and water supply systems were built as well as water tanks. Totally, there were about 400 houses 100 of which were residential buildings. In addition, the greenhouses and dairy farm were built and orchard was laid out.

During the war (1939–1945) the Livadia Palace complex suffered significant losses. A Small Palace was burnt down, a number of buildings were destroyed and a lot of parkland was cut down. In terms of Yalta Conference in spring of 1945 the repair and restoration and construction works were carried out within Big

Fig. 2 Tsar Nicolas II with family in the Livadia Palace

Fig. 3 The Livadia Palace construction

Fig. 4 *Left to right* W. Churchill, F. Roosevelt, I. Stalin in May 1945 in the Livadia Palace

Livadia Palace. Exhibition hall dedicated to Yalta Conference in the Livadia Palace was opened in 1974 (Fig. 4).

In Soviet times the Palace complex of buildings—Big Livadia Palace, church of the Exaltation of the Cross, the retinue building, Fredericks Palace were registered by the State as monuments of architecture and urban planning (Fig. 5). Livadia slope engineering protection partially was formed during the Palace construction (1911–1912) and consisted of collection galleries, rainwater drainages and retaining walls. After the Revolution in 1917 in Russia when the territory was passed to different USSR authorities the adjacent to the Livadia Palace sites were intensively constructed with buildings and structures. Apart from this, the reconstruction of water bearing structures was executed and the construction of new structures was carried out without taking into consideration the existing ones. In 1965–1968 coast-protection structures complex was constructed. It consisted of groynes and artificial beaches, hydro technical walls, cargo pier. The above structures have been distorted by the present days and as a result could not resist the soil slip and abrasive processes development. Intensification of landslides was facilitated by the lack of control over water-bearing and drainage structures the leakage from which along with the broken system of surface runoff led to watering of the Livadia Palace and Park Complex area.

In addition, a collection gallery around the Palace has adverse impact on the slope resistance (it creates groundwater dam, soil watering, reinforce hydrogeological power).

3 Central Livadia Landslide System and Its Activation

Lyvadia Palace and Park Complex is located within the upper part of the Central Livadia Landslide system (CLLS), its length is 1500 m, width in the top part is 200–500 m, while the bottom part is 700 m, the total surface gradient is 13°–15° (Fig. 6). The main (top) part of the system is located within 350 m of absolute marks, bottom is along the Black Sea present shore horizon. Ground waters are deposited allover at depths of 2.5 to 11.2–24.2 m. They practically do not have pressure and create separate waterlogged zones often with low hydraulic

Fig. 5 The Livadia Palace present day view

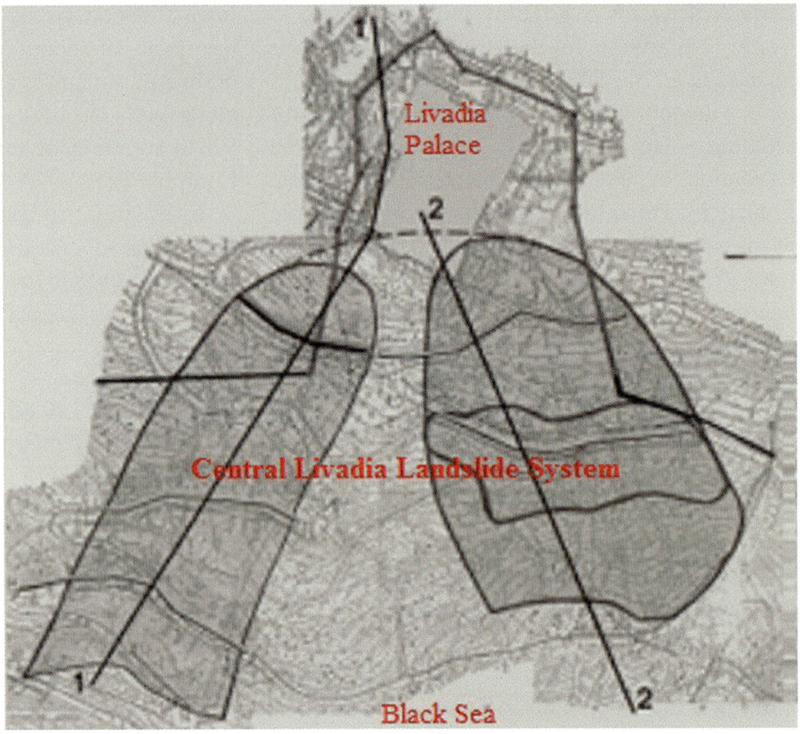

Fig. 6 Central Livadia landslide system (1–1: direction of the «West» landslide; 2–2: direction of the «East» landslide)

connection. The discharge of groundwater ranged from 0.008 to 1.8 l/s. Seismic force of the territory is 8.9 points according to Medvedev-Sponheuer-Karnik seismic scale depending on the soil condition. CLLS combined current landslides, land-slides epicenters and potentially dangerous sites in terms of land-slide of Livadia slope. It is a specific system of landslide bodies the dynamics and forming of which are predetermined by structural and tectonic, geologic and lithologic and geomorphologic peculiarities (Yerysh and Kulish 1998; Sheko et al. 1979). According to general hydrogeological concepts, the CLLS area prior to its development was in favorable conditions in terms of rock watering.

Rising over the adjacent areas, compartmentalization by relatively deep erosional cuttings created good conditions for drainage developed here aquifers. This led to the relative stability of the given site as compared to the adjacent ones. The territory development resulted in ecological disturbance of the slope and water regime change (due to irrigation, leakage from utility systems, etc.). In recent years, this watering process has been increased due to the lack of drainage and runoff organizing.

Geodynamic environment within the survey area also contributes to the landslides formation. The rocks forming the upper part of the slope deposits are extremely heterogeneous in terms of granulometric texture, physical and mechanical properties and water content and are in varying degree of strain caused by geodynamic processes. Within the active and constant ancient landslide bodies the rotation of the rock strain zones with zones of relative decompression is typical for the whole CLLS. And the strain zones are typical for areas adjacent to the walls of the active landslide slopes breakdown or discontinuity of stable sites. Natural and man-made seismic dynamic activity has been increased not only within CLLS but also the entire South Coast and the northern part of the Black Sea recent years. Every day the South Coast seismic stations record weak grade 1–2–3 earthquakes on the land and the sea and some cases of grade 3–4–5 earthquakes a year.

For over 100 years the slope territory within the Lyvadia Palace and Park Complex is constantly in state of stable (continuous) landslide with the transition in certain areas in such one that progresses, leads to the destruction of soil with formation of crown cracks landslides. Landslide speed in the cycles of increased landslide activation reaches 2–4 cm/year within the Complex and more than 6 cm/year in the park zone. Tendency of soil to shear deformations for similar SCC soils is confirmed by outcomes of rheological characteristics studies implemented by other researchers (Anosova et al. 1976). It was identified that within the weak areas (old and ancient slip planes) with loads up to 95–98% of the nominal momentary strength the continuous landslide with an approximate viscosity of 10^{14}–10^{16} poise is observed. The most dispersed and plastic samples of soil with a small (15%) content of debris inclusions show continuous landslide in the same range of viscosity values but at lower loads (about 65–75% of nominal momentary strength).

Landslide bodies' periodic activation is typical for CLLS. During 17 years (1980–1997) three cycles of activity were observed (1981–1983, 1986–1989 and 1995–1997). It was caused mainly by natural factors; however, man-made impact also plays a significant role in landslide bodies' activation. The capacity of landslide debris-clay deposits varying from 5–15 to 25 m is typical for the study area. There is a combination of rocks below these deposits, consisting of clay rock, sandstones and siltstones called Tauride Flick. The area of the territory under the survey is crossed by tectonic dislocations of the north-west and north-east course the role of which is not excluded in the landslide bodies' activation.

Due to intensive construction of the Yalta—Sevastopol highway and "Pogranichnik" health resort complex in the upper part of the Livadia Palace and Park Complex in 1960–70s the landslide processes started developing throughout the whole its territory. According to the above, currently the CLLS has rather high coefficient of man-made load that reaches 0.8 in the upper and up to 1.0 in the bottom parts. The whole territory has rather subdivided and dense water bearing structures network (old systems documents have been lost). During recent 10 years (Trofymchuk et al. 2013) a sustainable tendency of landslide bodies deformations intensity increasing has been observed. This has been resulted in local landslide bodies area expansion, destruction of retaining walls, drainage and water diversion structures, utilities; in expansion of existing cracks and cracking on the buildings of the Palace Complex and the Tsar's Palace as well.

At present, as a result of man-made factors integrated impact within Livadia Palace and Park Complex the spatially connected and interdependent dangerous geological processes are formed and developed: landslide, territory flooding, abrasion etc. During the "West" landslide investigation (Fig. 6) crown cracks and main scarps are observed along the access way with road base subsidence as well as numerous tension cracks in retaining walls surrounding the road (Malyuk 2006). Currently the transit across the road is impossible and destroyed sections of the road are used as soil and domestic waste landfill loading the head parts of active landslides. Under the threat of destruction there is a sewage collector that had been run along the lower destroyed retaining wall on the anchors within active landslide boundaries. The "West" landslide (Fig. 6) is evident on the slope, buildings and structures as cracks and scraps. Retaining walls, access road and footpaths destructions are observed on their crossing with landslide body flanks (Fig. 7). Visible deformations on the road asphalt coating westward of the Laurel alcove point at expansion of the "West" landslide borders. Access roads, retaining walls, footpaths, the Laurel alcove have been completely destroyed as a result of landslide movements (Fig. 8).

Main scarp of the "West" landslide has cut catchwater collector of the drainage gallery south branch (it was constructed at the same time with

Fig. 7 Cutting of the road with external retaining walls by the "West" landslide left border

Fig. 8 The Laurel alcove destruction by the "West" landslide right border

the Tsar's Palace in the early 20th century) that makes the slope landslide situation worse. Development of the west lower part of the landslide system and its sustainability entirely depends on the slope soils and landslide bodies watering by drainage and man-made water that mainly comes from destroyed water discharge pipe of the drainage gallery south branch (Fig. 9). Groundwater discharge is observed below the road in the "West" landslide toe part (near upper elevating equipment tunnel outlet).

Main factors of the "East" and "West" landslides formation are: significant gradient of natural ground contour and a top of bedrock of Tauride series (clay rock with sandstone bands), underlying quaternary deposits; high water-bearing nature of covering upper-quaternary and recent deposits (loamy and rubbly aggregations) due to the leakage from water-bearing structures including drainage gallery especially in the places where they are connected to the catchwater collectors (pipes); recent seismic and dynamic activation. The above landslides destroy Livadia Palace and Park Complex, various utilities (such as public sewer), elevator shafts (where cutting of tubing bolts takes place at a depth of 18 and 30 m), as well as coast protection structures in the toe parts of the "West" and "East"

Fig. 9 Broken down catchwater pipe of the drainage gallery south branch

landslides of the CLLS. Further development of landslide processes can lead to significant destructions of the Livadia Palace building and adjacent territory. Drainage system in the south-west part of the Palace is under threat of destruction. The CLLS global stability is rapidly decreased due to introduction of diverse man-made load against natural factors and as a result 10 local (rather developed) landslides have been formed within the Palace and adjacent area until present time.

Given that since 1995 the landslide processes activation within the "West" landslide area has been significantly intensified, the primary focus during the survey was on the slope site with the Big Livadia Palace and adjacent territory of the landslide slope almost to the Black Sea water edge. Significant landslide developments have been observed during recent years just at that spot (in toe part of landslides the coast protections are broken down and the coast abrasion is intensified (Malyuk 2006).

4 The Livadia Palace Engineering Survey

Currently the active evidences of landslide processes are observed in immediate vicinity of the Big Livadia Palace (Fig. 5) below which the recent landslides have been formed (the "West" and "East" Fig. 6). The marble staircase cutoff towards the sea coast is still in process. This staircase leads to the viewing platform over the yard in the north-east corner of the Palace. In addition, the deformations on the asphalt are observed 3–5 m off the south front of the Palace. As a result of visual investigation of the building structures of the Big Livadia Palace in 2012–2014 (Trofymchuk 2012) we have found the "East" landslide activity increasing, resulted in expansions of landslide deformations upper of its head part. Adjacent to the Palace waterproofed asphalt and concrete coatings contain numerous cracks that allow rainwater penetration and as a result this water wet the soils of its foundations. There is a visible crack in the basement wall that goes deep in the Palace foundation on its technical basement floor. There are also visible cracks with growth up to 55 mm and more at the junctions between the perron and staircase to the Palace building on the south-east front side. The cracks are observed on the south-east and south-west fronts of the Palace from the basement to the eave including arched constructions of the window apertures. During 2012–2014 the development of landslide cracks up to 6–12 cm has been in process within the Palace, the retinue building and church.

More than 30% of the CLLS investigated area has been unbalanced for the recent 80 years of Livadia Palace and Park Complex use. The territory flooding that causes the reduction of soil bearing capacity, increasing of hydrologic force and seismic magnitude is observed within the Palace and the park area. Seismic magnitude is increased due to degradation and water saturation of the CLLS soils. Local landslides are currently (progressively and regressively) expanded and spatially united that results in forming of global landslide on the whole territory. High season and year activity of local landslides within the Palace and Park Complex increases the social and economic risk level. At the present the Palace is located within 70–100 m distance from the main scraps of three local landslides as well as in the continuous rock creep zone. All the above promotes formation and development of the cracks on adjacent sites and palace walls and reduces its seismic resistance. The risk of collapse endangers the Palace as well as life of numerous tourists groups visiting its sightseeing every day.

5 The Monitoring and Early Warning System of the Livadia Palace Building Constructions

The "Program of Works on Livadia Palace and Park Complex Preservation and Protection from Geological Hazard" was put into effect with the aim of risk mitigation and avoidance for the further use of Livadia Palace and Park Complex (Kaliukh 1999). The above Program is a system of scientifically based and interrelated planning and surveying, protecting and lithomonitoring measures as local ones within the Palace and on the whole CLLS territory as well.

The monitoring and early warning system ZSUV (Trofymchuk 2012) was used for analysis of CLLS landslide hazardous geological processes. One of the fragments of this system is presented on Fig. 10. It has been measured the following parameters:

- Groundwater levels in **boreholes** identified by the geologists;
- Angles of gradient of selected elements (identified by the experts) inside the landslide slope under survey (angular deformations of the holed through the slope boreholes);
- Other parameters characterizing litho-dynamics of the CLLS landslides.

More detailed description of the ZSUV—the CLLS monitoring and early warning system is presented in the report by (Trofymchuk 2012) on the Conference in Curio city (Japan). On-line monitoring of the Big Livadia Palace building structures stress and strain state is carried out with the help of the new "Sensor" Program. It puts into practice the transformation of analog signal into digital code with the help of analog-to-digital converter and data insertion into computer in online mode as well as processing of measurements primary statistic data and saving of this data coming from high precision electric inclinometers set on different Palace building structures. One of the main elements of the "Sensor" monitoring system is high-precision sensors (inclinometers), the accuracy of which depends on the type of selected sensing elements, their sensitivity and errors both occasional and systematic ones. The inclinometers consisted of two orthogonal-oriented in horizontal plane high-precision accelerometers. Accelerometers sensitivity axes are oriented towards body roll and pitch axes. The principle of the signals proportional to the angles forming is based on gravitational vector **g** projection measurement on

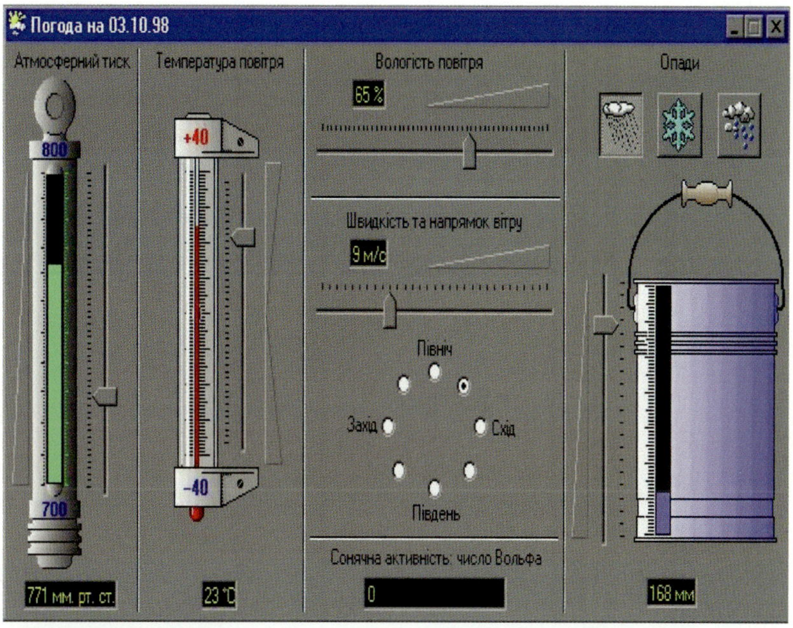

Fig. 10 ZSUV—CLLS monitoring and early warning system

Fig. 11 Location of one of the inclinometers in the south-east part of the Livadia Palace

accelerometers sensitivity axis. The accuracy of "Sensor" monitoring system high-precision sensors (inclinometers) is not more than 5 s of angle.

On-line measurements of dynamic vibrations of Livadia Palace spatial position have been carried out since mid-December 2001 with the help of "Sensor" (Kaliukh 1999) monitoring system. Evolution of spatial position is identified by measurement of the Palace fronts' gradient angles dynamics. One of the inclinometers has been set on marble window ledge (outside) of the corner room (empress's cabinet) on the 2 floor of the south-east aisle of the Palace (Fig. 11). The sensor has been fixed on the foot of marble column and oriented towards roll and pitch axes of Livadia Palace.

The Livadia Palace front part was taken for its roll axis (City of Yalta side). The signals as for angular position and vibrations of the Livadia Palace supporting constructions are formed within accelerometers high-precision sensors as volt alternating current proportional to gradient angles, amplitude and vibrations frequency. This signal enters the signal conditioner where it is amplified and converted into volt direct current. Amplified signals enter analog-to-digital convertor and further personal computer by special cable.

"Sensor" Program in real-time mode allows to register: the Palace building structures current gradient angles; seismic and man-made dynamic activity; implement primary statistic processing of the obtained data; implement archiving and storage of obtained data and results of primary statistic processing. "Sensor" program has been

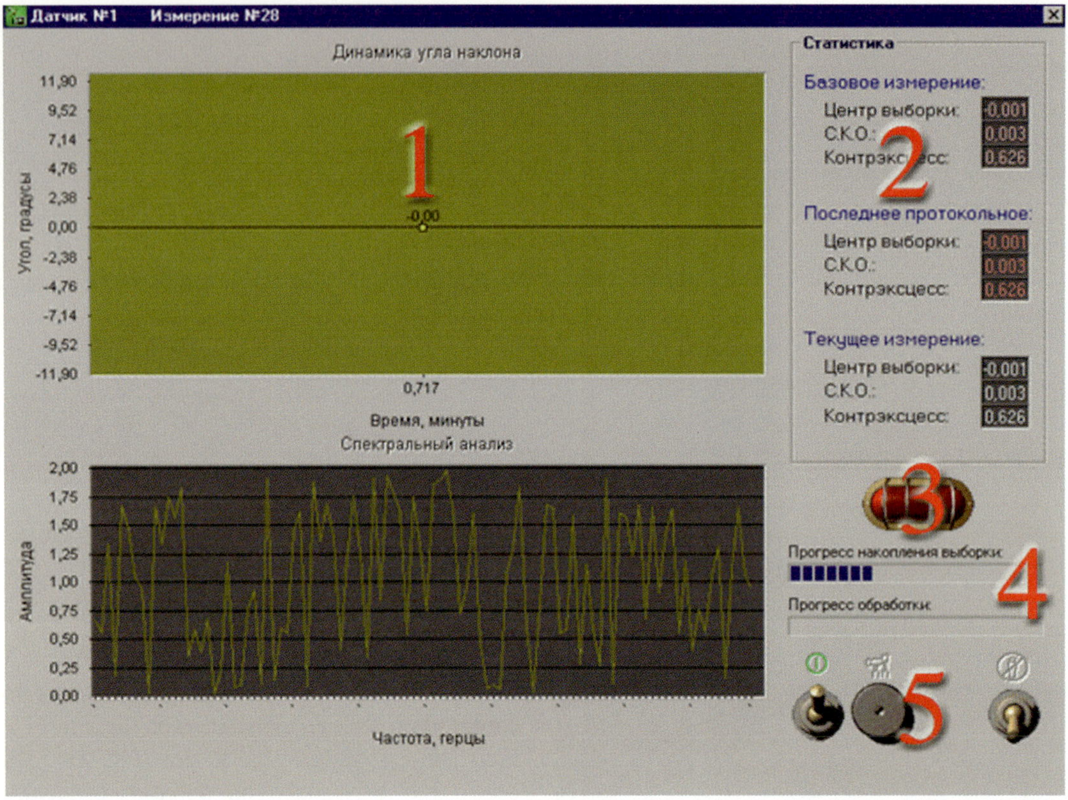

Fig. 12 "Sensor" program user shell

implemented in a view of several interrelated modules. "Sensor" Program user shell could be nominally separated into 5 parts (Fig. 12).

Here:

1. diagram presenting the dynamics of change of values under the survey (the diagram colour is changed depending on the high-precision sensor type ■:—for statistical data ■,—for dynamic data);
2. the results of statistic processing module operating;
3. indicator of value change under the survey (is followed by the audio signal in case of soundcard availability and specific file horn.wav);
4. indicator of data accumulation and processing;
5. "Sensor" Program controls.

"Sensor" software contains the following principles:

- Reasonable probability of measuring values;
- Performance reliability;
- Easy servicing;
- Possibility of operational evaluation of measurement data.

Information from inclinometers enters analog-to-digital convertor by 64 channels. The management of the above converter takes place within ADC module of the "Sensor" Program (Fig. 13).

Comments on some parameters of ADC module:

1. *Frequency of incoming signal discretization*, Hz—for "Sensor" Program given channel

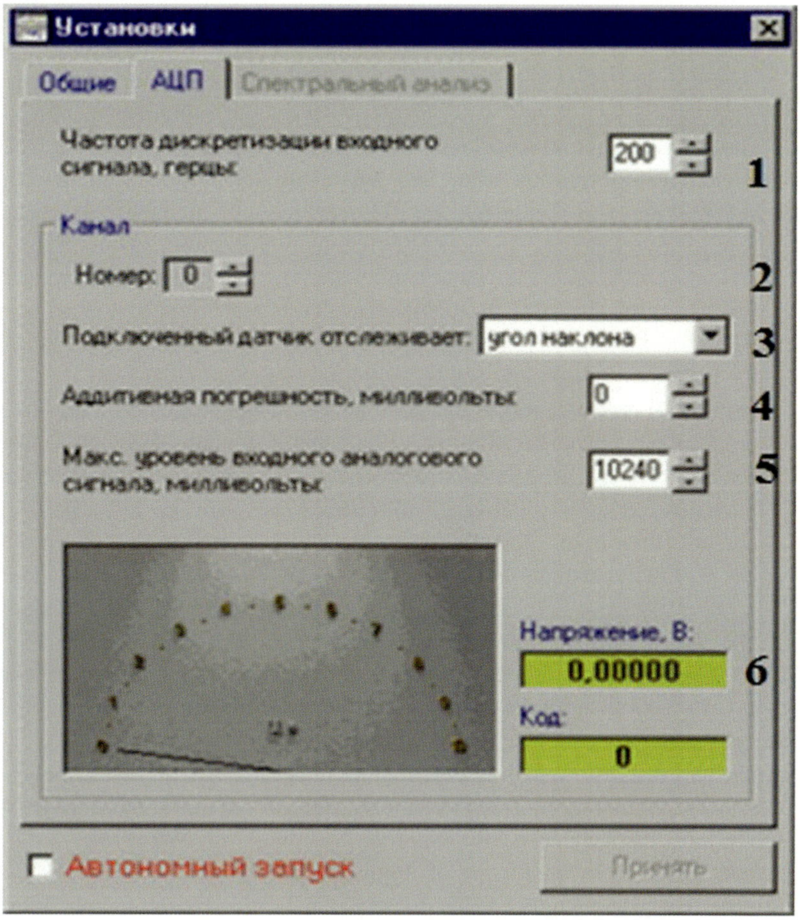

Fig. 13 Analog-to-digital conversion module

adaptation to the process under the survey. In terms of possible dynamic modes with frequency N Hz it should be set in the given analog-to-digital converter configuration window in accordance with Nyquist-Kotelnikov theorem, to establish frequency of signals registration in 2 N Hz.

2. *Channel, number*—number of channel of analog-to-digital convertor with high-precision sensor connected.
3. *Connected high-precision sensor tracks*—the type of measured physical quantity.
4. *Additive error, millivolt*—measurements additive error within the given channel.
5. *Maximum level of input analog signal, millivolt*—maximum voltage range coming from inclinometer sensor by the given measuring channel.
6. *Voltage*, B—instantaneous value of voltage within the given measuring channel.

Mathematical model of primary statistic processing of measurement data is developed in accordance with registration data requirements and designed for the following issues addressing: detailed presentation of measurements results in compliance with the State Standard of Ukraine; development of proximate confidential intervals of data spread by means of measurement system errors registrations; identification of minor but statistically valid data changes with the help of testing of statistic hypothesis method; prediction of reparameterization processes development:

1. Prediction as for change of the law on distribution of single measurements sample;

2. Prediction with use of measurement information results obtained during the finite time (monitoring time).

The present program on primary processing of measurement information puts into effect the confidential intervals computing algorithm with a probability $P = 0.90$:

1. Firstly, on the basis of PC simulation it was determined that there is a relationship among the error distribution law and required sample volume (number of single measurements) in order to get appropriate statistic assessment of specified relative integrated square error.
2. Secondly, change of error distribution law content often is an indication of some changes within experiment environment.

The above property can be used during the monitoring and prediction. Since, on the one hand, change of sample kurtosis and nonsymmetry is possible without change of the center of sample and dispersion, and on the other—change of statistic moments of lower orders (center of sample and dispersion) is not possible without change of statistic moments of higher order (nonsymmetry and kurtosis). Thus, they can be signals for further change of the sample center that can be significant for early hazardous processes prediction (evidences of landslides activation or crack formation and destruction of the Palace building structures).

Developed "Sensor" Program software allows solving following tasks:

1. To automatize the process of investigated parameters measurement (possibility of monitoring system long-term operation in real-time mode without the intervention of an operator).
2. To provide primary processing data in easy to learn way.

Detailed study of the distribution laws, their cataloguing and development of measurements primary statistic parameters base is essential primary information for timing series dynamics prediction program operating. This allows developing earlier and more reliable prediction of the Palace building structures strain stress state evolution, intensity of the CLLS landslide activity, carrying out correlation among geliogenous and lithogenous factors, etc.

6 Monitoring Studies Outcomes Analysis

Selected measurements results are presented on Fig. 14.

Figure 15 presents fragment of instantaneous information obtained from the one of the

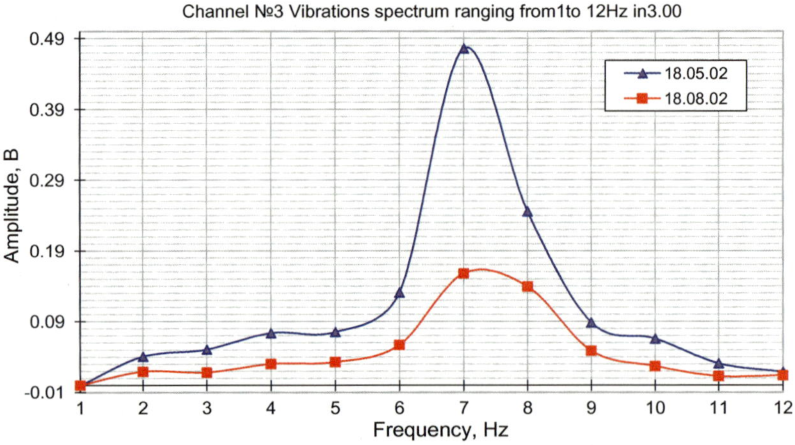

Fig. 14 The Livadia Palace monitoring measurements selected results

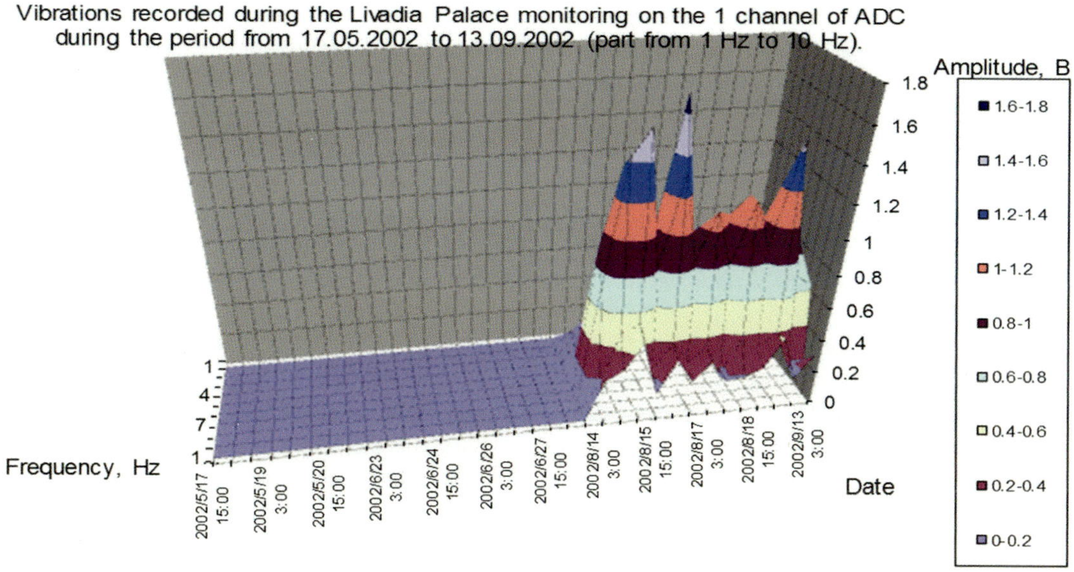

Fig. 15 The fragment of the instant information obtained from one of the numerous channels of high-precision sensors in 2002 during monitoring of the structural elements of the Livadia Palace

inclinometers during the Livadia Palace building structures monitoring in 2002.

On the basis of measurements results processing, it was determined:

1. The Livadia Palace south-east aisle executes continuous vibrations about some medium position. These vibrations are of clear periodic nature. And the period of vibrations is a period of 24 h. The diurnal vibrations amplitude is changed within approximately 1.5 s of angle, i.e. about 45 s of angle towards each side from medium position. Vibrations direction occurs towards building pitch axis. The Livadia Palace diurnal vibrations reasons could be the day and night temperature changes when differential heating and cooling of the building and adjacent territory takes place.
2. In certain cases, for example, such as on 13th–14th February, 26th–27th February and 22nd–27th March 2002 the Livadia Palace gradient angle increase was registered. In the above cases the amplitude increased up to 6 min of angle. Computational results have shown that regular diurnal vibrations are 1.9 mm towards each side from medium position, and on the above dates—about 4.2 mm towards each side from medium position. The causes of the Livadia Palace building front deviation amplitude increase could be the following circumstances and factors: heavy precipitation and temporary ground water level rise; consequent CLLS landslide activation; seismic and man-made dynamic impacts, etc.
3. We did not manage to identify precise correlation between the above factors and the Livadia Palace structural elements dynamics to the full extent due to frequent forced interruptions in monitoring system operation. These interruptions were caused by the necessity of meeting the Security Service of Ukraine requirements as for various official events of national (Ukrainian) and local (City of Yalta and Livadia) level carrying out and preparation within the Livadia Palace. Since January 2014 monitoring of the CLLS and Livadia Palace has been terminated in view of known reasons.
4. It is planned to create field Livadia landslide susceptibility map which will be supplement

by reviews of morphology derived from the available (2 × 2 m) digital elevation model (DEM) and by monitoring of recent movements from the period 2002–2014.

5. Future work is expected to utilize new data from: chemical and mineralogical analysis of rock samples and groundwater samples, hydrological, laboratory testing of mechanical properties and additional surface geophysical data to further improve understanding of the subsurface structure of the Central Livadia landslide system, including groundwater, soil and rock properties. The most important data will be collected by continuous monitoring of landslide movement and landslide causal factors, precipitation and earthquakes. Integration of real interpretation of landslide movement, the most important data from this period of investigation will be traces of sliding surfaces and vectors of displacements.

7 Conclusions

1. Until the present time no full-scale studies have been conducted on the whole territory of Livadia as unified nature and architectural-artistic complex. No precise territory zonation has been carried out with the aim of preservation of both ex-tsar's residence architectural and park complex as historic and artistic monument and adjacent environment reaching visual perception zone. In addition, up to the present moment, no historic and aesthetic evaluation of the given region and its potential as tourist and expositional center has been done. No complex program on unified use of all archeological, architectural, historical and natural objects present on Livadia territory has been developed.
2. Over the last years all the elementary CLLS landslides have been active and have had clear tendencies to spatial expansion and multiplexing. Especially last mentioned is typical for landslides that is a real threat to the Palace due to forming of horizontal tension cracks with the width of 2–5…50 mm within the Palace, retinue building, church and lower —on the footpaths. There are two sewage collector threads within active landslide processes zone (the "East" landslide Fig. 5). They are used for effluents discharge from health resorts located to the south-west of the Livadia Palace and Park Complex ("Livadia", "Glitzinnia", "Nizhnya Oreanda", "Pogranychnyk" health resorts), including Oreanda village. In practice, public sewage collector experiences landslide strain with real threat of outbreak that could lead to the whole Livadia park and the Black Sea waters contamination that could be considered as environmental disaster on the given site of the sea coast.
3. High season and year activity of local landslides within the Palace and Park Complex increases the social and economic risk level (now the Palace is located within 70–100 m distance from the main scraps of three local landslides as well as in the continuous rock creep zone that promotes formation and development of the cracks on adjacent sites and palace walls and reduces its seismic resistance). The risk of collapse endangers the Palace as well as life of numerous tourists groups visiting its sightseeing every day.
4. Proceeding from the tendencies of the three local landslides development located lower the Palace it is predicted that the Palace and adjacent south-east territory will be in intensive crack formation zone in 2–4 years and in active landslide dislocation zone in 2016–2018. It is possible that effect of the Levadia Palace and its territory flooding as well as developing destruction of coast protection structures could intensify landslide process. Local landslides will be easily (progressively and regressively) expanded and spatially united forming the global landslide on the whole CLLS territory within the "East" and "West" (Fig. 6) landslides in 2016–2018.
5. The local and central authorities should take urgent measures as for mitigation of risk of hazardous landslide processes activation that can cause destruction of the world-known historic and architectural monument—Tsar

Nicolas's II Livadia Palace. The above measures should be directed at the Livadia Palace and Park Complex preservation and its adaptation to the new environmental and social conditions.

6. In recent decades, the concept of cultural heritage (Margottini and Vilimek 2014; Migon 2013) has evolved into one that encompasses an understanding of the history of humanity, together with scientific knowledge and intellectual attitudes. This changing concept has prompted a subsequent reevaluation of what constitutes the outstanding universal values of World Heritage sites and the operational methods for implementing the UNESCO World Heritage Convention (The World Heritage Convention 1972). The scope has broadened from studying a single monument in isolation to one that values a multidimensional, multiregional, and interdisciplinary approach and encapsulates vast spans of human history, as demonstrated by the above "TXT-tool 2.380-1.1 Monitoring and Early Warning System of the Building Constructions of the Livadia Palace, Ukraine" materials.

Acknowledgements Results presented herein have been obtained with the financial support from private consulting firm "Center of Science-Engineering Service", Yalta, Crimea, Ukraine (owner Michail Rijii). "Center of Science-Engineering Service" have been financing preparatory and installation works in 1998–2008 years on the first stages of project. National Academy of Science of Ukraine, the Institute of Telecommunications and Global Information Space have been financing next stages of the project in 2009–2014 years. City Office for Construction of the City, Utility Services and Transport, City of Yalta enabled and financed supply of the system with electricity from public network. These supports are gratefully acknowledged.We would like to express our very great appreciation to the Ukrainian scientist from National Technical University "KPI" for consultations during the planning and development of the monitoring systems. Our special thanks are extended to our colleagues Mr. Iurii Gukovskii, Mr. Ruslan Litvinenko and Mr. Viktor Kochin for their support in the site and cabinet work.We wish to thank numerous citizens and the City of Yalta for their contribution to this project, by enabling use of private and City's land for measurement stations establishment.

References

Anosova L, Korobanova I, Kopylova A (1976) Common factors of landslide deposits properties formation. «Science». Moscow, p 182 c

Baum RL, Godt JP, Harp EL, McKenna JW, McMullen SR (2005) Early warning of landslides for rail traffic between Seattle and Everett, Washington, USA. In: Hungr O, Fell R, Couture R, Bernhard E (eds) Landslide risk management, Proceedings of the 2005 international conference on landslide risk management. A.A. Balkema, New York, pp 731–740

Highland L, Bobrowsky P (2008) The landslide handbook—a guide to understanding landslides: reston. U.S. Geological Survey Circular, Virginia, p 129

Huston D (2011) Structural sensing, health monitoring, and performance evaluation, p 619

Kaliukh I (1996) Experimental monitoring system of vibration of building constrictions of IY energetic block of Chernobyl atomic station. In: Vibrations in physical systems XYIth sump.-Poznan, Poland

Kaliukh I (1999) Application of modern information technologies, mathematical methods and measure instruments for registration and forecasting evolution of the process in the dangerous areas (in Russian). Society "Knowledge", Kiev, p 64

Karnawati D, Teuku F, Wahyu W, Syamsul M (2013) TXT-tool 4.062-1.1 A socio-technical approach for landslide mitigation and risk reduction [Online]. http://iplhq.org/icl/wp-content/uploads/2015/07/2-3-TXT-Tool-4.062-1.1-A-Socio-Technical-Approach-for-Landslide-Mitigation-and-Risk-Reduction.pdf

Lacasse S (2013) 8th Terzaghi oration protecting society from landslides—the role of the geotechnical engineer. In: Proceedings of the 18th international conference on soil mechanics and geotechnical engineering, Paris, pp 15–34

Lollino G (2014) In: Lollino G (ed) Engineering geology for society and territory IAEG XII congress volumes, vol 8643, p 8 volume-set

Malyuk Ю (2006) Development of hazardous geologic processes within the Central Livadia Landslide system. Protection and conservation of the Livadia Palace and Park Complex. In: Malyuk YA, Salomatin VN (eds) Construction and anthropogenic safety. Collection of scientific papers—Simferopol: NEPRD.—Issue 13–14, pp 205–212

Margottini C, Vilimek V (2014) The ICL network on "landslides and cultural & natural heritage (LACUNHEN)". Landslides 11:934–938. ISSN:1612-510X

Migon P (2013) Cultural heritage and natural hazards. In: Bobrovsky (ed) Encyclopedia of natural hazards. Springer Science + Business media, Dordrecht

Mihalic S, Mihalić A, Željko A, Martin K (2013) TXT-tool 2.385-1.2 A comprehensive landslide monitoring system: the Kostanjek landslide, Croatia [Online]. https://scholar.google.fr/citations?view_op=view_

citation&hl=ru&user=7eoSoYYAAAAJ&citation_for_view=7eoSoYYAAAAJ:D03iK_w7-QYC

Nagayama T, Spencer B (2007) Structural health monitoring using smart sensors, p 179

Reid M, LaHusen R (1998) Real-time monitoring of active landslides along highway 50, El Dorado County. Calif Geol 51(3):17–20

Sassa K (2005) Landslides: risk analysis and sustainable disaster management [Online]. http://www.ebook3000.com/Kyoji-Sassa-Landslides-Risk-Analysis-and-Sustainable-Disaster-Management-147509.html

Sheko A, Dvortzov P, Yerysh I, Korzhenevskiy I (1979) Analysis and prediction of the landslide processes. The South Coast of the Crimea. In: Sheko AI (ed) Prediction of exogenous geologic process on the USSR Black Sea coast. Nedra, Moscow, pp 140–146

The World Heritage Convention (1972) http://whc.unesco.org/en/convention/

Trofymchuk O (2012) IPL-153 Project "Landslide protection structures and their development in the Autonomous Republic of the Crimea, Ukraine"

Trofymchuk O, Kaliukh I, Hlebchuk H, Berchun V (2013) Experimental and analytical studies of landslides in the south of Ukraine under the action of natural seismic impacts. In: Earthquake-induced landslides. Proceedings of the international symposium on Earthquake-induced landslides. Kiryu, Japan, 2012. Springer, Berlin, Heidelberg, pp 883–890

Yerysh I, Kulish E (1998) About seismic factors role in landslides formation within the South Coast of the Crimea. Reports «Construction in seismic regions of Ukraine»—City of Yalta, State Construction Committee of Ukraine, pp 147–151

TXT-tool 2.039-4.2
LEWIS Project: An Integrated System for Landslides Early Warning

Pasquale Versace, Giovanna Capparelli
and Davide Luciano De Luca

Abstract

In the framework of the National Operational Programme 2007-13 "Research and Competitiveness", co-funded by the European Regional Development Fund, the Ministry of Research (MIUR) financed the project "AN INTEGRATED SYSTEMS FOR HYDROGEOLOGICAL RISK MONITORING, EARLY WARNING AND MITIGATION ALONG THE MAIN LIFELINES" with the acronym LEWIS (Landslides Early Warning Integrated System). The project aims to develop an integrated, innovative and efficient solution to manage risk issues associated with infrastructure, on landslide-prone slopes by developing and testing a system able to identify potentially dangerous landslides in a timely manner, and to activate all needed measures for impact mitigation, including information delivery. The system includes many components: standard criteria for evaluation and mapping landslides susceptibility: monitoring equipment for measuring the onset of landslide movement; telecommunication networks; mathematical models for both triggering and propagation of landslides induced by rainfall; models for risk scenario forecasting; a centre for data acquisition and processing; and a traffic control centre.

P. Versace (✉) · G. Capparelli · D.L. De Luca
Department of Informatics, Modelling, Electronics and System Engineering, University of Calabria—Ponte Pietro Bucci, 41B Building, 5th Floor, 87036 Arcavacata di Rende, CS, Italy
e-mail: pasquale.versace@unical.it

G. Capparelli
e-mail: giovanna.capparelli@unical.it

D.L. De Luca
e-mail: davide.deluca@unical.it

K. Sassa et al. (eds.), *Landslide Dynamics: ISDR-ICL Landslide Interactive Teaching Tools*, https://doi.org/10.1007/978-3-319-57774-6_38

Keywords
Landslide forecasting · Early warning systems · Risk scenarios

Contents

1 **Introduction** 510
2 **Landslide Susceptibility Assessment** 513
3 **Displacement Measurement** 516
3.1 Areal Monitoring Systems 516
3.2 Point Displacement Measurement Systems 517
4 **System for Displacement Forecasting** 519
5 **Data Transmission Network** 530
6 **Data Collecting and Processing Center (CAED)** 531
7 **Control Centre for Road Networks** 531
8 **The Intervention Model** 532
9 **Conclusions** 534
References 534

1 Introduction

In areas where landslide risk is very high and financial resources are scarce, an integrated approach that combines structural and non-structural measures and a complex strategy is necessary. It includes qualitative and quantitative risk analysis, monitoring, advanced early warning systems, mathematical modelling of rainfall-landslide relationships, decision-making procedures, a strategy for risk reduction measures, and plans for emergency management, works and maintenance.

In recent years, attention has been focused more and more on early warning systems, by developing both single components and integrated systems (Sassa and Yueping 2010). An early warning system aims to ensure the provision of timely and effective information that allow exposed people to make decisions in order to avoid or reduce damages and the loss of life (Intrieri et al. 2013). The safety actions are efficient when they are developed during the "lead-time", that is the time interval between the moment of the event prediction and the moment of the landslide impact.

A landslide early warning system, in its general configuration, should include landslide susceptibility maps for the investigated areas, scenarios for the impact of the events on exposed people and goods, monitoring of key parameters, real time data transmission, mathematical models and data processing for both current hazard evaluation and future hazard forecasting, a warning model, an emergency plan in order to avoid or reduce the damages and the loss of life, and a decision-making procedure.

In the design of these systems the velocity of the expected landslides plays a very important role; as it affects not only landslide destructiveness but also the procedures for risk mitigation, as velocities can range from some tens of meters per second to some millimetres per year.

From a general point of view, there are four crucial moments in landslide early warning systems: precursor forecasting (t_1), precursor occurrence (t_2), event onset (t_3) and impact on people and goods (t_4).

Early warning systems are quite efficient when the lead-time, corresponding to the time interval (t_4–t_3) is sufficiently long to make decisions and take actions such as evacuation or protection of structures and infrastructures. When the time between the event onset and its impact (t_4–t_3) is extremely short, the early warning procedure must be based on precursor measurements, when possible. This is the case for rapid landslides, as the time elapsing between the onset of slope failure and its impact on exposed goods is typically in the order of tens of seconds or a few minutes. When the time between precursor occurrence and event onset (t_3–t_2) is also short,

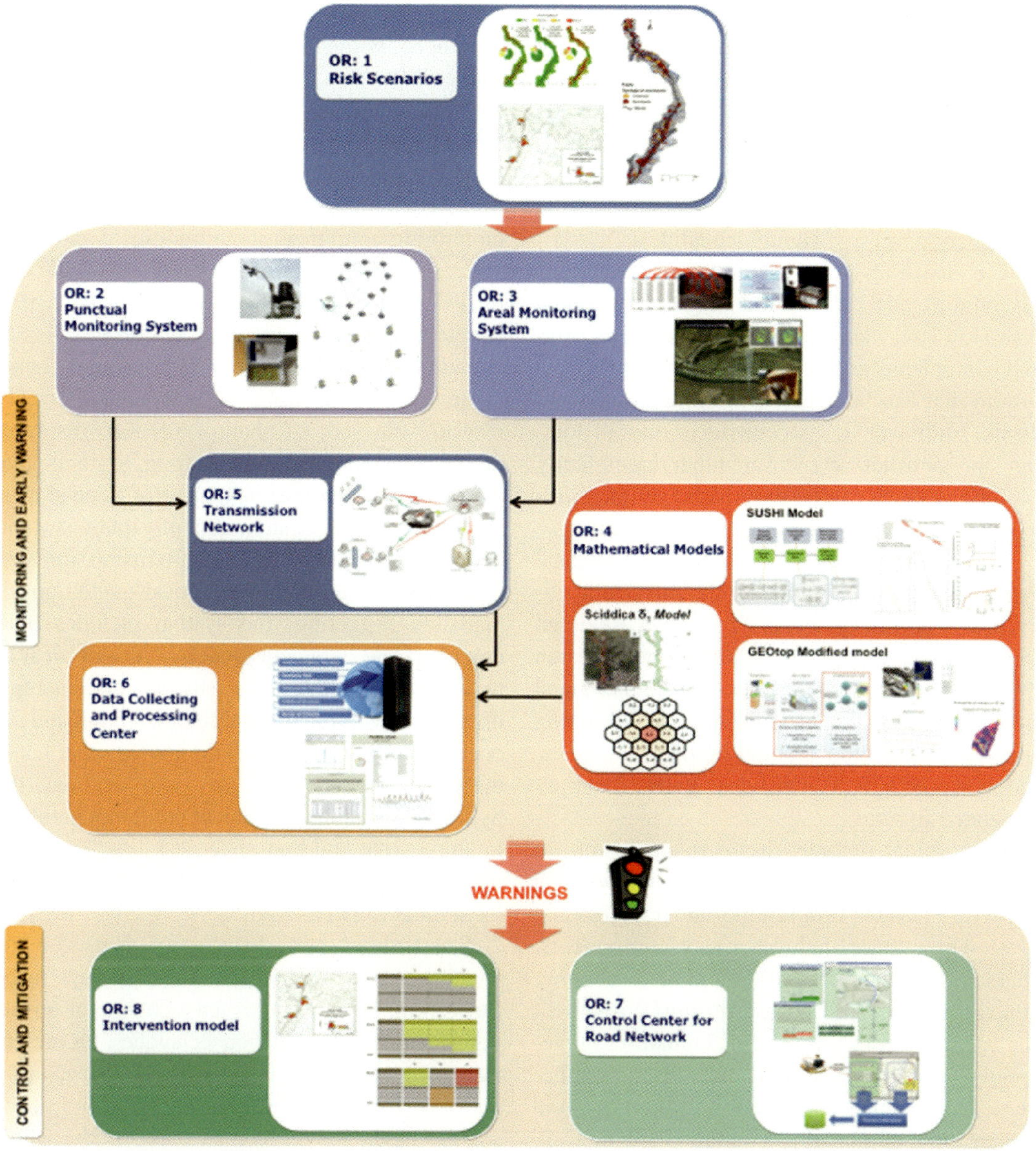

Fig. 1 Interrelation among the different research components

the forecasting of the precursor becomes essential. This is the case for shallow landslides, for which the difference (t_3–t_2) is in the order of tens of minutes or few hours. Then, depending on the landslide velocity, the system can address movement detection for "slow" landslides or forecasting, with mathematical models, the movement onset for "rapid" landslides.

This teaching tool describes an integrated system for landslide early warning that considers both slow and rapid movements, develops original components and integrates them in a flexible system that can be adapted to different environments. Moreover, it also considers rainfall forecast for landslide trigger. In other cases, early warning systems can be only based on measures of displacements, which cannot be forecasted.

In the framework of the National Operational Programme 2007-13 "Research and Competitiveness", co-funded by the European Regional Development Fund, the Ministry of Research (MIUR) financed the project "An Integrated System for Landslide Monitoring, Early Warning and Risk Mitigation along Lifelines", with the acronym LEWIS (Landslides Early Warning Integrated System).

The project includes industrial research, at site tests and training activities, with a two-year master programme at postgraduate level. The system includes many components: standard criteria for evaluation and mapping landslide susceptibility; monitoring equipment for measuring the onset of landslide movement, telecommunication networks, mathematical models for both triggering and propagation of landslides induced by rainfall, models for risk scenario forecasting, a centre for data acquisition and processing, and a traffic control centre.

In Fig. 1, a flow chart showing the interrelation among the different research components is outlined. The integration of the system allows for maximizing its operational flexibility, as each developed component provides different interchangeable technological solutions. Therefore, the final system may assume many different configurations, from the simplest to the most complex, to deal with different scenarios. The flexibility essentially depends on the wide range of monitoring equipment, both traditional and innovative, that have been considered and on the different kind of mathematical models that have been developed. In particular, six monitoring schemes are adopted here: three point-measuring systems, made up of a network of sensors that locally measure the start of shallow or deep displacements, and three area-measuring systems that remotely measure the movement of large slopes. All the monitoring systems are fully integrated and are connected to the same data transmission system. Concerning the mathematical modelling for landslide triggering, the system includes both empirical and complete models. The empirical or hydrological models are simple relationships, obtained by linking the antecedent rainfall and the occurrence time of a landslide. On the other hand, the complete models take into account the hydrological and geotechnical processes involved in slope scale and that affect stability. The complete models, adopted in this research, include local models and areal models.

Standardized and shared procedures for the identification of risk scenarios, for surveys to be carried out, for procedures for each type of on-site testing, for data assimilation techniques, for presentations of results, such as maps of risk along a highway, landslides susceptibility maps and so on, are the bases for the compactness of the whole system.

The setting up of the data acquisition and processing centre and of the traffic control centre are the core of the integrated system. The CAED ("Centro Acquisizione ed Elaborazione Dati"—Data Collecting and Processing Center) acquires

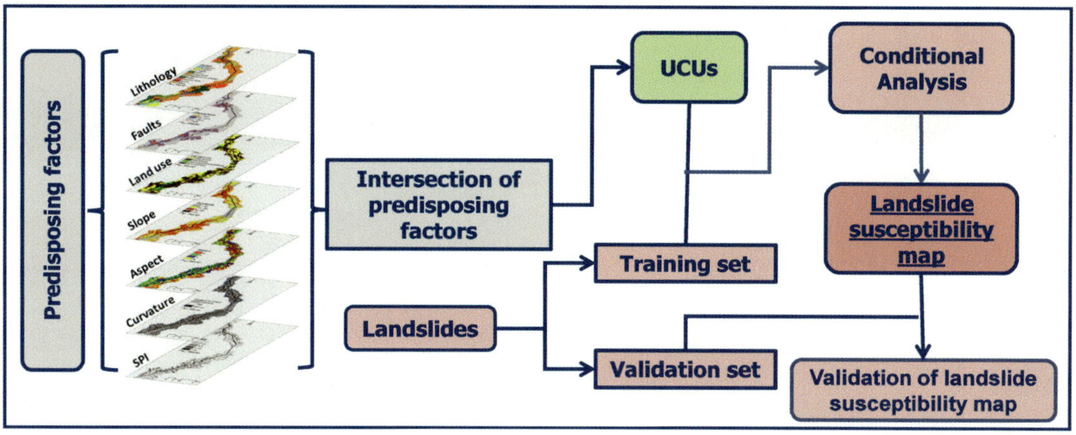

Fig. 2 Flow chart showing the landslide susceptibility method

and processes data that are tremendously variable in intensity, dimensions, characteristics and information content. The Control Center for Road Network (CCC) is meant to integrate the scientific and the management aspects of hydrological risk monitoring and early warning.

Tests for experimentation and validation of the system have been carried out in three highway segments, related to Campania, Puglia, Calabria and Sicily, which are Italian Regions interested in the Community Support Framework.

In the following chapters the LEWIS components will be described:

- Landslide susceptibility assessment;
- Displacement measurement;
- System for displacement forecasting;
- Data transmission network;
- Data Collecting and Processing Center (CAED);
- Control Center for Road Network (CCC);
- Intervention model.

The last topic is developed in greater detail (see also Versace et al. 2014). In fact, the intervention model is an important component for the correct functioning of an integrated system such as the one here described, which must combine the information coming from different sensors and models.

For the sake of brevity, experimental sites, and the related results, are not described in this teaching tool. For more details, please visit the website www.camilab.unical.it.

2 Landslide Susceptibility Assessment

In this study, the predisposing factors were selected from the most commonly used in literature (VanWestern et al. 2008) and based on the geological and geomorphological settings of the

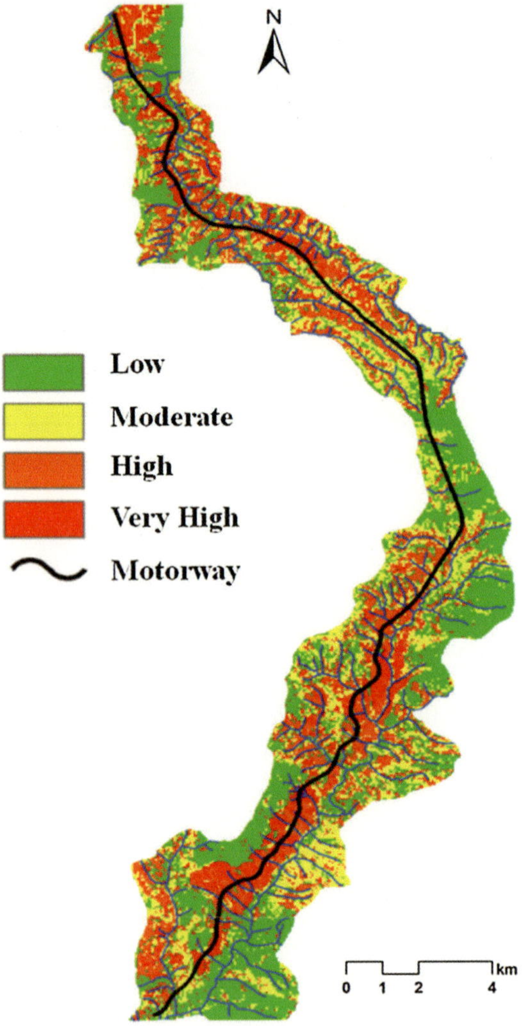

Fig. 3 Landslide susceptibility map of the study area

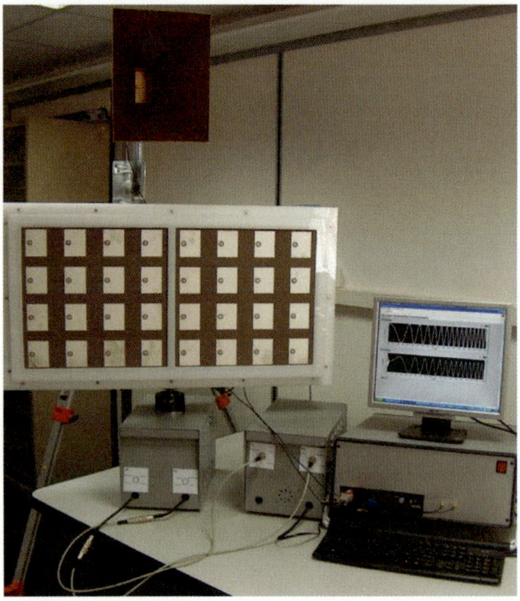

Fig. 4 L-Band radar system

study areas. The used data layers (Fig. 2) were transformed in raster format.

The land-slide susceptibility was performed on the basis of a 'Conditional Analysis', a statistical method, applied to subdivide territory into Unique Condition Units (Carrara et al. 1995). In this method, landslide susceptibility is expressed as a landslide density in correspondence with different combinations of predisposing factor classes (Clerici et al. 2006).

The thematic layers were combined in order to obtain all the possible combinations of the various classes of the different predisposing factors. Each specific combination represents a Unique Condition Unit (UCU). Their number and size depend on the criteria used in classifying the predisposing factors. Subsequently, the landslide presence, represented by the landslide area, is determined within each UCU and the landslide density is computed. Assuming that landslides are more likely to occur under those conditions which led to slope-failure in the past, the computed landslide density is equivalent to the future probability of occurrence. The conditional probability is given by:

$$P(L|UCU) = landslide\, area/UCU\, area \quad (1)$$

i.e., the probability of landslide occurrence (L), in an unique combination of factors (UCU), is

(a) (b)

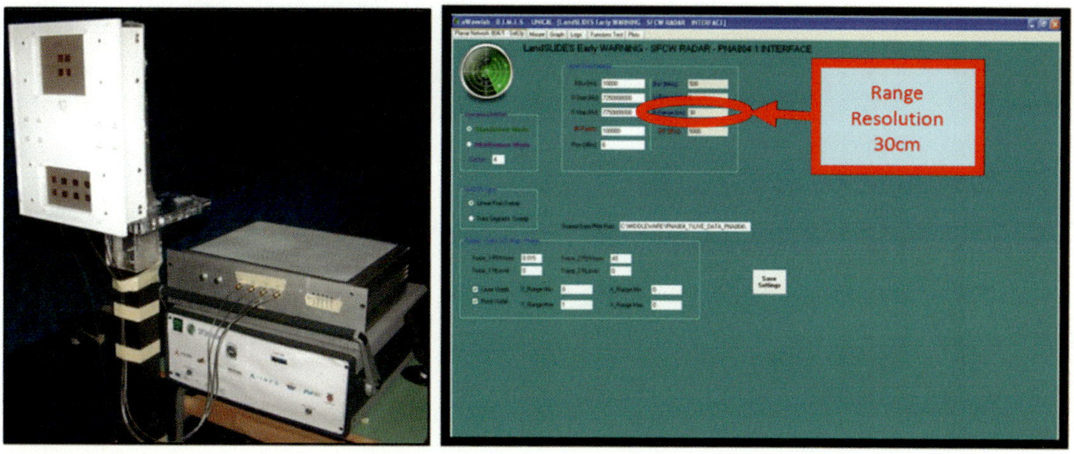

Fig. 5 **a** Radar system and **b** software interface

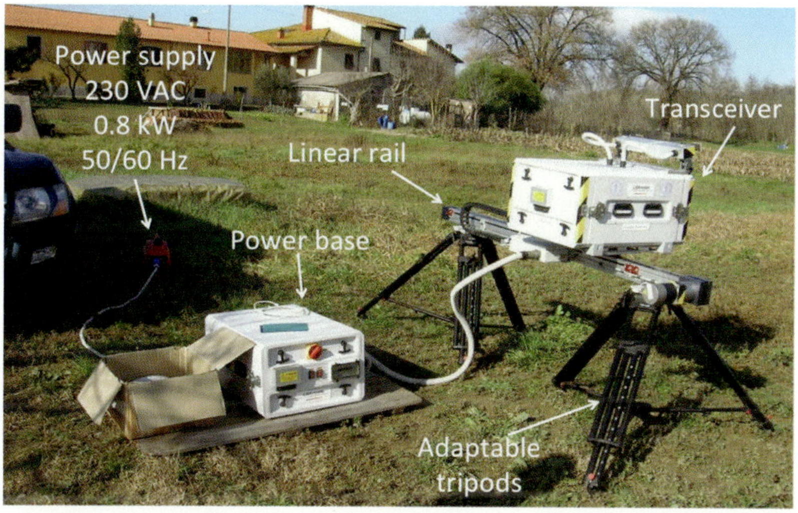

Fig. 6 Main components of the GB-InSAR system

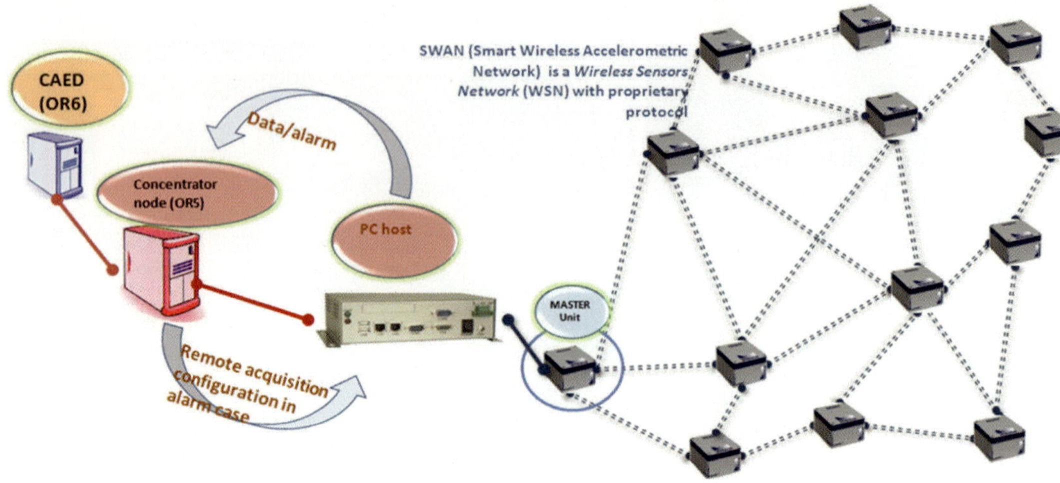

Fig. 7 SWAN system general scheme—communication with CAED via concentrator node

given by the landslide density in that specific UCU. Landslide density in each UCU was computed and the susceptibility map was realized. Of course, this approach only considers the occurred landslides in the inventory database.

As an example, an application of the described methodology is shown in Fig. 3 for a part of motorway "A3, Salerno-Reggio Calabria", between Cosenza Sud and Altilia, northern Calabria (Italy).

3 Displacement Measurement

3.1 Areal Monitoring Systems

The Microwave Lab at University of Calabria developed two compact and low-cost radar configurations (Costanzo et al. 2013), the first one based on the adoption of a software radio platform, the second one using a compact Vector Network Analyzer as a Stepped Frequency Continuous Wave (SFCW) scatterometer module. The Department of Earth Sciences at University of Firenze proposed a portable and versatile Ground-Based Interferometric Synthetic Aperture Radar (GB-InSAR).

The primary goal of the L-band software defined radar is the possibility for the radar signal to go over the possible vegetation layer on the mountain under observation; this justifies the choice of the L-band, able to utilize the wave penetration feature. The radar hardware is depicted in Fig. 4. However, the limit of this radar is the coarse azimuthal resolution, which can be improved by increasing the hardware and software features of the system.

A Stepped-Frequency Continuous-Wave (SFCW) radar has been constructed by adopting a compact Vector Network Analyzer, controlled by a C# .NET interface, and connected through a switch module and a Butler matrix to transmitting and receiving microstrip array antennas, in order to achieve an azimuthal scanning capability able to select a specific investigation area. To capture an entire scene, the scan of four different areas, illuminated by the same microstrip array antenna, is required. Since an operator is not constantly present at the installation site, a framework has been developed for the beam switching in reception and the remote switching on and off of the elements. This allowed us to create a system able to optimize a wide range of goals, such as reduced power consumption, size limits, performance, reliability and cost constraints. A picture of the radar system is illustrated in Fig. 5(a), while the software interface is shown in Fig. 5(b). Concerning the output, from vertical stripes of approximately 90 m, the system provides information on the presence or absence of azimuthal displacements (grouped into size classes of 30 cm), and on the rate of mobilized strip.

The Ground-Based Interferometric Synthetic Aperture RADAR (GB-InSAR, Fig. 6) is able to detect submillimetre displacements for an area up to a few hectares wide. The system is composed of a transceiver, a 2.1 m long linear rail and two antennas moving on it in order to obtain a maximum synthetic aperture of 1.8 m.

The instrument should be placed in front of the selected target at a distance ranging between a few dozen of meters and few kilometers. Spatial resolution is a function of this distance, while precision is a function of employed wavelength (Del Ventisette et al. 2011).

3.2 Point Displacement Measurement Systems

Displacements of terrain and/or other structures at selected points can be monitored, for landslide early warning, including underground troughs and surface point measurement systems. We can obtain significant information by using a wide range of instruments (total stations, Global Navigation Satellite System (GNSS) receivers, opto-mechanical systems, Micro-Electro-Mechanical Systems (MEMS) sensors); underground systems, in particular, imply the drilling of the soil and the installation of inclinometric tubes for every monitored point. The LEWIS project uses both conventional instruments (piezometers, inclinometers, stress cells, time-domain reflectometers (TDRs), meteorological sensors, etc.) and innovative techniques. In the following, the discussion will concern only the latter.

The STRAGO company optimized in the project framework a wireless network, named SWAN (Smart Wireless Accelerometric Network, Fig. 7), consisting of inclinometer units based on MEMS accelerometers (SMAMID unit) for continuous monitoring of landslide survey surface movements, for the purpose of warning/alarm in case of critical events. All system components (hardware, firmware and software) have been improved, increasing computing power and energy savings.

A middleware node concentrator permits data communication from the website (lab/field) to the CAED. The system acquires data with a specified frequency and duration, which are remotely

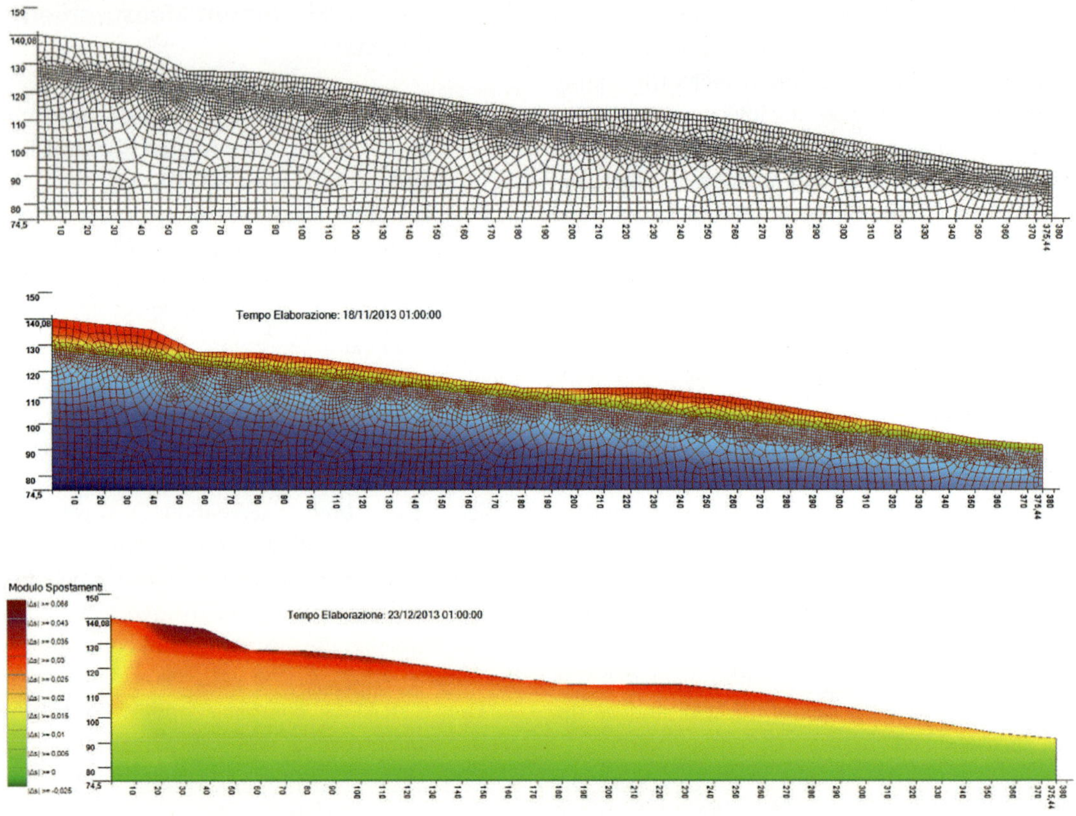

Fig. 8 Example of SUSHI output. Top to bottom, (1) image of the discretization of domain, (2) slope section with color bands indicating pore water pressure distribution, (3) slope section with color bands indicating displacement performed

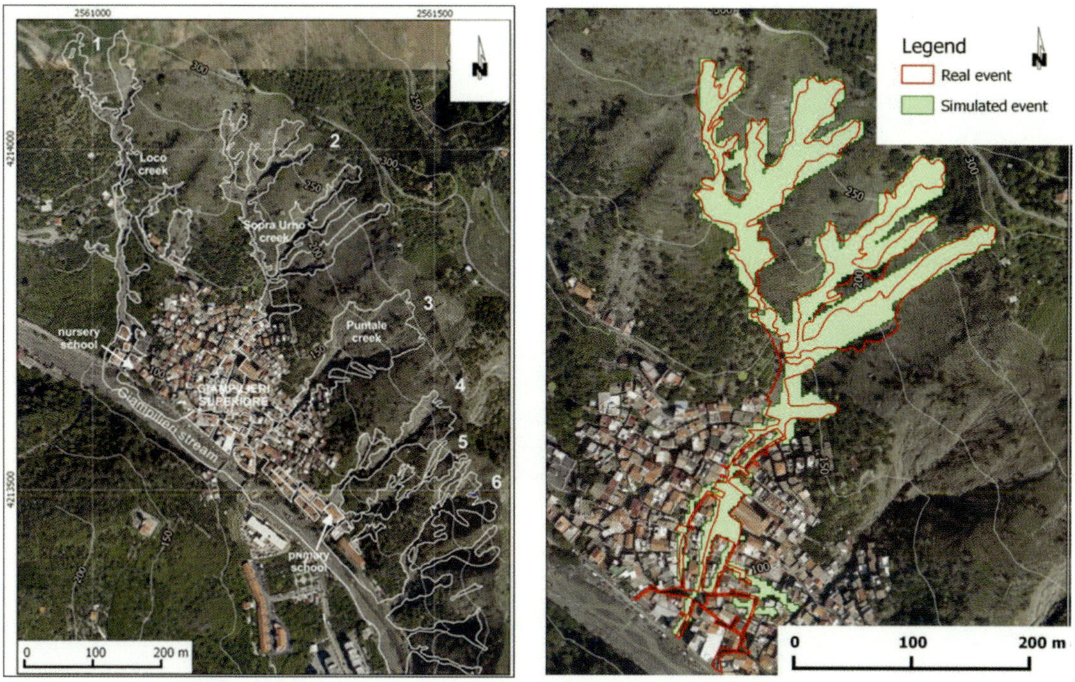

Fig. 9 Left—October 2009 debris flows occurred in Giampilieri Superiore, paths obtained by interpretation of aerial photos. Right—Comparison between Sopra Urno creek debris flows and a simulated event

configurable, and sends a data file containing information concerning each inclinometer measurement point and the landslide status, based on criteria editable by remote control.

The Geomatics Lab of University of Calabria has produced an integrated sensor for position and inclination measurement and monitoring (POsition and Inclination Sensor POIS), characterized by small size (Artese et al. 2015), low weight, low power consumption and low costs.

4 System for Displacement Forecasting

In the context of the modeling and forecasting displacement, a section of the project was devoted to the application of hydro-mechanical models. Three different types of models are used to deal with the analysis on different levels. Models are different for the structure and objectives of analysis.

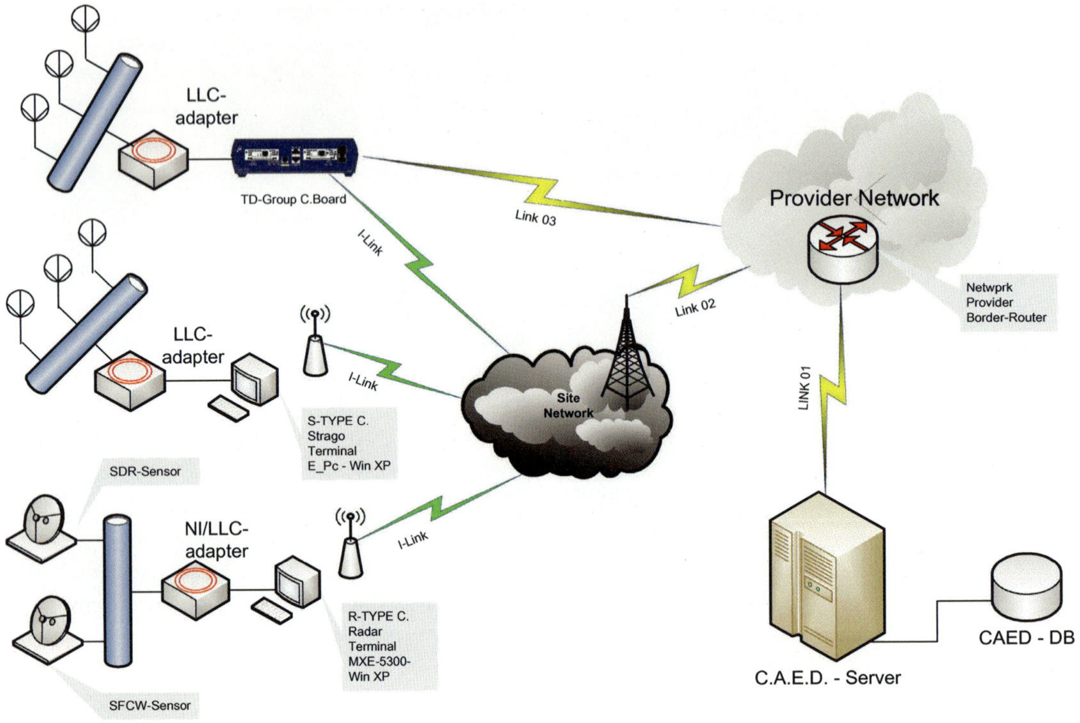

Fig. 10 Network architecture

The first model is named SUSHI (Saturated Unsaturated Simulation for Hillslope Instability), it is applied at slope scale and considers the relation between rainfall infiltration, pore water pressure and slope stability, taking into account several components, including specific site conditions, mechanical, hydraulic and physical soil properties, local seepage conditions, and their contribution to soil strength. Moreover, the model was developed in order to be suitable for cases with strongly heterogeneous soils, irregular domains and variable boundary conditions that vary in space and time. SUSHI is based on the combined use of two modules: an hydraulic module, to analyze the subsoil water circulation due to the rainfall infiltration under transient conditions, and a geotechnical module, which provides indications regarding the slope stability. With regard to the hydraulic module, variably saturated porous media flows were modeled by the classical nonlinear Richards equation:

$$\frac{d\theta(\psi)}{d\psi}\frac{\partial\psi}{\partial t} = C(\psi)\frac{\partial\psi}{\partial t} = \nabla[K(\psi)\nabla\psi] + \frac{\partial K}{\partial z} \tag{2}$$

and closed by constitutive relations describing the functional dependence of moisture content

C and hydraulic conductivity K on the pressure head ψ. Due to the high nonlinearity in the constitutive relations, analytical solutions of Richards equations rarely exist except for a limited number of simple configurations. For this reason, the Galerkin-type finite element method was used. In the geotechnical module the equations to be solved are the following:

$$\frac{\partial \sigma'_{ij}}{\partial x_j} + \frac{\partial u_w}{\partial x_i} + \gamma \delta_{iz} = 0 \quad (3)$$

where σ'_{ij} are the stress components, u_w is the pore pressure calculated from the hydraulic module, γ is the soil unit weight and δ_{iz} is the Kronecker symbol. The linear constitutive equations (plane stress) and strain-displacement relationship also must be taken into account. The displacement formulation is considered in the present approach in which stress and strain are eliminated, resulting in differential equations in which the displacements are the variables. Then, discretization and application of Galerkin's method leads to the stiffness equations for a typical element. To solve the equation over a generic domain an assembly strategy is chosen, leading to global algebraic linear equations $[K_G]\{U\} = \{F\}$ where $\{U\}$ and $\{F\}$ are the nodal variables (displacements) and known force components (gravity loading and seepage loads) respectively. The model takes into account material non linearity using constant stiffness iterations, in which non linearity is introduced by iteratively modifying the loads vector $\{F\}$: the loads vector at each iteration consists of externally applied loads and self-equilibrating loads that have the effect of redistributing stresses within the domain. The Mohr-Coulomb criterion was chosen to represent the yield function and associated flow was assumed. Also the self-equilibrating loads were calculated using an initial strain method. By means of the presented geotechnical module it is possible to assess:

- the safety factor of the slope subjected to gravity loading and to the pore pressure calculated from the hydraulic module;
- displacement, strain and stress under the effect of rainfall infiltration.

Another model is of areal type, named Geotop (Rigon et al. 2006), that couples a hydrological and a geotechnical module for the computation of the shallow landslide safety factor under the assumption of an infinite slope hypothesis. The hydrological module concerns a three-dimensional (3-D), physically based, spatially distributed model that performs water and energy budgets at pixel scale. It performs subsurface saturated and unsaturated flows, surface runoff, channel flows, and turbulent fluxes across the soil-atmosphere interface (e.g., latent and sensible heat fluxes, soil temperature, etc.). Tools for parameter calibration are used in order to estimate parameters of the soil water retention curve by comparing simulated suction and soil water content with those coming from the in situ sensors along test site of the project. The results are moisture, soil suction, and water table depth maps at different soil depths of the digital watershed model where it is applied. The geotechnical module uses these outputs and parameters such as soil friction angle and cohesion, root cohesion and local slope, in order to provide infinite slope safety factor raster maps for each soil layer for all the computational domain. The system's components are integrated by using the modelling framework Object Modeling System 3.0, which includes many model facilities such as tools for model output visualization, algorithms for model parameters calibration and sensitivity analysis.

Finally, a third model deals with the analysis of the debris flow propagation. It follows the innovative guidelines of the Cellular Automata (CA) methodology to develop efficient models for simulating complex dynamic systems, that evolve mainly on the base of local interactions of their constituent parts. Debris flows may be considered such a type of complex systems. Modeling such dangerous phenomena can supply new tools, using computer simulations to evaluate debris flow hazards and the effects of possible remedial works in the considered areas. In our application we use the Giampilieri zone (southern Italy, Peres and Cancelliere 2014), which was devastated in October 1st, 2009 by several catastrophic debris flows, triggered by high intensity rainfall concentrated in a few hours. In this context, SCIDDICA-SS3 (Simulation through Computational Innovative methods for the Detection of Debris flow path using Interactive Cellular Automata—both Subaerial and Subaqueous ones, Avolio et al. 2013) was used, which is a new version model of the SCIDDICA-SS family, that improves approximations regarding momentum conservation (Fig. 8).

The Giampilieri village is located on the eastern slopes of the Peloritani Mountains on the

Table 1 Thickness classification (THI)

Class	Type	Description (m)
THI1	Very shallow	≤ 0.5
THI2	Shallow	≤ 2
THI3	Medium	≤ 10
THI4	Deep	≤ 30
THI5	Very deep	>30

Table 2 Magnitude classification (MAG)

Class	Type
MAG1	Low
MAG2	Moderate
MAG3	High

Table 3 Geometric Index classification (GEI)

Class	Type	Description
GEI1	Very Small	All SUR, SCA, VOL, THI are equal to 1
GEI2	Small	At least a value is equal to 2 among SUR, SCA, VOL, THI
GEI3	Moderate	At least a value is equal to 3 among SUR, SCA, VOL, THI
GEI4	Large	At least a value is equal to 4 among SUR, SCA, VOL, THI
GEI5	Very Large	At least a value is equal to 5 among SUR, SCA, VOL, THI

Table 4 Velocity classification (VEL)

Class	Type	Description
VEL1	Slow	Slides, flows, spreads (<1 m/year)
VEL2	Moderate	Earth and debris flows, complex landslides (1 m/year to 1 m/h)
VEL3	Rapid	Falls and topples, rock slides and debris flows (>1 m/h)

Table 5 Matrix for magnitude estimation (MAG)

	VEL1	VEL2	VEL3
GEI1	MAG1	MAG1	MAG1
GEI2	MAG1	MAG1	MAG2
GEI3	MAG1	MAG2	MAG3
GEI4	MAG2	MAG3	MAG3
GEI5	MAG3	MAG3	MAG3

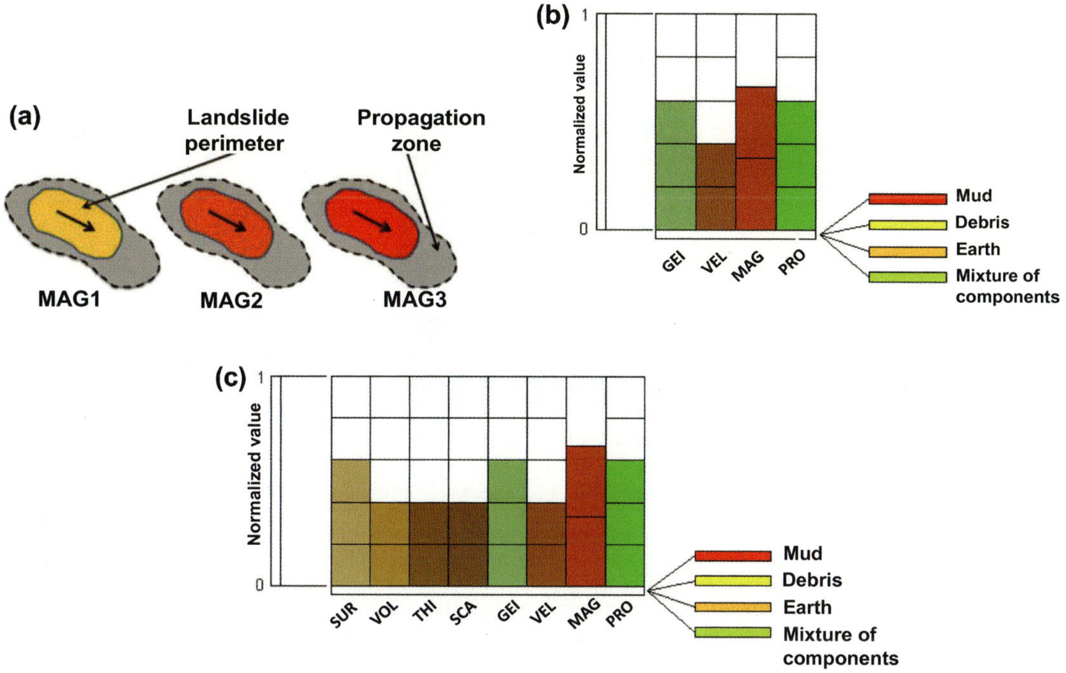

Fig. 11 Map of event scenarios: **a** level 1; **b** level 2; **c** level 3

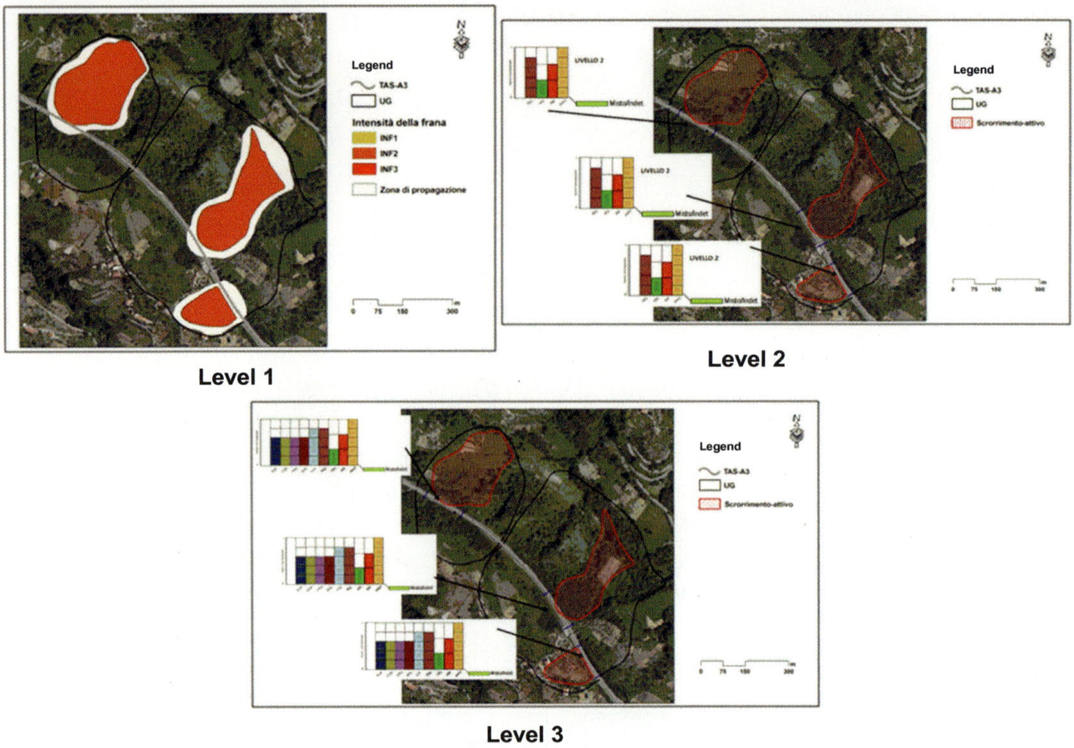

Fig. 12 Map of event scenarios: application for A3 motorway—Mancarelli and Fiego

left side of Giampilieri River. It is settled on ancient alluvial fan and is crossed by various creeks, tributaries of the Giampilieri River. During this 2009 alluvial event, several debris flows were mobilized from the basins behind the town (Fig. 9), and reached Giampilieri Superiore. The mutual interaction between different, nearly simultaneous, debris flows produced dramatic effects in terms of loss of human lives and damages to buildings close to the hill and along the principal streams that cross the town. Crossing the centre, the flows killed 19 people, destroyed houses and dragged away whatever was in their path.

Simulations were performed for the six debris flows that occurred in the Giampilieri area in 2009 (indicated with numbers from 1 to 6 in Fig. 9). In particular, the no. 2 debris flow was used in the model calibration phase (Fig. 9) while the other ones were used for model validation. The results show a good capability of the model to simulate the debris run-out, particularly in the upslope parts of the basins, while in the downslope urbanized area, the reproduction of the real events is less accurate, with significant differences due to lateral spreading. The model behaviour was satisfactory in terms of reproducing the global dynamic of the events, such as velocity, debris flow depth, thickness of deposit, and, in particular, the path of debris flows, that show a good correspondence with the real events. The program could be refined for the reproduction of debris flow propagation into highly urbanized areas, where streets are narrow. This improvement may be obtained by a better

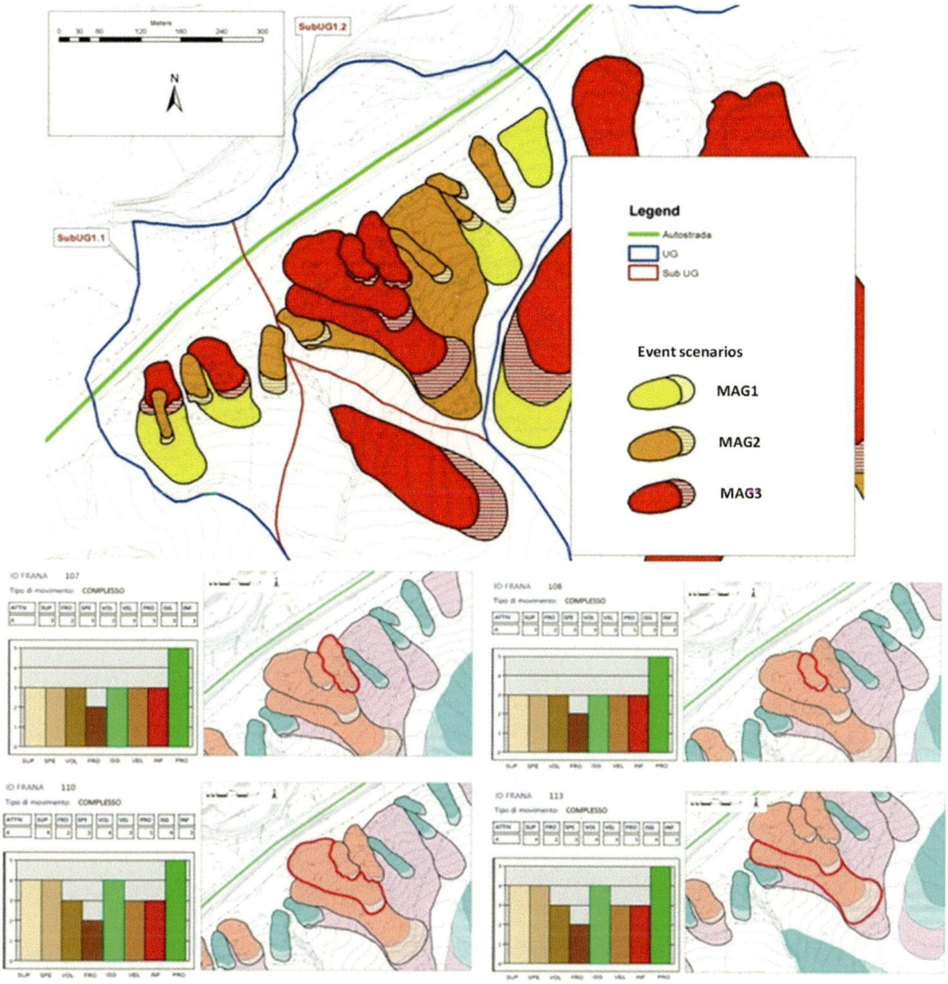

Fig. 13 Map of event scenarios: application for A16 motorway—km112+400

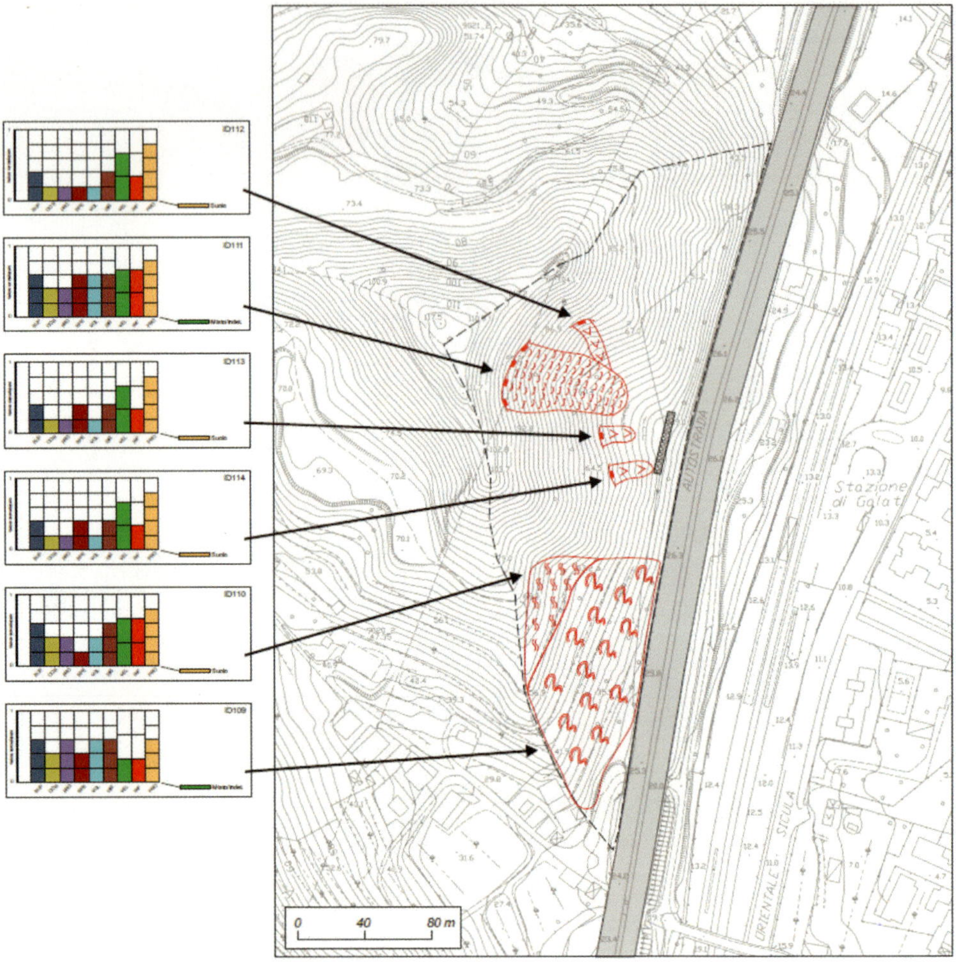

Fig. 14 Map of event scenarios: application for A18 motorway

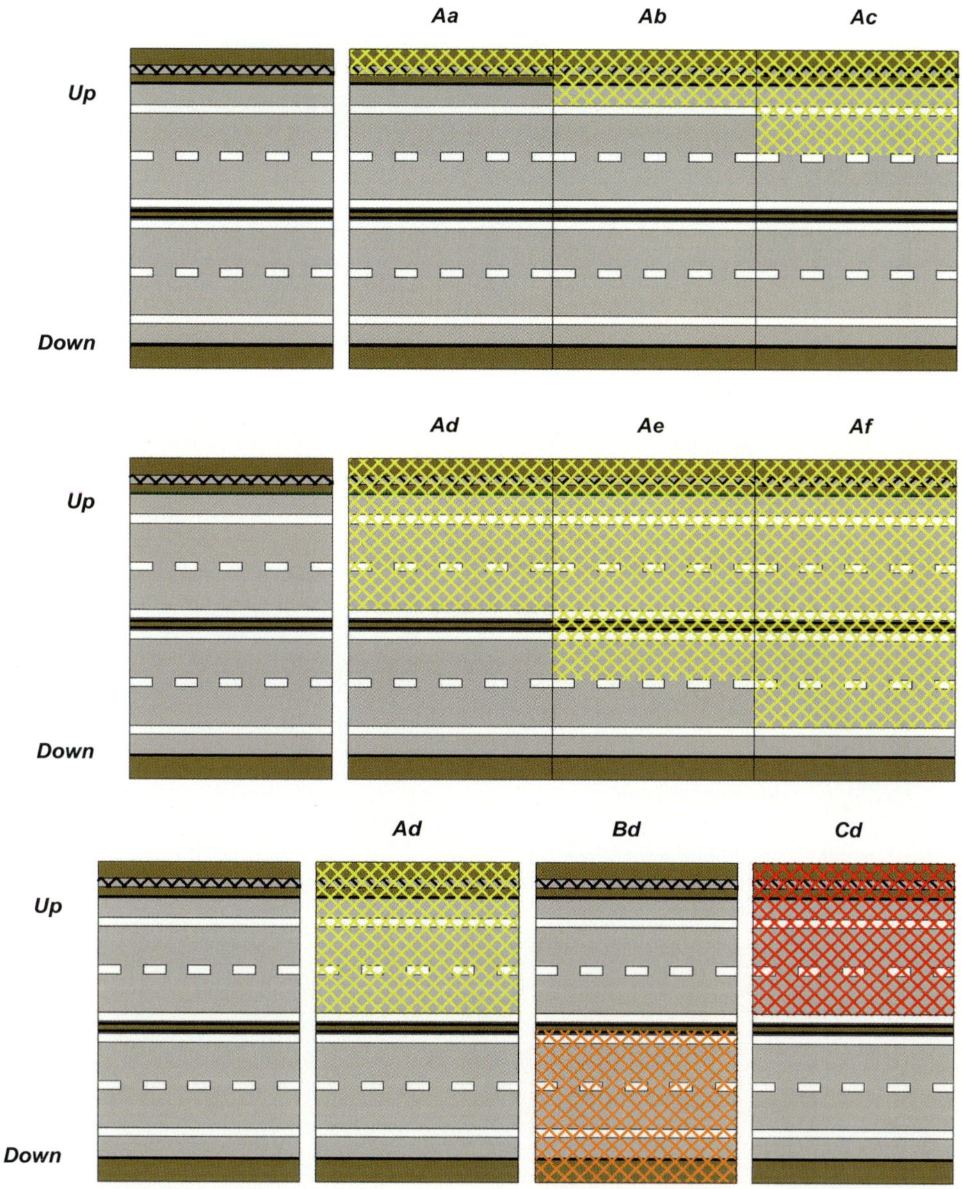

Fig. 15 Risk scenarios (class A: *dotted yellow lines*; class B: *dotted orange lines*; class C: *dotted red lines*)

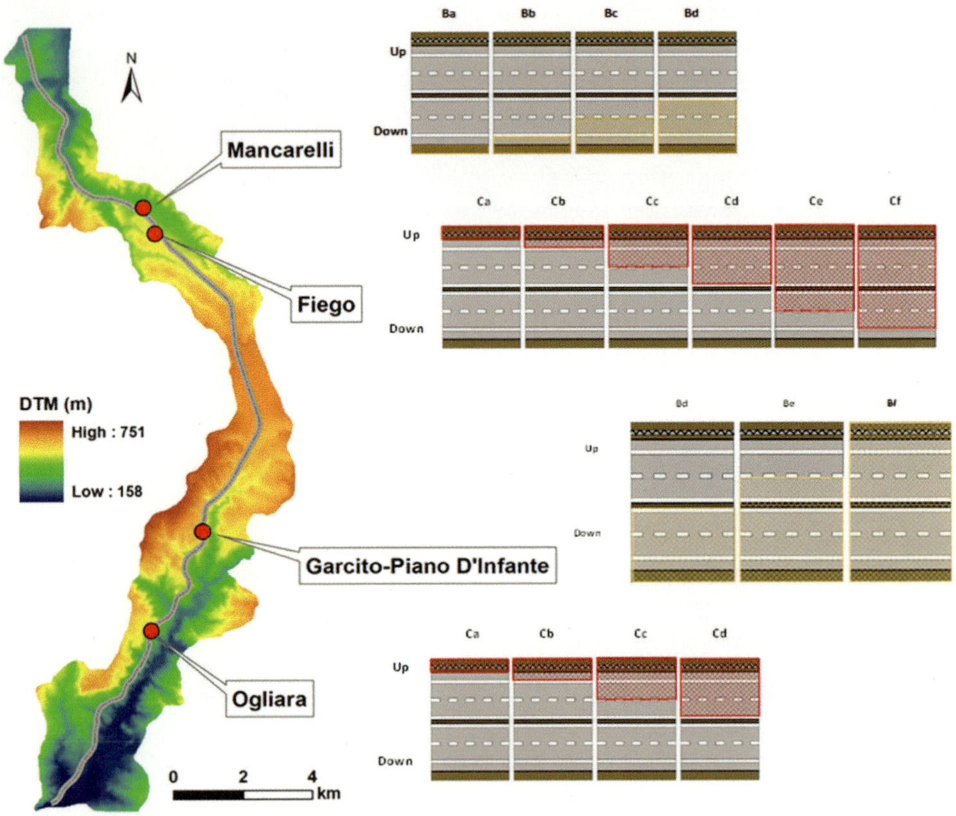

Fig. 16 Risk scenarios—application for A3 motorway

Table 6 CAED possible decisions

State of sensors and/or models	CAED decisions
All INDs and SENs are S0	0—no decision
At least one IND is S1 and all SENs are S0	1—activation of SOD (Sensors On Demand)
At least one SEN is S1	2—to intensify the presence up to 24 h/day
At least *n* SENs are S1 or at least one SEN is S2	3/1—to issue a notice of ordinary criticality (level 1)
At least *n* SENs are S2 or at least one SEN is S3	3/2—to issue a notice of moderate criticality (level 2)
At least *n* SENs are S3	3/3—to issue a notice of high or severe criticality (level 3)

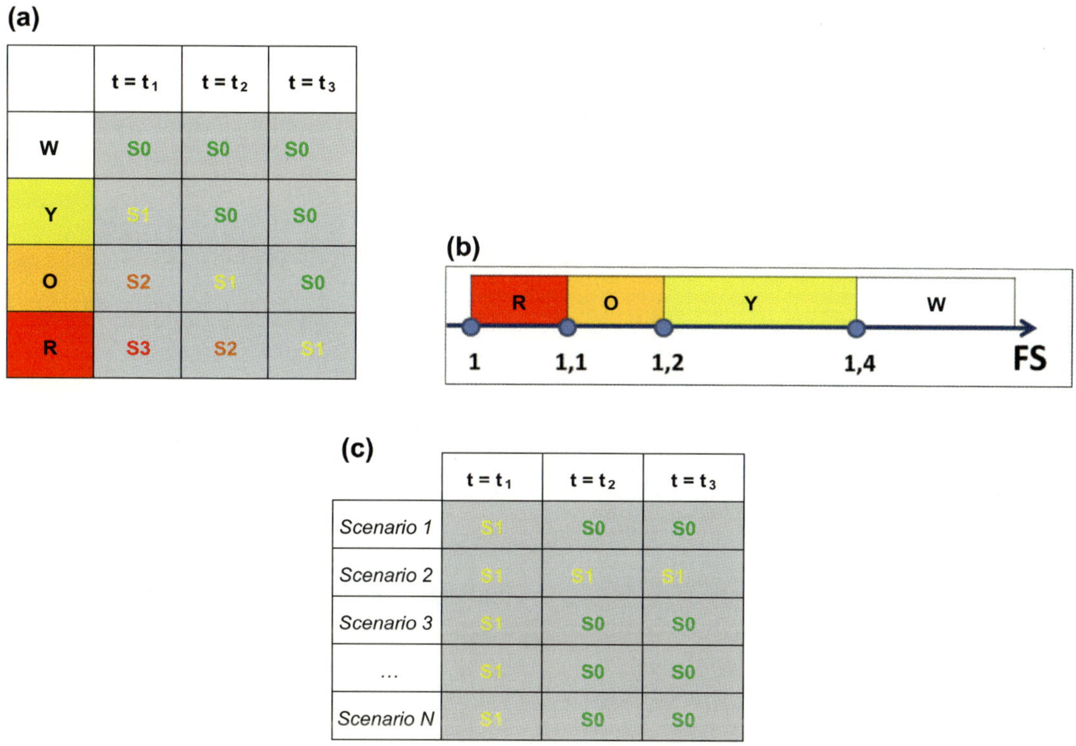

Fig. 17 **a** Matrix for definition of sensor state; **b** values of factor of safety; **c** state of sensor, related to several predicted scenarios

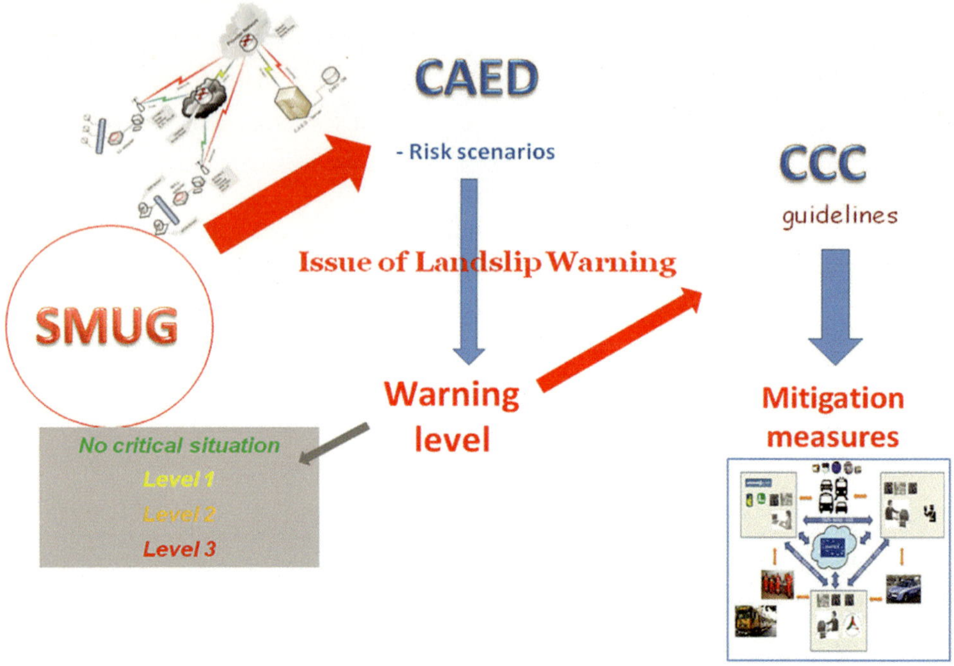

Fig. 18 Relationship among LEWIS components

cell discretization of the computation region, which enables accurate positioning of buildings.

5 Data Transmission Network

A network architecture with hybrid configuration, named LEWARNET ('Landscape Early WARning NETwork', Fig. 10) is used for data transmission from the various sensors involved in the monitoring (e.g., SDRadar, SFCW scatterometer, interferometer, SMAMID sensors) towards the CAED. The chosen topology can be represented by a three-level network, with a connection between the CAED and the monitoring sites through a direct link or by sink nodes.

The network framework includes: (1) a server component working as a software interface, designed in agreement with CAED specifications; (2) an own middleware, including all software components for the management of the sink nodes, namely the monitoring of the client process, the definition of data structures for the storage and the management of data messages, encryption utility and data compression utility; (3) the client software components for each sensor belonging to a sink node, namely the specific software interfaces established in agreement with the sensors designers.

The main task of the middleware implements the following functionalities: (i) 'Warm Up', consisting in the actions relative to the identification of CAED server and active sensors (network and sensor discovery); (ii) 'Messages Queue Manager', devoting to the management of messages queues in input/output from the sensors and towards the CAED; (iii) 'Activities Logger', devoting to the generation of log files reporting all developed activities and activated processes.

6 Data Collecting and Processing Center (CAED)

Management of information flows, telematic architecture and services for data management is entrusted to the Data Collecting and Processing Centre (CAED, namely "Centro Acquisizione ed Elaborazione Dati").

The CAED ensures the continuous exchange of information among monitoring networks, mathematical models and the Command and Control Centre (CCC) that is responsible for emergency management. A communication protocol, implemented by the CAED and named AqSERV, allows for the management of data flow from the monitoring network. AqSERV was designed by considering the heterogeneity of devices of monitoring and transmission networks (point and areal sensors) and the available hardware resources (microcontrollers and/or industrial computers). AqSERV was designed to link the CAED database (named LewisDB) and the monitoring networks, after validation for the authenticity of the node that connects to the centre. Data acquisition, before the storage in the database, is validated both syntactically and according to the information content. The procedures for extraction of the information content and validation were realized differently for point and areal sensors: the latter require a more complex validation, as they work in a 2D domain. The complete management of the monitoring networks by CAED is ensured by specific remote commands, sent to individual devices via AqSERV, to reconfigure the acquisition intervals or to activate any sensor, depending on the natural phenomena occurring in real time. The acquired and validated data are then accessible for the mathematical models through a further service, created ad hoc, which publishes all the acquisitions by sensors on a remote server for sharing. The configuration of the monitoring networks (composed of devices and sensors), the communication protocol used by each network, and the rules for extraction and validation of information content are carried out through a web application, that allows the users to manage the whole system.

7 Control Centre for Road Networks

The Centre, which is devoted to Road Infrastructure Monitoring and Management (CCC, namely "Centro di Comando e Controllo"), manages a monitoring and supervising system. It is responsible for integrating data from hydrogeological risk monitoring and traffic and road condition information, considering relevant alerts, and initiating road maintenance and traffic management operations based on global situation evaluation, which will act based on specifically designed management rules and procedures, derived from the emergency plan.

The CCC is able to activate communication channels with the operating and rescue teams in an automated way, after operator validation of the danger situations recognized by the systems. In addition, it may easily interact with other operating centers responsible for road management and safety (such as Authorities, Road Police, Civil Defence, Viability Patrols, etc.). Automatic Communication has been implemented by machine-generated messages, both on traditional communication means (such as e-mail, telefax, SMS, speech) and by a specific innovation in this framework, by means of DATEX protocol (www.datex2.eu), and up-to-date technology and machine readable protocol.

The functions performed by the CCC: Check the environmental conditions; Monitor traffic status; Detect eventual abnormal conditions and situations; Find out the intervention/information procedures to be applied against certain situations; Trigger the intervention/information procedures (whether directly or actively involved in the operation); Check the progress of the interventions; Check the information delivered on different media.

The organizational and procedural model provides a broad vision of the problem showing several operations centres (police, rescue organizations, road operators) that are able to interact with other operational centres through traditional and up-to-date technology communication means. CCC can operate by acting as a road operational centre that directly activates emergency rescue teams and patrols, and as well may implement the actions required to manage the critical situation. Otherwise, it can operate as a supervisory central, which gets information from other operating centres to oversee, and coordinate several collaborating operating centres, monitoring and triggering management risk and initiating specific rescue and management operations.

8 The Intervention Model

An intervention model is based on the following elements: Event scenarios, Risk scenarios, Levels of criticality, Levels of alert.

Event scenarios describe the properties of expected phenomena in terms of dimension, velocity, involved material and probability of occurrence. Occurrence probability depends on the associated time frame, which should be equal to a few hours at most for Early Warning Systems. Evaluation of occurrence probability is carried out by using information from monitoring systems and/or from outputs of adopted mathematical models for nowcasting. All the properties, to be analyzed for event scenarios, are listed below. A subdivision in classes is adopted for each one:

- Landslide velocity (3 classes from slow to rapid)—**VEL**;
- Landslide surface (5 classes from very small to very large)—**SUR**;
- Landslide scarp (5 classes from very small to very large)—**SCA**;
- Landslide volume (5 classes from extremely small to large)—**VOL**;
- Thickness (5 classes from very shallow to very deep)—**THI**;
- Magnitude (3 classes: low, moderate, high), which combines the previous information—**MAG**;
- Involved material (mud, debris, earth, rock, mixture of components)—**IM**;
- Occurrence probability (zero, low, moderate, high, very high, equal to 1)—**PRO**.

As an example, thickness classification is reported in Table 1.

The landslide magnitude is estimated by considering velocity and a Geometric Index (GEI), below defined, and three classes are identified (Table 2).

In details, GEI synthetizes the landslide dimensions and it is a combination of four parameters (SUR, SCA, VOL, THI); five classes are defined, as reported in Table 3.

Regarding landslide velocity (VEL), the classification of Cruden and Varnes (1996) was considered. In particular, three classes are detailed (Table 4).

Finally, landslide magnitude is computed by using the matrix given in Table 5.

Representation of event scenarios can be carried out by using in synthetic way the hazard or intensity of a phenomenon. In this work, three levels of representation (Fig. 11) are proposed: the first one is only related to intensity and, due to its simplicity, it allows to immediately identify which are the most dangerous phenomena. The other two levels also provide an indication of occurrence probability, and therefore the hazard of the phenomenon.

Some examples of application are reported in Figs. 12, 13 and 14.

Risk scenarios can be firstly grouped in the following three classes:

A. Mud and/or debris movements which could induce a friction reduction and facilitate slips (indicated with dotted yellow lines in Figs. 15 and 16);
B. Road subsidence induced by landslides that could drag or drop vehicles (indicated with dotted orange lines in Figs. 15 and 16);
C. Falls of significant volumes and/or boulders that could crush or cover vehicles and constitute an obstacle for others vehicles (indicated with dotted red lines in Figs. 15 and 16).

For each previous risk scenario, six sub-scenarios can be identified on the basis of the number of potentially involved infrastructure elements, carriageways and lanes: (a) hydraulic infrastructures and/or barriers, (b) only emergency lanes, (c) lanes, (d) fast lanes, (e) fast lanes of the opposite carriageway, (f) lanes of the opposite carriageway).

Thus, there are 18 possible risk scenarios, indicated by a couple of letters (Capital and small, Fig. 15). An application for A3 motorway is shown in Fig. 16.

For identifying the ***levels of criticality***, the CAED acquires measurements from sensors and model outputs. Moreover, CAED identifies four states for each of them: state 0 = no variation; state 1 = small variation; state 2 = moderate variation; state 3 = high variation. These four states were set with heuristic criteria and only the acquisition of experimental data will allow for a better definition of them.

Besides information from sensors and models, meteorological and hydrological models (named as indicators) give other information.

Indicators comprise weather forecasting and output of FLaIR model (Forecasting of Landslide induced by rainfall, Capparelli and Versace 2011) based on observed and predicted (for the successive six hours) rainfall heights.

Indicators define two states:

- state 0 = no variation or not significant;
- state 1 = significant variation.

To sum up, CAED has the following information in any moment:

- state (0, 1) of indicators (IND);
- state (0, 1, 2, 3) of sensors and models running for the specific highway segment (SEN);

and, on the basis of these states, four different decisions can be made by CAED, one of which with three options, that activate different level of criticality.

All the possible decisions are illustrated in Table 6, in which the weight of the several sensors is assumed to be the same.

As an example, identification of a state related to the SUSHI model is below described. SUSHI output is constituted by a series of Factor of Safety (FS) values, estimated on the basis on rainfall heights, which are measured and predicted with a stochastic model. In particular, the stochastic model provide N hyetographs for input in SUSHI, and the final result is a matrix of FS values, which are computed for several time instants (t_0, t_1, t_2, ...) and different rainfall inputs.

Based on FS, the state S is defined by using the matrix reported in Fig. 17a, that considers:

(i) four levels of FS (see also Fig. 17b), indicated as W (white), Y (yellow), O (orange) and R (red); (ii) the instant of prediction.

A greater weight is assigned to estimation very close to current instant t_0, as it is less uncertain.

Moreover:

- for each t, N values of S are provided (Fig. 17c), corresponding to N rainfall realizations;
- the highest value of S, which occurred in at least 10% of realizations, is assigned to each t;
- N is usually set equal to 100, and 3 time instants are fixed, for which $t_1–t_0 = t_2–t_1 = t_3–t_2 = 1$ h.

Thus, SUSHI model provides 3 values of S, related to t_1, t_2 and t_3, respectively.

Based on the notices of criticality levels provided by CAED, and on its own independent evaluations, the CCC issues the appropriate **level of alert** (Surveillance, Alert, Alarm and Warning) and makes decisions about the consequent actions (Fig. 18).

9 Conclusions

The example here proposed only provides a framework of a specific case of early warning system that was designed and realized for research and testing purposes. Obviously, given the huge variability of natural contexts, it cannot represent all the cases that might occur in reality. Depending on the specific site, it is necessary to evaluate and complete the best configuration.

Many problems still remain open, such as the definition of the threshold levels, the intervention model, the use of monitoring instruments or devices to spread alarms.

For such complex and so hyper-calibrated systems it is necessary to ensure an adequate validation period for better understanding the dynamics and for better defining all the implemented procedures.

References

Artese G, Perrelli M, Artese S. Meduri S, Brogno N (2015) POIS, a low cost tilt and position sensor: design and first tests, vol 5(5). Multidisciplinary Digital Publishing Institute

Avolio MV, Di Gregorio S, Lupiano V, Mazzanti P (2013) SCIDDICA-SS3: a new version of cellular automata model for simulating fast moving landslides. J Supercomput. doi:10.1007/s11227-013-0948-1

Capparelli G, Versace P (2011) FLaIR and SUSHI: two mathematical models for early warning systems for rainfall induced landslides. Landslides 8:67–79. doi:10.1007/s10346-010-0228-6

Carrara A, Cardinali M, Guzzetti F, Reichenbach P (1995) GIS technology in mapping landslide hazard. In: Carrara A, Guzzetti F (eds) Geographical information systems in assessing natural hazards. Kluwer Academic Publisher, Dordrecht, pp 135–175

Clerici A, Perego S, Tellini C, Vescovi P (2006) A GIS-based automated procedure for landslide susceptibility mapping by the conditional analysis method: the Baganza valley case study (Italian Northern Apennines). Env Geol 50:941–961

Costanzo S, Spadafora MOH, Scarcella F, DiMassa G (2013) Multiband software defined radar for soil discontinuities detection, J Electr Comput Eng. Article ID 379832

Cruden DM, Varnes DJ (1996) Landslide types and processes. Nat Acad Sci 247:36–75 (Special Report, Transportation Research Board)

Del Ventisette C, Intrieri E, Luzi G, Casagli N, Fanti R, Leva D (2011) Using ground based radar interferometry during emergency: the case of the A3 motorway (Calabria Region, Italy) threatened by a landslide. Nat Hazards Earth Syst Sci 11(9):2483–2495. doi:10.5194/nhess-11-2483-2011

Intrieri E, Gigli G, Casagli N, Nadim F (2013) Landslide Early Warning System: toolbox and general concepts. Nat Hazards Earth Syst Sci 13(85–90):2013. doi:10.5194/nhess-13-85-2013

Peres DJ, Cancelliere A (2014) Accounting for variability in rain-event intensity and initial conditions in landslide triggering return period mapping via a monte carlo approach. In: Landslide Science for a safer geoenvironment: volume 2: methods of landslide studies, pp 499–505

Rigon R, Bertoldi G, Over TM (2006) GEOtop: a distributed hydrological model with coupled water and energy budgets. J Hydromet 7:371–388

Sassa K, Yueping Y (2010) Early warning of landslides. Geological Publishing House, 166 p

VanWestern CJ, Castellanos Abella EA, Sekhar LK (2008) Spatial data for landslide susceptibility, hazards and vulnerability assessment: an overview. Eng Geol 102:112–131

Versace P, Artese G, Autiero M, Avolio MV, Bardi F, Borgia A, Cancelliere A, Capparelli G, Capuozzo M, Caruso A, Casagli N, Cavallaro L, Cianciosi O, Conforti M, Conte E, Costanzo A, Costanzo S, De Marinis M, Di Gregorio S, Di Massa G, De Luca DL, De Santis D, Donato A, Fanti R, Fidolini F, Formetta G, Foti E, Intrieri E, La Sala G, Luci A, Maletta D, Mannara G, Moreno D, Morrone L, Mungari T, Muto F, Paoletti F, Peres DJ, Raffo A, Rago V, Rigon R, Spadafora F, Spataro W, Troncone A, Trunfio GA, Vena M, Viggiani G (2014) An Integrated System for Landslide Monitoring, Early Warning and Risk Mitigation along Lifelines. In: Capparelli G, Greco R (eds) Aspetti idrologici e Idraulici per il controllo dei movimenti franosi, pp 1–68

TXT-tool 2.039-3.4 Methods to Improve the Reliability of Time of Slope Failure Predictions and to Setup Alarm Levels Based on the Inverse Velocity Method

Tommaso Carlà, Emanuele Intrieri, Federico Di Traglia, Giovanni Gigli and Nicola Casagli

Abstract

Estimating the time of slope failure is a topic of great importance in the field of landslide risk mitigation. Within this framework, time of failure forecasting methods based on the inverse velocity, typically intended as the extrapolation of linear trend lines of the inverse of velocity with time, are widely known as tools for early warning of slopes displaying accelerating trends of deformation rate. Although nominally simple, their correct application is actually tricky as many factors can influence displacement data and eventually heavily reduce the accuracy of the predictions. Such disturbing elements can be classified as noise caused by instrumental precision and as noise representing the diverging of a natural behavior with respect to an ideal inverse velocity trend. Hence correctly preparing the dataset is a pivotal and critical task. The present teaching tool describes how to filter displacement data by presenting three different approaches and discussing the results of their application to three large slope failure case histories in Italy, in order to improve, in retrospect, the reliability of the failure-time predictions. Procedures to automatically setup alarm levels of slope failure occurrence are consequently proposed for supporting the definition of landslide emergency response plans.

Keywords

Slope failure · Time of failure prediction · Inverse velocity
Alarm threshold · Early warning · Monitoring

T. Carlà (✉) · E. Intrieri · F. Di Traglia · G. Gigli · N. Casagli
Department of Earth Sciences, Università Degli Studi Di Firenze, Via Giorgio La Pira 4, 50121 Florence, Italy
e-mail: tommaso.carla@unifi.it

E. Intrieri
e-mail: emanuele.intrieri@unifi.it

F. Di Traglia
e-mail: federico.ditraglia@unifi.it

G. Gigli
e-mail: giovanni.gigli@unifi.it

N. Casagli
e-mail: nicola.casagli@unifi.it

K. Sassa et al. (eds.), *Landslide Dynamics: ISDR-ICL Landslide Interactive Teaching Tools*, https://doi.org/10.1007/978-3-319-57774-6_39

Contents

1 **Introduction** 538
2 **The Accelerating Creep Theory** 538
3 **Methods** 539
4 **Results** 540
5 **Setting up Alarm Levels Through Inverse Velocity Analyses** 545
6 **Conclusion** 549
References 550

1 Introduction

Quick failure of large landslides, which can involve both soil and rock materials, is one of the major forms of geo-hazard. Predictions of the time of failure are therefore essential to provide notice to the population exposed to the risk and define appropriate response action emergency plans. Progressive increase of the slope surface velocities has been frequently observed in anticipation of catastrophic events of slope failure. The analysis of the linear trend of the inverse of velocity with time is typically the most used tool for describing phases of slope acceleration and estimating when slope failure will occur. However, the approach implies several simplifications and assumptions that can hamper its practical use. Instrumental noise of the monitoring system can also impact the accuracy of the velocity measurements. In this paper the inverse velocity method (INV) is successfully applied to three large slope failure case histories in Italy: it is shown that data smoothing is fundamental in order to improve the reliability of failure-time predictions. Finally, procedures to set alarm levels of slope failure occurrence are introduced.

2 The Accelerating Creep Theory

The most diffused approaches to forecast the time of failure (T_f) are those based on the accelerating creep theory (Saito 1969; Fukuzono 1985; Cruden and Masoumzadeh 1987; Crosta and Agliardi 2003; Carlà et al. 2016), from which derives that the time of slope failure can be predicted by extrapolating the trend towards zero of the inverse velocity-time plot. The most notorious description of the topic was made by Voight (1988, 1989), who extended the theory to the behaviour of materials in terminal stages of failure and proposed a relation between displacement rate and acceleration, influenced by two dimensionless parameters (A and α):

$$\dot{\Omega}^{-\alpha}\ddot{\Omega} - A = 0 \quad (1)$$

where Ω is the observed displacement and the dot refers to differentiation with respect to time (i.e. $\dot{\Omega}$ is velocity and $\ddot{\Omega}$ acceleration). He then provided an analytical solution for the forecast of T_f that may be utilized especially when the inverse velocity plot approaches zero with a striking non-linear trend (perfect linearity is verified for $\alpha = 2$), which would make more uncertain the use of the graphical solution. A and α are not necessarily constants and can change if different time increments are considered (Voight 1988; Crosta and Agliardi 2003). Voight (1988), in fact, supposed that the mechanisms of deformation and the conditions of loading are time invariant, an assumption which often does not agree with reality. The accuracy of the analytical method is heavily influenced by the precision and frequency of the monitoring data (Voight 1989). Experience suggested that α is frequently nearly equal to 2 (Fukuzono 1985; Voight 1988, 1989). Petley et al. (2002) observed that the form of the inverse velocity plot is in fact predominantly linear for landslides where generation of a new shear surface and crack propagation are the dominant processes. Rose and Hungr (2007) also assessed that the inverse velocity plot often approaches linearity, especially during the final stages of failure and highlighted the need to seek consistent linear trends in the data. On this subject, Dick et al. (2015) suggested that the linear fitting procedure should only include data following the identification of Onset Of Acceleration (OOA) points; they also advised that additional fitting should be performed if a TU (Trend Update) point is detected.

The linear approximation of the inverse velocity method has thus become a widely applied tool for failure prediction and the simplicity of use is its most powerful feature. It can be used only for slopes which fail in accordance to the accelerating creep theory, excluding, for example, cases of structural (Mufundirwa et al. 2010) or brittle (Rose and Hungr 2007) failure. Rose and Hungr (2007) provided a list of elements that can cause variation of displacement rates and thus hamper the performances of INV. The theoretical method was in fact obtained from controlled laboratory conditions, which are extremely unlikely to be satisfied on natural slopes and field conditions. Such limitations can include measurement errors and random instrumental noise, local slope movements, variation of the displacement rates driven by periodically changing factors (e.g. rainfall, groundwater, snowmelt, human activities, etc.), α significantly diverging from 2, lack of clear OOA and TU points acceptably earlier than the occurrence of the failure event. Moreover, INV assumes that velocity at failure is infinite; this condition is obviously not verified and in fact velocity at failure varies for different rock types, volumes, mechanisms of failure and slope angles (Newcomen and Dick 2015). All these elements can decisively hamper the interpretation of inverse velocity plots, thus affecting the accuracy and reliability of T_f predictions and eventually discouraging users from using the method during the monitoring program.

For these reasons, it is of crucial importance to correctly process the velocity time series in order to remove as much as possible these disturbing effects, which here are generally termed "noise" for simplicity. Data smoothing is essential to help identifying OOA and TU and to increase the degree of fitting of the linear regression. Noise can be classified as:

- "instrumental noise" (IN), related to the accuracy of the monitoring instrument;
- "natural noise" (NN), that includes all the mentioned factors causing possible divergence of the slope movements from the ideal linear behaviour of the inverse velocity plot towards failure.

Different filters may generate different results depending on the shape of the rate acceleration curve and on the level of noise intensity. Therefore, testing should be performed on various types of case studies. In the following sections, data smoothing is executed on velocity data from three large slope failure case histories and the performances of INV are then evaluated. Each event displayed a noticeable accelerating trend towards failure. In particular, these are:

1. the Mt. Beni rockslide (December 2002), characterized by generally low-medium values of slope velocities approaching failure ($A \approx 0.1$; see later sections);
2. the catastrophic Vajont landslide (October 1963), which was anticipated by extremely high movement rates ($A \approx 0.04$; Voight 1988);
3. the roto-translational slide of the debris talus of Stromboli volcano (August 2014).

Finally we propose some standard procedures for the analysis of inverse velocity plots. These proposed guidelines address both the aspect of data handling and the issue of objectively establishing the alarm setup process for the emergency plan.

3 Methods

Two of the most common and easy-to-use smoothing algorithms are tested, i.e. the moving average and the exponential smoothing. Three forms of filter are considered:

1. a short-term simple moving average (SMA), where the smoothed velocity at time t is:

$$\overline{\mathrm{v}}_{\mathrm{t}} = \frac{\mathrm{v_t} + \mathrm{v_{t-1}} + \cdots + \mathrm{v_{t-(n-1)}}}{\mathrm{n}} \tag{2}$$

and $n = 3$;

2. a long-term simple moving average (LMA), where $n = 7$ in (2)
3. an exponential smoothing function (ESF), where

$$\bar{v}_t = \beta \cdot v_t + (1 - \beta) \cdot \bar{v}_{t-1} \quad (3)$$

and the smoothing factor is $\beta = 0.5$.

There is no formally correct rule or procedure in order to establish the order of the moving averages. In fact, this is highly dependent on data quality and temporal frequency. High acquisition rates, which in the field of slope monitoring are typical of advanced remote sensing techniques (e.g. ground-based radar, GPS and total station), usually require to perform smoothing over a greater number of measurements. Conversely, low acquisition rates will hide much of the background noise, causing on the other hand the inability to trace short-term movements and delaying the identification of eventual trend changes; in such instance, smoothing should be performed over relatively less measurements, compared to data obtained at high acquisition rates. It follows that the number of data points within the time series does not necessarily influence in a significant way the accuracy offailure predictions. It is difficult to determine a priori the minimum number of acquisitions, in addition to those necessary to apply the described filters (i.e. n), that is required to confidently extrapolate linear trends in the inverse velocity plot. Undoubtedly, a populated data set will help determine more clearly how well the time series is fitted by the linear regression line, thus giving an important indication about the reliability of the prediction. In any case, modern monitoring technologies, which can measure displacements several times per day, ensure that the former is no longer a common issue and typically provide users with a more than acceptable amount of data points in the time series, unless the monitoring activities are initiated extremely close to the failure event.

With reference to the failure events presented in this paper, which were mostly characterized by low acquisition rates of the displacements (about 1 measurement/day or less), using values of $n > 7$ in (2) seemed to level out excessively the trend changes which were actually due to OOAs. Moreover, the wider the temporal window over which smoothing is performed, the higher the lag that is introduced to the time series; consequently, in scenarios of near-real time monitoring, it is probably not practical or convenient to use windows of smoothing which span over too long time periods. A LMA with $n = 7$ seems in this case a reasonable compromise. Regarding the SMA, Eq. (2) with $n = 3$ is typically considered the most basic filter for smoothing a hypothetical outlier in a linear time series. This moving average responds quickly to trend changes and is sensitive to slight fluctuations in the data. For this reason it is selected for comparison to the "slower" and less sensitive LMA. The aim of the study is not determining if a certain value of n is the most ideal for treating displacement measurements, since this will certainly vary from case to case depending on several factors such as acquisition rate, landslide velocity, background noise, etc. As described more thoroughly in the following sections, both short- and long-term moving averages provide benefits and downsides in time series analysis. The intent is to evaluate how these properties influence the reliability of failure predictions and if short-term or long-term moving averages should be preferred for the application of INV.

Finally, in (3) a smoothing factor of 0.5 maintains a good balance between smoothing effect and sensitivity to trend changes in the data. An ESF has the form of a geometric progression and therefore, differently from a moving average, takes into account all past data, with recent observations having greater weight than older ones.

4 Results

Mt. Beni rockslide

On 28 December 2002 a landslide occurred on the Eastern flank of Mt. Beni (Central Italy), on a slope which had been previously exploited by quarrying activity until the 1980s. The failure mechanism, involving jointed massive basalts overlying ophiolitic breccias, was characterized by a volume of about 500,000 m^3 and has been classified as a rockslide/rock topple (Gigli et al. 2011). Several distometric bases were put in place along the perimetral crack and recorded cumulative displacements since April 2002. Increasing rates, ranging from few millimetres up to few centimeters per day, were detected by most of the devices starting from September 2002 until the collapse of 28 December. According to eyewitnesses, the event began at about 4:30 am local time. Gigli et al. (2011) made particular reference to distometric base 1–2, since this recorded the longest and most consistent progressive acceleration in the time series of displacements; they ultimately calculated a failure forecast according to the linear extrapolation of the trend in the inverse velocity plot, which could be distinguished since 2 August already (Fig. 1a). Although it provides a good estimate that failure is going to occur, fitting of these data gives a T_f which anticipates the actual failure (T_{af}) by 4 days and a half ($\Delta T_f = T_{af} - T_f = 4.5$), as is also shown by the related life expectancy plot in Fig. 2a. Life expectancy plots are very useful tools to examine the course with time of the T_f predictions, updated on an ongoing basis each time a new measurement is available (Mufundirwa et al. 2010; Dick et al. 2015).

Inverse velocity plots for data filtered by means of SMA, LMA and ESF are given in Fig. 1. In this case, short-term moving average smoothing eliminates the small steps in the raw data curve, yielding a significantly improved T_f ($\Delta T_f = 1.1$); on the other hand, long-term moving average, despite determining a slightly better fitting, generates a considerably negative ΔT_f (−4.3) and most importantly delays the identification of the OOA point by one month. The latter aspect is verified even more for the exponentially smoothed velocities. Their poor linearity would have also made their use in the emergency scenario quite troublesome, even if, in retrospect, it is found that the last T_f value is highly accurate. Respective life expectancy plots (Fig. 2) display additional, valuable information: all datasets, beside velocities filtered by means of ESF, converge parallel to the actual time to failure line. However, raw data consistently give anticipated forecasts of the time of failure, whereas SMA allows forecasts to get decisively closer to the actual failure line since November 21. In Fig. 2c, predictions approach the actual values already on late October, but finally diverge towards significantly negative ΔT_f, which is something to avoid in risk management of geological hazards (concept of "safe" and "unsafe" predictions, Mufundirwa et al. 2010). Exponential smoothing produces a reliable T_f in correspondence of the last measurement, but the predicted time to failure line does not have a regular pattern.

Summarizing, the Mt. Beni case study, as analyzed here, showed that linear fitting of unfiltered inverse velocity data, although surely constituting an indicator of the ongoing acceleration of the slope movements, produced T_f which anticipate by several days the actual time of collapse. Smoothing data with simple moving average algorithms allowed to benefit from improved qualities of fitting and of T_f predictions. Short-term moving average performed better when applied to the almost linear measurements registered by distometric base 1–2. Exponential smoothing did not produce satisfying improvements to these data.

Vajont landslide

The 1963 Vajont disaster in Northeast Italy has been one of the most catastrophic landslides in history. The collapse occurred at about 10:39 pm local time on 9 October 1963, when a 270 million m^3, mostly calcareous rock mass detached from the slope of Mt. Toc and slid at 30 m/s into the newly created Vajont reservoir. The consequent tsunami wave overtopped the dam and killed 2500 people in the villages downstream. The pre-failure movements of the unstable material were strongly controlled by the water level in the valley floor and creeping motions began to be observed immediately since the creation of the reservoir (Havaej et al. 2015). The final chain of events started with the April 1963 reservoir filling cycle; the final rupture followed 70 days of downslope accelerating movements (Helmstetter et al. 2004). According to measurements from four benchmarks installed at

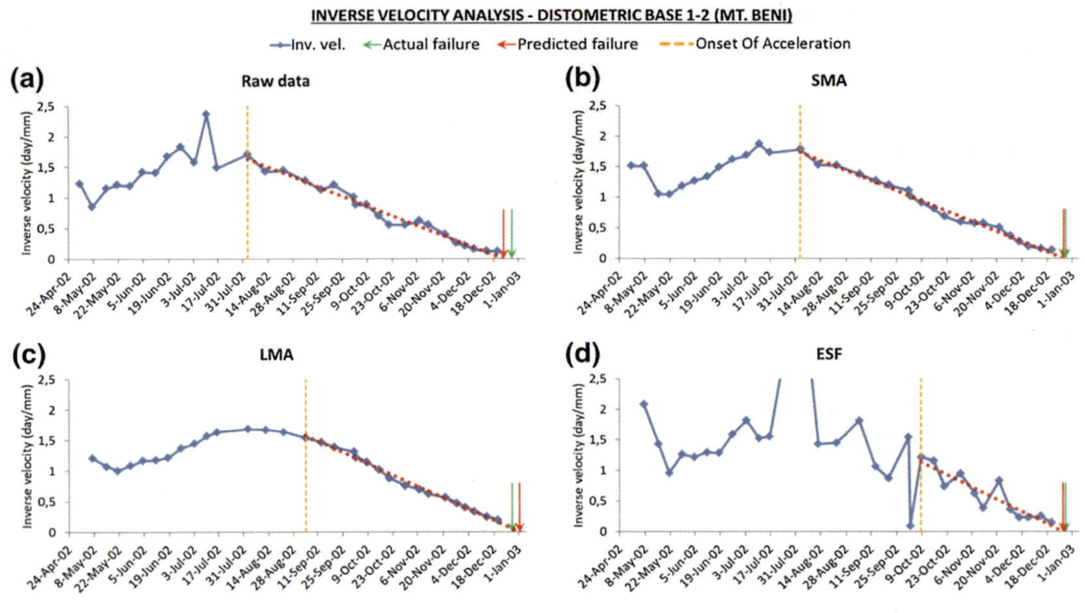

Fig. 1 Inverse velocity analysis for the distometric base 1–2 installed on the Eastern flank of Mt. Beni

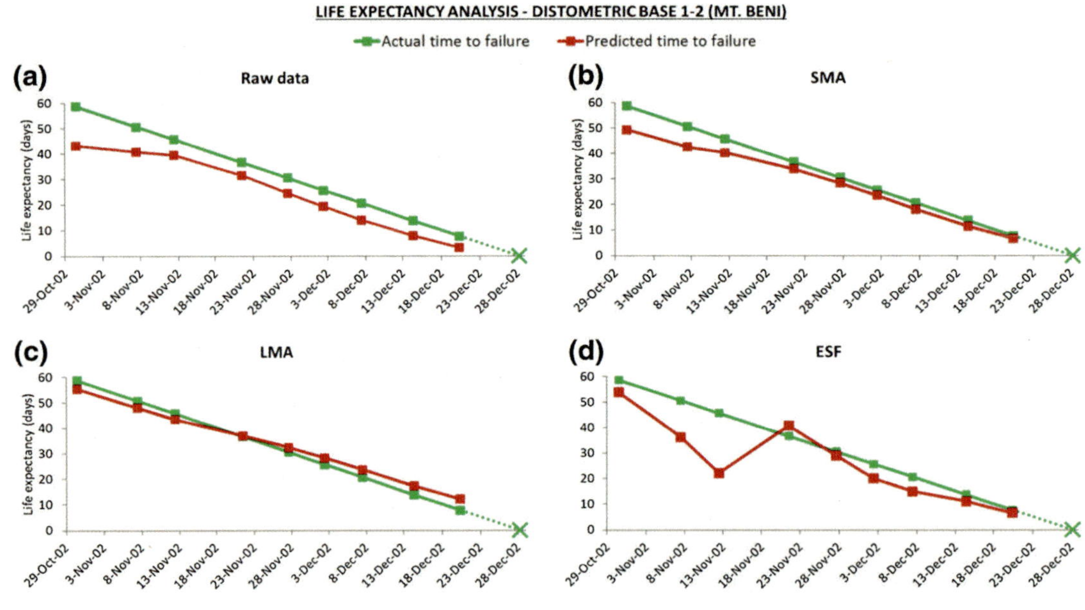

Fig. 2 Life expectancy analysis for the distometric base 1–2 installed on the Eastern flank of Mt. Beni

different positions on the mountain slope, pre-failure velocities were extremely high: among these, benchmarks 5 and 63 in particular were the ones which showed a clear state of accelerating creep, with displacement rates ranging from ≈5 mm/day up to over 20 cm/day and a total cumulative deformation of a few meters (Muller 1964).

In Fig. 3a, the inverse velocity plot for the unfiltered measurements of benchmark 63 are shown. Even though a slight step-like pattern can be noticed, the graph is remarkably linear as a whole and, most importantly, the linear trend line fits particularly well the last points leading up to failure. Therefore, if recorded velocities are regularly high since the initial stages of acceleration, i.e. A ≈ 0.04 (as in Vajont data; Voight 1988) or lower, and if the assumption of $\alpha = 2$ is consistently satisfied, it is suggested not to apply any strong data filtering, in order to avoid loss of sensitivity with regards to actual downward trends. At most, linear fitting of short-term averaged velocities may also be conducted in parallel to the fitting procedure of the original data. For benchmark 63, SMA yields a slightly better fitting and a slightly more accurate T_f ($\Delta Tf = 0.3$) with respect to results from the original data ($\Delta T_f = 1.1$, Fig. 3b). Conversely, the inverse velocity plot from benchmark 5 (Fig. 4a) seems to have a slightly different pattern: after the fourth point of the dataset, the trend appears to assume a certain concavity ($\alpha < 2$) and consequently, as opposed to benchmark 63, the last velocity points before failure are not well fitted by the linear regression line. This causes again a premature T_f prediction, which anticipates also the time of the last measurement on 8 October ($\Delta T_f = 2.7$, Fig. 4a). Excellent linearization is produced by LMA and ESF (Fig. 4), with the first one also determining an extremely accurate T_f ($\Delta T_f = 0.1$). From the relative life expectancy plot, it can be seen that a correct prediction of the time of failure could have been given about 10 days earlier than the actual event. In further support of their reliability, the forecasts of the time of failure remained consistent after each of the last six displacement measurements. On the other hand, life expectancies from the raw data are always some days off the actual time-to-failure line.

Again, as in the Mt. Beni case study, it is not argued that velocities from benchmarks at Mt. Toc were not indicating a clear accelerating trend towards slope collapse. Nevertheless it is found out that the accuracy of T_f predictions could be improved with appropriate data smoothing and thus linearization of the respective inverse velocity plots. Results indicate that, for fast-moving slopes, if $\alpha < 2$ it is suggested to filter data with a long-term moving average; instead, if $\alpha \approx 2$, it seems more convenient to proceed with the regular inverse velocity analysis performed on an ongoing basis or, at most, to filter data with a short-term moving average.

Stromboli debris talus roto-translational slide

On 7 August 2014 a debris talus located below the Northeastern crater (NEC) of Stromboli volcano was affected by a roto-translation slide, evolving into a rock avalanche on a 30°–45° steep slope called “Sciara del Fuoco”, and followed by the opening of an eruptive vent localized at ≈650 m a.s.l. (≈100 m below the NEC, Di Traglia et al. 2015; Rizzo et al. 2015; Zakšek et al. 2015). Debris cone material likely reached the sea but tsunami waves were not recorded. Growth stages of this debris talus at Stromboli are a common phenomenon related to the explosive activity, which produces accumulation

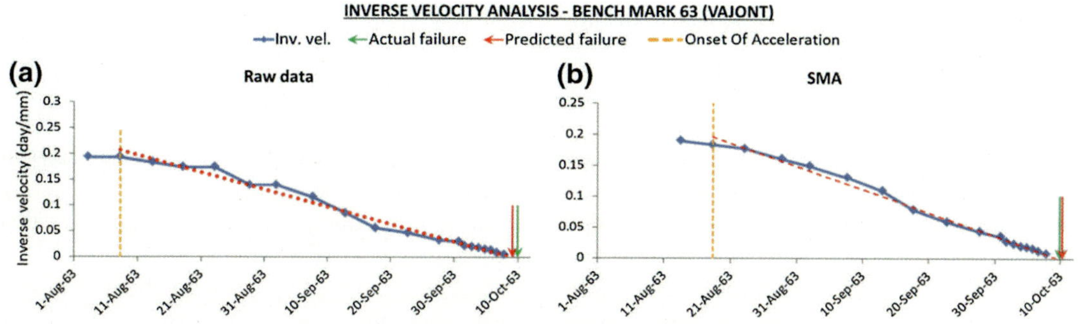

Fig. 3 **a–b** Inverse velocity analysis for benchmark 63 installed on Mt. Toc

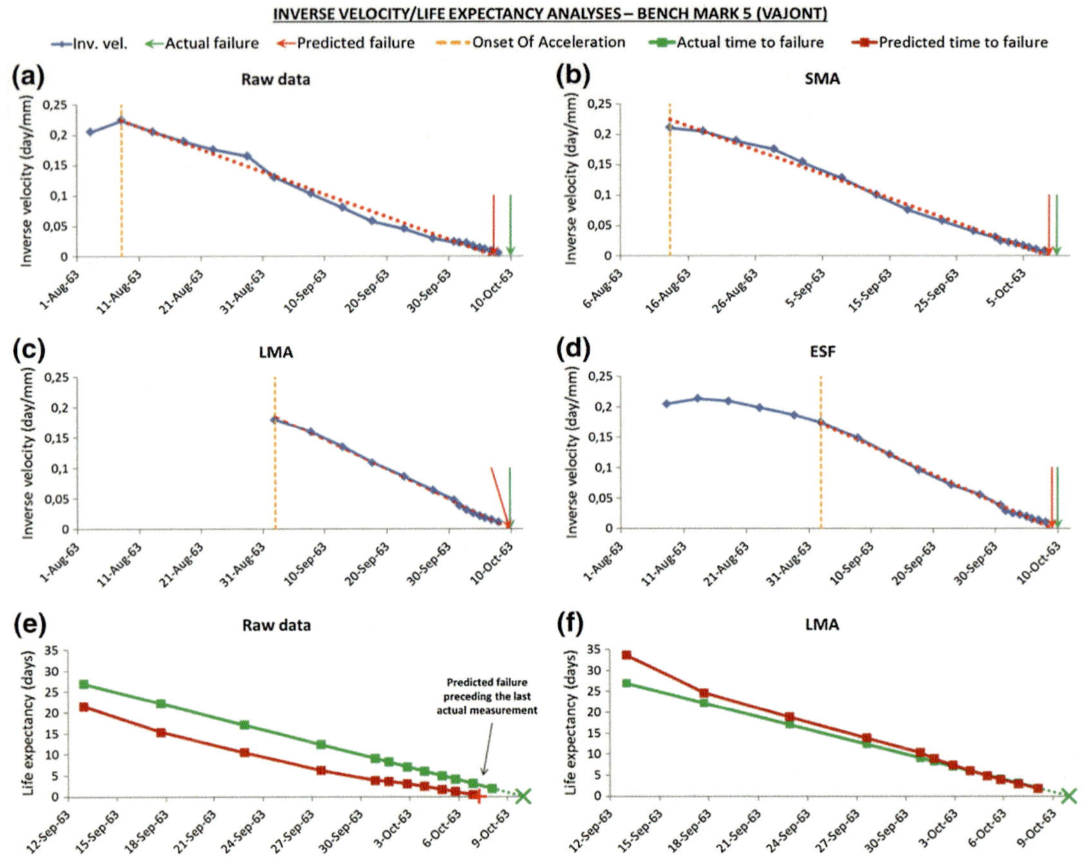

Fig. 4 Inverse velocity and life expectancy analyses for benchmark 5 installed on Mt. Toc

of volcaniclastic debris around the active craters (Calvari et al. 2014). Contrariwise, the roto-translational slides occurred at the onset of the 2002–03, 2007 and 2014 flank eruptions, characterized by magma propagation from the central conduit towards the NEC (Di Traglia et al. 2014, 2015; Intrieri et al. 2013). The debris talus is composed by loose volcaniclastic material, prevalently breccia alternating with tuff, lapillistones and thin (<2−3 m) lava flows (Apuani et al. 2005). The NEC area has been continuously monitored by a Ground-Based Interferometric Synthetic Aperture Radar (GBInSAR) system (Di Traglia et al. 2014). The latter recorded an increase in the line-of-sight displacement rate of the debris talus starting from 30 May 2014. The curve of deformation rate for this period presents a high degree of noise (Fig. 5a). This is mostly due to the dynamics of the magma within the volcanic edifice: at Stromboli, successive cycles of filling and emptying of the plumbing system can cause phases of inflation and deflation of the entire crater area, thus adding "natural noise" to the displacements of the debris talus. For this reason, strong negative velocities are displayed at times in Fig. 5a and a definite raise in rates can be observed only from late July. This final acceleration was made up of relatively low-medium rates, from few mm/day up to a peak velocity of 40.4 mm/day on 6 August 2014. Since 04:01 GMT on 7 August 2014, the GBInSAR recorded complete loss in coherence in the debris cone sector, consistently with a fast movement of the summit area (i.e. T_{af}).

In Fig. 5a, the OOA is found about only one week before the slope collapse. Hence, it was not possible to produce with confidence a failure

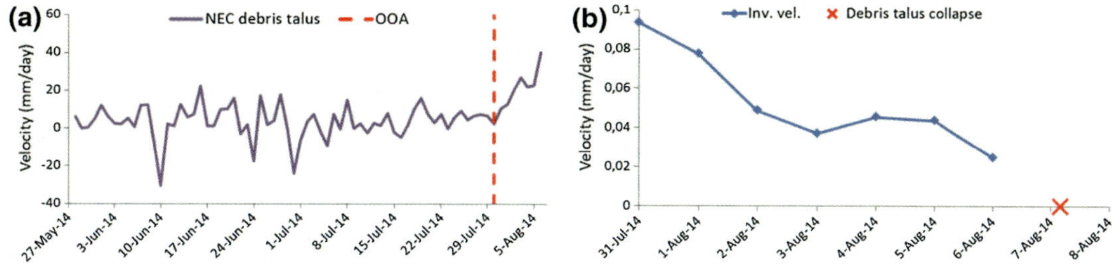

Fig. 5 Deformation rate (**a**) and inverse velocity plot after OOA point (**b**) for the GBInSAR time series of the debris talus below the NEC of Stromboli volcano

prediction, as only few measurements could be collected between OOA and T_{af} (Fig. 5b). Data filtering, eliminating short-term fluctuations and highlighting longer-term cycles, is here of invaluable help, as it may allow an earlier detection of the OOA point. With LMA (Fig. 6), the latter is found on 27 July, but the inverse velocity line appears to be excessively smoothed out, causing a late T_f prediction ($\Delta T_f = -1.7$); ESF generates the OOA on 24 July and provides a slightly better failure forecast ($\Delta T_f = -1.3$), despite a worse fitting ($R^2 = 0.77$). SMA undoubtedly produces, in this case, the best results overall (Fig. 6a, b): a trend towards the x-axis begins already on 23 July, the level of fitting is high ($R^2 = 0.88$) and T_f is extremely reliable ($\Delta T_f = -0.4$). In fact, from the life expectancy plot in Fig. 6b, it can be seen that a roughly correct time of failure forecast could have been made consistently since 3 August, approximately 4 days before the actual collapse of the debris talus. Similarly to the time series of distometric base 1–2 at Mt. Beni, it is found that, when accelerations towards failure are characterized by moderate velocity values, it is more reliable to filter data by means of SMA. This is even more important for the Stromboli case study, where failure is usually anticipated by a very brief final acceleration phase and therefore an early identification of the OOA is highly needed. A short-term moving average has higher sensitivity towards trend changes in the data and thus appears to be the most helpful mean of analysis, with regard to this specific issue.

5 Setting up Alarm Levels Through Inverse Velocity Analyses

INV can be a powerful tool for estimating the time of slope failure. The reliability and applicability of the method can be decisively improved by appropriately treating data. General guidelines for ideal data filtering of velocity data could be suggested:

- if movement rates are consistently high (i.e. $A \approx 0.4$ or lower), it is not advised to perform data smoothing, as it would generally yield no real improvements to the INV analysis and cause less sensitivity to actual trend changes. Only SMA could provide some benefits in terms of both OOA/TU identification and reliability of T_f. In general, focus should be put mostly on updating T_f predictions on an ongoing basis and on searching for new TU points in the data;
- if movement rates are low (i.e. $A \approx 0.1$ or higher) and noise has significant intensity, data smoothing is of great importance in order to perform INV analyses and could make the difference between being able to make effective T_f forecasts or not (e.g. Stromboli debris talus). IN mainly influences the initial stage of the acceleration, significantly delaying the identification of OOA/TU points. At the same time, NN may also disturb the linearity of the last stages of acceleration and thus hamper the accuracy of T_f predictions. It was observed that different filters could give better

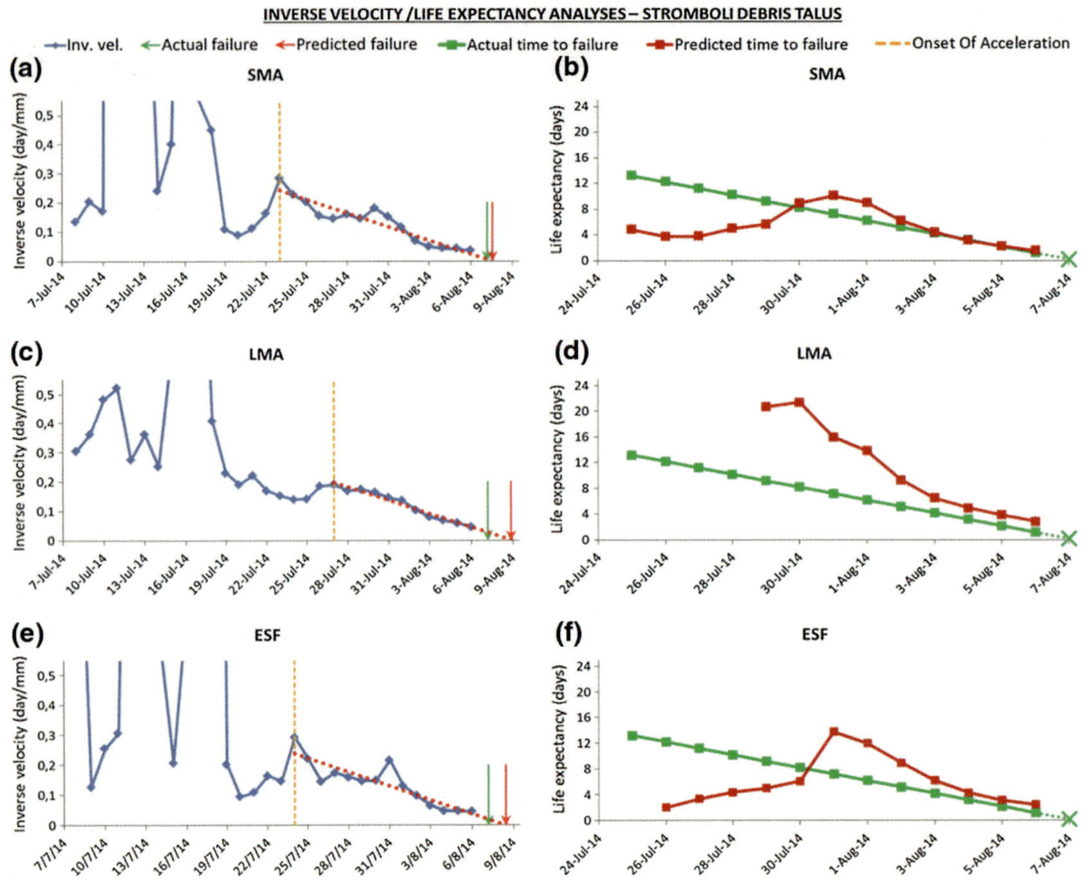

Fig. 6 Inverse velocity and life expectancy analyses for the smoothed GBInSAR time series of the debris talus below the NEC of Stromboli volcano

results in different case studies. Therefore, T_f predictions should be made while applying both short- and long-term filters and simultaneously studying the related trend lines. Exponential smoothing could also be run in parallel for comparison.

It should be noticed that studying trends of displacement rate remains in great part exposed to a high degree of subjectivity. In INV, much of the analysis is based on the interpretation and experience of the user. As shown in the case histories presented here and as hinted also by Dick et al. (2015), different techniques for data smoothing may be alternatively more suitable, depending on the specific characteristics of the velocity curves, on the properties of the monitoring device and on the requirements of the emergency plan. On top of that, it is obvious that the linear extrapolation of inverse velocity data trending toward the x-axis should not be considered as a precise forecast of the time of failure, but rather as a general estimation only.

We thus propose the introduction of some procedures for the analysis of inverse velocity plots; once established, these may be performed automatically and with little input from the user. We think that this would provide benefits especially when dealing simultaneously with multiple series of measurements, as in the case of point cloud data (e.g. ground-based radar, total station prisms, etc.). These procedures concern the identification of OOA and TU points and the setup of two different alarm levels, which would support the definition of the emergency plans. Like in mining operations, exceeding the thresholds between adjacent alarm levels could instantly prompt pre-determined actions of evacuation and/or slope remediation (Read and Stacey 2009).

1st alarm threshold: onset of acceleration

We consider the occurrence of an OOA as the first alarm threshold. Indicating the beginning of a progressive stage of failure (Zavodni and Broadbent 1980), such event would trigger the initiation of inverse velocity analyses, meaning that linear fitting may be applied only to velocity data from OOA onwards. Best-fits should be constantly revised to make sure that they successfully describe the most recent acquisitions. Predictions should be updated each time a new measurement is available and until failure occurs or, otherwise, an End Of Acceleration (EOA) point is detected (i.e. regressive stage of failure); an EOA signals the return to the condition of safety. Using a long time period in the calculations highlights longer-term trends, but can introduce more lag to data; on the other hand, shorter time periods generate higher responsiveness to slight trend changes, which in turn may just represent minor fluctuations.

In time series analysis, a crossover between a shorter-term and a longer-term average is typically one of the most basic signals suggesting the occurrence of a trend change in the raw data (Olson 2004; Shambora and Rossiter 2007). A crossover is defined as the point where a short-term moving average (c-SMA) line crosses through a long-term one (c-LMA). We consider that, if the c-SMA crosses above the c-LMA ("positive crossover"), the beginning of an uptrend and hence the occurrence of an OOA is signaled, while the opposite indicates the beginning of a downtrend ("negative crossover") and the occurrence of an EOA. Typically, in the context of moving average crossover analyses, the c-SMA is set so that it already smooths out significantly most of the slighter fluctuations, while the c-LMA is calculated over three or four times the c-SMA time period. The concept of c-SMA/c-LMA for the crossover analysis is therefore not necessarily related to the use of SMA/LMA for improving time of failure predictions, since a higher degree of smoothing is usually necessary in order to avoid false crossover points between the moving averages. This may cause to establish different types of c-SMA and of c-LMA for different case studies, depending on the properties and the noise of the rate curves.

The technique has been applied to the Stromboli debris talus and Mt. Beni movement series; those two only, among the examined case histories, have in fact enough points in the dataset in order to find a c-LMA. As already discussed, the GBInSAR series from Stromboli has a high degree of noise; therefore, a moving average calculated over 7 measurements (which is equivalent to the LMA in Sects. 2 and 3) was arbitrarily selected as the c-SMA. A moving average calculated over 21 measurements was consequently picked up as the c-LMA. The first positive crossover is found on 8 July, but is then followed by a negative crossover, which determines an EOA and the termination of the failure forecasts. A new positive crossover is detected on 18 July and leads up to the debris talus collapse on 7 August. Remarkably, the OOA is established even earlier than what had been done through visual examination of the data (Figs. 5 and 6). The other example is relative to the distometric base 1–2 at Mt. Beni (Fig. 7). The rate curve is in extreme accordance with the linear inverse velocity model, hence no substantial smoothing is required for the crossover analysis. A MA(3) and a MA(9) were selected as the c-SMA and c-LMA, respectively. The only positive crossover is found on 12 August and is not revoked until the slope collapse on 28 December, more than four months later. The OOA is detected roughly at the same time than through manual identification (Fig. 1).

Some key benefits may derive from the application of this type of approach. During the monitoring activity, an indication that a certain target has experienced a positive crossover (i.e. exceeded the first alarm threshold) could be issued automatically. In this way, the necessity to

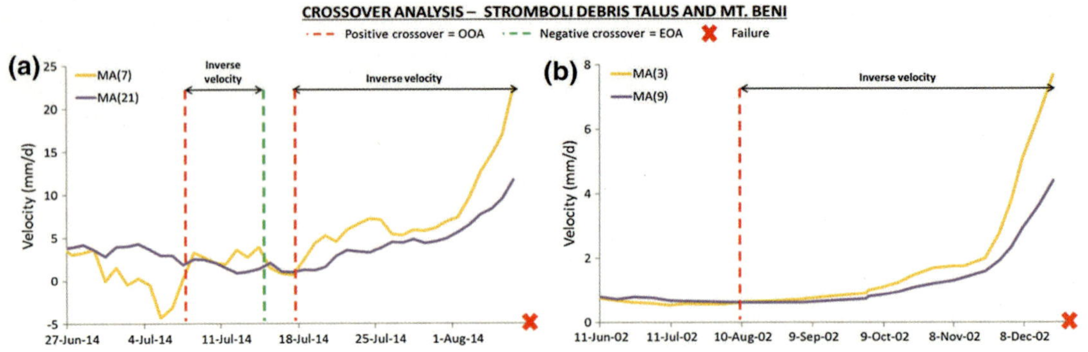

Fig. 7 Crossover analysis for Stromboli debris talus (**a**) and Mt. Beni (**b**) displacement rates

perform a subjective OOA analysis of the different rate curves would be avoided. This would be of great importance especially in radar applications, which can theoretically provide the user with thousands of time series of movement to evaluate at the same time. In terms of risk assessment on a slope, observing the spatial distribution of these alarms could allow to readily estimate the size of the instability and thus determine if the detected onset of acceleration is the expression of a local movement or instead the signal of an incoming partial/total failure event.

2nd alarm threshold: failure time window

Selecting the most appropriate smoothed time series can usually be done safely only in retrospect, during the post-event evaluation of the displacements which led up to failure. Therefore, the presence of a trend in the inverse velocity plot must be considered just as an indication that failure is presumably going to occur in proximity of its point of intersection with the *x*-axis. We propose a quick and reproducible procedure to define the time interval during which the occurrence of a failure event is considered highly probable ("failure window", T_{fw}). The failure window is obtained by projecting simultaneously the best-fits of both a SMA and LMA. This will yield two diverse T_f, with the difference between $T_{f(LMA)}$ and $T_{f(SMA)}$ being indicated as Δ; here, assuming $T_{f(SMA)} < T_{f(LMA)}$, T_{fw} was determined between:

$$\left[T_{f(SMA)} - \frac{\Delta}{2}; T_{f(LMA)} + \frac{\Delta}{2}\right]$$

The former corresponds to the period of activation of the second alarm level, i.e. the second alarm threshold is exceeded at time $T = T_{f(SMA)} - \Delta/2$. Understandably, there is no exact rule as to how use Δ for establishing the T_{fw} limits. These should be appropriately set based on the preferences of the user, on how well data seem to accommodate the INV and, most importantly, on the level of acceptable risk for the monitored case study. T_{fw} must be wide enough to account for the uncertainty implied with the failure predictions, while on the other hand excessive precaution might trigger too early activations of the second alarm threshold. The nature of the elements at risk and the characteristics of the emergency scenario should be carefully evaluated in order to assess the appropriate width of the failure window. Relatively to the case studies presented in this paper, the indicated T_{fw} limits were evaluated as the most suitable for successfully including T_{af} and not providing a too conservative alarm threshold (Fig. 8). The Stromboli debris talus and Mt. Beni slope failures are again used for reference to provide an overview of the methodology: the displayed failure windows have been calculated on the third last measurement of the rate curves (respectively 3 and 26 days before failure). For Stromboli (Fig. 8a), on 4 August the second alarm threshold is placed in the early morning of

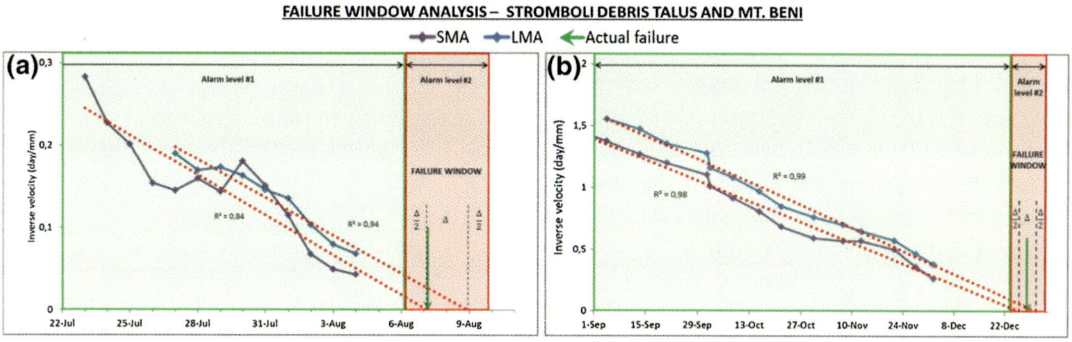

Fig. 8 Examples of failure window analysis for Stromboli debris talus (**a**) and Mt. Beni (**b**) data

6 August and T_{fw} lasts until 9 August, successfully including T_{af} with a convenient safety margin. The width of the failure window is about 3½ days. For Mt. Beni (Fig. 8b), the prediction performed on 2 December determines a wider failure window, spanning from late 23 December 2002 to early 2 January 2003 (≈10 days). T_{af} lies in the middle between $T_{f(SMA)}$ and $T_{f(LMA)}$ and thus also in the center of the T_{fw} interval. Equivalent results were obtained for the Vajont data.

The failure window must be continuously updated each time a new measurement is available and consequently new $T_{f(SMA)}$ and $T_{f(LMA)}$ predictions are performed. With this approach, it would be possible to recognize with a certain advance when the second alarm level will approximately be triggered and thus when the related emergency response actions will have to be taken. The eventual consistency over time of the T_{fw} boundaries can be considered as a strong element of support that failure will actually occur within the identified time frame. Lower Δ values suggest that measurements (and thus the forecasts) can be considered reliable, while higher Δ values may indicate that the downward trend in the inverse velocity plot is less clear, determining an increased width of the failure window and more uncertainty surrounding T_{af}. The value of Δ and the accuracy of the method in general is also influenced by the number and frequency of measurements available.

This simple and practical approach, once established within the monitoring program, could be extended to all the rate curves available and serve as a standard and automatic procedure to setup in real time the alarm level of each measuring point. Like for the moving average crossover approach, this may result of great advantage especially when simultaneously dealing with multiple targets. By estimating in advance when and on which points the second alarm threshold will be exceeded, more time would be provided to evaluate the risk in terms of size and probability of failure and promptly implement the pre-determined evacuation and/or remediation strategies.

6 Conclusion

The inverse velocity method is a powerful tool for predicting the time of failure of slopes displaying accelerating trends of movement. Because of the simplicity of use, its linear form ($\alpha = 2$) is arguably the most common mean of analysis of displacement data for this type of emergency scenarios. The assumed linear trend of inverse velocity toward infinite must be considered as an approximate estimation of the time of failure and its reliability is heavily dependent on the degree of instrumental and natural data noise. Noise can also decisively delay the identification of OOA and TU points. Smoothing measurements of velocity is thus a crucial process of INV analyses and possibly must rely on quick and easily applicable algorithms. In this study, two moving averages and an exponential smoothing function have been tested in order to filter velocities acquired at three large slope failures. In general, moving averages seemed to provide better fitting than exponential smoothing. It also resulted that filters have different performances depending on the shape of the inverse velocity curve and on the intensity of deformation rates. Following these findings, some general guidelines to suitably perform data smoothing were deduced. Because it may not be possible to safely carry out this interpretation process until the late stages of acceleration or even thereafter, some standard procedures for the setup of failure alarm levels were defined by employing both short- and long-term moving averages. The first alarm threshold identifies the OOA and is established following the implementation of the moving average crossover rule. The second alarm threshold corresponds to the time of activation of the failure window, which is calculated based on the forecasts of both SMA and LMA. The failure window defines the time frame during which the occurrence of the collapse event is considered most probable. The user would be able to anticipate when the thresholds will presumably be exceeded and, therefore, when the pre-determined response actions may have to be taken. Equally importantly, if several measuring points are included within the dataset, major benefit would derive from the ability to simultaneously and automatically assess their alarm level. The analysis of the spatial distribution of the alarms would determine a prompt estimation of the size of the instability and of the probability of an imminent failure.

References

Apuani T, Corazzato C, Cancelli A, Tibaldi A (2005) Physical and mechanical properties of rock masses at Stromboli: a dataset for volcano instability evaluation. Bull Eng Geol Environ 64:419–431

Calvari S, Bonaccorso A, Madonia P, Neri M, Liuzzo M, Salerno GG, Behncke B, Caltabiano T, Cristaldi A, Giuffrida G, La Spina A, Marotta E, Ricci T, Spampinato L (2014) Major eruptive style changes induced by structural modifications of a shallow conduit system: the 2007–2012 Stromboli case. Bull Volcanol 76(7):1–15

Carlà T, Intrieri E, Di Traglia F, Nolesini T, Gigli G, Casagli N (2016) Guidelines on the use of inverse velocity method as a tool for setting alarm thresholds and forecasting landslides and structure collapses. Landslides. doi:10.1007/s10346-016-0731-5

Crosta GB, Agliardi F (2003) Failure forecast for large rock slides by surface displacement measurements. Can Geotech J 40:176–191

Cruden DM, Masoumzadeh S (1987) Accelerating creep of the slopes of a coal mine. Rock Mech Rock Eng 20:123–135

Di Traglia F, Nolesini T, Intrieri E, Mugnai F, Leva D, Rosi M, Casagli N (2014) Review of ten years of volcano deformations recorded by the ground-based InSAR monitoring system at Stromboli volcano: a tool to mitigate volcano flank dynamics and intense volcanic activity. Earth Sci Rev 139:317–335

Di Traglia F, Battaglia M, Nolesini T, Lagomarsino D, Casagli N (2015) Shifts in the eruptive style at Stromboli in 2010–2014 revealed by ground-based InSAR data. Sci Rep 5. doi:10.1038/srep13569

Dick GJ, Eberhardt E, Cabrejo-Liévano AG, Stead D, Rose ND (2015) Development of an early-warning time-of-failure analysis methodology for open-pit mine slopes utilizing ground-based slope stability radar monitoring data. Can Geotech J 52:515–529

Fukuzono T (1985) A new method for predicting the failure time of slopes. In: Proceedings, 5th international conference & field workshop on landslides, pp 145–150

Gigli G, Fanti R, Canuti P, Casagli N (2011) Integration of advanced monitoring and numerical modeling techniques for the complete risk scenario analysis of rockslides: the case of Mt. Beni (Florence, Italy). Eng Geol 120:48–59

Havaej M, Wolter A, Stead D (2015) The possible role of brittle rock fracture in the 1963 Vajont Slide, Italy. Int J Rock Mech Min Sci 78:319–330

Helmstetter A, Sornette D, Grasso J-R, Andersen JV, Gluzman S, Pisarenko V (2004) Slider block friction model for landslides: application to Vaiont and La Clapière landslides. J Geophys Res 109. doi:10.1029/2002JB002160

Intrieri E, Di Traglia F, Del Ventisette C, Gigli G, Mugnai F, Luzi G, Casagli N (2013) Flank instability of Stromboli volcano (Aeolian Islands, Southern Italy): integration of GB-InSAR and geomorphological observations. Geomorphology 201:60–69

Mufundirwa A, Fujii Y, Kodama J (2010) A new practical method for prediction of geomechanical failure. Int J Rock Mech Min Sci 47:1079–1090

Muller L (1964) The rock slide in the Vaiont valley. Felsmech Ingenoirgeol 2:148–212

Newcomen W, Dick G (2015) An update to strain-based pit wall failure prediction method and a justification for slope monitoring. Proc Slope Stab 2015:139–150

Olson D (2004) Have trading rule profits in the currency markets declined over time? J Bank Financ 28:85–105

Petley DN, Bulmer MH, Murphy W (2002) Patterns of movement in rotational and translational landslides. Geology 30(8):719–722

Read J, Stacey P (2009) Guidelines for open pit slope design. CSIRO Publishing, Collingwood, VIC, Australia

Rizzo AL, Federico C, Inguaggiato S, Sollami A, Tantillo M, Vita F, Bellomo S, Longo M, Grassa F, Liuzzo M (2015) The 2014 effusive eruption at Stromboli volcano (Italy): Inferences from soil CO2 flux and 3He/4He ratio in thermal waters. Geophys Res Lett 42(7):2235–2243

Rose ND, Hungr O (2007) Forecasting potential rock slope failure in open pit mines using the inverse-velocity method. Int J Rock Mech Min Sci 44:308–320

Saito M (1969) Forecasting time of slope failure by tertiary creep. In: Proceedings, 7th international conference on soil mechanics and foundation engineering, pp 677–683

Shambora WE, Rossiter R (2007) Are there exploitable inefficiencies in the futures market for oil? Energ Econ 29:18–27

Voight B (1988) A method for prediction of volcanic eruption. Nature 332:125–130

Voight B (1989) A relation to describe rate-dependent material failure. Science 243:200–203

Zakšek K, Hort M, Lorenz E (2015) Satellite and ground based thermal observation of the 2014 effusive eruption at Stromboli volcano. Remote Sens 7(12): 17190–17211

Zavodni ZM, Broadbent CD (1980) Slope failure Kinematics. Bull Can Inst Min 73(16):69–74

TXT-tool 2.886-1.2 Guidelines for Landslide Monitoring Systems

An-Bin Huang and Wen-Jong Chang

Abstract

This report aims at providing guidelines for setting up field monitoring programs for landslides. Recognizing that many parameters may be involved in landslide monitoring, the report concentrates on ground displacement, rainfall, soil moisture and groundwater conditions as the key elements. The types of instrumentations involved and their field installations are presented. The options of using automated electrical or optical fiber sensor systems are described. A few cases of applying fully automated field monitoring schemes for slope stability monitoring are presented to demonstrate the capabilities of currently available techniques.

Keywords

Landslides · Rainfall · Groundwater · Slope stability · Field monitoring

Contents

1 **Introduction** 553
2 **What Is a Landslide?** 554
3 **What Triggers a Landslide?** 554
4 **Why Landslide Monitoring?** 554
5 **Key Elements in Landslides Monitoring** 555
5.1 Underground Lateral Displacement 555
5.2 Rainfall Monitoring 557
5.3 Groundwater Monitoring 557
6 **Automation of Field Monitoring** 557
7 **Field Monitoring Using Fiber-Optic Sensors** . 558
8 **Field Monitoring Using Integrated MEMS Sensor System** 562
9 **Examples of Landslide Monitoring** 562
9.1 Highway Slope at Five Turn Point 563
9.2 Hua-Fan University Campus Slope 564
10 **Techniques Under Development** 565
References 566

1 Introduction

This report will first present a series of questions and answers. The questions are those likely to be raised by readers who are not familiar with landslide monitoring. This arrangement enables the answers to be located easily by following the questions. The background information leading to

A.-B. Huang (✉)
Department of Civil Engineering, National Chiao Tong University, Hsin Chu 30010, Taiwan
e-mail: huanganbin283@gmail.com

W.-J. Chang
Department of Civil Engineering, National Cheng Kung University, Tainan 70101, Taiwan
e-mail: wjchang@mail.ncku.edu.tw

K. Sassa et al. (eds.), *Landslide Dynamics: ISDR-ICL Landslide Interactive Teaching Tools*, https://doi.org/10.1007/978-3-319-57774-6_40

the outline of how we conduct field landslide monitoring is presented by the following three questions.

2 What Is a Landslide?

According to Cruden (1991), the term landslide denotes "the movement of a mass of rock, debris or earth down a slope". The International Geotechnical Societies and United Nations Educational, Scientific and Cultural Organization (UNESCO) created a Working Party (WP) and published directory of World Landslide Inventory (WLI). A landslide can be classified (Varnes 1978; Cruden and Varnes 1996; WP/WLI 1990, 1993) by two criteria:

- The type of movement (falls, topples, slides spreads, flows), and
- The type of material involved in the movement (rock, debris, earth).

Combining movement and material type terms allows an appropriately descriptive landslide name to be formulated. This classification method is widely used throughout the world, and will be followed in this report.

3 What Triggers a Landslide?

Cruden and Varnes (1996) indicated three broad types of landslide processes or causes:

- increases in shear stress,
- contributions to low strength, and
- reduction in material strength.

The processes and characteristics that contribute to landslides fall into four categories: geological causes, morphological causes, physical causes and human causes. While there may be multiple causes for a landslide, there can be only one trigger (Varnes 1978). Wieczorek (1996) stated that a trigger is "an external stimulus that causes a near-immediate response in the form of a landslide by rapidly increasing the stress or reducing the strength of slope material". In some cases the landslide may occur without an apparent attributable trigger because of a variety or combination of causes, such as chemical or physical weathering of material, that gradually bring the slope to failure. The requisite short time frame of cause and effect is the critical element in the identification of a landslide trigger. According to Wieczorek (1996), the most common natural landslide triggers include intense rainfall, rapid snowmelt, water-level changes, volcanic eruptions, and earthquake shaking. Human activities such as excavation for road cuts and irrigation can also trigger landslides. This report however, will not be dealing with landslides triggered by volcanic eruptions or earthquake shaking.

4 Why Landslide Monitoring?

Intuitively, the goal of field monitoring is the identification of the cause(s) of an unstable slope or the forensic investigation of an earlier landslide. This goal can be difficult and sometimes impossible to accomplish, because of the time involved and sheer scale of the dimensions of a landslide. With this background in mind, various monitoring schemes have been used for the following:

- Determining the depth and shape of the sliding mass in a developing landslide.
- Determining the strength parameters and design of remedial measures.
- Determining the lateral and vertical movements, including rate of movements, within a sliding mass.
- Measuring groundwater table, soil moisture content or pore water pressure as a basis for effective stress analysis.
- Observing marginal natural or excavated slopes to evaluate the effects of construction or rainfall on slope stability.

- Observing and evaluating the effectiveness of control or slope stabilization measures, and
- Using the measurement data as a basis for warnings of impending danger.

5 Key Elements in Landslides Monitoring

The types of physical quantities involved in landslide monitoring can be many. This is especially true for slope surface measurements, such as ground cracking, tilting and heaving, in addition to displacement. The key elements that have a strong relationship with ground stability include ground displacement (especially underground displacement), rainfall, soil moisture and groundwater conditions. Monitoring of these parameters is likely to yield useful information as to the state of stability of a given slope. The reasons behind these measurements and procedures for their execution are now described.

5.1 Underground Lateral Displacement

Measurement of underground lateral displacement is useful because the results reveal the depth and thickness of slide shear zones, and the magnitude, rate and direction of landslide movement. A commonly used method for underground lateral displacement measurement is to use inclinometer casing. The method involves:

1. Installing the inclinometer casing in a vertical or near vertical position. The inclinometer casing should extend beyond the expected shear zone. One or multiple borings along the central axis of the landslide should be selected for inclinometer casing installation.
2. Use a traversing probe (inclinometer probe, IP) to measure the deflection of the inclinometer casing induced by ground displacement.

The inclinometer casing (shown in Fig. 1) can be made of ABS (Acrylonitrile Butadiene Styrene) or aluminum; their typical diameters range from 48 to 85 mm. The interior of the inclinometer casing has four longitudinal wheel-grooves spaced 90° apart to guide the inclinometer probe. Only one set of opposite grooves, in the anticipated direction of displacement, is used. For installation of the inclinometer casing, the following procedure is generally used:

1. Drill a borehole with a diameter large enough to insert the inclinometer casing.
2. Fill the casing with water to overcome buoyancy, if groundwater or drilling fluids are present in the borehole. Avoid applying forces to or twisting the top of the casing.

Fix the inclinometer casing in place by filling the space between the casing and borehole with cement grout or a granular material. In general, grouting is preferred. Grouting ensures that the annular space is reliably free of voids and soft zones. Use non-shrink, cement-bentonite grout with stiffness compatible to the surrounding material. Table 1 shows the recommended grout mix for various ground conditions. The grouting can be conducted using a tremie pipe, using a procedure shown in Fig. 2. Use sand or gravel as granular backfill, when monitoring is to be carried out immediately after casing installation and no time is available for the grout to cure; avoid bridging or voids in the granular backfill.

Figure 3 shows an inclinometer probe (IP) and its signal cable. Markers are engraved in the signal cable to indicate the depth of the probe when taking readings. The spring-loaded wheels are compatible with the wheel-grooves in the inclinometer casing (see Fig. 1). To take readings, the inclinometer probe is first inserted into the bottom of the inclinometer casing. The probe readings are taken in successive 50-cm increments (distance between the wheels) as the probe is pulled up. The weight of the probe and its cable assures that the signal cable is straight because of the pull and the marker properly

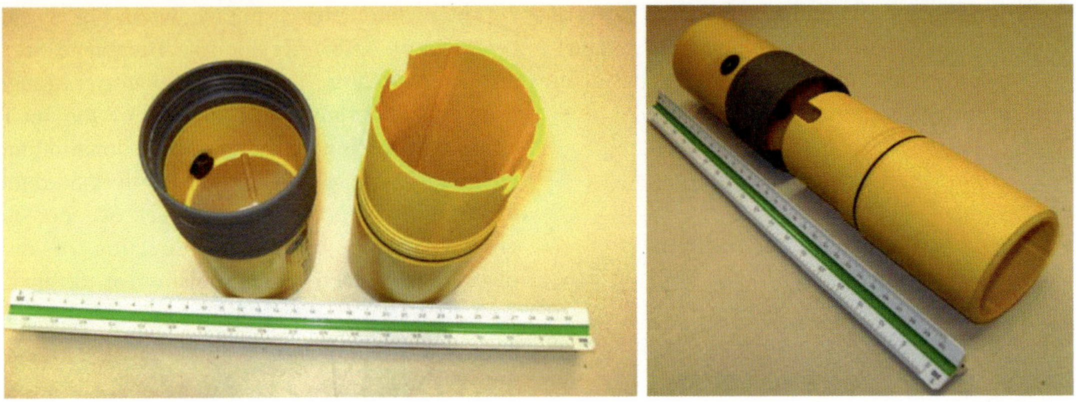

Fig. 1 The ABS inclinometer casing (Roctest)

Table 1 Cement-bentonite grout guidelines (slope indicator)

Materials	Soft soil environment	Medium to hard soil environment
Portland cement (kg)	43	43
Bentonite (powder) (kg)	18	11
Water (l)	284	114

Fig. 2 Casing insertion and grouting with a tremie pipe (slope indicator)

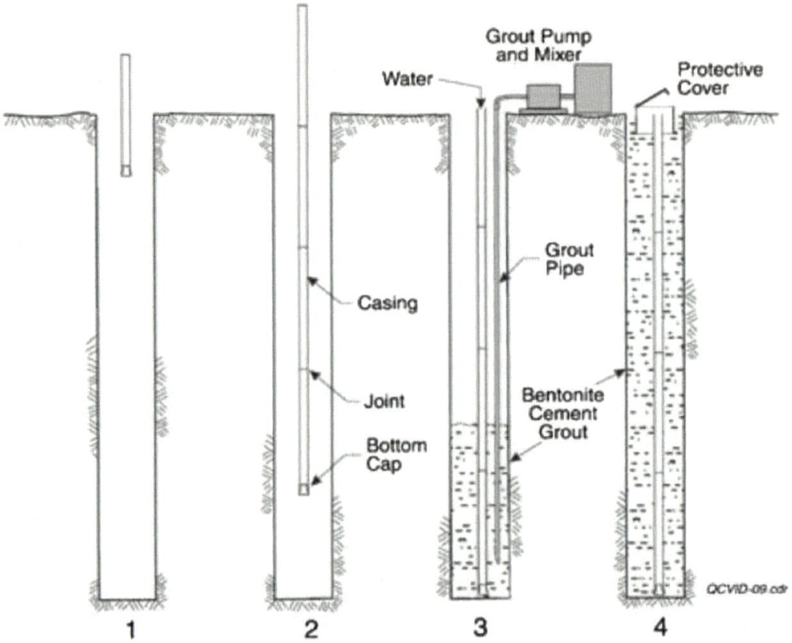

reflects the depth of the probe. The inclinometer probe reading shows the amount of tilt (θ) of the inclinometer probe in the casing. Figure 4 shows a person taking inclinometer probe readings in the field with a wireless readout/data logging unit. Figure 5 shows the principles of

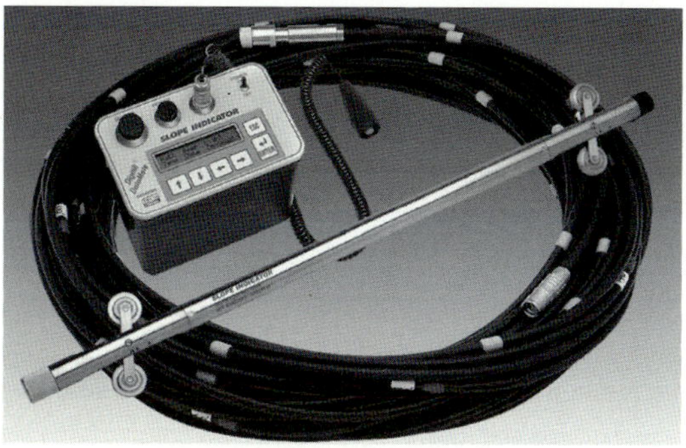

Fig. 3 The inclinometer probe (slope indicator)

Fig. 4 Taking inclinometer probe readings in the field (DECL, Taipei, Taiwan)

inclinometer operation and its data interpretation. In this figure, $\delta\theta$ = change of θ in comparison to the initial reading.

5.2 Rainfall Monitoring

Figure 6 shows a picture of a tipping bucket rain gauge often used to measure the amount of rainfall. The rain gauge consists of a funnel that collects and channels the precipitation into a small container. After a pre-set amount of precipitation falls, the lever tips, dumping the collected water and sending an electrical signal.

5.3 Groundwater Monitoring

Casagrande-type open-end standpipe piezometers, as schematically shown in Fig. 7, are commonly used to monitor groundwater level in a borehole. The open-end at the tip of a plastic pipe, typically made of a porous or slotted tube, is inserted to the bottom of a borehole. The space between the open-end and borehole is filled with sand as a filter. Groundwater enters the standpipe through the porous tube. A tape dropped into the plastic pipe is used to measure the water level in the pipe. A time lag may exist between the water level within the pipe and the surrounding groundwater. Alternatively, the effects of groundwater are directly measured as pore water pressure. A diaphragm type of pressure sensor is used to measure the water pressure. The amount of diaphragm deflection induced by water pressure is sensed by an electrical strain gage (e.g., vibrating wire strain gage). A vibrating wire piezometer is schematically shown in Fig. 8. The time lag is this case is minimal.

6 Automation of Field Monitoring

Automated data logging and transmission can be useful in many ways. For monitoring in remote areas, the automation can minimize both the cost and sometimes the danger of reaching the sensor locations and take readings manually. A fully automated field monitoring system (Fig. 9) may consist of:

- Electrical sensors, such as in-place-inclinometers, rain gauges and piezometers;
- Automated Data Analysis System (ADAS);
- Power supply (battery + solar panel or other forms of power-generating unit);
- Data transmission system.

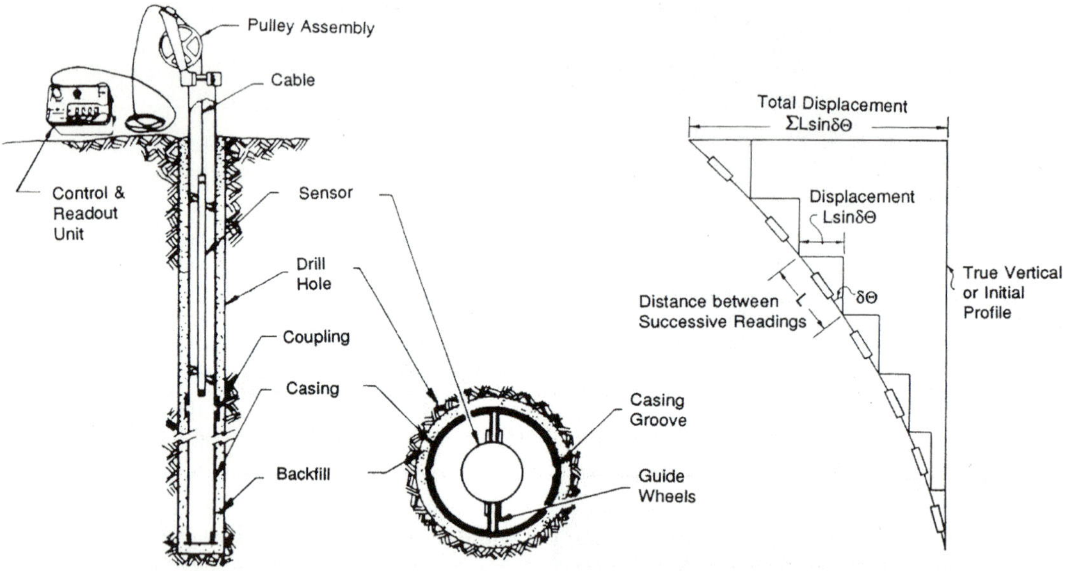

Fig. 5 Principles of inclinometer operation and its data interpretation (Mikkelsen 1996)

Fig. 6 A tipping bucket rain gauge (DECL, Taipei, Taiwan)

The ADAS serves the following functions:

- Regulates excitation voltage for sensors,
- Multiplexing installed sensors,
- Performs analog/digital conversion,
- Data storage,
- Transmits data via a communication system (i.e., internet, GSM, WiFi).

For automated ground displacement monitoring, electrical in-place inclinometers (IPI) as shown in Fig. 10 are often used. A string of in-place inclinometers are connected with rods and permanently inserted in the inclinometer casing. Readings can be taken with an ADAS. Vibrating wire piezometers can be used with the ADAS for automated pore water pressure measurements (Fig. 11).

7 Field Monitoring Using Fiber-Optic Sensors

Optical fibres are made of silica, with a diameter about the same as a human hair, and can transmit light over long distances with very little loss of fidelity. Optical fibres comprise two essential components: a core surrounded by an annular cladding. The core of the optical fibre serves to guide light along the length of the optical fibre.

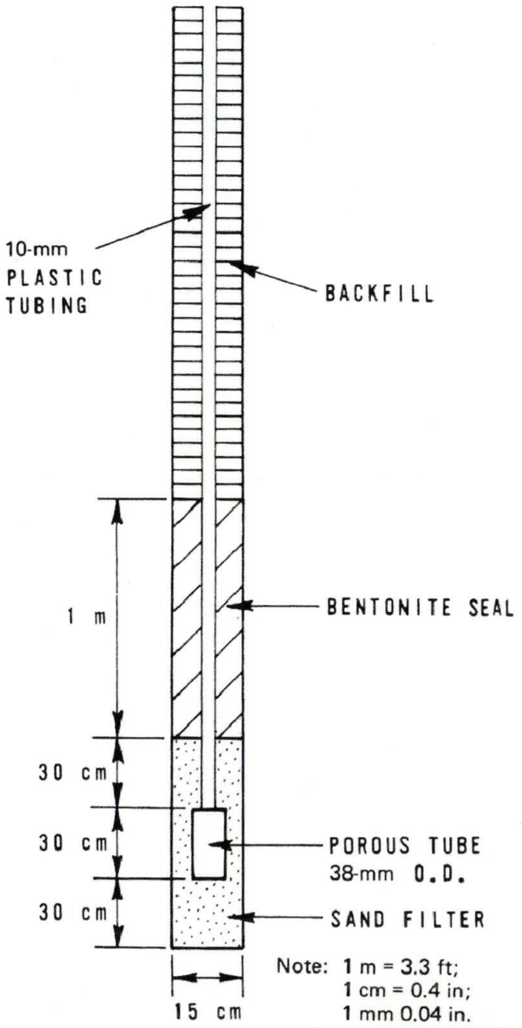

Fig. 7 Casagrande-type open-end standpipe piezometer (Mikkelsen 1996)

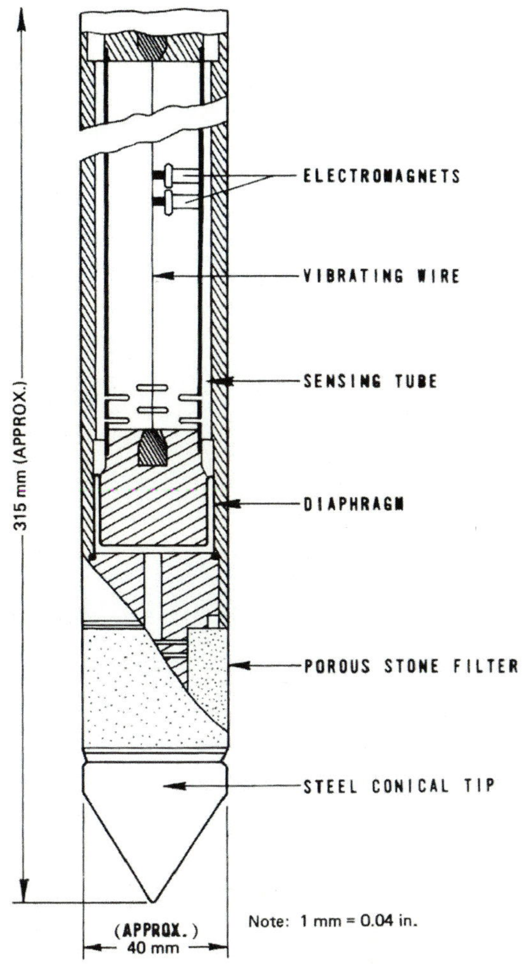

Fig. 8 A vibrating wire piezometer (Mikkelsen 1996)

The cladding has a slightly lower index of refraction than the core. Its primary function is to ensure total internal reflection within the core and that very little light is lost as it propagates along the core of the optical fibre. The typical combined diameter of core and cladding is 125 µm. The silica core/cladding is protected by an acrylic coating. The total outside diameter of an optical fibre with the acrylic coating is 250 µm. By adopting technologies from telecommunication, many fibre-optic-based sensing techniques have been developed. The fibre optic Bragg grating (FBG) is one of the many available forms of optical fibre sensors. An FBG is made by a periodic variation of fibre core refractive index. The typical length of an FBG is 1–20 mm long. When the FBG is illuminated by a wideband light source, a fraction of the light is reflected back due to interference by the FBG. The wavelength of the reflected light is linearly related to the longitudinal strains of the FBG, thus making FBG an ideal strain gage. The returned signal from every FBG carries a unique wavelength domain, making it possible to have multiple FBG elements on the same fibre. The multiplexing among various sensors on a single fibre can be accomplished by wavelength division addressing. The FBG is partially distributive

Fig. 9 A fully automated field monitoring system

Fig. 11 The vibrating wire piezometer (DECL. Taipei, Taiwan)

because only those parts of the optical fibre with FBG are used as strain sensors and these sensors can share the same optical fibre transmission line.

The FBG can be used directly as a strain gage, or, with the help of mechanical components, an FBG can be configured as displacement, pressure or inclination transducers (Ho et al. 2006, 2008). All advantages of the FBG can be inherited in FBG-based transducers for geotechnical engineering monitoring. These advantages can include partially distributive, high resolution; and good signal stability and immunity to electromagnetic interference (EMI), short circuits and lightning. The FBG-based sensors can be used as an individual sensor to reflect a physical quantity at a given location or connected into an array so that the profile of a given or multiple types of physical quantities can be monitored. FBG-based piezometers (Fig. 12), in-place inclinometers

Fig. 10 Field installation of in-place inclinometers (DECL, Taipei, Taiwan)

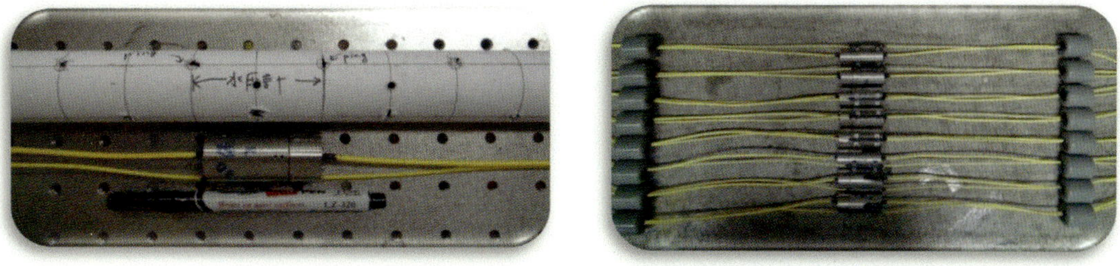

Fibre optic Bragg grating (FBG) pressure sensors.

String of FBG piezometers.

Fig. 12 Regularly spaced fibre optic Bragg grating (FBG) pressure sensors sealed in PVC pipes with filter protection to serve as a string of piezometers. Fibre optic Bragg grating (FBG) pressure sensors. String of FBG piezometers

Fig. 13 The fibre optic Bragg grating (FBG) in-place inclinometer

Fig. 14 The fibre optic Bragg grating (FBG) data logging system

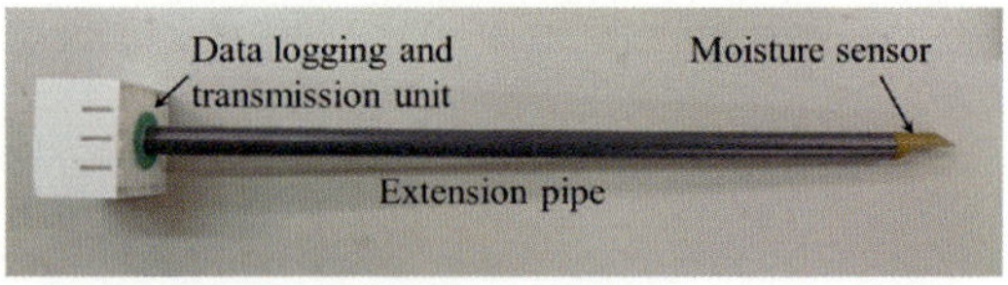

Fig. 15 The MEMS sensor stick

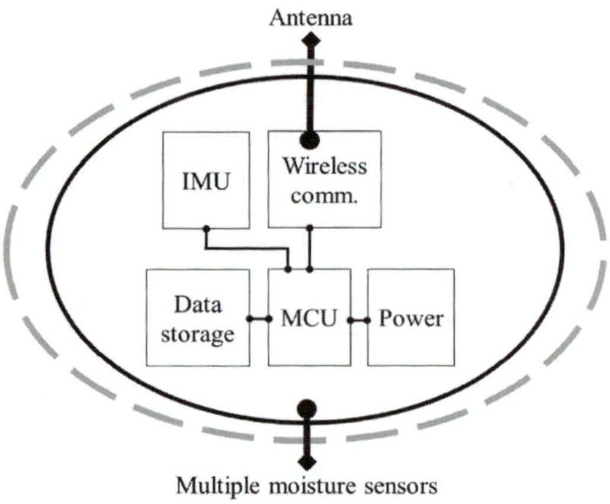

Fig. 16 Components of the MEMS sensor stick

(Fig. 13) and automated data logging systems (Fig. 14) (equivalent to ADAS) have been developed and are becoming available commercially.

8 Field Monitoring Using Integrated MEMS Sensor System

A slope failure can have its failure plane located above the ground water table where the soil is unsaturated. This type of slope failure is usually shallow, localized and involves less amounts of debris (shallow slope failure) but occurs much more frequently. Ideally, for shallow slope failure monitoring, we should measure the negative pore pressure in the field. The use of FBG piezometers for negative pore pressure measurements would require frequent maintenance to assure saturation which is rather impractical. The MEMS based sensor stick as shown in Fig. 15 is a low cost sensing system specifically developed for monitoring shallow slope failure. Major control and sensing components include a Microcontroller Unit (MCU), an inertial measurement unit (IMU) and a wireless communication unit as shown in Fig. 16. The IMU contains a gyroscope and a three axis G sensor. The IMU readings reveal tilting and linear motion of the sensor unit. Multiple soil moisture sensors can be placed on the extension pipe (see Fig. 15) at various depths up to 3 m below the ground surface. The moisture readings are converted to negative pore pressure (or matrix suction) via the soil water characteristics curve (SWCC). The system is controlled by the MCU and data are transmitted wirelessly to an aggregator located in the field. The system is compact, low cost and very energy efficient. The MEMS based sensor stick thus compensates well with the FBG sensing system.

9 Examples of Landslide Monitoring

The following section presents several examples of landslide monitoring, their field set up and available data.

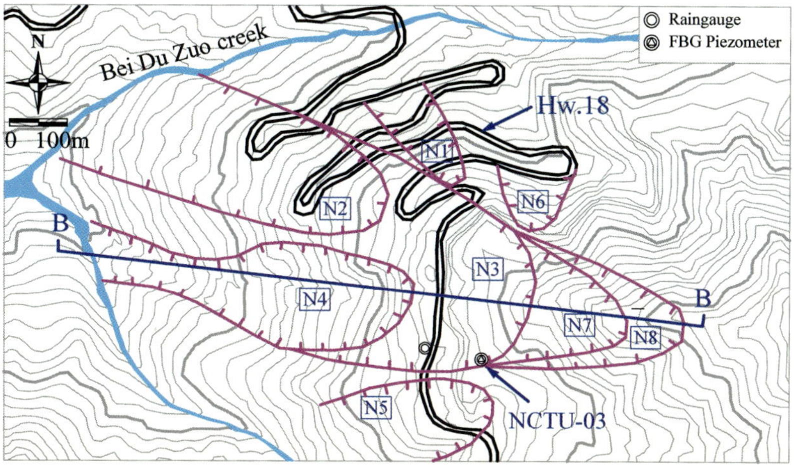

Fig. 17 Topographic map of the five turn point (Land Engineering Consultants, Co., Ltd. 2007)

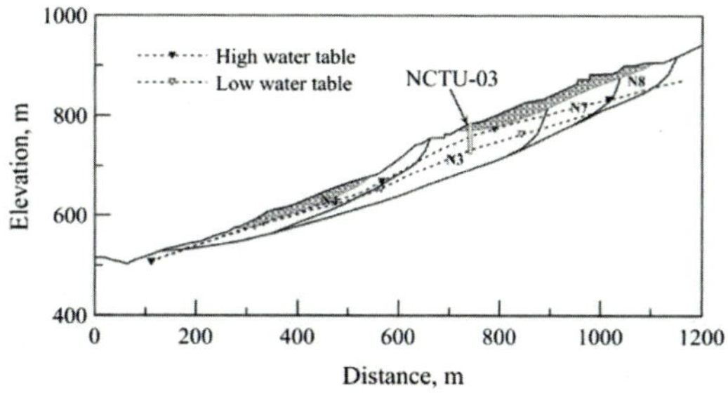

Fig. 18 Section B–B of the Five Turn Point (Land Engineering Consultants, Co., Ltd. 2007)

9.1 Highway Slope at Five Turn Point

A section of Highway 18 in Jia-Yi, Taiwan, referred to as the Five Turn Point, has been selected as the most dangerous highway in Southern Taiwan. At least eight sectors (designated as N1–N8 in Fig. 17) have been identified as having either previous slope failure or signs of continuous movement. Figure 18 shows a cross-sectional view of section B-B,, which has an average slope angle of 23°. Subsurface explorations revealed that the subject area was covered by 0–26 m of colluvial material underlain by interlayered fractured sandstone and shale extending from below the colluvium to over 200 m (deepest bore hole available) below the ground surface. The groundwater could rise from its low level by more than 20 m as a result of heavy rainfalls according to available the open-end piezometer data shown in Fig. 18. The sudden and significant change in groundwater table is believed to be a major cause for earlier slope failures in this area.

A 60 m deep borehole (marked as NCTU-03 in Figs. 17 and 18) was used to install the FBG piezometer array. The FBG pressure transducer, used as a piezometer, was fitted inside of a 28 mm (inner diameter) and 32 mm (outer diameter) PVC pipe. A comparison between an array of FBG piezometers installed in a single borehole to individual standpipes is depicted in Fig. 19. All sensors were connected to an on-site computer using optical fiber cables for optical signal interrogation and data logging/transmission. Field installation was completed in September, 2007.

The Five Turn Point slope endured three major typhoons (i.e., Sinlaku in 2008, Morakot in 2009, and Fanapi in 2010) since September 2008, and remained stable till now (January, 2013). Of the three typhoons, Morakot was the most damaging. A histogram of daily precipitation during typhoon Morakot at Five Turn Point is shown in Fig. 20. Figure 21 shows a set of representative pressure head (hp) profiles based on the FBG piezometer readings recorded from the beginning of typhoon Morakot to the time when the pressure head reached the maximum values.

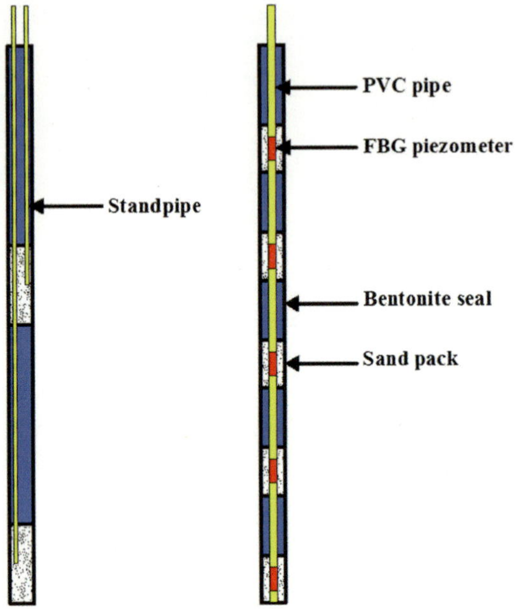

Fig. 19 Comparison between the individual standpipe and FBG piezometer array

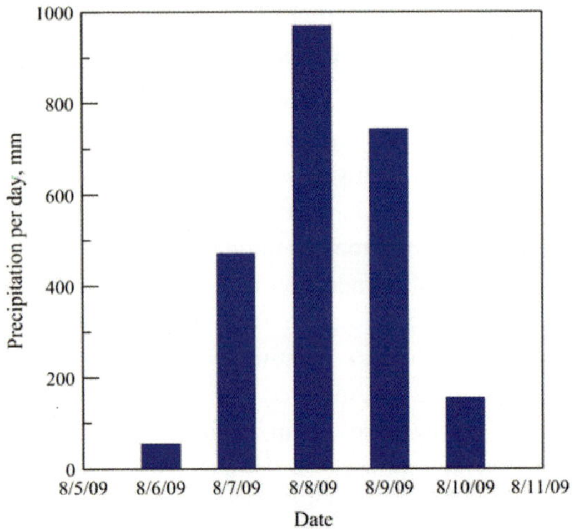

Fig. 20 Rainfall record during typhoon Morakot

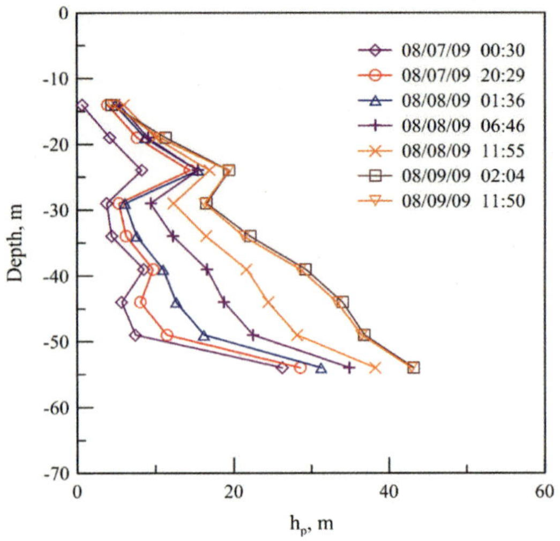

Fig. 21 Pressure head profiles from FBG piezometer reading (Huang et al. 2012)

Fig. 22 Field installation of an FBG in-place inclinometer on the Hua-Fan university slope

9.2 Hua-Fan University Campus Slope

The campus of Hua-Fan University is built on a hill side in New Taipei City, Taiwan. The surface of these steep slopes was mostly covered with colluvial material. Ground movement and the potential for landslides are major concerns for the university. Figure 22 shows the field installation of the FBG in-place inclinometer. The FBG in-place inclinometer readings and their

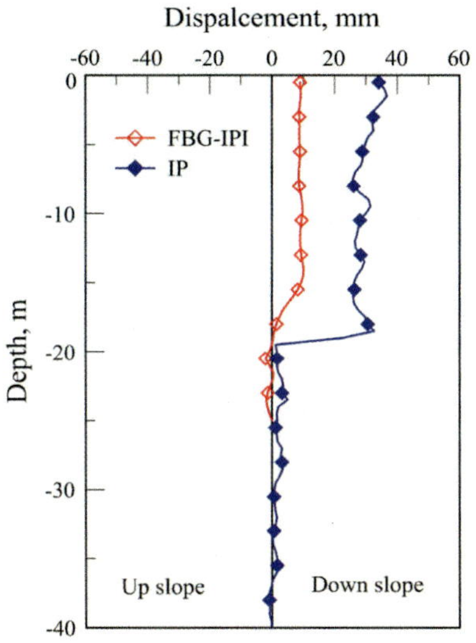

Fig. 23 Comparison between the FBG-IPI and IP readings

comparison with those of an inclinometer probe are shown in Fig. 23. Both types of readings, albeit taken in different nearby boreholes, showed a shear plane at approximately 20 m below the ground surface.

10 Techniques Under Development

By sharing data storage and computing with cloud computing, as conceptually described in Fig. 24, the efficiency and reliability of the data logging-communication chain can be much improved. The applications and capabilities of cloud computing in landslide monitoring are expanding rapidly.

The field monitoring performed in the field usually provides information at a given location. By coupling it with numerical analysis, the

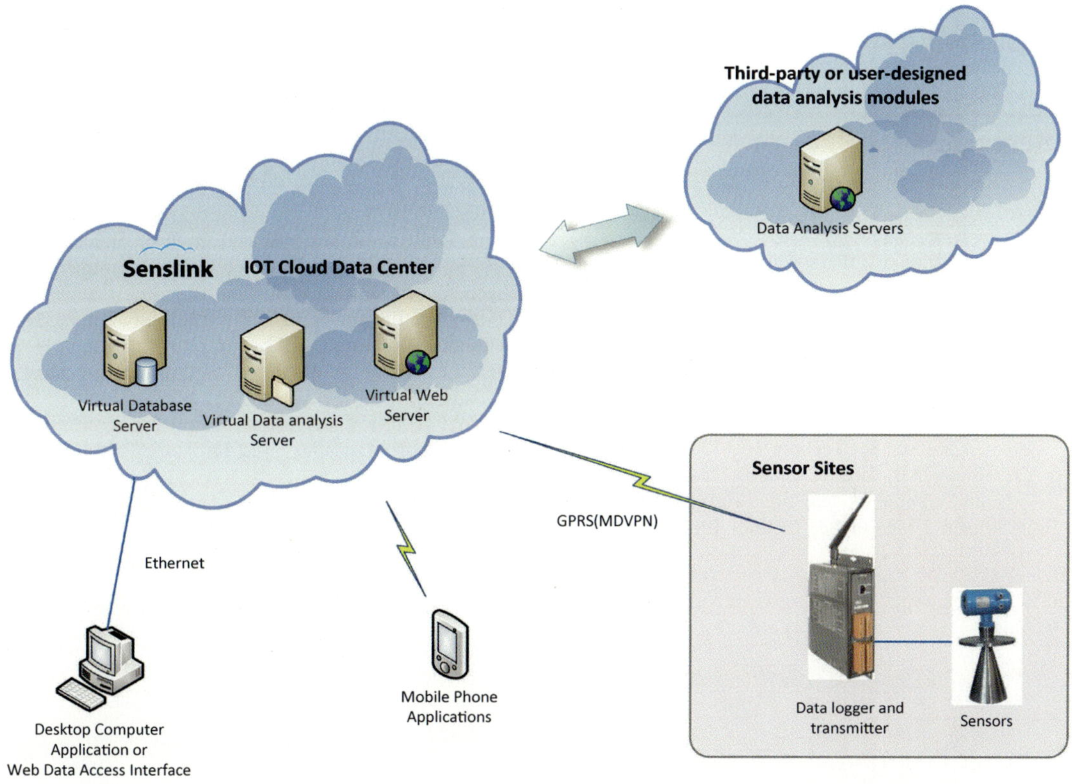

Fig. 24 Cloud computing based monitoring system (Senslink, Hsichu, Taiwan)

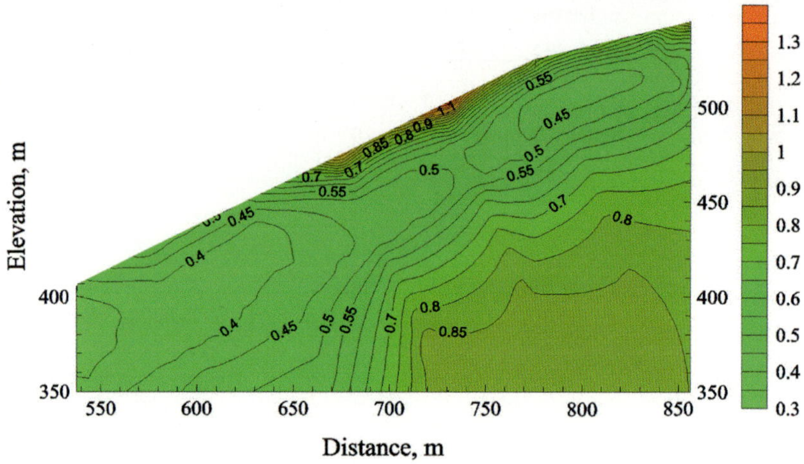

Fig. 25 Coupling real-time field monitoring with numerical analysis

information obtained at discrete locations can be extrapolated to cover a cross section of a slope as described in Fig. 25. By coupled data logging and numerical analysis on a real-time basis, an improved warning system for landslides that deals with the whole cross section of a slope can be established.

References

Cruden DM (1991) A simple definition of a landslide. Bull Int Assoc Eng Geol 43:27–29

Cruden DM, Varnes DJ (1996) Landslide types and processes. In: Turner AK, Schuster RL (eds) Landslides, investigation and mitigation, TRB special report 247. Transportation Research Board, National Research Council, Washington, DC, pp 36–75

Ho YT, Huang AB, Lee JT (2006) Development of a fiber Bragg grating sensored ground movement monitoring system. J Measur Sci Technol 16:1733–1740

Ho YT, Huang AB, Lee JT (2008) Development of a chirped/differential optical fiber Bragg grating pressure sensor. J Measur Sci Technol 19(4):6

Huang AB, Lee JT, Ho YT, Chiu YF, Cheng SY (2012) Stability monitoring of rainfall induced deep landslides through pore pressure profile measurements. Soils Found 52(4):737–747

Land Engineering Consultants, Co., Ltd. (2007) Highway 28.9 K ~ 31.5 K (Five Turn Point) landslide investigation, remediation planning and safety evaluation, 3rd overall report. Department of Highway Maintenance, Section V (in Chinese)

Mikkelsen PE (1996) Chapter 11, Field instrumentation. In: Turner AK, Schuster RL (eds) Landslides, investigation and mitigation. TRB special report 247. Transportation Research Board, National Research Council, Washington, DC, pp 278–316

Varnes DJ (1978) Slope movement types and processes. In: Landslides: analysis and control. Special report 176. Transportation Research Board, Washington, DC

Wieczorek GF (1996) Landslide triggering mechanisms: investigation and mitigation. Transportation Research Board. Washington, DC

WP/WLI (1990) A suggested method for reporting a landslide. Bull Int Assoc Eng Geol 41

WP/WLI (1993) A suggested method for describing the activity of a landslide. Bull Int Assoc Eng Geol 47

TXT-tool 2.386-1.2 Practice Guidelines on Monitoring and Warning Technology for Debris Flows

Johannes Hübl and Matjaž Mikoš

Abstract

This paper deals with practical guidelines on installing and running equipment for monitoring of debris flows, based on two decades of experiences in selected instrumented debris flow catchments in Europe. Starting with definitions and aims of any debris-flow monitoring system, especially those that are part of an early warning system, the chain of perception, monitoring and warning is presented. For measuring debris-flow signals, the paper focusses on issues regarding the positioning of measuring devices in the triggering, transport, and depositions areas. Different debris-flow data are recorded, communicated and archived, and all that needs a stable energy supply. The paper ends with a short review of installed debris-flow monitoring systems in Europe (with suggested further reading) and a short description of a human-based early warning system.

Keywords

Debris flows · Guidelines · Monitoring · Triggering

Contents

1 **Introduction** ... 568
2 **Definition and Aims of Monitoring** ... 568
3 **Chain of Perception, Monitoring and Warning** ... 568
4 **Signals** ... 569
5 **Positioning of Measuring Devices** ... 570
5.1 Debris-Flow Triggering Area ... 571
5.2 Debris-Flow Pathway ... 571
5.3 Debris-Flow Deposition Area ... 579
6 **Monitoring Data Recording** ... 579
7 **Monitoring Data Communication** ... 581
8 **Energy Supply** ... 581
9 **Monitoring Data Archiving** ... 581
10 **Debris-Flow Monitoring Systems in Europe** ... 581
11 **Debris-Flow Warning Systems** ... 582
12 **Conclusions** ... 583
References ... 583

J. Hübl
Institute of Mountain Risk Engineering, University of Natural Resources and Life Sciences, Peter-Jordan-Straße 82, 1190 Vienna, Austria
e-mail: johannes.huebl@boku.ac.at

M. Mikoš (✉)
Faculty of Civil and Geodetic Engineering, University of Ljubljana, Jamova c. 2, 1000 Ljubljana, Slovenia
e-mail: matjaz.mikos@fgg.uni-lj.si

K. Sassa et al. (eds.), *Landslide Dynamics: ISDR-ICL Landslide Interactive Teaching Tools*, https://doi.org/10.1007/978-3-319-57774-6_41

1 Introduction

In recent years, monitoring of landslides has improved through the development of new monitoring equipment, automatic measurements done by computerized equipment and an increase in equipment.

Within the last two decades several debris flow catchments in Europe were equipped with monitoring systems for a continuous measurement of important parameters describing debris flow initiation and flow dynamics. These systems are mainly operated by scientific institutions and therefore the main focus is on investigations of different aspects of the debris flow process itself. Further development of warning systems has taken place in recent years, because of the availability of cost-efficient and reliable sensors, the rapid enhancement of data transmission and the derivation of characteristic factors of already recorded debris flow events. Nowadays parameters of triggering (precipitation) and transport (flow depth, acoustic signals) are recorded, stored and transmitted in a more or less standardized way.

2 Definition and Aims of Monitoring

Monitoring (of the term "to monitor": to observe, to supervise) is a systematic observation of a process or a system. Depending on the objectives and applied technical means such an observation (surveillance) may be executed for certain limited periods of time or permanently continuous. For monitoring one can define different aims:

1. Observations to better understand the processes within the system, to obtain data to develop, test and calibrate models. Even though this goal is primarily of scientific interest, it also forms the basis for all further practical applications.
2. The detection of thresholds to be able to intervene in the process and cause changes to the process flow. For debris flows one can say that these processes in fact cannot be directly controlled; field applications for such a preventive goal for control of debris-flows can be, for example, performed via the construction of retention basins.
3. Recognition of the exceedance of thresholds to initiate further actions that cannot change the process itself. These applications can be found especially in the area of early warning or alarms, but also in surveillance of structures.

3 Chain of Perception, Monitoring and Warning

The systematic observation using technical means follows a chain of perception, monitoring and warning (Fig. 1) that is also typical for human beings. The signals (stimuli) are detected by appropriate sensors, converted into electrical signals (transduction or transformation) and processed (processing). In this case, we can just store the signals that are below a threshold of perception, and the signals above such a limit (perception) are used to determine further actions to be taken. In the latter case, the signals above the thresholds are compared with previous signal patterns (recognition). If a rule for the signal pattern recognition is present and the signal strength is large, fixed action steps can be executed.

The essence of systematic recording of the signals from debris flows is therefore the right choice of the needed "stimuli" to be detected as being transmitted from a debris flow, and the knowledge of the corresponding signal pattern. To obtain the longest possible response time, the required time interval (t_1–t_2) for perception and recognition should be short. After the time of the alert (t_2), the transmission paths of communication determine further the timing. When the warning according to the transmission protocol reaches the recipient(s) (t_3), it begins the time for action to achieve the desired protection goal (t_4).

During monitoring, different signals may be recorded at different time intervals. With rare debris flows, the measurement system can remain most of the time in a "sleep mode". In this mode,

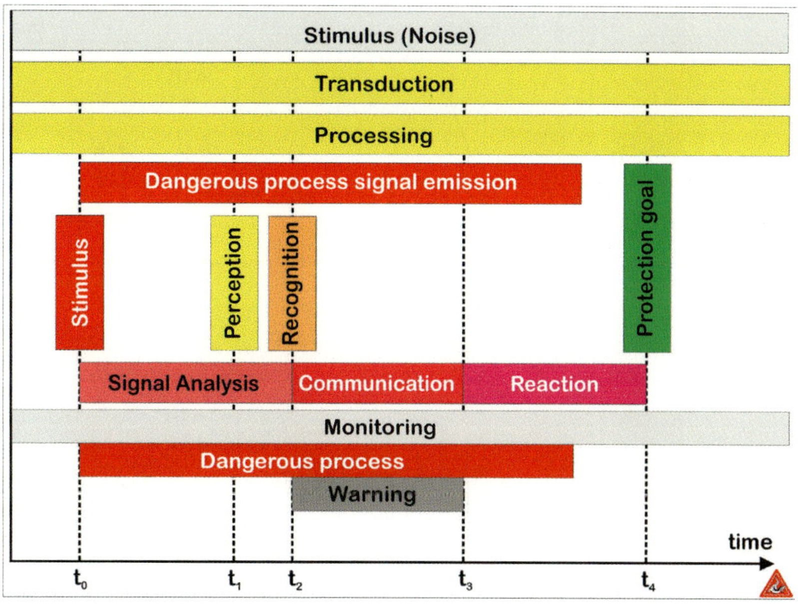

Fig. 1 From perception via monitoring to warning (From Hübl and Mikoš 2014)

the system should record signals at regular intervals (about 10–15 min) and should send messages by itself when the system detects a malfunction (maintenance messages).

Nevertheless, the system should, if a predefined signal pattern is detected, immediately "wake up", send messages, trigger actions (e.g., starting video recording) and save relevant signals with high temporal resolution. In this mode then the monitoring system works as a warning system.

4 Signals

To detect the signals (input variables, mostly physical parameters) of debris flows, different sensors can be used. In the measuring instruments, the signal is converted to a measured value, the output signal, which is presented mostly as mVs. On the basis of calibration functions, this measured value within the measurement range can be back-calculated into the desired measured value.

Signals can be divided into classes based on their properties (Rupprecht 2013). If the function values are defined only for discrete (usually equidistant) points in time, then it is called the signal in discrete time. If the function value for each time (t) defines a continuous time domain, then the signal is called a continuous signal. Accordingly, we referred to the signal as a discrete or continuous signal, if the value range for the signal is discrete or continuous.

Therefore, one can in principle differentiate four different signal types (Fig. 2):

(a) Continuous-time continuous-value signals
(b) Discrete-time continuous-value signals
(c) Continuous-time discrete-value signals
(d) Discrete-time discrete-value signals

Simplified, one can denote continuous-time continuous-value signals as analog signals (strip chart), and discrete-time discrete-value signals as digital signals.

The signal can include both real numbers, or can be made only of the numerical values 0 and 1. The latter can be compared with a switch that allows only one answer, with yes or no. These signals represent the ideal "wake-up signals" for a sleeping monitoring system.

The signals are usually measured in constant time intervals (synchronous). This facilitates evaluation, particularly when the measurement values of different sensors are compared with

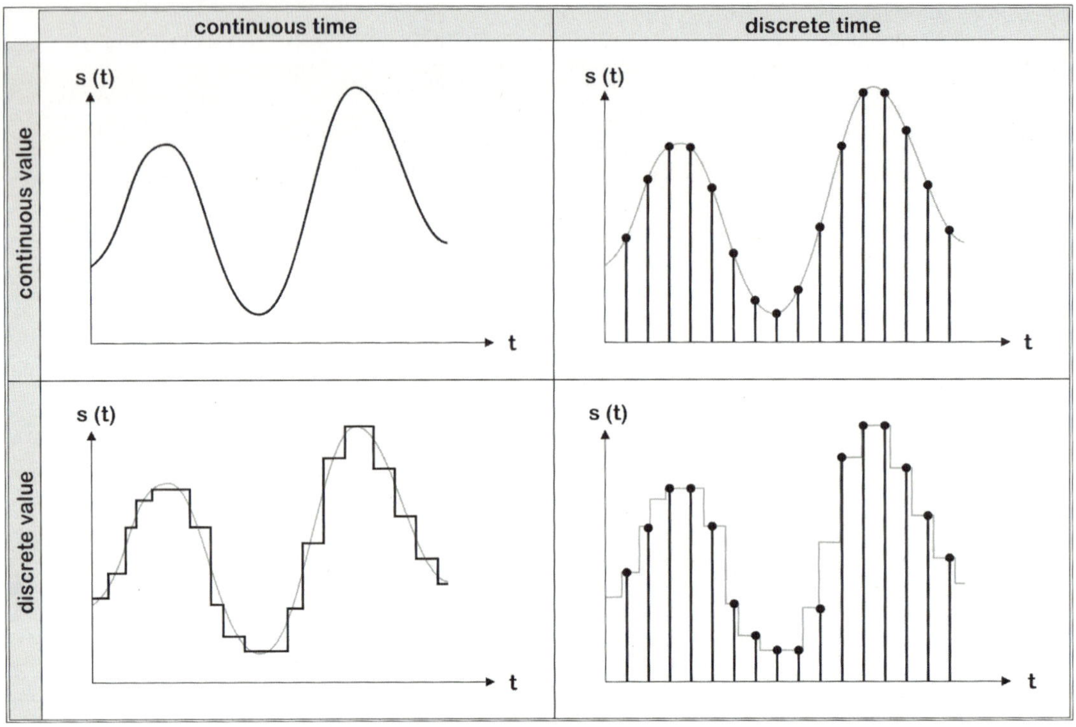

Fig. 2 Main characteristics of signals (adapted from Rupprecht 2013)

each other. For special applications, especially if one wants to achieve a higher resolution of the magnitude of the measured variable, the time can be detected when the measured variable undergoes a specified change: e.g., in rainfall detection when the precipitation increases by 0.1 mm. This is called event-driven or asynchronous recording.

5 Positioning of Measuring Devices

One can assess how likely a catchment is to produce debris flows by systematic observations of debris flows. If debris flows in a selected torrent channel do not often occur, in principle a debris-flow monitoring system monitors only the status I, in which fluvial processes predominate (Hübl 2009). If one wants to draw conclusions about the changes in a torrent system, especially conclusions about the changing tendencies with regard to debris flow initiation, of course, the parameters that reflect these tendencies are of paramount importance. A key parameter in this regard is soil moisture. To form conclusions about the change of the torrent system into the state II, characterized by debris-flow processes, of course, data on a triggering event is necessary. First and foremost, rainfall (precipitation) data collection is needed, which is measured locally, but also spatially by means of weather radar. After a debris flow has been already triggered, measurements that detect its movement are a focus. Debris flows typically show a sudden increase in their flow depths and they usually last only a short time. In case of debris-flow pulses, their duration can be limited to only a few seconds, somewhat longer pulses can last a few minutes, but very rarely a debris-flow event exceeds a period of half an hour.

If one wishes to detect a single debris-flow event in order to send this information to a decision point in a monitoring or warning system, or also to other gauges along the torrent channel in order to "wake them up", one is faced with the following problem (Fig. 3, from Kienholz 1998).

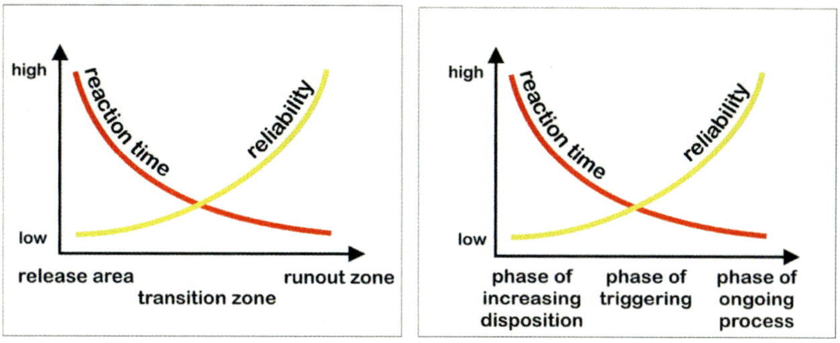

Fig. 3 Reliability and time for identification based on the localization of the alarm devices or on the phase of process (Kienholz 1998)

If the observation of the variable tendency for debris flows initiation is included into a debris-flow monitoring system, i.e., the time prior to the initiation of a debris flow, this will result in a high reaction time, but also the chance for "false alarms" is then very high. If a debris flow has already been triggered, any measurements (readings) can be recorded with high reliability, but then the reaction time is reduced. The next question with regard to a monitoring system is the question of where in the torrent catchment the debris-flow monitoring system should be placed. If it is situated in the upper part of the torrent catchment, i.e., close to or within the triggering/initiation area, one saves time for the reaction at the expense of the uncertainty of its perception and recognition. If the debris-flow monitoring system is placed at the apex of the torrent fan, we get very trustworthy signals, but the reaction time is extremely short.

Therefore, from the very beginning we should clearly communicate to the users of any debris-flow monitoring system (e.g., local residents) that the goal of debris-flow monitoring not to raise wrong hopes. You should also take as the rule the fact that the sooner you want to record a measurement, the greater the uncertainty of the signal is.

5.1 Debris-Flow Triggering Area

The determination of a debris-flow triggering precipitation event can best take place using rainfall gauges (ombrometers), if present in the torrent catchment area. It has been found that measuring devices that detect the precipitation using the weighing principle (a scale), are best suited, since they also record very high intensities reliably, provided that the collecting container is emptied regularly and the wind influence on the gauge is low. A major limitation is that the precipitation (rainfall) is measured only at certain points (locally), and that extrapolation of measured precipitation values to larger areas (to a spatial scale of the whole torrent catchment) is only limited (Fig. 4). Therefore, it is advisable to include spatial available precipitation (rainfall) data in the debris-flow monitoring. As an example, in Austria, the INCA data (ZAMG), based on the evaluation of the weather radar and their calibration by ground-based precipitation (rainfall) measurements, are available. Such data are currently available in Austria with a spatial resolution of 1 km^2 and a time resolution of 5–15 min (Fig. 5).

5.2 Debris-Flow Pathway

For the monitoring of debris flows, the debris-flow pathway is the most significant observation place, as in this reach the main parameters to describe the debris-flow process dynamics can be measured and monitored. These parameters include the debris-flow depth, its front and surface velocity, pressure waves in the air, seismic waves, density, shear stress at the channel wall, pressure, and others.

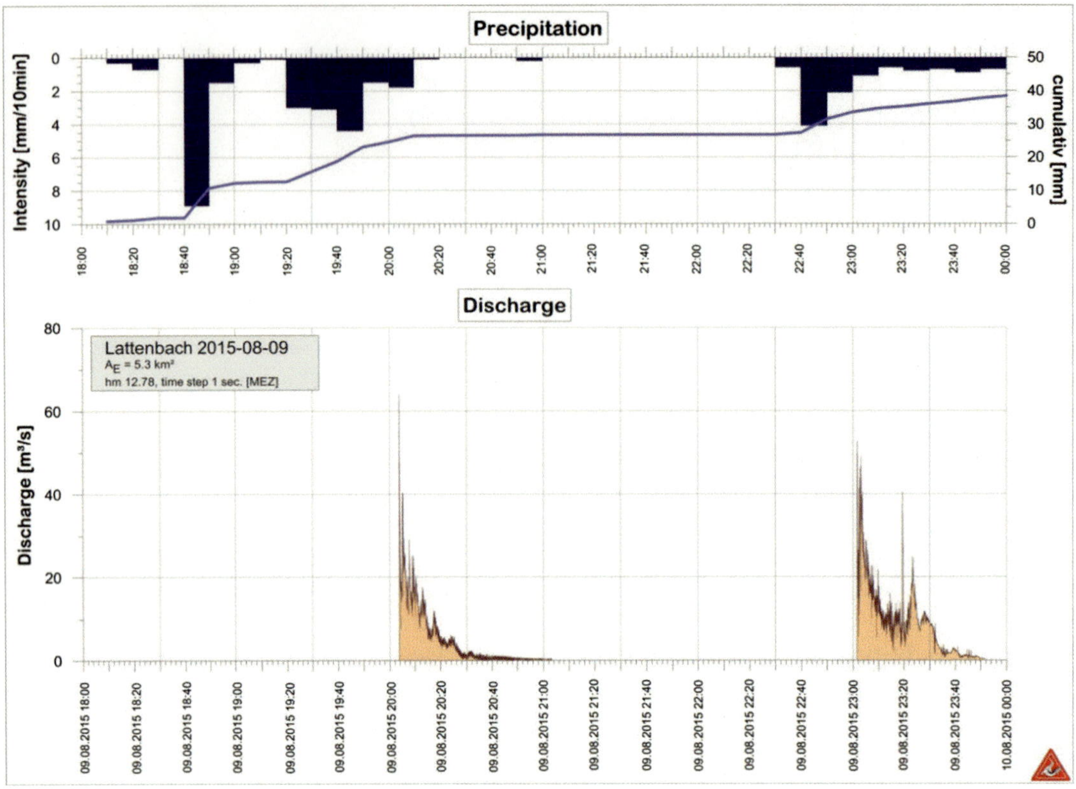

Fig. 4 Triggering precipitation for two subsequent debris flows at Lattenbach (rainfall station is located in the headwater, about 3.5 km away from monitoring site)

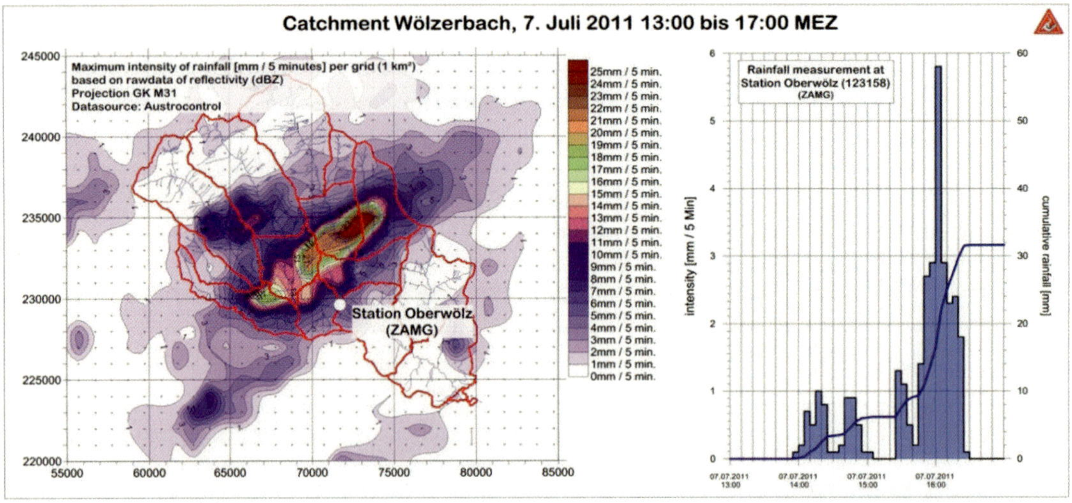

Fig. 5 Comparison of local and spatial distributed rainfall intensities, Wölzerbach, Styria 2011

5.2.1 Changes in Torrent Channel Morphology (Erosion)

The rate of erosion with respect to bedload (coarse sediment) mobilisation in the torrent channel can be measured or estimated continuously or after a debris-flow event by carrying out a field survey of the torrent channel (measuring cross sections or using TLS—Terrestrial Laser Scanning), comparing Airborne Laser Scanning (ALS) data (DTM; see Fig. 6), or by using a drone [UAV (DOM)]. It is also possible to install bendable elements (e.g., chain links) in the torrent channel bed and then to record their exposure by erosion.

5.2.2 Changes in Debris-Flow Behavior

In principle, two different groups of sensors can be used (Jingri 1989): those that come into contact with the debris flow (contact detection) and those that deliver measured data without a direct contact with the debris flow (contactless detection).

The contact detection of debris flows (Fig. 7) represents the beginning of the automated

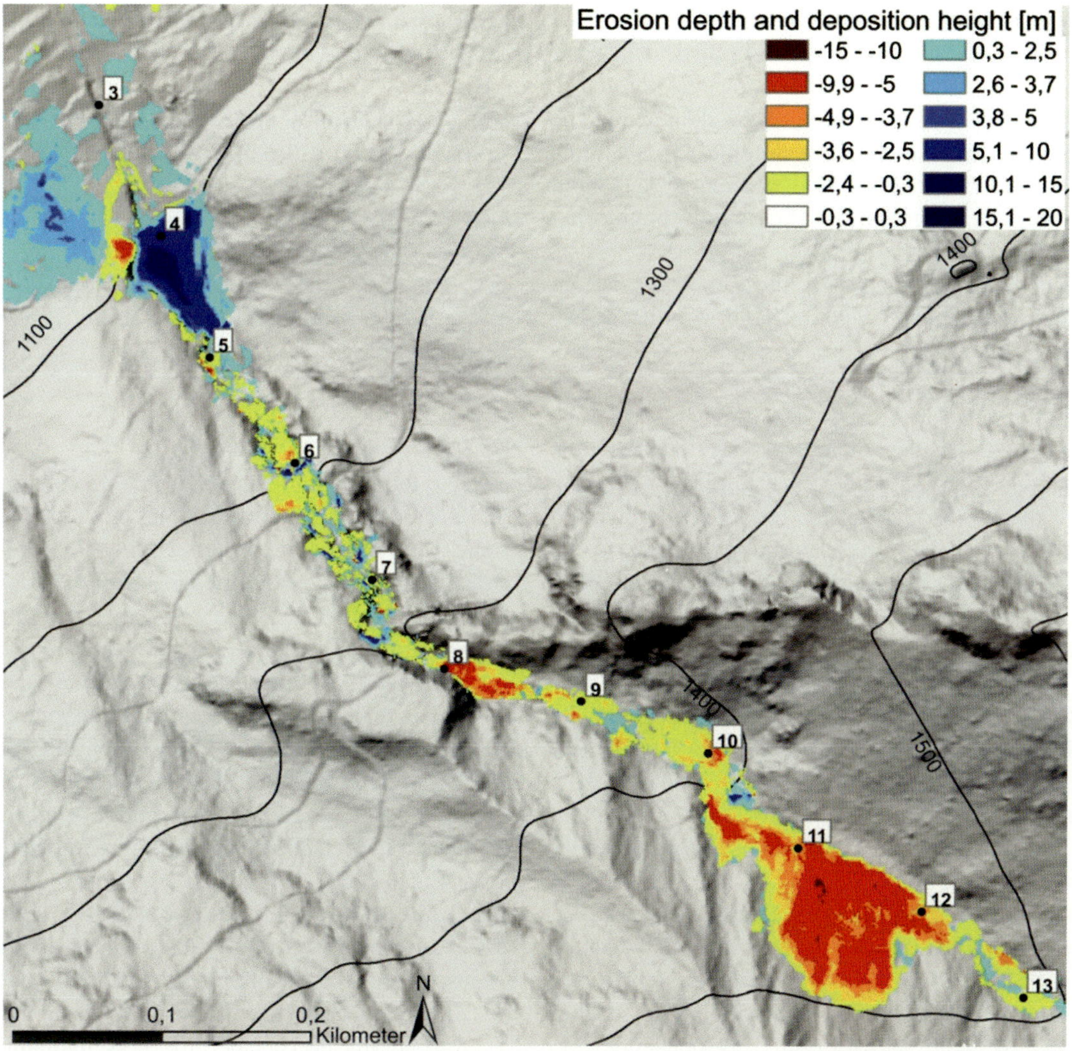

Fig. 6 Channel morphology changes after a debris flow at Schallerbach (2015), derived by ALS data with a 1 m spatial resolution

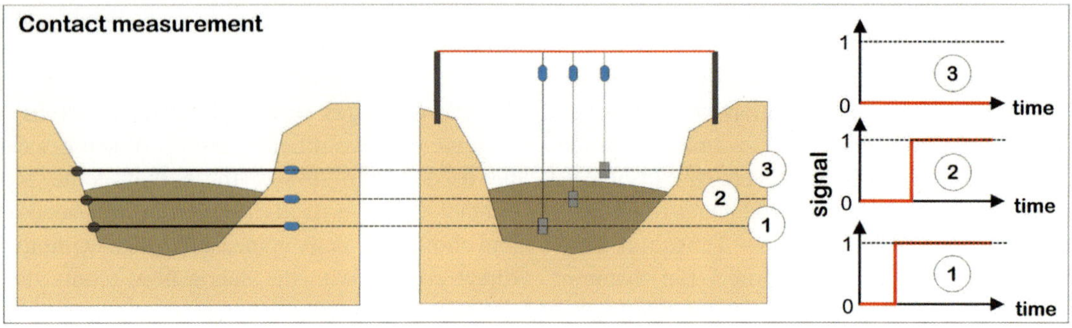

Fig. 7 Examples for contact monitoring by check-line and mercury inclinometer

detection of debris flows in the past, and has been used in many parts of China and Japan. But even today, this not technically sophisticated technique is an inexpensive and a practical possibility for the detection of debris-flow dynamics. The biggest benefit of this technique is its simple integration into debris-flow warning systems, but also in the triggering of debris-flow monitoring facilities.

The disadvantages of these systems are in their maintenance. Parts of the system needed to be replaced after each debris flow was detected. A continuous measurement is not possible; generally, only the first debris-flow pulse can be detected. The measured cross sections must be stabilized, because after each lowering or raising of the torrent channel the debris-flow triggering criteria are changed. In addition, damage to the measurement system by people (vandalism, playing) or animals (game) can be a problem, and a certain protection is therefore necessary.

5.2.3 Wire Sensors

In this system, one or more horizontal wire sensors are used as a triggering device. If a debris-flow surge exceeds the level of a certain control wire sensor at a given elevation, this wire will be dismantled, since a preset tensile force of an anchor (e.g., magnetic) has been reached. A wire pulled from its position triggers a change of the signal (0, 1). This system also works in a vertical position. If on a tether in a cross-section, multiple wire sensors (pull ropes) can be distributed in a vertical position, so wide cross-sections can be easily monitored. Counterweights are mounted at the lower end of each pull rope, possibly at different heights, so that they can be caught and pulled off when a debris flow flows through the control cross-section, thereby switching signals for each known elevation of the counterweight.

5.2.4 Inclinometers

The system with inclinometers works similarly to wire sensors. When an object undergoes a deflection by a debris-flow surge, a circuit will be closed by a switch (mercury switch, principle DLT).

5.2.5 Debris-Flows Depths

In stable torrent channel reaches, the debris-flow depths can be measured without coming into contact with the moving material. One may apply ultrasonic, radar and laser sensors, most of which are centered over the torrent channel (Fig. 8). During installation, one should ensure an appropriate working distance from the presumed maximum debris-flow surface to secure proper data acquisition. Currently applied sensors can detect without problems distances up to 30 m. In natural torrent channels with high channel roughness and small discharges, the measurement signal shows a high level of noise, since pure water runoff usually cannot be easily measured at the debris-flow measurement point because the effective torrent channel width and the current water flowline is constantly changing. When arranging the sensors, the flow hydraulic system should be taken into account; if possible, the debris-flow measurement should be made in a

Fig. 8 Examples on flow depth measurements at Lattenbach, Austria, and Illgraben, Switzerland)

cross-section where a debris flow will potentially have a smooth surface.

5.2.6 Ultrasound

Short ultrasonic pulses in the range of 30–70 kHz can be produced by an ultrasonic transducer (an electric signal causes a thin membrane to oscillate) and radiated to the medium (debris flow); pulses are reflected from the debris-flow surface and received again by the transducer. By the transit time method, the propagation of the pulse is carried out with supersonic velocity, from the time elapsed between emission of the sonic pulse and the reception of the echo signal, the distance to the debris-flow surface can be calculated. The disadvantage of this measuring principle is its large temperature dependence; a corresponding temperature compensation is therefore required. Put simply, the higher the temperature, the greater the absorption and vice versa.

5.2.7 Radar

Radar sensors have proven to be very efficient for measurement of the flow depth of debris-flows. With a measuring range of up to 20 m, an accuracy of 2 mm can be reached. The device sends high-frequency radar wave pulses (K-band, pulse duration ~ 1 ns) in the direction of the debris-flow surface. The surface reflects the radar waves, which are then received by the antenna system of the radar sensor as an echo signal. From the duration of the radar waves between being emitted and received, the device calculates the distance to the debris-flow surface. The time between emission and reception of the radar pulses is proportional to distance and thus debris-flow flow depth. The waves are emitted in the form of a cone, so the recorded measurement value is already an "edited" value. By appropriate parameterization of the device, one obtains from this measurement the corresponding flow depth of the debris-flow. The big advantage of the radar measurement technique is that environmental conditions such as temperature, wind or rain do not affect the measurement.

5.2.8 Laser

Distance measurement by laser is a type of optical method, the measurement is either performed using the transit time method, or by measuring the phase position. One can apply point as well as 2D laser devices (Fig. 9). The advantage of a 2D laser is that the entire cross section of a torrent channel can be measured (laser opening angle of 190°). Laser measurement applications are independent of weather conditions but fail when measuring pure water (no particles for reflection of waves).

5.2.9 Debris-Flow Velocities

The DF front velocity can be determined using various sensors and methods. An indirect measurement is, for example, the temporal comparison of the increase of seismic signals or of the debris-flow flow depths. The velocity is calculated from the time difference of the signal increases and the distance between the sensors, which are arranged along the torrent channel (Fig. 10). An analysis of video sequences can also be used to determine the debris-flow front or the debris-flow surface velocity; if video recording takes place from appropriately oriented positions (Arattano and Marchi 2000).

A direct measurement of the debris-flow front velocity, but also of the debris-flow surface velocity can be carried out using a high-frequency radar measurement (Fig. 11). The radar system transmits a short coded pulse and processes the echo signals at specific time intervals, which in turn correspond

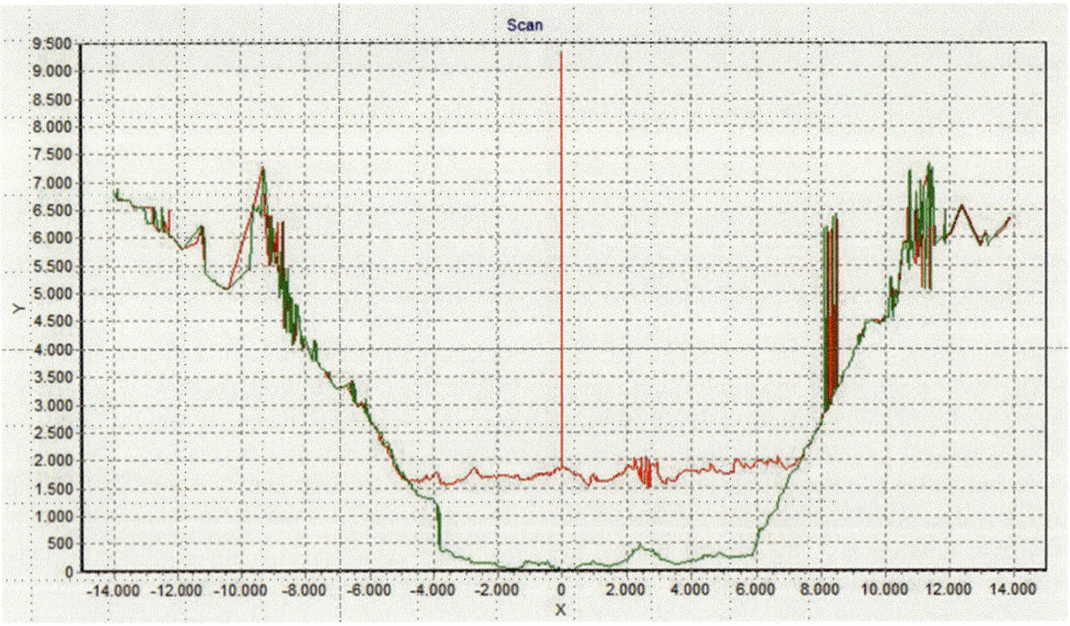

Fig. 9 2D Laserscan of a debris flow surge at Lattenbach, Austria, with a frame rate of 0.3 s (*green* antecedent channel surface; *red* flow surface)

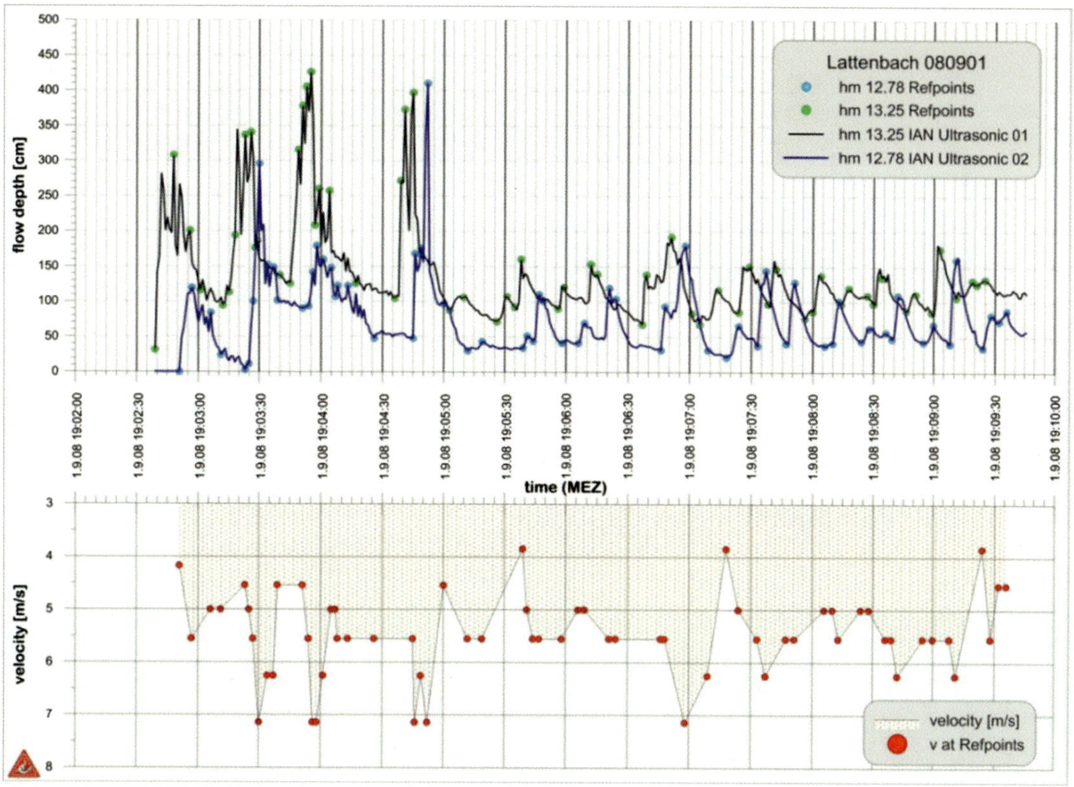

Fig. 10 Velocity estimation at two profiles with a distance of 47 m

to certain distance intervals (range gates). The measurement is based on a Doppler-frequency analysis, which represents the debris-flow front velocity and the debris-flow velocity spectrum. In principle, debris flows can be monitored to distances of up to 2 km with a high resolution.

5.2.10 Shear Stresses

The shear stress exerted by a debris flow can be detected directly or indirectly by means of a mechanical translation by force transducer, mounted in the torrent channel bottom or on its banks. Such a measuring device is currently installed in Illgraben, Switzerland.

5.2.11 Debris-Flow Densities

The debris-flow density can be determined either from a sample taken from a debris flow (Fig. 12) or by the combination of a measurement of the debris-flow depth and the debris-flow weight force by an installed weighing system.

5.2.12 Impact Forces

The measurement of impact forces on structures has so far only been carried out at Jiangjia Gou, China in a few debris-flow pulses (Hu et al. 2011). In Austria, field experiments at a scale 1:1 have been done in the Schesatobel.

5.2.13 Seismic Waves

Seismic waves are elastic waves of energy that spread through soils and rocks. Similar to sound waves, the speed at which seismic waves propagate depends on the medium in which they propagate. Therefore, there is a connection between the transition time of the seismic waves, the position of the source and the sensor, and the material properties of the Earth, through which they spread. The primary waves (longitudinal waves or pressure waves) propagate faster than the secondary waves (shear waves) and are therefore the first signal that can be seen from a wave emitting source in an elastic solid (Thorne

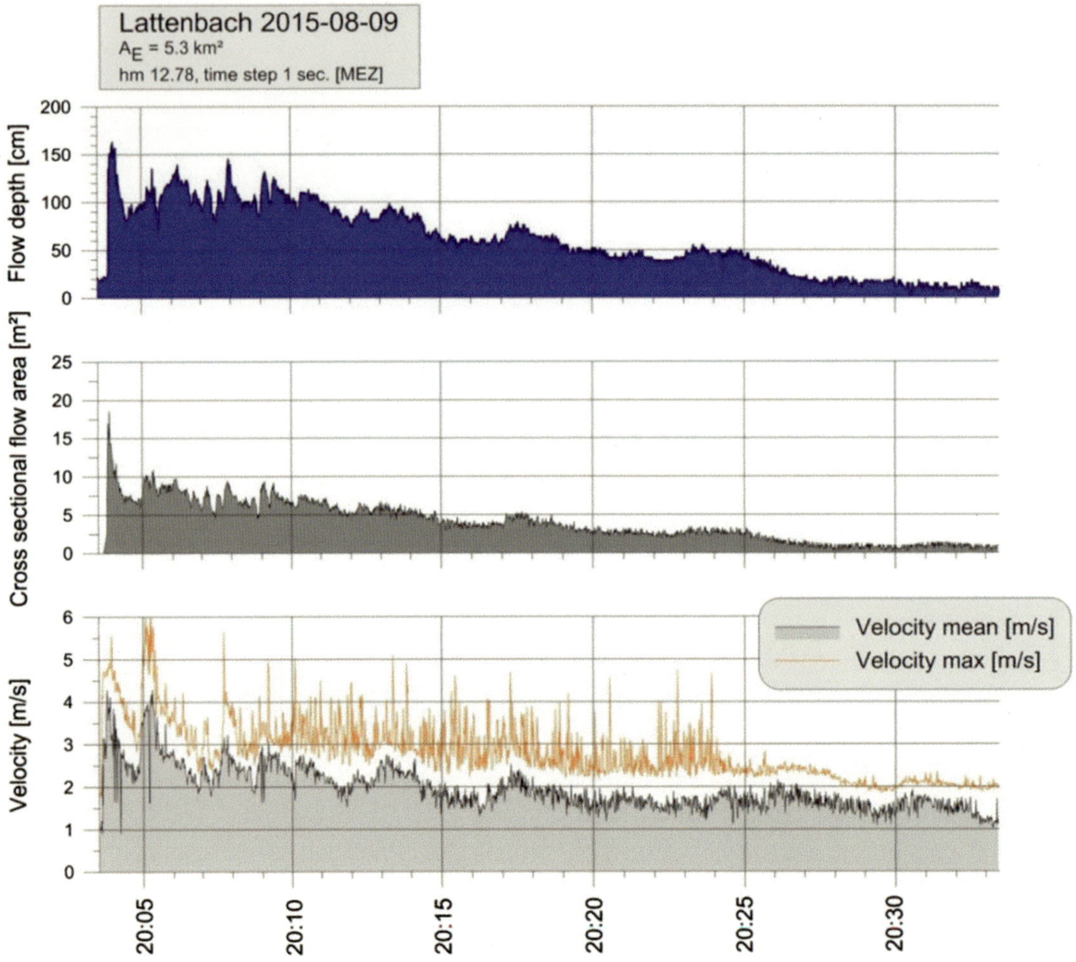

Fig. 11 Flow depth (Radar), wetted area (2D-Laser) and flow velocity within according range gate (high frequency radar) at Lattenbach (time step 1 s)

Fig. 12 Sampling device at Jiangjia Gou, China (density, concentration, rheological parameter)

and Wallace 1995). The recording of seismic waves has a long tradition in debris-flow monitoring (e.g., Okuda et al. 1979; Itakura et al. 1997; Lahusen 1996, 2005; Arattano 1999, 2003; Arattano et al. 2014; Suwa et al. 2000; Surinach et al. 2005; Abanco et al. 2012). Since the sensors can be easily installed outside of the torrent channel in the terrain, they are energy efficient, cost-effective and robust, and analysis routines are known. The latter allowed the detection of debris flows on the basis of a crossing over a selected threshold, but specific statements about the nature and the size of a debris flow cannot be made. This is possible only with the analysis of the raw signal (sampling rate >100 Hz).

5.2.14 Infrasound

Infrasonic waves are longitudinal pressure fluctuations in the air, with narrow-band low frequencies (0.001–20 Hz) and are thus are not perceptible to humans. The infrasound waves propagate at the velocity of 344 m/s, the same velocity as audible sound. The propagation velocity depends on the properties of the air (temperature, density) and not on the frequency or amplitude of the wave. They can propagate thousands of kilometers and can still be detected. If one monitors infrasound over short distances (<5 km) in particular, scattering losses must be taken into account, since the energy loss due to absorption in the atmosphere is relatively small (Albert and Orcutt 1990; Johnson 2003; Zhang 1993; Zhang et al. 2004; Hübl et al. 2008).

Due to the inter-collision of stones inside the debris-flow mass and the collision of the debris-flow mass with a torrent channel, debris flows produce both infrasound and seismic waves. The characteristic frequency range in which the largest amplitudes occur varies depending on the viscosity, from 5 to 10 Hz for debris flows with high coarse sediment content to 15–30 Hz for debris-flow like events with low viscosity. The fact that a debris flow emits in the frequency-time domain a characteristic pattern of infrasound and seismic signals permits an early detection of such events, even before a debris-flow pulse reaches the sensor location (Kogelnig et al. 2014). Figure 13 depicts the infrasonic and the seismic signal of a debris flow that occurred on September 1, 2008 at Lattenbach in Tirol, Austria.

5.2.15 Debris-Flow Dynamics

The dynamics of a debris flow in a torrent channel reach can be best seen (recognized) by means of a video camera (webcam) that was activated by a trigger. In order to make geometric statements about the debris flow, one should place markers in the video image area that are also visible during the debris-flow event (during video footage). The video time resolution should not be less than 10 frames per second so that still pictures will also contain enough image information. In order to assure video footage even at night, the use of day/night cameras is required. As an additional light source at night infrared emitters are preferable, because they cause no light pollution like a normal light. The recording time of the video footage can be process-based determined or by a timer.

5.3 Debris-Flow Deposition Area

In a debris-flow deposition area, spreading and two-dimensionally propagating debris flows are very difficult to measure. If video sequences are available, they can be evaluated with respect to the debris-flow propagation direction, debris-flow velocity and debris-flow depth. After a debris flow has deposited (after a debris-flow event), for monitoring purposes it is essential to determine the debris-flow deposition area, the debris-flow deposition heights, and also to sample the debris-flow deposits in order to determine the concentration of solids and the debris-flow granulometric composition.

6 Monitoring Data Recording

The central control and recording units of a debris-flow monitoring system are data acquisition systems, such as, e.g., Data Loggers. By programming the control unit the measured frequency of the sensors is determined.

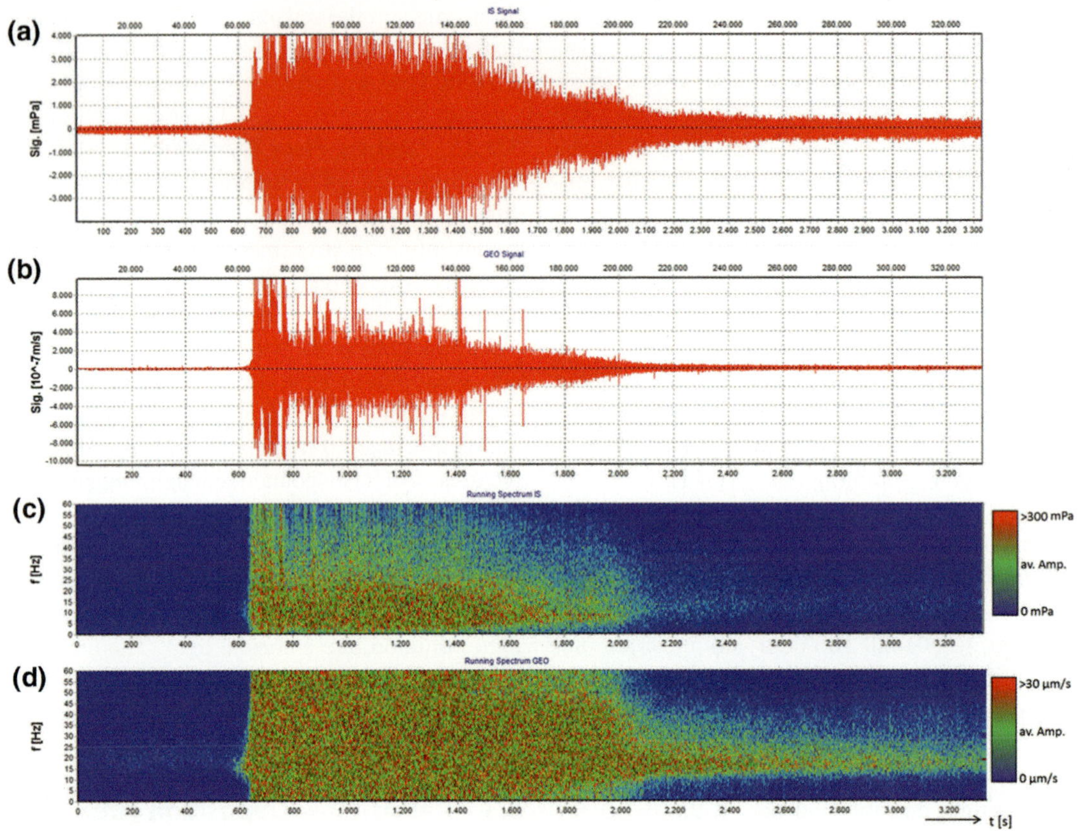

Fig. 13 Infrasound and seismic signal of a debris flow of 01.09.2008 at Lattenbach, Tyrol. **a** Infrasound signal, **b** seismic signal, **c** time-frequency spectrum of the infrasound signal, **d** time-frequency spectrum of the seismic signal

The measured signals are either stored in the selected measurement frequency, or stored as discrete time and discrete values such average value, maximum, minimum, current value or number of exceedances of a selected threshold value (Fig. 14). This data can be used to trigger further working steps, e.g., a warning system. For very high-frequency measuring signals, interposition of a microcontroller that can perform the initial analysis of the data by software on site is recommended (via, e.g., Fast Fourier Transform).

The data logger is also the communication interface to various transmission media. By default, ring buffers are located in the Data Loggers that can be combined with external storage media (such as SSD cards for example). Therefore, depending on the data volume of the measured values, measured data should be transmitted to a different storage medium (such as a database). This data transmission can in principle be done in two ways: either send data packets to the Data Logger at set time intervals (hourly, daily) or a server requests the data from the Data Logger in the desired time intervals. It is important to ensure that at the selected times, the communication between server and Data Logger is also guaranteed.

Regardless of the data recording, the Data Logger should be able to send service messages, such as when the battery level drops below a threshold value. Of course, a closed ground

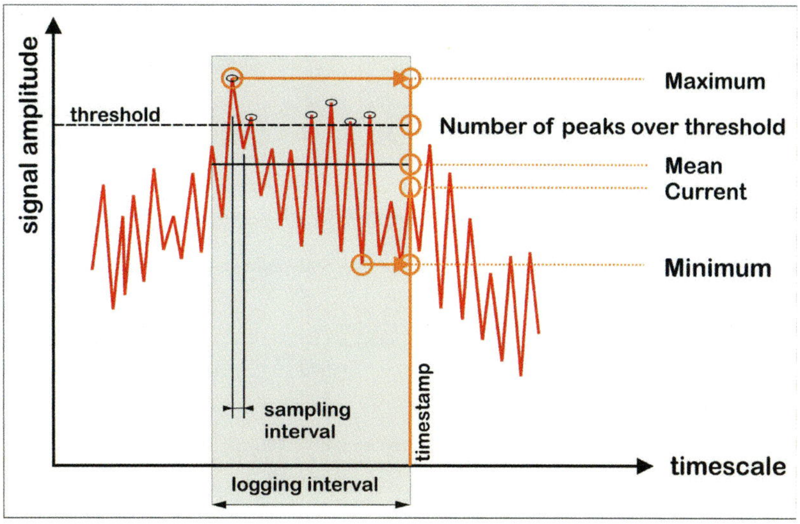

Fig. 14 Recording of data

system for the measurement system (sensors, data logger, and power supply) must be installed (with lightning protection).

7 Monitoring Data Communication

The communication units are the most energy-intensive component of debris-flow monitoring systems (Fig. 15). The transfer of data to the Data Logger is done on site mostly via cable; if this is not possible then through a radio link. Depending on the transmission power of the system, an official permit may be necessary. The transmission of data or information from the data logger to higher-level units (database, decision makers) depends on existing transmission paths (landline, broadband Internet, GSM, radio). If there is good coverage by mobile networks (3G, 4G), such a facility is recommended. For the transmission of short messages (SMS), the interposition of a tele-control unit is required.

8 Energy Supply

If a debris flow monitoring system is needed only for the summer months (or rainy season—monsoon season), the devices can be put into a hibernation status. The use a 230/240 V power supply (in Europe) proved very advantageous, at least for the central unit of the debris flow monitoring system. Nevertheless, additional buffering by a 12/24 V battery is required. If no electricity network connection can be established, the entire debris flow monitoring system must be operated via solar units or in combination with fuel cells. The (solar) supply unit must be designed and positioned in such a way that a full-day operation is guaranteed, if possible. The energy reserves should be able to efficiently bridge some days of "bad weather".

9 Monitoring Data Archiving

For a permanent debris-flow monitoring system, the archiving of data in a database is essential. In addition to storing the raw data, the data quality control must be carried out in order to identify faulty recordings in due time and to be able to replace defective components of the system.

10 Debris-Flow Monitoring Systems in Europe

Professional literature contains descriptions of catchments that are equipped with field facilities for long-term debris-flow monitoring. The main

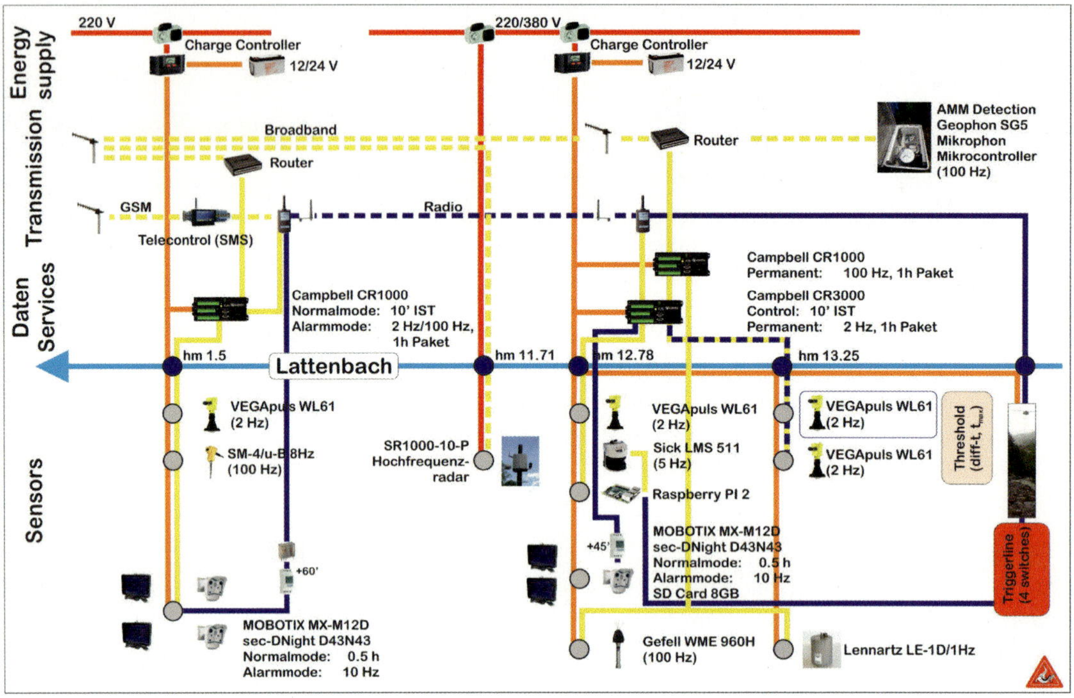

Fig. 15 Conceptual schema of the monitoring site at Lattenbach (sensors, communication, and energy supply)

criterion for selection of interesting debris-flow monitoring systems can be the frequency of debris-flow events in these areas. In Europe, debris-flow monitoring systems are installed in:

(a) Austria: Lattenbach, Wartschenbach, and others (e.g., Hübl and Moser 2006)
(b) Switzerland: Illgraben, Dorfbach, Spreitgraben, and others (e.g., Hürlimann et al. 2011, 2015)
(c) Italy: Rio Moscardo, Aquabona, Gadria (e.g., Genevois et al. 2000; Marchi et al. 2002; Comiti et al. 2014; Blasone et al. 2015)
(d) France: Manival Torrent, Torrent Réal and others (e.g., Theule et al. 2012; Navratil et al. 2012, 2013; Theule et al. 2015; Bel et al. 2014, 2015)
e) Spain: Rebaixader torrent, Portainé and others (e.g., Abancó et al. 2014)

11 Debris-Flow Warning Systems

An Early Warning System (EWS) is "the set of capacities needed to generate and disseminate timely and meaningful warning information to enable individuals, communities and organizations threatened by a hazard to prepare and to act appropriately and in sufficient time to reduce the possibility of harm or loss". There are five key elements of the human-centered EWS:

(a) Knowledge of the debris-flow risk;
(b) Monitoring, analysis, and forecasting of debris-flow hazards;
(c) Operational centers;
(d) Communication or dissemination of alerts and warnings;

(e) Local capabilities to respond to the warnings received.

The EWS are usually based on:

(a) Debris-flow hazard maps (hazard zones within the maximum run-out zone);
(b) Meteorological forecasts (rainfall forecasts, rain radars), and
(c) Monitoring data from hazard areas,
(d) Issuing debris-flow pre-trigger warnings (using different empirical thresholds) or
(e) Post-trigger warnings (event-triggered warnings).

12 Conclusions

A debris-flow monitoring system consists of different components, such as sensors, data acquisition units, a control unit, communications equipment, power supply and a storage unit for data archiving. Due to constant ongoing rapid technological development, new sensors, memory chips, etc. are available, so that the possibilities of debris-flow monitoring are steadily improving. In the meantime, industrial standard solutions (e.g., radar, ultrasonic) can be integrated into a debris-flow monitoring system, so that the equipment costs can be reduced. Experiences have shown that an optimal matched combination of the components of a debris-flow monitoring system is an essential criterion for success. A clear definition of debris-flow monitoring system objectives, an excellent knowledge of the characteristics of the debris-flow processes to be detected, and the acceptance of local and technical conditions are the main requirements for operation of a debris-flow monitoring system.

Acknowledgements The overview presented in this paper was prepared within the framework of the European Alpine Space project START_it_up (State-of-the-Art in Risk Management Technology: Implementation and Trial for Usability in Engineering Practice and Policy), financed by the European Commission.The main content of these practice guidelines was published as a paper in German language as Hübl and Mikoš (2014).

References

Abancó C, Hürlimann M, Fritschi B, Graf Ch, Moya J (2012) Transformation of ground vibration signal for debris-flow monitoring and detection in alarm systems. Sensors (Basel) 12(4):4870–4891

Abancó C, Hürlimann M, Moya J (2014) Analysis of the ground vibration produced by debris flows and other torrential processes at the Rebaixader monitoring site (Catalan Pyrenees, Spain). Nat Hazards Earth Syst Sci 14:929–943. doi:10.5194/nhess-14-929-2014

Albert D, Orcutt J (1990) Acoustic pulse propagation above grassland and snow: comparison of theoretical and experimental waveforms. J Acoust Soc Am 87 (1):93–100

Arattano M (1999) On the use of seismic detectors as monitoring and warning systems for debris flows. Nat Hazards Earth Syst Sci 20:197–213

Arattano M (2003) Monitoring the presence of the debris-flow front and its velocity through ground vibrations detectors. In: Proceedings of the 3rd international conference on debris-flow hazards mitigation: mechanics, prediction and assessment, vol 2. Millpress, Rotterdam, pp 731–743

Arattano M, Marchi L (2000) Video-derived velocity distribution along a debris flow surge. Phys Chem Earth Part B 25:781–784

Arattano M, Abancó C, Coviello V, Hürlimann M (2014) On the techniques for processing the ground velocity signal produced by debris flows: the methods of amplitude and impulses compared. Comput Geosci 73:17–27. doi:10.1016/j.cageo.2014.08.005

Bel C, Liébault F, Bellot H, Fontaine F, Laigle D, Navratil O (2014) Debris flow monitoring in French Alps. Proc Int Conf Fluvial Hydraulics RIVER FLOW 2014:1589–1595

Bel C, Navratil O, Liébault F, Fontaine F, Bellot H, Laigle D (2015) Monitoring debris flow propagation in steep erodible channels. In: Lollino G, Arattano M, Rinaldi M, Giustolisi O, Marechal J.-C, Grant GE (eds) Engineering geology for society and territory, vol. 3. Springer International Publishing, pp 103–107

Blasone G, Cavalli M, Cazorzi F (2015) Debris-flow monitoring and geomorphic change detection combining laser scanning and fast photogrammetric surveys in the moscardo catchment (Eastern Italian Alps). In: Lollino G, Arattano M, Rinaldi M, Giustolisi O., Marechal J, Grant GE (eds) Engineering geology for society and territory, Springer, Vol. 3, pp 51–54. doi:10.1007/978-3-319-09054-2_10

Comiti F, Marchi L, Macconi P, Arattano M, Bertoldi G, Borga M, Brardinoni F, Cavalli M, D'Agostino V, Penna D, Theule J (2014) A new monitoring station for debris flows in the European Alps: first observations in the Gadria basin. Nat Hazards 73:1175–1198. doi:10.1007/s11069-014-1088-5

Genevois R, Tecca P, Breti M, Simoni A (2000) Debris-flow in the dolomites: experimental data from a monitoring system. In: Wieczorek GF, Naeser ND (eds) Debris-flow hazards mitigation: mechanics, prediction, and assessment, Proceedings of the 2nd international conference, Balkema Press, Taipei, Taiwan, Rotterdam, pp 283–291, 16–18 Aug 2000

Hu K, Wei F, Li Y (2011) Real-time measurement and preliminary analysis of debris-flow impact forces at Jiangjia Ravine, China. Earth Surf Processes Land 36:1268–1278. doi:10.1002/esp.2155

Hübl J (2009) Hochwässer in Wildbacheinzugsgebieten. Wien Mitt Wasser Abwasser Gewässer 216:45–58

Hübl J, Moser M (2006) Risk management in Lattenbach: a case study from Austria. In: Lorenzi G, Brebbia C, Emmanouloudis D (eds) Monitoring. Simulation, prevention and remediation of dense and debris flows, WIT Press, Southamption, pp 333–342

Hübl J, Mikoš M (2014) Monitoring von Murgängen. Wildbach-und Lawinenverbau 78(173):50–66 (in German with English abstract)

Hübl J, Zhang S, Kogelnig A (2008) Infrasound measurements of debris flow. In: De Wrachien D, Brebbia CA, Lenzi MA (eds) WIT transactions on engineering sciences, 2nd international conference on monitoring, simulation, prevention and remediation of dense and debris flows II. vol 60, pp 3–12. doi:10.2495/DEB080011

Hürlimann M, Abancó C, Moya J, Raimat C, Luis-Fonseca R (2011) Debris-flow monitoring stations in the Eastern Pyrenees. Description of instrumentation, first experiences and preliminary results. In: Genevois R, Hamilton DL, Prestininzi A (eds) 5th international conference on debris-flow hazards mitigation: mechanics, prediction and assessment, Casa Editrice Universita` La Sapienza, Roma, pp 553–562

Hürlimann M, McArdell WB, Rickli C (2015) Field and laboratory analysis of the runout characteristics of hillslope debris flows in Switzerland. Geomorphology 232:20–32. doi:10.1016/j.geomorph.2014.11.030

Itakura Y, Koga Y, Takahama J, Nowa Y (1997) Acoustic detection sensor for debris-flow. In: Chen CL (ed) Debris-flow hazards mitigation: mechanics, prediction, and assessment, Proceedings of first international conference, San Francisco, 7–9 August 1997, USA, New York, ASCE, pp 747–756

Jingri C (1989) A study on debris flow warning in China. In: The Japan—China symposium on landslides and debris flows; Niigata, Tokyo, Japan; S 177–182

Johnson J (2003) Generation and propagation of infrasonic airwaves from volcanic explosions. J Volcanol Geoth Res 121:1–14

Kienholz H (1998) Early warning systems related to mountain hazards; manuscript hand-out; In: International conference on early warning systems for natural disaster reduction, Postdam

Kogelnig A, Hübl J, Suriñach E, Vilajosana I, McArdell B (2014) Infrasound produced by debris flow: propagation and frequency content evolution. Nat Hazards. 70 (3):1713–1733 doi:10.1007/s11069-011-9741-8

LaHusen R (1996) Detecting debris flows using ground vibrations. USGS Fact Sheet 236-96, USGS (ed)

LaHusen R (2005): Debris flow instrumentation. In: Jakob M, Hungr O (eds) Debris-flow hazards and related phenomena, Springer, Chichester, pp 291–304

Marchi L, Arattano M, Deganutti A (2002) Ten years of debris-flow monitoring in the Moscardo Torrent (Italian Alps). Geomorphology 46(1–2):1–17

Navratil O, Liébault F, Bellot H, Theule J, Travaglini E, Ravanat X, Ousset F, Laigle D, Segel V, Fiquet M (2012) High-frequency monitoring of debris flows in the French Alps. preliminary results of a starting program, Conference Proceedings Interpraevent

Navratil O, Liébault F, Bellot H, Travaglini E, Theule J, Chambon G, Laigle D (2013) High-frequency monitoring of debris-flow propagation along the Réal Torrent, Southern French Prealps. Geomorphology 201:157–171. doi:10.1016/j.geomorph.2013.06.017

Okuda S, Suwa H, Okunishi K, Yokoyama K, Ogawa K, Hamana S (1979) Synthetic observation on debris-flow, part 5, observation at valley Kamikamihorisawa of Mt. Yakedade in 1978. Annuals Disaster Prev Res Inst Kyoto Univ 22B–1:157–204

Rupprecht W (2013) Signale und Übertragungssysteme—Modelle und Verfahren; Skriptum zur Vorlesung Grundlagen der Informationsübertragung, Fachbereich Elektrotechnik und Informationstechnik, Universität Kaiserslautern (http://nt.eit.uni-kl.de/fileadmin/lehre/guet/skript)

Suriñach E, Vilajosana I, Khazaradze G, Biescas B, Furdada G, Vilaplana J (2005) Seismic detection and characterization of landslides and other mass movements. Nat Hazards Earth Syst Sci 5:791–798

Suwa H, Yamakoshi T, Sato K (2000) Relationship between debris-flow discharge and ground vibration. In: Wieczorek GF, Naeser ND (eds) Debris-flow hazards mitigation: mechanics, prediction, and assessment, Proceedings 2nd international conference, Balkema Press, Taipei, Taiwan, Rotterdam, pp 311–318, 16–18 Aug 2000

Theule JI, Liébault F, Loye A, Laigle D, Jaboyedoff M (2012) Sediment budget monitoring of debris-flow and bedload transport in the Manival Torrent, SE France. Nat Hazards Earth Syst Sci 12:731–749. doi:10.5194/nhess-12-731-2012

Theule JI, Liébault F, Laigle D, Loye A, Jaboyedoff M (2015) Channel scour and fill by debris flows and bedload transport. Geomorphology 243:92–105. doi:10.1016/j.geomorph.2015.05.003

Thorne L, Wallace T (1995) Modern Global Seismology. Academic Press, San Diego, USA

Zhang S (1993) A comprehensive approach to the observation and prevention of debris flow in China. Nat Hazards 7:1–23

Zhang S, Hong Y, Yu B (2004) Detecting infrasound emission of debris flow for warning purposes. International Symposium Interpraevent VII, Riva, Trient, pp 359–364

TXT-tool 2.886-1.3
Debris Flow Monitoring Guidelines

Hsiao-Yuan Yin and Yi-Min Huang

Abstract

In order to document debris flow events on-site, the Soil and Water Conservation Bureau (SWCB), Council of Agriculture, Taiwan, has developed a debris flow monitoring system since 2002. This paper introduces the technology of debris flow monitoring stations established by SWCB in Taiwan. At each on-site monitoring station, several observation instruments, including rain gauges, CCD cameras, wire sensors, geophones, and water level meters were installed to collect dynamic debris flow information that can be used as reference data for countermeasures for debris flow disaster mitigation. In addition, several meteorological sensors have also been adapted recently to record the effects of long-term climate change on the hill country of Taiwan. The framework of the debris flow monitoring system consists of monitoring sensors, instrumental cabin (vehicle platform for a mobile station), transmission systems and a web-based display system. An example of monitoring during Typhoon Kalmaegi in July 16–18, 2008 is provided in this guideline. On-site observation data, including the rainfall patterns, video images, wire sensor ruptures and ground vibrations caused by debris flows, are analyzed in detail.

H.-Y. Yin
Soil and Water Conservation Bureau, Council of Agriculture, 54044 Nantou, Taiwan
e-mail: sammya@mail.swcb.gov.tw

Y.-M. Huang (✉)
GIS Research Center, Feng Chia University, 407 Taichung, Taiwan
e-mail: niner@gis.tw

K. Sassa et al. (eds.), *Landslide Dynamics: ISDR-ICL Landslide Interactive Teaching Tools*, https://doi.org/10.1007/978-3-319-57774-6_42

Keywords
Debris flows • Debris flow monitoring system

Contents

1 **Introduction** 588
2 **Monitoring Scope** 588
2.1 Observing Environmental Changes in Potential Debris Flow Areas 589
2.2 Recording and Monitoring 589
2.3 Basin-Wide Coverage 589
2.4 Response to Disaster, Alerts, and Evacuation 589
3 **Monitoring Networks and Systems** 589
3.1 Instrument and Sensors 589
3.2 Communication 591
3.3 Monitoring Stations 591
3.4 Integrated Control and Management Systems 594
4 **Setup Procedures** 594
4.1 Determine the Areas of Potential Debris Flows 594
4.2 Conduct Site Investigations 594
4.3 Determine the Locations of Monitoring Stations, the Layout of Instruments and Sensors 594
4.4 Set up Housing, Lines, Power, and Satellite Communication 594
4.5 Test Communications 594
4.6 System Test 594
5 **Examples and Applications** 595
5.1 Debris Flow Disaster Prevention Website 595
5.2 Application: Jioufen-Ershan Monitoring Station 595
6 **Conclusion** 598
References 598

1 Introduction

With recent changes in severe weather, heavy rain and floods occur more frequently than before. Among natural hazards, debris flows have become a serious threat in recent years to the people who live near slopes or at the mouth of rivers. Debris flows usually carry a lot of rocks, driftwood, and other material from the upper stream area, and consequently usually cause destruction and damage to the environment. People and the government, therefore, need to be prepared for debris flow disasters and work together to reduce the impacts from debris flows.

So-called debris flows refer to a natural phenomenon—a mixture of mud, sand, gravel, rock and water that flows downhill along slopes, ditches or rivers by gravity. Figure 1 illustrates the generation of a debris flow. There are three main causes that contribute to debris flows:

- **Abundant sediments:** Enough loose debris provides the solid materials needed for a debris flow.
- **Sufficient water:** Sufficient water is needed as a lubricant that can reduce the friction between debris and help solid materials flow.
- **Adequate slope:** An adequate gradient is needed to provide the driving force for a debris flow, helping to decrease the friction between earth and rock so that the debris flow will keep moving to lower areas.

In order to be prepared for debris flow disasters, monitoring is an important approach. Given the major causes of debris flows, water is the key factor of their occurrence. Therefore, monitoring the water conditions, i.e., the rainfall (input) and water level of a river (output), becomes critical for a debris flow monitoring system.

2 Monitoring Scope

A debris flow monitoring network and management system should be established to provide necessary data and assistance in disaster response. The following outlines a debris flow monitoring network and management system.

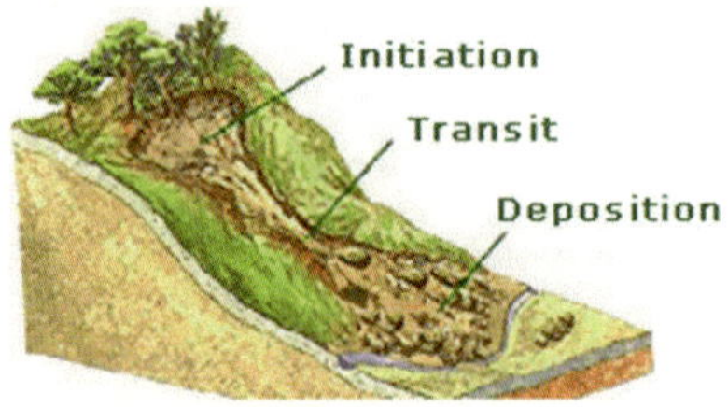

Fig. 1 The generation of a debris flow (courtesy of Prof. Hongey Chen and Soil and Water Conservation Bureau, Taiwan, http://246eng.swcb.gov.tw/)

2.1 Observing Environmental Changes in Potential Debris Flow Areas

The monitoring network can help observe the environmental changes at a potential debris flow area. The observations can include rainfall, soil moisture, water level, and CCD images.

2.2 Recording and Monitoring

The monitoring system can collect observation data and transmit the processed information to the emergency operation center. The system operates automatically and can control the local monitoring network through an integrated platform.

2.3 Basin-Wide Coverage

The concept of a basin-wide monitoring network comes from the fact that the natural hazards that cause debris flows usually affect a broad area, and so widely distributed monitoring sites are needed (Lee et al. 2010). On-site monitoring stations are commonly used for long-term data collection, usually at mid- and downstream locations, mobile stations for event monitoring, usually at mid- and upstream locations, and portable units are often used for event monitoring, usually at upstream and remote sites. All types of monitoring stations constitute the basin-wide monitoring network for debris flows. Figure 2 shows the concept of basin-wide monitoring.

2.4 Response to Disaster, Alerts, and Evacuation

The monitoring system can trigger debris flow warnings based on observations and predefined warning criteria. The warnings can be sent through different methods, e.g., cellphone message and emails, to disaster respondents. Evacuation then can be conducted accordingly.

3 Monitoring Networks and Systems

3.1 Instrument and Sensors

The instruments most commonly used for monitoring debris flows are rain gauges, geophones, water level meters, wire sensors, soil moisture sensors, and CCD cameras.

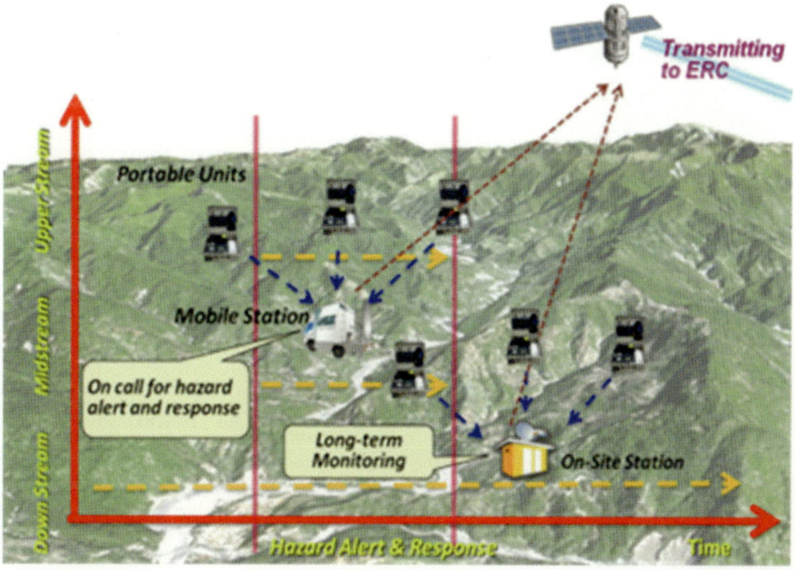

Fig. 2 Basin-wide monitoring network

3.1.1 Rain Gauges

The main purpose of a rain gauge (see Fig. 3) is to measure the amount of rainfall around the monitoring station. The recorded rainfall can be compared with the pre-defined warning value of rainfall to determine the potential for debris-flow occurrence. The rainfall intensity (i.e., hourly rainfall) and the cumulative rainfall (e.g., 24 h, 48 h, 7-day effective rainfall) are usually used for debris flow warning criteria. In order to ensure that the information is real-time, the frequency of rainfall data measurement is 10 min.

Fig. 4 Setup of the geophone

3.1.2 Geophones

Debris flows consist of a flowing body of mixed water, mud, sand, and gravel under the action of gravity. Therefore, when debris flows occur, a slight vibration of the ground surface can be detected. This kind of vibration wave, which is transmitted by the earth's crust, is called debris flow underground sound. When the debris flow occurs, the bed of the channel it flows down undergoes corrosion and attrition, with particle collision, thus producing vibration waves. The wave transmission in air is only a small portion of the energy and most of the energy is transmitted in the rock stratum of the channel bed. The primary reason behind the ground vibration is the large particles of gravel striking the bed, especially at the front end of debris flows.

Based on the results of research on debris flow seismic signals, the significant frequency produced by the flow of debris flow is around 10–40 Hz (Fang et al. 2008). In order to obtain the actual frequency that is characteristic of debris flows, the geophone (Fig. 4) is used to capture the underground signal from the three axes (along the river, transverse to the river, and in the vertical direction) (Fig. 5). To record the debris flow underground sound, the acquisition frequency is set to 500 Hz.

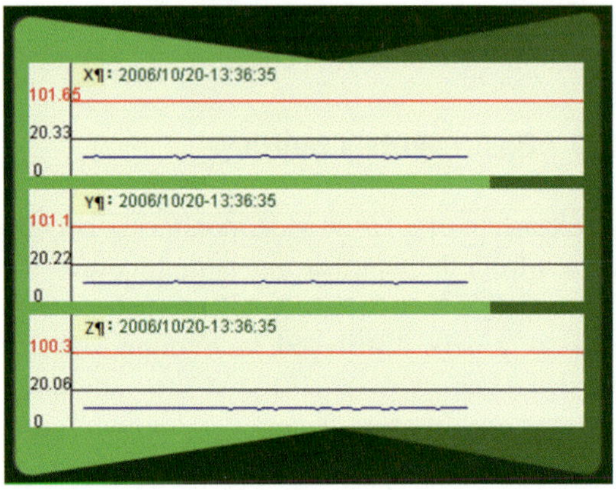

Fig. 5 Example of a geophone readout showing accumulated energy per second

Fig. 3 Setup of the rain gauge

3.1.3 Ultrasonic Water Level Transmitters

Ultrasonic water level transmitters (Fig. 6) may be placed on the spillways of Sabo dams, slit dams, and submerged dams. The principle of an ultrasonic water level transmitter is to obtain the water level in real-time from the reflected water position.

Fig. 6 Example of an ultrasonic water level transmitter

3.1.4 Wire Sensors

Wire sensors can be set up that will be broken by the advancing front of a debris flow so the circuit is disconnected. The locations chosen for positioning the wire sensors are the spillways of Sabo dams, slit dams, and submerged dams (Fig. 7). The height of the wire sensor depends on the design water level of normal floods and debris flows. In order to detect debris flows more effectively, two wire sensors often are employed at the same location at different heights.

3.1.5 Soil Moisture Sensors

Soil moisture sensors (Fig. 8) are used to measure the volumetric percentage of water in soil.

Fig. 7 Setup of wire sensors

The source of a debris flow is usually the shallow layer of soil along the slopes. When the soil moisture increases as rainfall continues, it is highly likely to cause shallow landslides along slopes and consequently generate the source of debris for debris flows. The soil moisture sensor usually is set up at a depth of 50 cm from the ground surface and in the region of river slopes.

3.1.6 CCD Camera and Infrared Spotlight

The debris flow image capture system employs both colour and black-and-white modes to capture the best images (the lowest sensitivity is 0.02Lux/54G1). These two modes can be switched automatically by a CCD camera. The light compensation device (including visible light and infrared light) is necessary to capture clear images (Fig. 9). Through the video server, the analog image can be transformed into a digital image. The digital image is then transmitted to the data acquisition center through the wireless VSAT or wired ADSL for handling and storage (Fig. 10).

3.2 Communication

In addition to the deployment of monitoring sensors, transmitting various types of real-time data to the control center is also an important factor. The primary communication network for debris flow warning systems is mainly via satellite during an event (e.g., a typhoon). Satellite transmission has the advantage of wider bandwidth and full coverage. Thus an enhanced transmission quality of the observation information is attained. In order to maintain the transmission of information to the Debris Flow Emergency Response Center, ADSL, GSM, GPRS or PSTN networks are used as backup transmission systems and operate in normal time.

3.3 Monitoring Stations

There are three types of monitoring stations: local permanent (on-site) stations, mobile stations, and portable stations (units).

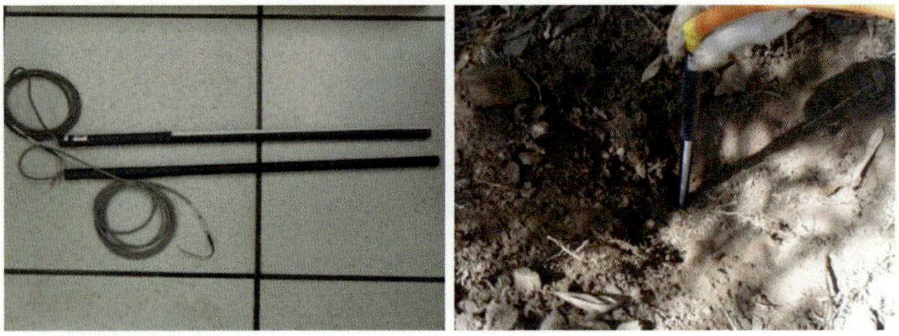

Fig. 8 Soil water moisture sensor

Fig. 9 CCD camera and infrared spotlight

Fig. 10 Illustration of a real-time image transmitted from a monitoring station

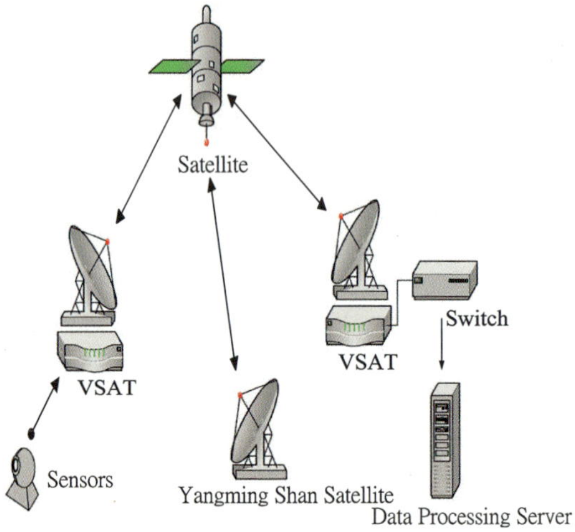

Fig. 11 Setup of the data acquisition center

Fig. 12 Example of an on-site monitoring station

3.3.1 On-Site Station (Local Data Acquisition Center)

Each on-site monitoring station is constructed as a permanent structure, and designed to provide long-term monitoring service. Each on-site station has a data acquisition center (Figs. 11 and 12) to collect and store the observation information from sensors. The acquisition frequency and format of each sensor are shown in Table 1.

Table 1 Example table shown in this template

Type of sensor	Capture frequency	Storage format	Extension name	Remark
CCD camera	30 images/s (share for the station)	Movie	VGX	All CCD of the station share the same resources (e.g. there are 2 sets of CCD and each CCD records in 15 images/s), and is stored in the information receive center
	Around 10 s/image	Ortrait	JPG	Transmit to back-end database instantly
Rain gauge	1 min	Text file	Txt	Accumulation of rainfall per min
Geophone	500 Hz/axis	Binary file	D00	
Ultrasonic level transmitter	1 Hz	Binary file	D01	Ultrasonic level transmitter and pore pressure transducer gauge share the same capture software
Pore pressure transducer gauge				
Wire sensor	Change of status	Assess database	Mdb	

3.3.2 Mobile Stations

Mobile stations are used to extend the coverage of the debris flow monitoring system. The mobile debris flow monitoring station (Fig. 13) is put to use only when a particular event occurs. The overall cost of a mobile station is much lower than that of an on-site station. It can be dispatched to a remote area when the location is accessible during an event. The instruments and sensors in the mobile station include a rain gauge, soil moisture sensor, geophone, and CCD camera. The satellite disk is installed to transmit data back to the operation center.

3.3.3 Portable Units

A newly developed monitoring device had been produced. The new equipment, called a portable unit, carries sensors to collect debris flow-related data at remote areas. The portable unit was designed to work with on-site monitoring stations and mobile stations. Figure 14 shows an example of a portable unit. The unit includes basic and necessary sensors, including a rain gauge, soil moisture sensor, and geophone. The collected data is transmitted through 3G or radio, depending on communication status at the sites.

Fig. 13 Examples of mobile stations

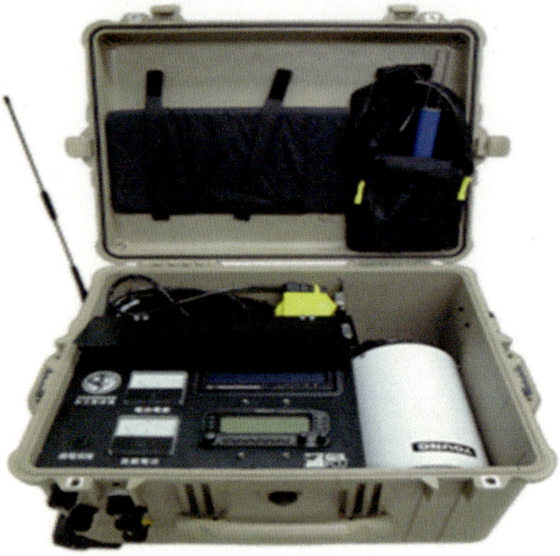

Fig. 14 A portable advanced unit

3.4 Integrated Control and Management Systems

Along with the monitoring sensors, an integrated system is necessary to control and manage the debris flow monitoring network (Lee et al. 2011, 2012, 2013, 2014). The monitoring system should provide easy access to control instruments and sensors, and function in data collection, exchange, and warning announcements. The system can display the locations of monitoring stations and instruments through a GIS map platform, or on the Google Earth platform. The system should operate automatically at all times and notify operators if anything unusual occurs.

4 Setup Procedures

4.1 Determine the Areas of Potential Debris Flows

The areas prone to debris flows can be initially determined by historical data, case histories, aerial photos, and local maps. The potential levels can be adjusted based on the needs of the operation. Usually three levels—high, medium, and low—are used to classify the potential for debris flows.

4.2 Conduct Site Investigations

After determining the potential debris flow areas, site investigations need to be conducted to determine the environmental status at the future locations of monitoring stations. The site investigation also can provide information for planning the instruments and sensors.

4.3 Determine the Locations of Monitoring Stations, the Layout of Instruments and Sensors

Based on the results of site investigation, the monitoring station should be placed at a safe location and be accessible for maintenance and repair. The layout of instruments and sensors is set up in a way that the monitoring network can observe both up- and down-stream.

4.4 Set up Housing, Lines, Power, and Satellite Communication

Construct the on-site monitoring station and set up connections to the instruments. Set up the power system and satellite disk.

4.5 Test Communications

Check the status and signal strength of ADSL, 3G, GPRS, and satellite while collecting observation data.

4.6 System Test

Conduct tests of the monitoring system, including data validation, a function test, and a warning system check-up.

5 Examples and Applications

5.1 Debris Flow Disaster Prevention Website

The Soil and Water Conservation Bureau (SWCB) in Taiwan has established a debris flow monitoring system online (Fig. 15, website: http://246eng.swcb.gov.tw), and has experience in debris flow monitoring and disaster response. The system updates the status of potential debris flows and announces warnings when critical rainfall criteria are reached. The public can access the website to obtain the real-time monitoring data. Also, the website provides knowledge and information about debris flows. Education is another key part in the Bureau's debris flow monitoring website. Currently there are 21 on-site debris flow monitoring stations, three mobile stations, and 17 portable units used in the monitoring network. The Geographic Information Systems Research Center at Feng Chia University Taiwan (GIS.FCU) helps the Soil and Water Conservation Bureau to develop and establish the monitoring network and system.

5.2 Application: Jioufen-Ershan Monitoring Station

Completed in 2002, Jioufen-Ershan monitoring station is equipped with a rain gauge, a water level meter, a CCD camera, a geophone, wire sensors, load cells and soil moisture sensors. Figure 16 shows the layout of instruments.

Case: Typhoon Kalmaegi

During the Typhoon Kalmaegi in July 16–18, 2008, the Jioufen-Ershan Monitoring Station provided observation data to the emergency operation center. Figures 17 and 18 show the path of the typhoon and the rainfall distribution of Taiwan, respectively. The accumulated rainfall measured at the monitoring station from July 16 to 23 was 625 mm. Figure 19 shows the rainfall intensity and accumulated rainfall. Figure 20 shows the real-time image recorded during the typhoon.

The geophone signal recorded during Typhoon Kalmaegi is shown in Fig. 21. Slight

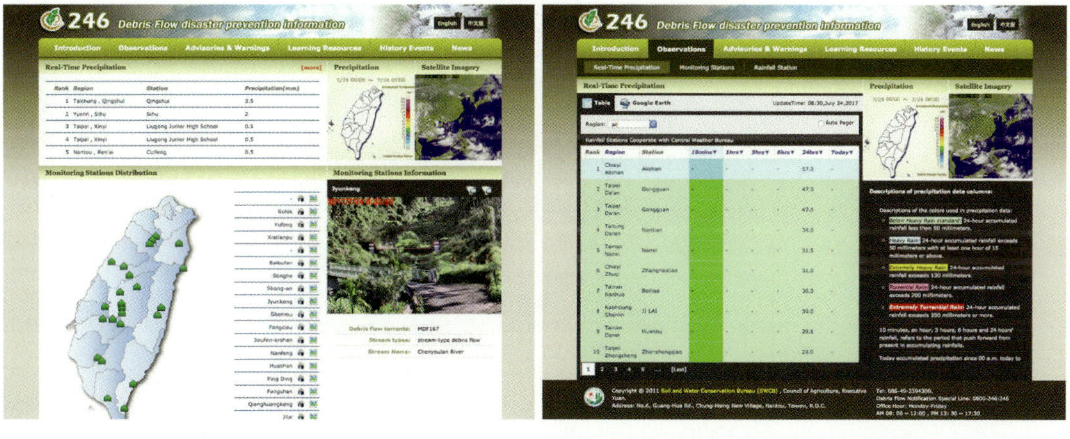

Fig. 15 The Taiwan Soil and Water Conservation Bureau (SWBC) debris flow monitoring website

Fig. 16 Layout of sensors in Jioufen-Ershan monitoring station

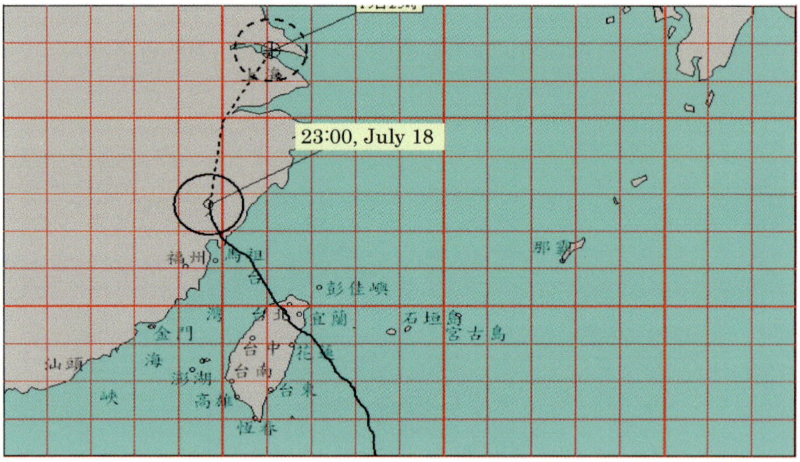

Fig. 17 Path of Typhoon Kalmaegi

fluctuations appear in both the time record and the wavelet transform. But the CCD images did not show any apparent debris flow; furthermore, it becomes clear from the wavelet transform results that there were no obvious peaks in low frequencies during this period. Therefore, it is determined that the wire sensor was broken by the flood.

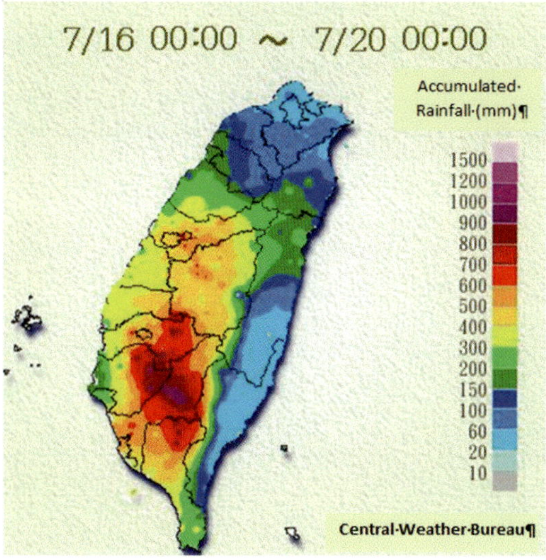

Fig. 18 Rainfall distribution in Taiwan during Typhoon Kalmaegi

Fig. 19 Rainfall measured from July 17 to 23

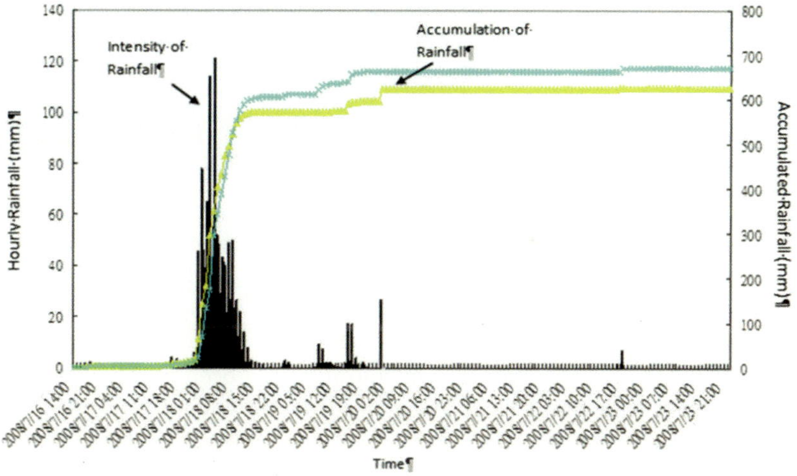

Fig. 20 The CCD camera images **a** Before the rainfall (2008/07/16 15:30). **b** Debris flows arrived (2008/07/18 13:56)

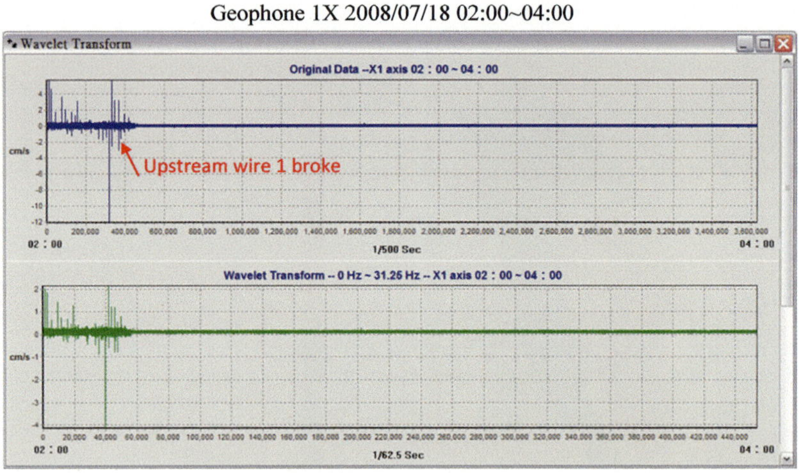

Fig. 21 Time record at Jioufen-Ershan station and wavelet transform analysis

Fig. 22 Damaged equipment and site condition after Typhoon Kalmaegi

Several instruments were damaged during the typhoon (Fig. 22) and the hydraulic structures in the river were no longer functional after the typhoon. Repair work has been conducted after the event to ensure that the monitoring at Jioufen-Ershan Station can be maintained on a regular basis.

6 Conclusion

One of the most important considerations of debris flow disaster monitoring is the understanding of causes. To better understand the mechanism of debris flow, observation and data collection are important for disaster prevention and response. A system is necessary for debris flow monitoring in order to manage sensors, equipment, and early warnings. The experience in Taiwan has shown that a monitoring network with different types of stations is effective in debris flow monitoring. Rainfall is commonly used to announce the early warning. The wire sensors and geophones provide useful evidence of debris flow occurrence. The whole system has been adopted in Taiwan for years and with successful outcomes. This approach may also be used to tackle similar disasters in other developing countries in Asia.

Acknowledgements Special thanks are directed to the Soil and Water Conservation Bureau, Taiwan, for their support for the monitoring program, as well as the Debris Flow Disaster Emergency Operation Center. We are grateful to the supporting team of GIS Research Center of Feng Chia University Taiwan. Thanks to all people providing help with this study.

References

Baum RL, Godt JW (2010) Early warning of rainfall-induced shallow landslides and debris flow in the USA. Landslides 7(3):259–272

Fang YM et al (2008) Analysis of debris flow underground sound by wavelet transform—a case study of events in Aiyuzih river. J Chin Soil Water Conserv CSWCS 39(1):27–44

Lee MH (2006) A rainfall-based debris flow warning analysis and its application. Ph.D. thesis. National Cheng Kung University, Taiwan (in Chinese)

Lee BJ et al (2010) Basin-wide monitoring network for debris flow at Laonong and Qishan rivers. Project report. Soil and Water Conservation Bureau, p 364 (in Chinese)

Lee BJ et al (2011) The 2011 on-site data gathering and monitoring station maintenance program. Project report. Soil and Water Conservation Bureau, p 364 (in Chinese)

Lee BJ et al (2012) The 2012 on-site data gathering and monitoring station maintenance program. Project report. Soil and Water Conservation Bureau, p 476 (in Chinese)

Lee BJ et al (2013) The 2013 on-site data gathering and monitoring station maintenance program. Project report. Soil and Water Conservation Bureau, p 446 (in Chinese)

Lee BJ et al (2014) The 2014 on-site data gathering and monitoring station maintenance program. Project report. Soil and Water Conservation Bureau, p 433 (in Chinese)

List of PPT-tools and PDF-tools

List of PPT tools & video tools

No.	Identifier	Title	Author	Email/website	Slides
1	PPT-tool 1.039-1.1	Remote Sensing data and methodology for event landslide recognition and mapping	Alessandro Mondini	alessandro.mondini@irpi.cnr.it	30
2	PPT-Tool 1.052-1.1	Logisnet Manual and Quick-start Tutorial	Gabriel Legorreta Paulín, Marcus I. Bursik	legorretag@hotmail.com	127
3	PPT-tool 2.039-1.1	Italian National Landslide Warning System	Mauro Rossi et. al.	mauro.rossi@irpi.cnr.it	29
4	PPT-tool 2.062-1.1	Landslide Monitoring and Early Warning System	Teuku Faisal Fathani, Dwikorita Karnawati	tfathani@ugm.ac.id, dwiko@ugm.ac.id	31
5	PPT-tool 2.062-1.2	Monitoring and Early Warning System for Debris Flows in Rivers on Volcanoes	Teuku Faisal Fathani, Djoko Legono	tfathani@ugm.ac.id, djokolegono@yahoo.com	37
6	PPT-tool 2.385-1.2	Landslide Comprehensive Monitoring System: The Grohovo Landslide Case Study, Croatia	Željko Arbanas, Snježana Mihalić Arbanas, Martina Vivoda Prodan, Josip Peranić, Sanja Dugonjić Jovančević, Vedran Jagodnik	zeljko.arbans@gradri.uniri.hr	28
7	PPT-Tool 3.385-1.3	Landslide Occurrence Prediction in the Rječina River Valley as a Base for an Early Warning System	Martina Vivoda Prodan, Sanja Dugonjić Jovančević, Željko Arbanas	martina.vivoda@gradri.uniri.hr	24
8	PPT-tool 2.886-1.2	Landslide Monitoring and Warning	An-Bin Huang, Wen-Jong Chang	huanganbin283@gmail.com	69
9	PPT-tool 3.039-1.1	Landslide Hazards and Risk Assessment	Fausto Guzzetti	F.Guzzetti@irpi.cnr.it	52
10	PPT-tool 3.039-1.2	Landslide-related WPS services	Ivan Marchesini	I.Marchesini@irpi.cnr.it	46

(continued)

K. Sassa et al. (eds.), *Landslide Dynamics: ISDR-ICL Landslide Interactive Teaching Tools*, https://doi.org/10.1007/978-3-319-57774-6

No.	Identifier	Title	Author	Email/website	Slides
11	PPT-tool 3.039-1.3	Probabilistic approach to physically based landslide modeling	Massimiliano Alvioli, Mauro Rossi, Fausto Guzzetti	Massimiliano.Alvioli@irpi.cnr.it	29
12	PPT-tool 3.039-1.4	Advanced 2D Slope stability Analysis by LEM by SSAP software: a full freeware tool for teaching and scientific community	Lorenzo Borselli	lborselli@gmail.com	52
13	PPT-tool 3.886-1.1	Debris-2D Tutorial	Ko-Fei Liu, Ying-Hsin Wu	kfliu@ntu.edu.tw, wu.ahsin@gmail.com	43
14	PPT-tool 4.039-1.1	Definition and Use of Empirical Rainfall Thresholds for Possible Landslide Occurrence	Maria Teresa Brunetti, Silvia Peruccacci	Maria.Teresa.Brunetti@irpi.cnr.it, Silvia.Peruccacci@irpi.cnr.it	39
15	PPT-tool 4.039-1.2	Landslide Risk to the Population of Italy	Paola Salvati, Cinzia Bianchi	P.Salvati@irpi.cnr.it	37
16	PPT-tool 4.062-1.1	Socio-Technical Approach for Landslide Mitigation and Risk Reduction	Dwikorita Karnawati, Teuku Faisal Fathani, Wahyu Wilopo, Budi Andayani	dwiko@ugm.ac.id	10
17	PPT-tool 4.062-1.2	Community Hazard Maps for Landslide Risk Reduction	Dwikorita Karnawati, Teuku Faisal Fathani, Wahyu Wilopo, Budi Andayani	dwiko@ugm.ac.id	10
18	PPT-tool 4.066-1.1	Course on Landslide Disaster Risk Reduction for Local Government Level Stakeholders	Asian Disaster Preparedness Centre	peeranan@adpc.net	416
19	PPT-tool 4.886-1.1	Typhoon Loss Assessment System (TLAS) Taiwan Web Tool	Hsin-Chi Li, Yi-Chen Chen, Mei-Chun Kuo	hsinchi@ncdr.nat.gov.tw	8
20	PPT-tool 4.886-1.2	Assessment Social Impact of debris flow disaster by Social Vulnerability Index	Ko-Fei Liu, Hsin-Chi Li, Mei-Chun Kuo, Hui-Hsuan Yang	kfliu@ntu.edu.tw, hsinchi@ncdr.nat.gov.tw	17
21	PPT-tool 4.886-1.3	Tutorial: Procedures for Constructing Disaster Evacuation Maps	Su-Chin Chen, Lien-Kuang Chen	scchen@dragon.nchu.edu.tw, steven_chen@ncdr.nat.gov.tw	56
22	Video-tool 3.084-2.1	Manual for undrained dynamic-loading ring shear apparatus	Lam Huu Quang	lhqlinh@yahoo.com	20 minutes

List of PDF tools

No.	Identifier	Title	Author	Email	Pages
1	PDF-tool 1.064-1.1	Field guide for the identification and assessment of Landslide and Erosion features and hazards affecting pipelines	Chris Massey, Graham Hancox, Mike Page	massey@gns.cri.nz	88
2	PDF-tool 3.081-1.1	An integrated model simulating the initiation and motion of earthquake and rain induced rapid landslides and its application to the 2006 Leyte landslide	Kyoji Sassa	sassa@iclhq.org	18
3	PDF-tool 3.081-1.3	A hypothesis of the Senoumi submarine megaslide in Suruga Bay in Japan -based on the undrained dynamic-loading ring shear tests and computer simulation	Kyoji Sassa, Bin He	sassa@iclhq.org	17
4	PDF-tool 3.081-1.5	Manual for the LS-RAPID software	Kyoji Sassa, Hendy Setiawan, Bin He, Karolina Gradiški, Khang Dang	sassa@iclhq.org	80
5	PDF-tool 3.081-1.6	Manual for undrained dynamic-loading ring shear apparatus	Hendy Setiawan et al.	sassa@iclhq.org	32
6	PDF-tool 3.081-1.7	Undrained dynamic- loading ring shear apparatus and its application to landslide dynamics	Kyoji Sassa, Hiroshi Fukuoka, Gonghui Wang and Naohide Ishikawa	sassa@iclhq.org	13
7	PDF-tool 3.081-1.9	Dynamic properties of earthquake induced large-scale rapid landslides within past landslide masses	Kyoji Sassa	sassa@iclhq.org	10
8	PDF-tool 4.091-1.2	Training Module on Comprehensive Landslide Risk Management	Surya Parkash	suryanidm@gmail.com	304
9	PDF-tool 4.064-1.1	Guidelines for assessing planning policy and consent requirements for landslide prone land	Wendy Saunders and Phillip J. Glassey	https://www.gns.cri.nz/	78
10	PDF-tool 4.064-1.2	Shut happens—Building hazard resilience for businesses in NZ	Resilient Organisations	www.resorgs.org.nz	16
11	PDF-tool 4.064-1.3	Working from the same page consistent messages for CDEM: PART B: Hazard-specific information—Landslides	Ministry of Civil Defence & Emergency Management, New Zealand	http:// www.civildefence.govt.nz	14
12	PDF-tool 4.007-2-1	Summer School guidebook	Alexander Strom	strom.alexandr@yandex.ru	131

International Consortium on Landslides

An international non-government and non-profit scientific organization promoting landslide research and capacity building for the benefit of society and the environment

President: Yueping Yin (China Geological Survey)
Vice Presidents: Irasema Alcantara-Ayara (UNAM), Mexico, Matjaz Mikos (University of Ljubljana), Slovenia Dwikorita Karnawati (Gadjah Mada University, Indonesia)
Executive Director: Kyoji Sassa (Prof. Emeritus, Kyoto University, Japan), Treasurer: Kaoru Takara (Kyoto University, Japan)

ICL Supporting Organizations:

The United Nations Educational, Scientific and Cultural Organization (UNESCO) / The World Meteorological Organization (WMO) / The Food and Agriculture Organization of the United Nations (FAO) / The United Nations International Strategy for Disaster Reduction Secretariat (UNISDR) / The United Nations University (UNU) / International Council for Science (ICSU) / World Federation of Engineering Organizations (WFEO) / International Union of Geological Sciences (IUGS) / International Union of Geodesy and Geophysics (IUGG) / Government of Japan

ICL Members:

Albania Geological Survey / The Geotechnical Society of Bosnia and Herzegovina / Geological Survey of Canada / Chinese Academy of Sciences, Institute of Mountain Hazards and Environment / Northeast Forestry University, Institute of Cold Regions Science and Engineering, China / China Geological Survey / Nanjing Institute of Geography and Limnology, Chinese Academy of Sciences / Tongji University, College of Surveying and Geo-Informatics, China / Universidad Nacional de Columbia, Columbia / City of Zagreb, Emergency Management Office, Croatia / Croatian Landslide Group (Faculty of Civil Engineering, University of Rijeka and Faculty of Mining, Geology and Petroleum Engineering, University of Zagreb) / Charles University, Faculty of Science, Czech Republic / Institute of Rock Structure and Mechanics, Department of Engineering Geology, Czech Republic / Cairo University, Egypt / Joint Research Centre (JRC), European Commission / Technische Universitat Darmstadt, Institute and Laboratory of Geotechnics, Germany / National Environmental Agency, Department of Geology, Georgia / Universidad Nacional Autónoma de Honduras, UNAH, Honduras / Amrita Vishwa Vidyapeetham, Amrita University / National Institute of Disaster Management, India / University of Gadjah Mada, Indonesia / Parahyangan Catholic University, Indonesia / Research Center for Geotechnology, Indonesian Institute of Sciences, Indonesia / Building & Housing Research Center, Iran / University of Calabria, DIMES, CAMILAB, Italy / University of Firenze, Earth Sciences Department, Italy / Istituto de Ricerca per la Protezione Idrogeologica (IRPI), CNR, Italy / Italian Institute for Environmental Protection and Research (ISPRA) - Dept. Geological Survey, Italy / DIA–Università degli Studi di Parma, Italy / Forestry and Forest Product Research Institute, Japan / Japan Landslide Society / Kyoto University, Disaster Prevention Research Institute, Japan / Korea Forest Research Institute, Korea / Korea Infrastructure Safety & Technology Corporation, Korea / Korea Institute of Civil Engineering and Building Technology / Korea Institute of Geoscience and Mineral Resources (KIGAM) / Korean Society of Forest Engineering / Slope Engineering Branch, Public Works Department of Malaysia /Institute of Geography, National Autonomous University of Mexico (UNAM) / International Centre for Integrated Mountain Development (ICIMOD), Nepal / University of Nigeria, Department of Geology, Nigeria / International Centre for Geohazards (ICG) in Oslo, Norway /Grudec Ayar, Peru / Moscow State University, Department of Engineering and Ecological Geology, Russia / JSC "Hydroproject Institute", Russia / University of Belgrade, Faculty of Mining and Geology, Serbia / Comenius University, Faculty of Natural Sciences, Department of Engineering Geology, Slovakia / University of Ljubljana, Faculty of Civil and Geodetic Engineering (UL FGG), Slovenia / University of Ljubljana, Faculty of Natural Sciences and Engineering (UL NTF), Slovenia / Geological Survey of Slovenia / Central Engineering Consultancy Bureau (CECB), Sri Lanka / National Building Research National Organization, Sri Lanka /Taiwan University, Department of Civil Engineering, Chinese Taipei / Landslide group in National Central University from Graduate Institute of Applied Geology, Department of Civil Engineering, Center for Environmental Studies, Chinese Taipei/ Asian Disaster Preparedness Center, Thailand / Ministry of Agriculture and Cooperative, Land Development Department, Thailand / Institute of Telecommunication and Global Information Space, Ukraine / California State University, Fullerton & Tribhuvan University, Institute of Engineering, USA & Nepal / Institute of Transport Science and Technology, Vietnam / Vietnam Institute of Geosciences and Mineral Resources (VIGMR), Vietnam

ICL Supporters:

Marui & Co., Ltd., Osaka, Japan / Okuyama Boring Co., Ltd., Yokote, Japan / GODAI Development Corp., Kanazawa, Japan / Japan Conservation Engineers & Co., Ltd, Tokyo / Kokusai Kogyo Co., Ltd., Tokyo, Japan / Ohta Geo-Research Co., Ltd., Nishinomiya, Japan / OSASI Technos Inc., Kochi, Japan / OYO Corporation, Tokyo, Japan / Sabo Technical Center, Tokyo, Japan / Sakata Denki Co., Ltd., Tokyo, Japan

Contact:

International Consortium on Landslides, 138-1 Tanaka Asukai-cho, Sakyo-ku, Kyoto 606-8226, Japan
Web: http://icl.iplhq.org/, E-mail: secretariat@iclhq.org
Tel: +81-774-38-4834, +81-75-723-0640, Fax: +81-774-38-4019, +81-75-950-0910